HANDBOOK OF MULTILEVEL METALLIZATION FOR INTEGRATED CIRCUITS

HANDBOOK OF MULTILEVEL METALLIZATION FOR INTEGRATED CIRCUITS
Materials, Technology, and Applications

Edited by

Syd R. Wilson
and
Clarence J. Tracy

Materials Research Organization
Motorola Semiconductor Products Sector
Mesa, Arizona

John L. Freeman, Jr.

Advanced Custom Technologies
Motorola Semiconductor Products Sector
Mesa, Arizona

np | **NOYES PUBLICATIONS**
Westwood, New Jersey, U.S.A.

Copyright © 1993 by Noyes Publications
    No part of this book may be reproduced or utilized in
    any form or by any means, electronic or mechanical,
    including photocopying, recording or by any
    information storage and retrieval system, without
    permission in writing from the Publisher.
Library of Congress Catalog Card Number: 93-26689
ISBN: 0-8155-1340-2
Printed in the United States

Published in the United States of America by
Noyes Publications
Fairview Avenue, Westwood, New Jersey 07675

10 9 8 7 6 5 4 3 2

Library of Congress Cataloging-in-Publication Data

Handbook of multilevel metallization for integrated circuits : materials,
technology, and applications / edited by Syd R. Wilson, Clarence J. Tracy,
John L. Freeman, Jr.
            p.   cm.
    Includes bibliographical references and index.
    ISBN 0-8155-1340-2
    1. Integrated circuits--Design and construction. 2. Metallizing.
    I. Wilson, Syd R. II. Tracy, Clarence J. III Freeman, John L.
  TK7874.H3493   1993
  621.3815--dc20   93-26689
                                                          CIP

MATERIALS SCIENCE AND PROCESS TECHNOLOGY SERIES

Editors

Rointan F. Bunshah, University of California, Los Angeles *(Series Editor)*
Gary E. McGuire, Microelectronics Center of North Carolina *(Series Editor)*
Stephen M. Rossnagel, IBM Thomas J. Watson Research Center
(Consulting Editor)

Electronic Materials and Process Technology

HANDBOOK OF DEPOSITION TECHNOLOGIES FOR FILMS AND COATINGS, 2nd Edition: edited by Rointan F. Bunshah

CHEMICAL VAPOR DEPOSITION FOR MICROELECTRONICS: by Arthur Sherman

SEMICONDUCTOR MATERIALS AND PROCESS TECHNOLOGY HANDBOOK: edited by Gary E. McGuire

HYBRID MICROCIRCUIT TECHNOLOGY HANDBOOK: by James J. Licari and Leonard R. Enlow

HANDBOOK OF THIN FILM DEPOSITION PROCESSES AND TECHNIQUES: edited by Klaus K. Schuegraf

IONIZED-CLUSTER BEAM DEPOSITION AND EPITAXY: by Toshinori Takagi

DIFFUSION PHENOMENA IN THIN FILMS AND MICROELECTRONIC MATERIALS: edited by Devendra Gupta and Paul S. Ho

HANDBOOK OF CONTAMINATION CONTROL IN MICROELECTRONICS: edited by Donald L. Tolliver

HANDBOOK OF ION BEAM PROCESSING TECHNOLOGY: edited by Jerome J. Cuomo, Stephen M. Rossnagel, and Harold R. Kaufman

CHARACTERIZATION OF SEMICONDUCTOR MATERIALS, Volume 1: edited by Gary E. McGuire

HANDBOOK OF PLASMA PROCESSING TECHNOLOGY: edited by Stephen M. Rossnagel, Jerome J. Cuomo, and William D. Westwood

HANDBOOK OF SEMICONDUCTOR SILICON TECHNOLOGY: edited by William C. O'Mara, Robert B. Herring, and Lee P. Hunt

HANDBOOK OF POLYMER COATINGS FOR ELECTRONICS, Second Edition: by James Licari and Laura A. Hughes

HANDBOOK OF SPUTTER DEPOSITION TECHNOLOGY: by Kiyotaka Wasa and Shigeru Hayakawa

HANDBOOK OF VLSI MICROLITHOGRAPHY: edited by William B. Glendinning and John N. Helbert

CHEMISTRY OF SUPERCONDUCTOR MATERIALS: edited by Terrell A. Vanderah

CHEMICAL VAPOR DEPOSITION OF TUNGSTEN AND TUNGSTEN SILICIDES: by John E. J. Schmitz

ELECTROCHEMISTRY OF SEMICONDUCTORS AND ELECTRONICS: edited by John McHardy and Frank Ludwig

v

HANDBOOK OF CHEMICAL VAPOR DEPOSITION: by Hugh O. Pierson

DIAMOND FILMS AND COATINGS: edited by Robert F. Davis

ELECTRODEPOSITION: by Jack W. Dini

HANDBOOK OF SEMICONDUCTOR WAFER CLEANING TECHNOLOGY: edited by Werner Kern

CONTACTS TO SEMICONDUCTORS: edited by Leonard J. Brillson

HANDBOOK OF MULTILEVEL METALLIZATION FOR INTEGRATED CIRCUITS: edited by Syd R. Wilson, Clarence J. Tracy, and John L. Freeman, Jr.

HANDBOOK OF CARBON, GRAPHITE, DIAMONDS AND FULLERENES: by Hugh O. Pierson

Ceramic and Other Materials—Processing and Technology

SOL-GEL TECHNOLOGY FOR THIN FILMS, FIBERS, PREFORMS, ELECTRONICS AND SPECIALTY SHAPES: edited by Lisa C. Klein

FIBER REINFORCED CERAMIC COMPOSITES: by K. S. Mazdiyasni

ADVANCED CERAMIC PROCESSING AND TECHNOLOGY, Volume 1: edited by Jon G. P. Binner

FRICTION AND WEAR TRANSITIONS OF MATERIALS: by Peter J. Blau

SHOCK WAVES FOR INDUSTRIAL APPLICATIONS: edited by Lawrence E. Murr

SPECIAL MELTING AND PROCESSING TECHNOLOGIES: edited by G. K. Bhat

CORROSION OF GLASS, CERAMICS AND CERAMIC SUPERCONDUCTORS: edited by David E. Clark and Bruce K. Zoitos

HANDBOOK OF INDUSTRIAL REFRACTORIES TECHNOLOGY: by Stephen C. Carniglia and Gordon L. Barna

CERAMIC FILMS AND COATINGS: edited by John B. Wachtman and Richard A. Haber

Related Titles

ADHESIVES TECHNOLOGY HANDBOOK: by Arthur H. Landrock

HANDBOOK OF THERMOSET PLASTICS: edited by Sidney H. Goodman

SURFACE PREPARATION TECHNIQUES FOR ADHESIVE BONDING: by Raymond F. Wegman

FORMULATING PLASTICS AND ELASTOMERS BY COMPUTER: by Ralph D. Hermansen

HANDBOOK OF ADHESIVE BONDED STRUCTURAL REPAIR: by Raymond F. Wegman and Thomas R. Tullos

CARBON–CARBON MATERIALS AND COMPOSITES: edited by John D. Buckley and Dan D. Edie

Preface

It is widely recognized that the successful design, development, and integration of metallization systems is and will continue to be key to current and future VLSI technologies. All major semiconductor companies have significant ongoing research teams focused in this area.

To be successful, such teams must view multilevel metallization as a system rather than a collection of isolated process modules. The interaction between unit processes and their materials pose special challenges and must be well understood.

However, by necessity MLM teams are usually composed of unit process specialists and, because of the large number of different process modules involved, it has been difficult for these team members to become generalists. Many different educational boundaries must be crossed since device, design, chemical, materials science, and analysis problems are encountered, to name a few. To date, the engineers composing these teams have been forced to accumulate and read large numbers of publications from each topical area to obtain a reasonably complete understanding of the numerous topics. In our personal experience, we have found this to be quite tedious and time consuming.

This *Handbook of Multilevel Metallization for Integrated Circuits* answers this important need by pulling together in one volume a thorough technical summary of each of the key areas that make up a multilevel metal system. Properly included are associated design, analysis, materials, and manufacturing topics. The book then serves three purposes: (1) It functions as a good learning tool for the engineer newly assigned to work in metallization. All important aspects of the fully integrated process are discussed in sufficient depth such that no additional literature searches are needed. (2) It serves as a reference text for any MLM engineer, new or experienced, who wishes to refresh his or her memory about the specifics of a concept or process. (3) For someone who wants to further specialize in one topical area, an extensive listing of references has been provided to simplify more in-depth study.

In Chapter 1 the terminology for an MLM system is defined, the trends in MLM are presented, and a methodology for developing an MLM system is discussed. In Chapter 2, the subject of contacts, silicides and barrier materials is covered in detail. The topic of metal-to-silicon contacts is reviewed. The properties of various silicides and barrier materials are discussed, and the process of creating a self-aligned silicide is explained. Chapter 3 provides a discussion of the various metals that may be used for interconnects in an MLM system. The emphasis is upon aluminum and aluminum alloys. The properties of the various alloys are listed. Chapter 4 covers all of the inorganic dielectrics that are used for ILD0 to separate the metal and the device structures as well as the intermetal dielectrics ILD1, 2, ... and the final passivation. The properties of these inorganic dielectrics are reviewed. In addition this chapter explains the chemistry and various processes for depositing these dielectrics. Chapter 5 discusses organic dielectrics. As in Chapter 4, the chemistry, properties and deposition processes associated with organic dielectrics are summarized. Chapter 6 covers the topic of planarization and explains not only why it is important but also the various approaches to achieve planarization. In Chapter 7, the two key unit processes of lithography and etch are described with an emphasis on special issues related to their use in building an MLM system. The areas of electromigration and stress migration are discussed in Chapter 8 which covers the issues associated with the reliability of the metallization layers. The theory as well as the experimental procedures associated with reliability are covered in detail. Chapter 9 talks about the design and layout of a MLM test vehicle. Various structures, how they are tested, and the information they provide are covered. Chapter 10 presents a discussion of manufacturing requirements associated with an MLM system. The concepts of statistical process control (SPC), design for manufacturability (DFM), and cost of ownership (CoO) are presented as well as a listing of currently available equipment. Chapter 11 gives a comprehensive discussion of the analytical tools and techniques for developing and evaluating the processes and materials in an MLM system. Chapter 12 covers the topic of packaging and how it interfaces with the chip metallization. Finally Chapter 13 takes a look at the future of multilevel metal and what will be required as devices continue to shrink and circuits increase in complexity.

We would like to thank all of the contributors for helping complete this book in a timely fashion. The assistance of Carol Wilson in the typing and layout of portions of the book is greatly appreciated. Finally we would like to thank our wives, Carol Wilson, Lorraine Tracy, and Sandee Freeman for their patience and support.

Mesa, Arizona   Syd R. Wilson
July, 1993   Clarence J. Tracy
                                                  John L. Freeman

Contributors

Harry K. Charles, Jr.
The Johns Hopkins University
Applied Physics Laboratory
Laurel, MD

Michael L. Dreyer
Motorola, Inc.
Mesa, AZ

John L. Freeman, Jr.
Motorola Semiconductor Products
Sector
Motorola, Inc.
Mesa, AZ

George E. Georgiou
AT&T Bell Laboratories
Murray Hill, NJ

Sumanta K. Ghosh
Rensselaer Polytechnic Institute
Troy, NY

Gregory W. Grynkewich
Advanced Custom Technologies
Motorola, Inc.
Mesa, AZ

John N. Helbert
Advanced Custom Technologies
Motorola, Inc.
Mesa, AZ

Paul S. Ho
Center for Materials Science and
Engineering
University of Texas at Austin
Austin, TX

Farhad Moghadam
Intel Corporation
Santa Clara, CA

Jeff C. Olsen
Applied Materials, Inc.
Santa Clara, CA

K. Ramkumar
Rensselaer Polytechnic Institute
Troy, NY

Laura B. Rothman
IBM
Yorktown Heights, NY

Arjun N. Saxena
International Science Company
Ballston Lake, NY

Dominic J. Schepis
IBM
East Fishkill, NY

Thomas Seidel
SEMATECH, Inc.
Austin, TX

Krishna Seshan
IBM
East Fishkill, NY

Simon Thomas
Core Technologies
Motorola, Inc.
Mesa, AZ

Chiu H. Ting
Intel Corporation
Santa Clara, CA

Clarence J. Tracy
Motorola Semiconductor Products
 Sector
Motorola, Inc.
Mesa, AZ

Charles J. Varker
Motorola Semiconductor Products
 Sector
Motorola, Inc.
Mesa, AZ

G. Donald Wagner
The Johns Hopkins University
Applied Physics Laboratory
Laurel, MD

Syd R. Wilson
Motorola Semiconductor Products
 Sector
Motorola, Inc.
Mesa, AZ

S. Simon Wong
Electrical Engineering Department
Stanford University
Stanford, CA

NOTICE

Table of Contents

1

INTRODUCTION

SYD R. WILSON, CLARENCE J . TRACY, And JOHN L. FREEMAN, JR.
Motorola Semiconductor Products Sector
Mesa, Arizona

Today's state of the art integrated circuits contain many active and passive elements including millions of transistors, capacitors, and resistors on a single chip.(1) These discrete elements must be connected with some form of wiring to form a circuit. Historically integration of the various elements was done by patterned lines of heavily doped polycrystalline silicon (polysilicon), polysilicon that had been silicided to lower its sheet resistance, and low resistivity metal lines (typically Al or Al alloys). As chips have become larger and more complex the requirements placed on the interconnect system have increased. With more devices, more layers of interconnect are required for efficient routing, resulting in multilevel metallization (MLM). As the distance across a chip becomes larger and the devices become faster, there is a need to improve the interconnect system so that it will not be the limiting factor in the overall performance of the circuit. This tends to be accomplished by reducing the interconnect resistance and capacitance. (Techniques for doing this are discussed throughout this book.) This need for lower interconnect resistance to improve speed has been a major driving force behind replacing many of the polysilicon interconnect lines with silicided polysilicon or metal. In addition, as the devices become smaller, the dimensions of the interconnect system must also scale down to prevent it from limiting the device packing density.

In this chapter we will introduce the structures and terminology for an MLM system, the uses of various components, why we use multiple layers, the current status, and some trends in the industry, and a methodology for choosing, and developing a particular interconnect system. Subsequent chapters will address, in much greater depth, the various components required to fabricate these multilevel metal interconnects.

1.0 MLM TERMINOLOGY AND STRUCTURE

Throughout the chapters of the book reference is made to the physical and electrical properties of the various components of the MLM interconnect system either as individual structures or integrated into the final system. It is the purpose of this section to establish a consistent terminology and identify the structures in a typical MLM system.

The definitions employed fall into three general categories: (a) the terminology associated with the MLM structures; (b) the terms associated with processing or fabrication sequences; and (c) those descriptors that identify electrical or physical properties or behaviors of the various MLM interconnect components.

1.1 Structure Nomenclature

A fully processed four layer MLM interconnect cross section is shown in Figure 1. The principal components of the structure beginning at the silicon or device interface are as follows:

[1] Contact This includes the contact opening and the various approaches used to ensure an ohmic contact from the first layer of metal to the silicon device or component. This structure is generally composed of various silicides, salicides, and doped polysilicon. Specific differentiation as to the type of contact is achieved by joining it with its purpose, i.e. base contact, source or drain contact, resistor contact, etc.

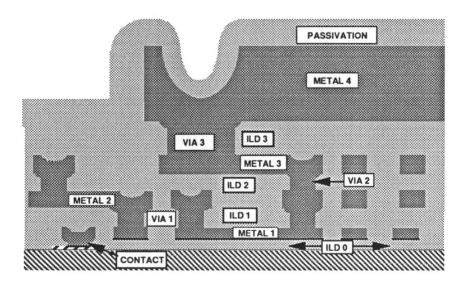

Figure 1. Schematic cross section of a four layer metal system.

[2] ILD and Passivation In Figure 1 there are five layers of dielectric shown for the four layer metal structure. The first four are referred to as

interlevel dielectric layers (ILDs) and serve to insulate the metal layers within the layer and from each other or the underlying silicon. The ILD separating the silicon or local interconnect lines from the initial or first metal interconnect layer is referred to as ILD0. The ILD separating the first layer of metal from the second is referred to as ILD1. ILD 2 and ILD3 then, are the dielectric layers between metal 2 and metal 3, and metal 3 and metal 4, respectively. It should be noted that each of these layers may be the product of processes which utilize several layers of deposition or etch processes in attempts to achieve planarized surfaces as described in Chapter 6 of this text. The final layer of dielectric is referred to as the 'passivating' layer or 'passivation.' The purpose of this layer is to provide physical and chemical protection to the underlying metal and device structures during the final assembly processes; and while the finished device is operating, in potentially hostile environments. Generally, this layer is composed of one or more layers of dielectric whose composition may act as mobile ion barriers or inhibit the diffusion of moisture and corrosive ions to metal surfaces, as discussed in section 8 of Chapter 4. Further, when bump or TAB assembly technologies are employed, the passivation is required to protect the metal layers from etchants or solder chemicals.

[3] Interconnects   Four metal interconnect layers are shown in the structure. They are simply identified as M1, M2, M3, and M4 for the first-through-fourth layers of metal interconnect, respectively. Again, as with the ILD structures it should be recognized that an individual metal interconnect layer may be composed of one or several metal films designed to enhance the reliability or performance of the interconnect system.

[4] Vias   The linkage from one layer of interconnect to another is provided by metallic conductors in holes between the dielectric layers referred to as vias. Identification of a specific via is achieved in the same fashion as the ILD nomenclature. V1, V2, and V3 identify the via linking metal 1 to metal 2, metal 2 to metal 3, and metal 3 to metal 4, respectively.

Table I summarizes the nomenclature for the MLM interconnect structures used in this text.

Other concepts have been employed elsewhere that utilize different numbering and/or lettering systems to identify and characterize the various components of VLSI/ULSI interconnect structures. One such system has been proposed by Saxena, with a rigorous labelling of the various layers.(2) It is based on the division of the interconnects into two types. Horizontal interconnects are those parallel to the wafer surface; vertical interconnects are perpendicular to the substrate surface (contacts and vias). The multiple layers which may comprise the layers of metal (e.g., under, main, and cap layers) or dielectric are tracked by the use of dual subscripts. Conductors (metals) are labelled as M_{ij}, where "i"is the layer number and "j" is the structure within the ith layer. Subsurface layers (e.g., diffusions) have i = 0; MLM systems have i = 1 to n. Similarly, dielectrics are denoted as D_{ij}, where "i" is the layer number and "j" is the structure within the ith layer. Again, i = 0 indicates subsurface dielectrics, whereas the MLM system dielectrics have i = 1 to n. Table II helps to clarify this system which, while not utilized in this book, may be useful to some readers.

Table I. Multilevel Metal Interconnect Nomenclature.

TERMINOLOGY	FUNCTION
ILD0	Provide dielectric insulation between silicon device components and the metal interconnect layers.
Contact	Provides electrical connection from the interconnect to silicon components through ILD0.
Metal one (M1)	First layer of metal interconnect.
Metal two (M2)	Second layer of metal interconnect.
Metal three (M3)	Third layer of metal interconnect.
Metal four (M4)	Fourth layer of metal interconnect.
ILD1	Dielectric layer between M1 and M2.
ILD2	Dielectric layer between M2 and M3.
ILD3	Dielectric layer between M3 and M4.
Via one (V1)	Provides electrical connection between M1 and M2 through ILD1.
Via two (V2)	Provides electrical connection between M2 and M3 through ILD2.
Via three (V3)	Provides electrical connection between M2 and M3 through ILD3.
Passivation	Final dielectric layer, provides protective physical, chemical barrier to the circuit.

Table II. An alternative multilevel metal nomenclature as proposed by Saxena, ref. 2.

LAYER DESIGNATION	USAGE	COMMENTS
M_{11}	"Barrier layer beneath metal one (e.g., TiW)"	
M_{12}	"Conducting layer (e.g., AlX)"	
M_{13}	"Cap layer for metal one (e.g., TiN)"	
D_{21}	Bottom barrier dielectric	These comprise the dielectrics separating metal one from metal two
D_{22}	Main dielectric	
D_{23}	Top barrier dielectric	

1.2 Process or Fabrication Nomenclature

Table III lists some of the more important terms used during the various process steps employed in creating MLM interconnect systems and their definitions. Where a common acronym exists it is included.

Table III. Multilevel metal processing nomenclature.

TERM	SYMBOL	DEFINITION
Alignment		Relative placement of one layer to the next in the lithography tool
Assembly		The part of the process that mates the chip to its package or board
Back end of the line	BEOL	Refers to that part of the process which builds the MLM system
Barrier		A thin underlayer used to prevent interactions with other materials
Border		Same as overlap; additional metal width in excess of the contact or via dimension
Cap		A thin additional film deposited on the surface of the conducting layer
Chemical Vapor Deposition	CVD	The deposition of thin films by reaction of gaseous sources
Chem-Mechanical Polish	CMP	Planarization with a polishing tool where the slurry chemistry plays a role
Collimated		PVD where material flux is directional
Critical Dimension	CD	Usually the minimum and most difficult to resolve feature size on any layer
Defect density	D_o	Process flaws/unit area that reduce yield
Etchback		An etch used without a resist pattern such that material is thinned everywhere
Hardmask		Material between the resist and the layer to be etched used to improve pattern transfer
Overlay		Relative placement of one layer to the next including all sources of error
Overlap		Same as border
Pad		A large metal area on which probes can be placed or bonding wires attached
Physical Vapor Deposition	PVD	The deposition of thin films by evaporation or sputtering
Pitch		Sum of metal linewidth and space
Plug		A metal element which completely fills a contact or via
Rapid Thermal Processing	RTP	Single wafer, short duration annealing with abrupt temperature rise and fall
Reactive Ion Etching	RIE	Plasma etching where the directional ion flux to the wafer leads to anisotropy
Sheet Resistivity	Rs	The ratio of resistivity to thickness of a conducting film (Ω/square)
Stacked		Refers to vias or contacts placed directly on top of one another

1.3 Properties / Behavior Nomenclature

Table IV presents terms used to describe various structures and properties of MLM interconnect systems.

Table IV. Terms used to describe structures and properties of MLM systems.

TERM	SYMBOL	DEFINITION
Capacitance (line)	C_1	The capacitance per unit length of an interconnect (fF/μm)
Chain		A test structure composed of a number of contacts or vias connected in series and used to evaluate continuity.
Coefficient of expansion	TCE	The fractional change in dimensions of a material with changing temperature (ppm/°C)
Comb		A test structure of interdigitated metal fingers used to test for intralayer shorting
Crosstalk		Signal induced on one metal line due to current in an adjacent metal line
Current density	"J, Jmax"	The value of current per unit area carried by a conductor; design limit is Jmax (A/sq. cm.)
Dielectric thickness	t_{ox}	Thickness of interlevel dielectrics separating metal layers (μm)
Electromigration	EM	Transport of metal atoms due to current flow, resulting in voids and line failure
Failures in ...	FIT	Number of failures in 10^9 hours of operation; used in reliability targets
Kelvin structure		Four terminal single contact structure used to measure specific contact resistance
Median time to fail	"MTTF, MTF"	The time at which half of a sample population have failed
Resistivity	ρ	Fundamental property; the resistance that a cubic cm. of material has to current flow
Snake/serpentine		A long metal line used on test chips to evaluate continuity
Stress migration		The transport of metal atoms due to mechanical stress
Stress voiding		Voids resulting from the transport of metal atoms due to mechanical stress
Specific contact resistance	R_c	The resistance of an ohmic contact to silicon from which the area dependence has been removed (Ω cm^2)

2.0   WHAT ARE THE USES OF METAL LAYERS IN AN INTEGRATED CIRCUIT?

2.1 Contact to the Devices and Other Elements

As indicated previously, once the individual devices and other elements are fabricated, they must be wired together to form specific circuits. This may involve connecting tens of millions of devices together in an area that is a few cm^2. In many flows, the first step is to form silicides on the source, drain, and gates of the MOS transistors (salicide process) as well as on portions of the doped polysilicon films that may be used as resistors or contacts in many bipolar circuits. The formation of silicides, contacts, and barrier layers is discussed in more detail in Chapter 2. In some flows, after the contact is etched in ILD0, the silicide is placed only in the floor of the contact. The next step in the process is to deposit a dielectric film(s), ILD0, to isolate the first metal layer from the device elements. (The various films that may be used for ILD0 are discussed in Chapter 4.) Depending on the ILD0 topography and the temperature limitations of the process at this point in the flow, ILD0 may be thermally flowed or planarized by some other technique such as etchback or chemical mechanical polishing. The ILD0 film is then patterned and etched to form contact holes through the oxide to underlying structures or silicide. The contact holes are filled by depositing a barrier film if there is not already one on the silicide and a conducting film (usually Al or W). If Al is used to fill the contact it is also patterned to form the first level of interconnects. If W is used it may be deposited by either a blanket or selective process. In the selective process, the W will deposit only in the contact hole. In the blanket process, an adhesion layer is deposited first (typically TiW, TiN or Ti), then the W is deposited. The W may be etched back to the top of the contact hole to form a plug or it may be patterned to use W as the metal one interconnect.

2.2 Connecting Devices Together To Create Functions

Once contact has been made to the device elements, the metal layers are used to connect various devices together to form simple functions. These functions will include gates, flip flops, memory cells, etc. This metal wiring is usually done at the lower metal levels and usually requires the minimum metal pitch (width of one metal line plus one space). The metal runs are usually not too long (with the exception of memory arrays) since they are wiring up the simplest cells; and, because long runs at the minimum pitch have the longest time constant as will be discussed in subsequent sections. Typically Al or Al alloys have been used for these metal runs as will be discussed in Chapter 3. In many earlier generations of MOS circuits, part or all of this type of interconnection was done with silicides or heavily doped polysilicon. However, the ratio of resistivity of polysilicon or silicides to Al is ~300 and ~6, respectively. In order to achieve faster circuits many of these higher resistivity interconnections have been replaced by Al. This trend has occurred even sooner in bipolar circuits where their primary advantage over MOS circuits has been speed, and the use of higher resistivity interconnects would have negated this advantage. In addition, bipolar circuits do not have the doped polysilicon films built into the flow in the same way that silicon gate MOS does.

2.3 Connecting Functions to Form Integrated Circuits

Once the fundamental cells have been formed, they must be wired together to form an integrated circuit. This is usually done at the second or third metal layer and may involve a larger metal pitch. The larger pitch is to help the speed (by reducing the line resistance), since the lengths of the metal runs are getting longer; and to help yield, since the topography gets more severe as the number of layers increases. Each of these metal layers are separated by an interlevel dielectric ILD1, 2, 3, etc. The composition of these films usually differs from the composition of ILD0 due to the different requirements and temperature constraints as discussed in Chapters 4 and 5. In order to deal with the increasing topography some form of planarization of the ILDs is usually required as discussed in Chapter 6. The wiring at the metal 2 and metal 3 layers may also be used to connect any peripheral circuitry that may be required to control or protect the rest of the chip. This wiring may also be used to connect hybrid technologies that now exist on the same chip such as a microcontroller and some associated memory.

2.4 Power Distribution and Interface to the Package

The top layer or two of metal are used for carrying the power distribution and other signals that must run the length of the chip. These metal layers may have the thickest metal and the widest pitch to reduce the resistance per unit length and thus, minimize any voltage drop in distributing the power around the chip. This metal layer also provides the bond pads that are used to form the interface to the package and the outside world. Therefore the metal used on this layer must be compatible with the packaging technology that will be used for the particular chip. The issues associated with packaging are discussed in Chapter 12.

3.0 WHY USE MULTIPLE LAYERS OF METAL

Historically each new generation of integrated circuits has involved a shrink of the device size. This has facilitated an improvement in the device speed, an increase in device packing density, and an increase in the number of functions that can be placed on a single chip. However in recent generations, the full benefits of shrinking the device have not been achievable without significant improvements in the metal interconnect system as well. As will be discussed in this section, the main approach in the interconnect system to realizing the full benefit of shrinking the device has been to use multiple layers of metal combined with a shrink of certain metal dimensions.

3.1 Allow a Tighter Packing of Devices

The first integrated circuits involved a relatively small number of devices, and these devices were quite large in size. Typically one layer of metal was used, and the minimum manufacturable metal pitch did not limit the device packing density. As the devices became smaller and the number of devices used in a circuit became larger, the metal system became a limiting factor in the number of devices that could be placed on a chip. As devices got smaller than the metal pitch, the pitch had to be reduced so that the leads required to make contact to the device did not cause an unnecessary increase in the device dimensions. A second problem that began to arise with larger circuits that used only one layer of metal is shown in Figure 2. It is obvious that the metal covers a large percentage of the circuit. It is understandable how this can impact both the packing density and circuit performance. The large power runs alone limit the area where contact can be made to devices. Eventually the point is reached where the area needed to route the interconnect lines exceeds the area required to build the devices, and the packing density is said to be interconnect limited. This problem is greatly minimized by moving some of the interconnect lines to other layers. This is especially true of power distribution lines which occupy more than 50% of the area of some circuits.

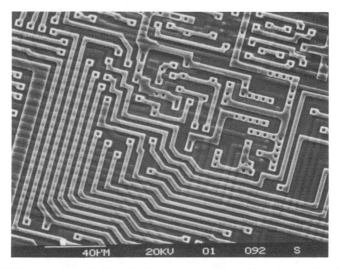

Figure 2. SEM micrograph showing the patterned metal layer on a circuit using only one layer of metal.

Another way in which the interconnect layer limits the packing density is the space required for routing the interconnect from one function to another as seen in Figure 2. Since metal lines can not touch each other, the path some metal runs have to take becomes quite long and winding in order to get from one point to another. These long paths require additional area and cause longer signal delays. This problem also can be minimized by using multiple layers of metal with the runs on one layer being orthogonal to the runs on the adjacent layers. This allows for the interconnects to follow a shorter path.

3.2 Reduce the Chip Size or Increase the Number of Functions Per Chip

The increase in packing density that was discussed in the preceding section can result in a smaller chip for a given circuit, and this in turn will result in more die per wafer. On the other hand, the desire for increased functionality on a chip leads designers to push chip sizes to be as large as possible. The maximum chip size then becomes limited by the field size of the stepper being used. If the packing density is increased, more functions can be placed within a given field size. This allows more functions or different functions to be combined on a chip rather than at the package or board level. This in turn may provide the end user a competitive advantage in cost, system speed or possibly weight. These potential advantages must be traded against the increased manufacturing complexity and the need for larger and more complex packages. Chapters 10 and 12 provide the reader some insight into the manufacturing and packaging issues.

3.3 Reduce the RC Time Constant of the Interconnect System

As will be discussed many times throughout this book, the impact of the interconnect on the circuit speed (performance) is related to the RC time constant. The resistance of the interconnect line is:

$$[1] \qquad R = \frac{\rho l}{w t_m}$$

where ρ=metal resistivity
l=length of the interconnect
w=width of the interconnect
t_m=metal thickness

To first order the capacitance of an interconnect line with a metal plate above and below the line is:

[2] $$C = 2\varepsilon\frac{lw}{t_{ILD}}$$

where ε=dielectric constant
    l=interconnect length
    w=metal width
    t_{ILD}=thickness of ILD

This equation assumes the ILD thickness above the line is equal to the thickness of the ILD below the line. This equation also ignores intraplane and fringing capacitance effects. (Historically these effects have been quite small. However, as the intraplane spacing between metal features has become comparable to the interplane spacing and the feature width, these effects can no longer be ignored as will be seen later in this chapter and in Chapter 13.) The product of R and C then becomes:

[3] $$RC = 2\varepsilon\rho\frac{l^2}{t_m t_{ILD}}$$

The intraplane capacitance can also be accounted for in the same fashion and the equation for RC becomes:

[4] $$RC = 2\varepsilon\rho l\left(\frac{l}{t_{ILD}t_m} + \frac{l}{ws}\right)$$

where s=intraplane distance between metal interconnects

Obviously with RC proportional to l^2, reducing l greatly improves RC. As we discussed in section 3.1, multiple layers of interconnect allow shorter metal runs.

4.0 TRENDS IN MULTILEVEL METALLIZATION

It is often beneficial to examine both some of the older and more recent literature to gain an idea of the directions the industry is taking in MLM and also to establish benchmarks to compare new plans and ideas against. The following two tables attempt to capture some of the essential elements of metal systems first as they existed in the mid-1980's (Table V) and as they exist now (Table VI). Shown are the number of metal layers, approximate minimum dimensions for contacts, vias, and metal runners (line / space [L / S] or pitch [P]), some information about the materials used, the company where the authors worked, and a reference number in parentheses. The following discussion will be based on these two tables and will consider what has changed in each of the major topical areas of MLM. It will soon become clear that in only six or seven years there have been significant changes in how the metal system is built but that there is a great deal of similarity in the approach that different companies are using.

Table V: Essential characteristics of MLM systems as described in the literature from the mid-1980's. L/S = Line / Space; (P) indicates pitch. Metal layers are listed starting closest to the silicon substrate.

NUMBER OF LAYERS	CONTACT	METAL ONE	VIA	METAL TWO, THREE, ...	ILD'S	COMPANY
	DIM.(μ) / BARRIER /FILL	L/S (μ)/ MATL	DIM(μ)/FILL	MIN L/S(μ)/ MAT'L.		
2	-- /none / none	4(P) / AlSi	1.5 / none	4(P) / AlSi	PEON	HP(3)
2	-- /none / none	3 / 4 /AlSi	6 / none	3 / 3 / TiW/AlSi	PEO	Philips(4)
2	3 / TiW /none	5 / - / AlSi	-- / none	-- / AlSi	ITO	DEC(5)
3	-- /none / none	2 / 6 / AlSiTi	3 x4 /none	3 / 8 / AlSiTi	Polyimide +PEN	Siemens(6)
2	1 / Ti/W	2.5(P) / W/AlSi/Ti/AlSi	1 / none	Ti/W/AlSi	PEO	TI(7)

The first and most obvious change over time has been in the number of layers and in the dimensions of those layers; and these, in turn, are the driving forces for all other changes in the materials and the way they are processed. In 1985, most MLM systems were limited to double level metal with dimensions usually significantly greater than one micron. Multilevel metal was just beginning to be an important topic; it was in 1984 that the first IEEE VLSI Multilevel Interconnection Conference was held. In 1993, three- and four-layer metal is common, and five layer metal systems exist. As well, all critical dimensions are now about 0.5 μm. These trends are expected to continue to even more layers of smaller dimensions; Chapter 13 of this book is devoted to the topic of the future of interconnects and suggests some of the limits that will be reached.

With more layers and submicron dimensions, reliability and processing constraints made straightforward evolution of the 1985 technology impossible. Hence, major changes occurred in the way contacts and vias were built and in the materials used for both the interconnections and the dielectrics.

Two primary issues occurred at the contact. First, shallow junctions and sometimes, the presence of silicides, combined with small contact area, ruled out AlSi directly on silicon because of silicon dissolution, unwanted reactions with the silicides, if present, and silicon precipitate formation (see Chapters 2 and 3 in this book). Secondly, the increased aspect ratio (contact depth / width) resulted in poor step coverage when standard sputtering processes were used for metal deposition. An enormous amount of work by many in the industry went into developing a usable CVD tungsten process; Table VI indicates its widespread acceptance as a solution wherever submicron contacts are involved.

Ti / TiN is commonly used as a barrier / adhesion / nucleation layer beneath the CVD tungsten to contact the silicon or silicide. Since standard sputtering will suffer from reduced step coverage for these layers, both collimated sputtering and CVD approaches are being developed to permit the extension of this approach past the 0.5 μm level. Chapter 2 of this book discusses contacts for ULSI; and in particular, sections 3 and 4 deal in depth with Ti/TiN, CVD tungsten, and other related metals.

Table VI: Essential characteristics of MLM systems as described in the literature from 1992. L/S = Line / Space; (P) indicates pitch. Metal layers are listed starting closest to the silicon substrate.

NUMBER OF LAYERS	CONTACT	METAL ONE	VIA	METAL TWO, THREE, ...	ILD'S	COMPANY
	Dim.(μ) / Barrier /Fill	L/S(μ)/ MAT'L.	Dim(μ) /Fill	MIN L/S(μ) / MAT'L.		
3	0.6 / Ti + TiN / W	0.4 / ? / W	-- / sel. W	0.4 / ?/ AlCu /W/TiN	APTEOS+ SOG	Mitsubishi (8)
3	≈0.5 / Ti + TiN / W	≈0.5 / ? / W	≈0.6 / W	≈0.5 / ? / Ti/TiN/AlX/W	Oxide	IBM(9)
2	0.5 / Ti + TiN / W	0.4 / 0.6/ W	0.5 / W	0.5/0.7 /TiW/W/AlCuSi	BPSG / PETEOS	TI(10)
5	0.6 / -- / W	1.4 (P)/ Ti/AlCuSi	0.9 / W	0.7/=1.1 / Ti/AlCuSi	PECVD Oxide	IBM(11)
4	0.6/ Ti + TiN / W	1.8 (P) / AlCu/TiW	0.6 / W	1.8(P) / AlCu/TiW	PETEOS	Motorola(1)
4	1.2 /-- / Al	1.2/1 .8/bar.+Al+cap	1.8 / Al	1.8/1 .8 / bar.+Al+cap	Polyimide	Vitesse(12)
2	0.5 / Ti + TiN / W	-- / W or AlCu(Si,Ti)	-- / W	-- / AlCu(Si,Ti)	-----	Fujitsu(13)

In the case of vias, just as for contacts, the shrinking design rules have led to widespread use of tungsten plugs. Unlike contacts, however, the similar depths of all vias permits the use of a selective tungsten process, and that is the direction some of the Japanese semiconductor companies are taking. In the United States blanket / etchback is the preferred approach, presumably because of the defect density associated with selectivity loss in manufacturing. With the blanket / etchback method, a combination nucleation and adhesion layer is first sputter deposited; both TiW and Ti/TiN are in use currently; and it remains to be seen if the industry will converge to a common solution.

For metal one, two basic approaches are currently in use. Some companies choose to pattern the CVD tungsten that is used to fill the contact as first metal. There is no question that fine tungsten lines can be processed, and that their reliability is excellent. The obvious drawback is the impact of the higher resistivity on line resistance, and it will be a circuit-design/performance-driven decision whether that is tolerable. The alternate but more complex and expensive approach is to etchback the tungsten to leave a plug in the contact followed by patterning and etching of a sputter deposited aluminum alloy to create the metal one lines.

It is also quickly apparent from comparing Tables V and VI that the preferred aluminum alloy, be it used for metal one, two, three or ..., is no longer AlSi as it was in the mid-1980's; but there is still much disagreement as to what the best aluminum alloy is. Almost everyone adds copper to the aluminum, although the percent varies greatly; and layered structures involving Ti, TiW, TiN, or W caps, intermediate layers, and sublayers are common. In some cases, titanium or silicon is added directly to the aluminum. For the most part, these choices are reliability driven; Chapter 8 in this book discusses the impact of various alloy additions on the electromigration and stressmigration failure times of the interconnects. Caps and sublayers may have additional benefits as antireflective coatings for improved lithography or as a hardmask or stopping layers for etching, respectively. Chapter 3 gives a detailed discussion of the issues and understanding of aluminum interconnects. Clearly this is an area of continuing research.

In the area of interlevel dielectrics (ILDs) the industry has used and apparently will continue to use silicon dioxide, with a rare paper describing a process based on polyimide. As the spacing between metal lines has shrunk, the step coverage associated with the oxide deposition process has had to improve to avoid gaps or voids which make subsequent processing difficult, hence the major changes have been in how the glass is deposited. Today PETEOS (Plasma Enhanced oxide from a TetraEthyl OrthoSilicate precursor) is commonly used, but it is unlikely that it can be extended beyond current 0.5 μm design rules. Several approaches (ECR oxide deposition, SOG, atmospheric pressure TEOS systems, other precursors, ...) are being vigorously investigated by the industry but no clear winner has yet emerged. Chapter 4 discusses these and other inorganic dielectric issues in detail. As for polymers, there is a growing consensus that their lower dielectric constants will dictate their increasing use in future MLM systems. To date, both real and imaginary process and reliability concerns, have restricted their application to a few products. Chapter 5 deals with organic ILDs and should be required reading for MLM engineers.

Planarization of the ILD has become a key problem driven by lithography requirements for near perfect global planarity on a structure with more layers and finer geometries. Chapter 6 is devoted to the topic of planarization and discusses the multitude of processes that are in use today and their associated problems.

The increasing technical importance of multilevel metallization means that it is critical from a business or competitive standpoint as well. A distinguishing difference between the various companies which is not apparent from Table VI is the method of implementation. Despite the fact that the MLM system materials and structure are similar and may eventually converge with time, the toolset and the efficiency with which it is operated in manufacturing will be a telling difference. This is such an important topic that Chapter 10 of this book has been devoted to it. In addition, the exact choice of design rules (linewidth, spaces, overlaps, thicknesses, etc.) based on a good understanding of the process, and materials capabilities balanced against the needs of the device and circuit technology will critically impact the yield; and therefore, the profitability of the fab line. This is one of the challenges that the MLM engineer faces and is the subject of the remainder of this chapter.

5.0 METHODOLOGY FOR DEVELOPING AN MLM SYSTEM

In this section we will discuss a methodology we have successfully used to determine preliminary design rules and processes for building a multilayer metal system.(14, 15)   The need for rationalization between performance, manufacturability, reliability, and packing density will be discussed. Since the processes required to improve performance and packing density (i.e. decreased feature sizes, more layers of metal, etc.) are usually contrary to increasing the manufacturability and the reliability (i.e. relaxed design rules), the significance of each of these requirements will vary depending upon the circuit technology. If the application is a high performance ASIC, where premium prices can be charged, the cost and manufacturability cannot be ignored, but often a more complex process must be used to achieve the required performance. However for commodity products, the interconnect must not impact performance to such an extent that the circuit is not competitive, but cost must (i.e. manufacturability) be heavily considered since pricing is often the key differentiator. There is no case where reliability can be compromised. The basic decision process can be flow diagrammed as shown in Figure 3. The boxes in the flow diagram are discussed in the following.

5.1 Definition of the Target Technology

The definition of the target technology requires obtaining inputs from a variety of other engineers. Information must be obtained from the device engineers on the device size and structure, layout of contacts, potential topography issues, etc. Information is gathered from the circuit engineers regarding the number of interconnect runs required, the minimum number of metal layers required to wire up the circuit, the cell size, performance requirements, electromigration requirements, etc. Similar information should be obtained from the end user/customer as well as information regarding acceptable costs for the MLM system. Care must be taken not to design a system that is not manufacturable or not reliable (see the discussion in Chapter 10 on design for manufacturability and the discussion in Chapter 8 for predicting reliability) even though it meets or exceeds the performance or packing density goals. (The customer in this case may be the business unit that will manufacture the circuit or a specific customer that is outside the company for certain custom products.) Finally information is needed from the packaging engineers to assure that the MLM system is compatible with the packaging and assembly process. (The packaging issues are discussed in Chapter 12.) Although every effort is made to meet these targets, it may be required that some goal has to be renegotiated if a fundamental roadblock is encountered.

After the overall circuit technology targets are determined, the initial MLM design rules can be established based on five key inputs or requirements: [1] The metal pitches are chosen to meet the cell size and circuit density targets. [2] The metal line widths and thicknesses as well as the interlayer dielectric thicknesses are chosen by computer modeling of the resistance and capacitance of the interconnect system to assure that it meets the performance requirements. [3] Processing capabilities are selected that are currently available or can reasonably be anticipated to be available in the manufacturing area where the circuits will be built. [4] The choice of metal line dimensions, ILD thicknesses and via dimensions are affected by reliability constraints. [5] The choice of available materials that are compatible with the processing and reliability goals also affect the choice of design rules.

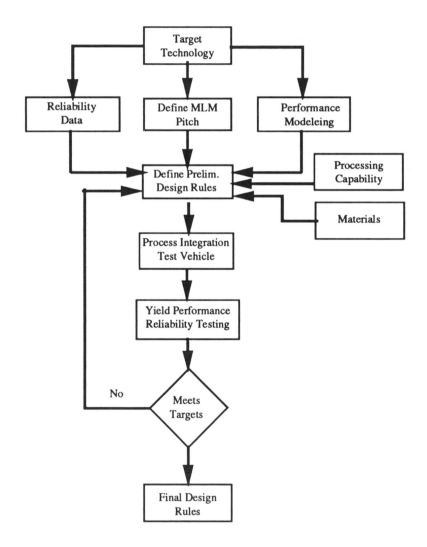

Figure 3. Flow diagram showing the methodology for developing a MLM system.

5.2 Choice of Metal Pitch

The choice of metal pitch is determined by the transistor size, the circuit layout, and the number of metal layers that can be used. As discussed in section 4.0 of this chapter, many circuits today use three or four layers of metal. In chapter 13 it is speculated that future circuits may use as many as six layers of metal interconnect. The following discussion is really relevant to three or four layers, but the reader can see how this will relate to more layers if required. The minimum metal 1 pitch is set by the minimum transistor size and lithography tolerances sufficient to insure that adjacent metal 1 lines at the minimum pitch can completely cover the contacts without shorting to each other. All misalignments, etch biases and other associated process variances must be designed into the overlap tolerances. The metal 2 and possibly metal 3 minimum pitches are set by the cell size and the number of routing tracks required per cell. This is done by simply dividing the cell size by the number of routing tracks required on each metal layer. In addition, if structures such as stacked vias or vias stacked over contacts are used (this is becoming more common with each generation to enhance performance and packing density) it is convenient if not essential that the pitch of each metal layer that is contacted with a stacked via is the same pitch as all other layers. In other words metal 1, 2, and 3 are all the same pitch where stacked vias are used between metal 1 and metal 3. This is required to facilitate the layout, since as was discussed in preceding sections the alternating layers tend to run orthogonal to each other. Without a constant or integer multiple pitch for the various layers, it would be very complicated to lay out a stacked via that would be in the center of a metal 1 and metal 3 line. The final metal layer (metal 3 or 4) must have a sufficiently large pitch such that power can be distributed around the chip without the voltage drop impacting the circuit operation. For this reason the pitch of this layer is usually much larger than the preceding layers.

5.3 Determination of the Metal Line and ILD Dimensions from Performance Modeling

After the pitch of the various metal layers has been determined from the density requirements, the line and space are defined in part by the circuit performance requirements. The performance of the interconnect system is determined by the RC analysis discussed in section 3.0 of this chapter. To minimize the effects of the product of the line resistance (R) and capacitance (C), we have used MOM (16) to model RC for a matrix of metal thicknesses, metal line widths, and ILD thicknesses for each of the metal pitches being considered. This computer code takes into account the intraplane capacitances and fringing effects as well as the interplane capacitances. Another code developed for this purpose is RAPHAEL.(17) If fringing can be ignored, (this is frequently the case for metal features and spaces greater than 1 μm) equation 3 can be modified to include intraplane as well as interplane capacitances as shown in equation 4 and the RC impacts can be modeled to first order fairly well with a desktop computer and a spread sheet. Although this simple approach can be used to gain insight into RC for submicron features, more sophisticated analysis is required to get accurate results. Typical results obtained using MOM are shown in Figure 4 where we have assumed the metal is AlCu with a resistivity of 3.3 $\mu\Omega$-cm and the ILD is SiO_2 with a dielectric constant of 3.9. The model calculates the capacitance/μm on a single line surrounded by grounded lines in the same metal layer and with a grounded metal plate above and below the line being modeled. This is the worst case situation for the capacitance. The model accounts for coupling to the surrounding lines and plates as well as fringing effects. The R is calculated from the metal resistivity and metal dimensions. The figure shows the calculated results for a series of 1000 μm long lines of different widths and thicknesses. The pitch was 3.0 μm, and the ILD thickness above and below the line of interest was 0.75 μm. As the linewidth decreases below 1 μm, the RC rises rapidly due to resistance increasing faster than capacitance is decreasing. As the space becomes <0.5 μm (i.e., linewidth >2.5 μm) the RC rises because the intraplane capacitance and plate capacitances are dominating. Increasing the metal thickness decreases R, and thus, RC; but increasing metal thickness increases the effect of the intraplane capacitance. Increasing the ILD thickness decreases RC. Since the ILD thickness mainly affects the plate capacitance term, this has essentially no effect on the choice of optimum linewidth. However as seen in Figure 5, decreasing the pitch from 3.0 μm to 2.0 μm causes the region of optimum RC to narrow and the minimum RC for a 0.75 μm thick metal line increases from 6.8 X 10^{-12} to 9.0 X 10^{-12} sec. It can be seen in Figure 5, that increasing the metal thickness causes the minimum RC to shift from a linewidth of 1.5 μm for a 0.5 μm thick line to 1.25 μm for a 1.0 μm thick line. This shift is due to the increasing contribution of the intraplane component of capacitance with increasing linewidth. Also, the minimum RC is lower because of the reduced metal resistance.

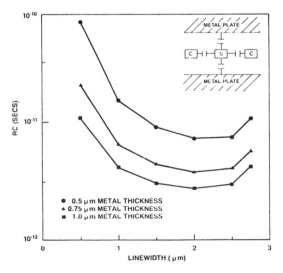

Figure 4. RC versus linewidth for a 3.0 μm metal pitch.

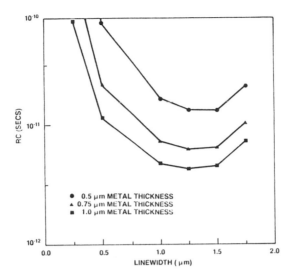

Figure 5. RC versus linewidth for a 2.0 μm metal pitch.

The modeling discussed in the previous paragraph is for an interconnect system where the load and driver device resistance and capacitance are not considered. In a real circuit, the device and load capacitances tend to be small relative to the interconnect capacitance, but the device resistance will be quite large relative to the interconnect resistance until the interconnect length is several millimeters or more. These effects have studied by defining a delay time (t) as the point at which an output signal at a load device reaches 90% of the input signal for an interconnect with a particular driver and load.(18) This calculation is dependent upon the type of device attached to the interconnect, because bipolar devices have somewhat lower device resistances than MOS devices. Using this type of model we have calculated the delay time as a function of length in μm for the different linewidths studied in Figure 4. The results are shown in Figure 6 for a device with a resistance of 100 Ω and a capacitance of a few fF (a high performance bipolar circuit), a metal thickness of 0.75 μm and a metal pitch of 3.0 μm. For interconnect lengths < 1500 μm, the t is dominated by interconnect capacitance and device resistance and the narrowest lines have the smallest delay. The differences in t at these lengths are small (~20%) and the delay time is small relative to the clock speed of today's circuits. For longer distances the interconnect resistance begins to exceed the device resistance for the interconnects studied and therefore the RC of the interconnect begins to dominate. For lengths of 5000 μm or greater the linewidths that produce the lowest t in Figure 6 are the same as the linewidths producing the lowest RC in Figure 4. As mentioned above, for MOS circuits the device resistances will be greater, but the speeds also tend to be slower than bipolar circuits. Therefore this calculation becomes very circuit dependent. For long interconnect distances the designer would need to consider lower resistance drivers (larger devices) and wider pitch lines or multiple lines to carry the signal.

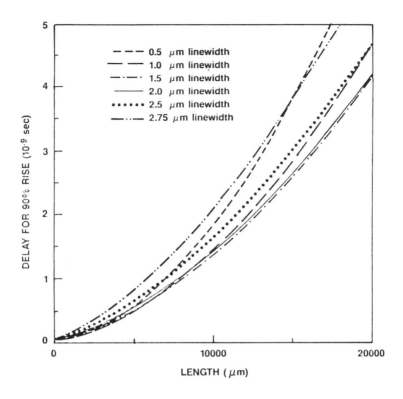

Figure 6. Delay time as a function of interconnect length and linewidth. Delay time is the time required for an output signal to reach 90 percent of the input signal for a 100 Ω device.

5.4 Taking Account of the Effects of Processing and Reliability on Metal Line Dimensions and ILD Thickness

The modeling results in the previous section indicate where the lowest RC is achieved theoretically, however this must be traded against the manufacturability and the reliability of the resultant metal system. It was clearly seen that the lowest RC is achieved with the thickest possible metal lines, the thickest possible ILD and intraplane spaces between 0.5 μm and 1.0 μm depending upon the pitch. In a multilevel metal system, both reflectivity and topography degrade the resolution of the lithography tools to the point that printing lines or spaces substantially less than 0.7 μm is only practical for a manufacturing operation with the most advanced steppers. In addition, the choice of the minimum metal linewidth cannot be less than the width determined by electromigration testing to be necessary to meet the reliability goals of the circuit under normal operating conditions. This linewidth is very dependent upon the particular metal alloy and the deposition condition. When the metal line must connect with a contact or a via the overlap tolerance must be taken into account. This is frequently accomplished by increasing the metal width as it passes over a via or a contact. Increasing the metal width is also used where a via is expected to land on a metal line. This is shown schematically in Figure 1. The maximum metal thickness must be limited to meet the lithography and metal etch requirements (discussed in Chapter 7) as well as ILD coverage (gap fill issues) between metal lines (discussed in Chapters 4 and 5). As a rule of thumb for today's technologies, the metal thickness is limited to approximately the minimum space between metal lines. This is because the most commonly used ILD, PETEOS can fill an aspect ratio (width / depth) of 1 with little or no voiding. The nominal ILD thickness is usually set by the minimum via size and the maximum aspect ratio that the via fill technology can achieve. For standard sputtered Al or Al alloys the useful aspect ratio tends to be between 0.5 and 1.0 depending on the minimum allowed step coverage in the contact or via. As discussed in Chapter 3 many enhanced sputtering techniques, such as heated substrates, collimated sputtering, etc. have been investigated to fill larger aspect ratio vias. In addition, considerable effort has gone into the development of CVD metal systems. As discussed in section 4 of this chapter, CVD W is becoming widely accepted for this purpose. CVD W allows vias with aspect ratios of 1 or greater to be used. Figure 7 shows how these processing and reliability effects must be overlaid on the modeling results presented in Figure 4. The lithography and etch processes limit the regions of minimum line and minimum space that can be used. Electromigration reliability also limits the minimum linewidth. The ILD coverage and via fill also limit the metal and ILD thickness as well as the minimum space.

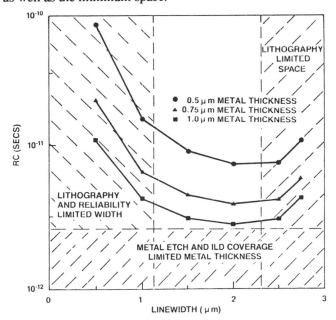

Figure 7. RC modeling results from Figure 4 with limitations from processing and reliability overlaid to define a usable linewidth and space.

5.5 Choice of Materials for Use in the MLM System

The various materials for use in the MLM system must be chosen based on their electrical and physical properties as well as their ease of processing which relates to their manufacturability.

The first material to be considered is the silicide that will be used to make contact to the silicon device structures. The choice of silicide depends on the circuit application. The characteristics of the most commonly used silicides are discussed in detail in Chapter 2. For most MOS applications, $TiSi_2$ is the silicide of choice. This is mainly due to: (a) its low resistivity, (b) the ability of Ti to form a barrier layer, TiN, simultaneously to forming $TiSi_2$, (c) the ability to form a self aligned silicide (salicide) and (d) the ability to withstand some thermal processing associated with the deposition of ILD0. $CoSi_2$ has received considerable interest for submicron applications. This is due in part to the fact that initially in the silicide formation, the metal is the diffusing species in the $CoSi_2$ reaction (as CoSi is forming) whereas Si is the diffusing species in the $TiSi_2$ reaction (see Chapter 2). When the metal is the diffusing species, the possibility of bridging of the salicide between the gate and source/drain region is reduced. For many bipolar and BiCMOS applications PtSi has been used as the silicide due to its ability to form both good ohmic and Schottky contacts to n-type Si depending upon the doping density of the Si. PtSi has a higher resistivity and cannot withstand as high temperature processing as $CoSi_2$ or $TiSi_2$ without Pt diffusion creating lifetime killers and causing excess junction leakage.

The ILD0 material choice is usually dictated by the thermal budget that has been allocated to the circuit fabrication at this point, as well as the topography that must be covered. In addition, it is usually required that ILD0 either getter or be a barrier to mobile ions such as Na. For these reasons BPSG is usually chosen for this layer. In the past Si_3N_4 and PSG have been used. However, nitride has a higher intrinsic stress as well as a greater dielectric constant than most oxides and therefore leads to a larger metal to substrate capacitance. PSG requires a significantly higher flow temperature than BPSG and therefore does not planarize topography as well. If the topography is severe and the thermal budget is limited such that sufficient thermal flow of ILD0 cannot be accomplished, some other form of planarization is required in the formation of ILD0.

The choice of metal barrier material is based on a number of different requirements: [1] Its ability to act as a barrier between the silicide and the contact fill material. [2] Its ability to achieve acceptable step coverage in high aspect ratio contact structures. [3] If CVD W is used the barrier material must act as a seed layer for W deposition. [4] If blanket W is used the barrier layer must also act as an adhesion layer between the W and the dielectric. These requirements and the characteristics of the various barrier material candidates are discussed in Chapters 2 and 3. As was reported in section 4 of this chapter, TiN and TiW are the most frequently used barriers today. In the future a CVD barrier will be required to handle the high aspect ratio contacts.

The material used for both contact and via fill depends upon the maximum aspect ratio. For aspect ratios <1.0, sputtered Al is used. As the aspect ratio approaches 0.5 many companies have adopted approaches such as elevated temperature deposition or bias sputter deposition to enhance the step coverage. For aspect ratios ≥ 1.0 CVD W is used in most companies. Other aproaches to enhance step coverage such as CVD Cu(19, 20) and CVD Al(21, 22) are being investigated. However, these are not currently manufacturable at this time.

The choice of interconnect metal depends upon the resistivity and reliability requirements of the circuit. AlCu alloys are used almost exclusively today. The electromigration reliability increases with increasing % Cu in the alloy. However, the resistivity also increases with increasing Cu content and for Cu content in excess of ~1% Al_2Cu precipitates can become quite large and can cause problems in processing metal lines (particularly during the lithography and etch steps).(23) These issues are discussed in Chapter 7. AlSi and AlCuSi alloys were used in prior generations. However, as contact dimensions have decreased below 1 μm and barrier layers have been required, there has no longer been a need for Si in the Al alloy. Some companies have left Si in the alloy longer than necessary due to the cost of requalifying a new metal alloy. Other metal alloys have been investigated, but are not widely used. In the future Al alloys will probably be replaced by Cu as discussed by Wong in Chapter 13. This is due to Cu's lower resistivity and greater resistance to electromigration. Before this can happen a number of processing problems associated with Cu must be overcome.

The choice of ILD material is influenced by a number of issues including its dielectric constant, its ability to fill high aspect ratio gaps between metal features, the maximum temperature required for deposition and its stress. Oxides are used rather than nitrides or oxynitrides due to the differences in dielectric constant. PETEOS is the most frequently used oxide due to its ability to fill high aspect ratio gaps at relatively low temperatures. In the future organic dielectrics are likely to be used due to their lower dielectric constant. The effect of a lower dielectric constant can be seen in Equations 3 and 4. Organic dielectrics are currently used on a limited basis, but many of the problems associated with organic dielectrics must be overcome for them to reach their full potential. The characteristics and issues associated with various inorganic and organic dielectrics are discussed in Chapters 4 and 5.

The final material choice that must be made is the passivation layer. This layer must provide mechanical protection from scratches as well as a barrier to mobile ions and moisture. Combinations of PSG and Si_3N_4 are frequently the choice for the passivation layer. They tend to be used in combination to balance the stress in the two films.

5.6 Defining the Preliminary Design Rules and Process Targets

After the information discussed in sections 5.1-5.5 has been obtained, it is used to create a preliminary set of design rules and process targets. As shown in the flow chart in Figure 3, these design rules will be used to build a multilevel metallization test vehicle. The test vehicle is then used to evaluate the metal system and the design rules to see if they will meet the targets. Figure 8 shows schematically the pieces of the metal system that need to be included in a preliminary set of design rules. This drawing is typical of the metal 1, poly and contact layers. Similar drawings would apply for other metal and via layers. The areas that will be discussed in the following paragraphs are labeled a-e in the drawing and the letters are indicated at the end of the title for each section. Additional items or rules may be required for special applications and should be added as needed. These rules are repeated for each layer although the size will often change for each layer. In addition to the design rules a set of process targets needs to be generated for each layer (film) that is deposited. Examples of these process parameters are also discussed.

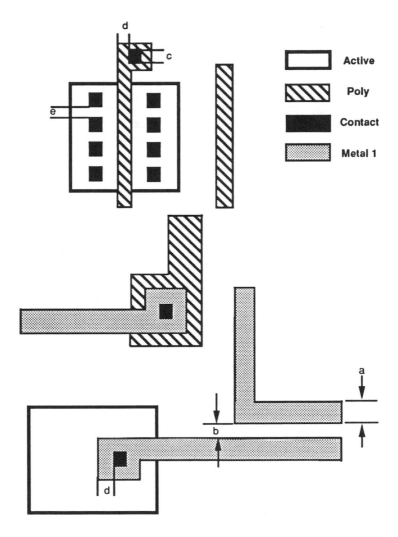

Figure 8. Schematic showing some of the elements of an MLM system that are included in the design rule definition.

Minimum metal linewidth (a): As discussed in the preceding paragraphs the choice of metal linewidth is determined by the ability to process a line of the minimum width, its reliability and its effect on RC. This width either remains the same or increases with increasing metal layers. As discussed in Chapter 9, a number of test structures are derived from the minimum metal linewidth. For example the minimum metal linewidth plus the overlap for a contact or via (discussed below) determine the minimum contacted metal linewidth.

Minimum metal space (b): This is the minimum space allowed between metal features on any given layer. The space is usually determined by the ability to pattern and etch the space as well as its effect on RC. The ability of the ILD process to fill the space should also be considered. As with the metal linewidth, the space remains the same or increases with increasing metal layers. Depending upon the lithography process used, some technologies add fat metal rules. These call for the space to be increased when the space is adjacent to a large metal feature.

Metal pitch: The minimum metal pitch is the sum of the minimum metal linewidth and minimum metal space. In addition to the issues that affect the minimum linewidth and space, the pitch is influenced by the cell size and the number of routing tracks that are required for each layer. The size of the minimum metal pitch may also be influenced by the width of the metal line at points where it intersects either a contact or via. At these points the linewidth may have to be increased to account for the overlap as is discussed later in this section.

Contact or via dimension (c): The contact / via dimension is set by the lithography and etch processes, the via fill technology, the current that each contact / via must carry, and the contact / via resistance. When large currents arc needed it must be decided whether to use one large contact / via or several small ones. Also if a blanket CVD metal plus etchback process is used, there must be a limit placed on the maximum contact / via dimension in at least one dimension. This is to ensure that the growth fronts meet in the center of the contact / via and close the seam in the feature. Otherwise, during the etchback the bottom of the feature will be etched away.

Overlap of contacts / vias (d): The overlap is the amount on each side that the metal exceeds the contact / via dimension if the alignment is perfect. This is to ensure that the contact / via is totally enclosed by the metal feature above and poly, active or metal feature below it. The amount of overlap should account for any variations in the metal and contact / via dimensions as well as the misalignment tolerance of the lithography tool for each layer. When a tapered contact / via is used the overlap may be larger for the top layer than the bottom. If the metal, active area or poly width is greater than the contact / via dimension plus the overlap no increase in the metal width is required. Otherwise the metal line width must be increased where it encounters a contact / via. This increase is referred to as a flag. Since the minimum space does not decrease where this occurs, this will cause the minimum contacted pitch to be greater than the noncontacted pitch. (Obviously contacts / vias that are not flagged will not result in a change in the pitch.) The difference in the two pitches will depend upon whether adjacent contacts / vias are allowed or if only staggered contacts / vias are used as shown in Figure 9. Staggered features result in a tighter contacted pitch at the expense of a less dense contact / via pattern.

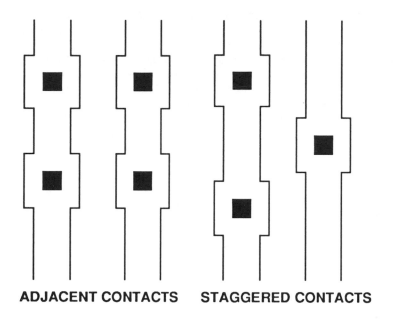

ADJACENT CONTACTS   STAGGERED CONTACTS

Figure 9. Schematic showing staggered and adjacent vias.

Minimum distance between contacts and vias (e): The minimum distance between adjacent contacts or adjacent vias varies depending upon whether they are contacting the same metal line or not. The minimum distance between structures on different metal lines is set by the overlap and metal pitch. When the features are contacting the same metal lines, the minimum distance is based on the spacing that can be resolved by the lithography process and if tapered vias are used, the distance between features must be large enough so that the top of the features will not touch. If the features are in two different layers such as via one and via two, the decision depends upon the misalignment tolerance, the resulting topology and the contact / via fill technology. This rule will not apply where stacked vias or vias stacked on contacts are used. This rule may also be expanded to include the spacing between contacts and underlying features such as gates. This is to prevent the possibility of the contact touching the gate and causing a short due to a misalignment or a tapering of the contact.

Silicide Thickness and Sheet Resistance: The silicide thickness is the first component of the MLM process target table. This thickness will determine the sheet resistance and perhaps the contact resistance. It must not be too thick relative to the junction depth such that it causes excess junction leakage as discussed in Chapter 2. The thickness must also be sufficient to stand up to the contact etch, especially when the contact etch reaches the silicide on the polysilicon gate well before it reaches the source drain region due to differences in contact depth. The silicide thickness required to stand up to this etch will depend upon the etch selectivity, the differences in contact depths and any lag differences in the etch (see Chapter 7 for more details).

Metal Layer Thicknesses: The minimum metal layer thickness is chosen to meet the sheet resistance and current carrying requirements for the particular layer. The maximum thickness is limited by the ability to pattern and etch the lines and the ability of the subsequent dielectric layers to fill the gaps between the metal lines. If a blanket CVD metal plus etchback process is used for a contact or via fill, the thickness of the film must be sufficient to ensure that the metal closes the contact or via opening.

Metal Step Coverage: This is the minimum thickness of metal on the walls or bottom of a contact / via to ensure that the electromigration design rules are met. This is a difficult requirement to measure electrically and is often done on representative test structures using a cross section SEM micrograph as explained in Chapter 11.

Dielectric Thicknesses: The thickness for each layer is set by the need to planarize the underlying topography, fill gaps between metal features with minimum formation of voids, and provide a low enough capacitance to meet the RC requirements of the circuit. The maximum thickness is set by the ability to pattern the films and the step coverage requirements of the contact / via fill.

5.7 Process Integration Test Vehicle

Using the design rules discussed in the preceding section a process integration test vehicle is created. This vehicle will be used to evaluate the various proposed processes and materials, the performance, robustness, yield and reliability of the selected MLM technology, and finally by the various process disciplines to develop or optimize new or alternative processes. The test vehicle may also be used to provide samples for the physical / chemical analysts who help evaluate the process technology as discussed in Chapter 11. It should include as a minimum the following test structures:

1. Snake and comb structures to evaluate the ability to pattern metal features.
2. Contact chains to evaluate the silicide, contact etch and contact fill processes. Chains should be included for contacts to all possible cases, i.e. single crystal silicon, polysilicon on single crystal silicon or gate oxide, and polysilicon on field oxide.
3. Via chains to evaluate the vertical interconnects between each of the various layers. If stacked vias are allowed they should be included also.
4. Plate capacitors for measuring the ILD thicknesses, to evaluate interlayer shorts and to measure the dielectric constants of the ILD.
5. Process control test structures for determining sheet resistance, electrical linewidths, alignments of one layer to the next, and contact / via resistance structures. These structures are usually of the form of linear bridges, van der Pauw structures or Kelvin structures.
6. Reliability test structures for evaluating the electromigration resistance as well as stress migration resistance.
7. Structures for physical and chemical evaluation. This will include large areas for depth profiling using techniques discussed in Chapter

11.   Arrays of contacts and vias for each layer should also be included for ease of evaluation with cross section SEM. This portion of the test vehicle should also include structures with various topographies to evaluate fill and planarization techniques.

8. If RC delays are critical, structures for evaluating line capacitances should also be included.

9. Other special structures as are allowed by the design rules.

Several MLM test vehicles have been described briefly in the literature.(24, 25) Chapter 9 of this book discusses the design and testing of these structures in great detail. That chapter also discuss the tradeoffs and choices that should be made as to when one electrical test structure should be used over another. The tools and techniques used for the physical / chemical evaluation of the test structures are presented in Chapter 11.

5.8 Final Design Rules

After the various processes have been optimized and the test vehicle has been exercised a sufficient number of times, it must be decided if the yield, performance and reliability of the MLM system meet the required targets for the technology. If the system meets the targets, the design rules are made final and the transfer to production can begin. If the system fails to meet the targets, the engineer must go back to the step of defining preliminary design rules. At this point it should be determined why the preliminary rules failed and see if they can be met by a change in process or materials or if the requirements are too stringent for currently available options. If the targets are too difficult, then the targets need to be renegotiated. Clearly the methodology can be repeated until the technology meets the targets. A key goal though should be to minimize the number of iterations through this flow due to the cost and timeliness of getting the product to market on time.

6.0 SUMMARY

In this chapter we have defined the terminology, in section 1, for an MLM system that will be used throughout this book. We have discussed what are the uses of a multiple layer interconnect system in section 2.0 and why it is important in section 3.0. In section 4.0 we present the trends for what was used in the past and what is being used currently to fabricate an MLM system based on papers in the literature. In section 5.0 we discuss the process for gathering information and developing an MLM system. The system is then processed and evaluated using a test vehicle. If the integrated processes meet the targets of the technology the design rules are made final and the transfer to manufacturing begins.

The following chapters provide the basic information and necessary references for the MLM process integration engineer to develop a MLM system.

These chapters may also serve as a good starting point or reference point for the process engineer working in each of these areas. In Chapter 2 the subject of contacts, silicides and barrier materials is covered in detail. Chapter 3 provides a discussion of the various metals that maybe used for interconnects in an MLM system. Chapter 4 covers all of the inorganic dielectrics that are used for ILD0 to separate the metal and the device structures as well as the intermetal dielectrics ILD1, 2,... and the final passivation. Chapter 5 discusses organic dielectrics. Chapter 6 covers the topic of planarization, why it is important and explains the various approaches to achieve planarization. In Chapter 7 the two key unit process steps of lithography and etch are presented as well as special issues that apply to the MLM system. The areas of electromigration and stress migration are discussed in Chapter 8 which covers the issues associated with the reliability of an MLM system. Chapter 9 talks about the design and layout of a MLM test vehicle. Various structures, how they are tested and the information they provide is covered. Chapter 10 presents a discussion of manufacturing requirements associated with an MLM system. The concepts of statistical process control (SPC) and Design for Manufacturability (DFM) are presented as well as a listing of currently available equipment. Chapter 11 gives a comprehensive discussion of the analytical tools and techniques for developing and evaluating the processes and materials in an MLM system. Chapter 12 covers the topic of packaging and how it interfaces with the MLM system. Finally Chapter 13 takes a look at the future and what will be required of MLM systems as devices continue to shrink and circuits increase in complexity.

REFERENCES

1.   J. Kirchgessner, J. Teplik, V. Ilderem, D. Morgan, R. Parmar, S.R. Wilson, J. Freeman, C. Tracy, and S. Cosentino, "An Advanced 0.4 μm BiCMOS Technology for High Performance ASIC Applications," IEDM Technical Digest (1991), p. 97.

2.   A. J. Saxena, "The Need for Multilevel Interconnection and Developments to Date," VLSI Multilevel Interconnection Tutorial (June 14, 1987), p. 3.

3.   D. Barton and C. Maze, "A Two-level Metal CMOS Process for VLSI Circuits," Proc. IEEE VLSI Multilevel Interconnection Conf. (1984), p. 268.

4.   J.M.F. van Dijk and R.A.M. Wolters, "A Two-level Metallization System with Oversized Vias and a TiW Etch Barrier," Proc. IEEE VLSI Multilevel Interconnection Conf. (1985), p. 123.

5.   A.L. Wu, "The Manufacturability of a Dual Layer Metal MOS Process for VLSI Circuits," Proc. IEEE VLSI Multilevel Interconnection Conf. (1985), p. 145.

6.   H. Eggers, H. Fritzsche, and A. Glasl, "A Polyimide-isolated Three-layer Metallization System for Bipolar Gate Arrays," Proc. IEEE VLSI Multilevel Interconnection Conf. (1985), p. 163.

7.   T.D. Bonifield, R.J. Gale, B.W.Shen, G.C. Smith, and C.H. Huffman, "A One Micron Design Rule Double Level Metallization Process," Proc. IEEE VLSI Multilevel Interconnection Conf. (1986), p. 71.

8.   H. Kotani, Y. Takata, Y. Hayashide, A. Ohsaki, M. Iwasaki, T. Twutsumi, M. Matsuura, A. Ishii, Y. Maekawa, I. Tottori, K. Mori, Y. Ii, Y. Yamaguchi, and T. Katayama, "An Advanced Multilevel Interconnection Technology for 0.4-μm High Performance Devices," Proc. IEEE VLSI Multilevel Interconnection Conf. (1992), p. 15.

9.   S. Roehl, L. Camilletti, W. Cote, D. Cote, E. Eckstein, K.H. Froehner, P.I. Lee, D. Restaino, G. Roeska, V. Vynorius, S. Wolff, and B. Vollmer, "High Density Damascene Wiring and Borderless Contacts for 64 M DRAM," Proc. IEEE VLSI Multilevel Interconnection Conf. (1992), p. 22.

10.   P.S. Ying, R.A. Chapman, P.J. Wright, D.A. Prinslow, W.F. Richardson, A.R. Peterson, S.W. Huang, C.M. Garza, M.M. Moslehi, and D.W. Reed, "Scalable 0.4 μm DLM for Fast-flow Single Wafer Processing," Proc. IEEE VLSI Multilevel Interconnection Conf. (1992), p. 51.

11.   S. Luce and S. Pennington, "Interconnect Technology for 16-Mbit DRAM and 0.5-μm CMOS Logic," Proc. IEEE VLSI Multilevel Interconnection Conf. (1992), p. 55.

12.   M. R. Schneider, " A Multilayer Interconnect Process for VLSI GaAs ICs Employing Polyimide Interlayer Dielectrics," Proc. IEEE VLSI Multilevel Interconnection Conf. (1992), p. 93.

13.   T. Ohba, "Multilevel Metallization Trends in Japan," Proc. Advanced Metallization for ULSI Applications Conf. (1991), P. 25.

14.   S. R. Wilson, J. L. Freeman and C. J. Tracy, "A High Performance, Four Metal Layer Interconnect System for Bipolar and BiCMOS Circuits," Proc. IEEE VLSI Multilevel Interconnection Conf. (1990), p. 42

15.   S. R. Wilson, C. J. Tracy, and J. L. Freeman, "A Four-Metal Layer, High Performance Interconnect System for Bipolar and BiCMOS Circuits," Solid State Technology 36 (11) 67 (1991).

16.   M. Scheinfein, et al., computer code MOM, University of Arizona, Tucson, AZ, 1986.

17.   RAPHAEL, Technology Modeling Associates, Inc., Palo Alto, CA.

18.   D. S. Gardner, J. D. Meindel and K. C. Saraswat, "Interconnection and Electromigration Scaling Theory," IEEE Trans. Electron. Dev. ED34, 633 (1987).

19.   J. Norman, B. Muratore, P. Dyer and D. Roberts, "New OMCVD Precursors for Selective Copper Metallization," Proc. IEEE VLSI Multilevel Interconnection Conf. (1992) p. 123.

20   Y. Arita, N. Awaya, T. Amazawa, and T. Matsuda, "Deep Submicron Cu Planar Interconnection Technology Using Cu Selective Chemical Vapor Deposition," IEDM Technical Digest, (1989) p. 893.

21.   R. A. Levy, and M. L. Green, "Low Pressure Chemical Vapor Deposition of Tungsten and Aluminum for VLSI Applications," J. Electrochem. Soc. 134, 37C (1987).

22.   H. W. Piekaar, L. F. T. Kwakman, , and E. H. A. Granneman, "LPCVD of Aluminum in a Batch-Type Load-Locked Multi-Chamber Processing

System," Proc. IEEE VLSI Multilevel Interconnection Conf. (1989) p. 122.

23.   D. Weston, S. R. Wilson and M. Kottke, "Microcorrosion of Al-Cu and Al-Cu-Si Alloys: Interaction of the Metallization with Subsequent Aqueous Photolithographic Processing," J. Vac. Sci. Technol. A8, 2025 (1990).

24.   Thomas E. Wade, "Proposed Comprehensive Test Vehicle for Monitoring Multilevel Interconnection Process Variabilities, Misalignment, Parametrics and Defect Density," Proc. IEEE VLSI Multilevel Interconnection Conf. (1986), p. 354.

25.   D. J. Radack, J. C. Swartz, L. W. Linholm, and M. W. Cresswell, "A Comprehensive Test Chip for the Characterization of Multi-Level Interconnect Processes," Proc. IEEE VLSI Multilevel Interconnection Conf. (1987), p. 238.

2

SILICIDES AND CONTACTS FOR ULSI

GEORGE E. GEORGIOU
AT&T Bell Laboratories
Murray Hill, New Jersey

1.0   INTRODUCTION

Two main driving forces for continuing to decrease device dimensions into the submicron ULSI regime are economics and performance. The metallization technology required for improving performance is the subject of this chapter. Here, metallization includes contact technology and involves concepts of silicides, barrier materials, and local interconnects. In addition, multilevel metallization is needed to effectively connect the device to the external world without significantly increasing the interconnect resistance and capacitance and degrading the device performance or speed. Below we will discuss the materials and process integration issues with silicides and metallization. Of course, there are circuit design and layout tradeoffs which emphasize one technology and de-emphasize another. For example, additional levels of metal can remove the need for silicides. Such alternate design considerations are beyond the scope of this chapter.

1.1   Metallization and Device Performance

The speed of a device is directly related to the resistance and capacitance in the equivalent circuit. Capacitance is due to the metallization and to the device processes. In general, capacitance is a complicated topic which involves the interaction between device design, metal layout and dielectric processing. The device component is related to the specific device implantation and diffusion front end processes. The component due to metallization is related to the separation between the metal levels and the metal lines.   The intralevel capacitance is set by the submicron design rules for the metal pitch of a given ULSI technology. In principle, the interlevel capacitance is minimized by using thicker interlevel dielectrics and by optimizing the circuit layout. Interlevel dielectrics are discussed in detail, in Chapters 4 and 5. However, it should be noted that real limitations exist in the deposition and patterning of thicker dielectric films. Additionally, as discussed below, metal coverage into deeper contacts and vias of a thicker interlevel dielectric significantly complicates the metallization process.

The resistance components are illustrated for the MOSFET cross-section in Figure 1. (1) Figure 1a shows the cross-section with a contact to Si source and poly-Si gate. Figure 1b shows the case of a self-aligned "salicide" process where a metal (commonly Ti or Co) silicide is sintered on the source/drain and gate. A variation (not shown) is a "polycide" process where reactive sputter etching is used to define a multilayer gate of doped poly-Si under a metal (commonly W, Mo or Ta) silicide. Here, the metallization is responsible for the contact resistance (R_{CO}), the series resistance of the metal runner to the source (R_{AL}) and the series resistance of the gate (R_G) which is in series with the gate oxide capacitance C_{OX} (=$\varepsilon_{ox}A_G/t_{ox}$ where ε_{ox} and t_{ox} are the gate oxide permittivity and thickness and A_G is the gate area) to the device current channel.

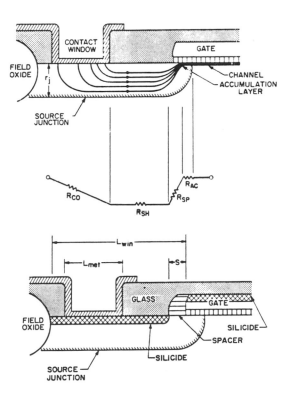

Figure 1. The series resistance components of a MOSFET with salicide (1a) and without salicide (silicided source/drain and gate) (1b). After Ref. 1, reprinted with permission of the publisher.

The device process and geometry are responsible for the source sheet resistance (R_{SH}) and the spreading and accumulation resistances (R_{SP} and R_{AC}) where the source current crowds into the narrow channel under the gate. Similar series resistance components exist at the drain connecting to the external power supply. For all practical purposes, the metal resistance R_{AL} is small compared to all the other resistance values.

The gate runner connecting the gate to the external power supply is usually patterned over a field oxide of several-thousand angstroms. The runner is usually long and has a significant capacitance C_{BL} (of order $\varepsilon_{ox}A_{BL}/t_{FOX}$ where ε_{ox} and t_{FOX} are the field oxide permittivity and thickness, and A_{BL} is the gate runner area) to the ground plane. In most cases, the device capacitance is small compared to C_{BL} and, to first order, the device switching speed is approximately $R \times C_{BL}$ where R is the equivalent device resistance.

Figure 2 (2) shows a typical calculated scaling (the exact scaling is process specific) of R_{CO} and $R_{SP} + R_{AC}$ with the effective gate length L_C. Two curves are shown for R_{CO} of 10^{-6} and 10^{-7} Ω-cm^2. Most metal to heavily doped Si contacts are in this range. (3) The two curves for $R_{SP} + R_{AC}$ correspond the unsilicided (K const) and silicided (K scaled) source/drain (S/D) cases. Also shown is R_{SH} for the unsilicided case. (As we will see later, the silicide R_{SH} is significantly lower.) For L_C below .5 μm, the device resistance is significantly lower with silicided S/D. Also, R_{CO} becomes an important component. The gate runner resistance is reduced in order to decrease the $R_G \times C_{BL}$ time constant. Therefore, high performance devices will have silicided gate and for dimensions below ~0.5 μm, silicided S/D. In addition, for CMOS circuits, the contact metallurgy must have a low contact resistance (R_{CO}) to both n$^+$ and p$^+$ Si.

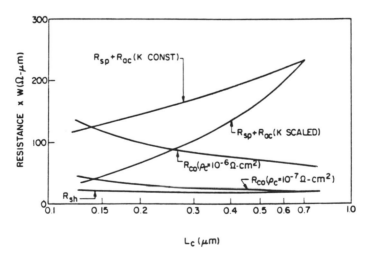

Figure 2. Scaling of each MOSFET resistance component without (K const) and with salicide (K scaled). After Ref. 2, reprinted with permission of the publisher.

1.2   Al-Si Reaction, Device Scaling and Barrier Materials

There is a direct correlation between short channel effects and junction depth x_j measured from the gate oxide/Si interface, for "surface" channel devices (n-poly gate on n-channel and p-poly gate on p-channel) and a less direct correlation for "buried" channel devices (n-poly on p-channel). (1, 4) Thus, decreasing the gate length requires shallower junctions to avoid short channel effects. To preserve the integrity of these junctions and the integrity of the contact resistance, the contact metallization must not interact with Si during any of the subsequent heat cycles. These anneal cycles of ~400°C-500°C, for ~60 min., are needed to remove any gate oxide damage due to deposition and reactive sputter etching, to grow Al metal grains to improve electromigration, to process the interlevel dielectrics needed for multilevel metallization and finally, to package the integrated circuit.

The pure aluminum metallization on silicon is unstable when the contact is annealed at 450°C. Si has a high diffusivity (~10^{-8} cm^2/sec primarily along the Al grain boundaries) and a high solubility (~0.5 wt.% (5)) in Al. Therefore, the contact failure mechanism is Si dissolving in the Al and Al spiking into Si. (6,7,8) The depth of the Al spikes is related to the cleanliness of the Al/Si interface. (7) A typical scanning electron (SEM) micrograph after the Al is chemically removed, of Al spiking in Si, is shown in Figure 3a. (8) A solution to the spiking problem is to satisfy the Si solubility in Al at the maximum post-metal anneal temperature, by sputter-depositing an Al~1%Si alloy metallization. However, upon cooling, the Al-Si alloy becomes supersaturated and the excess Si epitaxially precipitates at the Al/Si interface. A typical SEM micrograph after the Al-Si is chemically removed, of these Si precipitates, is shown in Figure 3b. (8) These epitaxial Si mesas are doped p-type with ~10^{18} cm^{-3} Al. As the contact diameter decreases, these Si precipitates can completely fill the contact and give very high contact resistance. Therefore, Al-Si-Cu (Cu is added to improve the Al electromigration lifetime) cannot be used to contact Si in ULSI technologies.

Aluminum alloy contacts to silicide (on Si) are also unstable for relatively low temperature anneal cycles. However, the contact failure mechanisms are neither spiking nor Si precipitates. Figure 4 (9) shows a Rutherford backscattering spectrum (RBS) of Al/CoSi$_2$/Si before and after annealing at 400-420°C, 30-60 min. Clearly seen is the reaction of Al with Co and the subsequent penetration of Al to Si. Al reacts with Ti in TiSi$_2$ at ~450°C. In addition, Al and Si interdiffuse primarily through the TiSi$_2$ grain boundaries, at 400°C. (10) Contact failure is also observed by Al and Si interdiffusion through other (e.g. Pd, Pt) silicides. (11) Thus, since silicides are generally not barriers to the Al-Si interaction, Al atoms quickly reach the silicide/Si interface after annealing at >400°C. In this case, the important depth for the Al-Si metallurgical reaction is the depth of the junction x_j' measured from the silicide/Si interface and not the deeper junction depth x_j measured from the metallization/silicide interface.

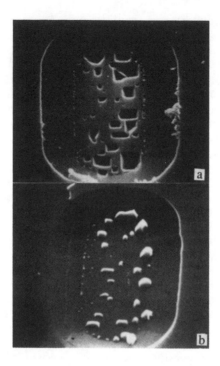

Figure 3. Scanning electron micrographs of at top, Al on Si showing spiking and at bottom, Al-Si alloy on Si showing Si precipitates. Al removed after annealing at 400°C, 60 min. After Ref. 8, reprinted with permission of the publisher.

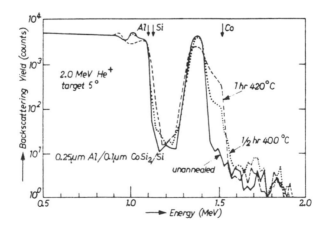

Figure 4. 2 MeV He+ RBS spectrum of Al/CoSi$_2$/Si annealed at 400°C, 30 min or 420°C, 60 min. After Ref. 9, reprinted with permission of the publisher.

Gross Al alloy interaction with Si in general will degrade the device electrical properties. (Reverse bias diode leakage increases because impurities diffuse into the junction depletion region. Contact resistance may increase because the metal/Si interaction may affect the barrier height ϕ_B.) Low level Al/Si interdiffusion through silicide, may be tolerated for deep junctions but cannot be tolerated for the <2000 Å total depth junctions common in submicron device designs. Therefore, other contact metallurgies using barrier materials are required for ULSI.

Barrier materials (12) retard the reaction between the metallization and the device. The barrier X inserted between conductor A and substrate B allows little or no transport of A and B across it. In reality, there will always be a temperature at which X fails; i.e., X reacts with A and B. However, before total barrier failure allowing A and B to interdiffuse, the device electrical properties may not degrade during a small interaction of A and X. For metallization of shallow junctions, X should not interact with substrate B. Again, it is important to note that atomic diffusion at low temperature is primarily grain boundary diffusion; and is thereby, affected by barrier material microstructure. Therefore, the exact failure temperature of a barrier material is sensitive to the particular deposition processes and equipment, both of which affect the barrier film microstructure. The common sputter deposited barrier materials between Al and Si or silicide are Ti, Ti-W composites, and TiN. (Ti beneath TiN is the preferred barrier structure which combines the lower contact resistivity of Ti with the higher stability of the TiN barrier.) A thin TiN (or other refractory metal) layer over the Al alloy is sometimes used to further improve the electromigration resistance of the Al alloy. Finally, note that Si is no longer needed in the Al alloy to prevent Al spiking when the Al alloy is sputtered on a barrier layer.

1.3   Sputtering Step Coverage and Need for CVD Metallization

Current metal deposition is by high rate magnetron sputtering achieved by having the substrate close to the target. In this case, the large accepted-incident angle (Figure 5 (13) top) on the substrate surface compared to a much smaller accepted-incident angle (Figure 5 center) in the contact results in a buildup of material at the feature corners. This corner buildup further shadows the physical vapor deposition. Thus, the high aspect ratio contacts in ULSI circuits have very poor sputtering step coverage. An example of the coverage of 1 μm thick Al deposited by magnetron sputtering into a contact with aspect ratio (height/diameter) ~2, is shown in Figure 6. Note that the self-shadowing at the contact corners reduces the Al thickness to <1000 Å at the bottom of the contact. A thinner Al metal film would have less buildup of material at the contact corner and better coverage.

Increasing the target to substrate distance would decrease the difference in the accepted angle between the substrate surface and contact bottom. This case (Figure 5 bottom) of "collimated" sputtering can also be obtained by introducing a collimator element between the target and substrate to reduce the accepted angle in a high rate sputtering system. Collimation by either technique, increases the film thickness at the bottom of the feature but obviously results in worse sidewall coverage. Therefore, the collimated deposition must be followed by another deposition process to thicken the film thickness on the sidewall and provide continuity. Thus, the improved coverage multistep sputtering process has a lower effective deposition rate and a lower throughput for the metallization system.

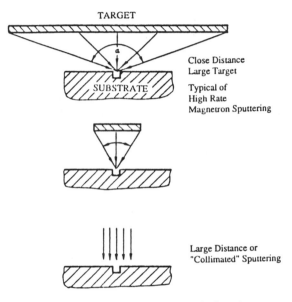

Figure 5. Sputtering geometry showing the relation between target-aspect ratio (target-to-substrate distance divided by target diameter) and the angle a of the atomic flux accepted by a feature patterned in the substrate. Originally presented at the Fall Meeting of the Electrochemical Society, Inc. After Ref. 13, reprinted with permission of the publisher.

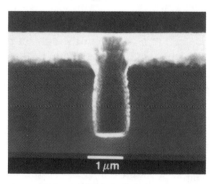

Figure 6. Al-Cu deposited 1 µm thick, by magnetron sputtering at 300°C, into a contact with aspect ratio (height/width) = 2.

At a given target to substrate distance increasing the substrate temperature during the deposition can improve step coverage by increasing the mobility of the sputtered atoms on the surface. This is shown schematically in Figure 7. (14) However, a significant improvement in step coverage is not observed when the substrate temperature is less than half the melting temperature of the metal film. Further, the increase in mobility depends on the substrate surface composition and cleanliness which in turn is affected by the vacuum base pressure, and the vacuum oxygen and water vapor partial pressures.

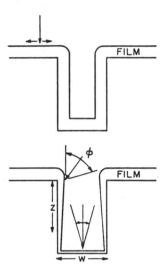

Figure 7. Schematic diagram showing (a) good step coverage resulting from surface migration and (b) poor step coverage due to line-of-sight deposition. After Ref. 14, reprinted with permission of the publisher.

Increasing the substrate temperature to increase the surface mobility improves the sputtering step coverage of materials such as Al. (15, 16) Increasing the substrate temperature within an acceptable range (<500°C) does not improve the step coverage of sputtered refractory metal barrier materials. However, ~1000 Å thin barrier layers have better coverage than ~0.5-1 μm thick Al films. Figure 8 (17) shows an SEM micrograph of the contact coverage of 1500 Å thin magnetron sputtered Ti(10 wt%)-W. The thickness at the bottom of the contact is ~500 Å or ~30%. Again, collimation would increase the thickness at the bottom of the contact.

An attractive alternative to improving step coverage is provided by chemical vapor deposition (CVD). Here, surface mobility; and thereby, coverage into contacts is additionally controlled by the deposition chemistry. Tungsten deposited by CVD is common in development applications. CVD TiN is now being developed. Plasma-enhanced CVD or PECVD is being studied for improving the film purity and step coverage with a higher deposition rate and at a lower deposition temperature.

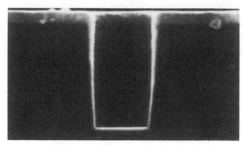

WINDOW 0.85µm diameter x 1.3µm deep

1µm

Figure 8. Ti-W sputtered 1500 Å thin into a contact with aspect ratio ~2. After Ref. 17, reprinted with permission of the publisher.

1.4 Planarization and Metallization

Multilevel metallization and metal reliability further complicate the metallization requirements. Topography becomes more severe as an IC technology uses more layers. Contact (to S/D or gate) or via (between the various metal levels) reflow or tapering will improve metallization coverage. However, reflow has limited application at the contact level and no application at the via levels because the temperature required to reflow or round the contact corners in the deposited dielectric can be high (~800°C). As we will see later, this adversely affects dopant diffusion; and silicide stability. Tapering or increasing the diameter at the top of the contact or via during the etching cannot be used with submicron design rules because of geometry considerations. Reducing topography improves dielectric coverage but does not "planarize" in the usual interpretation. Therefore, the topography must be reduced and each interlevel dielectric surface must be "planarized" before proceeding to the deposition of the next level of metal. See Chapter 6 for discussion of planarization techniques.

Planarization adds to the complexity of the metallization. The interlevel dielectric must be thicker than a reflowed dielectric to planarize the underlying topography and to account for the deposited dielectric thickness nonuniformity and for the RSE nonuniformity. This increases the contact aspect ratio (height/diameter) and worsens sputtering step coverage. Additionally, contacts have different depths to the various device features and metal levels. CVD metal plugs reduce these problems.

The first level contacting the device is shown in Figure 9 (18). Here, a selective plug process is used to partially plug or overfill the various contacts. Selective CVD W is widely considered for this application. A selective process adds only one more step to the metallization process. However, in the case of W, the selective process has a variety of difficulties. (19) A metal process using blanket CVD W is shown in Figure 10. Here, CVD W is deposited everywhere on a thin barrier layer. Since W has poor adhesion to SiO_2, the barrier layer is also an adhesion layer. The metallization (e.g. Ti/TiN/Al/TiN -- here Ti, being listed first by convention is the layer closest to the substrate) is then sputter deposited on the blanket W or on the W plugs (formed when the blanket W is blanket etched back by RSE).

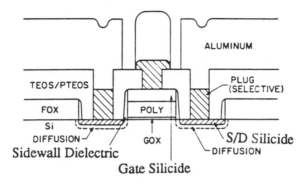

Figure 9. Selective contact plug process after metallization with Al. After Ref. 18, reprinted with permission of the publisher.

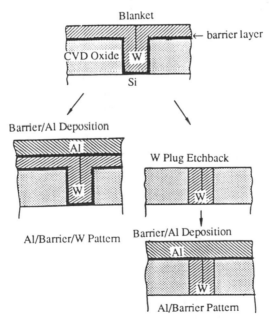

Figure 10. Blanket CVD W metallization showing at left, Al/barrier layer deposited and patterned on W; and, at right Al/barrier layer deposited and patterned on W plugs formed by etchback.

1.5   Chapter Outline

The current understanding of silicides, and Al alloy metallization on thin barriers (deposited by sputtering or CVD) and thick plugs (formed selectively or by RSE etchback) is described below. The silicides include $TiSi_2$ or $CoSi_2$ for the salicide approach and WSi_x, $MoSi_x$ or $TaSi_x$ for the polycide approach. Thin barrier materials include sputtered Ti, Ti-W, TiN, or Ti/TiN and CVD W, or TiN. (Note that the cleanliness of the interface between the barrier and Si is important for low contact resistance. In many cases, the interface is improved after the contacting layer is sintered before sputtering the Al alloy.) The thick plugs are selective or blanket CVD W. Subsequent sections will discuss the ideas of and materials used for local interconnects and multilevel metallization. The final section will summarize the large variety of materials discussed in the previous sections. The following discussions will concentrate on aspects of silicides and metallization intimately related to ULSI process integration. The materials science aspects are discussed within this context.

The design of semiconductor processing equipment is beyond the scope of this chapter. However, it should be noted that ULSI processing requires a lower particle defect density and better control of interlayer contamination. Both improvements are obtained from the minimum handling of the wafers in a clean room environment. Therefore, the trend for ULSI processing equipment is toward single-wafer cold-wall deposition and RSE modules which are interconnected with a central transfer chamber containing a robot arm. The resulting cluster tool performs a certain combination of process steps which form silicide or deposit the metal film structure without breaking vacuum and contaminating the surface or interface. For example, to obtain the right side of Figure 10, a cassette of wafers with contacts etched and chemically precleaned is loaded into the central transfer chamber. After the central chamber is evacuated, the robot arm transfers the wafers one at a time into separate chambers to backsputter clean, to deposit the barrier / adhesion layer, to deposit CVD W, to RSE etchback CVD W, to deposit the barrier layer, and to finally sputter Al. After all the wafers are processed, the cassette is removed and taken to a track which applies photoresist before lithography and metal patterning.

2.0 SILICIDES FOR APPLICATION TO SALICIDE AND POLYCIDE

The basic applications for silicide in advanced MOS processes are for salicide, polycide and contact. For the salicide process, metal is sputter deposited and is thermally reacted or sintered only on the Si areas by limiting the sintering temperature below that needed for the metal to react on oxide. The unreacted metal is then wet etched leaving behind the self-aligned silicide on S/D and gate. The polycide process is conceptually a simple extension of the usual poly-Si only gate process. Here, a silicide is sputter deposited from usually a composite target, on the gate poly-Si. The silicide/poly-Si gate stack is then exposed and patterned with reactive sputter etching. For the contact, a thin metal layer is deposited after contacts are patterned in a deposited oxide. The metal is then heated to a temperature greater than that required for silicidation. Reaction of metal with Si or silicide in the contacts reduces the negative effect of metal-Si interface residues deposited at the bottom of the contacts patterned with reactive sputter etching.

A wide variety of near-noble and refractory metals react with Si to form silicides. Table I lists selected electrical and kinetics properties of the metal silicides (MSi_x) most frequently used for salicide, polycide, and contact processes.

Low series resistance partly results from using materials with a low resistivity. $TiSi_2$ and $CoSi_2$ have the lowest resistivity. The Ti silicide sintering kinetics involve Si diffusing into Ti. The Co silicide sintering kinetics involve primarily Co diffusing into Si during the Co_2Si phase which immediately precedes the CoSi phase and Si diffusing into CoSi during the $CoSi_2$ reaction. However, the kinetics of both Ti and Co silicide can be engineered to give a selective salicide process.

Silicide deposition ease is an additional parameter for polycide. Here, the availability of a composite target is an important consideration. WSi_x, $MoSi_x$ and $TaSi_x$ are available. However, the target composition must be controlled to give a thermally stable, near-stoichiometric MSi_x film on poly-Si. (As will be discussed later, the silicide is slightly Si rich with x~2.5. This excess Si is not alloyed with the metal and is free to oxidize during the subsequent device processing, without affecting the mechanical stability of the polycide gate and interconnect. The composition of a film deposited from a composite target varies across the wafer and the mean across the wafer value of the Si/M ratio must be chosen high enough to provide this oxidation margin everywhere on the wafer.)

The silicide to Si barrier height is a fundamental property which, for a clean contact interface, determines the (Ohmic or Schottky) contact resistivity. (3) Therefore, the silicide/Si barrier height is important for selecting the contact and S/D silicide.

Figure 11 (3) shows the specific contact resistance as a function of Si n-type doping concentration, N_D, and for various barrier heights, ϕ_B. R_C decreases rapidly for high $N_D > 10^{19}$ cm^{-3} and decreasing ϕ_B. Similarly, R_C to p-type Si decreases rapidly for higher doping $N_A > 10^{19}$ cm^{-3} and for increasing ϕ_B. Therefore, low contact resistance between silicide and n-Si or p-Si is obtained from mid-band gap $TiSi_2$ or $CoSi_2$. These silicides are the choice for ohmic contacts to CMOS circuits. On the other hand, PtSi has a higher contact resistance to n-Si and makes a good Schottky contact to n-Si. The PtSi Schottky contact is used for the metallization of bipolar or BICMOS circuits. Below, each silicide in Table I is discussed in conjunction with its common application.

Table I: Selected electrical and kinetics properties of selected metal silicides. (20)

Silicide	Resistivity ($\mu\Omega$-cm)	Moving Species	Reaction Temp. ('C,undoped Si)	Barrier Height (ϕ_B eV)
CoSi		Co	400-450	0.65
CoSi$_2$	18-25	Si	>550	0.65
MoSi$_2$	80-250	Si	>600	0.55
Pd$_2$Si	30-35	Pd	>400	0.75
PtSi	28-35	Pt	600-800	0.87
TaSi$_2$	30-45	Si	>600	0.59
TiSi$_2$ (C-47)	60-80	Si	600-700	
TiSi$_2$ (C-54)	14-18	Si	>700	0.6
WSi$_2$	30-70	Si	>600	0.65

2.1   Salicide — TiSi$_2$ and CoSi$_2$

The salicide process involves metal deposited on Si and thermally reacted to selectively form silicide. This thermal reaction is sensitive to the cleanliness of the metal/Si interface. The reaction is also a function of the dopant type and (high) substrate doping concentration. (20)   The salicide process flow is schematically shown in Figure 12. Figure 12 shows the source and gate and drain cross-section after the poly-Si gate is patterned, a sidewall oxide spacer is formed, and the S/D are implanted and activated. (Note that a lightly doped drain or LDD extension is commonly used to improve device reliability.) After a pre-clean, Ti or Co is sputter deposited.   The metal is reacted with Si at a temperature below that which is needed for the metal to react with oxide and, additionally for Ti, below the temperature/time needed for the Si to move enough to form the Ti-Si compound on the sidewall spacer and bridge S/D to gate. The unreacted metal is chemically removed without etching the Ti-Si or Co-Si compounds. A final anneal is needed to convert the S/D and gate silicide to its final low resistivity disilicide phase.

There is a volume change during the silicidation reaction associated with a density difference between the original metal-Si structure and the final metal silicide. Calculations indicate that 1 Å Ti + 2.27 Å Si gives 2.51 Å TiSi$_2$ while 1 Å Co + 3.64 Å Si gives 3.52 Å CoSi$_2$. This volume change gives a rough silicide surface and silicide/Si interface. In addition, the silicide surface is not at the original substrate surface.

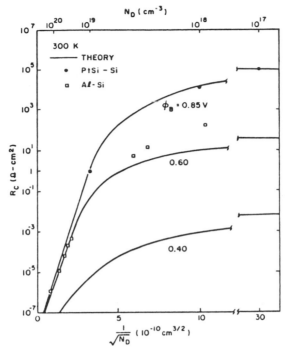

Figure 11. Theoretical and experimental values of specific contact resistance of materials with various barrier heights ϕ_B to n-type Si. After Ref. 3, reprinted with permission of the publisher.

Figure 13 (21) schematically shows the effect of the cleanliness of the metal/Si interface on silicide roughness. The left side of the figure shows the presence of a native oxide between Ti and Si. The Ti/Si reaction will first begin in the area of pinholes in the native oxide/interface residue. The Ti/Si reaction will begin elsewhere only after the native oxide is consumed or bridged by the reaction. This additional reaction nonuniformity will add to the silicide surface and silicide/Si interface roughness. The right side of the figure compares the Ti/Si reaction for the case where the native oxide is broken up by ion beam mixing. (Typically, >10^{15} cm^{-2} Si dose is implanted with an energy of 10 keV/100 Å Ti, to give an implant range at the Ti/Si interface. Arsenic implanted with a lower dose and higher energy can also be used particularly for application to NMOS.) Here, the implantation creates more pinholes in the native oxide, and the reaction proceeds more uniformly. Note that in an actual integrated circuit process, the native oxide is only one component affecting interface cleanliness. The effect of dopant type and concentration on oxide thickness and the presence of reactive sputter etch residues are additional concerns. Whereas the thin native oxide increases silicide roughness, these additional RSE residues may prevent the silicide reaction. Therefore, ion beam mixing of the Ti/Si interface is important to the silicide reaction on processed substrates. Ion beam mixing of the Co/Si interface also reduces the CoSi$_2$ roughness.

- Pattern Gate and Implant LDD
- Form Sidewall Spacer

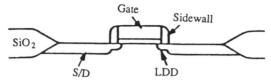

- Preclean and Deposit Metal

- Ion Mix Interface
- React (C-47 $TiSi_2$ or CoSi) at Lower Temperature
- Wet Etch Metal from Oxide
- Anneal at Higher Temperature (C-54 $TiSi_2$ or $CoSi_2$)

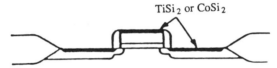

Figure 12. Ti salicide process sequence.

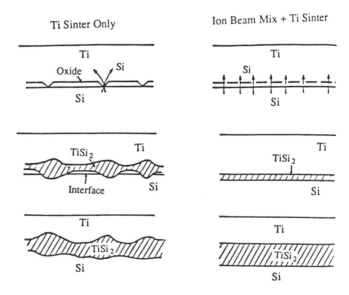

Figure 13. The effect of Ti-Si interface quality with (right) and without (left) ion beam mixing, on $TiSi_2$ roughness.

Figure 14 shows the source-sidewall spacer-gate area for the cases where, at left, metal or at right, Si is the dominant diffusing species during the silicidation reaction. The two distinct cases are discussed below for the Ti and Co silicide sintering processes.

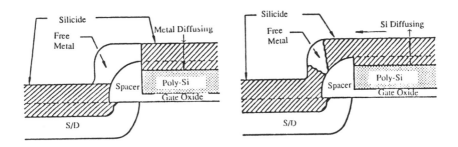

Figure 14. Source-spacer-gate geometry showing the different silicide bridging obtained when metal diffusion (left) or Si diffusion (right) controls silicidation.

TiSi$_2$. Si is the dominant diffusing species during the formation of TiSi$_2$. Therefore, Si from the substrate and from the poly-Si gate can diffuse into the metal on the sidewall oxide. The silicide reaction can creep over the sidewall and bridge or short S/D to gate. Figure 15 (22) plots the lateral growth of TiSi$_2$ over SiO$_2$ because of the diffusion of Si in Ti, as a function of time at temperature. The TiSi$_2$ overgrowth is several microns after sintering for several minutes at 750°C. (This is compared to a typical poly-Si gate height <0.5 μm.) Thus, the short time at low temperature needed to avoid TiSi$_2$ bridging necessarily requires rapid thermal processing (RTA or RTP) to form TiSi$_2$. The higher resistivity C-47 phase of TiSi$_2$ forms during this low temperature (~650°C), short time (~60 sec) RTA1 cycle. After etching (e.g., in a mixture of H$_2$O$_2$ and H$_2$SO$_4$) the unreacted Ti from the oxide, the low resistivity 16-18 μΩ-cm C-54 phase is formed again using a rapid thermal process RTA2 at 800-900°C for <60 sec. Both RTA1 and RTA2 are in flowing N$_2$ or NH$_3$ ambients so as to minimize the oxidation of Ti during the silicidation reactions.

Figure 16 shows the TiSi$_2$ sheet resistance R$_S$ (which can be converted to an approximate TiSi$_2$ thickness using ρ=18 μΩ-cm) as a function of sputtered Ti thickness. Here, Ti is sputtered on undoped or As doped Si wafers after a dilute HF acid clean. Si ion mixing is used before a 625°C RTA1 + Ti etch + 900°C RTA2 sequence which avoids bridging. Note that for the thicker Ti depositions, R$_S$ is weak function of Ti thickness. This implies that all the Ti metal is not consumed during this RTA1 cycle. Therefore, the across-the-wafer TiSi$_2$ thickness uniformity is not necessarily that of the sputtered Ti uniformity but is also related to the temperature uniformity of the rapid thermal processing (RTP) equipment. Typically, since higher temperature RTA processes have better spatial uniformity, RTA1 is at the highest temperature which avoids bridging.

Note that for the thinner Ti depositions, R_S is a strong function of Ti thickness. This implies that all the Ti thickness is consumed either by the silicide reaction or by Ti surface oxidation/nitridation during RTA1. Ti oxynitride is a chemically preferred compound which is not converted to Ti silicide. Thus, the thinner $TiSi_2$ across-the-wafer uniformity is additionally affected by the competing oxidation and nitridation of the Ti surface.

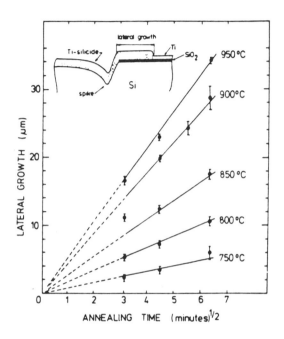

Figure 15. Lateral growth of $TiSi_2$ over SiO_2 for various temperature/time Ti sintering cycles. The inset shows the lateral growth test structure. After Ref. 22, reprinted with permission of the publisher.

The junction doping/anneal process is related to the choice of silicide. Dopant atoms (As, P, and B) have a high solubility in and form stable compounds (23) with $TiSi_2$. Since dopants cannot be implanted into and reliably outdiffused from $TiSi_2$ sintered on undoped Si, the Ti salicide process involves Ti sintering on already doped poly-Si gates and on already formed S/D junctions. Thus dopants from the underlying Si are incorporated into $TiSi_2$ during silicidation; and, for a given RTA1/RTA2 sequence, affect the resulting $TiSi_2$ thickness. Figure 16 also shows the $TiSi_2$ sheet resistance as a function of Ti thickness for Ti sintered on As implanted (2×10^{15} cm^{-2}) and activated (950°C) Si with the same RTA1/RTA2 sequence used to sinter Ti on undoped Si. Note that $TiSi_2$ has a higher sheet resistance and is thinner on As doped Si than on undoped Si. This thickness difference is a function of the particular dopant type, dose, and activation cycle. In general, thinner $TiSi_2$ is sintered on more heavily doped Si substrates. For all Ti thicknesses, heavy As doping greater than a few times 10^{15} cm^{-2}, produces a thinner and less uniform $TiSi_2$.

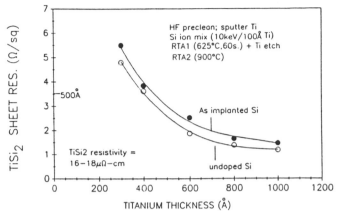

Figure 16. $TiSi_2$ sheet resistance obtained from sintering various Ti thickness on undoped or As doped Si, with a particular rapid thermal process (RTP).

(Although junction formation is beyond the scope of this chapter, it should be noted that dopant diffusion does not simply scale with temperature and time. Enhanced dopant diffusivity has been measured and related to residual implant damage. The implant damage is not effectively removed by a lower diffusion temperature. Thus, interstitial dopant diffusion results in a junction depth which is deeper than that expected (24, 25) from classical dopant diffusion in Si. Therefore, scaling of junction depth with $TiSi_2$ will be sensitive to damage enhanced diffusion. RTP may be needed for a shallow junction process to remove implant damage before significant transient dopant diffusion.)

A minimum junction depth ~1000 Å below $TiSi_2$ is needed to avoid excessive reverse bias diode leakage. (26) Therefore, shallower junctions require thinner $TiSi_2$. Figure 17 shows transmission electron micrographs (TEM) of nominally 450Å and 750Å thick $TiSi_2$ sintered on undoped Si with the previous RTA1/RTA2 sequence. The grain size is ~5000 Å and is not limited by the $TiSi_2$ thickness. Figure 18 shows the same silicide thicknesses after a post-silicide anneal of 850°C, 30 min. Note that the roughness of the thicker silicide significantly increases while the thinner silicide shows islanding behavior. Therefore, the thermal stability of $TiSi_2$ thin films is an important issue with negative device effects. Since $TiSi_2$ is sintered on a higher resistivity poly-Si gate, a thermally unstable $TiSi_2$ gives a higher gate sheet resistance (higher RC time constant) and a resulting decrease in device speed. Since dopants have a high solubility and are incorporated in $TiSi_2$, thermally unstable $TiSi_2$ gives a higher reverse bias diode leakage, a higher transistor off current, and a lower transistor on current. To first order, the maximum temperature that $TiSi_2$ can be annealed at before it agglomerates, is related to silicide thickness and microstructure (grain size). Dopant effects impact thermal stability primarily through their effect on silicide thickness. The cleanliness of the Ti/Si interface influences thermal stability through its effect on the as-sintered $TiSi_2$ roughness and grain size.

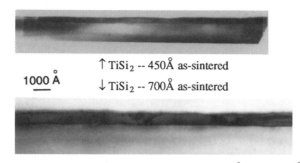

1000 Å   ↑ $TiSi_2$ -- 450Å as-sintered
          ↓ $TiSi_2$ -- 700Å as-sintered

Figure 17. Cross-section TEM micrographs of 450 Å or 750 Å thick $TiSi_2$ as-sintered.

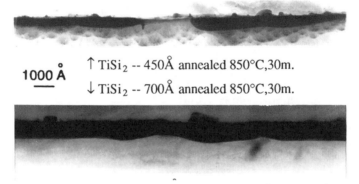

1000 Å   ↑ $TiSi_2$ -- 450Å annealed 850°C,30m.
          ↓ $TiSi_2$ -- 700Å annealed 850°C,30m.

Figure 18. Cross-section TEM micrographs of 450 Å or 750 Å thick $TiSi_2$ after annealing at 850°C, 30 min.

CoSi2. Co metal is the dominant diffusing species in the Co-Si reaction kinetics of the CoSi phase. This basic difference is advantageous in avoiding silicide bridging S/D and gate. The Co salicide process therefore proceeds as follows.(27, 28) Co is sputter deposited after a preclean of the Si substrate. CoSi is formed by heating to 400-450°C. This temperature is low enough to avoid reaction of Co with SiO_2. Since bridging by Si diffusion,does not limit time at temperature; a furnace sinter in a nonoxidizing ambient, such as H_2, can be used to sinter CoSi. Therefore, the CoSi sinter time is chosen to be long enough to completely react the entire Co thickness with Si. After CoSi sinter, the unreacted Co is selectively wet etched for example, in an $H_2SO_4 + H_2O_2$ mixture. The CoSi is then converted to the low resistivity $CoSi_2$ after a 600-700°C furnace anneal, again in a nonoxidizing ambient (Ar or N_2). It should be noted that since the entire Co thickness reacts, the across-the-wafer $CoSi_2$ uniformity is that of the Co metal uniformity. Also note that similar to $TiSi_2$,

RTP at 700-800°C can be used to sinter $CoSi_2$. In this case, the uniformity tradeoff between furnace and RTP processing must again be addressed.

Figure 19 shows a cross-section TEM micrograph of the resulting Co salicide process on P doped poly-Si gate and undoped S/D, from 200 Å Co sintered to $CoSi_2$ with 450°C, 90 min in H_2 + Co etch + 700°C, 30 min. in Ar. Note that bridging of S/D and gate is not observed even for this excessive sinter. However, microbridging may be seen. Even though Co is the dominant diffuser during the sintering of CoSi, a finite Si diffusivity results in the creep of silicide over the oxide spacer. This creep is related to the temperature/time cycle used for sintering CoSi.

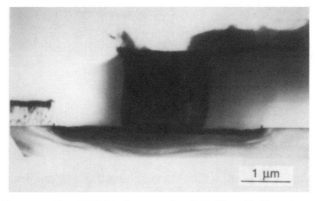

Figure 19. Cross-section TEM micrograph of $CoSi_2$ salicide process with an Al/TiN contact to Si.

The salicide process in Figure 19 shows $CoSi_2$ sintered on undoped S/D. The $CoSi_2$ thickness sintered by a given temperature/time sequence is a function of dopant type and dose. (29) However, shallow n^+/p or p^+/n junctions can be formed by implanting As and/or P or BF_2 and/or B in $CoSi_2$ and outdiffusing either by rapid thermal or by furnace anneal. (30) (The higher temperature, shorter time RTA process has the advantage of higher dopant activation in Si and a lower silicide/silicon contact resistance.) The implant energy is chosen low enough to confine the implant in $CoSi_2$. Therefore, the outdiffusion cycle is determined by the required junction depth and not by annealing implant damage. The cross-section TEM micrograph at the top of Figure 20 shows the microstructure of 150 Å Co sintered to ~600 Å $CoSi_2$ on undoped Si with a minimum sinter cycle of 400°C followed by 600°C. $CoSi_2$ has a grain size of ~1000 Å and a roughness of ~200 Å for this case where ion mixing was not used after Co deposition. The bottom TEM micrograph of Figure 20 shows the $CoSi_2$ microstructure after 5×10^{15} cm^{-2}, 40 keV BF_2 is implanted into $CoSi_2$ and outdiffused at 800°C, 90 min. Note that the $CoSi_2$ roughness does not increase and that there is no implant damage in the underlying Si. Secondary ion mass spectroscopy (SIMS) shows the 10^{17} cm^{-3} B concentration is ~500 Å below the $CoSi_2$/Si interface. Similar junction depths are obtained by outdiffusion of As or P. The $CoSi_2/n^+$ or p^+ (with 10^{20} cm^{-3} surface concentration) contact resistivity was measured (16) to be $\sim 2 \times 10^{-7}$ Ω-cm^2 and was much larger than the Al/CoSi$_2$ contact resistivity $\sim 10^{-8}$ Ω-cm^2.

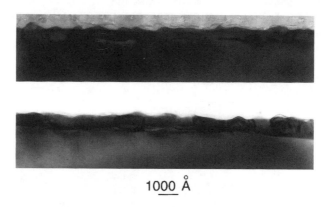

1000 Å

Figure 20. Cross-section TEM micrograph of ~500 Å CoSi$_2$ as-sintered (top) and after being implanted with BF$_2$ and annealed at 800°C, 90 min (bottom).

Finally, several points on the manufacturability of CoSi$_2$ should be noted. (a) Co is a magnetic material which is not compatible with standard magnetron sputtering systems designed primarily for sputtering nonmagnetic materials such as Al alloys and barrier materials. Therefore, special provisions involving magnetron guns with magnetic coils rather than permanent magnets, may be needed to sputter Co in a manufacturing development environment. (b) Co is a midgap impurity in Si and will affect MOS properties if allowed to cross-contaminate processing equipment. One obvious source of cross-contamination is using the same RSE system to pattern contacts and diffusion or gate levels. Since Co does not form volatile fluoride or chloride compounds in typical RSE systems used to etch SiO$_2$ or poly-Si with fluorine containing gases or chlorine, the Co sputtered from the bottom of the contacts during the contact overetch, will deposit on the walls of the RSE system. Using the same RSE system to then pattern the diffusion or gate levels will then redeposit low levels of Co from the system wall onto the Si surface. (c) As was previously discussed, dopants diffuse rapidly and have a low solubility in CoSi$_2$. This is advantageous for fabricating ultra shallow junctions. However, dopants implanted into CoSi$_2$ will diffuse in the CoSi$_2$ overlayer from n-poly to connected p-poly gates (lateral diffusion) and segregate into the poly-Si. Depending on the distance between the NMOS and PMOS device, this lateral diffusion may influence the poly-Si / gate oxide doping concentration and the device threshold voltage for a symmetric Co salicide CMOS process.

For comparison, Ti is not magnetic and can be deposited with conventional magnetron equipment. TiSi$_2$ etches with fluorine and chlorine RSE and pre-clean chemistries and may not have the cross-contamination concerns of CoSi$_2$.

Dopants have a high solubility in $TiSi_2$. This limits the minimum junction depth (and gate height) to ~2000 Å but also reduces the impact of lateral diffusion on the connected NMOS and PMOS gates in a symmetric CMOS process.

2.2 Polycide — WSi_x or $TaSi_x$

The polycide gate stack consists of a sputter deposited metal silicide on poly-Si. (31) The dominant metal silicides for polycide application are WSi_x and $TaSi_x$. These 10-20 Ω/sq silicides have a sheet resistance which is ~5-20 times lower than that of heavily PBr_3 diffused or implanted and activated poly-Si. (Note that $MoSi_x$ is also used for polycide applications with results similar to those discussed below for WSi_x. However, a disadvantage for $MoSi_x$ in ULSI applications may be it's significantly higher resistivity.) Figure 21 shows a cross-section TEM micrograph of a transistor with Ta silicide polycide gate. (28) In this case, $CoSi_2$ was sintered on the S/D after the Ta silicide polycide gate was patterned with RSE.

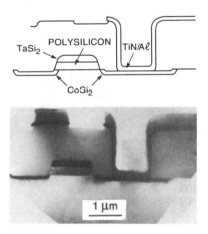

Figure 21. Cross-section TEM micrograph of MOSFET structure with $CoSi_2$ S/D, $TaSi_2$ polycide gate and Al/TiN contact metallization. After Ref. 28, reprinted with permission of the publisher.

Decreasing the gate stack height reduces topography. This topography consideration becomes important as device dimensions decrease and the number of metal levels increases, for improving process margins in lithography (depth of focus for optical steppers), reactive sputter etching (etch selectivity related to say, etching different height contacts), and metal coverage (contact and via aspect ratio). Polycides can reduce the gate stack height by using outdiffusion of dopants implanted into the silicide to dope the poly-Si gate down to the gate oxide surface. The poly-Si gate without polycide is usually limited to a thickness greater than that with polycide by either the channeling of S/D or gate dopant implants through the gate thickness or by the salicide process giving a gate silicide too close to the gate oxide. The first affects the device threshold voltage V_T, by changing the designed substrate doping under-the-gate oxide. The second affects V_T by changing the gate work function.

Microstructure and Composition Stability. Figure 22 shows cross-section TEM micrographs of a nominally 1200 Å W-Si$_x$ thick film with Si to W ratio x = 2.5 sputtered from a composite target on Si. The top micrograph shows the W-Si$_{2.5}$ as deposited. The bottom micrograph results after annealing at 900°C, 30 min. in flowing N$_2$. Rutherford backscattering spectroscopy (RBS) data interpreted with the thickness measurement by cross-section TEM, give a mass density for this W-Si$_{2.5}$ film of ~7 gm/cm^3. This mass density is that of the W-Si atomic mixture and is lower than the ~10 gm/cm^3 measured for WSi$_2$ formed by reacting W with Si. (20)

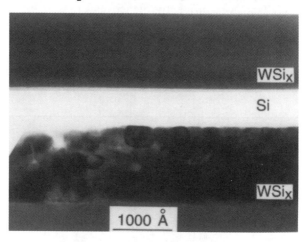

Figure 22. Cross-section TEM micrographs of sputtered WSi$_x$ as-deposited (top) and after annealing at 900°C, 30 min. (bottom).

The as-sputtered W-Si$_x$ is amorphous. Grains grow to ~500 Å after annealing at 900°C. Grains grow to 2000-3000Å after annealing at 1000°C. (32) A corresponding decrease in resistivity (ρ) is measured and correlates with WSi$_x$ grain size. The as-sputtered resistivity is ~800 $\mu\Omega$-cm and decreases to ~150 $\mu\Omega$-cm or ~50 $\mu\Omega$-cm after respectively annealing at 900°C or 1000°C. Thus, a WSi$_{2.5}$ film with thickness t = 1000 Å annealed at ~900°C, has a sheet resistance ($R_s = \rho/t$) of ~15 Ω/sq. The resistivity does not significantly change with annealing time at temperature, longer than ~5 min. Thus, the WSi$_x$ microstructure equilibrates after several minutes at temperature. (Similar results are obtained with TaSi$_x$. The as sputtered film is amorphous and grows grains after annealing. The resistivity again decreases to a value comparable to that for

WSi$_x$.) Finally note that the annealed W-Si$_{2.5}$ film (Figure 22 bottom) forms a top surface SiO$_2$ cap layer and epitaxial precipitates at the interface with the Si wafer.

Nonstoichiometric Si rich (x > 2) sputtered W or Ta silicides processed for long enough time at temperatures >650°C, where W or Ta disilicide are the stable compounds, equilibrate to approximately stoichiometric silicides (33, 34) by rejecting Si to the top surface exposed to the annealing ambient or to the Si interface. Typically, oxygen unintentionally backstreams into nonevacuated furnaces. Therefore, annealing in a flowing inert ambient oxidizes the Si rejected to the top surface and gives the Si oxide cap. The Si rejected to an interface with single crystal Si shows epitaxial regrowth. The Si rejected to an interface with a poly-Si adds to the poly-Si thickness.

In general, Si rejection is by Si diffusing along grain boundaries, to the interface. At a given processing temperature, each dopant has a solubility limit in the silicide or poly-Si. Dopants in excess of this solubility limit must diffuse or precipitate into the grain boundaries. This grain boundary stuffing will limit the diffusivity of excess Si.

Experimental analysis by RBS and TEM verifies that the entire silicide thickness does not become stoichiometric after annealing at <900°C. In this case, the silicide under the oxide cap and close to the top surface remains at near its original Si/W ratio while that at the Si interface has a locally lower Si/W ratio. Annealing at >950°C, 60 min equilibrates the entire silicide thickness by rejecting the excess Si primarily to the interface with Si. The measured thickness of rejected Si agrees well with that calculated from the final equilibrium composition. For example, 3200 Å TaSi$_{3.2}$ sputtered on 4000 Å undoped poly-Si becomes ~2500 Å TaSi$_{2.2}$ on ~4700 Å poly-Si. The equilibrium disilicide is slightly Si rich.

Metal rich nonstoichiometric sputtered silicides equilibrate by consuming Si from the underlying Si or poly-Si substrate. The Si diffuses up into the W-Si$_x$ or Ta-Si$_x$ during this silicidation reaction, at a rate which is influenced by substrate doping. Since silicidation has a lower activation energy than Si grain boundary diffusion, the temperature and time needed to equilibrate the metal rich sputtered silicides are lower than those needed to equilibrate the Si rich sputtered silicide films. The measured amount of Si consumption again agrees well with that calculated from the final silicide composition. For example, 2500 Å Ta-Si$_{1.3}$ equilibrates at <850°C by consuming 900 Å Si to become TaSi$_{2.1}$. Again, the final silicide is slightly Si rich.

Since the equilibrium disilicides are slightly Si rich, polycide applications begin with the as-sputtered disilicides being slightly Si rich. This avoids possible processing and device problems associated with changing silicide composition. For example, the volume and stress changes and possible oxidation during further processing can result in loss of adhesion between the silicide and Si. In addition, the precipitation of an undoped Si layer between silicide and Si will result in a higher contact resistance to doped Si in the case where silicides are used as a local interconnect to the doped Si S/D. The excess Si also allows margin for annealing in oxygen without affecting the Si under silicide. Such oxidation properties are essential in most integrated ULSI processes since; e.g., a thin thermal oxide is grown before implanting the S/D or before depositing the CVD oxide for the sidewall spacer.

Finally, it should be noted that the sputtering rates of W or Ta and Si are different enough to require a target that is more Si rich than the sputtered silicide film. Additionally, silicides can be deposited by chemical vapor deposition. For example, WSi_2 can be deposited (32) using a variation of the $WF_6 + SiH_4$ chemistry used to deposit blanket CVD W in Sec. 4.3 However, the improved step coverage of CVD is probably not needed at the gate level, and the control of the Si-to-metal ratio needed to avoid processing problems is more difficult for CVD metal silicides. CVD silicides are more appropriate for application to local interconnect (discussed in Sec. 5) where coverage into high aspect ratio contacts is important.

Dopant Outdiffusion. As was the case of $TiSi_2$, B implanted into WSi_x or $TaSi_x$ has a higher solubility in the silicide than in Si. Additionally, B reacts with the metal to form localized metal boride compounds. (35) Arsenic more readily outdiffuses from Ta (36) or W (37) disilicides. However, the outdiffusion temperature is high (>950°C) and a majority of the As dopant remains in the silicide. Thus, B and As do not sufficiently outdiffuse at reasonable ULSI process temperatures, to degenerately dope the underlying poly-Si/gate oxide interface.

On the other hand, P readily outdiffuses from W or Ta silicide. Figure 23 shows a secondary ion mass spectroscopy (SIMS) depth profile of P implanted into $WSi_{2.5}$ and outdiffused into the underlying poly-Si deposited undoped on SiO_2, with an 850°C, 30 min furnace anneal or 1050°C, 10 sec RTA cycle. Both RTA and furnace give similar P profiles. (Note that SIMS measures the physical profile and does not distinguish between P which is electrically active in the Si grains and P which is electrically inactive in the grain boundaries. Therefore, depending on the poly-Si microstructure, the apparently identical P doping can give different MOS capacitor depletion effects.) The P is relatively uniformly distributed in the WSi_x and poly-Si away from the interfaces. SIMS profiling artifacts due to the change in materials, accounts for some of the decrease in P doping at the interfaces. However, some of the decrease is related to P evaporating from the silicide surface and to Si precipitating at the WSi_x/poly-Si interface.

2.3   CMOS and Bipolar Contact Silicides

As was previously discussed, the cleanliness of the metal/Si interface is important in determining the microscopic and macroscopic thickness uniformity and roughness of the sintered salicide. An additional interface residue exists at the bottom of the contacts which are reactive sputter etched in a deposited oxide, to Si or silicide. This residue must be removed before proceeding to the metallization process since the cleanliness of the final metal interface contacting Si (usually the barrier metal) will affect the mean contact resistance to S/D and gate and its uniformity across the wafer and from wafer-to-wafer.

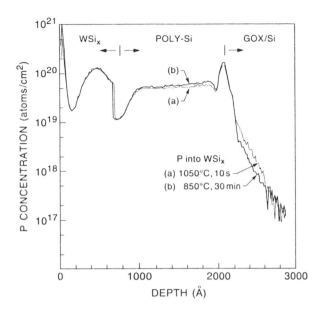

Figure 23. SIMS profile of P implanted into sputtered WSi_x on poly-Si and outdiffused at 850°C, 30 min or 1050°C, 10 sec.

The exact condition of the contact interface remaining after contact RSE is related to the specific surface composition (type of silicide) and to the specific reactive sputter overetch chemistry. Figure 24 (18) quantifies with x-ray photoemission spectroscopy (XPS), the condition of a large area As implanted $CoSi_2$ surface exposed to a CHF_3/CO_2 overetch with a self-induced bias of ~200 V. The chemical composition of the RSE residue is that of an ~10 Å thin teflon-type layer on an ~50 Å thick suboxide which is implanted with C during the RSE overetch. A clean $CoSi_2$ surface begins ~70 Å below the surface. Selective chemical cleaning of this residue may be difficult because the cleaning solution (usually a mixture of H_2SO_4 with H_2O_2 followed by dilute HF) will etch the various silicides at the bottom of the contact, to various degrees. An added concern for ULSI with submicron design rules is that the contact clean will etch the deposited oxide and increase the contact diameter.

Processing margin can be improved by using a milder wet clean and by consuming the remaining residue by reacting a suitable metal in the contact. Ti metal can be used effectively here since Ti silicides at a temperature low enough not to redistribute the dopants in and affect the electrical characteristics of the underlying device structure. As will be discussed in the next section, if the reaction does not consume the entire Ti thickness, the remaining Ti will provide some barrier properties between Al and Si. W can also be used effectively. However, the adhesion of W to oxide is usually improved with a Ti adhesion layer between W and oxide. Ti and W and their silicides have midgap barrier heights (see Table I) to both n- and p-Si and are therefore good choices for ohmic contacts to CMOS circuits with typical doping concentration $>10^{19}$ cm^{-3}.

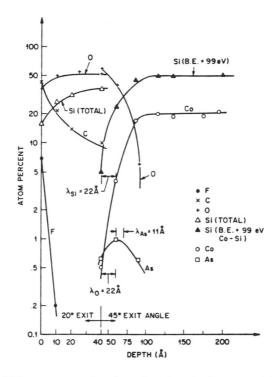

Figure 24. XPS spectrum showing the chemical composition of a CoSi$_2$ surface after a contact RSE overetch with CHF$_3$/CO$_2$. Reprinted by permission from the publisher, The Electrochemical Society, Inc. After Ref. 18.

PtSi and Pd$_2$Si (see Table I) have barrier heights to n-type Si which are above the midgap energy level. Thus, (see Figure 11). Pt and Pd can be sintered to obtain Schottky contacts to n-Si (38). Since ϕ_B(n-Si) + ϕ_B(p-Si) is the band gap of Si, PtSi and Pd$_2$Si have barrier heights to p-type Si which are below the midgap energy level. Therefore, Pt and Pd silicides can be sintered to obtain even for low doping concentration, ohmic contacts to p-Si. Note that PtSi is preferred because it has a higher barrier height to n-Si, gives better (lower reverse bias diode current) Schottky junctions with n-Si and gives better (lower resistivity) ohmic contacts to p-Si, than Pd$_2$Si. PtSi also has better thermal stability than Pd$_2$Si. Of course, to minimize the impact of silicide thermal instability, PtSi is sintered at a point in the fabrication process where the subsequent processing can be done at <800°C.

Both PtSi and Pd$_2$Si are used for bipolar and BICMOS circuit contacts. Some applications follow: (a) In self-aligned bipolar devices with poly-Si base electrodes, the poly-Si resistance adds to the total base resistance. Thus, the base resistance increases (and the base transit time and bipolar performance decrease) as the poly-Si thickness is reduced say to reduce topography. Here, the thinner poly-Si is silicided to reduce the base resistance; (39) (b) siliciding a thin poly-Si emitter can increase the injection efficiency and bipolar gain by consuming any oxide interface between the n^+ poly-Si (which is also consumed during silicidation) on n^+ Si emitter; (c) a Schottky diode shunting the base and collector is simply obtained in a planar integrated structure by sintering say, PtSi in a contact overlapping the base and collector. This increases the switching speed of a n^+pn bipolar device by allowing the inherently faster majority carrier Schottky diode to switch the inherently slower minority carrier bipolar transistor through saturation; and, (d) low-resistance ohmic contacts can be made by directly contacting the p-type base with PtSi. This eliminates the need for an extra p^+ implant used when contacting the p-type base with a materials such as Ti, $TiSi_2$ or Al which have a midgap barrier height.

Finally, it should be noted that the sintering of the contact silicide consumes some underlying Si and may not be appropriate for use with very shallow junctions. Bridging is not a concern here since any metal silicide layer on the oxide is etched during metal patterning.

3.0 SPUTTERED BARRIER MATERIALS BETWEEN Al AND Si OR SILICIDE

A metallization structure with low contact resistance and without degradation of the shallow junction device integrity during sequent thermal cycles, necessarily requires a barrier layer between Al and Si or silicide. In the previous section, it was seen that refractory metals such as Ti or W, do not react with Si to form silicide until ~600°C. Therefore, the requirement that the barrier material not react with the Si substrate is satisfied by Ti and W for temperatures less than ~600°C. Since W does not adhere well to oxide, Ti is usually used as an adhesion layer below W. To simplify the deposition sequence, a small amount (~10 wt.%) of Ti is pressed into a composite target and a Ti-W barrier is deposited. Ti and W are sputtered from the Ti-W target with Ar ions. Again, since Ti and W have different sputtering yields, the target has a different Ti/W ratio than that of the deposited Ti-W. (For a basic discussion of sputtering fundamentals, see Refs. 40 and 41.)

Ti or W react with Al at a lower temperature ~450°C. Therefore, barrier failure is initiated by reaction with Al. In general, the temperature for reaction of the barrier material with Al is higher when the barrier is sputter deposited in a chamber with a partial nitrogen ambient. Here, the sputtered atoms react with the nitrogen atoms as they travel from the target to the substrate and deposit a refractory metal nitride is on the substrate.

The barrier material adds to the total series resistance (the sum of the contact and vertical bulk resistances of each material in the layered metallization) of the contact. However, the contact resistivity of a metal with a midband gap barrier height (e.g., Ti or Co silicide, Ti, or W) has a contact resistivity ~10^{-7} Ω-cm^2 to doped Si and accounts for 20-50Ω series resistance for a single 0.5 μm diameter contact. The highest bulk resistivity is usually ρ_{BM} of the barrier material and the barrier material bulk resistance is $\rho_{BM} t_{BM}/A_C$ where t_{BM} is the barrier thickness and A_C is the contact area. In the usual case where the barrier is <1000 Å thick, the barrier bulk resistance does not significantly add to the total contact resistance until ρ_{BM} approaches the high value of 1 m Ω-cm. Thus, barrier materials with moderate resistivity, can be used provided they have the appropriate metal to metal barrier height needed for ohmic contacts with low specific contact resistivity.

The sections below describe the temperature stability of Ti, Ti-W, and TiN between Al and Si or silicide. Physical analysis by RBS and TEM is correlated with shallow junction reverse bias diode leakage and contact resistance.

3.1   Ti and Ti-W

Figure 25 (42) shows the RBS spectrum (counts/channel plotted against backscattering energy) of 1500 Å Al on 500 Å Ti on Si as deposited (solid line) and after annealing at 400°C, 30 min (light dotted line) or 450°C, 30 min (heavy dotted line). Indicated on the plot are the backscattering energy corresponding to finding Ti, Si or Al at the surface. As deposited, Al is at the surface and Ti and Si are displaced from the surface by an energy corresponding respectively to the thickness of Al and Al + Ti. (Note that RBS spectra are quantitative. (43) The backscattering energy is converted to depth by knowing the energy loss dE/dx, for the He$^+$ or α-particle, in the given material. The backscattering yield or counts/channel is converted to an atomic concentration by knowing the scattering cross-section of the α-particle with the particular atom in the given material.) The Ti and Al react after annealing at 400°C. This is seen by the broadening of the leading edge of the Ti spectrum and the trailing edge of the Al spectrum. After annealing at 450°C, Ti is seen at the Al surface. In addition, Al has penetrated the entire Ti thickness and, as is seen from the distortion of the leading edge of the Si spectrum, into the underlying Si. Al and Ti react to form Al$_3$Ti.

Figure 26 (11) shows the RBS spectrum of 2000 Å Al on 2000 Å Ti(10 wt.%)-W on Si as deposited and after annealing at 500°C, 30 min (solid line) or 550°C, 30 min (dotted line) or 550°C, 60 min (dash-dot line). The arrows indicate the backscattering energy for W, Ti, Si, and Al at the surface. (Note that to increase the visibility of the interactions in Figure 26, the counts scale for the Si, Al and Ti spectra is 5x that of the W spectrum.) Ti-W does not interact with Si or Al at 500°C. Al and Ti-W interact at 550°C, a temperature higher than that for Ti-Al interaction. However, since the height of the trailing half of Ti and W spectra of the annealed sample is not significantly reduced from the height of the Ti and W spectra of the as-deposited sample, the Al does not react over the entire Ti-W thickness. The broadening of the trailing edge of the Ti and W spectra and of the leading edge of the Si spectrum indicate a Ti-Si and W-Si reaction starting at 550°C. Therefore, the 2000 Å thick Ti-W film is a barrier to the Al-Si reaction up to 550°C. The failure mechanism for thinner Ti-

W is Al completely reacting with Ti and W (to form Al_3Ti and $Al_{12}W$). The silicidation reaction adds to the failure mechanism when annealing at >550°C.

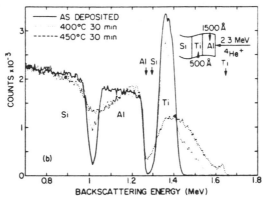

Figure 25. Rutherford backscattering spectra of 2.3 MeV He⁺ ions from a Al/Ti/Si layered structure as deposited (solid line) and after annealing at 400°C, 30 min (light dotted line) or at 450°C, 30 min (heavy dotted line). Reprinted by permission from the publisher, The Electrochemical Society, Inc. After Ref. 42.

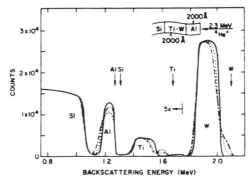

Figure 26. RBS spectra of 2.7 MeV He⁺ ions from an Al/Ti-W/Si layered structure as deposited and after annealing at 500°C, 30 min (solid line), at 550°C, 30 min (dotted line) or at 550°C, 60 min (dash-dotted line). After Ref. 11, reprinted with permission of the publisher.

Barrier properties are more accurately quantified for a specific device process, by noting the degradation of electrical properties. Figure 27 (44) uses reverse bias (5V) diode leakage to study 5000 Å Al on 900 Å Ti (10 wt.%)-W metallization on $CoSi_2$. The effect of the anneal temperature (for 30 min) on the mean leakage current and its across the wafer standard deviation are plotted for 500 μm x 500 μm square n⁺/p diodes. The shallow junctions are ~1000 Å below ~700 Å $CoSi_2$ and are made by outdiffusing P from $CoSi_2$ at 800°C. Note that thin $CoSi_2$ is not a barrier to the Al-Si interaction (see Figure 4). Figure 27 shows the effect of the Al/Ti-W metallization on ~1000 Å deep junctions. The initial decrease in the diode leakage is due to the annealing of process induced damage. The mean leakage and standard deviation remain low up to 450°C, 30 min. After 475°C, a significant "sport" population associated with nonuniform across the wafer barrier failure is indicated by the large increase in standard deviation with a low mean value. Complete 900 Å Ti-W barrier failure is noted after these shallow junctions are annealed at 500°C.

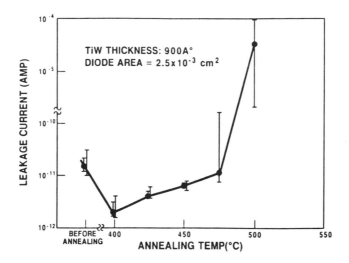

Figure 27. Mean reverse bias (5V) diode leakage and its standard deviation for $n^+(P)/p$ junctions 1000 Å below the 700 Å $CoSi_2$ and metallized with Al/900Å Ti-W and annealed at various temperatures for 30 min 10^{-11} A equivalent to 4 nA/cm². After Ref. 44, reprinted with permission of the publisher.

For the same thickness of Ti-W, the Ti-W deposition conditions (Ar pressure, target power, and substrate temperature) do not significantly affect the degradation of shallow junction leakage. (17, 44) Therefore, the Ti-W deposition condition can be used to minimize stress and particle defect density.

The barrier properties of Ti-W inferred from Figures 26 and 27 do not indicate the true Ti-W barrier properties at the bottom of the contact. Because of sputtering step coverage, Ti-W sputtered ~1000 Å thick, as measured on a large area, is <500 Å thin at the bottom of the contact with aspect ratio >1. Figure 28 (45) shows a cross-section TEM micrograph of a contact (aspect ratio ~1) with Al (alloy with 1% Si) sputtered on "1000" Å Ti-W after the sample is annealed at 450°C, 30 min. The thinner Ti-W in the contact is not a barrier to Al spiking at 450°C. Similar Ti-W barrier failure in contact patterns is observed during high temperature (~500°C, 5 min) deposition of Al. (15)

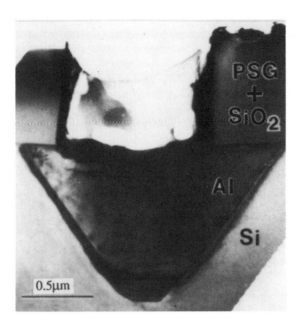

Figure 28. Cross-section TEM micrograph of Al/Ti-W sputter deposited into a contact and annealed at 450°C, 30 min Ti-W is 1000 Å thick as measured on the open areas. After Ref. 45, reprinted with permission of the publisher.

3.2 TiN and TiN on Ti

Figure 29 (46) shows the nitrogen to Ti ratio and the deposited film resistivity as functions of the percentage of nitrogen in the Ar-N_2 sputtering gas mixture. For this s-gun deposition system, a stoichiometric TiN is deposited with >20% nitrogen. The N/Ti ratio increases slowly for a higher percentage of nitrogen. Note that the resistivity increases from ~90 $\mu\Omega$-cm characteristic of Ti sputtered in 100% Ar, to a maximum of ~200 $\mu\Omega$-cm for films sputtered with 15% N_2. A minimum resistivity of 70 $\mu\Omega$-cm is measured for TiN sputtered in 80% Ar- 20% N_2. The resistivity again increases with a higher percentage of nitrogen but the N/Ti ratio remains <1.2. (Note that increasing the % N_2 as required to deposit TiN, also causes nitridation of the Ti target. Since the sputtering yield from the target depends on the target material as well as the sputtering gas, the deposition rate also sharply decreases by a factor of 3-4 for sputtering with >20% N_2.)

In general, less background oxygen is incorporated into films sputtered with higher bias and lower system pressure. The resistivity of the TiN decreases with decreasing total system pressure and with increasing negative-substrate bias (up to a certain system-dependent value, after which resistivity begins to increase). (47, 48) Thus, the TiN resistivity correlates well with the amount of oxygen incorporated into the sputtered TiN film. Resistivity also correlates well with the TiN microstructure.

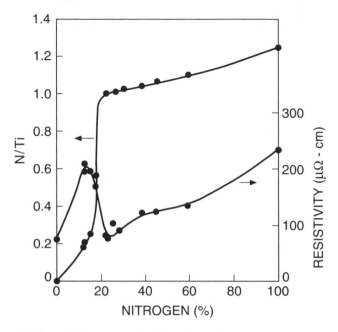

Figure 29. Ratio of nitrogen to titanium in (left axis) and resistivity of (right axis) TiN films sputtered with an S-gun as a function of the percentage of nitrogen in the Ar + N_2 sputtering gas mixture. After Ref. 46, reprinted with permission of the publisher.

Figure 30 (48) shows cross-section TEM micrographs of the microstructure of stoichiometric TiN obtained from sputtering with different total system pressure and different negative substrate bias. In this case, the negative substrate bias is self-induced and is varied by varying the sputtering power. The stoichiometry of Ti/N = 1 is maintained by adjusting the Ar/N_2 ratio of the sputtering gas mixture. The substrate temperature is maintained at 80°C during deposition.

Sputtering with high pressure and lower bias (~20 mT and -150 V in this deposition system) results in the columnar TiN structure in Figure 30a. The structured TiN is also indicated by the electron diffraction pattern shown in the inset. The columns with ~200 Å diameter, extend from the Si substrate to the top surface. The resistivity is high (300 μΩ-cm) but the compressive stress is low (-5 x 10⁹ dynes/cm²). Sputtering with a lower pressure and higher negative bias (~5 mT and -250 V for this deposition system) results in the closely packed, nearly amorphous TiN structure in Figure 30b. The amorphous TiN is also indicated by the electron diffraction pattern shown in the inset. The resistivity is minimized at ~70 μΩ-cm but the compressive stress is high (-2 x 10¹⁰ dynes/cm²). The TiN microstructure is between amorphous and columnar for intermediate values of pressure and bias. As a result, intermediate values of compressive stress and resistivity are measured.

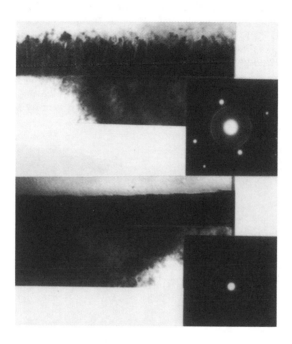

Figure 30. Cross-section TEM micrographs of RF bias sputtered TiN with (a) lower stress, higher resistivity and columnar microstructure deposited with lower negative bias and higher pressure and (b) higher stress, lower resistivity and very fine grain microstructure deposited with higher negative bias and lower pressure. After Ref. 48, reprinted with permission of the publisher.

Since the barrier thickness is typically <1000 Å, a moderate barrier resistivity of 100-200 $\mu\Omega$-cm does not add significantly to the Al/barrier series resistance. Therefore, in reality, a compromise TiN deposition condition can be obtained with moderate stress and moderate resistivity. In addition, the TiN stress can be controlled with a higher substrate temperature during deposition.

Figure 31 (49) shows the RBS spectrum (counts/channel plotted against backscattering energy) of 3000 Å Al on 1000 Å TiN on Si as deposited (solid line) and after annealing at 550°C, 30 min (dotted line). Here, the Al and TiN film thicknesses are such that the Al and Ti spectra partially overlap. The theoretical Al spectrum (calculated for the 3000 Å thick as-deposited Al) is shown by the dashed line. TiN and Si do not interact at 550°C, 30 min since the leading edge of the Si spectrum remains unchanged. However, the small change in the trailing edge of the Ti + Al spectrum indicates a small Al/TiN interaction. Similar Al/TiN stability is measured by RBS, on $TiSi_2$, and $CoSi_2$. (49, 50)

Barrier properties are more accurately quantified for a specific device process, by noting the degradation of reverse bias diode leakage. Figure 32 shows the probability plot of diode leakage (% of diodes measured with less than a given value of reverse bias diode current at -5 V) for 900 Å deep n^+/p junctions (below ~1500 Å thick $CoSi_2$). The diodes are metallized over their entire 500 µm x 500 µm square area with Al on 500 Å thin TiN and annealed at 330°C-550°C, 45-60 min. The diode leakage distribution does not change up to annealing at 500°C, 60 min. However, both the mean leakage and the sport population increase after annealing at 550°C, 60 min. The mean leakage current density increase to <100 nA/cm^2 indicates low average levels of Al penetrating 500 Å TiN. However, the unacceptably high sport population indicates barrier failure at many sites across the wafer. Therefore, the annealing temperature for failure of 500 Å TiN between Al and $CoSi_2/Si$, as measured by shallow junction leakage, is ~500-550°C.

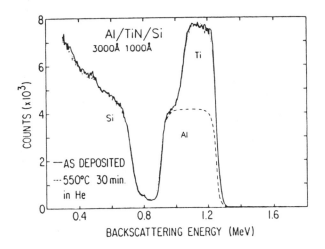

Figure 31. Rutherford backscattering spectra of 2.3 MeV He+ ions from a Al/TiN/Si layered structure as deposited and after annealing at 550°C, 30 min. After Ref. 49, reprinted with permission of the publisher.

The annealing temperature for failure of a TiN barrier between Al and Si or silicide is modified by the diffusion of Al along the TiN grain boundaries. Since the TiN microstructure is significantly different for different deposition conditions, the exact TiN barrier properties have a second order dependence on the deposition system. However, in general, 500 Å TiN is stable up to annealing at 550°C. Note that this temperature limit is ~100°C above the ~450°C stability limit seen in Figures 27 and 28, for a similar thickness of Ti-W.

An important property of the metallization structure is its contact resistance which is usually dominated by the contact resistance of the final contact layer to doped Si. (18)   For the same dopant concentration in Si, the contact resistivity of Ti to Si is significantly lower than that of TiN to Si. (11) Therefore, Ti below TiN can be used to lower contact resistance. Since Ti does not silicide below ~600°C; and TiN is chemically stable even above 600°C, the barrier properties of the Ti/TiN structure are determined by the Al/TiN reaction.

Therefore, Ti/TiN is the barrier structure of choice between Al and Si (without or with silicide) and is also stable to annealing at 500°C-550°C.

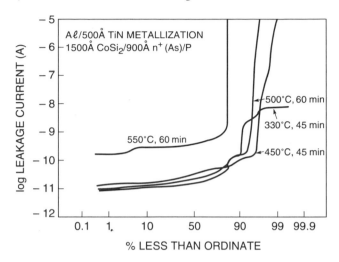

Figure 32. Probability plot of reverse bias (5V) diode leakage of $n^+(As)/p$ junctions 900Å below $CoSi_2$. Metallized with Al/TiN and annealed with various temperature/time cycles, in H_2.

4.0 CVD BARRIER MATERIALS

The poor coverage of sputtering into high aspect ratio contacts reduces the thickness of the barrier material at the bottom of the contact and hence, the useful temperature/time before barrier failure in the contact. Chemical vapor deposition, usually at low pressure (CVD or LPCVD), depending on the deposition chemistry, has excellent step coverage. CVD can be applied to the deposition of a thin barrier (predominantly TiN) followed by sputtered Al. To planarize each metal level, CVD can be applied to the deposition on a thin barrier layer of a thick nonselective plug material (predominantly W), which is either etched back or patterned with the subsequent sputtered Al conductor (see Figure 10). CVD can be applied to a selective deposition (predominantly selective W) on Si or silicide followed by a sputtered Al conductor. The three sections below describe CVD TiN, blanket CVD W on a TiW or TiN barrier / adhesion layer, and selective CVD W.

In general, Al on the CVD barrier materials described below, has an equal or higher thermal stability than that of the Al on the sputtered barrier layers previously discussed. In other words, the barrier properties of CVD TiN are similar to those of an equal thickness of sputtered TiN, as previously discussed. However, the better step coverage of CVD, compared to that of sputtering, gives a thicker TiN barrier at the bottom of the contact. The blanket W thickness is usually greater than half the contact diameter to provide a contact plug. Therefore, as shown schematically in Figures 9 and 10, the Al must completely react through its sputtered barrier layer, through the height of the W plug (~1μm), and through any barrier / adhesion layer beneath the CVD W plug, before it can interact with Si S/D.

(It should be noted that depending on the conductor resistance needed for the speed of a particular circuit design, the ~3 times higher resistance W can be patterned without a subsequent Al metal. Such a W-only metallization can be used especially with multilevel metallization where the conductor length can be minimized at all levels except at the final bus level. Therefore, the W only metallization would be stable to the higher temperature ~600°C, needed for the W reaction with Si through the barrier / adhesion layer.)

Since thermal stability is not a worse problem for CVD metallization, the sections below will concentrate on the various deposition chemistries and processes. An understanding of the deposition is important since the degradation of device properties can now result from the improper control of the deposition chemistry and process.

4.1   CVD TiN

CVD TiN is deposited by reacting $TiCl_4$ with NH_3 (or N_2+H_2)on a substrate heated to >500°C. (51-54)

$$[1] \qquad 6TiCl_4 + 8NH_3 \rightarrow 6TiN \downarrow + N_2 + 24HCl \uparrow$$

The reactants are diluted in a N_2 or a N_2+Ar carrier gas to obtain the desired pressure with the desired flow rates. The free H (from the decomposition of NH_3) reduces the $TiCl_4$ and the resulting "free" Ti reacts with N to deposit TiN. A large amount of vapor HCl results from the deposition. (The thermal energy required to give atomic hydrogen from the decomposition of NH_3 is lower than that needed for molecular hydrogen. Thus, NH_3 is more reactive than molecular H_2, and the NH_3 reduction chemistry may give a more complete reaction at a given deposition temperature.) In general, the flow rate ratio of NH_3 to $TiCl_4$ is large, and the total system pressure is several-hundred mT. Below is discussed the effect of the reactant flow rates and the total system pressure on the chlorine contamination, resistivity, and step coverage of the deposited TiN film.

Figure 33 (51) shows a typical cold wall single wafer reactor used for the thermally activated deposition of TiN. Here, the reaction occurs on the substrate heated to ~550°C-600°C. Note that $TiCl_4$ is a liquid with a low vapor pressure. Therefore, the Ti source vapor is introduced through a gas line heated to ~50°C. To avoid reaction in the heated gas line producing TiN particulates which may clog the gas line and deposit on the substrate, $TiCl_4$ and NH_3 are introduced separately into the reactor. Alternatively, the reactants are mixed at the entrance to the reactor. Figure 34 (51) shows a cross-section SEM of CVD

TiN deposited into a contact with aspect ratio ~1. The step coverage is nearly conformal. However, increasing the $TiCl_4$ flow rate to increase the deposition rate results in an enhanced gas phase reaction and in a less than conformal deposition.

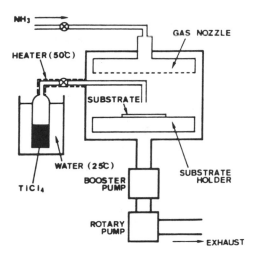

Figure 33. Schematic diagram of a single wafer system for thermal CVD deposition of TiN. Reprinted by permission from the publisher, The Electrochemical Society, Inc. After Ref. 51.

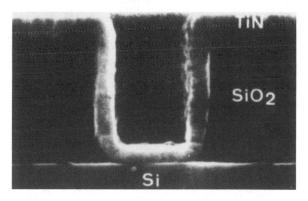

Figure 34. Cross-section SEM micrograph of thermal CVD TiN deposited into a 0.9 μm diameter contact, showing step coverage. Reprinted by permission from the publisher, The Electrochemical Society, Inc. After Ref. 51.

The level of chlorine contamination of the CVD TiN film depends on the deposition temperature, reactant flow rates and total system pressure. Figure 35 (52) shows the atomic % Cl incorporated into the thermal CVD TiN as a function of substrate temperature and for two system pressures, both with a NH_3 to $TiCl_4$ flow rate ratio of 20 and $TiCl_4$ flow rate of 2.2 sccm. The % Cl contamination decreases with increasing substrate temperature and system pressure. This is expected since a higher temperature and longer residence time in the reactor allow the hydrogen to more completely break the Ti-Cl bonds.

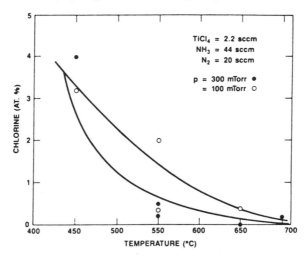

Figure 35. Atomic percent of chlorine incorporated into thermal CVD TiN as a function of deposition temperature and pressure. Reprinted by permission from the publisher, The Electrochemical Society, Inc. After Ref. 52.

The resistivity of TiN also decreases with higher temperature and higher pressure. Again for 2.2 sccm $TiCl_4$ and 44sccm NH_3 flow rate, CVD TiN resistivity decreases to <100 $\mu\Omega$-cm typical of low resistivity sputtered TiN, when the substrate temperature is >550°C-600°C with a total system pressure >100 mT. Figure 36 (53) shows a strong correlation between CVD TiN resistivity and atomic % Cl contamination. Higher resistivity is measured with higher % Cl content.

Again, the higher bulk resistivity of the thin CVD TiN barrier resulting from a higher % Cl content, does not significantly add to the total contact series resistance. However, chlorine in TiN if it is not chemically bound to Ti, will diffuse into the overlying Al during the subsequent heat cycles needed for multilevel metallization and packaging, and result in corrosion of the Al. Therefore, to reduce the % Cl content and the reliability concerns for the Al/CVD TiN structure, a substrate temperature of >600°C is not uncommon for thermal CVD TiN. This high deposition temperature limits the use of thermal CVD TiN to contact to S/D and gate.

The deposition rate for LPCVD TiN with 1.65 sccm $TiCl_4$, 33 sccm NH_3 and 20 sccm N_2 at 300 mT and 650°C is ~600 Å/m. Keeping the $TiCl_4$ and NH_3 flow rates and the deposition temperature and pressure constant, the deposition rate increases with increasing the N_2 flow rate. The deposition rate also increases with increasing $TiCl_4$ flow rate. However, this will increase the amount of Cl incorporated in the TiN film if, at the given temperature, the NH_3 or N_2+H_2 flow rate is insufficient to reduce the additional $TiCl_4$. Therefore, higher TiCl$_4$ flow rate will require higher NH$_3$ or N$_2$+H$_2$ flow rate. Of course, the maximum total flow rate with a given system pressure is limited by the pumping capacity associated with a particular reactor design.

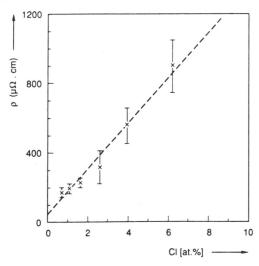

Figure 36. Thermal CVD TiN resistivity as a function of the percent chlorine incorporation. Reprinted by permission from the publisher, The Electrochemical Society, Inc. After Ref. 53.

Additionally, an incubation time which can exceed 1 min, is observed for deposition on oxide. (52) This is too long for a reasonable throughput in a single wafer cold wall system. The incubation time is reduced by increasing the system pressure or by increasing the NH$_3$ flow rate. The incubation time is also a function of the composition of the surface onto which CVD TiN is deposited. A sputtered TiN surface immediately catalyzes the surface reaction and does not show an incubation time.

Ionizing the reactants into the atomic components, increases the efficiency of the chemical reactions and affects the incubation time needed to activate the surface reaction. The increase in reaction efficiency can also reduce the substrate temperature needed to obtain film properties e.g., resistivity and % Cl content, comparable to those obtained with higher temperature thermal CVD.

Figure 37 shows a schematic diagram of a plasma enhanced CVD or PECVD TiN deposition system which also uses electron cyclotron resonance (ECR) heating with microwaves and a confining magnetic field to further increase the plasma density. N$_2$ is introduced into the plasma chamber and ionized by the usual RF excitation and the microwaves which are introduced through the attached waveguide. The ionized nitrogen plasma leaves the plasma chamber and enters the reaction chamber which contains the NH$_3$ or H$_2$ and TiCl$_4$ reactants. Downstream deposition occurs on the heated substrate. Since the plasma ionization now affects the downstream reaction, the TiN film properties (resistivity, % Cl content and step coverage) are additionally affected by the power used to generate the plasma. Usually, the RF power is fixed and the microwave power is varied to affect the deposition.

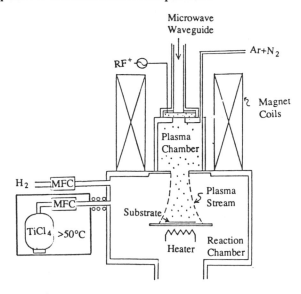

Figure 37. Schematic diagram of a single wafer system for ECR PECVD deposition of TiN.

ECR TiN deposition is in the early development stages and has not yet been optimized. The deposition temperature required for low Cl incorporation (<1%) is still ~600°C. However, further optimization of the deposition conditions to increase the plasma ionization, can decrease the deposition temperature to 400°C. Early results indicate low TiN resistivity ~100 $\mu\Omega$-cm can be obtained with a deposition rate >500 Å/min. The higher deposition rate obtained by increasing the $TiCl_4$ flow rate, is limited by the Cl incorporation with a given microwave power. The % Cl incorporated in the TiN film, can be reduced by increasing the microwave power. However, increasing power can increase plasma potential and, depending on the design of the deposition system, can cause a higher substrate bias. This can cause current flow through and damage to the gate oxide. Such gate oxide damage would result in a lower gate oxide breakdown voltage and in variability of the MOSFET threshold voltage V_T.

The step coverage degrades if the deposition rate is too high to be smoothed by surface mobility. Figure 38 shows deposition into features etched in a deposited phosphorus doped TEOS (tetraethyloxysilane) glass. The deposition is made thicker than that of the normal barrier layer to clearly illustrate the overhang at the top corner of the feature. This overhang indicates that the reaction may also have a gas phase component which shadows deposition into high aspect ratio features. Since the surface mobility is controlled by the material composition and surface cleanliness, the step coverage is also a function of the particular type of deposited oxide. However, the typical plasma enhanced

CVD TiN thickness at the bottom of the contact is ~50% of the top thickness. This coverage is worse than that of thermal LPCVD TiN but better than of sputtered TiN.

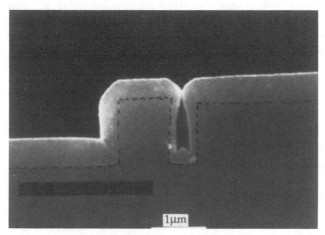

1μm

Figure 38. Cross-section SEM micrograph of PECVD TiN deposited on CVD oxide patterned on Si, showing step coverage difference between a small and large opening.

4.2 Selective CVD W

CVD W is deposited selectively on a heated Si substrate and not on SiO_2, with a two-step reaction process using tungsten hexafluoride (WF_6) + H_2 chemistry diluted with argon. In the first step, Si from the substrate reduces the WF_6 gas in the contact. This provides a W seed layer onto which hydrogen adsorbs. Hydrogen reduction of WF_6 on the W surface adds to the W thickness in the contact. SiO_2 is a thermodynamically stable compound which does not have any free Si atoms to reduce WF_6 on the oxide surface. However, the selectivity of the CVD W is finite and only a certain thickness of W can be deposited selectively before the CVD W reaction byproducts activate the oxide to W deposition.

Si reduces WF_6 heated to greater than 300°C, to deposit W on Si through the reduction reaction

[2] $$2WF_6 + 3Si(+ > 300°C) \rightarrow 2W \downarrow + 3SiF_4 \uparrow$$

with SiF_4 as a gaseous byproduct. The Si reduction reaction is self-limiting in thickness. However, ~2 Å Si are consumed for 1 Å W deposited in the contact. The self-limiting W thickness of the Si reduction reaction is shown in Figure 39, as a function of the deposition temperature, on an undoped or on an As doped

Si substrate. (55) Note that the self-limiting thickness is a function of the deposition temperature. The self-limiting W thickness is ~100 Å up to a certain temperature, then reaches a maximum value ~1000Å, before again decreasing at a higher deposition temperature, to ~100 Å. Dopant in the Si substrate translates the temperature function of the self-limiting W thickness. In particular, an improperly chosen deposition temperature results in a self-limiting W thickness, which consumes a large fraction of a shallow junction depth in the contact, and increases shallow junction leakage.

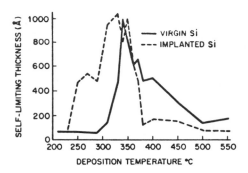

Figure 39. Self-limiting CVD tungsten thickness on undoped or As doped Si from the WF$_6$/Si reaction as a function of substrate temperature. After Ref. 55, reprinted with permission of the publisher.

Additional W thickness is added by the subsequent hydrogen reduction reaction

[3] $$WF_6 + 3H_2(+ > 300°C) \rightarrow W \downarrow + 6HF \uparrow$$

with HF as a gaseous byproduct. It should be noted that the free energy for the Si reduction reaction is comparable to that of the H$_2$ reduction reaction. Thus, the Si reduction can provide the self-limiting W thickness even in the presence of H$_2$, and selective CVD W can be deposited with a one-step process using "only" hydrogen reduction, or with a two-step process using Si reduction followed by hydrogen reduction.

Selective W had several early problems, as shown in the cross-section TEM micrograph in Figure 40. (56) Here, W was deposited in an opening in thermally grown SiO$_2$ on Si. W grows laterally, or encroaches, under the oxide by several tenths of microns. Tunnels, or wormholes, also extend under the oxide edges by several tenths of microns. These tunnels are empty of material but have a W (or W silicide) particle at their end.

The quality of the selective W deposition critically depends on the cleanliness of the Si surface. A thin "native" oxide is not continuous on Si. The selective W chemistry can effectively nucleate through the numerous and closely spaced, pinholes in the native oxide. Thus, W surrounds the closely spaced islands of native oxide, and the W/Si interface and the W surface are smooth. Thicker residues (not only Si oxides but also as shown in Figure 24, polymers resulting from the contact RSE process), are usually more continuous. As in the case of silicide sintering with a thicker metal/Si interface shown in the left side of Figure 13, the W chemistry preferentially consumes Si from the fewer existing pinholes for a longer time before forming the continuous W layer needed to uniformly nucleate additional W thickness. Thus, selective W deposited on improperly cleaned Si has a rougher W/Si interface (56) and a rougher surface which can be recessed from the original Si surface by much more than the expected ~100 Å.

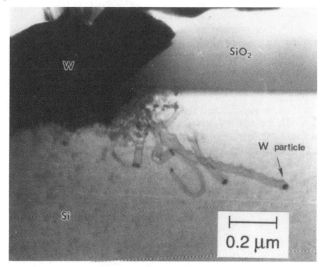

Figure 40. Selective CVD tungsten deposited into patterns in thermal SiO_2 on Si using a two-step deposition chemistry. Reprinted by permission from the publisher, The Electrochemical Society, Inc. After Ref. 56.

Encroachment and tunneling are empirically related to the type of oxide mask. Significantly fewer problems are seen with selective W deposited into contacts etched in a CVD deposited oxide with lower tensile stress, than into contacts etched in a thermally grown SiO_2 with a higher compressive stress. Additionally, the Si surface can be effectively cleaned with a H_2SO_4 + dilute HF sequence (dilute HF for a controlled time so as not to excessively increase the diameter of the contacts in the deposited oxide). Figure 41 (57) shows a cross-section TEM micrograph of 1000Å thick selective one-step W deposition into contacts RSE etched in a phosphorus doped on undoped deposited oxide. The W is deposited at 300°C, on As-doped Si. A thick Al overlayer is sputter deposited with poor step coverage on the W. Note that smooth W is deposited without obvious encroachment or wormholes.

W can be selectively deposited on silicides with the above chemistries. Figure 42 (57) shows a cross-section TEM micrograph of selective W deposited with a one-step process at 300°C on $CoSi_2$ on As-doped Si. Note that the WF_6

chemistry reacts in filaments with $CoSi_2$. Additionally, the WF_6 diffuses through the $CoSi_2$ grain boundaries to react with the underlying Si. The silicide grain boundaries can be preferentially etched by the contact RSE chemistry. In this case, the larger grain boundaries can result in missing silicide grains and severe W/Si reaction. The W/Si reaction, at the enlarged silicide grain boundaries, can consume a large fraction of the junction depth below the silicide and may be enhanced by the high tensile stress (high 10^9 dynes/cm^2) of the silicide.

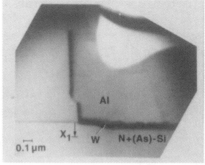

Figure 41. Selective CVD tungsten deposited into patterns in deposited CVD oxide on Si using a one-step deposition chemistry after sputter depositing Al. After Ref. 57, reprinted with permission of the publisher.

Figure 42. Selective CVD tungsten deposited on $CoSi_2$ using a one-step deposition chemistry. After Ref. 57, reprinted with permission of the publisher.

The TEM micrograph in Figure 41 shows a smooth W deposition without encroachment and wormholes. However, even such high-quality deposits can degrade the reverse bias diode leakage of shallow junctions. Figure 43 (57) shows the mean reverse bias (5V) diode leakage (measured for 130 diodes) as a function of unsilicided n$^+$(As)/p junction depth. The unmetallized large area diodes are directly probed and compared with the same diodes after 1000 Å selective W deposition with two deposition chemistries. The two-step deposition (Si reduction followed by hydrogen reduction) severely degrades junctions <3000 Å deep. The one-step deposition (hydrogen reduction only) degrades junctions <2000 Å deep. However, an increase in the high leakage sport population is measured even for 2500 Å deep junctions.

Selective W deposition conditions can be found to successfully deposit W on <2000 Å shallow junctions. (58) However, the "correct" selective W deposition conditions may be different for each particular device process and deposition system design.

Thicker selective W depositions eventually loose their selectivity to oxide. The W reaction byproducts (SiF_x) and the WF_x molecules (59-61) will adsorb on the oxide surface near where W deposits in the opening to Si. (Sub-fluorides with x=4 or 5 exist in the reactor because of incomplete reduction of the majority of the WF_6 flow which does not participate in the W deposition.) This adsorption, given enough time, will nucleate W deposition. The adsorption depends on the type of oxide. Selectivity is lost to thermal oxide before it is lost on boron and phosphorus doped deposited oxides.

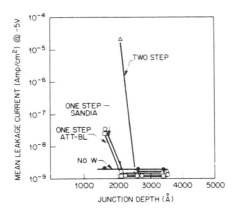

Figure 43. Mean reverse bias (5V) diode current for various $n^+(As)/p$ junction depths in Si, unmetallized, and after one-step or two-step selective CVD tungsten deposition. After Ref. 57, reprinted with permission of the publisher.

Since selectivity loss is related to the adsorption of reaction byproducts, a reactor which reduces the volume of these byproducts will provide better selectivity. The quartz tube of a hot wall reactor is easily activated to W deposition. Thus, W deposition on the large area of a hot wall reactor generates a relatively large volume of reaction byproducts. A cold wall reactor has only the much smaller wafer surface area heated. W deposits on a much smaller wafer area and produces a much smaller volume of reaction byproducts than in a hot wall reactor. Additionally, the flow of WF_6 is lower and more localized than that in a hot wall reactor. Therefore, cold wall reactors result in better selectivity and are the systems of choice for selective W (and, as discussed below, for blanket W) deposition.

Figure 44 (62) shows ~1µm thick selective W plug deposition in a cold wall system into contacts etched in a deposited oxide. Note the good selectivity. Finally, it should be noted that most ULSI process technologies will use planarization of all contact and via dielectrics. Such a highly planar topography will have contacts of different depth to S/D and gate. Therefore, a compromise

W plug thickness must be chosen to avoid overfilling the shallower contacts and shorting between the closely spaced metal lines.

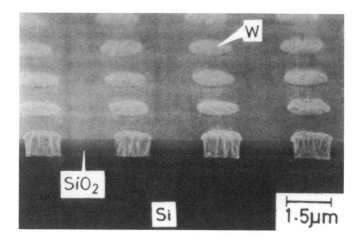

Figure 44. Thick selective CVD W deposited in a cold wall reactor showing the filling of contact holes in a deposited oxide without losing selectivity. After Ref. 61, reprinted with permission of the publisher.

4.3   Blanket CVD W

The sensitivity of device degradation and the loss of selectivity to the type of oxide, and to the selective W deposition conditions and reactors, can be avoided by a more complex blanket W process. The selective CVD W deposition becomes a blanket deposition on an appropriate surface which is say, sputter deposited after contact RSE. Blanket CVD W is deposited on a barrier / adhesion layer by the reacting WF_6 source gas with hydrogen (H_2) or with silane (SiH_4) on a substrate heated to >300°C. Argon is usually used to dilute the reactants and to obtain the required reactor pressure with the required component flow rates.

Again, cold wall reactors are preferred to deposit blanket CVD W. In a hot wall reactor, which is just a pumped tube furnace, CVD W deposits on all heated surfaces are exposed to the reaction. Thus, the relatively expensive high purity WF_6 gas deposits W on the hot furnace tube wall. Since the surface area of the tube wall is much greater than that of the Si wafer, W deposition on the hot wall reduces the W deposition rate on the Si wafer, to <100 Å/min. Additionally, the buildup of stressed W thickness results in W peeling from the hot wall and in W particles being deposited on the wafer. Since the entire wafer is heated in a tube furnace, CVD W also deposits on the backside of the Si wafer. Typically, the barrier / adhesion layer is only sputtered on the device front side of the wafer, while the backside has all the films grown during prior processing, the last of which is the deposited oxide dielectric. Since W does not adhere well to oxide, the CVD W deposited on the wafer backside would lift during subsequent handling and add to the defect density problem during later steps in the processing.

Cold wall reactors are designed with nozzles to introduce the reactants close to the heated wafer surface in a low pressure Ar ambient. Additionally, in a cold wall reactor, only the wafer front is exposed to the deposition chemistry, and CVD W does not deposit on the oxide backside of the wafer. Therefore, the particle problem is significantly reduced. It should be noted that the particle problem must still be addressed because of possible W buildup on the nozzles and other unintentionally heated surfaces close to the substrate.

Silane (SiH_4) or H_2 can are used to reduce WF_6 and deposit blanket W on the heated wafer surface. W deposited by the hydrogen reduction of WF_6, as before, produces six HF molecules for each deposited W atom. Such a large volume of HF byproduct will etch oxide and may etch the barrier / adhesion layer. Possible attack of the underlying Si can then result in device degradation before enough W is deposited to "seal" the contact. On the other hand, W deposited by the SiH_4 reduction of WF_6 has both SiF_4 and HF as thermodynamically possible reaction byproducts. Therefore, less HF and reduced etching of the oxide and barrier / adhesion layer are expected during blanket W deposition with SiF_4 reduction of WF_6. Thus, silane reduction is the preferred chemistry to deposit blanket W in a cold wall reactor with a deposition rate ~0.5-1 μm/min. A possible negative aspect of silane reduction is that WF_6 can react in the gas phase, with Si from the thermal decomposition of SiH_4. This gas phase component of the silane reduction chemistry results in a directional component of the deposition and in a slightly worse step coverage than that resulting from H_2 reduction of WF_6. However, the very high CVD W surface mobility on most barrier materials (62) overwhelms any directional component and results in nearly 100% step coverage.

Figure 45 (55) shows a cross-section SEM micrograph for the H_2 reduction of WF_6 onto a sputtered TiN barrier / adhesion layer. The very good filling of the contact with aspect ratio ~1.5, with an ~1 μm thick W deposition, clearly shows the good mobility of the surface activated reaction. (The small keyhole is probably due to the step coverage of the sputtered TiN underlayer.) Therefore, the H_2 or SiH_4 must first adsorb on the surface before the reduction of WF_6 can proceed. The adsorption time depends on the surface to be activated. Deposition on TiN shows an incubation time which can be as long as several minutes. (55) On the other hand, W is autocatalytic, and no incubation time is measured when blanket W is deposited on a very thin sputtered W layer.

Figure 46 shows a planar TEM micrograph of the thick blanket CVD W deposit. The CVD W grain size is 0.1-0.2 μm. This grain size is consistent with the resistivity ~7-8 $\mu\Omega$-cm for the CVD W film being higher that the ~6 $\mu\Omega$-cm value for bulk W. Figure 47 shows a cross-section TEM micrograph of the Si/sputtered TiN/ CVD W/sputtered Al metallization. Note the columnar

TiN microstructure which can allow WF_6 to diffuse through the barrier grain boundaries during the early stages of the deposition.

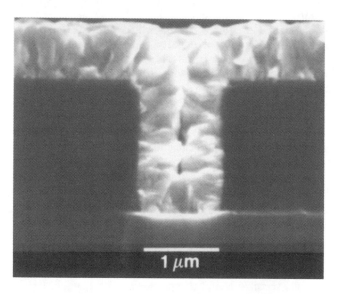

Figure 45. Cross-section SEM micrograph of blanket CVD tungsten deposited on TiN into a contact etched in a deposited oxide. After Ref. 55, reprinted with permission of the publisher.

Figure 46. Plan view TEM micrograph showing grain size of blanket CVD W deposited on TiN.

WF_6 diffusing through the barrier / adhesion layer grain boundaries would react with the underlying Si and increase the shallow junction leakage after CVD W deposition. Figure 48 (55) shows the effect of deposition temperature on reverse bias (5V) diode leakage. The n+(As)/p junctions are made by outdiffusing As implanted into ~700 Å $CoSi_2$ to a depth ~500 Å below the silicide. The data shows the probability plot of the percentage of diode measurements with less than a certain leakage current (10^{-11}A is equivalent to 4 nA/cm^2). W is deposited on a 900 Å thin sputtered TiN barrier, at 325°C, 375°C or 425°C. A very high sport population ~80% develops after W is deposited at 325°C. The sport population decreases with higher deposition temperature and finally no sport population is measured after W deposition at 425°C. Note that the 425°C deposition temperature corresponds to the minimum self-limiting W thickness on As doped Si, for the selective W-Si reduction reaction (see Figure 39).

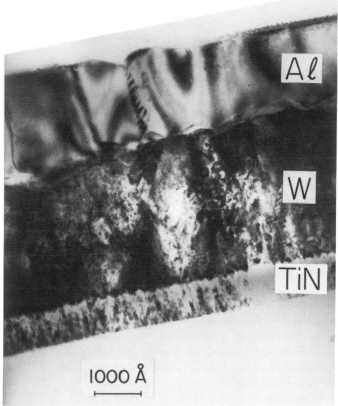

Figure 47. Cross-section TEM micrograph of the sputtered Al on blanket CVD W on sputtered TiN on Si layered metallization structure.

It should be noted that deeper junctions are less sensitive to the CVD W deposition conditions. Also, sputtered TiW is less effective than TiN as a barrier to the CVD W chemistry. (64) Thus, as before, the preferred barrier / adhesion layer for blanket CVD W is TiN on Ti. An additional thin sputtered W layer may be needed on the TiN to avoid the incubation time associated with the W reaction on TiN. To avoid the keyhole due to poor sputtering step coverage, the CVD W deposition conditions must be consistent with a minimum barrier layer thickness. Reducing the size of the keyhole and choosing an appropriate barrier is also important to avoid completely etching the W in the contact during the reactive sputter etchback step to form the W plug. (65, 66)

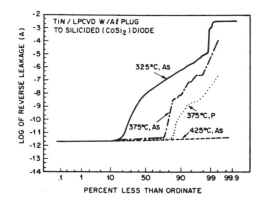

Figure 48. Probability plot of reverse bias (5 V) diode leakage of n$^+$(As)/p junctions 500 Å below 700 Å thick CoSi$_2$ after metallization with blanket CVD W on sputtered TiN as a function of the W deposition temperature. After Ref. 55, reprinted with permission of the publisher.

5.0 LOCAL INTERCONNECTS AND MULTILEVEL METALLIZATION

A large fraction of the integrated circuit area is occupied by metallization. The fraction of the area used for interconnects, will increase as more and smaller devices are included in ULSI circuits, unless a vertical dimension is added to the metallization process. This concept of multilevel metallization using local and global interconnects is illustrated in Figure 49. (67) The vertical metallization adds to the complexity of the ULSI process integration but also adds significant flexibility to the circuit layout. Typically, advanced circuit layout uses several levels of "poly" for local interconnects and several levels of "metal" for global interconnects.

Local interconnects connect parts of devices on a local scale. For example, a CMOS inverter has a common gate for the NMOS and PMOS device. Thus, the poly-Si gate is a local interconnect. A local interconnect can also be used to connect the PMOS drain of a CMOS inverter to the NMOS source. Local interconnects have a length usually measured in microns or tens of microns. Thus, the bulk material resistivity for local interconnects, is an important but not a critical consideration. A moderately high resistivity ~100 $\mu\Omega$-cm, will not significantly add to the much larger contact and device series resistances.

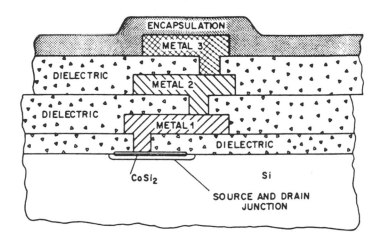

Figure 49. Schematic cross-section diagram of a multilevel metallization structure. After Ref. 67, reprinted with permission of the publisher.

Global interconnects connect subcircuits to each other and to external power. Bus lines and bit lines for large DRAM and SRAM memories are examples of global interconnects. Here, the line length is measured in hundreds or thousands of microns. Thus, the bulk material resistivity must be very low, and global interconnects are usually patterned in Al (or in the future, possibly Cu or Ag for the final bus level).

In general, materials for local interconnect must have a relatively low resistivity and must be compatible with the deposition of good quality interlevel dielectrics (usually CVD oxide at >700°C). Poly-Si and silicides are stable materials at such high temperature. Thus, both heavily doped poly-Si and silicides can be used to contact the junction in Si.

The poly-Si resistivity is related to the doping concentration and the annealing temperature at which the poly-Si grains are grown, and the doping is made electrically active. The activation temperature, <900°C, needed so as not to degrade the intrinsic short channel device performance by redistributing the device dopants, will not give a sufficiently low implanted poly-Si resistivity. Thus, the implanted poly-Si must be silicided with either a polycide or a salicide process. The contact resistance is determined by the cleanliness of the interface between poly-Si and Si. The CVD poly-Si deposition can add to the interface condition, by slightly oxidizing the Si in the contact. Outdiffusion of the dopants from poly-Si into the Si in the contact (68, 69) during the activation, will reduce the interface effects. Note that to avoid a diode contact, the dopant type in the poly-Si local interconnect must be the same as the dopant type in the S/D junction.

Sputtered and annealed silicide has a resistivity <100 $\mu\Omega$-cm. However, the silicide stoichiometry is important. As seen before in Section 2.2, Si rich silicides equilibrate by precipitating Si in the contact. A high contact resistance will result if this Si is undoped. Thus, a "contact wash" implant into and outdiffusion from the silicide local interconnect may be needed to lower contact resistance. Note that WSi_x is a good choice for local interconnect contact to n^+ S/D since phosphorus readily outdiffuses from WSi_x when annealed at $<900°C$.

Tungsten can be used as a local interconnect. However, to avoid W silicide formation and degradation of shallow junctions, W cannot contact S/D when the next level dielectric is deposited at above $600°C$. Thus, W is safely used above the first level metallization. For example, if in Figure 49, the local interconnect is labeled as "metal 1," W is used beginning with the metal level labeled "metal 2".

Ideally, the multilevel metal process (except for the local interconnect to S/D and gate and maybe for the final bus line) is a repeating sequence similar to that used for the first level of metal. Thus, the materials are the same as those previously discussed. However, the previous concern for the metal-Si interaction is now replaced by a concern for the metal-metal interaction which can also give high contact resistance. When step coverage is not an issue, the sputter deposition of a barrier material (say TiN on Ti) is followed by the sputter deposition of the Al alloy. A refractory metal cap (W or TiN) can be deposited over the Al alloy (say Al -.5% Cu) to improve the electromigration resistance. (70, 71)

The design rules for multilevel metal of ULSI circuits are the same as those for metal 1. Thus, the vias between metal levels have a high aspect ratio, and step coverage is an important issue. Here, the deposition of a barrier / adhesion layer is followed with blanket CVD W. If the circuit design allows the use of a metal interconnect with the higher W resistivity, a blanket CVD W with low enough surface roughness can be directly patterned. Otherwise, the blanket CVD W must be then etched back to form a via plug. The barrier/Al alloy/cap structure is sputter deposited and patterned.

Selective CVD W can also be applied to via plugs. W can be selectively deposited in vias to Al or W. The WF_6 reduction on Al leaves a fluoride interface between W and Al. (72) This interface results in a high contact resistance between CVD W and Al. This AlF_3 interface becomes less of a problem when the CVD W is deposited on Al at a higher temperature. (73) However, the temperature at which the Al fluoride evaporates before W deposition is ~$600°C$, close to the melting point of Al. CVD W deposited on W, Ti-W, or TiN does not leave a fluoride interface since the W and Ti fluorides are gaseous at low temperature, and the TiN is chemically stable. Thus, a selective W via plug will probably be on an Al alloy with a capping layer.

Inorganic and organic interlevel dielectric materials are discussed in detail in Chapters 4 and 5, respectively. However, for completeness, a brief discussion of the interlevel dielectric issues impacting metallization reliability is included below.

The interlevel dielectric (ILD, usually CVD oxide) is deposited on Al alloy lines and must therefore be deposited at a low temperature sufficiently below the Al melting point. To further avoid potential barrier failure at any Al contact to S/D or gate, the ILD should be ideally deposited below ~450°C. Low temperature CVD (usually plasma enhanced -- PECVD or ECR) (74) oxides are deposited using silane (SiH_4) or tetraethyloxysilane (TEOS) as the Si source, O_2, N_2O or ozone (O_3) as the oxidizer, and Ar or N_2 as the inert carrier gas at 100mT to atmospheric pressure. However, the oxide quality and step coverage usually degrade at a lower deposition temperature of 400°C-450°C.

Low oxide quality is due to the incomplete reaction depositing an oxide which is not dense SiO_2. The unsaturated Si-O-H bonds can potentially cause a reliability problem because they allow for diffusion of impurities such as sodium or chlorine during subsequent processing. Note that additional ionization of the reactants in an ECR-CVD (75) system similar to that shown in Figure 37 may improve the oxide quality, and possibly with a lower deposition temperature.

Step coverage is important for depositing the ILD without a void in the gap between Al lines. Gap filling becomes more difficult with increasing metal wall angle or aspect ratio. High packing density requires near vertical metal walls. (To improve step coverage, the metal is slightly tapered with a wall angle of ~85°, and the top is slightly smaller than the bottom of the metal feature.) Thus, the aspect ratio of the gap between metal features is ~1 for ULSI applications with a metal thickness ~0.5μm and the minimum design rule spacing ~0.5μm.

The top of Figure 50 (74) shows the poor gap filling of a low temperature PECVD oxide. The large gas phase component and poor surface mobility result in the breadloaf profile over metal steps. Applying a bias voltage to the substrate adds a sputtering component with its preferred sputtering angle. Thus, bias sputtering during the deposition preferentially removes the corner buildup. This is shown at the bottom of Figure 50. The gap filling is improved, but material is simultaneously removed and deposited at the expense of a lower effective deposition rate. A variation of this bias deposition is a sequential dep-etch-dep process where an ILD thickness is deposited, etched back, and deposited again to its final thickness. Finally, it should be noted that the biased plasma deposition process must again be consistent with low radiation damage of the gate oxide and tight V_T control.

Spin-on materials are also used as low temperature ILDs. Siloxanes or polymers containing Si-O bonds can be spun-on and baked at <400°C to deposit oxides. Polyimides can also be spun on and cured at <400°C to deposit a hard dielectric material which can withstand subsequent processing up to the polyimide glass transition temperature ≥450°C. However, spin-on materials still have several problems. Spin-on glasses with low mobile ion contamination are available. In the case of polyimides, the mobile ion concentration must be further reduced by several orders of magnitude from the current parts per billion range. The high thermal expansion coefficient of the spin-on material, compared to that of Si, causes high stress in, and cracking or buckling of the spin-on material during subsequent processing. Polyimides additionally have a lower dielectric constant and can absorb moisture (possible corrosion of the Al metal lines). Therefore, spin-on dielectrics, as shown in Figure 51, are used in sandwich ILD structures. Here, a thin PECVD oxide is deposited on the metal pattern. The spin-on oxide (76) or polyimide (77) is applied to smooth the metal topography and is then capped with another thin layer of PECVD oxide which provides the moisture barrier and controls the composite ILD thermal expansion. However, the larger thermal expansion of polyimide can be difficult to accommodate, even for the deposition of a low temperature CVD oxide cap. Finally, it should be noted that some CVD oxides such as ozone TEOS, deposited at a very low temperature, also absorb moisture and should also be used in a sandwich structure.

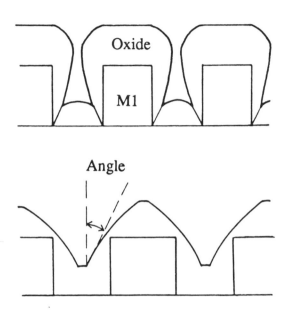

Figure 50. Schematic representation of low temperature PECVD oxide deposition on metal steps at top, without bias; and at bottom, with bias. Reprinted by permission from the publisher, The Electrochemical Society, Inc. After Ref. 74.

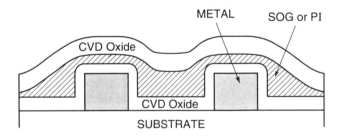

Figure 51. Schematic diagram showing the use of lower quality spin-on dielectrics sandwiched between higher quality low temperature CVD oxide films to smooth metal patterns.

6.0 CONCLUSION AND FUTURE TRENDS

ULSI circuit performance improves with the downward scaling of device dimensions. However, in order not to lose circuit performance, the circuit layout must be optimized, and the correct metallization materials and processes must be chosen to minimize the interconnect resistance and capacitance.   This metallization process must be compatible with narrower metal lines (higher current density) and have good coverage of high aspect ratio contacts and vias (CVD replaces PVD or sputtering). Longer, narrower, and thinner (to reduce the device topography seen by subsequent processes such as CVD deposition of the interlevel dielectric) metal interconnects required for ULSI will add to the total series resistance. Thus, local interconnects and multilevel metallization are necessary.

Selecting the correct metallization requires an understanding of the material, contact,and thermal stability properties and the deposition processes which will affect these properties. An additional and important process step which, if inadequate, will increase the contact resistance is the cleaning of the submicron contacts and vias before metallization.

In many cases, the zero level of metallization is the silicide used in a polycide or salicide process. Silicides reduce the resistance of the poly-Si gate local interconnect and the Si S/D. Additionally, silicides are used to engineer the contact resistance to doped Si. Previous sections described various aspects of silicides. WSi_x (x ~2.5) polycide has desirable dopant diffusion properties. However, since P most readily outdiffuses form WSi_x, this process is probably limited to CMOS circuits with buried channel PMOS devices (n-poly on p-channel) and may be difficult to scale to <0.35 μm gate length with low Vt ~-0.5 V.

$TiSi_2$ or $CoSi_2$ salicides are attractive self-aligned processes. To be consistent with the shallow junctions needed for ULSI, the silicide must be nominally ~500 Å thin. Thermal stability of the thin silicide is an important concern which, depending on the silicide thickness, may limit the back end process annealing to ~800°C. The silicide stability can be improved by: (a) using as thick a silicide as is consistent with low junction leakage; (b) degrading (but in a controlled manner) the cleanliness (78) of the metal/Si interface across which the silicide reaction occurs; (c) engineering the silicide grain size (interface cleanliness and sintering temperature); or, (d) implanting dopants which amorphize the silicide and segregate to and "stuff" the silicide grain boundaries during silicide grain regrowth. Increasing the silicide thickness particularly for $TiSi_2$, where the junction is formed by conventional implant and anneal processes before silicidation, may not be consistent with low junction leakage; or degrading the silicide interface may not be consistent with reproducible silicide sintering. In this case, the silicide can be raised above the original Si surface by selectively depositing and completely siliciding an epitaxial Si or poly-Si plug. (79, 80)

It should be noted that dopant outdiffusion from $CoSi_2$ gives shallower junctions than those made by sintering $TiSi_2$ on already formed junctions.

Lateral diffusion is a major difficulty for a symmetric $CoSi_2$ salicide process. Dopant segregation from the poly-Si into the silicide and an increase in silicide roughness due to thermal processing may be a problem particularly for $TiSi_2$ salicide. Therefore, so as not to effect the gate doping at the poly-Si/gate oxide interface and thereby effect the gate work function, the salicided gate height will probably be larger than the polycide gate height.

Shorter channel devices must have <0.25 μm shallow junctions. The metallization process must not degrade either the reverse bias diode leakage of, or contact resistance to, these shallow junctions. Here, barrier materials are needed between Al and Si, to defer to higher temperatures, the catastrophic Al interaction with Si. Thin Ti, W or Ti-W barrier materials fail after annealing at ~450°C, because of the Ti or W interaction with Al. TiN barriers in contact with Al are stable up to annealing at ~550°C. Ti under TiN is the preferred barrier material structure which combines the low contact resistance of Ti to n^+ or p^+ Si and the superior barrier properties of TiN to Al. Note that barrier failure correlates with grain boundary diffusion. Thus, the exact failure temperature depends on the deposition equipment and conditions which affect material microstructure.

The poor sputtering step coverage reduces the thickness and hence the effectiveness of the barrier material at the bottom of high aspect ratio contacts and vias. Chemical vapor deposition (CVD, PECVD, and ECR CVD), in a cold wall reactor, has better contact coverage and is used to deposit W and TiN barrier materials. CVD W is being widely developed. CVD TiN is entering the development phase. However, collimation increases the thickness of the barrier material at the bottom of the contact and is extending the usefulness of reactive sputtered TiN for high aspect ratio contacts.

Tungsten on Si or metal silicide is stable to annealing up to 600°C-650°C. However, an Al overlayer interacts with Si through 1000 Å CVD W at 450°C. Al spiking occurs where the grain boundaries of Al, W, and silicide align to provide a direct path for the Al to the underlying Si. Al on a thicker CVD W plug can be annealed at a higher temperature or longer time, simply because the Al is that much further removed from Si.

Thick CVD W can be deposited selectively into contacts in a deposited oxide to either Si or silicide. However, selectivity results from Si reduction of WF_6 and Si consumption. Thus, shallow junction leakage depending on the deposition conditions and can increase during selective CVD W deposition. Blanket CVD W deposited on a barrier/adhesion layer (again, TiN on Ti is preferred because of its superior contact and barrier properties), and etched back by RSE etchback to form contact plugs, reduces damage to shallow junctions. However, the W chemistry can still affect the Si through the grain boundaries of the barrier/adhesion layer, and the CVD W deposition conditions can still impact shallow junction leakage. Additionally, the CVD W chemistry can react with the barrier material (particularly Ti-W). Thus, the barrier thickness in the contact under blanket W is important. Again, contact coverage must be considered particularly when the barrier/adhesion layer is sputter deposited.

CVD TiN has significantly better, although not perfect, contact coverage. For lower temperature depositions, and depending on pressure and flow rate parameter values, the resistivity of CVD TiN can be higher than that of sputtered

TiN. However, even this higher CVD TiN resistivity can be used for thin barrier layers. Unfortunately, the higher resistivity correlates with chlorine contamination, which can corrode the Al overlayer. Recently, TiN has been deposited using a cleaner metalorganic CVD chemistry which does not contain chlorine. (54, 81) TiB_2 (82) can be deposited by CVD with comparable resistivity, good (but not perfect) contact coverage with a cleaner chemistry, and with better temperature stability (83) than TiN. Again it should be noted that an attractive shorter term approach for increasing the barrier thickness at the bottom of the contact or via (at the expense of worse sidewall coverage and the need for a two-step sputtering process) is to sputter TiN in a PVD system with "collimation".

Local interconnects relax the requirements on thermal stability of barrier materials to the reaction with Al simply because the metal contact can be moved from the S/D to over oxide. Multilevel metallization is needed to decrease the total series resistance by shortening the length of the metal interconnects. Conceptually, the metal levels are similar in structure and materials (from top to bottom, barrier material/Al/W plug/barrier material). One additional requirement is a low temperature, good quality interlevel dielectric CVD oxide with good step coverage. Reducing the metal thickness will reduce the aspect ratio of the spaces between the metal lines and reduce the difficulty of filling the spaces with a CVD oxide. Spin-on oxide or polyimide effectively smooth the metal patterns. However, the spin-on materials are of lower quality and are used over a thin CVD oxide to improve the coverage of a subsequent CVD oxide capping layer. ECR oxides are being deposited with a low temperature dep-etch-dep process with high density and good step coverage.

Multilevel metallization shortens the length of the metal runners. Therefore, a higher resistivity interconnect such as W, with significantly better electromigration and stress voiding reliability, is a reasonable alternative to a lower resistivity material such as Al, with lesser reliability. However, the longer, and usually slightly wider, final bus metal interconnect must have a lower resistivity. If the resistivity of sputtered Al alloys is not low enough, copper may be necessary. Polycrystalline copper has lower resistivity and significantly better electromigration lifetime than polycrystalline Al but has major compatibility (corrosion, drift migration, and spiking through oxide to Si, poor adhesion to oxide), and processing (patterning by RSE) problems. (84, 85)

Interconnect reliability is discussed elsewhere in this book. However, it should be finally noted that the higher current density carried by the metal interconnects in ULSI will decrease the electromigration lifetime of the metal lines. Currently, most of the electromigration failures are correlated with Al narrowing (due to the depth of focus of the lithography tools) and thinning (due to sputtering step coverage) over device topography. Electromigration is the displacement of metal atoms along grain boundaries because of metal atom collisions with electrons. If the polycrystalline Al has a barrier underlayer and overlayer which do not readily electromigrate, the electromigration of the Al will not result in an open interconnect. Thus, barrier materials improve the metal reliability. Al sputtering coverage is not an issue when the high aspect ratio contacts and vias are filled with a CVD W plug. The Al/W plug/Al via contact is more reliable than the Al/Al (86) via contact. Thus, contact and via plugs improve reliability. Unfortunately, multilevel metallization with barrier materials may decrease the stress voiding lifetime of the metal interconnect. Stress voiding is the movement of material to relax mechanical stress. An interconnect with nonzero stress will, in time, stress void without the flow of current. Thus, the metallization structure (metal layers, interlevel dielectrics, and cap material) must have near zero sum stress; (87) i.e., the stress of the metallization layers must be minimized or compensated.

In conclusion, this chapter discussed certain advanced materials and processes which will be used for 0.5 µm and 0.35 µm design rules. Reliable metallization is a critical part of scaling dimensions into the ULSI regime. Silicides and barrier materials are needed to engineer and maintain the integrity of the various contacts. However, thermal instability of a thin salicide process is a limiting factor. Therefore, any circuit layout which can replace salicide with additional levels of metal, allows more freedom in the choice of interlevel dielectric materials and process temperatures before Al metal deposition. CVD metallization is needed to improve coverage into the high aspect ratio contacts. Metallization is further complicated by the need for multilevel wiring. Additionally, each interlevel dielectric must be planarized to allow optical lithography to achieve dimension control in the sub-half micron regime.

Processing with the 0.25 µm design rules is a challenge which will require further evolutionary improvements in the metallization process and layout to further increase interconnect reliability, to moderate the increase in interconnect series resistance, and to compliment further improvements in the front-to-middle end device process. CVD W will be needed (probably blanket on collimated TiN on Ti); and, either RSE etched back to form contact or via plugs, or patterned for short metal interconnects. In particular, layout improvements will be needed to minimize the interconnect lengths and to allow the use of refractory metals and their silicides for local interconnects. These higher resistivity interconnects of CVD W or WSi_x (deposited by sputtering or CVD as contact and via coverage requires) have a significantly better reliability than Al, for the typical operating ULSI current density ~10^6 A/cm^2 and temperature ~200°C. Cu at the bus level is a better theoretical choice than Al. However, the processing experience with Al, and the processing uncertainties with Cu, make Al a better practical choice. In this case, the bus level design rules must be (and typically are) relaxed,and an appropriate cap material must be used to satisfy the current density and stress limits imposed by interconnect reliability.

REFERENCES

1.   J. Brews, "The Submicron MOSFET," in "High Speed Semiconductor Devices," S. M. Sze, ed., (Wiley-Interscience, New York, 1990) p. 141-143.
2.   K. K. Ng and W. T. Lynch, "Impact of Series Resistance on MOSFET Scaling," IEEE Trans. Electron Devices, **ED-34**, 503 (1987).

3. S. M. Sze, "Physics of Semiconductor Devices," 2nd ed., (Wiley-Interscience, New York, 1981) p. 305.
4. K. M. Cham, S.-Y. Oh, D. Chin, J. L. Moll, K. Lee, P. V. Voorde, "Computer-Aided Design and VLSI Device Development," 2nd ed., (Kluwer Academic, Boston, 1988) p. 285.
5. M. Hansen and A. Anderson, "Constitution of Binary Alloys," (McGraw-Hill, New York, 1958).
6. D. Pramanik and A. Saxena, "VLSI Metallization Using Aluminum and its Alloys," Solid State Technol., 26, 127 (Jan. 1983).
7. D. Pramanik and A. Saxena, "VLSI Metallization Using Aluminum and it Alloys - Part II" Solid State Technol., 26, 131 (Mar. 1983).
8. R. Rosenberg, M. J. Sullivan and J. K. Howard, "Effect of Thin Film Interactions on Silicon Device Technology," in "Thin Films-Interdiffusion and Reaction," J. M. Poate, K. N. Tu and J. W. Mayer,eds. (Wiley-Interscience, New York, 1978) p. 13.
9. G. J. Van Gurp, J. L. C. Daams, A. van Oostrom, et.al., "Aluminum-Silicide Reactions. I. Diffusion, Compound Formation and Microstructure," J. Appl. Phys., 50, 6915 (1979).
10. C. Y. Ting and M. Wittmer, "Investigation of the $Al/TiSi_2/Si$ Contact System," J. Appl. Phys., 54, 937 (1983).
11. C. Y. Ting and M. Wittmer, "The Use of Titanium-Based Contact Barrier Layers in Silicon Technology," Thin Solid Films, 96, 327 (1982).
12. M.-A. Nicolet, "Diffusion Barriers in Thin Films," Thin Solid Films, 52, 415 (1978).
13. H. Bader, M. A. Lardon, and K. J. Hoefler, "Topographical Aspects of Physical Deposition Methods for VLSI - Metallization," in "Multilevel Metallization, Interconnection and Contact Technologies," L. Rothman and T. Herndon, eds. (Electrochemical Society, Princeton, N.J., 1987) p. 185.
14. A. C. Adams, "Dielectric and Polysilicon Film Deposition," in "VLSI Technology," 1st ed., S. M. Sze, ed. (McGraw-Hill, New York, 1983), Ch. 3.
15. G. E. Georgiou, K. P. Cheung and R. Liu, "Planarized Aluminum Deposition on TiW and TiN Layers by High Temperature Evaporation," Proc. VLSI Multilevel Interconnection Conf. (1989) p. 315.
16. M. Inoue, H. Hashizume and H. Tsushikawa, "Properties of Aluminum Thin Films Sputter Deposited at Elevated Temperatures," J. Vac. Sci. Technol. A, 6, 1636 (1988).
17. G. E .Georgiou, R. M. Baker and S. A. Eshraghi, "Effect of Sputtered TiW Deposition Conditions on Barrier Properties for Submicron Metallization," Proc. IEEE VLSI Multilevel Interconnection Conf. (1991) p. 420.
18. G. E. Georgiou, F. A. Baiocchi, H. S. Luftman, et.al., "Thick Selective Electroless-Plated Cobalt-Nickel Alloy Contacts to $CoSi_2$," J. Electrochem. Soc., 138, 2061 (1991).
19. T. Ohba, "Multilevel Metallization Trends in Japan," in "Advanced Metallization for ULSI Applications," V. V. S. Rana, R. V. Joshi and I. Ohdomari, eds., (Materials Research Soc., Pittsburgh, Pa., 1992) p. 25.

20.   S. P. Murarka, "Silicides for VLSI Application," (Academic Press, New York, 1983), pp. 15-20, 30-43 and 78-97.

21.   D. Pramanik, A. N. Saxena, O. K. Wu, et.al., "Influence of the Interfacial Oxide on Titanium Silicide Formation by Rapid Thermal Annealing," J. Vac. Sci. Technol. B, **2**, 785 (1984).

22.   P. Revesz, J. Gyimesi, L. Pogany and G. Peto, "Lateral Growth of Titanium Silicide Over a Silicon Dioxide Layer," J. Appl. Phys., **54**, 2114 (1983).

23.   V. Probst, H. Schabber and P. Lippens, "Limitations of $TiSi_2$ as a Diffusion Source," Appl. Phys. Lett., **52**,1803(1988).

24.   C. M. Osborn, "Formation of Silicided Ultra-Shallow Junctions Using Low Thermal Budget Processing," J. Elect. Mat., **19**, 67 (1990).

25.   Y. Kim, H .Z. Massoud, and R. B. Fair, "The Effect of Ion Implantation Damage on Dopant Diffusion in Silicon During Shallow-Junction Formation," J. Elect. Mat., **12**, 143 (1989).

26.   D. C. Chen, T. Cass, J. E. Turner, et.al., "$TiSi_2$ Thickness Limitations for Use with Shallow Junctions and SWAMI and LOCOS Isolation," IEEE Trans. Electron Devices, **ED-33**, 1463 (1986).

27.   H. J. Levinstein, S. P. Murarka, and A. K. Sinha, "Cobalt Silicide Metallization for Semiconductor Integrated Circuits," U. S. Patent 4,378,628, dated April 5, 1983.

28.   S.J.Hillenius, R.Liu, G.E.Georgiou, et.al., "A Symmetric Submicron CMOS Technology," IEDM Tech. Digest, (1986) p. 256.

29.   A. Sitaram and S. P.Murarka, "Formation of Cobalt Silicides in Arsenic Implanted Cobalt on Silicon System," in "Advanced Metallizations for Microelectronics," A. Katz, S. P .Murarka, and A. Applebaum, eds., (Materials Research Soc., Pittsburgh, Pa., 1990), p. 97.

30.   R.Liu, D.S.Williams, and W.T.Lynch, "A Study of the Leakage Mechanisms of Silicided n^+p Junctions," J. Appl. Phys., **63**, 1990 (1988).

31.   S. P. Murarka, D. B. Fraser, A. K. Sinha, and H. J. Levinstein, "Refractory Silicides of Titanium and Tantalum for Low-Resistivity Gates and Interconnects," IEEE Trans.Electron Devices, **ED-27**, 1409 (1980).

32.   D. Lrors, J. A. Fair, K. A. Monning, and K. C. Saraswat, "Properties of Low Pressure CVD Tungsten Silicide as Related to IC Process Requirements," Solid State Technol., **26**, 183 (April 1983).

33.   F. A. Baiocchi, N. Lifshitz, T. T. Sheng, and S. P. Murarka, "Equilibration of Nonstoichiometric Ta-Si Deposits on Polycrystalline Silicon at High Temperatures," J. Appl. Phys., **64**, 6490 (1988).

34.   M. Y. Tsai, F. M. d'Heurle, C. S. Peterson, and R. W. Johnson, "Properties of Tungsten Silicide Film on Polycrystalline Silicon," J. Appl Phys., **52**, 5350 (1981).

35.   P. Eichinger, E. Frenzel, and F. Neppl, "Diffusion of As, B and P in $TaSi_2$," in "Thin Films and Interfaces II," J. E. E. Baglin, D. R. Cambell and W. K. Chu, eds., (North-Holland, New York, 1984) p. 165.

36.   N. Lifshitz, F. A. Baiocchi, and D. Malm, "Study of Implanted Dopant Distribution in Polysilicon-Tantalum Silicide Gate Structure," Proc. 1st

International Conf. on ULSI Science and Technology, S. Broydo and C. M. Osborn, eds., (Electrochemical Soc., Princeton, N. J., 1987) p. 465.

37.   T. Hara, H. Takahashi, and S.-C. Chen, "Ion Implantation of Arsenic in Chemical Vapor Deposited Tungsten Silicide," J. Vac. Sci. Technol. B, **3**, 1664 (1985).

38.   M. P. Lepselter and J. M. Andrews, "Ohmic Contacts to Silicon, in "Ohmic Contacts to Semiconductors," B. Schwartz, ed., (Electrochem. Soc., Princeton, N.J., 1969) p. 159.

39.   T. Tashiro, H. Takemura, T. Kamiya, et.al., "An 80 ps ECL Circuit with High Current Density," IEDM Tech. Digest, (1984) p. 686.

40.   L. I. Maissel and R. Glang, eds.,"Handbook of Thin Film Technology," (McGraw-Hill, New York, 1970).

41.   D. B. Fraser, "Metallization," in "VLSI Technology," 1st ed., S. M. Sze, ed., (McGraw-Hill, New York, 1983), Ch. 9.

42.   C.Y.Ting and B.L.Crowder, "Electrical Properties of the Al/Ti Contact Metallurgy for VLSI Application," J.Electrochem.Soc., **129**, 2590 (1982).

43.   W. -K. Chu, J. W. Mayer, and M.-A. Nicolet, "Backscattering Spectroscopy," (Academic Press, New York, 1978) Ch. 4.

44.   S. A. Eshraghi, G. E. Georgiou, R. Liu, et.al., "Electrical Degradation of Al/TiW/CoSi$_2$ Shallow Junctions," J. Vac. Sci. Technol. B, **9**, 69 (1991).

45.   P.-H. Chang, R. Hawkins, T. Donifield, et.al., "Aluminum Spiking at Contact Windows in Al/Ti-W/Si," Appl. Phys. Lett., **52**, 272 (1988).

46.   M. Wittmer, "Properties and Microelectronic Applications of Thin Films of Refractory Metal Nitrides," J. Vac. Sci. Technol. A, **3**, 1797 (1985).

47.   K. Ahn, M. Wittmer, and C. Y. Ting, "Investigation of TiN Films Reactively Sputtered using a Sputter Gun," Thin Solid Films, **107**, 45 (1983).

48.   D. S. Williams, F. A. Baiocchi, R.C. Beairsto, et.al., "Nitrogen, Oxygen, and Argon Incorporation during Reactive Sputter Deposition of Titanium Nitride," J. Vac. Sci. Technol. B, **5**, 1723 (1987).

49.   C. Y. Ting, "TiN formed by Evaporation as a Diffusion Barrier between Al and Si," J. Vac. Sci. Technol., **21**, 14 (1982).

50.   R. J. Schutz, "TiN as a Diffusion Barrier between CoSi$_2$ or PtSi and Aluminum," Thin Solid Films, **104**, 89 (1983).

51.   N. Yokayama, K. Hinode, and Y. Homma, "LPCVD TiN as a Barrier Layer in VLSI," J. Electrochem. Soc., **136**, 882 (1989).

52.   A. Sherman, "Growth and Properties of LPCVD Titanium Nitride as a Diffusion Barrier for Silicon Device Technology," J. Electrochem. Soc., **137**, 1892 (1990).

53.   M. Butting, A. F. Otterloo, and A. H. Montree, "Kinetical Aspects of the LPCVD of Titanium Nitride from Titanium Tetrachloride and Ammonia," J. Electrochem. Soc., **138**, 500 (1991).

54.   J. T. Hillman, M. J. Rice, Jr., D. W. Studiner, and R. F. Foster, "Comparison of Titanium Nitride Barrier Layers Produced by Inorganic and Organic CVD," Proc. IEEE VLSI Multilevel Interconnection Conf. (1992) p. 246.

55. N. Lifshitz, J. M. Andrews, and R. V. Knoell, "Shallow Silicided Diodes with LPCVD Tungsten Plug," in "Tungsten and Other Refractory Metals for VLSI Application III," V. A. Wells, ed., (Materials Research Soc., Pittsburgh,Pa., 1988) p. 225.
56. W. T. Stacy, E. K. Broadbent, and M. H. Norcott, J. Electrochem. Soc., 32, 444 (1985); R. S. Blewer and M. E. Tracy, "Detrimental Effects of Residual Silicon Oxides on LPCVD Tungsten Depositions in Shallow Junction Devices," in "Tungsten and Other Refractory Metals for VLSI Application II," E.K.Broadbent, ed. (Materials Research Soc., Pittsburgh, Pa., 1987) p. 235.
57. G. E. Georgiou, J. M. Brown, M. L. Green, et.al., "The Influence of Selective Tungsten Deposition on Shallow Junction Leakage," in "Tungsten and Other Refractory Metals for VLSI Application II," E. K. Broadbent, ed., (Materials Research Soc., Pittsburgh, Pa., 1987) p. 227.
58. T. Moriya and H. Itoh, "Selective CVD of Tungsten and its Application to MOS VLSI," in "Tungsten and Other Refractory Metals for VLSI Application," R. S. Blewer, ed., (Materials Research Soc., Pittsburgh,Pa., 1986) p. 21.
59. C. McConica and K. Cooper, "A Model for Tungsten Nucleation on Oxide," in "Tungsten and Other Refractory Metals for VLSI Applications II," E. K. Broadbent, ed., (Materials Research Soc., Pittsburgh, Pa, 1987) p. 51.
60. N.Lifshitz, "Nature of the Self-Limiting Effect in the Low Pressure Chemical Vapor Deposition of Tungsten," Appl. Phys. Lett., 51, 967 (1987).
61. I. Hirase, T. Sumiya, M. Schack, et.al., "The Effects of Impurities and Byproducts on Selective W Deposition," in "Tungsten and Other Refractory Metals for VLSI Application III," V. A. Wells, ed., (Materials Research Soc., Pittsburgh, Pa., 1988) p. 133.
62. H. Itoh, T. Moriya ,and M. Kashigawi, "Tungsten CVD: Application to Submicron VLSICs," Solid State Technol., 30, 83 (Nov. 1987).
63. J. E. J. Schmitz, R. C. Ellwanger, and A. J. M. van Dijk "Characterization of Process Parameters for Blanket Tungsten Contact Fill," in "Tungsten and Other Refractory Metals for VLSI Application III," V. A. Wells, ed., (Materials Research Soc., Pittsburgh,Pa.,1988), p. 55.
64. S. A. Eshraghi, G. E. Georgiou, and R. Liu, "The Effect of the Chemical Vapor Deposition of Tungsten on Shallow n^+p and p^+n Junctions using Titanium-Tungsten as a Barrier," J. Appl. Phys., 68, 2839 (1990).
65. R. J. Saia, B. Gorowitz, D. Woodruff, and D. M. Brown, "Plasma Etching Methods for the Formation of Planarized Tungsten Plugs used in Multilevel VLSI Metallization," J. Electrochem. Soc., 135, 936 (1988).
66. J. M. F. G. van Laarhoven, H. J. W. van Houtum, and L. De Bruin, "A Novel Tungsten Etchback Scheme," Proc. 6th IEEE VLSI Multilevel Interconnection Conf. (1989) p. 129.
67. S. P. Murarka, "Metallization," in "VLSI Technology" 2nd ed., S. M. Sze, ed. (McGraw-Hill, New York, 1988), p. 415.

68. G. E. Georgiou, T. T. Sheng, F. A. Baiocchi, et. al., "Shallow Junctions by Outdiffusion from As Implanted Polycrystalline Silicon," J. Appl. Phys., **68**, 3714 (1990).

69. G. E. Georgiou, T. T. Sheng, J. Kovalchick, et.al., "Shallow Junctions by Outdiffusion from BF_2 Implanted Polycrystalline Silicon," J. Appl. Phys., **68**, 3707 (1990).

70. H .P. W. Hey, A. K. Sinha, S. D. Steenwyck, et.al., "Selective Tungsten on Aluminum for Improved VLSI Interconnects," IEDM Tech. Digest, (1986), p. 50.

71. H. H. Hoang and J. M. McDavid, "Electromigration in Multilayer Metallization Systems," Solid State Technol., **30**, 121 (Oct. 1987).

72. V. V. S. Rana, J. A. Taylor, L. H. Holschwandner, et.al., "Thin Layers of TiN and Al as Glue Layers for Blanket W Deposition" in "Tungsten and Other Refractory Metals for VLSI Application II," E. K. Broadbent, ed., (Materials Research Soc., Pittsburgh, Pa, 1987) p. 187.

73. R. V. Joshi, S. Brodsky, T. Bucelot, et.al., "Low-Resistance Submicron CVD W Interlevel Via Plugs on Al-Cu-Si," Proc. IEEE VLSI Multilevel Interconnection Conf. (1989) p. 113.

74. G. C. Smith and A. J. Purdes, "Sidewall-Tapered Oxide by Plasma Enhanced Chemical Vapor Deposition," J. Electrochem. Soc., **132**, 2172 (1985); C.-P. Chang, C. S. Pai, and J. J. Hseih, "Ion and Chemical Radical Effects on Step Coverage of Plasma Enhanced Chemical Vapor Deposition Tetraethylorthosilicate Film," J. Appl. Phys., **67**, 2119 (1990).

75. C. S. Pai, J. F. Miner, and P. D. Foo, "Electron Cyclotron Resonance Microwave Discharge for Oxide Deposition Using Tetraethoxysilane," J. Electrochem. Soc., **139**, 850 (1992).

76. H.W.M.Chung, S.K.Gupta, and T.Aaldwin, "Fabrication of CMOS Circuits Using Non-Etchback SOG," Proc. IEEE VLSI Multilevel Interconnection Conf. (1989) p. 373.

77. H.Eggers, H.Fritzsche, and A.Glasl, "A Polyimide-Isolated Three-Layer Metallization System for Bipolar Gate Arrays," Proc. IEEE VLSI Multilevel Interconnection Conf. (1985) p. 163.

78. H.Sumi, T.Nishihara, Y.Sagano, et.al., "New Silicidation Technology by SITOX (Silicidation Through Oxide) and its Impact on Sub-half Micron MOS Devices," IEDM Tech. Digest, (1990), p. 249.

79. C.-S. Wei, V. Murali, M. Lawrence, et.al., "The Use of Selective Silicide Plugs for Submicron Contact Fill," Proc. IEEE VLSI Interconnection Conf. (1989) p. 136.

80. T.Iijima, A.Sishiyama, Y.Ushiku, et.al., "A Novel Selective Ni_3Si Contact Plug Technique for Deep-Submicron ULSIs," Symposium on VLSI Technol. Tech. Digest, (1992), p. 70.

81. I. V. Raaijmakers, R. N. Vrtis, G. S. Sandhu, et.al, "Conformal Deposition of TiN at Low Temperature by Metal Organic CVD," Proc. IEEE VLSI Multilevel Interconnection Conf. (1992) p. 260.

82. L.M.Williams, "Plasma Enhanced Chemical Vapor Deposition of Titanium Diboride Films," Appl. Phys. Lett., **46**, 43 (1985).

83.   J.Shapiro, J.J.Finnegan, and R.A.Lux, "Diboride Diffusion Barriers in Silicon and Ga As Technology," J. Vac. Sci. Technol. B, **4**, 1409 (1986).
84.   D.S.Gardner, J.Onuki, K.Kudoo, and Y.Misawa, "Encapsulated Coppers Interconnection Devices Using Sidewall Barriers," Proc. IEEE VLSI Multilevel Interconnection Conf. (1991) p. 99.
85.   P.-L.Pai and C.H.Ting, "Copper as the Future Interconnection Material," Proc. IEEE VLSI Multilevel Interconnection Conf. (1989) p. 258.
86.   F. Matsuoko, K. Hama, H. Itoh, et.al., "An Electromigration and Related Resistance Increase Phenomenon on a Tungsten Filled Via Hole Structure," Proc. IEEE VLSI Multilevel Interconnection Conf. (1988) p. 491.
87.   A.Isobe, Y.Numazawa, and M.Sakamoto, "Increase in EM Resistance by Planarizing Dielectric Film over Al Wirings," Proc. IEEE VLSI Multilevel Interconnection Conf. (1989) p. 161.

3

ALUMINUM BASED MULTILEVEL METALLIZATIONS IN VLSI/ULSICs

K. RAMKUMAR And SUMANTA K. GHOSH
Rensselaer Polytechnic Institute
Troy, New York
And
ARJUN N. SAXENA
International Science Company
Ballston Lake, New York
And
Rensselaer Polytechnic Institute
Troy, New York

1.0   INTRODUCTION

The purpose of this chapter is to inform the reader about the aluminum (Al) based multilevel metallizations in VLSIC/ULSIC's. They are used predominantly in today's microelectronics manufacturing, and are expected to continue to be used for years to come despite several limitations of Al and its alloys. The reader is assumed to be familiar with the various microelectronics manufacturing technologies. However, a detailed discussion of the fundamentals of the properties and processes of aluminum based multilevel metallization materials and technologies, current status, problems and approaches to their solutions are given in addition to the future directions. Key process, reliability and equipment issues are also discussed.

It is obvious that no device nor an IC would function without making contacts of the right kind to the different regions of the devices, interconnecting them in a desired configuration reliably on the chip, and eventually bringing out the selected inputs/outputs through the package to interface with the rest of the electronic system. These contacts are essentially ohmic contacts to the various n^+ and p^+ junctions in bulk silicon, gates over oxides in MOST's, and rectifying Sckottky contacts in bulk silicon. The contacts interconnecting various levels of metallization separated by interlevel dielectrics are referred to as vias, which obviously, are ohmic also. Pure aluminum, Al, hasbeen used as the contact and the interconnect metal, however, it is unsatisfactory in today's IC's. Various problems are encountered when pure Al is used; its alloys (AlX) are used to overcome some of these problems where X is usually Si and/or Cu. Even homogeneous AlX alloys are not always satisfactory, so various layers of AlX with other metals/alloys (Y/AlX/Z) are used. Y is termed as the underlayer and usually Ti, TiN or TiW is used for this layer. Z is termed as the cap layer and TiN or TiW is used for it. These are explained in Section 4.4.

In many of the IC's, the devices are laid out on a chip in such a configuration that some interconnects in a single level metallization either cannot be made (due to crossings causing shorts), or they have to meander through long zig-zag paths, causing performance degradation due to large RC time constants. To reduce the chip size while increasing the number and the type of devices per chip (for ever higher levels of integration and improvement in performance), the interconnects cannot be made in a single level metallization. Therefore, multilevel metallizations in which several levels of metallizations separated by dielectric layers are needed. The speed performance, chip size, reliability and yield of VLSI/ULSIC's are heavily impacted by multilevel metallizations, not just by device scaling anymore. This is even more so as the technologies are being scaled down to sub-0.5 µm. A simple equation for the RC time constant of an interconnect is,

[1]
$$RC = \rho \frac{l}{dw} \frac{lwk_{ox}\varepsilon_o}{t_{ox}}$$

$$= \rho_\square \frac{l^2}{t_{ox}} \kappa_{ox}\varepsilon_0$$

where,

ρ = thin film resistivity of the interconnect

$\rho_\square$ = sheet resistivity of the interconnect

l = length of the interconnect

w = width of the interconnect

d = thickness of the metal/alloy film comprising the interconnect

t_{ox} = thickness of dielectric between adjacent metallization levels

κ_{ox} = dielectric constant of the oxide (or insulator)

ε_o = permittivity constant.

The term l^2 strongly suggests the need to shorten the length of the interconnects to reduce the RC time constant, and enhance the speed performance of the IC's. Reduction of l can be best achieved by the use of multilevel metallizations which allow shortest paths to connect various interconnect levels through vias. Thus, multilevel metallizations are a necessity in today's and the future IC's. Further enhancing the speed performance requires lowering the $\rho_\square$ of the interconnect from that of Al, and the κ of the interlevel dielectric from that of SiO_2. These new approaches, however, are far from being production worthy yet. Therefore, Al based multilevel metallizations with SiO_2 based interlevel dielectrics are expected to continue to be used for the foreseeable future in microelectronics manufacturing.

As discussed in Section 5.3, the need for the planarization of both the metals and the dielectrics in multilevel metallizations is of paramount importance.   Planarization technologies are in use in microelectronics manufacturing now, and they shall be used increasingly in the more advanced microelectronics manufacturing technologies of the future.   This, viz., planarization, essentially groups the interconnects and the dielectrics into two categories:

1. Horizontal, viz, parallel to the wafer surface.

2. Vertical , viz, perpendicular to the wafer surface.

Further, at each level, the interconnect comprises several layers of metal. As an example, the interconnect at the first level makes contacts to the shallow n^+ and p^+ junctions through a silicide and a barrier metal layer; further it generally has an under-layer, main conductor layer, and a cap layer of another barrier metal. This interconnect system becomes even more complicated when contact planarization is used employing metals other than Al.   Similar complications are encountered as we follow the subsequent second, third, fourth and more levels of interconnects. Also, the inter-level dielectrics, planarized or not, sometimes consist of more than one type of material. For example, the dielectric layer between the first and second level of interconnects can consist of PECVD-TEOS-SiO_2 (*Plasma Enhanced Chemical Vapor Deposition of SiO_2 using the liquid precursor Tetra Ethyl Ortho Silicate*) followed by O_3-TEOS SiO_2 and etch-back, and finally, a top layer of PECVD-TEOS-SiO_2. Keeping track of all of these various layers of materials, both metals and dielectrics, in a multilevel metallization scheme becomes a difficult task.   A variety of designations for each layer in various levels of metals and dielectrics have been used in the literature, which sometimes create confusion. A nomenclature which helps in designating the various layers in different levels in a systematic manner will be helpful in R & D as well as in manufacturing. Such a nomenclature has been presented in Reference 1.

The focus of this chapter is on the Al based multilevel metallizations, so the properties and technologies of pure Al, homogeneous alloys AlX and layered structures Y/AlX/Z are discussed primarily.   Associated technologies of interlevel dielectrics, which are mainly SiO_2 based, and of a few other metals needed for Al based metallizations are also discussed for a complete multilevel metallization system. Future directions of metals with resistivities lower than Al, and dielectrics with dielectric constants lower than SiO_2, are also discussed briefly in Section 7.2.

A brief comment on the use of the expression *multilevel metallizations* is that it is somewhat more generic and all-encompassing than *multilevel interconnections*. An interconnection at any level can, and usually does, consist of several layers of metals/alloys which have already been patterned by the lithographic / etching / planarization technologies. So the discussion of an interconnection may only be limited to the materials used rather than the additional information needed for a more complete discussion of the fundamental properties of materials, deposition, interface, lithographic, etching and planarization technologies of a metallization. However, the expressions *metallizations* and *interconnections* are *synonymous* and used interchangeably in this chapter depending on the technology step being discussed. The term multilevel in both metallizations or interconnections implies the use of interlevel dielectrics to insulate and isolate each metallization level. Examples are interlevel dielectrics and sidewall spacers around the gates and interconnects.

In any microelectronics manufacturing, the technologies of multilevel metallizations are important in determining the bottom line, viz, the profitability, in addition to determining other crucial technical factors such as the chip size, its speed performance and reliability. Since all the manufacturing technologies of current multilevel metallizations are Al based, it is of paramount importance to understand and improve them.

2.0   REQUIREMENTS IN MULTILEVEL METALLIZATIONS

Multilevel metallizations require, by definition, the use of metals/conductors and dielectrics/insulators. Their usage and key requirements of their properties are given below.

2.1   Conductors/Metals

As indicated earlier, conductors are used in multilevel metallizations for many applications in IC.'s. The important ones among these are:

1. Ohmic contacts to various devices
2. Schottky contacts
3. Horizontal interconnects at different levels
4. Vertical interconnects (vias) between various horizontal interconnect levels
5. Barrier metals for both vertical and horizontal interconnects

The desirable properties of conductors for these applications can be grouped into three categories, viz, electrical, physical/metallurgical and process/chemical. These are summarized in Table I (1).

Table I: Desirable properties of conductor materials

ELECTRICAL	PHYSICAL / METALLURGICAL	PROCESS / CHEMICAL
1. Low sheet resistivity, $\rho_\square$. 2. Low specific contact resistance (ρ_c) to n^+ and p^+ Si. 3. High electromigration resistance at contacts and in interconnects; good lifetime. 4. Good MOS properties (reliable interface with SiO_2; no mobile ions).	1. Reliable shallow junction contacts (no spiking). 2. No hillock formation; surface smoothness. 3. Low stress - as deposited and after thermal cycling 4. No stress voiding 5. Good adhesion. 6. Ability to withstand post metallization processes and high alloying / sintering temperatures. 7. No intermetallic reactions with adjacent layers.	1. Deposition -process uniformity, reproducibility, cost (for large volume) -Good step coverage -Selective deposition. 2. Patterning -Fine line lithography -Good dry etch processes. 3. Structuring -Planarization -Lift-off. -Chem-mechanical polishing 4. Reliability -No mobile ions -No corrosion. 5. Bonding

2.2   Dielectrics/Insulators

Dielectrics are used in a multilevel metallization system for the following applications:

1. Interlevel dielectric,
2. Passivation layer,
3. Sidewall spacer, and
4. Isolation.

The desirable properties of dielectrics for these applications are summarized in Table II.

Table II. Desirable properties of dielectric materials

ELECTRICAL	PHYSICAL / METALLURGICAL	PROCESS / CHEMICAL
1. Low dielectric constant (K).	1. Low defect density.	1. Deposition -process uniformity, reproducibility, cost (for large volume)
2. High dielectric breakdown strength.	2. Low reflow temperature.	-Good step coverage on metal lines
3. Low leakage	3. Low stress - as deposited and after thermal cycling; no stress voiding.	-Selective deposition.
4. Good MOS properties (reliable interface with metallization, low charge density; no mobile ions).	4. Good adhesion.	2. Patterning -Fine line lithography -Good dry etch processes.
	5. Planarization.	3. Structuring -Planarization -Lift-off. -Chem-mechanical polishing
	6. No interactions with adjacent metal layers.	4. Reliability -No mobile ions -Good moisture resistance.

More detailed discussion on the properties of dielectrics used in multilevel metallizations can be found in Chapters 4 and 5 of this book.

3.0 APPLICATIONS OF AlX IN MULTILEVEL METALLIZATIONS AND KEY PROBLEMS

The most important applications of pure Al, its homogeneous alloys (AlX), and layered structures (Y/AlX/Z) in multilevel metallizations and key problems associated with these are discussed below. Explanation of the causes of the problems and approaches to their solutions are discussed in detail in section 6. The most common additives to Al for homogeneous alloys, AlX, are Si and Cu which overcome several problems. Typical weight percentage of Si and Cu added to Al are 0.8-1.2 and 0.5-4.0 respectively. Most common layered Al alloys use Ti, TiW or TiN for under-layers, and TiW or TiN for cap layers. A cross-section of a multilevel metallization scheme with various contacts, interconnecting vias, and dielectrics is shown in Figure 1(a). Figure 1(b) shows the cross section of any one level of the metallization which shows the under-layer and the cap layer.

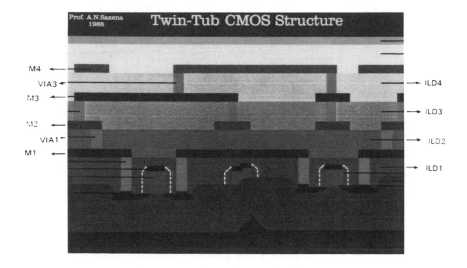

a

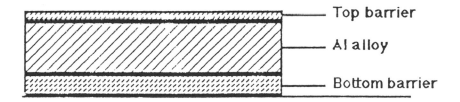

b

Figure 1. Cross-section of (a) multilevel metallization scheme, and (b) one level of metallization.

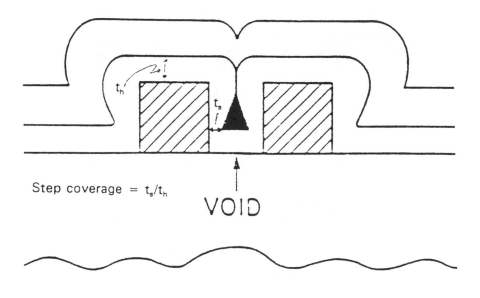

Step coverage = t_s/t_h

VOID

Figure 2. Definition of step coverage.

3.1   Contacts

In the earlier technologies, pure Al was used for making ohmic contacts to p^+ and n^+ junctions in Si directly, and to make rectifying Schottky contacts (2). In the present technologies, pure Al is no longer used either for ohmic or for Schottky contacts. With the shrinking device geometries at higher levels of integration, which also require shallow junctions, several problems become more severe in pure Al based contacts. These are given below.

Junction Spiking: This failure is associated with the dissolution of Si from the contacts into Al, when pure Al is used, due to high solubility of Si in Al. This causes Al penetration into the contact through the junction, causing shorts (3). This problem gets worse as the junctions get shallower with device scaling.

Step Coverage: The step coverage is defined as the ratio of the film thickness on the walls of the contact cut or via, to the thickness on planar region (any one or more of T_1/T_o, T_2/T_o, and T_3/T_o), as shown in Figure 2.   Al deposited by conventional sputtering has poor and inadequate step coverage of sub micron contacts/vias with high aspect ratios. Ideally, complete filling of such contacts/vias, viz., planarization, is highly desirable.

High Contact Resistance: The inadequate filling of Al in sub-micron contacts with high aspect ratio (height/width), because of poor step coverage, results in a high contact resistance. Inadequate cleaning of contacts by wet chemical or sputter-etch (due to re-deposition of SiO_2 from the walls on the contacts) processes also cause high contact resistance. Another contribution to the high contact resistance comes from the epitaxial re-deposition of Si in the contacts following its dissolution in Al at higher temperatures.

To overcome these problems due to pure Al in contacts, AlSi alloys or an underlying barrier metal (also known as underlayer) have been used. However, even this has not proved to be adequate for sub-micron contacts with high aspect ratios. The reliability problems in these contacts have led to the use of alternate metallizations such as tungsten, for the vertical interconnects, in the present VLSI technologies. In addition, new methods of Al deposition are also being developed for improving the contact filling for lower contact resistance and higher reliability. These will be discussed in detail in section 5.1.

3.2 Horizontal Interconnects

Pure Al interconnects were used for horizontal interconnects widely in the earlier technologies. However, in VLSI technologies, they posed a variety of reliability problems. The most significant ones are listed below.

Electromigration: Electromigration refers to the mass transport in Al as a result of current flow through it. This mass transport within a horizontal interconnect made of pure Al results in formation of voids or opens in the interconnect, leading to its failure. This failure mechanism becomes very severe with the shrinking of interconnect dimensions because of the increase in current density. Also, this failure is more likely to occur, in particular, in the interconnect regions with poor step coverage where the metal is thinner and hence the current density gets increased. Small quantities of additives such as Cu into Al helps slow the electromigration process, thereby enhancing the useful life of the interconnect.

Hillock Formation / Lateral Extrusion: This is another major cause of failure in the horizontal interconnects. Stresses developed in the Al interconnect due to the difference in the thermal expansion coefficients of Al and Si/SiO_2 result in Al protruding out from the interconnect in the vertical (hillock) or horizontal (lateral extrusion) directions. This leads to shorting between adjacent metallizations in different levels, and between adjacent lines of metal in the same level. Electromigration can also affect the hillock formation and lateral extrusion in the Al interconnects.

Stress Voiding: This failure mode is brought about by the stresses imposed on the Al interconnect by the underlying and overlying dielectrics, and the intrinsic stress and the microstructure of the interconnect itself. It can lead to the formation of voids in the Al interconnect when it is subjected to high temperature excursions either during subsequent processing, storage or operation. It can also accelerate failure due to electromigration.

Corrosion: Al interconnects get corroded subsequent to dry etching with Cl_2 based chemistry and photoresist removal processes. They can also get corroded by interaction with the external package, primarily by the moisture. The moisture reacts with Cl_2 remaining on AlX interconnects after RIE to produce HCl, and with phosphorous in the PSG to produce H_3PO_4, which can etch AlX in an irregular manner causing notching (5). This gives rise to increases in the interconnect resistance and accelerates its failure due to electromigration.

Etching: The RIE of AlX containing Cu poses problems as Cu does not have volatile compounds at normal processing temperatures, and hence, it does not etch easily. This gives rise to precipitates and can also accelerate the corrosion of AlX lines.

Passivation: The top most AlX level in a multilevel metallization needs to be covered with a passivation layer such as SiO_2, Si_3N_4, polyimide, or a combination of these layers. This is done for scratch protection and hermeticity of the underlying interconnects. The passivation can cause stress voiding in the AlX interconnect due to thermal mismatch and defects in the passivation. Also, if the passivation does not provide good hermeticity, moisture can cause corrosion of the interconnects and lead to other instabilities.

3.3   Vertical Interconnects (Via)

A via connects two horizontal interconnects at different levels. It is similar to a contact except that the metal in the via makes contact with AlX or Y/AlX/Z on both sides while in a contact the metal makes contact with Si or silicide on one side. The problems of AlX based vias are similar to those of contacts, viz, inadequate filling and high via resistance, although Si spiking is not a problem. Further, since vias are basically interconnects, electromigration and stress voiding problems also occur and are important.

3.4   Interlevel Dielectrics

Interlevel dielectrics (ILD's) are used to insulate adjacent levels of metallization in a multilevel metallization and have a very significant impact on the reliability of the underlying/overlying interconnects, and hence the multilevel metallization itself. The electromigration and stress voiding are both affected by the stresses in the ILD's and their variations during thermal cycling. The planarization of the ILD over an interconnect pattern is also very important to minimize, if not eliminate, the step coverage problems of the next level metallization. The filling of narrow gaps between AlX lines without voids is also essential for planarization because voids can lead to reliability problems. The thermal conductivity, k, of the ILD is another crucial parameter. It is found that the thin films of SiO_2 have lower k than its value for bulk quartz. This can lead to a larger temperature rise in the interconnects than that obtained in the thermal modelling of the ICs which assume bulk k values. This in turn can further accelerate the failure due to electromigration.

Some aspects of dielectric planarization and its impact on the reliability of metallization are covered later in section 5. More detailed discussion on planarization and reliability can be found in other chapters of this book.

4.0   PROPERTIES   OF Al, AlX AND RELATED MATERIALS

Aluminum has been and continues to be the most widely used material for the IC metallizations, including the present day multilevel metallizations, although its alloys and layered combination with other metals are used instead of pure Al. The features that make Al so popular have been discussed in several references (6, 7). The important ones are as follows:

1. Low sheet resistance,
2. Good adherence to SiO_2 and Si_3N_4,
3. Ease of deposition,
4. Patternability,
5. Good bondability, and
6. Easy availability and low cost.

To be able to have the most reliable multilevel metallization with Al and its alloys, it is important to understand their key properties and interactions with Si, SiO_2 and many other materials of interest to the microelectronics industry .

In the following sections, the important properties of Al, AlX and other related materials used in multilevel metallizations are discussed. These properties are grouped under two categories, viz, material and electrical.

4.1   Material Properties

The most significant material properties of Al for its application in multilevel metallization refer to its interaction with Si, SiO_2 and other materials such as TiN and TiW. Some of the key properties are discussed below.

AlX Phase Diagrams: The Al-X phase diagrams shown in Figure 3 (8) summarize the metallurgical interactions between Al and alloying elements such as Si and Cu, over a wide range of temperatures. As shown in Figure 3(a), the eutectic temperature of the AlSi system can be identified as 577°C. This limits the maximum temperature to which an IC/Si wafer can be subjected to after Al deposition. In multilevel metallization technologies, the post metal (after the first level metal, AlX, deposition) processing temperatures are usually limited to about 400°C-450°C.

Figure 3(a) also shows that no compound (silicide) is formed between Al and Si, unlike other metals such as Pt, Ti, W, Ni etc. that are also used in microelectronic fabrication technologies.

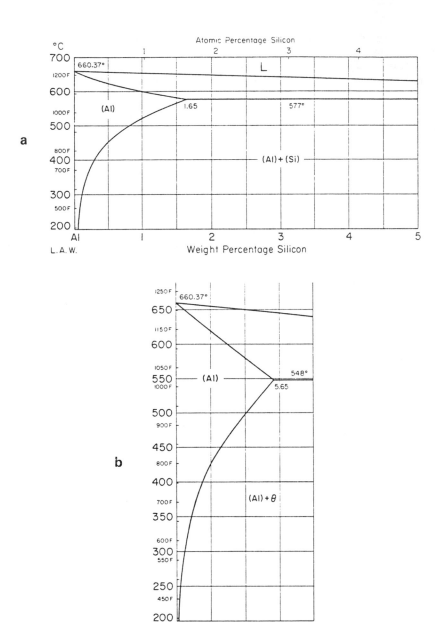

Figure 3. Phase diagram of (a) AlSi and (b) AlCu. <u>Metals Handbook, 8th Edition, Vol. 8,</u> (American Society for Metals, Metals Park, Ohio, 1973), pps. 259-263. Reprinted with permission.

An appreciable fraction of Si goes into solution in Al depending on temperature/time as shown by the solubility curve in Figure 3(a). This is because of the high solubility of Si in Al. For example, at 500°C, the solubility of Si in Al is as much as 0.8wt%. This means that Si from the substrate will dissolve in Al metallization during annealing, whenever the two are in contact. Such dissolution of Si depends on the annealing temperature, time, and "volumes" of Si and Al interacting, until solid solubility limits are reached. This leads to Al spiking through contacts, epitaxial re-deposition and re-configuration of polysilicon. These are discussed in detail in reference 7. The AlCu phase diagram is shown in Figure 3(b). Formation of various compounds such as $CuAl_2$ and Cu_2Al, have been indicated in the phase diagram.

Diffusion of Si in Al: The diffusion coefficient, D, of Si in Al determines the distance that Si travels in Al during an anneal and hence, determines the volume of Al which will be subjected to saturation. The values of D for Si in Al are shown as a function of temperature in Figure 4. for both bulk and thin film Al. The diffusion coefficient in thin film Al is higher by a factor of about 40 than in bulk Al in the temperature range 400-500°C. This is due to the fact that the diffusion in thin films occurs mainly along grain boundaries. Thus, the consumption of Si in pure Al films is enhanced by the polycrystalline nature of the Al film.

Al Reactivity: Al reacts readily with oxygen to form Al_2O_3. The heat of formation of Al_2O_3 is 399 Kcal/mol. as compared to 205 Kcal/mol of SiO_2. Therefore Al can reduce SiO_2 to Si by the reaction

[2] $$3SiO_2 + 4Al \rightarrow 3Si + 2Al_2O_3$$

This reduction of SiO_2 by Al is very significant because it removes the native SiO_2 existing on Si, leading to a better contact between Al and Si (lower contact resistance). This is also an important factor for the good adhesion of Al to SiO_2 layers, which is a key property needed for metallization. Although this reduction reaction can take place theoretically even at room temperature, in actual processing, however, annealing is done at 400 -450°C to ensure that complete reduction takes place.

Stress in Al Films: Mechanical stresses are developed in thin metal films because of the mismatch in the coefficients of thermal expansion (CTE) of the film and its underlying & overlying materials. In addition, there is usually an intrinsic stress in the deposited metal film on account of its microstructure. These two components of stress in the metal film result in a significant net as-deposited stress. The magnitude of stress is measured in MPa or dynes/cm^2 (1MPa = 10^7 dynes/cm^2), and its sign follows the convention of tensile being positive and compressive being negative. On heat treatment (thermal cycling), this stress undergoes changes, and its value after thermal cycling is generally different from the as-deposited value. For Al films, the as deposited stress is about 100-200 MPa, tensile. After thermal cycling to about 400°C, the stress at room temperature increases to about 300 MPa, tensile. During thermal cycling, the stress variation shows a hysteresis behavior as shown in Figure 5(a) (9) . The variation of stress can be understood as follows. The stress in as deposited films is mainly thermal stress caused by the difference in coefficients of thermal expansion (CTE's) of Al (23.5 x 10^{-6} /$^{\circ}$K) and Si (2.5 x 10^{-6} /$^{\circ}$K). The thermal stress, σ_{th}, is given by

$$[3] \qquad \sigma_{th} \cong \left(\frac{E_f}{1-v_f}\right)\int_{T_A}^{T_D}(\alpha_{Al}-\alpha_{Si})dT$$

$$\approx \left(\frac{E_f}{1-v_f}\right)(\alpha_{Al}-\alpha_{Si})(T_D-T_A)$$

where E_f is the Young's modulus of Al
           v_f is the Poisson's ratio of Al
           α_{Al} and α_{Si} are the CTEs of Al and Si respectively, a
           T_D and T_A are the temperatures of deposition and stress
           measurement respectively.

For $T_D > T_A$, the resulting stress is positive, i.e., tensile. With increase in T_A, σ_{th} decreases and crosses zero into the compressive regime at about 50-100°C. With further increase in T_A, σ_{th} becomes more and more compressive. In this temperature range, referred to as the elastic region, the stress variation is reversible. Beyond about 100°C, the plastic flow of Al occurs with increase in temperature and hence, the compressive stress gets relaxed, leading to a decrease in total stress. On cooling, the compressive stress is initially reduced elastically. The stress eventually becomes tensile and continues to increase. Ultimately, the yield strength of the material is reached and hence plastic flow deformation occurs for the relaxation of the tensile stress. This is more clearly illustrated in Figure 5(b), which shows the stress temperature behavior of AlSi and AlCu alloys. The mechanisms of stress relaxation have been extensively investigated. Grain boundary sliding, diffusional creep, and grain growth are some of the important compressive stress relaxation mechanisms proposed while dislocation slip generation and grain boundary sliding have been suggested as tensile stress relaxation mechanisms. Depending on the structure of the film, different stress relaxation mechanisms may operate in different regions of the stress temperature curve. A combination of different mechanisms is also likely to operate in the boundaries of certain regions in the stress-temperature curve. The stress behavior changes with the addition of alloying elements (9, 10) as shown in Figure 5(b). The causes for such changes are the changes in the CTE and the microstructure of AlX with different alloying elements. The CTE of AlX decreases by 1.07% per weight percent of Si added in the alloy. In the case of Cu or Ti addition, the decrease is 0.33% and 1.0% per weight percent respectively. As will be explained in section 6.1, the stress variations in the Al and the underlying/overlying dielectric layers can lead to voids in the interconnects giving rise to reliability problems in the interconnect.

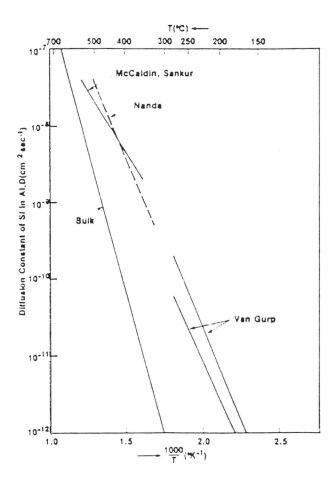

Figure 4. Variation of diffusion coefficient of Si in Al with temperature. Ref. 7, reprinted with permission.

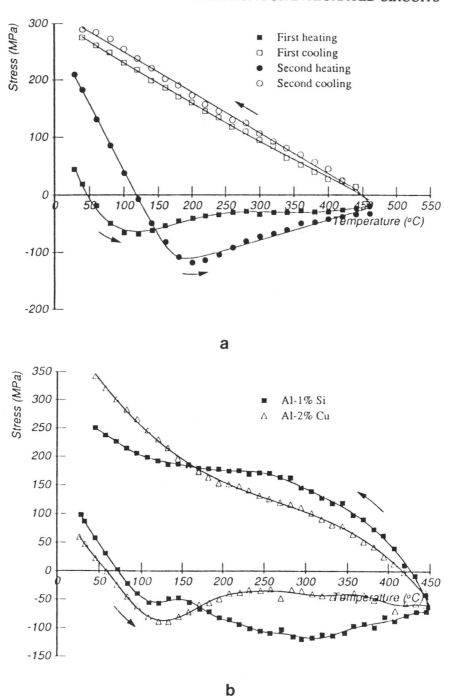

Figure 5. (a) Temperature vs. stress of Al films during thermal cycling and (b) Stress variation in AlSi and AlCu alloys during thermal cycling. Ref. 9, reprinted with permission.

In addition to the above properties, there are other process related properties which have significant impact on multilevel metallizations. The important ones are:

Microstructure of Al and AlX Films: Thin films of pure Al and AlX deposited by conventional PVD processes are polycrystalline in nature. The dominant orientations and the grain sizes and their distributions are dependent on the method of deposition and the process conditions used. The annealing of the Al and AlX films generally increases the average grain size. However, after certain duration of annealing, the average grain size tends to saturate. The percentage of small grains also tends to saturate and these cannot be eliminated under reasonable (normal) annealing conditions. This is attributed to the segregation of oxygen, present in the film, in the grain boundaries which reduces their mobilities.

The grain sizes obtained for different AlX films, obtained by SEM/TEM analyses, are summarized in Table III for different process conditions (11).

Table III: Dependence of grain size of Al on deposition conditions

Material	Deposition Method	Substrate Temperature	Average Grain Size (µm)	
			As-deposited	After 450°C 15 min. anneal in N_2
Al	In-S (a)	Room temp.	0.53	0.9
Al	In-S	200°C	0.96	1.4
Al	Magnetron Sputtering	Room Temp.	0.30	0.7
Al-2%Cu	In-S	Room Temp.	0.23	1.0
Al-2%Cu	In-S	200°C	0.67	2.3
Al-2%Cu	Magnetron Sputtering	Room Temp.	0.22	1.2
Al-0.7%Si-4.0%Cu	In-S	Room Temp.	0.05	0.2
Al-0.7%Si-4.0%Cu	In-S	200°C	0.07	0.2
Al-0.7%Si-4.0%Cu	Magnetron Sputtering	Room Temp.	0.07	0.1

a. In-S = In - Source Evaporation

From this table, it can be seen that the addition of Si and Cu to Aluminum reduces the average grain size in the AlX films as compared to that of pure Al film. However, the grain size distributions, in as-deposited films and after anneal are narrower in AlCu and AlSiCu films as compared to pure Al films.

The orientations of the Al/AlX films obtained by different methods of deposition have also been studied in detail. The most dominant orientation in Al, AlCu and AlSiCu films deposited on Si or SiO_2 is <111>. Small amounts of <200> orientation are also observed (8). The relative amounts of <111> and <200> orientations are dependent on deposition/annealing conditions and grain growth/suppression due to residual gases.

TEM analysis of AlCu and AlSiCu films have shown that when the Cu content is in the range 0.8 to 2.0 wt%, precipitates are found to be distributed in the films after annealing, predominantly at the grain boundaries. Electron probe microanalysis has been used to identify these precipitates as $CuAl_2$ or Cu-rich Al. The sizes of these precipitates and their numbers increase with increase in Cu content (11, 12).

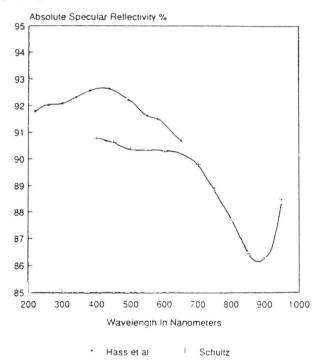

Figure 6. Specular reflectivity of a smooth Al surface. Ref. 8, reprinted with permission.

Reflectivity: The specular reflectivity of Al and its alloy films is used as a measure of the film quality. The reflectivity of the deposited film depends on the method of deposition and the deposition conditions. The specular reflectivity of a smooth Al surface for incident wavelengths between 200 and 1000 nm is shown in Figure 6 (8). The high reflectivity of a deposited AlX alloy film can affect the subsequent photolithographic processes. If the surface below the AlX film has topography, and it is replicated by the AlX film, the extraneous reflections from certain regions of the AlX can expose the photoresist in undesired areas. This can lead to notching of the Al lines or bridging of adjacent lines after patterning. An anti-reflection coating (ARC), either as a part of the photolithography process or as a layered metallization Y/AlX/Z process, is sometimes used on top of the Al to overcome these problems associated with the reflectivity of the Al alloy film. As an alternative to this, dyed photoresists are also used sometimes to minimize the effect of the extraneous reflections from Al or AlX.

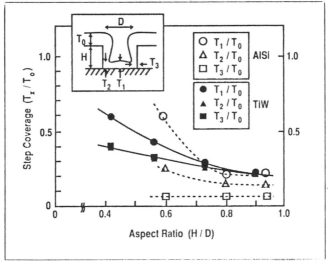

Figure 7. Step coverage of AlX in vias. Reference 20, reprinted with permission.

Step Coverage: Step coverage of a film becomes an important issue when it is deposited on a surface which has topography (non-planarized surface). For example, in a CMOS process, the first metallization layer needs to make contacts to the source, drain and gate regions through small holes in the dielectric, and provide the first interconnect layer which may be over severe topography variations due to field oxide and gate regions. If the dielectric and contact planarizations are not used, step coverage over the edges, sidewalls and the bottom of the contacts, and over other topographies could be quite poor, viz., the thickness of the metal film is thinner over steps compared to the flat regions. As defined earlier in section 3.1, the step coverage is the ratio of the minimum thickness deposited on the vertical walls of a step to the thickness on the horizontal portion of the step. In the case of a via, the ratio of the minimum thickness at the bottom to the thickness on the horizontal portion is also often used to describe the step coverage in addition to that indicated above. Ideally, the step coverage at any step or via should be 100%. In practice, however, it is less and sometimes the coverage at the bottom of a via can be as low as only 10%. Figure 7 shows a scanning electron micrograph of a poor step coverage with Al metallization. Step coverage of Al or any other metal being used depends on the method of deposition. It also depends on the topography of the substrate (13). The figure clearly shows poor step coverage for vias with high aspect ratio, which is defined as the height to width ratio of the via. Novel methods of deposition have been proposed to improve the step coverage of Al alloys in submicron contacts and vias. These are discussed in section 5.1 of this chapter.

4.2   Electrical Properties

The important electrical properties of Al are its electrical resistivity, specific contact resistivity and electromigration. These are discussed below.

Resistivity: The room temperature resistivity of bulk pure Al (2.7 $\mu\Omega$-cm) is one of the lowest among all the metals. Thin Al films can be deposited easily with a resistivity within 10% of this value. This makes Al highly suitable as an interconnect material. The resistivity becomes crucial for both horizontal and vertical interconnects, as we approach VLSI where the cross sectional areas of the interconnect lines become smaller and lengths become larger. The addition of alloying elements to Al to improve other key properties, as discussed in section 4.3, increases its resistivity too. In AlSi alloy, for each weight percent of Si added, the resistivity increases by about 0.7 $\mu\Omega$-cm. In AlCu alloys, the addition of each percent of Cu increases the resistivity by 0.3 $\mu\Omega$-cm. Thus, the resistivity of Al-1%Si is about 3.4 $\mu\Omega$-cm while that of Al-1%Cu is about 3 $\mu\Omega$-cm. When two or more alloying elements are added (eg. AlSiCu), the approximate resistivity of the Al alloy is obtained by adding the resistivity increases for each individual alloying element (8), Therefore, the resistivity of a Al-0.5%Si-1%Cu will be about 3.35 $\mu\Omega$-cm. In the present VLSI technologies, Al-1%Si-0.5%Cu and Al-0.5%Cu are typically the most widely used conductors, although slight variations in the weight percentages of Si and Cu depend on the proprietary processes of different corporations. Figure 8 shows the typical resistivity vs. temperature plot for a particular AlSiCu alloy (14). The temperature coefficient of resistivity (TCR) of the Al alloys is found to be about 4×10^{-3} /$^{\circ}$K. As the technologies move into 0.25 micron regime and beyond, the resistivity of Al or its alloys becomes a limiting factor for getting the required speed performance in the IC. At that point, alternate metallizations based on metals like Cu, Ag and Au, whose resistivity is lower than Al, need to be considered.

Contact Resistance to n^+ and p^+ Si: Al forms good ohmic contacts to both n^+ and p^+ junctions in Si, even to lightly doped p-Si. With n-Si, however, the contact is ohmic only for $N_D > 10^{19}$ cm^{-3}. The contact resistance is characterized by a parameter called " specific contact resistance (ρ_c) ", which is defined as

[4] $$\rho_c = R_c \times A$$

where R_c is the actual resistance across the contact and A is the contact area.

ρ_c is independent of the contact area and hence is an index of the quality of the contact for a given metallurgy.

The ρ_c of Al to both n+ and p+ Si is about 10^{-6} Ω-cm^2. When the contact dimensions become very small, this value of ρ_c may prove to be too large. For example the resistance of a contact of area 0.25 μm^2 can be as high as 400 Ω! The variation of specific contact resistance of Al with doping concentration in n and p type Si is shown in Figure 9. This Figure clearly shows that ρ_c can be reduced by increasing the surface dopant concentration in Si. In the present technologies with shallow junctions, Al or AlX does not make a direct contact with Si for ohmic contacts. These contacts are silicided, followed by a barrier metal deposition on the silicide, to prevent junction spiking. Al makes contact with the barrier metal. Therefore, the contact resistance between Al, the barrier metal and the silicide determines the net contact resistance. Typical values of ρ_c for Al/barrier metal/silicide on n+ and p+ Si are about $10^{-7}\Omega$-cm^2.

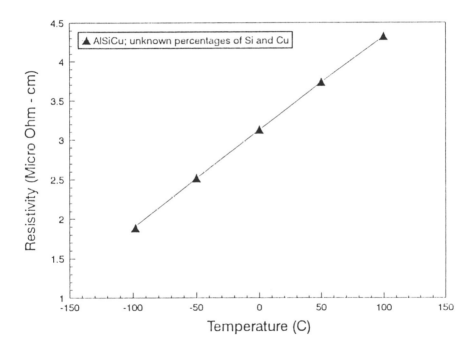

Figure 8. Resistivity vs. temperature for AlSiCu alloy.

Electromigration: As indicated earlier, electromigration is a phenomenon which causes failure of an interconnect. This failure mechanism is characterized by the "median time to failure (MTF)" of the interconnect. Typical values of the MTFs under a current density 10^6 A/cm^2 at 80°C are in the range

10^5 to 10^6 hours (15) for AlX interconnects. Table-IV lists the electromigration lifetimes of different AlX alloys deposited under different conditions.

Electrical Properties of Other Materials: The other conductor materials that are often used in AlX based multilevel metallizations are W, Ti, TiW and TiN. The resistivity of W films deposited by CVD is in the range of 7 to 12 $\mu\Omega$-cm depending on the deposition conditions. Thin film resistivities of TiW and TiN, on the other hand, are much higher. In general, the resistivity ranges from 50 to 80 $\mu\Omega$-cm for TiW films and 50-100 $\mu\Omega$-cm for TiN.

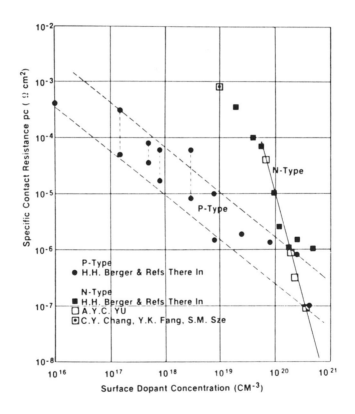

Figure 9. Variation of specific contact resistance of Al with doping in Si. Ref. 7, reprinted with permission.

Table IV: Electromigration aging results

Film Comp.	Dep. Tech.	Linewidth w (μm)	Aging Conditions		Extrapolated Lifetimes	
			Temp. (°C)	Current Density $(10^6 A/cm^2)$	MTF $(10^5 hr)$	Std. Dev. s [a]
Al-0.5%Cu	e-gun	2.6 1.0	250 250	2.0 2.0	12 >80	0.45
Al-0.5%Cu	In-S [b] (4700 Å/min)	7.3 3.1 1.2	220 240 235	1.4 2.0 2.0	6.1 1.2 1.9	0.39 0.55 0.26
Al-0.5%Cu	In-S (1500 Å/min)	3.8 1.8	215 215	1.5 1.5	0.91 0.45	0.5 0.2
Al-0.5%Cu	In-S (300 Å/min)	2.9 1.9	220 210	2.0 2.0	0.76 0.52	0.2 0.4
Al-0.5%Cu	S-gun (330°C)	2.3 1.25	200 200	1.8 1.8	3.2 2.1	0.5 0.5
Al-0.5%Cu	S-gun (cold)	3.5 1.7	200 200	1.8 1.8	3.3 1.8	0.2 0.5
Al-1%Si- 0.25%Cu	S-gun (cold)	3.7 1.7	205 220	1.8 1.8	5.2 1.4	0.2 0.3
Al-2%Si	S-gun (cold)	3.5 1.4	200 195	1.6 1.6	4.3 2.9	0.3 0.2

a. For log-normal distributions
b. In-S=In-source evaporation.

4.3   Aluminum Alloys

To improve the reliability of the Al based interconnects, Al alloys containing small amounts (0.5% to 4%) of elements like Si and Cu are used in the current microelectronic manufacturing. The features of these alloys are discussed below.

AlSi: This is the first alloy of Al that got its popularity as an interconnect material in the semiconductor industry. The addition of Si to Al was used to minimize, if not eliminate, the Si diffusion from the substrate to the Al, thereby reducing the probability of failure by junction spiking (16). The amount of Si in Al is such that the solubility of Si in Al is satisfied at the highest temperatures encountered in processing. Al-1%Si is a widely used alloy. Although the addition of Si suppresses spiking, new problems arise because of reprecipitation of Si. This occurs because the additional Si which dissolves in Al at high temperatures comes out of solution on cooling. This poses serious problems in very small contact windows. In the case of Al-polysilicon contacts, the Si from regions around the grain boundaries in polysilicon dissolves in Al which, in turn, is precipitated back on the polysilicon grains. The dissolution and precipitation of Si progresses with the growth of large grains of Si at the expense of the smaller grains (Ostwald ripening). Ultimately, a continuous film of polysilicon can be restructured into isolated crystalline Si particles by this process. Since the restructuring of the polysilicon is caused by the transport of Si in Al, it occurs even when AlSi alloy is used for the contact. To prevent the dissolution of Si in Al, a thin layer of a diffusion barrier material is often used to separate the two at the contacts, as will be described later in this chapter.

There are conflicting reports on the electromigration performance of AlSi interconnects (17).

Some results suggest that addition of Si improves the lifetime while others indicate very little change. However, this is no longer a very popular metallization in VLSI/ULSI technologies because of the development of newer alloys of Al with significantly improved electromigration resistance.

AlCu: Cu is used to improve the electromigration resistance of the Al based interconnect. Cu enters the grain boundaries in Al and occupies several vacant sites. This reduces the grain boundary diffusion and hence the electromigration resistance is increased. One of the key parameters to describe electromigration performance is the activation energy. The alloying components increase the activation energy for electromigration. Typical values of activation energies for pure Al and for AlCu are 0.6 eV and 0.71 eV respectively.

The effect of Cu content in Al on the electromigration lifetime has been investigated (18) over a range of 0.5 to 19 wt.%. Figure 10 shows the electromigration lifetime of AlCu alloys in comparison with those of other Al alloys.The lifetime increases with increases in Cu content up to 4.0 percent Cu. This improvement is attributed to the $CuAl_2$ precipitates in the grain boundaries. With further increase in Cu concentration, there is little improvement in the lifetime. This is because of retardation of the grain growth by the precipitate phase at high concentrations of Cu. In addition, the presence of Cu in the AlX metallizations makes the plasma etch process more difficult because the reaction products of Cu are not volatile at typical processing temperatures. Further, the AlCu alloy has increased susceptibility to corrosion in the presence of chlorine. Therefore, the Cu content is limited usually to about 0.5% in the presently used AlX alloys.

When AlCu metallization is used, especially at contacts, a suitable barrier layer such as TiW or TiN is needed between the contact metallization and the Si substrate.

AlSiCu: This is the most widely used Al alloy for contacts and interconnects in present day technologies. This alloy offers the advantages of both AlSi and AlCu alloys, as described above.

AlTi and AlSiTi: These alloys obtained by both simultaneous sputtering of Al or Al/Si and Ti and by layering of Al/Si and Ti have been characterized with respect to composition, resistivity, hillock formation and other film characteristics (19). It is found that Al/Ti alloy does not yield smooth films while Al/Si/Ti does. The presence of Si yields to significant reduction in hillock formation. The addition of Ti, however, increases the resistivity of the alloys to about 4.5 - 5.5 $\mu\Omega$ - cm.

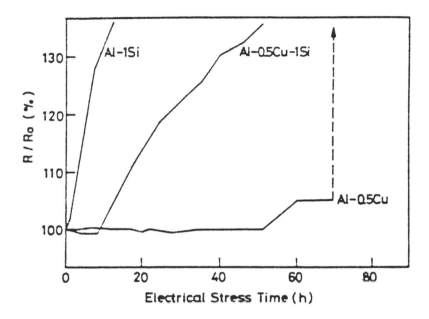

Figure 10. Effect of addition of Cu on electromigration lifetime of AlX. Ref. 17, reprinted with permission.

4.4   Layered Interconnects With Barrier Metals

With the continued shrinking of interconnect geometries in the VLSI technologies with device scaling, it was realized that even AlX is not good enough for obtaining interconnects of high reliability.   This led to the development of "barrier layers" to act as underlayer and cap layer for the AlX interconnect, viz, Y/AlX/Z. These layered interconnects of AlX are most widely used in the present technologies. In these, the underlayer (Y) and the cap layer

(Z) are conductors whose resistivities are not as low as that of the Al alloy; but they are more resilient to electromigration and stress voiding. Typically, the underlayer and cap layer are refractory metals/compounds--the most widely used being Ti, TiN, and TiW. The layered interconnects, Y/AlX/Z, exhibit very desirable characteristics and significantly higher reliability than AlX. The important features of the layered interconnects are:

1. The under layer acts as a barrier layer and prevents any interaction between Al and Si - no junction spiking, no Si precipitation.
2. The underlayer provides a clean, uniform nucleating layer for Al which improves its step coverage and electromigration performance.
3. The under/over layer provides a redundant conducting path for current flow.
4. The interactions between the under/cap layers and AlX form intermetallic compounds which suppress the void growth in Al - improved electromigration resistance.
5. The cap layer suppresses hillock formation.
6. The cap layer acts as an antireflection coating (ARC) for subsequent lithography processes.
7. The under layer and the cap layer reduce the effects of stresses in the underlying and overlying dielectrics on the interconnect.

The following requirements (20) are important to be satisfied by the barrier layers to act as under or cap layers. Obviously, those materials which meet most of these requirements are preferred. These layers should:

1. provide a good barrier against interactions between Al and Si/SiO_2 and their thickness preferably should be less than 2000 Å;
2. not react excessively with Al so as to lead to a large increase in resistance;
3. not degrade the electromigration characteristics of the interconnect; if any, the electromigration resistance should increase;
4. adhere well to SiO_2 or Si_3N_4 films;
5. have good adherence of AlX on them;
6. have good step coverage in high aspect ratio vias in ILDs
7. be easily dry etched along with the Al conductor;
8. have deposition processes compatible with those used in IC technologies; and,
9. have low stress.

The under/over layers that are most commonly used are TiN and TiW refractory materials. It is found from various investigations that TiN and TiW are good barriers against Si diffusion into Al. These barrier layers are used for both horizontal and vertical interconnects. The important properties of these barrier metals and their interactions with other materials used in multilevel metallizations are described below.

TiW: This is an alloy of Ti in W. The commonly used TiW contains about 10 wt% of Ti in the TiW film. The TiW films usually consist of columnar grains of W without any free Ti phase. The Ti is in solid solution with some W atoms distributed at the grain boundaries. The distribution of Ti is important for improving the barrier properties of TiW. The resistivity of TiW films is in the range of 50 to 80 $\mu\Omega$-cm.

The step coverage of TiW in a via compared with that of AlX is shown in Figure 11. It can be seen from the figure that for low aspect ratios, the step coverage of AlX is better than that of TiW. However, the step coverage of TiW does not decrease as rapidly as that of AlX at high aspect ratios. The TiW film profile is different from that of AlX, indicating a different mechanism controlling the step coverage. This is attributed to a lower sticking coefficient of Ti and W which causes the depositing atoms to scatter off the surface of the substrate several times before sticking to it, thereby giving a better step coverage.

TiW makes good contact to n^+Si but the contact resistance to p^+Si is large. A high temperature anneal at 600-700°C has been shown to decrease the contact resistance to p^+Si. This high contact resistance is caused by a shallow surface damaged region in the Si. If this layer is removed by an isotropic etch, contact resistances as low as 30 Ω can be obtained with p^+ Si. Direct contact with TiW to n^+ and p^+ junctions in Si is generally not used in the industry.

TiN: Stoichiometric TiN has 50 at.% each of N and Ti. For N concentrations more than 50%, the excess N exists in solid solution in the stoichiometric TiN. TiN films have columnar grains with resistivity in the range 50-100 $\mu\Omega$-cm. The step coverage of TiN is similar to that of TiW. The contact resistance of TiN with p^+ and n^+ Si is large. To reduce the contact resistance, a thin Ti layer is deposited prior to TiN deposition.

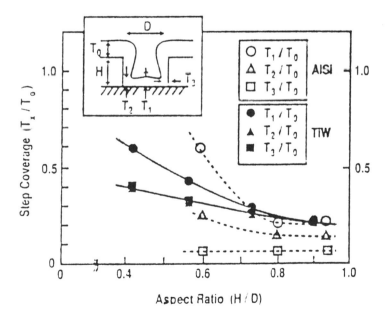

Figure 11. Step coverage of TiW in a via as compared to that of AlSi. Ref. 13, reprinted with permission.

Ti: Pure Ti is often used as an underlayer and for contact metallization on top of which a TiN layer is either formed in-situ, or deposited. Ti helps in reducing the interfacial oxide layer, improving the step coverage and adhesion of AlX layers.

4.5   Tungsten Clad AlX Interconnects

This is another type of layered Al interconnect that has been proposed in which tungsten selectively deposited by CVD (discussed in Section 5.1) covers the Al interconnect (21). This type of encapsulated interconnect using AlSiCu is found to exhibit improved step coverage in vias and increased resistance to electromigration. The W layer also helps in reducing the reflection problems during photolithography. In this type of W-clad interconnect, fluorine gets incorporated between Al and W during the W deposition, through the reaction of WF_6 with Al. The thin layer of AlF_6 formed is found to increase the resistance of vias/contacts. To prevent this incorporation of F, the use of sputtered TiW on top of Al prior to encapsulation by CVD W has been proposed (22). This layer prevents fluorine interaction with Al and also greatly reduces hillock formation during selective W deposition.

4.6   Multilayer Interconnects

In these interconnects, several layers of Al alloy and Ti or TiN are stacked alternately (23). These interconnects have been found to exhibit superior electromigration and stress migration resistance (24). It has also been found that Al/AlX films layered with Ti or W show no hillocks after heat treatment. In the case of AlSi/Ti multilayer system, on annealing, Si and Al redistribute between AlSi and Ti. A new phase, $TiAl_3$, enriched with Si forms locally between the AlSi and Ti layers. This phase minimizes hillock growth. Precipitation of Si and $TiAl_3$ also suppresses hillock growth. When this AlSi/Ti is compared to the Al/Ti system, there is a more uniform distribution of $TiAl_3$ in the former which indicates that the presence of Si in AlSi reduces the migration of Ti into Al. Similar improvement in the characteristics have been obtained in layered Al-1%Si-0.5%Cu based interconnects with intermediate TiN layers.

4.7   AlX Interactions With Barrier Metals

Interactions between the barrier layer and AlX restricts the highest temperature that may be used in subsequent processing. On the other hand, limited interactions at the interface often improves the quality of the layered interconnect.

In the case of TiW, above 400°C, Ti diffuses into Al and the Ti concentration increases with time (25). No new phase formation is observed but Ti remains in solution in Al and gives rise to an increase in resistance. At about 450°C, Al reacts with TiW to form WAl_{12}. At higher temperatures around 550°C all the Al is converted primarily into WAl_{12} and some WAl_5 is also formed. The reaction is nonuniform and initially occurs near the grain boundaries (26). Even at 600°C, the dominant phase is WAl_{12}. Surface passivation of the TiW by exposure to air retards the reaction with Al which improves the barrier properties. Compounds of Ti and Al are formed only above 550°C although such compounds can form even below 450°C when Al reacts with pure Ti.

In the case of TiN, the interface reactions are different for AlCu as compared to those for AlSi and AlSiCu. In AlCu/TiN interconnects, at an annealing temperature of 400°C, there is some diffusion of Ti into the Al matrix as well as the reaction between Al and TiN. Annealing at 500°C results in the formation of $TiAl_3$ which acts as a barrier against any further interaction. However, in the case of AlSi/TiN and AlSiCu/TiN, after anneal at 500°C, neither Al and Ti compounds involving Si nor $TiAl_3$ are seen prominently. Localized $TiAl_3$, which contains dissolved Si, is seen. This shows that in this case, formation of a new phase $Ti_xAl_ySi_z$ is preferred over the formation of $TiAl_3$ and this phase acts as a barrier to further interaction between Al and TiN. Thus the presence of Si appears to reduce the migration of Ti into Al (23). A TiN layer also suppresses the hillock formation in AlX due to the formation of $TiAl_3$ or $Ti_xAl_ySi_z$ phase and their precipitation in AlX.

As already indicated, in contacts, AlX makes contact to a silicide through a barrier metal like TiN. In one of the techniques reported, Ti deposited on Si in the contact is sintered in nitrogen at 650°C which forms a $TiN_x/TiSi_y$ interface. Significant dopant redistribution is observed during the sintering process. However, this barrier technology shows no significant leakage in shallow junctions after thermal stressing at 400°C for 8 hours (27).

4.8 AlX Interactions With ILDs

The interactions of the AlX interconnects with the underlying/overlying interlevel dielectrics (ILDs) also affect the reliability of a multilevel metallization. The most significant interaction is through the stress variations in the interconnect and ILDs during thermal cycling. The ILDs used in the present technologies are deposited by CVD and these develop significant intrinsic stress on thermal cycling. This stress can cause material movement in the underlying/overlying interconnects. For instance, a high compressive stress in the overlying ILD causes void formation in the interconnect.

4.9   AlX Interactions With Silicides

In most of the current ohmic contact technologies, AlX makes contact to silicides through a barrier metal. So, the interaction of AlX with the silicides is not significant, except to recognize the need for a barrier metal to be interposed between them to make a reliable ohmic contact. The extent of interaction between the AlX and the underlying silicide depends on the temperature. The lowest temperature at which Al interacts with the commonly used silicides are given in Table V (28).

Table V: Interactions of Al with silicides

Silicide	Lowest temperature of Interaction	Silicide	Lowest temperature
$MoSi_2$	500°C	$CoSi_2$	400°C
$TaSi_2$	500°C	NiSi	350°C
$TiSi_2$	450 - 600°C	PtSi	250 - 300°C
WSi_2	500°C	Pd_2Si	300°C

The interactions of some of the above mentioned silicides with Al have been studied in detail (29, 30). Reactions between Al and the silicide starts with interdiffusion of Al into silicide and Si dissolution in Al. The diffusion is prominent at the grain boundaries. The next step in the interaction is intermetallic compound formation and Si precipitation. Compounds that are most Al rich in the phase diagrams are formed first . In the case of Pd_2Si, reactions start at 300 °C and intermetallics such as $PdAl_3$ are formed. In the case of PtSi, beyond 400°C, $PtAl_2$ gets formed. The interaction of $MoSi_2$ at 540°C and beyond leads to the formation of $MoAl_{12}$. In the case of $CoSi_2$, $CoAl_2$ is formed as a result of interaction with Al at 400°C.

5.0 CURRENT STATUS OF Al BASED MULTILEVEL METALLIZATIONS

Two level and three level metallizations are currently used widely in the manufacturing of VLSICs. In these metallizations, layered AlX based conductors are used for the horizontal interconnects at most of the levels. These consist of an underlayer (diffusion barrier) such as Ti/TiN or TiW, the main conductor (Al-1%Si-0.5%Cu or Al-0.5%Cu) and a cap layer (TiN or TiW). The vertical interconnects usually consist of W on a barrier layer (Ti/TiN or TiW). Various technologies used for depositing, etching and planarizing these conducting layers are discussed below.

5.1   Deposition Technology

In this section the different technologies used for the deposition of different types of conducting films for multilevel metallizations will be discussed briefly. Since Al is the most dominant metal in these metallizations most of the discussion is on Al and AlX.

Deposition of Al and its Alloys: Thin films of Al or AlX can be deposited primarily by two techniques, viz., physical vapor deposition (PVD) and chemical vapor deposition (CVD). In PVD, Al or AlX in bulk form is used as the source (also called the "target"). From this source, atoms/molecules of Al or AlX are transported to the substrate where they deposit on the substrate surface layer by layer to form a thin film. Two of the widely used PVD technologies are electron beam (e-beam) evaporation and sputtering. In the evaporation technique, the Al or AlX atoms are produced by evaporation from the source by heating it with an electron beam. In sputtering, the Al or AlX atoms are produced by dislodging them from the source with high energy ion bombardment. Of the two, sputtering is the preferred technique for Al and AlX deposition in VLSI technologies because of the following features.

(1) The composition (stoichiometry) of the target is closely reproduced in the deposited film.
(2) Deposition rates are higher than in evaporation.
(3) Step coverage is better than that in evaporation because the target is large and the depositing atoms normally experience many collisions and arrive at a wide range of angles.
(4) Refractory metals like W, Ti and dielectric films having high melting temperatures can be sputter deposited which is difficult with evaporation.
(5) X-ray generation due to electron bombardment, which is a problem n e-beam evaporation, is absent.

In CVD, the source is usually a compound containing Al. The most common form of CVD is MOCVD where the MO stands for metalorganic, meaning the source is organometallic in nature. The molecules of this compound are transported in vapor phase to the surface of the substrate where they undergo chemical reactions, at elevated temperature and/or under plasma ambient. Al atoms, which are generated in these reactions, get deposited on the substrate to form the thin film. Laser assisted CVD or photo CVD of Al is also of interest because of its selective deposition capability, as will be discussed in Section 5.2. However, CVD of Al is still in the *R&D* phase and hence all the current VLSI manufacturing technologies use PVD for depositing AlX films for multilevel metallizations. Limitations of CVD to date are obtaining high quality Al, AlX (Si, Cu) and layered Y/AlX/Z films.

Other Conductor Materials Such as TiN, Ti, TiW, W, etc: As already described earlier, Al based interconnects in the present technologies usually have conducting underlayers and cap layers. Ti, TiN and TiW are some of the popular materials for these applications. Thin films of these materials can be deposited by either of the above PVD or CVD techniques. However, sputtering is most commonly used technique for depositing Ti and TiW while TiN is deposited by reactive sputtering of Ti in a nitrogen ambient at elevated temperature (~300°C). CVD of TiN films is also being evaluated as a manufacturing technology to exploit its superior step coverage properties as will be described later in this section. When W is used as interconnect metal, the preferred deposition technique is CVD using SiH_4 and/or H_2 reduction of WF_6.

Some of the important features of sputtering and CVD relevant to film deposition for VLSI multilevel metallizations will be discussed below.

Al/AlX Film Deposition by Sputtering: In a sputter deposition system, Al/AlX atoms/molecules are dislodged (sputtered) from a solid Al/AlX source (target) and these atoms/molecules get deposited on the substrate. The high energy ions that cause sputtering are those of an inert gas like argon generated in a plasma. Depending on the method by which the plasma is generated, different sputtering techniques are possible. In all these techniques, sputtering is done in a chamber which is evacuated to about 10^{-7} Torr before being filled with argon to a pressure of a few mTorr. In D.C. diode sputtering, a d.c. voltage is applied between the Al/AlX target and the substrate holder, to generate the plasma. In R.F. diode sputtering, r.f. voltage, typically at 13.56 MHz, is used to generate the plasma. In magnetron sputtering (D.C. or R.F.), stationary or rotating magnetic fields are used to confine the electrons that are knocked off the target surface, thereby confining the plasma for higher sputter rates and uniform erosion of targets. Figure 12 shows the schematic (31), of diode and magnetron sputtering systems.

D.C. sputtering is suitable for targets which are electrical conductors. It cannot be used for sputtering from insulating targets because the plasma cannot be sustained with insulator covered electrodes by applying d.c. voltage. Therefore, for depositing insulating films, r.f. sputtering is used. Different types of sputtering systems as stand alone or cluster tools are commercially available that offer various types of magnetrons, planetary motion, vacuum system options, multiple targets, substrate heating, etc.

The step coverage characteristic of sputtering, however, is usually not adequate for depositing metal films into contacts/vias of sub-micron and sub-half micron dimensions having aspect ratios>1. Sputter deposition results in incomplete filling of contacts/vias, which can lead to an increase in contact/via resistance and reliability problems.

Several variations of the sputtering process have been used to improve the step coverage of the deposited Al/AlX films.

Bias Sputtering: In bias sputtering, the metal film is deposited on the wafer by sputtering from a target with the wafer being held at a negative potential with respect to the plasma. Two processes occur simultaneously in bias-sputtering, viz, the deposition due to sputtering from the target, and in-situ etching (due to the sputtering by the impinging positive ions on account of the negative potential on the substrate) of the depositing film. For most materials the sputter etch rate depends on the angle of incidence of the ion with the surface being etched (32). This angular dependence of etch rate causes the material to be preferentially removed from the edges when the film is being deposited on a surface with a topography such as contacts/vias. Several other phenomena such as heating can occur during the bias induced ion bombardment which affects the microstructure, but the sputtering and redeposition of the Al atoms from the depositing film improves the step coverage. This technique can also be used for contact/via planarization, although the net deposition rate is lower. With bias sputtering, the grain size is greatly reduced in the Al.

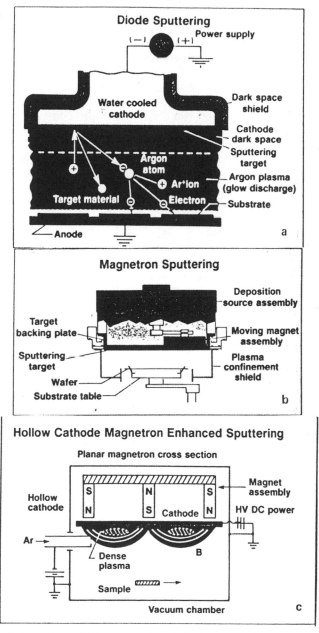

Figure 12. Schematics of (a) conventional diode sputtering system; (b) single wafer magnetron sputtering systems; and (c) hollow cathode magnetron enhanced sputtering system. Ref. 31, reprinted with permission.

Sputtering at High Substrate Temperatures: Sputtering at substrate temperatures in the range 500°C - 550°C has been used for deposition of Al/AlX into submicron vias/contact holes. The high substrate temperatures enhance the surface mobility of Al/AlX, and hence allow the deposited film to smooth out the topography. However, at high temperatures, even though the step coverage is improved, several problems arise. The Al reacts with impurities at high temperatures which degrades the film properties. Also, the initial nuclei deposited at high temperatures, tend to be large, and if they are formed at the top corners of vias, will shadow the walls below, thereby preventing deposition in these regions. Further, at high temperatures, the nuclei tend to grow larger and are widely spaced. These factors can result in the films being broken over walls of some vias but may have excellent step coverage over others. To overcome this difficulty, a thin nucleating layer of Al is deposited at lower temperature and is followed by high temperature deposition of the remainder of the film.

If the bias sputtering is done at high temperature with low bias voltages, improved step coverage is obtained without a substantial reduction in net deposition rate. The improvement in step coverage is due to the higher mobility of Al atoms owing to higher temperature and the increased vacancy concentration caused by the ion bombardment.

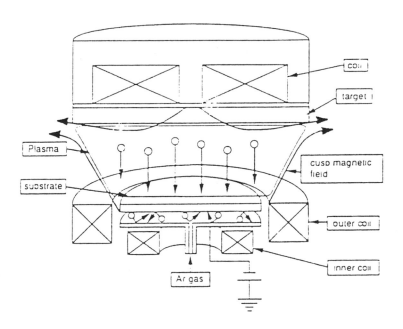

Figure 13. Cusp magnetic field electrode used in low energy ion bias sputtering. Ref. 34, reprinted with permission.

Low Energy Ion Bias Sputtering: A low energy, high density ion bombardment bias sputtering method at substrate temperatures of about 400°C has been reported (33) by Ohmi and his co-workers. In this method, as shown in Figure 13, a cusp magnetic field electrode is used. Under appropriate substrate bias conditions, there is a continuous low-energy ion bombardment on the growing Al film. This ion bombardment is also used for cleaning the surface of the substrate prior to Al deposition. Figure 14 illustrates the effect of substrate bias (voltage and current density) on the Al film characteristics. This method has been found to yield, under optimized conditions, Al films with a high degree of <111> preferred orientation, and low intrinsic stress. Hillock suppression, excellent step coverage (Figure 15), and high electromigration resistance are also achieved (34) with this process.

Collimated Sputtering: This is a simple but effective technique proposed recently for depositing films on high aspect ratio contacts and vias that could not normally be filled using conventional sputtering. In this, a metal grid is placed between the target and the substrate, as shown in Figure 16, to angularly collimate the velocity distribution of the sputtered beam. It is found that the filling in the contacts and vias is improved substantially (35) with such collimated sputtering. This is because the angular collimation removes the atoms/molecules arriving at high incidence angles which typically degrade the bottom coverage. This technique of collimated sputtering is undergoing a lot of development because it appears to be a promising technique for contact/via planarization with AlX, in particular for submicron and subhalf micron contacts/vias having high aspect ratios.

Sputtering of Other Conductor Materials: To deposit the underlayers and overlayers of Ti, TiN and TiW for Al based interconnects, the conventional sputtering technique is adequate and results in reasonably good and reliable metallizations. TiW films are deposited by sputtering from a TiW target. TiN films are deposited by reactive sputtering from a Ti target in a N_2 ambient or by rapid thermal annealing (RTA) of a Ti film in N_2 ambient (the former is the more widely used technique). For depositing TiN films to act as barrier layers in vias with high aspect ratios, however, conventional sputtering is not adequate because the step coverage, especially at the bottom of the vias is not adequate. To overcome this problem, newer sputtering techniques are being developed. These include: (i) high temperature sputtering (ii) sputtering with low energy ion bombardment (iii) collimated sputtering. The collimated sputtering of TiN has been studied extensively in the last one-to-two years to determine the effects of process parameters on the quality of contact/via fill obtained. These studies have shown that the aspect ratio of the collimator, i.e., the ratio of the thickness to grid opening has a significant effect on the TiN film properties. The deposition rate decreases substantially with increasing aspect ratios as shown in Figure 17, while the stress in the film becomes more compressive (36). The step coverage in the trenches and vias, improves significantly with increasing aspect ratio, as shown in Figure 18. Collimated sputtering has been used successfully for deposition of thin films of TiN and Ti as a first step for planarizing contacts and vias. Typical step coverage obtained in contacts is shown in Figure 19. Bottom coverage up to 50% and sidewall coverage upto 40% have been achieved in contacts 0.2 μm in size with an aspect ratio close to 7.0!

For contacts/vias, Ti/TiN barrier layers are used rather than just TiN alone. This is preferred because Ti acts as an adhesion layer and also, in the case of contacts, Ti on the silicide gives a lower contact resistance. To deposit such a sandwich of Ti/TiN layers, collimated sputtering is preferred (45).

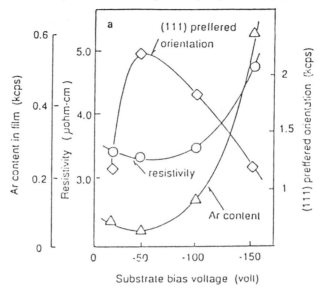

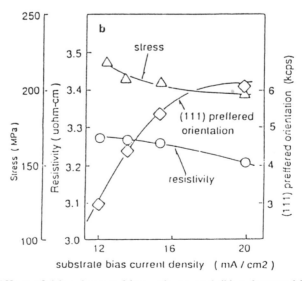

Figure 14. Effect of (a) substrate bias voltage and (b) substrate bias current density on Al film characteristics. Ref. 34, reprinted with permission.

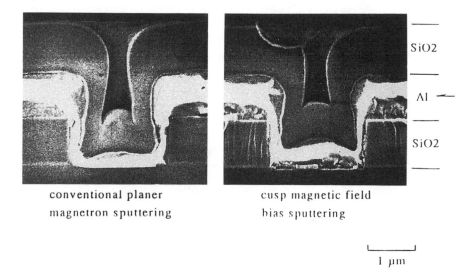

conventional planer
magnetron sputtering

cusp magnetic field
bias sputtering

SiO2

Al

SiO2

1 μm

Figure 15. Step coverage of Al film deposited by low energy ion bias sputtering. Ref. 34, reprinted with permission.

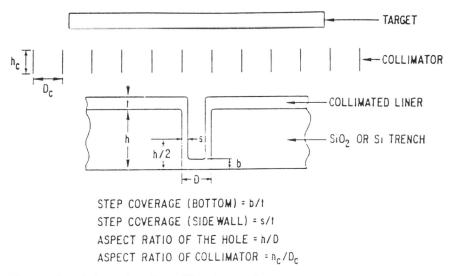

TARGET

COLLIMATOR

COLLIMATED LINER

SiO₂ OR Si TRENCH

STEP COVERAGE (BOTTOM) = b/t
STEP COVERAGE (SIDE WALL) = s/t
ASPECT RATIO OF THE HOLE = h/D
ASPECT RATIO OF COLLIMATOR = n_c/D_c

Figure 16. Schematic of a collimating grid.

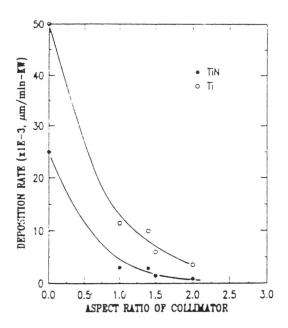

Figure 17. Variation of deposition rate with aspect ratio of collimator. Ref. 35, reprinted with permission.

Chemical Vapor Deposition of Al: As mentioned earlier, CVD of Al is still in the R&D stage. It has yet to be developed as a production worthy microelectronics manufacturing technology. All the CVD techniques proposed so far are based on organometallic precursors. The main features of CVD which make it attractive for VLSI/ULSI technologies are better step coverage and selective deposition capability. Both of these features are essential for filling of contacts and vias of submicron dimensions and high aspect ratios.

The most widely used precursor for CVD of Al is tri-isobutyl aluminum (TIBA). This is a liquid precursor having a boiling point of 40°C. This compound decomposes to di-isobutyl aluminum hydride in the temperature range 50-150°C (37). In the temperature range of 200-300°C, the di-isobutyl aluminum hydride thermally dissociates into pure Al by releasing H_2 and isobutylene gas

[5] $Al(C_4H_9)_3 -----> AlH(C_4H_9)_2 + i-C_4H_8$

[6] $AlH(C_4H_9)_2 -----> AlH_3 + 2i-C_4H_8$

[7] $AlH_3 ---> Al + 3/2\ H_2$

The LPCVD process carried out at pressure of 150 mTorr and a temperature of 250°C yields a growth rate of 200 A/min (38).

Other chemistries that have been proposed very recently for CVD of Al are based on trimethylamine complex of alane (TMAA) (39) and dimethyl aluminum hydride (DMAH).

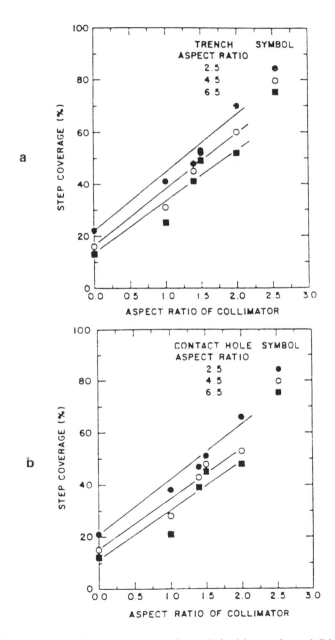

Figure 18. Variation of step coverage of metal, in (a) trench; and (b) contacts, with aspect ratio of collimator. Ref. 35, reprinted with permission.

COLLIMATED SPUTTERING

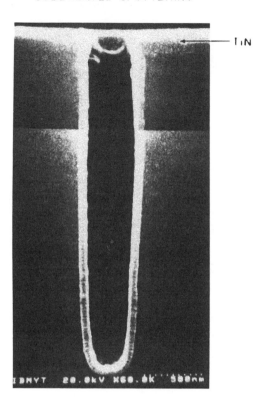

Figure 19. SEM of a high aspect ratio contact fill with collimated sputtering. Ref. 35, reprinted with permission.

Blanket and Selective Deposition: Both blanket and selective depositions of Al have been achieved with CVD. For selective deposition, thermal decomposition of TIBA is used. Initial work on this showed that the surface of the selectively deposited Al was rough. A new double wall CVD system has been proposed which deposits smooth Al films selectively on conducting surfaces (40). This system is shown in Figure 20. It has two heaters, one at the front and the other at the back, which simultaneously maintain the wafer at the desired temperature. The minimum temperature for Al deposition is around 230°C. Depending on the relative values of the temperatures T_f and T_b of the front and back heaters respectively, different conditions of deposition arise. Figure 21 shows these deposition conditions in this system. By carrying out the deposition in the hot wall region, smooth films are obtained which is attributed to the high density of nuclei. Selective deposition also occurs in the super hot wall region shown as "A" in the Figure.

Under this condition, Al gets deposited on Si, poly-Si, W, Ti, Mo and silicides. It does not deposit on insulators like SiO_2 and Si_3N_4. The selective deposition has been used for filling submicron contacts/vias with aspect ratios greater than 1.0, using $TiSi_2$ as the nucleating layer (40). The selectivity is mainly attributed to the nucleation process, which is surface catalyzed. The surface energy of metal is larger than that of an insulator, which produces a difference in nuclei density during film deposition and gives rise to selectivity. If the substrate temperature becomes high the selectivity is lost because of the homogeneous reaction on both the conducting and insulating surfaces. The selectivity is also critically dependent on the cleaning prior to via filling. Chamber contamination can lead to random nucleation (41) thereby causing loss of selectivity also.

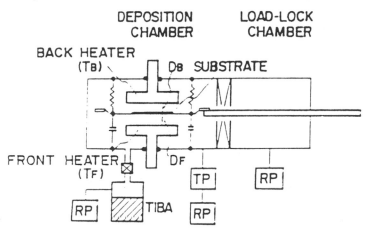

Figure 20. Schematic of the double wall CVD system for Al film deposition. Ref. 40, reprinted with permission.

Another selective CVD process proposed recently for Al is based on dimethylaluminum (DMAH) (42). This is claimed to be a controllable process in which selective or blanket deposition of Al can be obtained. The Al films are selectively deposited by thermal decomposition of DMAH with H_2. Blanket deposition is obtained by exciting a plasma at 13.56 MHz. The deposition temperature is 270°C. The resistivity of the Al obtained by this technique is about 3 µΩ-cm. This process has also been successful in filling vias. However, it is being evaluated further for use in IC. manufacturing. In the CVD techniques reported so far, incorporation of alloying elements such as Cu, Si and Ti has not been demonstrated satisfactorily. There is a great need to develop CVD processes to deposit various AlX films, because they have a potential for being more production worthy than sputtering as the Si wafer sizes are increasing from 8" and larger, the device/interconnect dimensions are being scaled down to 0.5 µm and lower, and the need for multilevel metallizations is ever increasing with larger number of metallization levels.

Laser assisted selective CVD Al process has been reported to achieve resistless patterning of Al. This technique will minimize, if not eliminate, the number of photolithography steps in the metallization process. A light assisted process activates the adsorbed metalorganic precursor, dimethyl aluminum hydride (DMAlH) (43), onto the substrate surface, to delineate the pattern as a seed layer for subsequent selective CVD. The substrate temperature is raised after this pattern delineation for the selective CVD process. The patterned Al achieved by this method has been reported to be of good quality having a low contact resistivity (0.2 $\mu\Omega$-cm^2) without any pre- or post-deposition treatments such as annealing.

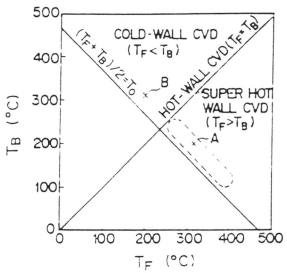

Figure 21. Deposition conditions in double wall CVD system. Ref. 40, reprinted with permission.

CVD Al Properties: Chemical analysis of the CVD Al films deposited from TIBA showed that the carbon content in the film is less than 0.3% (37). No other impurities including O_2 are detected. X-ray diffraction spectra indicate that the films show <111> structure, both on mono-crystalline Si and amorphous SiO_2 substrates. As a consequence of this, CVD Al films do not show any hillocking when heated to 400°C for 30 min (38). Step coverage and via filling by CVD Al is superior to sputtered Al. As-deposited films exhibit a tensile stress of around 170 MPa at room temperature. These films show good adherence to a variety of substrates including Si, SiO_2, TiN and TiW. The resistivity of the as-deposited CVD Al is about 2.8 to 3.0 $\mu\Omega$-cm.

CVD of Other Conductor Materials: As mentioned previously, conventional sputtering of underlayers and cap layers such as W, TiN and TiW does not yield adequate step coverage in sub-micron contacts/vias. CVD of TiN and W have been used successfully to overcome this problem. These techniques are discussed below.

TiN: Two CVD processes based on different chemistries are used commonly in the present technologies. The precursors used in these are $TiCl_4$ (inorganic) and tetrakis (dimethylamino) titanium, TDMAD, (organic). The

$TiCl_4$ based process needs higher temperature, around 650°C, as compared to TDMAD process that needs around 400°C. However, the resistivity of TiN deposited by MOCVD is very high (500-1000 $\mu\Omega$-cm) as compared to that deposited by $TiCl_4$ processes (70 $\mu\Omega$-cm), because of the higher impurity (C, O) content. Also the $TiCl_4$ based process provides films with higher density and better conformality of step coverage (44).The lower deposition temperature of the metalorganic CVD makes it attractive for use in the second level and beyond. Attempts to reduce the deposition temperature of $TiCl_4$ based CVD have yielded encouraging results. The TiN films deposited at 450°C have a higher resistivity (200 $\mu\Omega$-cm) than those deposited at 650°C but have 100% step coverage.

Tungsten: CVD tungsten has been used as a first level interconnect, but it is used more extensively for filling vias of high aspect ratios. The CVD of W overcomes the problems of step coverage of sputter deposited Al/AlX into such vias and contacts. The processing temperature in CVD of W is in the range of 300 to 500°C which is compatible with most VLSI processing. Both blanket and selective deposition of W are possible, although the latter needs further development before it can be used in manufacturing. In the selective deposition process, W deposits only on Si and other conducting surfaces, and not on insulating surfaces such as SiO_2.

The precursor most commonly used for CVD W is tungsten hexafluoride, WF_6. Three chemistries are possible depending on the reduction reaction used:

[8]   (i) Hydrogen reduction : $WF_6 + 3H_2 \rightarrow W + 6HF$

[9]   (ii) Silane reduction : $4WF_6 + 3SiH_4 \rightarrow 4W + 3SiF_4 + 12HF$
or, $2WF_6 + 3SiH_4 \rightarrow 2W + 3SiF_4 + 6H_2$

[10]   (iii) Silicon reduction : $4WF_6 + 6Si \rightarrow 4W + 6SiF_4$

Of these, the hydrogen and silane reduction processes can give either selective or blanket deposition of W. The Si reduction process is selective, i.e., it occurs only in regions where the Si is exposed. Nevertheless, this reaction consumes Si, can cause undercut of the oxide and penetrate Si. This leads to reliability problems in the multilevel metallizations. Presently, the SiH_4 reduction process is the most widely used CVD process for W deposition. This reaction offers a higher deposition rate as compared to the H_2 reduction reaction. The main feature of the H_2 chemistry is that it yields excellent step coverage. The deposition temperature for both silane based and H_2 based processes are in the 350-450°C range. The variation of deposition rate with temperature is shown in Figure 22. The deposition rates are around 2000 Å/min and 1000 Å/min for silane reduction and H_2 reduction respectively (45). Using the W CVD, contacts and vias of high aspect ratios are filled without voids. To act as a nucleating layer and improve the adhesion, a Ti liner is normally used. The WF_6 used in the CVD process tends to interact with the Ti layer. To prevent this interaction, a thin layer of a barrier metal like TiW or TiN is deposited prior to the CVD of W. The step coverage of this layer, especially at the bottom, becomes a crucial issue for the overall reliability of the via interconnect. The via or contact resistance depends on several factors such as the quality of the barrier layer, the via/contact filling by the W and its resistivity. The contact resistance is found to depend strongly on the deposition temperature (46). Figure 23 shows the via resistance of W/AlCuSi structure with TiN/Ti barrier as a function of the temperature of the substrate holder (note that the actual wafer temperature is 120-130°C lower).

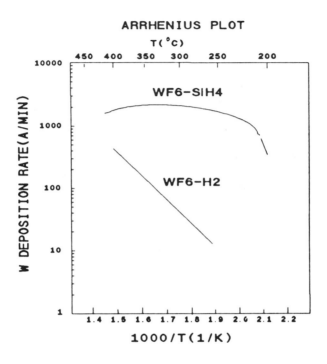

Figure 22. Variation of deposition rate of W with temperature for silane and hydrogen reduction reactions. Ref. 48, reprinted with permission.

Selective deposition of W by CVD has been successfully demonstrated (47) for contact fills. A typical two-step selective process is to first react WF_6 with the exposed Si layer in the contact. This reaction is self limiting in that continued deposition requires the WF_6 to diffuse through the newly formed W layer before reaching the Si. Thus a thin film of about 200 Å formed by this reaction acts as a seed for the subsequent reduction reaction with H_2 for the deposition of W. As already mentioned, a problem associated with the Si reduction of WF_6 is the under cutting of Si and the formation of defects such as worm holes.

One major problem with selective CVD of W is the loss of selectivity because of which W deposits on dielectrics outside the contact/via. Another, more serious problem, is when a few of the vias are not completely filled with W. The reasons for these are not yet fully understood. However, several mechanisms have been proposed recently to explain the occurrence of these problems (48). The important ones are:

(1) Nucleation provided by the metallic contaminants settling down on the dielectric adjacent to the contact/via hole. These metallic particles come from the contact metal and are back sputtered on to the resist mask during contact etching. When the resist is stripped with an O_2 plasma, while the organic resist gets removed, the metal contaminants settle on the dielectric surface in the form of metallic oxide. This oxide can act as a nucleation site for W during subsequent selective deposition.

(2) Nucleation provided by the OH bonds on the dielectric surface. During the resist strip, the ILD gets exposed to the plasma which changes its surface stoichiometry leading to an increase in dangling bonds. When the dielectric is exposed to air, moisture causes dangling OH bonds on the surface which act as nucleation sites for W deposition.

(3) Nucleation provided by other defect sites on the surface of SiO_2.

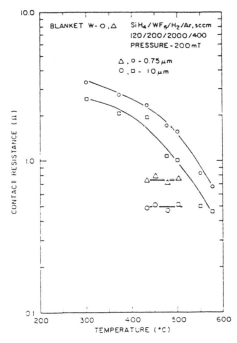

Figure 23. Variation of tungsten via resistance with temperature. Ref. 46, reprinted with permission.

One of the solutions to the first two problems of selectivity loss is to eliminate the plasma stripping of resist and instead use organic solvent for its removal. Figure 24 shows the surface of the dielectric after selective W deposition following resist stripping by two processes, viz., in O_2 plasma and in organic solvent. It is clear that in the case of resist removal by organic solvent, the surface of the dielectric is free of W nuclei.

The general approach to solving loss of selectivity problems is to passivate the defects on the surface of SiO_2 and minimize the movement of particles/species which cause this problem (49). It also requires designing-in system level purity by selection of appropriate materials for the CVD reactor, by incorporating in-situ wafer cleaning processes and by maintaining high purity of gases (moisture free) at the point-of-use. Modifications have been proposed to the selective deposition processes for preventing loss of selectivity . One of these is the Dep/Etch technique. In this, during the selective W deposition, a mid-process etch is carried out to remove the few nucleation sites that would be forming on the dielectric surface. The time at which this etching is to be done is determined by the rate of increase of the nuclei density. Once these nucleii are removed, the selectivity of the deposition is restored, at least partially. This process is schematically shown in Figure 25 with reference to the nuclei density.

a

b

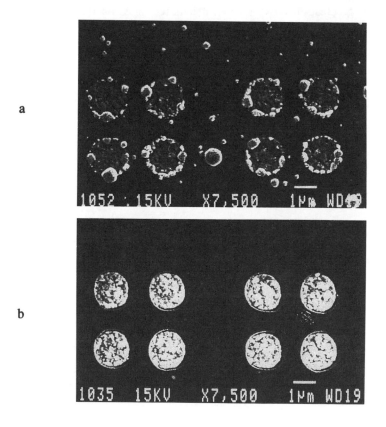

Figure 24. Surface of dielectric after selective W deposition following resist stripping in (a) oxygen plasma and (b) organic solvent. Ref. 48, reprinted with permission.

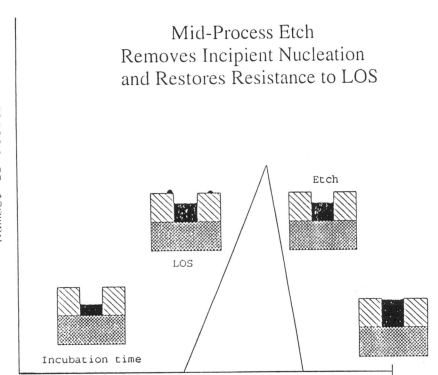

Figure 25. Selective W deposition with a mid-process etch step. Ref. 49, reprinted with permission.

5.2   Etching of Al and Its Alloys

Wet etching: Wet etching has been widely used for etching Al and its alloys from the very beginning in the semiconductor industry. The etch chemistry uses various ratios of phosphoric acid, nitric acid and DI water, depending on the etch rate and control desired. However, because of the isotropic nature of these processes, wet etching is absolutely unsuitable for etching fine line geometries used in present day technologies. Therefore, its use has almost been eliminated in recent years in advanced microelectronics manufacturing.

Dry etching: Dry etching can be done either in the RIE (Reactive Ion Etching) or plasma etching modes. The RIE processes for metal etching have become very popular in VLSI technologies and will be indispensable with further decrease in linewidths and spaces in the future ULSI technologies. The reason for this is that the high degree of anisotropy of RIE processes permit the use of smaller metal pitches that are a must for etching in sub-micron geometries. The anisotropy also gets rid of the *undercutting* problem that is common in wet etching processes, since wet etching is isotropic.

Anisotropy, however, is not the only criterion for etching in metallizations. Etch rate, selectivity with respect to the photoresist, and uniformity of etching are also important issues, along with good film stack profiles with no residues and very low film defect density, particularly for etching multi-layer metallizations like TiN/AlSiCu/TiW on any level.

Dry etching chemistry uses halides of various elements. In general, etch processes used for etching Al alloys are chlorine based in today's technologies. A mixture of $BCl_3/Cl_2/\{CHCl_3$ or $CCl_4\}/N_2$ etches Al and Al-based alloys with adequate etch rate and anisotropy (50). Incorporation of $CHCl_3$ or CCL_4 helps passivate the sidewalls. Chlorine chemistry works better than other halide chemistries with AlX because the etch product of etching AlX with chlorine based etchants have sufficiently high vapor pressure. This is not the case with fluorine based etchants to etch AlX, where the etch products have low vapor pressure. Fluorine based chemistry, however, gives good results in etching W and TiW (in addition to etching Si and SiO_2). Bromine based chemistry for Al etch has also been reported (51) where a small amount of a Br_2 containing chemicals such as HBr added to an optimized chlorine based chemistry results in a significantly improved selectivity with respect to the photoresist. In some cases, a combination of different etch gases can be used to optimize etch rate, anisotropy and selectivity. Multistep processing involving some 10-15 steps are being used to etch 2 or 3 layers of materials, as will be discussed later in this section, where each step is optimized for high selectivity, etch rate, etch uniformity, etc.

Al alloy etching processes, using various combinations of BCl_3, Cl_2 and N_2 and a passivating agent like CCl_4 or $CHCl_3$, can vary significantly by altering the etch conditions. Important among them are the pressure in the etch chamber and the RF power. At low pressure (20-200 mTorr) the etching is more anisotropic than that at higher pressures. At such low pressures, collisions in the plasma sheath can be better controlled (52). The gas species present in the etch reaction chamber dissociate more readily and completely at such low pressures. This helps ease the selection of gases for the RIE process. This is an important issue since many of the gases used for Al alloy etching are either carcinogenic or highly toxic.

Gas combinations of $Cl_2:BCl_3:CH_4$ have also been attempted (53) to avoid the use of such strong carcinogens as CCl_4 and $CHCl_3$ in the RIE of Al alloys. The etch rates of these processes involving CH_4 can be controlled easily. A photoresist or polyimide mask is used along with CH_4 to achieve good, anisotropic etching of the Al alloy. In reality, however, CH_4 in the gas flow leads to deposition of a polymer layer on the surface and sidewalls of the sample. If the CH_4 flow is relatively small, the thin polymer layer that is deposited on the metal surface, is sputtered off the surface, allowing etching to occur while the polymer film remains on the sidewalls, preventing undercutting.

Etching of AlX films sometimes results in the formation of a carbon rich film on the sidewalls of the patterned lines because of the presence of the organic photoresist (54). N_2 acts as the carrier gas for the other reactant gases in Al alloy etching. The presence of N_2 in the plasma also influences the structure, thickness and composition of the carbon rich film. In addition, it also helps remove the nonvolatile byproducts or species such as CuCl formed by Cu from AlSiCu alloys. For higher Cu percentage in AlSiCu alloys, a higher flow rate of N_2 is required to avoid any residues. Sometimes, sacrificial Al is placed in the etch chamber to produce $AlCl_3$ which helps remove the Cu residue that is left behind upon etching of AlSiCu.

The etch rate is primarily dependent upon the Cl_2 concentration in the RIE chamber (55). Figure 26 shows the etch rate, selectivity and uniformity of etch for Al-1%Si with varying r.f. power and varying chlorine flow rate. It is clear that increasing Cl_2 concentrations increases the etch rate and selectivity with respect to the photoresist, but reduces the etch uniformity. Figure 27 shows the effect of r.f. power and chamber pressure on these three parameters. There is a minimum in the uniformity (most uniform etch) and maximum in selectivity to photoresist in the low pressure and power range. An increase in pressure by increasing the etchant species resulting in higher etch rate, also results in an increase in the residence time which lowers the replacement rate of the reactive by-products and results in lower etch-rate. These two opposite effects of pressure on etch rate tend to balance each other. The net effect, however, is an increase in etch rates at increased pressures. Passivating agents, such as $CHCl_3$, are added to the etching gas mixture to provide sidewall protection. However, this incorporation reduces the etch rate as shown in Figure 28. Increasing the BCl_3 flow rate, on the other hand, increases the etch rate and also improves the etch selectivity over photoresist.

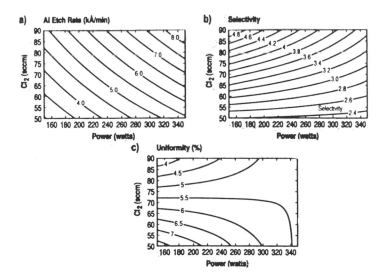

Figure 26. Variation of etch rate, selectivity and uniformity of Al-1%Si with r.f power and chlorine flow rate. Ref. 50, reprinted with permission.

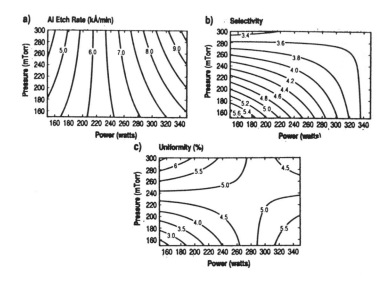

Figure 27. Effect of chamber pressure on etch rate, selectivity and uniformity. Ref. 50, reprinted with permission.

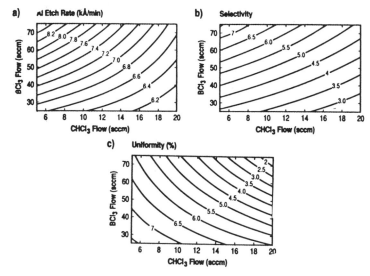

Figure 28. Effect of passivating agent on etch rate, selectivity and uniformity. Ref. 50, reprinted with permission.

Figure 29 shows the etch profiles corresponding to various combinations of all the etch parameters where 90° refers to a perfect vertical sidewall. An increase in the flow rates of $CHCl_3$ and BCl_3 result in more vertical profiles, i.e., better edge acuity. An even better edge acuity, one that is very close to 90°, can be made in the profiles by increasing the r.f. power as can be seen from Figure 29.

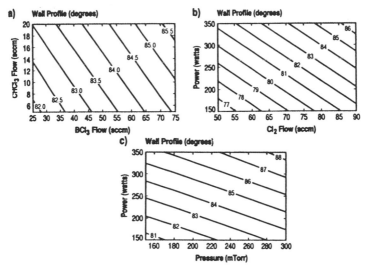

Figure 29. Etch profiles of various combinations of etch parameters. Ref. 50, reprinted with permission.

Multistep etching techniques are in use in the microelectronics industry today for dry etching of layered Al alloys. For example, layered structures of TiW/AlSi/TiW, where TiW is an alloy of W with a variable percentage of Ti, which acts as an underlayer as well as a cap layer on the AlSi metallization, can be etched using a BCl_3/Cl_2 chemistry in conjunction with CF_4. The CF_4 etches the TiW layer. Addition of O_2 helps increase the etch rate of TiW. The various endpoints can be detected by using laser interferometry (56), as shown in Figure 30. Indication of endpoint by this technique works well because most alloys of Al are highly reflective while most underlayers and cap layers such as TiW are only slightly reflective.

Corrosion and Postetch Treatments: Al alloy films patterned by RIE using chlorine chemistry tend to get corroded when exposed to humid atmosphere, giving rise to aluminum hydroxide or oxide on the surface and below the surface. This is not a problem when patterning with wet etching using phosphoric acid/nitric acid solutions. To minimize and prevent post-etch corrosion in dry etching, therefore, many post-etch treatments such as H_2O rinse, CF_4/O_2 plasma and O_2 plasma are used. In one study, a correlation between post-etch corrosion of Al-0.9%Si-0.4%Cu film and the residual chlorine after etching was done (57). Figure 31 shows the ratio of corroded area to observed metal surface area with respect to the residual Cl_2 intensity (measured by Auger electron spectroscopy, AES) for various types of post-etch treatments. It is clear from this figure that post etch corrosion does take place for the samples as-etched and post-treated with O_2 plasma at 40°C and 180°C. Corrosion is not observed in the other samples where the post etch treatment included H_2O rinse and/or

CF_4/O_2 plasma. The reason for this is that the H_2O rinse removes chlorine bound to the Al and not the chlorine bound to carbon, while the CF_4/O_2 plasma removes chlorine bound to Al as well as the chlorine bound to carbon. O_2 plasma, on the other hand, is effective in eliminating only the chlorine bound to carbon and not the ones bound to Al, causing some corrosion. The sidewalls of conductors, post treated with a fluorine containing plasma, such as CF_4 plasma, are covered with a passivation film consisting of F, Al, C and O atoms which works as an excellent protective mask against Al corrosion.

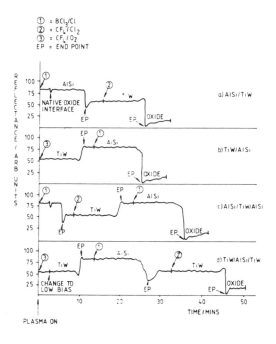

Figure 30. End point traces from laser interferometry. Ref. 56, reprinted with permission.

Prevention of Sidewall Etching: For high selectivity and high throughput in dry etching Al alloys, relatively high pressures (>100 mTorr) are used in the RIE chamber (58). However, several problems such as sidewall roughness, linewidth reduction and negatively tapered shapes occur as a result of this high pressure in the etch chamber, and can lead to yield loss and reliability problems in the interconnect and eventual failure of the devices/interconnect. The etchant species, particularly Cl_2, and the reactive oxygen-containing species from the oxide underlayer can etch the sidewall polymers to some extent. To remove metal residues from the oxide underlayer, particularly for the steep and high aspect ratio steps/openings, some overetch of the metal is very commonly used in practice. The sidewall polymers are susceptible to such overetching processes and, therefore, stable sidewall polymers are required to sustain the metal overetch process. Fluorinated polymer films have good stability against such attacks and can be used for the sidewall protection. Plasma treatment with CHF_3 between main etch and overetch processes (59) can give such sidewall protection. A comparison of profiles from conventional dry etching and the process using a fluorinated polymer sidewall protection layer has been shown in Figure 32.

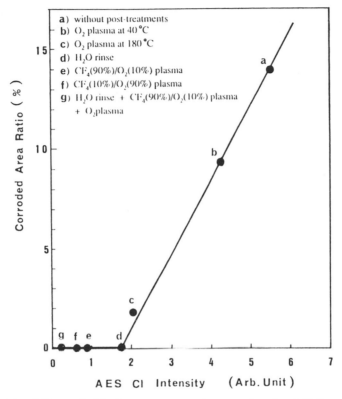

Figure 31. Effect of chlorine on post-etch corrosion in AlSiCu. Ref. 57, reprinted with permission.

Tapered etching: Good step coverage of the dielectric layers in any multilevel metallization scheme is of supreme importance because a poor step coverage is prone to result in reliability problems at the step in the form of dielectric voids, intra-level shorts by lateral extrusion and so on. These problems become more acute as the linewidths and pitches become smaller. To achieve good step coverage of the overlayer dielectric on the patterned AlX, many techniques have been attempted. One such technique is to etch the pattern such that the AlX lines have tapered walls, with a positive slope, rather than a vertical profile. The process may involve sidewall polymer formation technique as mentioned before, or, the photoresist erosion technique (60) during etching. In the first method, a polymeric sidewall is grown with the help of polymer generating species such as $CHCl_3$ and CHF_3 added in small quantities to BCL_3/Cl_2 processes. These processes help protect the Al sidewall from Cl_2

attack and to yield anisotropic profiles (61). In the other method, photoresist erosion technique, sloped profiles are generated in the resist and then transferred on to the metal by controlling the resist erosion and the Al alloy etch rate. This technique has some limitations, however, in terms of its ability to control narrow linewidths, severe topography, and varying photoresist thicknesses. Figure 33 shows the SEM of taper etched Al features with and without the photoresist and the sidewall.

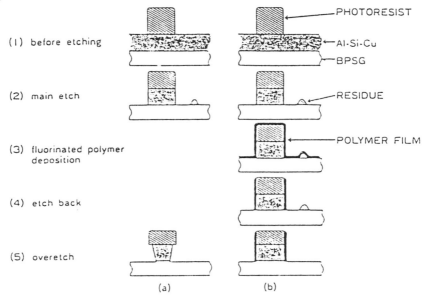

Figure 32. Etch profiles from conventional dry etching and the process using a fluorinated polymer sidewall protection. Ref. 59, reprinted with permission.

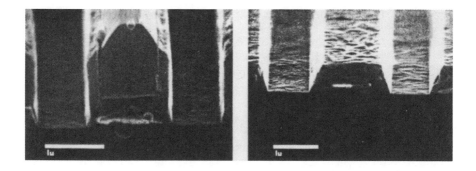

Figure 33. SEM of taper etched Al features with and without the photoresist and the sidewall. Ref. 61, reprinted with permission.

Laser Etching: Laser etching of fine patterned Al lines for CMOS test circuits has been reported (62) in which a tunable dye laser with wavelengths ranging from 2850 to 5730 Å has been used. It is found that a lower wavelength results in better etch selectivity. With the increasing need for cluster processing in microelectronic manufacturing to eliminate contamination caused by handling the wafers outside a vacuum chamber, processes like laser CVD and laser assisted etching are of interest because they cause less contamination as compared to the conventional CVD and etching processes.

5.3   Planarization for Multilevel Metallization

Planarization of both the metal layers (to "plug" contacts and vias and possibly even trenches) and the ILDs is critical to the success of multilevel metallization. The subject of ILD planarization is covered in detail in chapter 6 of this book. However, its salient features are given here briefly in addition to a more complete discussion of metal planarization to facilitate the discussion of Al based multilevel metallizations.

In a multilevel metallization, patterned conductors and dielectric layers at any level give rise to severe topography for the layers of the next level. This can result in opens in these interconnects or voids in the dielectric layers on account of the step coverage problems of the deposited metals or dielectrics. Shorts between adjacent lines can be caused by lithography problems due to the topography on the wafer. Such problems can cause severe yield losses at a point where the wafer has been through the major steps of the total process. For example, in a three level metallization scheme, metal lines at the third metallization level (M3) have to traverse the topographies caused by patterned lines of the first dielectric (ILD0), metal lines of the first level (M1) covered by the second dielectric layer (ILD1) and the metal lines at the second level (M2) as well as various stacked combination of these. Also, it has to pass over vias connecting the first level metallization to the second level metallization (V1) and cover vias (V2) connecting it (i.e., the third level metallization) to the second level metallization. These topographies can lead to opens in the third level metallizations due to step coverage problems. Figure 34 shows problems that can arise in a two level metallization due to topography. Planarization of layers in a multilevel metallization attempts to eliminate or smooth out the topographic undulations caused by conductors, dielectrics, contacts and vias. Planarization ensures that metal, at any level, is deposited on a planar surface which means that step coverage problems of the metal do not arise at all. It also aids the photolithography with steppers using high numerical aperture lenses and short wavelength light sources, which limit the depth of focus. Therefore the pattern definition all across the wafer remains good when complete planarization is used. Thus, the advantages of both ILD and metal planarization are (63):

1.   Problems of step coverage of metal over dielectric steps and into contacts/vias can be avoided.
2.   Reflection problems in lithography can be minimized, if not eliminated.

3   The walls of contacts/vias may need not be sloped; either by reflow or slope etching, depending upon the metal planbarization techniques.

4. Design rules can be tightened to allow denser packing. Conductors from various levels can overlap instead of requiring a minimum distance between them.

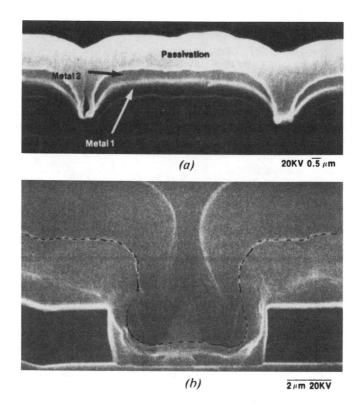

Figure 34. Problems in multilevel metallizations due to topography: (a) opens in second Al layer owing to small spaces between lines of first Al; (b) poor step coverage of Al in a contact. Ref. 63, reprinted with permission.

Planarization is desirable in double metal technology but is almost essential in triple and 4-level technologies and beyond. Various degrees of planarization are shown schematically in Figure 35. Referring to this Figure, if t^f_{step} is the step in the dielectric after planarization, then, in the case of local planarization, t^f_{step} is zero only in certain localized regions on the wafer while in the case of global planarization, t^f_{step} is zero everywhere on the wafer. Ideally, global planarization is desired. However, many of the planarization techniques yield "local" planarization where good step coverage, gap filling and conformality are achieved. Planarization can be divided into three groups:

1. Planarization of vertical interconnects (contacts/vias),
2. Planarization of horizontal interconnects, and
3. Planarization of dielectrics.

The technologies used for planarization can be classified into two broad groups, viz. structuring and deposition. Each of these include a variety of techniques, shown schematically in Figure 36, and are briefly discussed below. A variety of combinations of techniques from these two categories are also used.

Structuring: In this category, planarization of the dielectric or metal layer is achieved by changing or modifying certain characteristics of the layer. The techniques that come under this category are thermal reflow, etchback after spin-on or LPCVD/PECVD, lift-off, chem-mechanical polishing and material modification.

Deposition: In this category of techniques, planarization is achieved during the deposition of the layer. This category includes spin on and bias sputtering.

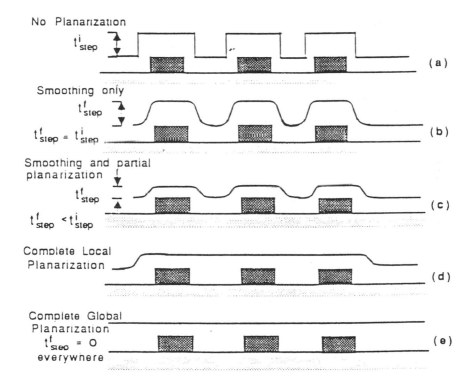

Figure 35. Degrees of planarization. (F. Moghadam in VMIC State-of-the-Art Seminar, p. 319 (1992), reprinted with permission.

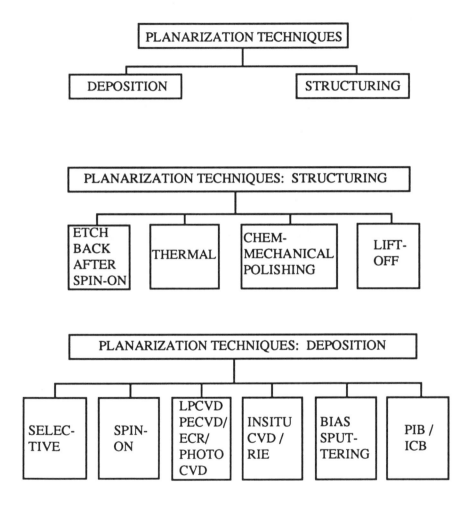

Figure 36. Classification of planarization techniques.

The following sections discuss the features of the different technologies used for planarization of vertical interconnects, horizontal interconnects and dielectrics.

Planarization of Vertical Interconnects (Contact/Via):
Planarization by structuring:

Thermal Reflow : In this technique, the metal in the interconnect is raised to a temperature such that it flows and smoothes out the topography below it. This can be achieved either by raising the entire wafer to the flow temperature in a conventional furnace, or in a rapid thermal anneal equipment for a short time, or by heating locally using lasers and electron/ion beams, without raising the temperature of the entire wafer. Conductors must be raised to their melting points before flow occurs. Heating the entire wafer either in a conventional furnace or by RTA is undesirable because of interaction of AlX with other materials at melting temperatures for relatively long periods of time. However, pulsed lasers allow flowing of Al locally and thus have been extensively investigated. So far, excimer lasers such as XeCl laser (λ=308 nm) have been most widely used to demonstrate AlX planarization. One of the commercially available excimer laser planarization systems (64) is shown schematically in Figure 37. It consists of a pulsed XeCl excimer laser source and a process vacuum chamber with substrate heating arrangement. The laser fluence range in this system is between 2 and 5 Joules/cm^2. The planarization is carried out under a vacuum of 7.5E10^{-7} Torr. The substrate temperature can be maintained between 250 and 450°C. The planarization of a variety of Al alloys such as Al-1%Si, Al-1%Si-0.5%Cu, Al-1%Si-0.1%Ti by the excimer laser shows (65) excellent planarization for all topologies including sub-micron contact holes. However, for the AlSi alloys, spot ablation occurs resulting in the removal of metal (see Figure 38. This is not seen in AlSiCu and AlSiTi alloys. The minimum laser fluence required for planarizing vias in a two metal level test structure increases as the via size increases. It decreases with increases in substrate temperature (66). Also, in the case of contact/via with high aspect ratio, the as-deposited metal profile is re-entrant. Therefore, upon laser irradiation, the surface tension effects result in planarized metal surface at the top with a void inside (67). For sloped or tapered contacts/vias, however, the opening is larger and hence the molten metal can flow into the contact/via, resulting in a voidless fill. Several other approaches have been developed for via/contact planarization using excimer laser reflow technique. In one of the techniques, a thin metal cap is patterned to cover the entire via and melted with the laser. The molten metal cap gets drawn into the via by surface tension forces resulting in the formation of the metal plug (69). Chem-mechanical polishing is used to remove the Al left on the surface (70).

Insufficient laser energy causes grain boundary separation and metal cracking as a result of grain boundary scattering and preferential melting of Al at the grain boundary. A threshold energy of greater than 3 J/cm^2 at a substrate temperature of 400°C is required to completely melt and flow Al in the entire laser spot without grain boundary separation (71) and metal cracking. No damage is done to the underlying structures at these elevated temperatures.

The barrier metal in the interconnect affects the planarization. It impacts the wetting of the overlying AlX and hence its reflow and contact filling during laser irradiation. A Ti barrier layer results in better planarization, compared to either TiN or a TiW barrier (72). The sputtering conditions of Al film deposition also affect the via planarization. A significant improvement in the process window, viz., the range of laser fluence and substrate temperature, has been observed by using high temperature and/or bias sputtering (73).

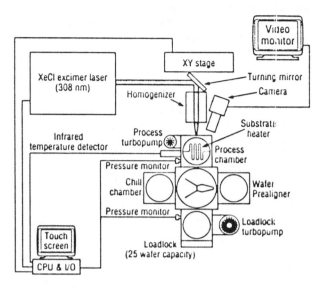

Figure 37. A commercial Al planarization system that uses excimer laser. Ref. 64, reprinted with permission.

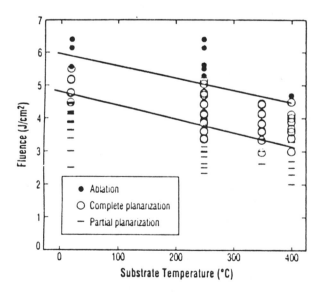

Figure 38. Excimer laser fluence vs substrate temperature effects. Ref. 64, reprinted with permission.

The grain size of the Al increases with laser fluence and substrate temperature. Typical grain size after the irradiation ranges from 3 to 7 micron for Al-1%Cu film on a TiW barrier layer. Stresses in AlSi/TiW change from slightly compressive to slightly tensile after processing. Further, no hillocks are observed on the laser reflowed samples after post-reflow annealing at 450°C for 30 minutes. This is attributed to the stress relaxation and out diffusion of O_2 and other impurities during the laser reflow process. A significant improvement in the contact/via resistance (74) and the electromigration resistance (75) has been observed with the laser planarization process.

Etchback After Blanket CVD: In this technique, a thick blanket metal is deposited and subsequently etched back such that it fills in all the spaces/gaps, yielding essentially a planar surface. Even though it can be used for either dielectric or metal planarization, it is mostly used for planarizing contacts/vias in multilevel metallizations. A blanket W layer is deposited by CVD and subsequently etched back to give a planar surface with W plugs in the vias. To avoid interaction between the CVD reaction species (WF_6) and the SiO_2, barrier layers such as TiW are used as liners in the vias. The W deposition has already been discussed in section 5.1. The etch back of W is usually done in fluorine plasma since fluorine reacts with W to form volatile WF_6. The W and the TiW liner are etched back in one single step till SiO_2 is reached in areas outside the contact/via. The end point can be detected automatically by monitoring the F intensity or the pressure. One of the problems in this process is that after the etch back, the W plug can be below the top of the contact hole, by as much as 0.4 µm, because of the loading effects (76). This recession in the via holes reduces the planarity of the surface of the via which can give rise to step coverage problems in the next level Al/AlX metallization. The surface roughness can also affect the lithography in the subsequent patterning.The loading effects responsible for this recession in W in vias are caused by the excess F becoming available in the holes after the SiO_2 is reached in the outside regions, which results in deeper etching in the holes. The high etch selectivity between W and SiO_2 is responsible for the excess F to be available. In one solution, a Si_xN_y sacrificial layer is sandwiched between SiO_2 and W. The etch rates of Si_xN_y and W are nearly same and hence consume F at the same rate. Thus when W is completely etched from the regions outside the via, excess F is not made available and hence the loading effects are minimized. The etch back sequence using the Si_xN_y sacrificial layer is schematically shown in Figure 39.

In all CVD W based planarization techniques, TiW or Ti/TiN liners are first deposited into the contact/via before the W deposition. The TiN acts as an adhesion promoter and a diffusion barrier between the W CVD reaction species, especially WF_6 and Ti. Ti is used, especially in contacts, to reduce the contact resistance. The step coverage of the Ti and TiN in contacts/vias of high aspect ratios also becomes an important issue for planarization. Novel techniques like collimated sputtering and CVD are used to achieve good step coverage. These aspects have been discussed in preceeding sections.

The blanket CVD of W and its etchback is the most widely used technique for metal planarization in the present VLSI manufacturing.

CMP after CVD: In this technique, planarization of the wafer surface is achieved by co-polishing with chemical and mechanical processes (CMP).   This method provides global planarization.   The polishing medium used consists of abrasive particles as well as appropriate chemicals (depending on the surface to be planarized).   The CMP occurs due to a combined action of chemicals, abrasive particles and the fibers of the polishing pad in presence of applied pressure/lateral motion.   To planarize contacts/vias after CVD of W, the W layer is removed from the regions outside the contact/via by CMP.   In the chemistry that has been developed for the polishing of W (77), the mechanical action removes a passivating layer continuously and the chemical action dissolves the exposed W and reforms the passivating layer.   In this way, only the high spots get removed while the lower spots are protected.

1. SiO$_2$ and SiN deposition.
    CO definition and etch.

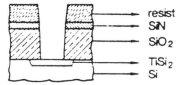

2. resist strip, TiW and W deposition.

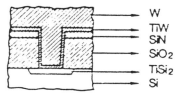

3. W and TiW etchback (stop in SiN).

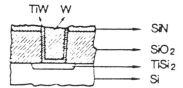

4. SiN removal, (TiW⁺) AlSiCu deposition.

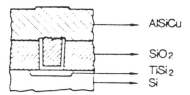

Figure 39. Etch back planarization with a sacrificial SiN layer. Ref. 76, reprinted with permission.

Planarization by Deposition

Bias Sputtering and Other Sputtering Techniques: In bias sputtering, the material to be deposited is sputtered from a target onto the wafer held at a negative potential relative to the plasma, causing simultaneous deposition of the material from the target and sputter-removal (etch) of the material from the wafer. At low bias voltages, better step coverage is obtained because of the resputtering of the film material from the base of the groove to the sidewalls and corners. As the bias voltage is increased, the sputter-etch rate is increased and the high points of the topography are levelled out at the same time as the material is being deposited. The final deposited layer planarizes the surface. Planarizing by this technique is also influenced by other factors such as the temperature of the substrate and the relative size and location of the target and the substrate.

Bias sputtering has been used to obtain planarized Al films. Sputtering with certain modifications is also used for filling vias/contacts with metals like Al. The modifications proposed are (i) the use of high substrate temperature (500°C-550°C) during the deposition (78) and (ii) the use of a collimator in front of the substrate to direct the Al atoms. By using these modifications, vias with high aspect ratios are filled with Al alloys for contact planarization. Among these sputtering techniques, collimated sputtering seems to be the most promising for filling of high aspect ratio contacts and vias.

Selective Deposition In this, a metal is deposited in contacts/vias over Si, silicides or Al without any deposition taking place over the dielectric. At present, processes for selective deposition of Al, W and Cu are under development (42, 47). In the case of W, the problems associated with loss of selectivity are yet to be fully solved.

Pillar Interconnects: In this method, metallic pillars are used for interlevel connections instead of the conventional vias. The pillar is formed by sputter deposition and consists of a layered structure made of refractory metals like Ti and W and the main Al alloy. The Ti and W act as both an etch stop during Al etching and a diffusion barrier between Al and the Si substrate. Next, the pillars are defined by lithography and multistep etching processes. A CVD SiO_2 layer is then deposited to such a thickness that its lowest point is above the top of the pillars. The oxide surface is planarized by using a sacrificial spin-on photoresist layer. The photoresist and SiO_2 layers are then etched back till the top surfaces of the pillars are exposed (79). The process sequence is shown in Figure 40.

Planarization of Horizontal Interconnects: Planarized horizontal interconnects can be obtained by exploiting the concepts of "buried interconnects", wherein groves in the ILD are filled with metal to create the interconnections at a given layer. The buried interconnect technology provides a planarized surface which means that there is no need for planarization of the next dielectric layer. In addition to that, this technology has other key advantages. It obviates the need for the metal etching which can be a crucial issue, particularly with new metallizations such as Cu. To fabricate buried interconnects, trenches are first etched in a dielectric layer . These trenches, when filled with metal up to the surface, result in the formation of buried interconnects. The techniques that are used for achieving planarized interconnects presently are briefly discussed below.

Structuring--chem-mechanical   polishing   after   blanket deposition : In this technique, metal is deposited on the patterned oxide so as to fill the trenches. The metal on top of the planarized dielectric is subsequently removed by chem-mechanical polishing (CMP), explained in detail in a preceeding section, or any other etchback technique to get the planarized interconnects. This is also called the "recessed metal" approach for planarized interconnects.

In an alternate technique, the metal interconnects are first formed on an insulator. These are then conformally covered with an insulator film. This is followed by a planarizing of the dielectric by etchback or CMP to expose the surface of the metal interconnects. Both of these approaches to achieve buried interconnect by structuring process are illustrated in Figure 41.

Deposition : In this technique, the metal is selectively deposited in the trenches in a dielectric layer to a thickness equal to the trench height. This technique has been used for getting buried Cu interconnects. Selective deposition of Cu is achieved by electroless plating technique (80) The sequence of processes is shown in the schematic in Figure 42.

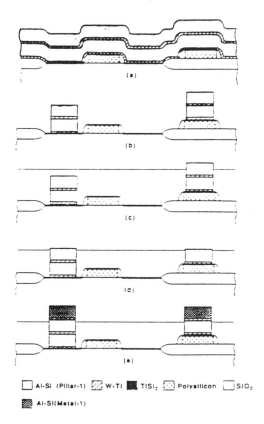

Figure 40.   Process sequence for getting pillar interconnects. Ref. 79, reprinted with permission.

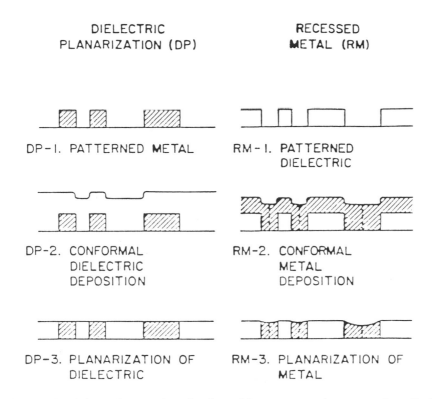

Figure 41. Approaches to planarization of interconnects by structuring. Ref. 77, reprinted with permission.

A reverse pillar process, as shown in Figure 43, has been suggested (81) for getting self aligned contacts and buried interconnects. In this, trenches are partially etched in the inter level dielectric (ILD), followed by contact etching of the dielectric at the desired locations. The contacts and the trenches are then filled with Al or AlX and etched back to achieve recessed metal in the trench (buried interconnects) and in the contacts (plugs) as shown in the figure.

Planarized Interconnects by Material Modification: In this technique, planarized Al interconnects are formed by selectively converting certain portions of it into insulator. This automatically produces a planarized interconnect pattern. In one approach proposed for this (82) nitrogen is implanted selectively into Al to convert it into insulating aluminum nitride. The patterned photoresist which would normally be used for etching Al will serve as the mask for the implantation. Nitrogen doses in the range 6×10^{17}-8×10^{17} cm^{-2} at 80 to 100 KeV are needed to produce insulating layers in 0.3 µm thick Al. The process steps involved in fabricating multilevel interconnect structures using this technique are schematically shown in Figure 44. This technique, however, is not expected to be production worthy because of the high cost of the hi-dose implantation, and the problems with the properties of the modified metal to give good insulator films.

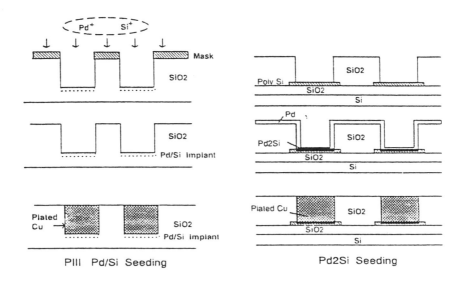

PIII Pd/Si Seeding   Pd2Si Seeding

Figure 42. Interconnect planarization by selective deposition of Cu. Ref. 80, reprinted with permission.

Planarization of Dielectrics: As in the case of interconnects, the various techniques of planarization of dielectrics can be discussed under two broad categories, viz, structuring and deposition. The techniques that are most widely used presently in manufacturing, and some of the promising techniques are discussed below. Details of the dielectric planarization techniques can be found in Chapter 6.

Planarization by structuring

Chem-mechanical Polishing (CMP): CMP technique for planarization of dielectrics has been investigated recently due to its potential ability to yield "global" planarization. This technique eliminates step coverage problems of the metal deposited at the next level. Also, as discussed earlier, it provides almost planar surfaces for subsequent photolithography steps which eliminates the problems due to the depth of focus limitations of today's steppers. In the technique reported (83), the SiO_2 films having topography are subjected to polishing in a commercial polishing machine. In this, the wafer surface with the oxide is actively rotated against a rotating polishing pad made out of cast, filled polyurethane, in the presence of a polishing slurry. The slurry contains fused silica particles (30-40 Å) colloidally suspended in an alkaline medium (KOH ; PH = 12-13). CMP removal rates of isolated elevated features is very high compared to large flat areas, with the rate inversely proportional to the area of the feature.

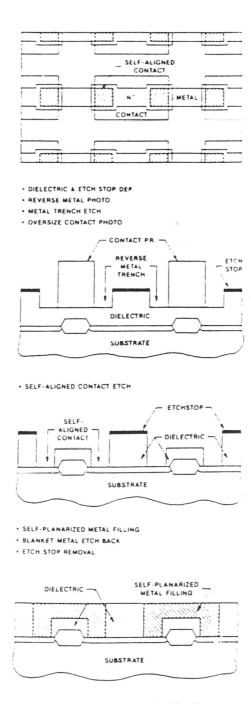

Figure 43. Reverse pillar process for self aligned contacts and buried interconnects. Ref. 81, reprinted with permission.

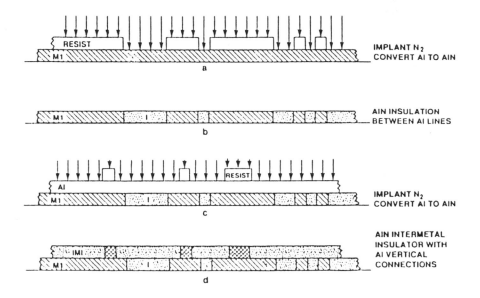

Figure 44. Process sequence for interconnect planarization by material modification. Ref. 82, reprinted with permission.

Presence of water in the polishing medium is very critical to the CMP. However, H_2O incorporation in SiO_2 occurs, which affects its mechanical (stress) and electrical (dielectric breakdown strength) properties. The SiO_2 surface after CMP is found to have significant amounts of SiOH.

CMP looks very promising as a global planarization technique in VLSI/ULSI manufacturing. Therefore, it is being evaluated extensively for various types of ILDs.

Etchback After Dielectric Deposition: This is the most widely used planarization technique in the present multilevel metallization processing. In this technique, a spin-on dielectric such as photo-resist or spin-on-glass (SOG) or SiO_2 is deposited and subsequently etched down (blanket etch) partially or fully. When the dielectric is fully etched away then it is termed as a "sacrificial layer". A typical process sequence used for this purpose is shown in Figure 45. Photoresist is the most commonly used material for sacrificial layer. The important criterion in this planarization technique is that the etch rates of the sacrificial layer and the dielectric film must be equalized by adjusting the parameters of the dry etch process (gas chemistry, flow rates, R.F power). An example of how this is achieved in a typical dry etch process is shown in Figure 46. Planarization is also achieved by depositing a thick layer of the actual dielectric such as SOG, and etching it back partially (no sacrificial layer). The thick dielectric layer creates a smoothened topography. An anisotropic etch back preserves this smooth surface as shown in Figure 47.

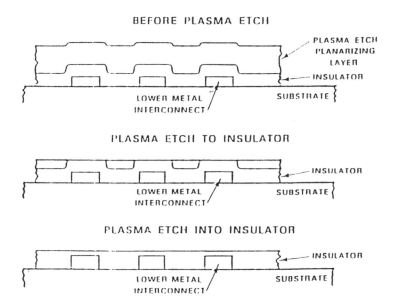

Figure 45. Dielectric planarization by etch-back process. (Moghadam, F., VMIC State-of-the -art Seminar, p. 319 (1992), reprinted with permission.

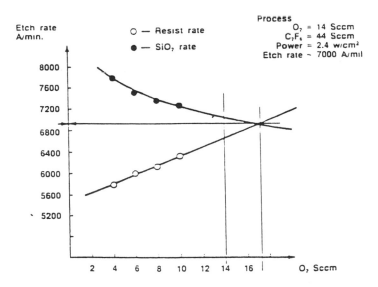

Figure 46. Etch rates of resist and SiO_2 can be equalized by varying the gas mixture used in the plasma etching process. (Moghadam, F. , VMIC State-of-the -art Seminar, p. 319 (1992).

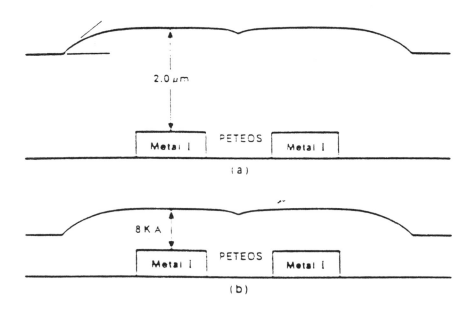

Figure 47. Dielectric planarization without using any sacrificial layer, by partial etch back. (Moghadam, F. , VMIC State-of-the -art Seminar, p. 319 (1992), reprinted with permission.

Several alternate approaches have been used in the last few years which do not use SOG for gap filling. These use TEOS (Tetra Ethyl Ortho Silicate) based CVD processes for deposition of SiO_2 films which have superior step coverage than the silane based SiO_2 films. Plasma based (PECVD) and O_3 based (ThCVD) processes are both used for this purpose. A five step Dep/Etch process is the most widely used in the present technologies and is shown schematically in Figure 48. This process uses the ThCVD SiO_2 for filling the narrow gaps , just like the SOG in the previous case.

Thermal Reflow: As mentioned earlier, in this technique, the temperature of the layer to be planarized is raised to sufficiently high values to enable the atoms to move around locally (material flow) and smooth out the topography. Pure SiO_2, which is the most widely used interlevel dielectric, flows at 1160°C which is too high a temperature for most IC applications. However, the flow temperature can be reduced by addition of various impurities to SiO_2 such as P, B, As, Ge etc. BPSG films are now widely used in which flow temperatures as low as 920°C can be achieved in N_2 ambient with 5% P and 3% B. This temperature can be lowered further by 25-30°C by using a steam ambient. This technique has been used extensively in planarizing/reflowing the pre-metal-1 dielectric layer.

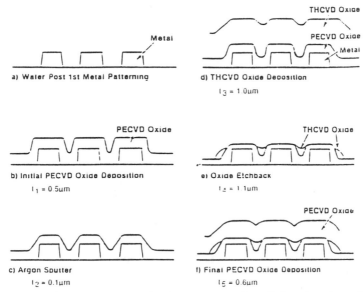

Figure 48. Five step dep-etch-dep process for dielectric planarization. (Moghadam, F. , VMIC State-of-the -art Seminar, p. 319 (1992).

Planarization by Deposition

Bias Sputtering and Other Sputtering Techniques: The preferential resputtering of the deposited film from certain angles in bias sputtering is exploited to smooth out the topography of the deposited dielectric layers and hence achieve planarization. For SiO_2 deposition by sputtering with Ar, the etch rate is low at incidence angles of $0°$ and of $90°$ while it is maximum at about $45°$. On account of this, when the SiO_2 is being deposited by bias sputtering on metal lines, it is preferentially removed at a higher rate from the curved portions over the corners of the metal lines as compared to the flat portions on the metal as well as the space between adjacent lines. Thus the net deposition of the dielectric in these regions is much less than at the flat portions. Under appropriate process conditions, the deposition and etch rates are such that the SiO_2 layer builds up at a faster rate in the flat regions than on the sidewalls and fills up the spaces between metal lines to produce a planarized surface without any voids (84, 85).

Selective Deposition for Planarization: Selective deposition of SiO_2 for planarization has also been proposed (86). The selectivity is achieved by liquid phase deposition (LPD) using a supersaturated hydrofluorosilicic acid (H_2SiF_6) aqueous solution. The oxide deposited by this technique has been characterized and its properties have been reported to be superior to those of CVD SiO_2. It is found that good selectivity of the LPD SiO_2 is obtained on photoresist patterns. This selectivity is caused by the wettability difference between the oxide and the photoresist. Full planarization of the interlevel dielectric has been reported using the LPD SiO_2 deposition with the photoresist mask.

5.4   Four-Level Metallization Systems

Two-level and three-level metallization schemes are currently used in IC manufacturing. During the past few years, various four level metallization schemes have been attempted in fab lines to manufacture reliable ICs. Many of these schemes use some of the very new materials for metallization and ILDs as well as new techniques of deposition and planarization. The important issues which need to be considered in a four level metallization and beyond are:

1. The metals/alloys used for the different levels - Although AlX is the most widely used metal, some schemes have used W as the metal in the first level. Presently, AlCu and AlSiCu alloys are the most widely used in the multilevel metallizations. Layered interconnects, with Ti/TiW or Ti/TiN underlayers and cap layers are used in most of the present technologies. Sputtering is the most widely used technique for metal/alloy deposition.

2. The dielectrics used at different levels - SiO_2 films deposited by TEOS based CVD processes are the most commonly used ILDs. Combinations of PECVD and ThCVD deposited oxides are generally used for planarizing Al interconnects with narrow, high aspect ratio gaps. Doped TEOS based SiO_2 films are used as pre-metal-1 dielectrics. SOG films in conjunction with PECVD SiO_2 were widely used in earlier technologies on account of their ability to planarize severe topographies.   These films are still used in some corporations.

3. ILD planarization - The most popular dielectric planarization technique is the dep-etch-dep process in which a combination of PECVD and ThCVD (both TEOS based) SiO_2 films with intermediate etching steps are used to planarize interconnect layers with gaps down to sub-half micron. Global planarization is achieved by depositing a sacrificial layer like photoresist and etching it back. Chem-mechanical polishing has also been successfully employed by several corporations in their four level metallization technologies.

4. Contacts/Vias and their planarization - W is the most widely used metal for contact/via metallization. Blanket CVD W followed by an etch-back process is the most common technique for planarization. To improve the reliability of these interconnects, TiN liners are deposited which minimize the interaction between the reactant species of W CVD and SiO_2. Novel sputtering and CVD techniques are used to get good step coverage of TiN in vias of very high aspect ratios. CVD of Al for via filling is still under development. CMP is also employed in some technologies for contact/via planarization.

5. Design rule - The various four level metallization schemes reported in the literature have progressively tighter design rules. In the most recently reported scheme, the pitch of the metal lines in the first level of metallization is 1.4 μm.

Some of these attempts to make multilevel metallization with four and more number of metal levels work in practice are summarized in Table VI

Table VI: Features of different four level metallization schemes

YEAR	COMPANY	DESIGN RULE	METAL	CONTACT / VIA	ILDs	ILD PLANAR-IZATION	COMMENTS
1989	Hitachi (87)	width = 0.8 µm, pitch =1.4 µm	M1 : Sputtered W; M2-M4 : Al/TiN	Al/TiN	D1 : PSG/SOG/PSG: D2: to planarize Al/TiN steps, double sandwich P-SiO / SOG/ P-SiO / SOG/ P-SiO is used	Annealing, SOG	Applied in CMOS and BICMOS VLSIs
1990	Bell Labs (88)	w = 1.2 µm; pitch = 2.2 µm	M1 thru M4: Al-Cu/TiW	CVD W plugs with sputtered Ti:W barrier	TEOS/BPTEOS and PTEOS dep-etch-dep process	Resist etchback planarizaiton at all dielectric levels	Constant design rule (0.9 µm maintained throughout the 4 levels
1990	Motorola (89)	M1 pitch ≤3.0 µm	Ti:W/Al-1.5%Cu/Ti:W	PtSi for contact metallization; multi-step hot Al process or selective CVD W process for via filling	PTEOS	"N" layer photoresist process(90)and etchback	MOSAIC IV and V processes; MOSAIC IV MLM process applied successfully to a 50,000 gate array
1991	IBM(91)	M1pitch=2.0 µm	Ti/Al-0.5%Cu/TiN (TiN for ARC)	Sputtered Ti/TiN layer + CVD W studs; use of CMP	PECVD SiO$_2$; dep-etch-dep	CMP	Functional 300K circuit ASIC logic test sites fabricated
1992	IBM (92)	M1pitch=1.4 µm	Ti/Al-Cu-Si	W	PECVD SiO$_2$	CMP	Offers upto 5 levels of metal for 0.5 µm CMOS logic designs

6.0   KEY RELIABILITY ISSUES IN AIX BASED MULTILEVEL METALLIZATIONS AND SOME SOLUTIONS:

The development of newer technologies for deposition, etching and planarization of metal and dielectric layers have all been aimed at improving the quality and reliability of Al based multilevel metallizations. The key reliability issues and solutions for some of these that are being currently used are discussed below.

6.1   Horizontal Interconnects

The reliability issues of importance for horizontal interconnects in multilevel metallizations are electromigration, stress migration/voiding, hillock formation and corrosion. These phenomena and their dependence on the interconnect structures and fabrication processes are discussed below.

Electromigration: As mentioned in sections 3.2 and 4.1, electromigration in Al interconnects is caused by the mass transport due to current flow.   This is the most significant failure mechanism in IC interconnects. The impact of this on the IC reliability has become more severe because of the shrinking of the interconnect dimensions in present day VLSICs.

Electromigration failure involves the displacement of the atoms of the interconnect as a result of a direct momentum transfer from conduction electrons in the direction of their motion. This causes removal of material in some locations, which generates voids, and accumulation of material in other locations. The former can result in open circuits while the latter can cause a short circuit with an adjacent interconnect. Both these problems cause failure of the interconnects. As indicated earlier, the electromigration failure is characterized by the "Median time to Failure (MTF)" which is the time needed for 50% of the interconnects to fail. This failure time depends on the current density, the temperature and the diffusivity of the atoms of the interconnect. An empirical equation for MTF which is extensively used for reliability evaluations, is (93).

$$[11] \qquad MTF = \left(\frac{F}{J^n}\right)\exp\left(\frac{E_a}{kT}\right)$$

where F is a factor dependent on the geometry of the interconnect and the diffusivity of the Al atoms, J is the current density, E_a is the activation energy, and T is the temperature. E_a is a parameter which depends on the grain structure of the film, defects, alloy compositions, etc. A higher E_a signifies a higher reliability of the interconnect with respect to electromigration.

The MTF of a pure Al interconnect is generally low, on account of the grain boundary dominated diffusion of atoms. The failure time is also determined by the grain size which in turn depends on the deposition method, rate and the substrate temperature. For small grain size Al, E_a is in the range 0.48 to 0.6eV . In textured Al it is about 0.73 eV and in large grain (1-8 μm) Al is about 0.9 eV (94).

Studies on single crystal Al lines with submicron width have shown that these have extremely high EM resistance as compared to polycrystalline Al (95) due to the large E_a resulting from lattice diffusion. It is found that in fine Al lines, as the ratio of the linewidth to grain size decreases, MTF decreases to a minimum and then increases exponentially (96, 97).

Certain types of surface treatment of the Al lines improves the EM lifetime. For instance, O_2 plasma treatment and a H_2O_2 dip after Al patterning have been shown to improve the MTF by a factor of 2 to 3 depending on the treatment conditions. This treatment can be explained as due to suppression of diffusivities of Al in the surface and/or grain boundary because of the partial oxidation of Al surfaces (98)

In the case of bias sputtering, which is preferred for depositing planarized Al, the MTF is found to be less than that for Al obtained by normal sputtering (99). This is attributed to the very small grain size and its distribution in Al deposited by bias sputtering.

The addition of Cu to Al (a few weight percent) improves the electromigration lifetime by a factor of 10 to 100. The same effect is seen with Mg also (100). Both Cu and Mg segregate at the grain boundaries to form $CuAl_2$ and Al_3Mg_2 respectively and hence reduce the grain boundary diffusivity by about two orders of magnitude. AlCuSi alloy is widely used in the present day VLSI technologies to improve electromigration performance of Al interconnects (Si is present in the alloy to prevent Si diffusion from the substrate). Unfortunately, Cu itself is found to migrate towards the anode and its depletion near the cathode causes failure. This requires replenishment of Cu into regions near cathode. Up to 4% Cu has been used to improve the EM resistance. Al-1%Si-0.5%Cu is a very widely used Al alloy. AlTi and AlCuTi alloys also show an improved EM resistance.

Layered Al based interconnects also show improved EM resistance. For example, sandwiching of the Al film between Ti and TiN layers greatly improves EM lifetime. The improved stability against EM is related to the formation of a continuous barrier of intermetallic compound like $TiAl_3$, which prevents voids from propagating across the film. The annealing temperature has been found to have significant effect on the intermetallic compound formation and hence on the EM lifetime (101). As mentioned earlier, the EM resistance of bias sputtered Al films is inferior to the films obtained by normal sputtering. However, improved performance has been achieved in Al-2%Cu /Ti layered films in which Al-Cu layer is deposited at high temperature (500°C) and with substrate bias (-600V) (102). Also, TiN and TiW under and cap layers provide an additional conducting path. So, even if AlX fails, TiW and TiN are intact and maintain the current flow. One study on layered interconnects with TiN, W over/under layers, however, has reported that layering degrades the electromigration immunity of Al layer because they tend to suppress Al grain growth and crystal orientation (103).

The EM lifetime is also found to increase, in certain cases, if the metallization is covered with a passivation layer. This is attributed to the compressive stress imposed by the passivating layer (104). Systematic studies on different passivation layers have shown that Al films passivated with dielectric deposited at low temperature has longer electromigration lifetime than the unpassivated ones. However, the passivating layer can lead to stress induced voiding in the interconnect.

Planarization of the interlevel dielectric layer increases the electromigration resistance of the underlying interconnect (105). This improvement is attributed to the reinforcement and crack suppression at the interlayer sidewalls.

Stress Migration/Voiding: This phenomenon refers to the mass transport in the interconnect due to the mechanical stress imposed by the underlying and/or overlying dielectric layers. This also may result in the formation of voids in the interconnect, ultimately leading to its failure. This phenomenon can occur in the absence of any current flow in the interconnect. Stress voiding occurs mainly because of the thermal mismatches between the different layers in a multilevel metallization. Therefore the failures due to voids can occur during processing itself (because of subsequent heat treatments) or because of high temperature storage. Although stress voiding has been known for quite some time, its impact on the reliability of interconnects has become significant in recent years because of the shrinking of dimensions. While stress migration/voiding by itself can cause failures in the interconnects, it can also accelerate failures due to electromigration.

Stress migration has been observed in all the Al alloys that are commonly used in VLSI technologies, both in sputtered and evaporated metal. For Al-(1-3%)Si films, passivated with SiO_2 or Si_3N_4, voids can form below 200°C. Opens result from the extension of dendritic voids completely across the interconnect. Such defects are more likely in narrower lines (106). The void growth due to this phenomenon is observed to continue even after the open circuit in the interconnect. The activation energy associated with the stress voiding is found to be in the range 0.46 eV to 0.64 eV. The failure of the interconnect due to stress voiding depends on the line width. The cooling rate after a high temperature excursion is also found to have a significant effect on stress voiding. The extent of stress voiding in the interconnect depends on the stresses in the underlying and overlying (passivation) dielectric layers, Systematic studies have been done on the effect of the stress in the passivation layer, σ_p , on the stress voiding (107). The void density is found to increase with σ_p, when it is compressive, showing a power law dependence. No voids are found when the stress in the passivation layer is tensile (108). Void formation is also suppressed by the addition of Cu to Al and also in layered interconnects.

New Al alloys have been proposed which have higher stress migration resistance than the conventional AlCuSi. A systematic study of the effect of adding different elements on the tensile strength of Al showed that Li, Mg, Mn, Fe, Co, Ni, Cu and Pd are effective in enhancing the high temperature tensile strength. However, Li and Mg are too active for sputter deposition and Fe,Co,Ni and Mn increase the resistivity of Al substantially. The most effective among these is the AlPdSi. Onuki et al. (109) have correlated higher creep strength with higher electromigration resistance. In Figure 49. they show electromigration failures with time for AlSiCu and Al-1%Si-0.3%Pd alloys passivated by successively depositing PECVD SiN and bias sputtered SiO_2, after annealing at 450°C for 30 min. They conclude that the stress migration resistance of AlSiPd is much higher than that of AlSiCu. This improvement is attributed to the higher creep strength of this alloy.

There have been a large number of studies to model stress voiding theoretically. Most of these are based on grain boundary or lattice diffusion in Al due to stress. There have been some contradictory views on the role of stress in the passivation layer on the voiding in interconnects.

In multilevel metallizations, stress voiding can be a severe reliability problem at all metallization levels. The metal layers in all levels have underlying and overlying dielectric layers. These layers are, in general, made of different dielectrics such as BPSG, PECVD SiO_2, SiN etc. The stress (as-deposited) and its variation in these layers are different. Further, during processing, different metal layers undergo different thermal cycles. For example, the first level metal goes through all subsequent thermal cycles (depositions and annealing cycles of subsequent layers) whereas the last level metal goes through only one cycle during the deposition of the passivation layer. In view of these, the stress voiding in the interconnects at different levels and their impact on the overall reliability of the multilevel metallization is very difficult to predict.

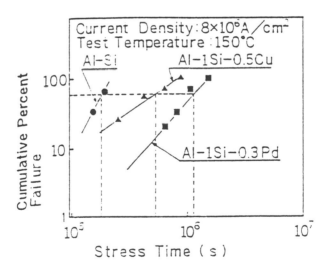

Figure 49. Electromigration failure distribution of AlSiCu and AlSiPd alloys. Ref. 109, reprinted with permission.

Hillock Formation: This is yet another reliability problem in interconnects arising out of the mismatch in the coefficient of thermal expansion of the metal and the underlying dielectric layers. Hillocks are found to develop when Al films are heated to 400°C or thermally cycled from room temperature to about 200°C. At these temperatures irreversible plastic flow occurs to relieve the large compressive stress in the metal, resulting in the formation of hillocks. The hillocks can grow to 2 μm or more in height. In general, the hillock height is proportional to the temperature excursion for a fixed heating time and alloy composition. In Al interconnects, the regions which supply material for the hillocks are depleted of Al. This causes inter-metal shorts predominantly, however, it can also cause open circuits in the interconnect. The extent of hillock growth is related to the self diffusion rate of Al as well as that of grain boundary diffusion. Therefore, the techniques adopted for improving the electromigration and stress migration resistance (alloying with Cu, Pd ; layered interconnects) should also help in reducing hillock formation. Other techniques that have been used to reduce hillock formation are surface hardening by ion implantation, forming anodic surface oxide and capping with PECVD SiO_2 layer (110). Also, as indicated in section 4.6 the use of multilayered structures, such as Ti/AlX/Ti/AlX/Ti, minimizes the hillock formation.

Corrosion: Al alloy based interconnects are affected by corrosion. The most dominant corrosion process is by chlorine during the dry etching of Al for interconnect fabrication. The chlorine in presence of moisture forms HCl which corrodes the Al. The presence of Cu in the Al alloy enhances the corrosion process (111). The corrosion due to chlorine can also occur after fabrication due to the transport of Cl and moisture from either chlorine containing residues or from the external package regions. One model which has been proposed to describe the time-to-failure of the interconnects under accelerated conditions (high humidity and elevated temperature) is (112,113)

$$(12) \qquad TF = A[Cl]^{-1} \exp[-a(\%RH)] \exp\left(\frac{Q}{kT}\right)$$

where, TF = time-to-failure due to corrosion, T is the temperature, %RH is the relative humidity, k is the Boltzmann constant, A is a constant, [Cl] represents the concentration of chlorine or other corrosive ions, Q is the activation energy which is in the range 0.7 to 0.9 eV for transport of moisture into plastic packages and 0.3 to 0.4 eV for liquid state diffusion of corroding species and a has a value in the range 0.1 to 0.3 [%RH]$^{-1.}$

Leaching of phosphorous from P-doped oxides such as PSG or BPSG, typically used as ILD or passivation layer on the metallizations, is yet another cause of corrosion. Reaction of this P with adsorbed moisture forms phosphoric acid, which then attacks the Al alloy.

The adverse impact of Cu on the corrosion susceptibility in AlCu and AlSiCu alloys is observed at 0.3% of Cu and increases sharply when the Cu content rises beyond 1%. However addition of 1%Si to the AlCu lowers the onset of corrosion by several hours but cannot change the corrosion rate once initiated. Annealing of the AlCu alloys for 30 min at 450°C is also found to improve the corrosion resistance. This corrosion behavior of AlCu alloys can be understood by considering a defective native oxide present on the Al which exposes the Al to corrosive attack. The Cu doped alloys seem to have more of such defective sites. Corrosion due to galvanic action has been observed in AlCu and AlSiCu alloys after conventional photolithographic processing (114). Al-0.5%Cu-0.5%Si had the least corrosion while Al-1.5%Cu had the most, indicating that Cu incorporation in the Al alloy enhances the corrosion due to galvanic action. During the cleaning process of the vias, as described in section 6.2, precipitates containing Cu are often left in the vias. These precipitates form a galvanic cell with Al, giving rise to corrosion by galvanic action. A post-deposition baking at 200-450°C dramatically reduces this precipitation and hence, the corrosion.

The other mechanisms of corrosion are: (a) chemical corrosion caused during cleaning processes, (b) electrolytic corrosion, and (c) corrosion from passivation/overlying dielectric layer.

Effect of contamination: Al films deposited by either sputtering or e-beam evaporation are contaminated by small amounts of oxygen and water vapor that are present in the deposition ambient. Depending on the deposition temperature and partial pressure, these contaminants have significant effect on the microstructure and electrical properties of the Al films (115, 116). Figures 50 and 51 show the effect of partial pressures of O_2 and H_2O on the resistivity and grain size of the Al films deposited by e-beam evaporation at room temperature as well as at higher deposition temperatures.

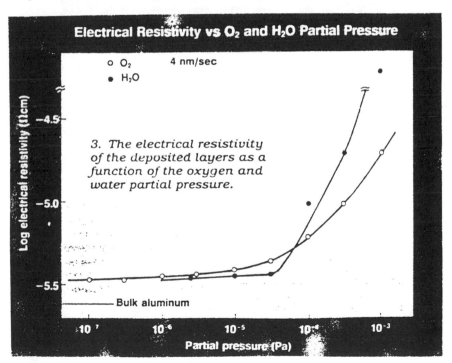

Figure 50. The electrical resistivity of the deposited layer as a function of the oxygen and water partial pressure; room temperature deposition. Ref. 115, reprinted with permission.

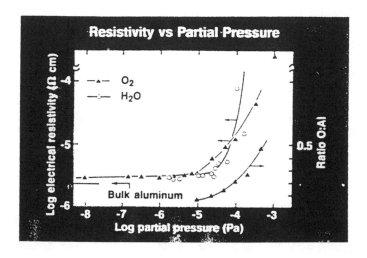

Figure 51. The electrical resistivity of the deposited layer as a function of the oxygen and water partial pressure; deposition at 250°C. Ref. 115, reprinted with permission.

6.2   Vertical Interconnects (Contacts and Vias)

The reliability issues that are significant in contacts and vias are electromigration, stress voiding, and Si precipitation. These issues and their dependence on processes are discussed below.

Contact/Via Electromigration: The electromigration in contacts and vias has been studied systematically only in the last few years. Two situations arise in these vertical interconnects depending on the type of via fill used. In one type, an Al alloy, with a barrier metal is used for the via fill. In the other type, CVD W, with a barrier metal is used for the vertical interconnect. The failure due to electromigration can occur either in the vertical interconnect itself or in the interfaces of the vertical interconnect with the horizontal metallizations or contacts. The electromigration in contacts and vias is strongly influenced by the current crowding effects. The current crowding is caused by various factors such as poor step coverage of the metal and Si precipitation in contacts. Depending on the direction of current flow, vias can have different failure modes (117). Referring to Figure 52, in the case where electron flow is from M2 to M1, a build up of Al would occur at the via and Al depletion occurs in M1 due to its transport away from the via interface. In the case where the electron flow is from M1 to M2, an accumulation of Al occurs at the via interface with M1, while Al is transported away from the via interface with M2. The presence of barrier layers like TiW at the via interface with M1 and/or M2 affects the electromigration (118). The barrier prevents the material transport at the via interface and hence improves the electromigration lifetime. The via itself having refractory metal doesn't fail even when the current density is much higher than that in the interconnect. This is because the refractory metals used for the vias have very low atomic mobilities in the temperature range of electromigration tests. Therefore, in most cases, the horizontal interconnects feeding the vias fail much earlier than the via itself.

In the case of vias filled with Al alloys, it has been found that via lift off is the mechanism for electromigration (119) failure which results in an open circuit. The failure occurs at the via interface where the upper level Al makes contact with the lower level Al. This is explained by the sharp current divergence at the via because of the dimensional differences between the two metal layers. In a two level metallization, the second level metal is normally wider and thicker. This asymmetry enhances the migration of Al away from the via interface when the electron flow is from M2 to M1 while Al accumulation occurs for electron flow from M1 to M2. The presence of a barrier metal below the top level Al alloy is found to improve the electromigration lifetime, showing an increase by nearly one order of magnitude for a 1000 Å Ti layer (120). This is attributed to the lowering of localized flux divergence at the via interface. The electromigration activation energy increases with increase in grain size, suggesting that different diffusion mechanisms prevail for different microstructure of the Al in the via. The electromigration lifetime is also strongly dependent on the cleanliness of the via before top metal deposition. A clean via interface gives a significant improvement in lifetime.

Figure 53 shows the electromigration lifetime of three different types of vias interconnecting AlX metallizations. It clearly shows that when the via is filled with the Al alloy itself, the highest lifetime is obtained.

In the case of contacts, the vertical interconnect links an Al based interconnect to an ohmic contact to Si (e.g. silicide). In this case, the electromigration can influence the material movement which can cause either an increase in contact resistance or accelerate phenomena such as junction spiking. For instance, the Si transport into Al is aided by the momentum exchange between conducting electrons and Si atoms. This diffusion of Si can result in (i) an accumulation of Si in the contact window which increases the contact resistance and (ii) depletion of Si from the contact window and junction leakage due to Al spiking. The electromigration in contacts due to these two effects has been studied. The activation energies for the two cases are found to be 1.1eV and 1.55eV respectively (121) for silicided contacts and 0.88eV and 0.84eV respectively for non-silicided contacts. This clearly indicates that silicided contacts show better electromigration resistance.

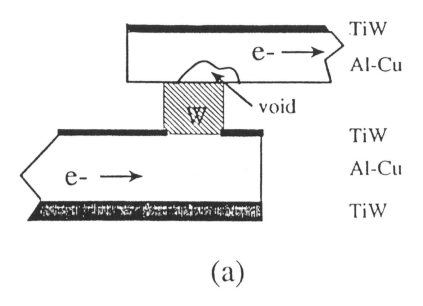

TiW

e- ⟶

Al-Cu

void

TiW

e- ⟶

Al-Cu

TiW

(a)

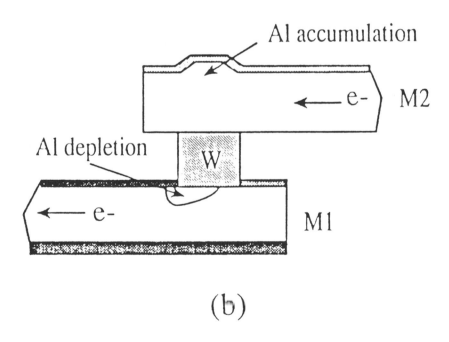

Al accumulation

⟵ e- M2

Al depletion

W

⟵ e- M1

(b)

Figure 52. Via electromigration. Ref. 118, reprinted with permission.

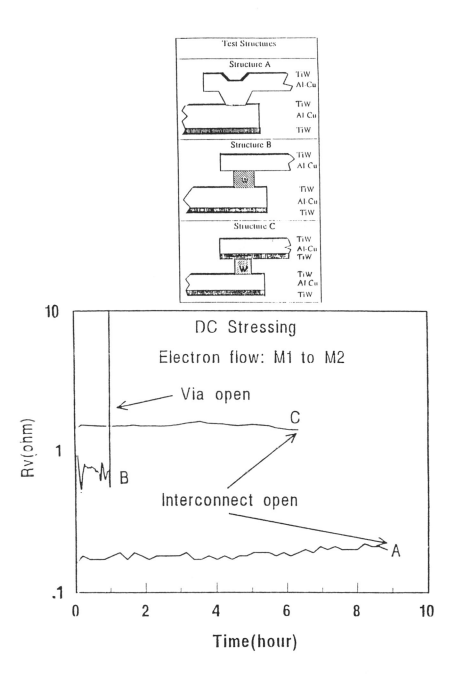

Figure 53. Electromigration lifetime of different types of vias between AlX interconnects. Ref. 118, reprinted with permission.

Stress Migration in Contacts/Vias: There have been very few reported studies on the stress migration/voiding in vertical interconnects. Open failures have been observed in vias with Al/Al interface during high temperature storage (122). This has been attributed to the stresses in the passivation layers on the metallization level over the via. As the via diameter decreases, open failures increase. This could be due to the poor step coverage of sputter deposited films and the crystal discontinuity of Al grain at the via interface. The layered structure of AlTiCu/Ti is found to be effective in preventing stress induced void formation and hence improving the stress migration resistance.

Problems Due to Si Dissolution: Silicon dissolution in Al has a significant effect on the reliability of contacts made by AlX directly or through barrier metals to the n^+ and p^+ junctions. The main reliability problem is the junction spiking. For contacts made by Al/AlX directly, this essentially results in Al penetrating into the Si as "spikes" in certain regions. If the spike length which can be 1 μm or more, is larger than the junction depth (which is the case for shallow junctions), then, the junction gets shorted. During the heat treatment after Al deposition, certain sites in the native SiO_2 on Si get reduced earlier than the rest, thereby exposing local sites in the underlying Si to Al. Si from these regions dissolves in Al and diffuses rapidly through the Al film. By the time SiO_2 in other regions is reduced, the Al above them is already saturated and no dissolution of Si takes place at these points. Thus the Si dissolution in Al takes place in only a few sites and Al spikes form only at these sites. The extent of spiking is strongly dependent on the thickness of the SiO_2 between Al and Si. For thinner oxides (<10 Å), the Al breaks through at a number of places and the spikes are relatively shallower and irregularly shaped. For thicker oxides (> 20 Å), the Al penetrates only at a few points and the spikes are deeper (16).

The effect of a diffusion barrier, TiW, on Al spiking has been investigated (123) to study contacts made through a diffusion barrier metal. This has shown that for thicknesses less than 1000 Å, the diffusion barrier cannot prevent the spikes from forming in both AlCu and Al-1%Si alloys for the aspect ratios greater than one used in the current technologies.

The problem of spiking is avoided by adding Si to the Al itself so that it is saturated even at the maximum temperature of processing. However, during cooling from a high temperature, the dissolved Si comes out of solution in two forms (6) as: (a) Silicon precipitates on nucleation sites such as preexisting Si particles, grain boundaries in Al or steps in the dielectric and (b) epitaxial growth on exposed areas of the Si substrate.

The epitaxial layer is p-type because of the small but finite solubility of Al in Si. When this layer is formed in contacts to n^{++} regions, the p^- - n^{++} junction formed can lead to a significant increase in contact resistance, especially in small area contacts, as shown in Figure 54. This is because in small area contacts, the Si precipitates cover a large fraction of the contact area. The contacts no longer are ohmic.

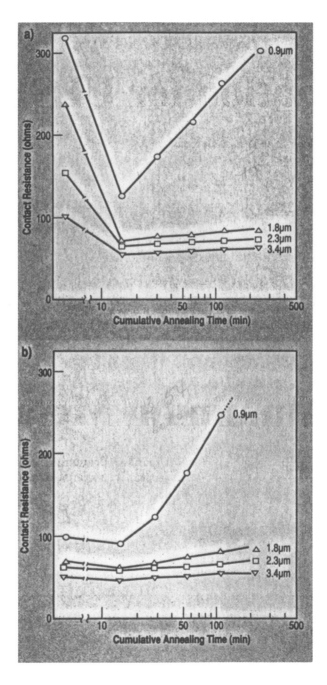

Figure 54. Contact resistance variation with cumulative annealing time for different contact sizes.

Epitaxial growth of Si can be minimized by
1. Cooling the wafer rapidly which reduces the precipitation. Although this can be achieved in a rapid thermal annealing process, the gain is limited because of the additional conventional heat/cool cycles for dielectric depositions.
2. Making the Si precipitate on existing Si nuclei rather than on the contacts. This calls for adding more Si to the Al than is required to satisfy the solubility requirements so that Si particles are left to act as nucleation sites during cooling. In addition, the Si particles and their distributions must be such that the contact area is not reduced substantially. The most desirable condition, a uniform distribution of medium size particles after the completion of all heat cycles, is best achieved by optimizing the deposition temperature. A temperature of 300°C appears to be optimum.
3. Interposing an additional metallic layer between the Al film and the contact. This layer should minimize the interaction between Al and Si. At the same time it should have low contact resistance with both n^{++} and p^{++} regions. Silicides meet both of these requirements and hence are universally used for contacts in present technologies.

Cleaning of Vias: Reliability problems in vias as a consequence of sputter etch cleaning have been observed (124). Sputter etch cleaning removes the natural oxide from the Al surface but leaves a thin insulating layer due to re-deposition of atoms sputtered out of ILDs. This layer causes contact failure during storage or operation.

Failure of Diffusion Barrier: Several failure mechanisms have been observed in the diffusion barriers such as TiW and TiN used with AlX. These are caused by structural defects in the barrier metal film itself. These defects are more likely to occur at the corners of small area contacts because of the columnar grain structure of TiN and TiW. Once a path for Si diffusion is established at these weak points there is a large diffusion of Al through these and large Al spikes are formed.

Diffusion barriers like TiN are deposited in contacts to prevent interaction between the reacting species of CVD of tungsten (mainly WF_6) and the Ti liners used for improving the contact resistance with the silicide. The WF_6 has a strong tendency to attack Ti. The diffusion rate of the WF_6 is very high near the corners of the contacts as compared to the flat surface. To prevent the interaction between WF_6 and Ti, a diffusion barrier, TiN, is deposited prior to W deposition. If there is unreacted Ti in this TiN, it is attacked by WF_6, especially at the corners (125). This leads to a breaking up and peeling of TiN. Tungsten deposited on this contact forms a mound as shown in Figure 55. This contact can have severe reliability problems.

6.3   Interlevel Dielectrics

The ILDs contribute significantly to the reliability of the AlX interconnects. Large stress variations in the ILD can lead to significant stress migration/voiding, change microstructure, and deteriorate electromigration resistance of the interconnect. As discussed in section 6.1, the nature and magnitude of the stress in the overlying dielectric determines the extent of stress voiding in the interconnect. The stress in the ILD, in turn, depends on the material, the method of deposition and the annealing temperature.

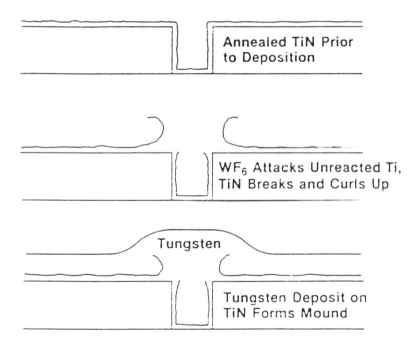

Figure 55. Schematic of the series of steps in the failure of a TiN diffusion barrier. Ref. 39, reprinted with permission.

7.0   FUTURE   DIRECTIONS

7.1   New   Dopants/Alloys

Several new Al alloys and different treatments of alloys have been suggested in the last few years to overcome various reliability problems of Al based metallizations. The features of the important ones are discussed below. It must however be remembered that in most of these cases the improvements claimed are only in certain characteristics and complete characterizations have not been done yet .

AlSiPd: This alloy has been proposed as an alternative to AlSiCu (126). AlPd alloys have a high creep strength which leads to lower grain boundary diffusion of metal atoms. It also has high corrosion resistance. The role of Si in AlSiPd is to prevent junction spiking, as in the case of AlSiCu. The AlSiPd shows little decrease in linewidth due to overetching or irregularities of sidewalls compared to Al-Si-0.5%Cu. Electromigration and stress migration performance of AlSiPd is also found to be superior to that of Al-Si-0.5%Cu.

AlSiCu Alloys Doped with Hf and B: Addition of a few hundred ppm of Hf and B has been found to improve the performance of AlCuSi alloy interconnect material, without changing the manufacturing feasibility (same etching characteristics). The hillock formation is suppressed considerably. The electromigration lifetime improved by more than two orders of magnitude by addition of Hf and B. The void formation due to stress migration is also found to be greatly suppressed.

Al-Samarium   (Sm)   Alloys: AlSm alloys are attractive for metallization because of the low solid solubility of Sm in Al which leads to low resistivity (about 3 $\mu\Omega$-cm after anneal at 450°C). Al-1%Sm alloys also exhibited low hillock growth (127).

Al-Yttrium   (Y)   Alloys: Just as in the case of AlSm alloy, the AlY alloy is also found to be attractive from the point of its resistivity. Upon anneal at temperatures above 300°C, the resistivity of Al-0.7%Y alloy film is found to decrease to values very close to that for pure Al (~ 2.9 $\mu\Omega$-cm). The addition of small amounts of Y is also sufficient to minimize the generation and growth of hillocks.

C-Doped Al: Al films deposited by magnetron-plasma CVD and doped with carbon have shown superior characteristics (128) in some respects. The in-situ C doping suppresses the growth of Al crystal grains, hillocks and spikes. Electromigration lifetimes are one order of magnitude higher than that of pure Al. The resistivity is significantly higher than pure Al in as-deposited films. After annealing at 450°C for 30 mins, it decreases considerably. Beyond a C concentration of 10%, the resistivity increases steeply due to the formation of Al_4C_3. For C content below 10%, hillock free Al film with resistivity of 3.8 $\mu\Omega$-cm was obtained after annealing at 600°C. The electromigration characteristics of C doped Al films are compared with pure Al films in Figure 56.

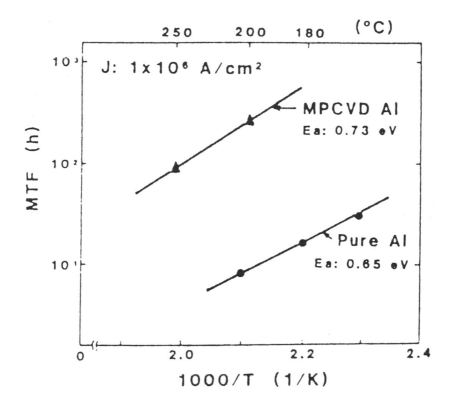

Figure 56. Arhenius plot of MTF. The C concentration of MPCVD Al is about 12%. The thickness of the Al line covered with PSG was 1 μm and the width was 4 μm. Ref. 128.

Fluorine Incorporation in Al: Incorporation of very small amounts of F (<0.1 atomic %) in Al and AlCu films is found to reduce the hillock formation considerably (129). The F incorporation is achieved by ion implantation with a dosage of about 10^{15} per cm^2, followed by annealing. The hillock density and resistivity of F incorporated Al films are compared with those with Cu in Figure 57. At high dosages F incorporation gives rise to a significant reduction in hillock density. This is attributed to the reactive nature of F. The highly electronegative F forms much stronger chemical bond than Al with itself or with Cu.

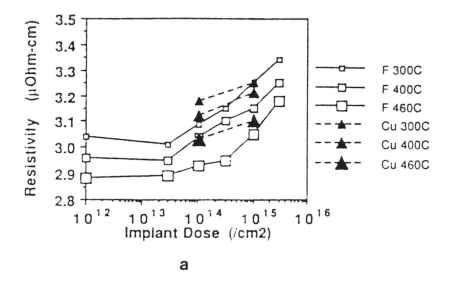

a

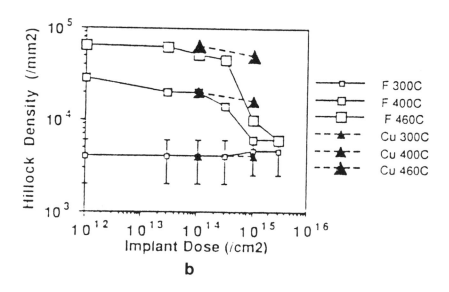

b

Figure 57. (a) Resistivity and (b) hillock density, as function of implant dose, species, and anneal temperature. Ref. 129, reprinted with permission.

Al-Scandium (Sc) Alloy: This new Al alloy containing 0.15% Sc has been proposed recently and characterized with respect to stress migration and electromigration (130). It is found that Sc addition to Al gives rise to a much lower grain size as compared to AlSiCu, even after annealing. However, the stress migration resistance is far superior to that of AlSiCu. The electromigration lifetime is also better by about one order of magnitude. It is known that, in general, smaller grain size is good for avoiding stress migration but gives poor electromigration resistance. But, it is not clear how the AlSc alloy with smaller grain size has good electromigration resistance also. It may be due to the formation of Al_3Sc or Sc precipitates in the grain boundaries.

7.2   Alternate Metallizations

As mentioned earlier, when the technologies move to 0.25 µm regime and beyond, alternate metallizations based on metals having resistivities lower than that of Al will be needed to minimize the contributions of the interconnects to the speed performance in the ICs. The candidates for the alternate metallizations are Cu, Ag and Au whose resistivities are lower than that of Al. In the last few years extensive research has been carried out on Cu based metallizations. The main features of Cu films and their suitability for multilevel metallizations are briefly discussed below. More detailed description of these can be found in other chapters.

Cu-Based Metallizations: Most of the research on Cu in the last few years has been on the development of interconnect structures with appropriate barrier metals/adhesion promoters, and deposition techniques for these. The reliability of Cu interconnects has also been studied. The following sections discuss the recent developments in these areas.

Deposition of Cu: Conventional sputtering is the PVD technique that has been used so far for depositing Cu films. Step coverage of sputtered Cu is found to be better than that of Al, especially at the bottom of vias (131). However, in the last few years, novel methods such as CVD and low energy ion bias sputtering have been proposed. These appear to be promising for contact metallization and via filling because of their better step coverage. CVD has the added advantage of giving selective deposition. The CVD process most widely studied uses bishexafluro acetylacetonate copper $[Cu(HFA)_2]$ as the precursor (132). The reaction involved is the hydrogen reduction of $Cu(HFA)_2$. Both thermal and plasma based processes have been reported. The substrate temperatures in the thermal process are in the 300-350°C range while it is in the 150-200°C range for the plasma enhanced process. The deposition rates achieved so far are 10-20 nm/min. The resistivity of Cu obtained by the CVD is about 2.0 µΩ-cm.   Selective deposition in contacts and vias has also been demonstrated (133). The low kinetic energy ion bias sputtering process has also been applied to the deposition of Cu (134). It has been claimed that Cu films deposited on SiO_2 by this technique exhibits excellent adhesion to SiO_2 without requiring any glue layer. Cu deposition by electroless plating has also been investigated (135). This method requires a conducting surface for deposition. Pd activation is used to activate the surface of insulators for subsequent Cu deposition. Another scheme reported for selective electroless deposition of Cu uses $PdSi_2$ as the activating layer. Pd layer deposited on Si or poly-Si patterned by SiO_2 is converted to $PdSi_2$ by annealing. The Pd remaining on SiO_2 is etched off to leave the silicide only in selected regions. Cu gets deposited only on these regions by electroless plating. Using this technique, via filling down to 0.5 μm with an aspect ratio of 6 has been demonstrated (136).

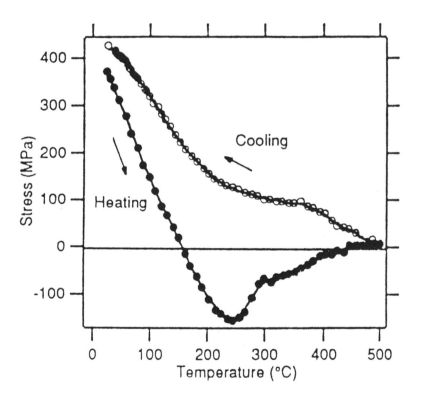

Figure 58. Stress behavior of sputtered Cu films Ref. 140, reprinted with permission.

Diffusion barriers/Adhesion promoters for Cu: Cu diffuses rapidly through SiO_2. It also has relatively poor adhesion to SiO_2. To be able to use Cu for multilevel metallizations with SiO_2 as the ILD, these problems have to be overcome. For this, diffusion barriers and adhesion promoters are needed. Extensive research has been reported on the barrier properties of a variety of materials against Cu diffusion. The prominent among these are Ti, TiN and TiW. Ti acts as a diffusion barrier as well as adhesion promoter (137). TiN is effective as a barrier layer up to 500°C. With some oxygen incorporated, TiN is found to be more effective as a barrier (138) TiW is also effective as a barrier up to 550°C (139).

Characteristics of Cu interconnects: The resistivity of Cu films obtained by the various PVD and CVD techniques is about $2\mu\Omega$-cm. The typical stress behavior of a sputtered Cu film on SiO_2 shown in Figure 58 is similar to that of Al (140). Cu interconnects are known to have a higher electromigration resistance than AlX interconnects. Cu films deposited by low energy kinetic energy ion bias sputtering technique are found to exhibit 3 to 5 orders of magnitude larger electromigration lifetime than AlSi (141).

Gold-Based Metallizations: Gold based metallizations are used extensively for ohmic and Schottky contacts to compound semiconductor devices. The suitability of Au for Si VLSI technologies is yet to be evaluated. Au is attractive for metallizations because of low resistivity and superior electromigration performance. However, these films also require a barrier layer to prevent diffusion into the underlying layers or devices and to prevent void formation in Au at high temperatures. A variety of materials such as Ti, V, Cu, Ni, etc. have been studied to determine their suitability (142) .

7.3   Equipment

From the descriptions of the processes involved in Al based multilevel metallizations and their impact on the reliability of these metallizations, the key requirements of equipment for VLSI/ULSI metallizations (deposition, etching and planarization) can be identified as:

High repeatability from run to run and from wafer to wafer within a run.

High uniformity of deposition or etch.

High throughput

Flexibility for changing process parameters, preferably through software.

Larger wafer capability.

Low levels of contamination.

In the last few years, there have been continuous improvements in the equipment for metal deposition and etching in meeting many of the above requirements. Novel concepts such as "the cluster tool approach" have been proposed to meet some of the requirements. The following sections discuss some of the features of the present day metallization equipment.

Sputtering Equipment: As pointed out in section 5, sputtering is the most widely used process for the deposition of AlX and other materials such as

TiW and TiN. Some of the key characteristics which are desirable for a ULSI metallization sputtering equipment include (6, 143):

1. Base pressure in the low 10^{-8} to mid 10^{-9} range in the sputtering chambers.
2. Deposition temperatures up to 500°C.
3. Backsputter etch capability offering minimal damage to thin gate oxides.
4. Sputter chambers completely isolated from all other processes.
5. Ability to outgas wafers prior to deposition.
6. Complete control over wafer temperature during all stages of deposition.
7. Control over partial pressures of OH, N, O in the 10^{-10} torr range.
8. High uniformity of deposition.
9. Good control over step coverage, hillocks, and stress.

Several sputtering systems with novel features are now commercially available. Some of these features are discussed below.

Single Wafer Systems With Cassette-to-Cassette Processing: This yields greater reproducibility and better control over the deposition ambient because the distance from source to wafer is fixed and the chamber is much smaller. However, these systems are relatively more expensive.

Multiple Target Systems: This allows sequential deposition of different materials like Ti, AlX, etc. without exposing the wafers to the ambient which inturn minimizes the oxidation of the surface layers of the deposited films.

Large (10"-12") diameter targets for greater uniformity of film thickness: As the wafer diameter approached 8" in the VLSI/ULSI technologies, the need for large diameter sputtering systems has increased in order to deposit highly uniform film. Such targets are now being manufactured with uniform stoichiometry across the entire area.

CVD Equipment: As explained in earlier sections, CVD of Al is yet to become a production worthy technique. No commercial CVD system is yet available for Al deposition. From the reported work in the literature, the following requirements are important for CVD equipment:

1. Precursor delivery system
2. Facility for plasma enhanced and thermal CVD processes
3. Large diameter wafer capability
4. Ability to change process recipes easily
5. Wafer handling system
6. Single wafer reactor.

CVD of W is used in present day VLSI manufacturing. Blanket deposition is the widely used technique. Commercial systems are now available for CVD of W. Most of these are single wafer reactors. They use different methods of substrate heating, viz, radiant heating from either the front or the back of the wafer, rapid thermal heating and resistive heating. Some of the tungsten CVD equipment are also capable of selective deposition.

Cluster Tools: The cluster tool has been accepted as the key concept in *advanced microelectronics manufacturing.* A cluster tool generally refers to a modular, multichamber, integrated processing system. It consists of a central wafer handling module and a number of peripheral process stations. The wafers go through a set of process steps sequentially in the process stations without being exposed to the ambient conditions. The transfer of the wafers for the processes is managed by the wafer handling module. Cluster tools offer potential advantages in improved overall throughput, better process control and significantly higher yields on account of lower defect densities. Different types of cluster tools (linear or radial) with different types of architecture (open or defined) are possible.

The fundamental equipment issue for multilevel metallization technologies is that of a cluster tool capable of performing several processes in-situ versus a stand alone single process equipment. As discussed in previous sections of this chapter, a typical AlX based multilevel metallization production process consists of the following for a planarized vertical interconnect (contact/via) followed by a horizontal interconnect:

1. In-situ clean (sputter etch; RIE)
2. Deposition of underlayers for the vertical interconnect (Ti, TiN or TiW)
3. Blanket deposition of WSi_x / W
4. In-situ etch back to form vertical interconnects
5. Deposition of underlayers for horizontal interconnects (Ti, TiN or TiW)
6. Deposition of the main conductor (AlX)
7. Deposition of the cap layer (TiN or TiW)
8. Deposition of an interlevel dielectric layer
9. Planarization of the ILD.

Clustering of a number of the above processes is desirable for improving the quality of multilevel metallizations. In the current technologies, cluster tool concepts have been applied for dielectric planarization (dielectric deposition and etch back) and for contact/via fill with metal (metal deposition and etch back) (144). A typical processing sequence for contact fill and the appropriate cluster tool configuration for this sequence are shown in Figure 59. It is clear that cluster tool will be a dominant factor in the future multilevel metallization technologies.

8.0   SUMMARY

From the discussions in the various sections of this chapter, it is clear that multilevel metallizations are the key to achieving VLSI/ULSICs with superior performance and higher packing density. It is also clear that Al based multilevel metallizations are here to stay for many more years as the most widely used for VLSI / ULSI technologies. It is the interconnect metal that is best understood so far in terms of fundamental properties and limitations. Both of these have been investigated extensively over the last three decades and solutions have been developed for many of the limitations. These solutions are being successively employed in the present multilevel metallization technologies. The problems of electromigration, Si dissolution and hillock formation are minimized in todays metallizations by using Al alloys and layering them with suitable materials. As the interconnect dimensions shrink and multilevel metallizations become mandatory, newer failure mechanisms such as stress voiding have become significant in the last few years. Extensive research is in progress on this phenomenon to characterize it and find solutions for it. The processes for deposition and etching of AlX for fabricating horizontal interconnects are well established. However, the technologies for Al based vertical interconnects are not yet available for manufacturing. Thus W based vertical interconnects are used even in the present technologies. Research is being done on novel sputtering methods for the fabrication of Al based vertical interconnects. Novel deposition techniques like collimated sputtering and CVD show considerable promise for the deposition of TiN or AlX into high aspect ratio contacts/vias. These are under careful evaluation for determining their suitability for manufacturing. Cluster tool processing for metallizations is another area which is under extensive research and development. These new developments in materials, processes and equipment are expected to yield highly reliable Al based multilevel metallization technologies which can be used profitably in the sub 0.5 μm ULSI regime. However, this requires equally good technologies for deposition and planarization of ILDs.

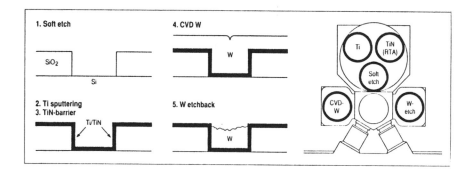

Figure 59. Integrated tungsten plug process sequence for contact hole filling using one multiple process module and two single process stations. Ref. 144, reprinted with permission

REFERENCES

1. Saxena, A.N., Keynote address, in Proc. First International IEEE Conf. on VLSI Multilevel Interconnections, June 1984.
2. Hamilton, D.J. and Howard, W.G., Basic Integrated Circuit Engineering, (McGraw Hill, New York 1975).
3. Fraser, D.B., in VLSI Technology, Ed. Sze, S.M., (McGraw Hill 1983)
4   P.K. Chatterjee, in Handbook fo Advanced Semiconductor Technology and Computer Systems, (Van Nostrand Reinhold Co. New York,1988) pp. 279-282.
5. Wolf, S. and Tauber, R.N., Si Processing for the VLSI Era, Vol. 1: Process Technology, (Lattice Press, Sunset Beach, CA, 1986).
6. Pramanik, D. and Saxena, A.N., "Aluminum Metallizations for ULSI," Solid State Technol. 33(3), 73-79 (1990).
7. Pramanik, D. and Saxena, A.N., "VLSI Metallization Using Aluminum and its Alloys," Solid State Technol. 26(1), 127-133 (1983).
8. Wilson, R.J. and Weiss, B.L., "A Review of the Properties of Aluminum Alloy Films Used During Silicon Device Fabrication," Vacuum, 42(12), 719-729 (1991).
9. Gardner, D.S. and Flinn, P.A., "Mechanical Stress as a Function of Temperature in Aluminum Films," IEEE Trans. Elec. Dev., 35(12), 2160-2169 (1988).
10 Gardner, D.S. and Flinn, P.A., "Mechanical Stress as a Function of Temperature in Aluminum Alloy Films," J. Appl. Phys., 67(4), 1831-1844 (1990).
11. Ghate, P.B., "Electromigration Testing of Al-Alloy Films," 19th Annual Proceedings of the IRPS, pp. 243-252 (1981).
12. Mayumi, S, Umemoto, T., Shishino, M., Nanatsue, H., Ueda, S., and Inoue, M., "The Effect of Cu addition to Al-Si Interconnects on Stress Induced Open-Circuit Failures," 25th Annual Proceedings of the IRPS, pp.15-21 (1987).
13. Pramanik, D. and Jain, V., "Barrier Metals for ULSI; Processing and Reliability," Solid State Technol. 34(5), 97-102 (1991).
14. Ghosh, S.K., Ramkumar, K., and Saxena, A.N. (to be published)
15. Pramanik, D., in: VMI Tutorial Short Course, Santa Clara, CA, pp. 66-112 (1992).
16. Pramanik, D. and Saxena, A.N., "VLSI Metallization Using Aluminum and its Alloys," Solid State Technol. 26(3), 131 (1983).
17. Learn, A.J., "Electromigration Effects in Al-Alloy Metallization," J. Elect. Mat. 3(2), 531 (1974).
18. Learn, A.J., "Electromigration Effects in Aluminum Alloy Metallizations," J. Elect. Mat. 3(2), 531 (1974).
19. Gardner, D.S., Michalka, T.L., Saraswat, K.C., Barbee, Jr, T.W., McVittie, J.P. and Meindl, J.D., "Layered and Homogeneous Films of Aluminum and Aluminum/Silicon with Titanium and Tungsten for

Multilevel Interconnects," IEEE Trans. Electron. Dev., 32 (2), 174-183 (1985).

20.   Pramanik, D., and Jain, V., "Barrier Metals for ULSI: Processing and Reliability," Solid State Technol., 34(5), 97 (1991).

21.   Hey, H.P.W., Sinha, A.K., Steenwyk, S.D., Rana, V.V.S., and Yeh, J.L., "Selective Tungsten on Aluminum for Improved VLSI Interconnects," IEDM Tech. Dig., pp. 50-53 (1986).

22.   Dalton, C.M., "Enhanced Selective Tungsten Encapsulation of TiW Capped Al Interconnect," Proceedings 7th Intl. IEEE VMIC, Santa Clara, CA, pp. 289-295 (1990).

23.   Joshi, A., Hu, H.S., Yaney, D.L., Gardner, D., and Saraswat, K., "Fundamental Factors Governing Improved Performance of Al-Si/Ti Multilayer Metallization for Very Large Scale Integration," J. Vac. Sci. Technol. A7(3), 1497 (1989).

24.   Kikkawa, T, Aoki, H, Ikawa, E and Drynan, J, "A Quarter- Micron Interconnection Technology Using AlSiCu/TiN Alternated Layers," IEDM Tech. Dig., pp. 281-285 (1991).

25.   Olowolafe, J.O., Palmstorm, C.J., Colgan, E.G., and Mayer, J.W., "Al/TiW Reaction Kinetics; Influence of Cu and Interface Oxides," J. Appl. Phys., 58(9), 3440 (1985).

26.   Palmstorm, C.J., Mayer, J.W., Cunningham, B., Campbell, D.R., and Totta, P.A., "Thin Film Interactions of Al and Al(Cu) on TiW," J. Appl. Phys., 58(9), 3444 (1985).

27.   Sun, S.W., Lee, J.J., Boeck, B. and Hance, R.L., "Al/W/TiN$_x$/TiSi$_y$/Si Barrier Technology for 1.0 μm Contacts," IEEE Electron. Dev. Lett., 9(2), 71 (1988).

28.   Saraswat, K.C., in VMI Tutorial Short Course , Santa Clara, CA, pp. 78-125 (1985).

29.   Van Gurp, G.J, Daams, J.L.C., Van Oostrom, A., Augustus, L.J.M., and Tamminga, Y., "Aluminum-Silicide Reactions. I. Diffusion, Compound Formation, and Microstructure," J. Appl. Phys, 50(11), 6915 (1979).

30.   Grinolds, H, and Robinson, G.Y, "Study of Al/Pd$_2$Si Contacts on Si," J. Vac. Sci. Technol., 14(1), 75 (1977).

31.   Skidmore, K., "Sputtering for Deposition and Etch," Semiconductor International, 11(6), 74 (1988).

32.   Ting, C.Y., Vivalda, V.J., and Schaefer, H.G.,"Study of Planarized Sputter Deposited SiO$_2$," J. Vac. Sci. Technol., 15(3), 1105 (1978).

33.   Ohmi, T., Kuwabara, H., Saitoh, S., and Shibata, T., "Formation of High Quality Pure Aluminum Films by Low Kinetic Energy Particle Bombardment," J. Electrochem. Soc., 137(3), 1008 (1990).

34.   Okutani, K., Horiuchi, M., Kurogome, M., Tateishi, H., and Sasaki, S., "Properties of Aluminum Films Deposited by Low Energy and High Density Ion Bias Sputtering Method Using Cusp Magnetic Field Electrode," Proceedings 7th Intl. IEEE VMIC, Santa Clara, CA, pp. 296-302 (1990).

35.   Joshi, R.V., and Brodsky, S., "Collimated Sputtering of TiN/Ti Liners into Sub-Half Micron High Aspect Ratio Contacts/Lines," Proceedings 9th Intl. IEEE VMIC, Santa Clara, CA, pp. 253-259 (1992).

36.   Meikle, S., Kim, S. and Doan, T., "Semiconductor Process Considerations for Collimated Source Sputtering of Ti Films," Proceedings 9th Intl. IEEE VMIC, Santa Clara, CA, pp. 289-291 (1992).

37.   Levy, R.A. and Green, M.L., "Low Pressure Chemical Vapor Deposition of Tungsten and Aluminum for VLSI Applications," J. Electrochem. Soc., 134(2), 37C (1987).

38.   Piekaar, H.W., Kwakman, L.F.T., and Granneman, E.H.A., "LPCVD of Aluminum in a Batch-Type Load-Locked Multi-Chamber Processing System," Proceedings 6th Intl. IEEE VMIC, Santa Clara, CA, pp. 122-128 (1989).

39.   Blewer, R., in Advanced Metal Systems, VMIC State-of-the-Art Seminar, Santa Clara, CA pp. 17-104 (1992).

40.   Amazawa, T., Nakamura, H., and Arita, Y., "Selective Growth of Aluminum Using A Novel CVD System," IEDM Tech. Dig., pp. 442-445 (1988).

41.   Fleming, C.G., Blonder, G.E., and Higashi, G.S., in Laser and Particle-Beam Chemical Processing for Microelectronics Symp., .MRS., p. 183, 1987.

42.   Tsubouchi, K., Masu, K., Shigeeda, N., Matano, T., Hiura, Y., and Mikoshiba, N., "Complete Planarization of Via Holes With Al By Selective and Nonselective CVD," Appl. Phys. Lett., 57, 1221 (1990).

43.   Cacouris, T., Scarmozzino, R., and Osgood, R.M., "Resistless Patterning of Aluminum," Proceedings 7th Intl. IEEE VMIC, Santa Clara, CA, pp. 268-274 (1990).

44.   Hillman, J.T., Rice, M.J., Studiner, D.W., Foster, R.F., and Fiordalice, R.W., "Comparison of Titanium Nitride Barrier Layers Produced by Inorganic and Organic CVD," Proceedings 9th Intl. IEEE VMIC, Santa Clara, CA, pp. 246-252 (1992).

45.   Small, M.B., and Hu, C.K., VMI Tutorial, pp. 59-109, June 1990.

46.   Joshi, R.V., Brodsky, S.B., Bucelot, T., Jaso, M.A., and Uttecht, R., "Low Resistance Sub Micron CVD W Interlevel Via Plugs on AlCuSi," Proceedings 6th Intl. IEEE VMIC, Santa Clara, CA, pp. 113-121 (1989).

47.   Singer, P.H., "Selective Deposition Nears Production," Semiconductor International, 13(3), 46 (1990).

48.   Bradbury, D.R., Turner, J.E., Nauka, K., and Chiu, K.Y., "Selective CVD Tungsten as an Alternative to Blanket Tungsten for Submicron Plug Applications on VLSI Circuits," IEDM Tech. Dig., pp. 273 -276 (1991).

49.   Blewer, R.S., in VMIC State-of-the -art Seminar, pp. 17-103 (1992).

50.   Colombo, L. and Illuzzi, F., "Plasma Etching of Aluminum Alloys for Submicron Technologies," Solid State Technol., 33(2), 95 (1990).

51.   O. Krogh, "Bromine Based Aluminum Etching," Semiconductor International, 11(6), 276 (1988).

52. L. Peters, "Plasma Etch Chemistry: The Untold Story," Semiconductor International, 15(6), 66 (1992).

53. Lutze, J.W., Perera, A.H., and Krusius, J.P., "Anisotropic Reactive Ion Etching of Aluminum Using Cl_2, BCl_3, and CH_4 Gases," J. Electrochem. Soc., 137(1), 249-252 (1990).

54. Torrisi, R.L., Vasquez, P., and Viscuso, O., "Surface Characterization of The Al/Si-Ti/W Metallization After Chlorinated Plasma Treatments," J. Electrochem Soc., 138(4), 1171-1174 (1991).

55. Chen, C.-H., DeOrnellas, S. and Burke, B., Microelectron. Manuf. Test. (USA), 11(11), p. 11, Oct. 1988.

56. May, P., and Spiers, A.I., "Dry Etching of Al/TiW Layers for Multilevel Metallization in VLSI," J. Electrochem. Soc., 135(6), 1592-1594 (1988).

57. Mayumi, S., Hata, Y., Hujiwara, K., and Ueda, S., "Post Treatments for Reactive Ion Etching of Al-Si-Cu Alloys," J. Electrochem. Soc., 137(8) 2534-2538 (1990).

58. Hu, C.K., Canney, B., Pearson, D.J., and Small, M.B., "A Process for Improved Al(Cu) Reactive Ion Etching," J. Vac. Sci. Technol., A7, 682-685 (1989).

59. Dohmae, S., Mayumi, S., and Ueda, S., "A New Etching Process of Aluminum Alloy for Submicron Multilevel Metallization," Proceedings 7th Intl. IEEE VMIC, Santa Clara, CA, pp. 275-281 (1990).

60. Abraham, T., "Sidewall Tapering of Plasma Etched Metal Interconnects," Proceedings 3th Intl. IEEE VMIC, Santa Clara, CA, pp. 198-204 (1986).

61. Selamoglu, N., Bredbenner, C.N., Giniecki, T.A., and Stocker, H.J., "Tapered Etching of Aluminum with $CHF_3/Cl_2/BCl_3$ and its Impact on step Coverage of Plasma Deposited Silicon Oxide from Tetraethoxysilane," J. Vac. Sci. Technol. B9, 2530-2535 (1991).

62. Contolini, R.J., and Alspector, J., "Fine Line Aluminum Etching in Air and in Solution Using a Tunable Dye Laser," J. Appl. Phys., 65(4), 1759-1765 (1989).

63. Saxena, A.N., and Pramanik, D., "Planarization Techniques for Multilevel Metallization," Solid State Technol., 29(10), 95-100 (1986).

64. Ong, E., Chu, H., and Chen, S., "Metal Planarization with an Excimer Laser," Solid State Technol., 34(8), 63-68 (1991).

65 Woratschek, B., Carey, P., Stolz, M., and Bachman, F., "Excimer Laser Planarization of AlSi, AlSiTi and AlSiCu Alloys," Appl. Surf. Sci., 43, 264-270 (1989).

66. Baseman, R.J., Andreshak, J.C., Schnitzel, R.H., and Cronin, J.E., "Excimer Laser Planarization of Patterned Metal Features," J. Vac. Sci. Technol. B8, 1158-1160 (1990).

67. Marella, P.F., Tuckerman, D.B., and Pease, R.F., "Void Formation in Pulsed Laser Induced Via Contact Hole Filling," Appl. Phys. Lett., 56, 2625-2627 (1990).

68. Ong, E., Chu, H., and Chen, S.A., "Metal Planarization With an Excimer Laser," Solid State Technology, 34(8), 63-68 (1991).

69. Mukai, R., Tizuka, M., Kudo, H., and Nakano, M, "Metal Plugs Produced by Excimer Laser Melting for Submicron Interconnection:

Mechanism, Electrical Properties," Proceedings 8th Intl. IEEE VMIC, Santa Clara, CA, pp. 192-198 (1991).

70. Yu, C., Doan, T.T., and Grief, M., "A Novel Submicron Al Contact Filling Technology for ULSI Metallization," Proceedings 8th Intl. IEEE VMIC, Santa Clara, CA, pp. 199-205 (1991).

71. Chen, S. and Ong, E., Proc. SPIE, 1190, p. 207, 1990.

72. Yu, C., Doan, T.T., and Kim, S., "Laser Planarization of AlSiCu on Various Barrier Metals for VLSI," Proceedings 7th Intl. IEEE VMIC, Santa Clara, CA, pp. 444-446 (1990).

73. Yu, C. et al, Symp. on Electronic Packaging (MRS), p. 357, 1990.

74. Pramanik, D., and Chen, S., "Characterization of Laser Planarized Aluminum for Submicron Double Level Metal CMOS Circuits," IEDM Tech. Dig., pp. 673-676 (1989).

75. Boeck, B.A., Fu, K.Y., Pintchovsky, F., Crain, N., Chen, S., and Chu, S., "Evaluation of Laser Planarized Second Aluminum for Semiconductor Devices," Proceedings 7th Intl. IEEE VMIC, Santa Clara, CA, pp. 90-96 (1990).

76. Van Laarhoven, J.M.F.G., Van Houtum, H.J.W., and de Bruin, L., "A Novel Blanket Tungsten Etchback Scheme," Proceedings 6th Intl. IEEE VMIC, Santa Clara, CA, pp. 129-135 (1989).

77. Kaufman, F. B., Thompson, D.B., Broadie, R.E., Jaso, M.A., Guthrie, W.L., Pearson, D.J., and Small, M.B., "Chemical-Mechanical Polishing for Fabricating Patterned W Metal Features as Chip Interconnects," J. Electrochem. Soc., 138(11) 3460-3465 (1991).

78. Nishimura, H., Yamada, T., and Ogawa, S.,"Reliable Submicron Vias Using Al alloy High Temperature Sputter Filling," Proceedings 8th Intl. IEEE VMIC, Santa Clara, CA, pp. 170-176 (1991).

79. Castel, E.D., Kulkarni, V.D., and Riley, P.E., "A Scalable Multilevel Metallization with Pillar Interconnections and Interlevel Dielectric Planarization," J. Electrochem.Soc, 137(2) 609-613 (1990).

80. Cheung, N.W., "Reliability Issues of Multilevel Metallizations," VMIC State of the Art Seminar pp. 283-313 (1992).

81. Yeh, J.L., Hills, G.W., and Cochran, W.T., "Reverse Pillar and Maskless Contact - Two Novel Recessed Metal Schmes and Their Comparisons to Conventional VLSI Metallization Schemes," Proceedings 5th Intl. IEEE VMIC, Santa Clara, CA, pp. 95-100 (1988).

82. Herndon, T.O., "Nitrogen-Implanted Aluminum for Planarized Insulation," J.Electrochem. Soc, 138(10), 3107-3111 (1991).

83. Webb, D., Sivaram, S., Stark, D., Bath, H., Draina, J., Leggett, R., and Tolles, R., "Complete Intermetal Planarization using ECR Oxide and Chem-mechanical Polish," Proceedings 9th Intl. IEEE VMIC, Santa Clara, CA, pp. 141-148 (1992).

84. Singh, B., Mesker, O., and Devlin, B., "Deposition of Planarized Dielectric Layers by Bias Sputter Deposition," J. Vac. Sci. Technol. B5, 567-574 (1987).

85.   Ting, C.Y., Vuvalda, V.J., and Schaefer, H.G., "Study of Planarized Sputter Deposited SiO_2," J. Vac. Sci. Technol., 51, 1105 (1978).

86.   Homma, T. et al, NEC Res. Dev., 32, p. 315, 1991.

87.   Nishida, T., Saito, M., Iijima, S., Kuro, T., Sasaki, E., and Yagi, K, "Multilevel Interconnection for Half-micron ULSIs," Proceedings 6th Intl. IEEE VMIC, Santa Clara, CA, pp. 19-25 (1989).

88.   Bollinger, C.A., Grube, D., Lytle, S.A., Martin, E.P., Shimer, J.A., and Siddiqui, H.R., "An Advanced Four Level Interconnect Enhancement Module for 0.9 Micron CMOS," Proceedings 7th Intl. IEEE VMIC, Santa Clara, CA, pp. 21-27 (1990).

89.   Wilson, S.R., Freeman, J.L., and Tracy, C.J., "A High Performance, Four Metal Layer Interconnect System for Bipolar and BiCMOS Circuits," Proceedings 7th Intl. IEEE VMIC, Santa Clara, CA, pp. 42-48 (1990).

90.   Sheldon, D.J., Gruenschlaeger, C.W., Kammerdiner, L., Henis, N.B., Kelleher, P., and Hayden, J.D., "Application of a Two Layer Planarization Process to VLSI Intermetal Dielectric and Trench Isolation Processes," IEEE Trans. on Semiconductor Manufacturing, 1(4), 140-146 (1988).

91.   Uttecht, R.R., and Geffken, R.M., "A Four Level Metal Fully Planarized Interconnect Technology for Dense High Performance Logic and SRAM Applications," Proceedings 8th Intl. IEEE VMIC, Santa Clara, CA, pp. 20-26 (1991).

92.   Luce, S., and Pennington, S., "Interconnect Technology for 16-Mbit DRAM and 0.5-µm CMOS Logic," Proceedings 9th Intl. IEEE VMIC, Santa Clara, CA, pp. 55-58 (1992).

93.   Black, J.R., "Electromigration - A Brief Survey and Some Recent Results," IEEE Trans. ED, 16, 338-347 (1969).

94.   Howes, M.J., and Morgan, D.V., Reliability and Degradation, John Wiley (1981).

95.   Shingubara, S., Nakasaki, Y., and Kaneko, H., "Electromigration in a Single Crystalline Submicron Width Aluminum Interconnection," Appl. Phys. Lett., 58, 42-44 (1991).

96.   Kwok, T., "Effect of Metal Line Geometry on Electromigration Lifetime in Al-Cu Submicron Interconnects," 26th Annual Proceedings of the IRPS, pp. 185-191 (1988).

97.   Cho, J., and Thompson, C.V., "Grain Size Dependence of Electromigration Induced failures in Narrow Interconnects," Appl. Phys. Lett., 54, 2577-2579 (1989).

98.   Wada, T., Sugimoto, M., and Ajiki, T., "Effect of Surface Treatment on Electromigration in Aluminum Films," IEEE Trans. Reliability, 38, 565-570 (1989).

99.   Kim, M.J., Skelly, D.W., and Brown, D.M., "Electromigration of Bias Sputtered Al and Comparison With Others," 25th Annual Proceedings of the IRPS, pp. 126-129 (1987).

100. Gangulee, A., and d'Heurle, F.M., "Mass Transport During electromigration in Al-Mg Thin Films," Thin Solid Films, 25, 317-325 (1975).

101. Hong, H.H., "Effects of Annealing Temperature on Electromigration Performance of Multilayer Metallization Systems," 26th Annual Proceedings of the IRPS, pp. 173-178 (1988).

102. Hariu, T., Watanabe, K., Inoue, M., Takada, T., and Tsuchikawa, H., "The Properties of Al-Cu/Ti Films Sputter Deposited at Elevated Temperatures and High D.C. Bias," 27th Annual Proceedings of the IRPS, pp. 210-214 (1989).

103. Hinode, K., and Homma, Y., "Improvement of Electromigration Resistance of Layered Aluminum Conductors," 28th Annual Proceedings of the IRPS, pp. 25-36 (1990).

104. Shingubara, S. et al, Conf. on Sol. State Dev. and Materials, p. 455, 1987.

105. Isobe, A., Numazawa, Y., and Sakamoto, M., "Increase in EM Resistance by Planarizing Dielectric Film Over Al Wirings," Proceedings 6th Intl. IEEE VMIC, Santa Clara, CA, pp. 161-167 (1989).

106. Tice, W., and Slusser, G., "Relationship of Ambient Deposition Conditions to Formations of Thermally Activated Voids in AlSi Interconnects," J. Vac. Sci. Technol. B8 106-107 (1990).

107. Hinode, K., Asano, I., and Homma, Y., "Mechanism of Stress Induced Migration in VLSI Aluminum Metallization," Proceedings 5th Intl. IEEE VMIC, Santa Clara, CA, pp. 429-435 (1988).

108. Sullivan, T.D., "Stress Induced Voiding in Microelectronic Chips," IRPS Tutorials, pp. 3.1-3.24 (1992).

109. Onuki, J., Koubuchi, Y., Fukada, S., Suya, M., Misawa, Y., and Itagaki, T., "Development of Highly Reliable AlSiPd alloy Interconnection for VLSI," IEDM Tech. Dig. pp. 454-455 (1988).

110. Minkiewicz, V.J., Moore, J.O., and Eldridge, J.M., "Some Factors Affecting Hillock Formation Due to PECVD Processing of Sputtered Al-4%Cu-1%Si Films," J.Electrochem.Soc, 139(1), 271-275 (1992).

111. Lawrence, J.D., and McPherson, J.W., "Corrosion Susceptibility of Al-Cu and Al-Cu-Si Films," 29th Annual Proceedings of the IRPS, pp. 102-106 (1991).

112. Striny, K.M., and Schelling, A.W., "Reliability Evaluation of Aluminum Metallized MOS Dynamic RAMs in Plastic Packages in High Humidity and Temperature Environments," IEEE Trans. CHMT, 4, 476-481 (1981).

113. Dunn, C.F., and Mcpherson, J.W., "Recent Observations on VLSI Bond Pand Corrosion Kinetics," J.Electrochem Soc, 135(3), 661-665 (1988).

114. Weston, D., Wilson, S.R., and Kottke, M., "Microcorrosion of Al-Cu and Al-Cu-Si Alloys: Interaction of the Metallization with Subsequent Aqueous Photolithographic Processing," J. Vac. Sci. Technol. A8(3), 2025-2032 (1990).

115. Van der kolk, G.J., Verkerk, M.J., and Brankaert, W.A.M.C., "Effects of Contamination on Aluminum Films. Part I: Room temperature deposition," Semiconductor International, 11(6), 224-227 (1988).
116. ibid 11(7), 106-111 (1988).
117. Martin, C.A., and McPhereson, J.W., "Via Electromigration Performance of Ti/W/Al-Cu(2%) Multilayered Metallization," Proceedings 6th Intl. IEEE VMIC, Santa Clara, CA, pp. 168-175 (1989).
118. Cheung, N.W., VMIC 1992, State-of-the art Seminar.
119. Bui, N.D., Pham, V.H., Yue, J.T., and Wollesen, D.L., "A Via Failure Mode in Electromigration of Multilevel Interconnect," Proceedings 7th Intl. IEEE VMIC, Santa Clara, CA, pp. 142-148 (1990).
120. Bui, N.D., Pham, V.H., and Yue, J.T., "Effect of Barrier Metal, Grain Size, And Interface Cleanliness on Electromigration Performance of Via Chain," Proceedings 9th Intl. IEEE VMIC, Santa Clara, CA, pp. 344-351 (1992).
121. Bonifield, T., "Reliability Issues in Multilevel Interconnect," VMIC Tutorial, pp. 279-392, June 1992.
122. Kanazawa, M., Shishino, M., Hata, Y., and Umemoto, T., "Stress Induced Void Formation of the Vias in the Al-Based Multilevel Interconnection System," Proceedings 8th Intl. IEEE VMIC, Santa Clara, CA, pp. 221-227 (1991).
123. Chang, P.H., Hawkins, R., Bonifeield, T.D., and Melton, L.A., "Aluminum Spiking at Contact windows in Al/Ti - W/Si," Appl. Phys. Lett., 52, 272-274 (1988).
124. Tomioka, H., Tanabe, S.-i, and Mizukami, K., "A New Reliability Problem Associated with Ar ion Sputter Cleaning of Interconnect Vias," 27th Annual Proceedings of the IRPS, pp. 53-56 (1989).
125. Blewer, R.S., Advanced Metal Systems, VLSI State-of-the-art Seminar, pp. 19-103, June 1992.
126. Koubuchi, Y., Onuki, J., Suwa, M., and Fukada, S., "Stress Migration Resistance of Al-Si-Pd Alloy Interconnects," Proceedings 6th Intl. IEEE VMIC, Santa Clara, CA, pp. 419-425 (1989).
127. Joshi, A., Gardner, D., Hu, H.S., Mardinly, A.J., and Nieh, T.G., "Aluminum Samarium Alloy for Interconnections in Integrated Circuits," J. Vac. Sci. Technol., A8, 1480-1483 (1990).
128. Kato, T., Ito, T., and Ishikawa, H., "In-Situ Carbon Doped Aluminum Metallization for VLSI/ULSI Interconnects," IEDM Tech. Dig. pp. 458-461 (1988).
129. MacWilliams, K.P., "Improved Yield and Reliability in Aluminum Interconnects Through Fluorine Incorporation," VLSI Technol.Sym, pp. 33-34 (1990).
130 Ogawa, S., and Nishimura, H., "A Novel Al-Sc (Scandium) Alloy For Future LSI Interconnection," IEDM Tech. Dig., pp. 277-280 (1991).
131. Park, Y.H., Chung, A.H., and Ward, M.A., "Step Coverage Evaluation of Copper Films Prepared by Magnetron Sputtering," Proceedings 8th Intl. IEEE VMIC, Santa Clara, CA, pp. 295-297 (1991).

132. Arita, Y., Awaya, N., Amazawa, T., and Matsuda, T., "Deep Submicron Cu Planar Interconnection Technology Using Cu Selective Chemical vapor Deposition," IEDM Tech. Dig., pp. 893-895 (1989).

133. Awaya, N., Ohono, K., Sato, M., and Arita, Y., "Double Level Copper Interconnections Using Selective Copper CVD," Proceedings 7th Intl. IEEE VMIC, Santa Clara, CA, pp. 254-257 (1990).

134, Ohmi, T., Saito, T., Shibata, T., and Nitta, T, "Room Temperature Copper ULSI Metallization by Low Kinetic Energy Particle Process," Proceedings 5th Intl. IEEE VMIC, Santa Clara, CA, pp. 135-141 (1988).

135. Pai, P.L., and C.H.Ting, "Copper as the Future Interconnection Material," Proceedings 6th Intl. IEEE VMIC, Santa Clara, CA, pp. 258-264 (1989).

136. Mak, C.Y., Miller, B., Feldman, L.C., Weir, B.E., Higashi, G.S., Fitzgerald, E.A., Boone, T., Doherty, C.J., and Vandover, K.B., "Selective Electroless Copper Metallization of Palladium Silicide on Silicon Substrates," Appl.Phys.Lett, 59, 3449-3451 (1991).

137. Shacham-Diamand, Y., Dedhia, A., Hoffstetter, D., and Oldham, W.G., "Reliability of Copper Metallization on SiO2," Proceedings 8th Intl. IEEE VMIC, Santa Clara, CA, pp. 109-115 (1991).

138. Olowolafe. J.O, Li. J.A, Mayer. J.W and Colgan. E.G, "Effects of Oxygen on the Diffusion of Cu in Cu/TiN/Al and Cu/TiNX/Si Structures," Appl.Phys.Lett, 58, 469 - 471, (1991).

139. Li, J., Shacham-Diamand, Y., Mayer, J.W., and Colgan, E.G., "Thermal Stability Issues in Copper Based Metallizations," Proceedings 8th Intl. IEEE VMIC, Santa Clara, CA, pp. 153-159 (1991).

140. Gardner, D.S., Onuki, J., Kudoo, K., and Misawa, Y., "Encapsulated Copper Interconnection Devices Using Sidewall Barriers," Proceedings 8th Intl. IEEE VMIC, Santa Clara, CA, pp. 99-108 (1991).

141. Otsuki, M. et al., in Int. Conf. Sol. Stat. Dev and Materials, p.186 (1991).

142 Kim. J.Y, and Hummel. R.E, "Hole and Hillock Formation in Gold Metallizations at Elevated Temperatures Deposited on Titanium, Vanadium and Other Barrier Layers," Phys. Stat. Solidi A, 122 (12), 255 - 273 (1990).

143. Burggraaf, P, "Sputtering: Wafer and Disk Challanges," Semiconductor International, 14(13), 38-41 (1991).

144. Bader, M.E., Hall, R.P., and Strasser, G., "Integrated Processing Equipment," Solid State Technology, 33(5), 149-154 (1990).

4

INORGANIC DIELECTRICS

CHIU H. TING
Intel Corporation
Santa Clara, California

1.0   INTRODUCTION

Inorganic dielectric films are used extensively in the fabrication of very large scale integrated (VLSI) circuits. The most commonly used dielectric materials are: silicon dioxide (SiO_2), phosphosilicate glass (PSG), borophosphosilicate glass (BPSG), silicon nitride (SiN), and silicon oxynitride (SiON). With the exception of thermally grown SiO_2, these dielectric films must be deposited onto the wafer surface as a part of the VLSI circuit fabrication process. The purpose of this chapter is to discuss the deposition processES, material properties, and applications of these deposited dielectric films for multilevel metallization.   More exotic dielectric films used for special applications, such as tantalum oxide, ferroelectric films, silicon carbide, etc., will not be discussed.

Thin films can be deposited either by physical vapor deposition (PVD) or chemical vapor deposition (CVD) processes. Physical deposition processes, such as evaporation and sputtering, are used extensively in depositing metal layers and are discussed in Chapter 3. With the exception of bias sputtered quartz (BSQ), physical deposition processes are seldom used in depositing dielectric films; instead, CVD processes are used extensively. Most of the modern sophisticated CVD technology was developed in order to meet the stringent VLSI circuit production requirements such as quality, uniformity, composition reproducibility, and defect free and through put requirements. Therefore, this chapter concentrates on the dielectric films deposited by CVD processes and plasma-enhanced CVD (PECVD) processes. In addition, spin on dielectric films, such as spin on glass (SOG), are discussed briefly because they are also used in multilayer metallization technology.

Many excellent reviews covering all aspects of CVD (1-5), and PECVD (6-8) are available. A large compilation of original research results are available in symposia proceedings (9, 10). This chapter provides a broad outline of the fundamental and practical aspects of the dielectric films used in modern VLSI multilayer metallization technology. The emphasis is on the dielectric film characteristics, deposition techniques and the effect of deposition conditions on film characteristics. Different applications, such as dielectric films used under the first metal (ILD0), intermetal layer dielectric (ILD1, ILD2, ...), and top passivation dielectric layer, are discussed in separate sections since their requirements are quite different from each other.

2.0   CVD PROCESSES

CVD is a very complicated technology. Understanding it requires knowledge from many different disciplines such as chemistry, thermodynamics, gas transport, heat transfer, and film growth kinetics. In addition, process variables such as temperature, pressure, reactant concentrations, reactor geometries and substrate surface conditions must also be considered. Although a detailed understanding of each reaction step is difficult and beyond the scope of this chapter, a good understanding of the basic principles is essential for the development of sophisticated CVD processes used to satisfy the stringent requirements of today's VLSI technology. The purpose of this section is to provide a broad, but brief, review of basic CVD dielectric deposition principles.

2.1 Deposition Chemistry

Many different chemical reactions are used to deposit CVD dielectric layers for VLSI circuit fabrication. Some typical reactions are listed in Table I. The criteria for choosing a particular reaction are:

1) deposition temperature - must be compatible with materials already on the wafer, such as aluminum metallization;

2) film properties - must have proper density, stress and step coverage;

3) manufacturing requirements - must give satisfactory throughput, cleanliness, and safety, etc.

The most common reactions used for depositing SiO_2 is oxidizing silane with oxygen (11,12). This reaction gives a high quality SiO_2 film at a temperature low enough for Al metallization. Doped oxide can be obtained by adding appropriate doping gases, such as phosphine (PH_3) or diborane (B_2H_6), to the deposition reactions.

The silane/oxygen reaction has some serious drawbacks. First, it is a major safety concern because silane is a pyrophoric gas which is capable of spontaneous combustion when exposed to air. Many industrial accidents have been caused by the improper handling of silane. To reduce the hazardous safety issues, silane is usually supplied in a diluted form, often with N_2. Another drawback is that oxide films deposited from the silane process usually have insufficient step coverage characteristics. The step coverage requirement becomes more critical as devices are scaled down to ever smaller dimensions.

Table I. Dielectric Deposition Processes

APCVD (Atmospheric Pressure Chemical
Vapor Deposition)

$SiH_4 + O_2 \rightarrow SiO_2$ (380-450 °C)
$TEOS + O_3 \rightarrow SiO_2$ (350-450°C)

LPCVD (Low Pressure Chemical Vapor
Deposition)

$SiH_4 + O_2 \rightarrow SiO_2$ (380-450 °C)
$TEOS + O_2 \rightarrow SiO_2$ (680-750 °C)
$TEOS + O_3 \rightarrow SiO_2$ (350-450 °C)
$SiH_2Cl_2 + O_2 \rightarrow SiO_2$ (850-950 °C)
$SiH_2Cl_2 + NH_3 \rightarrow Si3N_4$ (800-850 °C)

PECVD (Plasma Enhanced Chemical
Vapor Deposition)

$SiH_4 + N_2O \rightarrow SiO_2$ (150-400 °C)
$TEOS + O_2 \rightarrow SiO_2$ (300-400 °C)
$SiH_4 + NH_3 + N_2 \rightarrow SiN(H)$ (200-400 °C)
$SiH_4 + NH_3 + N_2O \rightarrow SiON$ (200-400 °C)

TEOS or tetraethylorthosilicate, $Si(OC_2H_5)_4$, has gained great popularity as an alternative Si source material because it is a non-toxic organic compound and is a liquid at room temperature with reasonable vapor pressure (13-15). Furthermore, the oxide films deposited with TEOS reactions have better step coverage than silane oxide (14, 16, 17). The oxidation of TEOS by oxygen, however, requires a temperature above 700°C which is too high for Al metallization.

Significant effort has been invested to develop alternate organosilicon source materials capable of depositing oxide at lower temperatures. Some of these new organosilicon precursors include tetra-methyl-cyclo-tetra-siloxane (TMCTS), octa-methyl-cyclo-tetra-siloxane (OMCTS), di-acetoxy-ditertiary-butoxy-silane (DABS), hexa-methyl-di-siloxane (HMDS), and di-ethyl-silane (DES). These alternative Si sources have not yet gained common use because their film qualities are not well characterized

There are other methods to reduce the TEOS deposition temperature. The most common method to lower the deposition temperature is to use plasma enhanced CVD, or PECVD, processes (18-22). Another method is to use ozone instead of oxygen to oxidize TEOS. With the TEOS-ozone reaction, oxide films can be deposited at temperatures of 400 C or lower (23, 24). An additional advantage of the TEOS-ozone process is the deposited films have superb step coverage (23), an important property for ILD applications.

CVD reactions for depositing silicon nitride and silicon oxynitride also require temperatures too high for Al metallization, thus are limited to applications for layers underneath the metallization (25). The lower temperature PECVD processes are used to deposit SiN and SiON over the Al layers for ILD and passivation layer applications (26, 27).

2.2 Fundamental Reactions

The deposition of a solid film from the vapor of reactants involves many complicated steps. These steps are given in Table II. These sequence of events can be classified into three categories:

1) Gas phase transport
2) Surface reactions
3) Surface desorption

The important features of each category are discussed next.

Table II. Events Occurring During CVD Deposition

1)	Transport of reactants to the reaction chamber
2)	Transport of reactants to the edge of boundary layer
3)	Transport of reactants across the boundary layer
4)	Gas phase reactions
5)	Adsorption of reactants
6)	Surface re-emission or desorption
7)	Surface reaction
8)	Surface migrations
9)	Incorporation at growth site
10)	Byproduct migration and reactions
11)	Byproduct desorption
12)	Transport of byproducts across the boundary layer
13)	Transport of byproducts out of reaction chamber

Gas Phase Transport: In order for film deposition to occur, the reactants must first be transported to the substrate surface. The gas phase transport process involves a complicated sequence of events, including gas phase reactions, convection of reactant gases, boundary layer formation, and diffusion of reactants across the boundary layer, etc.

Gas Phase Reactions: The transport of reactants to the substrate surface is governed by gas flow dynamics which is a complicated function of temperature, pressure, gas or vapor characteristics, and the reaction chamber geometries. Homogeneous gas phase reactions can occur during the transporting and mixing of the reactant gases. In general, homogeneous gas phase reactions are not desirable because they do not contribute to film growth and are a source of particle generation. Problems associated with gas phase homogeneous reactions can be minimized by using cold wall reactors. In a cold wall reactor, the reactant gases remain cool until reaching the vicinity of the substrate surface. Hot wall reactors tend to generate more particles not only from gas phase reactions but also from loosely bonded films deposited on hot chamber walls. Homogeneous gas phase reactions can also be reduced by using a low pressure deposition system. The following equation gives the approximate value of the mean free path (average distance between gas phase collisions), L, of an ideal gas as a function of pressure, P. In a low pressure system the mean free path is greatly increased, reducing the chances of a gas phase reaction.

$$[1] \quad L \, (\mu m) = \frac{50}{P \, (Torr)}$$

Convection: Deposited film thickness uniformity is dependent on the uniform delivery of reactants to the substrate surface. Mixing of reactant gases can be achieved by either free convection or forced convection. Free convection is established by temperature gradients in the reactor - gas generally flows from the hot substrate surface to the cool chamber walls. Forced convection is achieved by high gas flow velocity. At high velocities, forced convection dominates and free convection can be neglected. In the forced convection regime, gas flow is characterized by Reynold's number, R_e,

$$[2] \quad R_e = \frac{D_t \, v \, p}{n}$$

where D_t is a characteristic dimension of the reactor chamber, v is the linear gas velocity, p is the gas density and n is the gas viscosity. Low Reynold's number corresponds to laminar flow while high Reynold's number corresponds to turbulent flow.

The Boundary Layer: Because of the viscous nature of gases, gas velocity is not uniform in the vicinity of walls even when the flow is dominated by forced convection. Stationary chamber walls and the substrate surface create a drag on the gas flow so the gas velocity is lower near the stationary surfaces. This low velocity region, separating the stationary surface from the free-flow region, is called the "boundary layer". The concept of a low velocity boundary layer is important to both low pressure and atmospheric-pressure CVD processes.

For example, the boundary layer for a flat substrate surface parallel to the gas flow is illustrated in Figure 1. The gas velocity increases from zero at the substrate interface to its full velocity at the edge of the boundary layer. This increase is often assumed to be parabolic as illustrated. In simplified calculations, the velocity throughout the boundary layer is sometimes assumed to be zero. Under this assumption, the boundary layer is also called a "stagnant layer". Since the boundary layer is developed along the length of the stationary surface, the gas velocity is not only a function of the perpendicular distance, y, from the plate but also the longitudinal position, x.

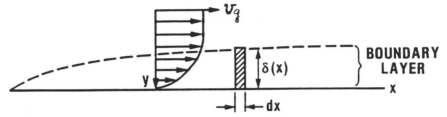

Figure 1. Schematic diagram of a boundary layer separating the fast moving gas region from the stationary deposition surface.

Boundary Layer Diffusion: Reactants must diffuse across the boundary layer in order to reach the substrate surface. The diffusion flux across the boundary layer is given by Fick's law:

$$[3] \quad F = D \; \frac{dC}{dy}$$

where D is the gas phase diffusion coefficient of the reactant gas, C is its concentration, and y is the perpendicular distance from the surface within the boundary layer. Assuming the concentration varies linearly with the position in the boundary layer, then

$$[4] \quad F = D \; \frac{C_g - C_s}{d}$$

where C_g and C_s are the concentrations of the reactant gas at the edge of the boundary layer and substrate surface respectively, and d is the thickness of the boundary layer as illustrated in Figure 2. Thus, the concentration gradient, $(C_g - C_s)/d$, provides the driving force for the transport of reactants from the gas flow across the boundary layer to the substrate surface. While C_g is fixed for a given gas flow condition, the surface concentration, C_s, is a function of the consumption rate (deposition rate) of the reactant at the substrate surface. For steady state, the surface reaction rate, R, is a strong function of temperature and is expressed as:

$$[5] \quad R = R_o \; EXP \left(- \frac{E_a}{kT} \right)$$

where k is the Boltzmann constant, T is temperature in degrees Kelvin, and E_a is the activation energy of the reaction.

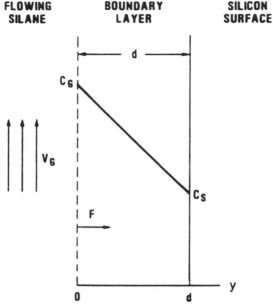

Figure 2.   Schematic diagram of the reactant concentration gradient in the boundary layer.

Surface Reactions: The reactant species diffused across the boundary layer are consumed at the substrate surface by the deposition process which changes the reactants into a deposited film. Some of the important events that take place at the substrate surface are discussed below.

Effect of Deposition Temperature: The reactant concentration at the substrate surface, C_s, depends on the balance between the supply rate of reactants diffusion across the boundary layer and the consumption rate at the surface due to the deposition reaction. At low temperatures, the reaction rate is small as given by Equation (5). Therefore, the reactant concentration at the surface, C_s, is large, approaching the value of C_g to give a very small concentration gradient, or driving force, across the boundary layer. Under these conditions, the overall deposition reaction is said to be reaction rate limited and the deposition rate has a strong temperature dependence with an activation energy, E_a, given in Equation (5). In Figure 3, the deposition rate is shown as a function of the deposition temperature with the reaction rate limited region labelled as region II. Conversely, at high temperatures, the reaction rate is large and the reactant is consumed rapidly at the surface. Thus, the surface concentration, C_s, decreases with increasing temperature. At the limit, the reactant is consumed immediately after it reaches the surface and the value of C_s approaches zero to give the maximum driving force across the boundary layer. Under these conditions, the reaction rate at the surface is limited by the rate at which the reactant can be transported across the boundary layer. The overall reaction under these conditions is mass transport limited. This situation is labelled region I in Figure 3. In this region, the deposition rate is high but not a strong function of substrate temperature. The weak increase in deposition rate is primarily due to the increase of the gas phase diffusion coefficient with temperature.

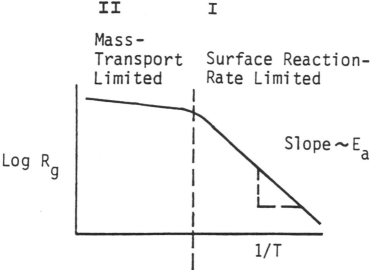

Figure 3.   The deposition rate as a function of temperature for I) Mass transport limited region and II) Surface reaction rate limited region.

It may seem desirable to operate in the mass transport region for the highest deposition rate; however, other issues must be considered. For example, the thickness uniformity in the deposited film may be difficult to control for a deposition process operating in the mass transport region. The thickness uniformity in this region is determined by the uniformity of the reactant in the gas phase which, in turn, is determined by the gas flow dynamics of the system. Controlling the uniformity of gas flow could be difficult because of the complex reactor geometries. Furthermore, operating in the mass transport region may require a temperature too high for aluminum metallization. Therefore, dielectric layer deposition processes generally operate in the surface reaction rate limited regime where the temperature must be carefully controlled. For uniform and reproducible depositions, the substrate heater must be carefully designed to give accurate temperature control and uniformity over the entire substrate surface. In designing the substrate heater, the effect of heat loss at the edge of the wafer must be taken into account in order to achieve uniform temperature distribution across the entire wafer.

Adsorption: The reactant species must first be adsorbed at the substrate surface for the deposition reaction to take place. For SiO_2 deposition from silane (SiH_4 and oxygen (O_2), the silicon sources, SiH_4 or its intermediates, SiH_3 or SiH_2, etc., are adsorbed on the surface according to the reactions:

[6a] $SiH_4 + * \longrightarrow > SiH_4^*$

[6b] $SiH_2 + * \longrightarrow > SiH_2^*$

where * represents a free surface state, and SiH_4^* and SiH_2^* are the adsorbed species. The reaction given by Equation (6) requires the presence of both a free reactant species and a free surface site to form an adsorbed reactant species. Different reactant species have different adsorption probabilities at the free surface sites. For example, SiH_2 is more strongly adsorbed at Si and SiO_2 surfaces than SiH_4. Thus, even though SiH_4 is more abundant in the gas phase than SiH_2, SiH_2 may still dominate the deposition reaction because it is strongly adsorbed (has a high sticking coefficient) at the substrate surface (28).

Surface migration: After adsorbing onto the surface, the reactive species can either remain fixed at the surface sites until the reactions are complete or, more likely, migrate from surface site to surface site before they are consumed by the deposition reactions. Surface migration can be accomplished either by surface diffusion or by desorption and re-adsorption of the reactive species at a different location on the surface. Diffusion on the surface is expressed in terms of the surface diffusion coefficient, D_s, of the adsorbed species,

[7] $D_s = D_0 \theta \, Exp \left(- \dfrac{E_a}{kT} \right)$

where D_0 is a constant, θ is the fraction of unoccupied surface sites, and E_a is the activation energy of surface diffusion. For example, surface diffusion is an essential part of Si epitaxial growth where the atoms must move to the most energetically favorable sites, i.e. lattice sites at the ledge (step) of a crystalline plane, to form epitaxial layer deposition. Surface diffusion and surface desorption/re-adsorption are also essential in improving the step coverage of CVD dielectric films over surface topography.

Oxygen to Silane Ratio: The heterogeneous reaction between adsorbed silane and oxygen on the substrate surface to form a solid SiO_2 film involves a complicated series of events. The deposition rate depends not only on the deposition temperature but also on the gas flow rates. At low silane partial pressures, the deposition rate is proportional to the silane flow rate because only a small fraction of available surface sites are covered by the silane species. Therefore, the deposition rate can be increased by increasing the silane partial pressure. Increasing the silane flow has a practical limit because when the silane partial pressure becomes too high, hazy deposits of SiO_2 are often observed which are probably due to gas phase homogeneous reactions.

The deposition rate also increases with oxygen flow, reaches a maximum and then decreases at higher oxygen partial pressures. At the maximum deposition rate, the oxygen to silane ratio increases with higher deposition temperatures, as illustrated in Figure 4 (29). The reduction in deposition rate at very high O_2/SiH_4 ratios is attributed to the depletion of available surface sites by too much adsorbed oxygen on the surface, thus retarding the adsorption of silane species needed for the reaction. This retarding effect is also observed when other gas additives, such as doping gas, are used.

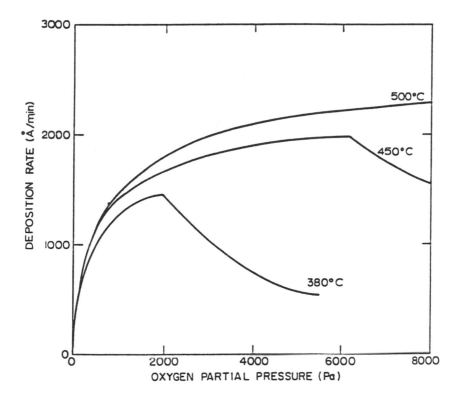

Figure 4.   The deposition rate of SiO_2 for different O_2 concentrations (29), reprinted with permission of the publisher.

Substrate Surface Conditions Since the deposition process is a surface activated reaction, the composition and the condition of the substrate surface can have a profound effect on the deposition process. Chemical impurities on the substrate surface can either act as catalysts or inhibitors to the growth rate. Different surface properties, such as hydrophobic or hydrophilic surfaces, can significantly alter the deposition rates under some circumstances. In general, surface contaminants cause defects in the deposited film such as particles, pinholes, pits, poor adhesion, cloudy deposits, and local thickness variations, etc. A cleaner surface helps to achieve better uniformity and reproducibility. In addition, substrate surface topography can change local gas flow patterns resulting in local film quality variations.

Surface Desorption: For the SiO_2 deposition reaction,

[8a] $SiH_4^* + O_2^* \longrightarrow SiO_2^* + 4\,H^*$

[8b] $4\,H^* \longrightarrow 2H_2$

adsorbed hydrogen atoms, H^*, and free hydrogen, H_2 are byproducts of the deposition process. The adsorbed hydrogen atom will diffuse on the substrate surface until it encounters another hydrogen atom to form an adsorbed hydrogen molecule, H_2. The hydrogen molecule can then be desorbed, diffused through the boundary layer and removed from the reaction chamber. If these by-products are not removed, they can become a limiting factor for further deposition reactions. Since the above reaction produces many hydrogen atoms at nearby sites, formation and removal of hydrogen molecules should be rapid. However, at low deposition temperatures the desorption of hydrogen can be slow and the entire deposition process will be limited by the hydrogen desorption process. When organic compounds, such as TEOS, are used as a Si source the byproducts consist of complicated organic molecules which must also be desorbed from the surface. Incomplete removal of organic byproducts results in the incorporation of carbon or other contaminants in the deposited SiO_2 film.

2.3   PECVD

In the previous section, only conventional CVD processes activated by thermal energy were considered. Other energy sources, such as electric energy or photon energy, can also be used to enhance the CVD process. The most common technique is to use electrical energy in an electrical discharge, or plasma, to enhance the CVD deposition processes. This is called plasma enhanced CVD or PECVD. Even though PECVD is a relatively new technology, it is used extensively for depositing dielectric layers. The purpose of this section is to briefly summarize the principles and practice of PECVD for dielectric deposition. A great amount of literature has been published on the subject already; for more details, refer to several excellent review papers (6-8).

Deposition Process: The primary motivation for using PECVD is to reduce the deposition temperature. The energetic electrons in the plasma can activate many reactions among the gases to produce excited neutral and ionized species. With excited species, reactions can proceed rapidly at a lower temperature than reactions excited by thermal energy alone. Although the plasma produces energetic electrons and ions, the bulk of the gas and the substrate do not reach equilibrium with the more energetic electrons. With the glow discharge in a non-equilibrium state, films can be deposited with compositions that are not at thermal equilibrium with the deposition temperature. With PECVD, many dielectric materials can be deposited with high rates at temperatures less than 400°C. Some typical PECVD reactions used to deposit silicon dioxide and silicon nitride are listed in Table I. The fundamental aspects of CVD deposition discussed previously also apply to the PECVD process. However, PECVD processes are much more complex than thermal CVD process because of the variety of reactive species in the plasma. As a result, PECVD film properties depend not only on the standard CVD parameters, such as substrate temperature and gas flow conditions, but also vary significantly with plasma conditions such as frequency, power, substrate bias, electrode spacing, reactor geometry, etc. (30, 31). This makes comparing various PECVD films to one another very difficult without the detailed knowledge of equipment configurations and operating conditions.

Equipment Requirements PECVD systems are different from thermal CVD systems because additional components must be added to generate and maintain a glow discharge. A schematic diagram of a typical PECVD equipment is shown in Figure 5 (8), and some of the important components are listed below.

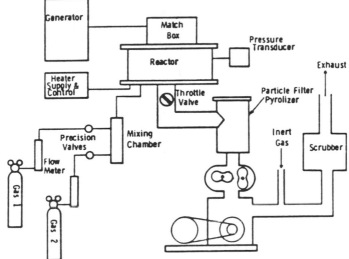

Figure 5.   Schematic diagram of a PECVD system (ref.8), reprinted with permission of the publisher.

Power Supply: For dielectric deposition, a DC discharge cannot be used because the deposited insulating film will block the conduction path of the discharge. An rf (radio frequency) power supply at 13.6 mHz is often used to generate the discharge. Microwaves can also be used to generate the plasma but are less commonly used because of the complication of an additional resonator. A good example of a microwave discharge is the ECR, or electron cyclotron resonance system, which combines a microwave field with an appropriate magnetic field creating an electron resonance condition to generate an intense plasma at very low pressures. Sometimes, additional rf or low frequency (kHz range) power supplies are used to bias the substrate electrode in order to increase ion bombardment during the film deposition to give desired film properties.

Match Network: The purpose of the match network is to pass the rf power from the power supply to the plasma load. The match network introduces variable inductance and capacitance between the generator and the plasma load so when adjusted properly a resonant circuit is formed. Even though high circulating current flows within the resonant circuit, the power supply sees only the resistive component of the load, thus passing the rf power from the generator to the plasma load.

Electrodes: Many electrode designs are possible, such as tube type, parallel plate type, inductively coupled or capacitively coupled. The simplest and the most common type is the parallel plate electrodes with radial gas flow. This design allows the wafers to lie flat on the electrode for uniform heating and temperature control. An example of a parallel plate system is shown in Figure 6 (31). In this design, the power electrode also serves as a "shower head" for gas distribution.

Vacuum System: A PECVD system requires a pressure of several Torr to facilitate the generation and maintenance of the glow discharge. The system is often pumped down to high vacuum level before turning on the gas flow to ensure against leaks or other contaminations that could affect the deposition process. In general, high pumping speeds are used to achieve low pressures at high gas flow rates. Safety precautions are needed for pumps and exhausts to handle flammable and toxic gases.

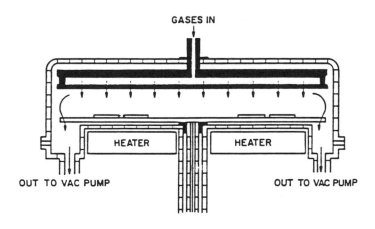

Figure 6.   Schematic diagram of a parallel plate electrode system (31), reprinted with permission of the publisher.

3.0   DEPOSITION EQUIPMENT

CVD reactions can take place either at atmospheric or at low pressures. A wide variety of reactor designs have been developed and are available commercially, such as horizontal reactors, vertical reactors, hot wall reactors, cold wall reactors, etc. The choice of reactor design depends on the nature of the reactions. These reactors are incorporated into systems with different wafer handling capability such as batch processing, single wafer processing, continuous processing systems, modular systems, etc. The choice of wafer handling system depends on the uniformity and throughput requirements. It is beyond the scope of this chapter to discuss the details of reactor and system designs for different CVD applications. A few commonly used CVD systems are briefly discussed below to illustrate their important features.

3.1   LPCVD Systems

LPCVD (low pressure) systems of many different designs are used in depositing dielectric films for device fabrication. The most common configuration is the hot wall horizontal tubular furnace (32). A schematic diagram of a LPCVD furnace is shown in Figure 7. The reactant gases are introduced at one end and pumped out at the other end. The wafers are placed vertically and separated from each other by a narrow spacing. The small spacing between wafers makes the system capable of very high throughput but makes the diffusion of reactant gases to the center of the wafer difficult. In order to obtain uniform film thickness across the wafer, it is necessary to operate the system in the reaction limited region. Because the gases are introduced at one end, a portion of the reactant gas is consumed along the length of the reaction tube resulting in decreased deposition rate along the length of the tube. This gas depletion effect must be compensated in order to obtain uniform film thickness for all wafers. For deposition processes operating in the reaction limited region, the reaction rate changes exponentially with the temperature as given in Equation (5). Therefore, a slight increase in temperature along the length of the tube can be used to compensate for moderate gas depletion to improve the thickness uniformity among the wafers. In addition, the gas depletion problem can also be minimized by using a high capacity pump to achieve higher gas flow rate and higher gas velocity at a given pressure to allow more reactant gas to reach down stream wafers.

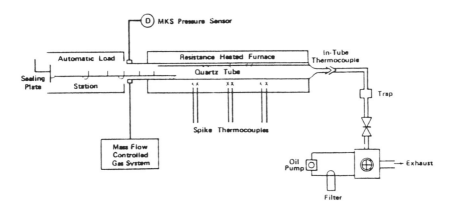

Figure 7.   Schematic diagram of a LPCVD system (32), reprinted with permission of the publisher.

3.2   APCVD Continuous Processing System

In a APCVD system, wafers are often placed horizontally on a heated susceptor. The reactor walls remain relatively cool. The reactant gases usually are introduced at one end of the reactor at slightly higher than atmospheric pressure and pushed out at the other end. The gases flow in the reactor primarily by forced convection. With cool reactor walls, gases do not react until they diffuse across the boundary layer to reach the hot wafer surface. An innovative APCVD system is represented by the Watkins Johnson 999 continuous processing system (33). This system uses a specially designed nozzles to introduce reactant gases over the wafers which are transported on a heated conveyer belt. The reactant gases are separated by a nitrogen gas curtain to prevent homogenous reactions before they reach the wafer surface. A schematic diagram of the system and the gas injector are shown in Figure 8 (a) and (b), respectively. With multiple injectors, different films can be deposited in a single pass. To minimize particles, the conveyer belt is continuously cleaned to remove deposited films. The system is capable of very high throughput and is widely used for SiO_2 and doped glass deposition.

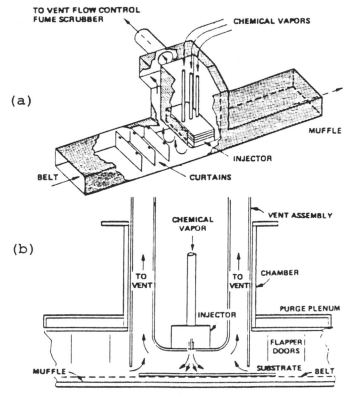

Figure 8. Schematic diagram of Watkins-Johnson APCVD system (a) Overall system, (b) Gas injector (33), reprinted with permission of the publisher.

3.3   PECVD Continuous Processing System

Many different types of PECVD systems have been developed for dielectric film deposition. As in CVD systems, a variety of reactor designs and wafer handling systems are available. In addition, there are different designs in rf power coupling, electrode configurations, match network, and bias power to the electrodes, etc. An example of a commonly used PECVD system is the Novellus continuous processing system (34). The schematic diagram of the system is shown in Figure 9. In this system, the wafers are fed into the reaction chamber through a vacuum loadlock to eliminate the pressure and temperature cycling of the process chamber. In the process chamber, there are six individual processing stations and wafers are processed sequentially at multiple deposition stations to give an averaging effect for better uniformity. The multiple station approach allows the flexibility of maintaining reasonable throughput even for processes with relatively low deposition rates. The system also has a dual rf generator so low frequency rf bias can be applied to the substrate electrode to control the amount of ion bombardment during film deposition. The effect of ion bombardment on the properties of the deposited films will be discussed later.

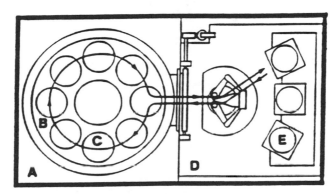

A = Process Chamber
B = RF Powered Al Susceptor
C = Deposition Stations
D = Vacuum Loadlock
E = Cassettes (3)

Figure 9. Schematic diagram of a Novellus continuous processing plasma reactor (34), reprinted with permission of the publisher.

3.4   Multichamber System

Perhaps the most flexible system is a multichamber system that is capable of handling different processes in different chambers for flexibility or the same process for high throughput. An example of such systems is the Applied

Material Precision-5000 system (35). A schematic diagram is shown in Figure 10. This system can have up to four processing chambers. Wafers are fed into each chamber through a central vacuum loadlocked load station. Each chamber can be fitted with the same process or different processes including LPCVD, PECVD, or etching. Multichamber systems are especially suitable for integrated processes that require multiple layers of dielectric films and for etchback of deposited films for planarization (36, 37). For the P-5000 system, the reactor chamber is cleaned by an etching cycle following each deposition cycle to remove any unwanted deposition on the chamber walls to minimize particulates.

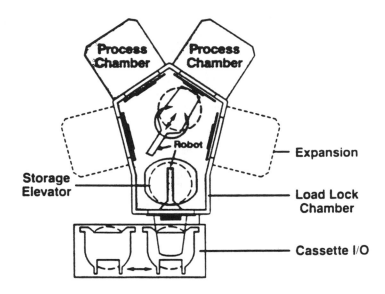

Figure 10. Schematic diagram of a Applied Materials Precision 5000 multichamber CVD reactor (35), reprinted with permission of the publisher.

4.0   SPIN ON DIELECTRICS

Due to the severe surface topographies generated in the multilevel metallization process, it is necessary to planarize the inter level dielectric (ILD) in order to avoid processing problems such as poor step coverage of the deposited metal, void formation in the deposited dielectric films and limited depth of focus of high resolution lithography. Many different processing techniques have been developed to planarize the dielectric surface (38). One of the most common technique used to avoid the problems mentioned above is to use a thin layer of spin on glass (SOG) film to fill in the gaps to give a locally planarized surface (39-41). The SOG layer, which can be deposited at low temperatures, can provide the dielectric properties of SiO_2 and the planarizing properties of a liquid.

SOG represents a family of materials that consists of a slilicon-oxygen backbone to which are attached different side groups. If the attached side groups are inorganic, such silicon, oxygen and hydroxyl groups, then they are called silicates. If the side groups are organic, such as methyl, ethyl and phenyl, then they are called siloxanes. SOG materials can also be doped with P and B to give film properties similar to PSG or BPSG. These materials are dissolved in solvents, such as alcohols and ketones, to form various SOG compositions. The SOG material is spin coated onto the wafer and then thermally annealed to evaporate the casting solvent and to densify the film. With high enough annealing temperatures, such as 900°C or higher, SOG film can be converted into amorphous SiO_2. An SOG film formed with an annealing temperature of 450°C or less, a limitation of Al metallization, is composed of silicon-oxygen back bone with some side groups attached to it. The physical and the dielectric properties of the SOG films cured at relatively low annealing temperatures changes not only with the type of SOG material but also with its preparation conditions.

4.1   Silicate SOG

Silicate SOG materials contain no organic side groups. It is prepared in the solution form by hydrolysis of organic silicon source materials such as TEOS,

$$[9] \quad Si(OEt)_4 + 4\ H_2O \xrightarrow{\ H^+\ } 4EtOH + Si(OH)_4$$

In practice, various degrees of hydrolysis can take place to give intermediate products such as $Si(OEt)_3OH$, $Si(OEt)_2(OH)_2$, etc. After spin coating, an annealing cycle will change it into a three dimensional polymer of SiO_2, with essentially every Si-atom bonded to four O-atoms,

$$[10] \quad \text{-Si-OH + HO-Si-} \longrightarrow \text{-Si-O-Si- + } H_2O$$

With high purity and careful preparation, silicate SOG films are essentially free of any metallic impurity, with Na contamination below the parts per million range. The main contaminants for films cured at low temperatures are residual organic materials such as ethoxy groups (OEt) and hydroxyl groups (OH). A schematic diagram of the three dimensional net work formed after a 425°C cure is shown in Figure 11, together with possible broken bonds and attached (OH) groups. The silicate SOG can also be doped with P to give phosphosilicate like films (42). The integrity of the silicate SOG films varies greatly with the preparation conditions. The first encountered problem in using SOG is likely to be particulates because gel or SiO_2 particles can easily form in the SOG solution according to the reaction given in Equation (10), even at room temperature. The silicate SOG solutions, even in sealed bottles, have a finite shelf life due to gel or particle formation. The easiest way to prolong the shelf life of SOG is to keep it refrigerated. After opening the SOG bottle, solvent evaporation will enhance particle formation, particularly at the mouth of the bottle. This problem can be minimized by keeping the solution covered and using a small amount of SOG so that the bottle is replaced frequently. A more troublesome source of particle generation happens during the spin coating operation. The excess SOG solution spun onto the wall of the spinner bowl dries quickly to generate particles. This problem is solved by using a special SOG spinner that has a built-in mechanism to rinse the tip of the SOG dispensing nozzle and the surface of spinner bowl with a solvent after each spin coating so that the excess material is rinsed away before it can be dried to form particles.

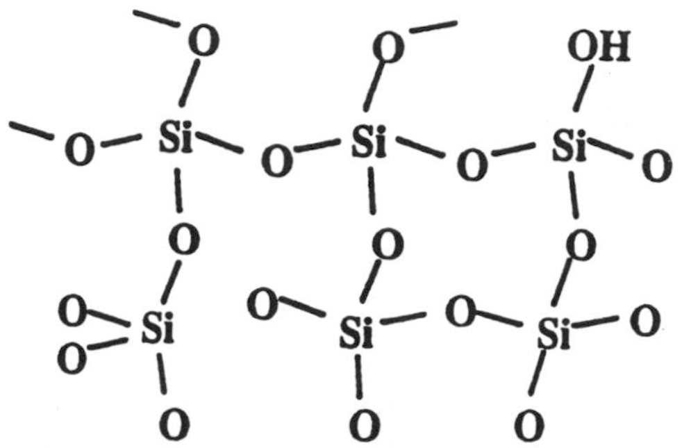

Figure 11. Schematic representation of silicate SOG material after 425°C cure.

In order to get a dense film, the coated SOG must be properly cured. This is usually done with a series of baking steps at approximately 100 C, 250 C and 425°C on hot plates. These baking temperatures are chosen to bake out the solvent, moisture, and ethoxy groups respectively. These hot plates are usually an integral part of the SOG spinner track so that the spin-bake sequence can be carried out automatically. Other forms of energy such as UV curing, plasma curing, and rapid thermal curing have been tried to cure SOG films. The merits of other curing methodologies over properly designed conventional hot plate baking are not yet clearly established.

Even with proper spinning and curing, the silicate SOG films still have many limitations. One of the most important is film cracking which limits the use of silicate SOG to only a very thin layer, often 100 nm or less. Film cracking gets worse when SOG is coated over severe surface topography. Film cracking can be reduced by adding P dopant in SOG to reduce the film stress. Another common method to achieve a thicker film is to use multiple coatings where a very thin layer, typically 100 nm thick or less, is first coated, cured, then followed by second layer coating and curing, and so on. The multiple coating process, although straightforward, adds a lot of process steps with only limited improvements. A better way to get thicker SOG films is to formulate SOG materials that are inherently more cracking resistant. The siloxane materials, which will be discussed next, were developed to give better coating properties. Due to its thickness limitation, the silicate SOG layer is almost always used in conjunction with other CVD or PECVD dielectric layers for ILD applications (43-44).

The major reason why the silicate SOG films crack easily is that the film is rather porous. The density and hardness of silicate SOG films are poor compared to SiO_2 grown at high temperatures. Porous SiO_2 films also tends to absorb a large amount of moisture which leads to device stability problems. Moisture content in SOG films is a serious problem and will be discussed separately in the section on ILD.

4.2   Siloxane SOG

Various organic side groups can be attached to the Si-O backbone in order to alter the properties of the SOG material. The common organic side groups, R, are methyl ($-CH_3$), ethyl ($-C_2H_5$), or phenyl ($-C_6H_5$) groups in various amounts. The purpose of adding these side groups is to reduce the film stress, particle generation and moisture absorption. For example, when methyl side groups are added in moderate amount to the starting material, the reaction can be represented by,

[11] $Si(OEt)_4 + MeSi(OEt)_3 + H_2O \longrightarrow EtOH +$
$Si(OEt)_3OH + MeSi(OEt)_2OH + MeSi(OEt)(OH)_2$
$+ \ ....$

When every starting molecule of the Si source material has an organic radical attached to it then the reaction can be represented by,

[12] $MeSi(OEt)_3 + H_2O \longrightarrow [MeSiO_{1.5}]_n$

This material is schematically represented in Figure 12. It is a form of silicon polymer often called silsesquioxane. Similar to the silicate material, the siloxane materials can also be doped with phosphorus to give phospho-siloxanes. The selection criteria for the type and amount of attached organic radicals to optimize the siloxane material for a particular application is beyond the scope of this chapter. Major manufacturers of SOG materials usually offer a variety of formulations for different applications. Some formulations of siloxane material have lower dielectric constants (45-47), a very attractive feature for high speed circuits. Siloxane materials can give thicker film coatings without cracking than silicate materials. Spin and bake conditions are similar to those used for the silicate materials. As in the case of silicate materials, multiple coatings are often used to prepare thicker siloxane films. Siloxane films usually contain less water than silicate films, however, the integrity of the siloxane material can be damaged by subsequent processing. Excessively high temperatures or oxygen plasma can destroy the carbon structures in the siloxane film resulting in a very porous structure. The damaged siloxane film is prone to film cracking and moisture absorption. For example, it has been reported that the carbon content in the siloxane film leads to device instability (48), but the real cause is more likely to be the destruction of the carbon structure in the siloxane film that leads to high moisture absorption. The issue of water absorption will be discussed in the section on ILD because siloxane SOG films have been used extensively as a part of ILD dielectric layers. Cracking resistance of many siloxane materials are still marginal, particularly when coated over severe surface topography. It is possible to improve cracking resistance by additional treatments such as sylilation (49) or fluorination (50).

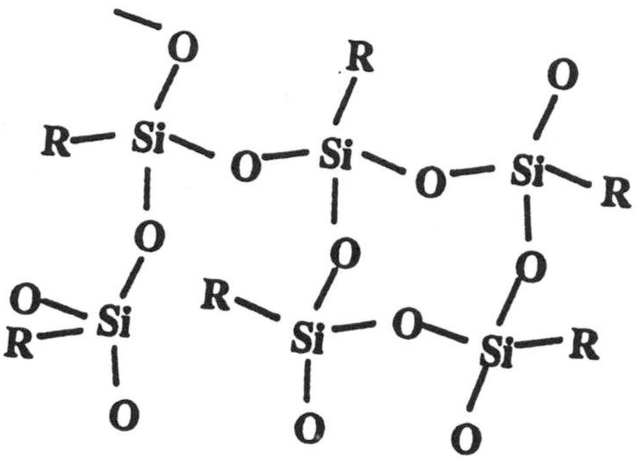

R = CH₃-(methyl),
C₆H₅-(phenyl)

$$R = CH_3\text{-(methyl)},$$
$$C_6H_5\text{-(phenyl)}$$

Figure 12. Schematic representation of a siloxane (silsesquioxane) SOG material after 425°C cure.

The SOG films, whether silicates or siloxanes, often have to be etched in device processing with other oxide films deposited by CVD or PECVD process. The wet etch rate of SOG films in HF solutions is usually considerably different from the CVD oxide films. The plasma etch rate of various SOG films is, however, approximately the same as CVD oxide films when proper plasma chemistry and conditions are used. With siloxane materials the subsequent processing conditions need to be carefully controlled in order to preserve its integrity.

SOG materials, whether silicates or siloxanes, do not possess all required properties of a stand alone ILD layer. They are often used as a smoothing layer between two PECVD oxide layers. Because of its high moisture content, the SOG layer is frequently used in an etchback process leaving SOG material only in isolated pockets for gap filling to minimize its impact on device performance (39, 51). With proper care, non-etchback SOG process have been developed as a part of multilayer ILD structures (52, 53).

5.0   FILM CHARACTERIZATION

The deposited dielectric film must satisfy the stringent requirements of VLSI technology, such as thickness control, dopant concentration control, reproducibility, low moisture content and good step coverage, etc. A variety of measurement techniques are used to qualify the deposited films properties. A list of commonly monitored film properties and their measurement techniques are listed in Table III. Standard analytic techniques are discussed in Chapter 11. In this section we will discuss the origin, measurements and the effect of dielectric film stress and moisture content. Both stress and moisture are often overlooked by standard testing procedures even though they have serious impacts on the device reliability. With high purity source chemicals and proper processing conditions, very little metallic or organic impurities are in the deposited films. The most common impurities in the deposited dielectric films are the ubiquitous moisture and related hydrogenous species. Hydrogenous species are difficult to measure and are highly mobile even at relatively low temperatures. The mobile hydrogenous species have a significant effect on device operations. Film stress and moisture content are interrelated because highly stressed films tend to absorb moisture from the ambient in order to relax the film stress. Both film stress and moisture content are a function of the thermal history of these films.

5.1   Stress Measurements

Mechanical stresses in deposited thin films have been studied extensively (54-57). Stress is a major failure mechanism of deposited thin films since high stresses cause film cracking and delamination. High stresses in the dielectric films also cause reliability failures in the underlying aluminum conductors such as stress induced voids and stress migration failures (58-60). In extreme cases, high film stresses can even cause failures in the single crystalline silicon substrate, such as dislocations, slip lines, and cracking. Many methods have been developed for measuring stress. The most common technique is to use an optical level to measure the wafer curvature (54). The stress of the thin film deposited on a wafer will cause the wafer to bend. The stress in the deposited thin film, σt, can be determined by measuring the change in wafer curvature before and after the film deposition,

$$[13] \quad \sigma t = \frac{E_s}{6(1 - V_s)} \frac{T_s}{R\, T_f}$$

where T_f is the film thickness, R is the radius of curvature, T_s is the Si substrate thickness, E_s is the Young's modulus of Si (with a value of $1.689*10$ N/m), and V_s is the Poisson's ratio of the substrate (with a value of 0.064). The stress is tensile if the curvature on the film side is concave and compressive if convex. A schematic diagram of the laser cantilever beam system for stress measurement is shown in Figure 13 (61), where the radius of curvature, R, is obtained by measuring the translation in the detected signal, dD, for a given wafer translation, dW,

$$[14] \quad R = 2L \; \frac{dW}{dD}$$

If the wafer is placed in an enclosed heating chamber, the stress measurements can be carried out in-situ during thermal cycling either in air or inert N_2 ambient.

Table III. Film Characterization Techniques

Thickness
  Step Height
  Ellipsometer
  Optical Interference

Dopant Concentration
  IR Absorption Spectroscopy
  X-ray emission (e⁻ probe)
  X-ray fluorescence

Surface Roughness
  Optical Microscopy
  SEM

Step Coverage
  Cross sectional SEM and TEM

Impurities
  SIMS (Secondary Ion Mass Spect.)
  Auger e⁻ analysis
  Wet Chemical analysis

Stress
  Wafer curvature
  X-ray diffraction

Moisture
  IR Spectroscopy
  MEA (Moisture Evolution Analysis)
  UHV Thermal Desorption

STRESS MEASUREMENT

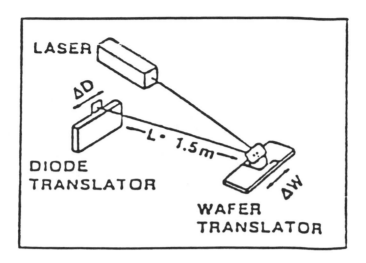

Figure 13. Schematic of a canterlever laser beam system for film stress measurement (61), reprinted with permission of the publisher.

A typical stress versus temperature measurement of a CVD oxide is shown in Figure 14 (62). During the first heating cycle, the stress increasingly becomes tensile due to the difference in the thermal coefficient of expansion (TCE) of the oxide film and the silicon substrate. During cooling, the stress decreases fairly linearly with temperature. The final stress is considerably more tensile than at the start of the temperature cycle producing a hysteresis in the stress versus temperature curve. The size of the hysteresis is an indication of the amount of film densification by annealing out either the vacancies or the moisture in the film. Since oxide films can not flow in this low temperature range to eliminate vacancies, it must be the moisture in the film being annealed out. The size of the hysteresis is, therefore, an indication of the moisture content and the stability of the oxide film. The second cycle of the stress versus temperature measurement has no hysteresis because the time between the two measurements is too short for the incorporation of moisture from the ambient into the film.

As deposited CVD oxide films are usually dry, moisture is incorporated into the film by absorption from ambient after the film deposition. The measured stress of an oxide film decreases with the exposure time in the room environment because the absorbed moisture provides the volume expansion needed to decrease the tensile stress of the film. A compressive film becoms more compressive. This is illustrated in Figure 15 (61). The speed and the magnitude of stress change after exposing the deposited oxide to the ambient is a good indication of the oxide quality.

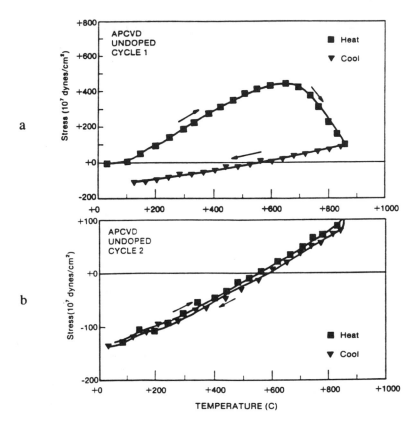

Figure 14. Stress versus temperature measurements of an APCVD undoped SiO_2 film, a) First annealing cycle, b) Second Annealing cycle (62)., reprinted with permission of the publisher.

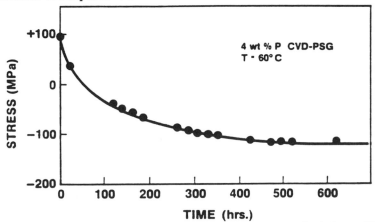

Figure 15. Stress versus storage time measurement of a typical CVD SiO_2 film (61), reprinted with permission of the publisher.

The wafer curvature method is only capable of measuring stress for blanket films deposited over flat wafers. It is not suitable for situations where stress concentrations generated by surface topographies plays a major role, such as stress induced void formation in patterned Al conductors. In this case, X-ray diffraction techniques are used to measure the stress in metal lines under the passivation layer (63, 64), or numerical calculation using finite element methods can be used (65, 66).

5.2 Moisture Measurements

It is well known that all CVD oxide films contain some moisture. The amount of moisture in the film, however, varies significantly with the deposition conditions. Moisture in the oxide films contributes not only to dielectric film failures such as film cracking and delamination but also to device instability such as trapped charge in gate oxide and at the oxide-Si interface (67). Therefore, moisture content in the deposited dielectric film is a major reliability concern. It should be carefully measured and minimized by optimizing the dielectric structures. Some of the common techniques for moisture measurements are described below.

FTIR Measurements: Chemical bonding structures in the deposited dielectric films can be determined by using a FTIR spectrometer. In FTIR measurements, the IR intensity spectrum after passing through the film are measured to determine the characteristic absorption of a specific chemical bonding structure at a particular wavelength. SiO_2 gives a sharp absorption peak at approximately wave number 1070 cm^{-1} plus two other smaller peaks located at wave numbers 810 and 445 (68). Moisture gives an absorption peak at 3650 for the Si-OH bond and a very broad peak stretching from wave number 3700 to 2700 for loosely bonded water molecules (68). Figure 16 is a FTIR spectrum of a wet CVD oxide film with about 5 wt % moisture content. The relative amount of moisture in the film is obtained by graphically integrating the areas under the moisture peak and the SiO_2 peak. Quantitative measurement can be achieved by calibrating the FTIR results to a direct measuring method, such as Moisture Evolution Analyzer (MEA) or thermal desorption spectroscopy (TDS). With multiple measurements, accurate calibration and computer data analysis, moisture content as low as 0.2 wt % in one micron thick oxide films can be determined with FTIR measurements.

Moisture Evolution Analyzer: Quantitative measurement of the moisture content in thin films is possible with a moisture evolution analyzer (69). A schematic diagram of the measurement system, based on a commercial DuPont 902H Moisture Evolution Analyzer system, is shown in Figure 17 (a). The sample is placed in a heated chamber with dry nitrogen flow; moisture in the film is baked out at elevated temperatures and swept out by the dry nitrogen carrier gas into an electrolytic cell detector. The detector has a pair of helical electrodes wound on a phosphorous penta-oxide coated surface. This surface will absorb the moisture from the carrier gas and then electrolyze it to give a current flow between the helical electrodes. This current is easily monitored and integrated to give the total amount of water electrolyzed. Integration is necessary because the absorbed moisture is consumed during the electrolyzing process resulting in a current that changes with time. An expanded view of the detector is shown in Figure 17 (b). The system is interfaced with a personal computer so the moisture evolution rate, temperature and elapsed time are monitored simultaneously. The temperature ramping rate is controlled by a rheostat and temperatures in excess of 600 C can be readily reached. With this setup, moisture content of approximately 10 ppm in one micron thick oxide film can be measured quantitatively.

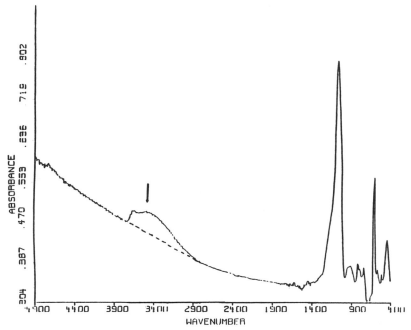

Figure 16. FTIR spectrum of a very wet CVD oxide film.

UHV Thermal Desorption Spectroscopy: The measurement of moisture content in deposited thin dielectric films can be carried out using thermal desorption mass spectroscopy (TDS) in a UHV chamber (70-72). In principle, any vacuum chamber equipped with a heating stage and mass spectrometer can be used for TDS measurement. However, the background moisture, which is present ubiquitously on all chamber surfaces, must be minimized in order to get good accuracy and sensitivity. Therefore, a UHV chamber with a vacuum loadlock is often needed for this purpose. In addition, the sample holder should remain cool or be thoroughly baked out to prevent excessive out-gassing during the measurement. Many research laboratories have custom built UHV desorption systems to carried out these measurements; one such system is shown schematically in Figure 18 (72). This is a sophisticated system equipped with turbomolecular pumps capable of 10^{-9} Torr working pressure with a compact type double focused mass spectrometer located very close to the sample. A computer controls the heating unit and the mass spectrometer for dynamic thermal desorption studies. Even though the sensitivity of UHV TDS is approximately the same as MEA, it has the advantage of being able to detect moisture and other volatile contaminants, such as hydrocarbons, in the deposited film. However, for quantitative measurements, elaborate calibration procedures are needed.

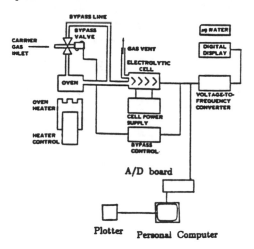

(a)

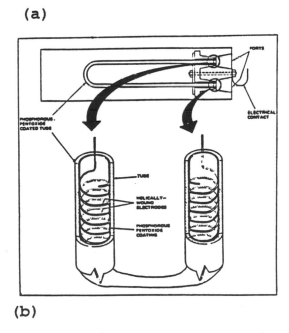

(b)

Figure 17. Schematic diagram of a Moisture Evolution Analysis (MEA) system. a) Overall system, b) Expanded view of the detector cell (69), reprinted with permission of the publisher.

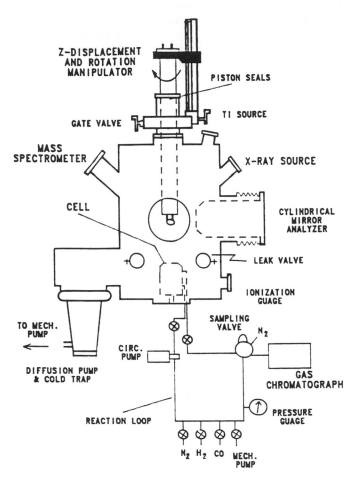

Figure 18. Schematic diagram of an UHV Thermal Desorption Mass Spectroscopy (TDS) system (72), reprinted with permission of the publisher.

6.0   DOPED GLASSES

The first deposited dielectric layer after the formation of transistors (ILD0) is a thick doped SiO_2 or doped glass layer. It is used as an insulator between polysilicon gates and the metallization levels. Contact holes are opened in this dielectric layer in order for the metallization pattern to contact the transistor structures below. Phosphorous dopant is incorporated into the deposited oxide to give phosphosilicate glass (PSG). PSG is favored for this application because of its capability to getter mobile alkali ions (73) and because the softening temperature of PSG is significantly lower than pure SiO_2. By using a high temperature annealing cycle, the dielectric surface and contact holes can be smoothed by the thermal flow process (74,75) in order to facilitate subsequent processing steps. For applications that require even lower flow temperatures, boron is added to PSG to give borophosphosilicate glass (BPSG)(76-78). Although SiO_2 doped heavily with boron (BSG), has even lower flowing temperatures, it is seldom used because it lacks gettering capability of mobile alkali ions. The thermal reflow process is covered in Chapter 6 as a part of the planarization process. This section focuses on the incorporation of phosphorous and boron dopant into the oxide films and their effect on the film properties. Some typical doped glass compositions and their flow temperatures are listed in Table IV.

Table IV. Typical Doped Silicate Glass Compositionsand Flow Temperatures

	WT %		FLOW TEMPERATURE (°C)	
GLASS	P	B	DRY N2	STEAM
SiO_2	0	0	1725	1600
PSG	8.5	0	1000-1050	950-1000
BSG	0	6.5	825-850	800-825
BPSG A	4	4	925-950	900-950
BPSG B	4	5	825-850	800-825
BPSG C	3	5.7	800-825	750-825

6.1   PSG

Deposition Process: The chemical reaction for depositing PSG is the same as CVD oxide deposition with an additional phosphorus doping gas,

$$[15] \quad 4 \, PH_3 + 5 \, O_2 \longrightarrow 2 \, P_2O_5 + 6 \, H_2$$

This can be carried out in APCVD reactors, LPCVD reactors or PECVD reactors. The P concentration in the deposited film can be adjusted by changing the PH_3/SiH_4 ratio in the gas flow. The result is an approximately linear relationship as shown in Figure 19 (79).

The deposition rate of PSG depends on the silane, oxygen and phosphine flow rates. It increases initially with the oxygen flow rate, reaches a maximum, and then deceases at higher oxygen flow rates (80). The initial increase in the deposition rate is caused by the oxidation enhancement of silane in the presence of oxygen. Similar to the oxide deposition case discussed in Section 2.2, the decrease in deposition rate at high oxygen flow is due to an overabundance of adsorbed oxygen on the substrate surface which retards the adsorption of Si reaction species. The deposition rate first decreases and then increases with the phosphine flow rate (79). The decrease in deposition rate with phosphine flow rate can also be explained with the same retardation theory discussed earlier.

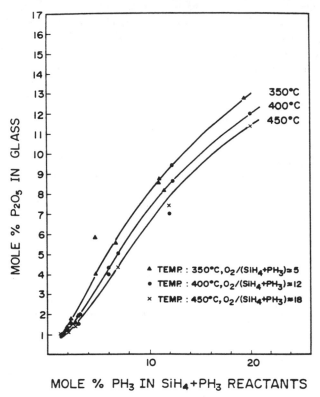

Figure 19. Composition of PSG films as a function of hydride concentration and the hydride/silane mole ratio (79), reprinted with permission of the publisher.

Measurement of P Content: The phosphorous content in the PSG films can be determined by FTIR measurements. The P=O bonds give an absorption peak at wave number 1330 cm^{-1} (81) as shown in Figure 20. The absorption ratio

[16] $R = \log (I_p / I_{po}) / \log (I_{si} / I_{sio})$

where I_p and I_{po} are the phosphorous bond absorption peak and baseline located at 810cm^{-1}; I_{si} and I_{sio} are the silicon bond absorption peak and baseline located at 1330 cm^{-1} as indicated in Figure 21 (a) The absorption ratio can be calibrated to give the phosphorous concentration in the film as indicated in Figure 21 (b). The value of this absorption ratio, R, has been calibrated (82) to give the mole % of P_2O_5 in the oxide film as shown in Figure 21. It must be cautioned that the P absorption peak at 1330 cm^{-1} is for P=O bonds only. For films deposited at low temperatures not all P in the film is in the penta-valance state, for example it could be in the four-valance state.

```
         O
  |    |     |
-Si-O-P-O-Si-
  |    ||    |
         O
```

A high temperature annealing step, typically at 900°C, is often used to convert all phosphorous (those in tri-valent or four valence states) into phosphorous penta-oxide. Since phosphorous penta-oxide in the film is a good desiccant, undensified PSG film can absorb water molecules from the ambient changing phosphorous penta-oxide to hydride according to the reaction:

```
         O                           O O-H
         |                           | /
[17]  -O-P=O + H₂O ------------> -O-P
         |                           | \
         O                           O O-H
```

$$[17] \quad -O\text{-}P{=}O + H_2O \longrightarrow -O\text{-}P\begin{array}{l} O \; O\text{-}H \\ \\ O \; O\text{-}H \end{array}$$

with a corresponding decreasing in the P=O bond absorption peak height at 1330 cm^{-1} in FTIR measurements. Therefore, phosphorous concentrations should be measured immediately after the high temperature annealing step to avoid the possibility of significant change in the P=O bond concentration cause by water absorption. Water absorption of PSG films will be discussed in the next section.

Another common method for measuring P content is X-ray fluorescence analysis (83). With this method, the sample is irradiated by an intense X-ray beam and the secondary emission from the sample is analyzed for the characteristic phosphorous emission. The advantages of method are that it is a non-destructive method and only a small sample area is needed. It can measure the total P content regardless of the chemical bonding state. Commercial instruments are available and easy to use. The disadvantages are that suitable substrates that contain no P must be used and calibration standards are needed to determine the amount of P in the film.

Water Absorption: Since the softening temperature of the doped glass film is lower for higher P concentration, as shown in Table IV, the tendency is to use the highest P doping concentration possible. However, there is a practical limit of P concentration because too high a P content will result in a hygroscopic glass, which readily absorbs moisture from ambient. Absorption of water will cause phosphorous oxide to change into phosphoric acid which can cause leakage current and aluminum corrosion. Therefore, the stability of PSG in a humid environment is an important reliability concern.

Hydration of PSG films was studied by measuring the phosphorous content of PSG films with variable amounts of P content before and after moisture testing (84). To accelerate the testing time, the moisture test was conducted by exposing the wafers to the saturated water vapor in a steam pot, at 120 C, with a steam pressure of 2 atmospheres, for 4 hours. It was found that for PSG films with P concentration above a certain value the P content was mostly leached away after the moisture test. This concentration value is called the critical concentration. The critical concentration was about 8 mol % for as deposited films and increased to about 12 mol % for PSG films annealed at 900 C, for 15 minutes, in N_2. This is shown in Figure 22. The P concentration in the PSG film should be kept below the critical concentration in order to maintain stability. The critical concentration value is not fixed but rather depends on the thermal history of the film such as the deposition temperature and annealing temperature. Different deposition reactions that can effect the P bonding states in SiO_2 matrix will also give different values of phosphorous critical concentration.

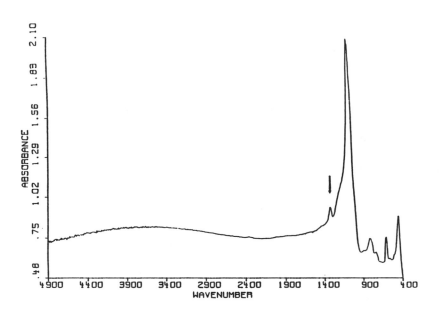

Figure 20. FTIR spectrum of PSG showing the P absorption peak at wave number 1330 cm^{-1}.

6.2   BPSG

The continuing decrease in device dimensions requires a corresponding decrease in the thermal budget of the fabrication processes. The PSG flow temperature is approximately 950°C for P concentration near the critical concentration. This is still too high a thermal budget for submicron CMOS fabrication. To achieve even lower flow temperatures, boron doping is added to the PSG deposition process to get borophosphosilicate glass or BPSG films with a flow temperature of 850°C or lower.

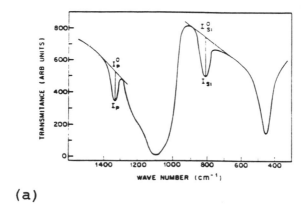

(a)

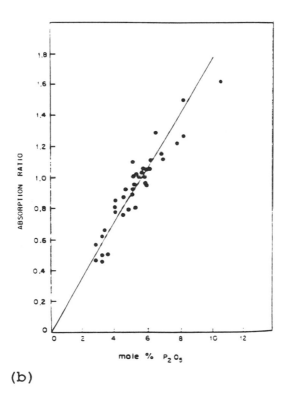

(b)

Figure 21. P concentration in PSG film as a function of absorption ratio of P peak and SiO_2 peak at wave number 810 cm^{-1}. a) Absorption peaks, b) Phosphorous content vs. absorption ratio.

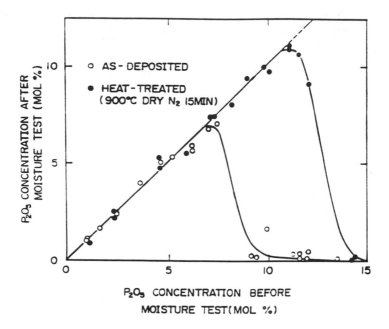

Figure 22. P_2O_5 concentration in PSG film before and after moisture test (steam pot at 120 C, 4 hours) for as deposited film and annealed film (900 C, for 15 minutes, in N_2) (84), reprinted with permission of the publisher.

Deposition Process. Deposition of BPSG can be achieved by the co-oxidation of silane (SiH_4), phosphine (PH_3) and diborane (B_2H_6) with oxygen and an inert diluent gas, usually N_2. The reactions are similar to the PSG deposition process, with the addition of the diborane oxidation reaction:

[18] $2 B_2H_6 + 3 O_2 \longrightarrow 2 B_2O_3 + 6 H_2$

The overall deposition rate is lowered significantly with the addition of B_2H_6, indicating a strong retardation effect on SiH_4 oxidation by B_2H_6. The boron concentration is approximately proportional to the fraction of B_2H_6 in the gas mixture. The deposition can be carried out in APCVD, LPCVD or PECVD reactors. Infrared studies have shown that the boron is incorporated into the silicate matrix as boron oxide (85),

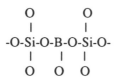

The flow temperature of the BPSG depends on the concentration of both B and P in the film, but is enhanced more strongly by the B concentration than the P concentration as indicated in Table IV.

Measurement of B Content: Infrared absorption spectroscopy can be used to determine the composition of BPSG. The B-O bond has a strong absorption peak at 1370 cm^{-1} (86). Boron concentration in BPSG can be determined by measuring the absorption ratio of the B-O to Si-O peaks. Similarly, the P concentration in BPSG can be determined by measuring the P absorption peak at 1330 cm^{-1}. This is illustrated in Figure 23. FTIR measurements should be carried out after densification of the BPSG film since the high temperature densification cycle will improve the sharpness of the P=O peak as discussed in the PSG section. The time delay between densification and FTIR measurement should be minimized to avoid the possibility of moisture absorption. The dominant B peak is located very close to the P peak and there is considerable overlap of these two peaks. This makes accurate measurement of P concentration more difficult. While FTIR measurement may be good enough for determining B concentration, more accurate measurements of P concentration in BPSG can be obtained using the X-ray fluorescence spectrum as discussed earlier.

Phase Relationship: BPSG films were developed in order to achieve a lower flow temperature than attainable with PSG. Actually, heavily doped boro-silicate glass or BSG can give an even lower flow temperature than BPSG. BSG, however, does not have the gettering capability for alkali ions, and heavily doped BSG is highly hygroscopic (87). Therefore, BSG is seldom used; instead, BPSG is carefully formulated to retain the advantages of both BSG and PSG. Optimization of BPSG requires the understanding of the tertiary B_2O_3-P_2O_5-SiO_2 system. Figure 24 is the tertiary phase diagram together with the flow or fusion temperatures of a few specific BPSG compositions (88). For comparison, the fusion temperature of one BSG and one PSG composition are also shown. Note that the flow of BPSG or any other doped glasses depends not only on the temperature but also on the flow ambient. For example, the flow temperature can be lowered about 30 C if flowed in a steam ambient, or 60 C if flowed under 10 atm of pressure. In addition, the amount of glass flow also depends on the length of time at the flow temperature as well as the pattern geometries on the wafer.

Organic Doping Sources: Even after replacing the hazardous SiH_4 with organosilicon precursors, such as TEOS and TMCTS, the doped glass deposition process is still a safety concern because the doping gases, phosphine (PH_3) and diborane (B_2H_6), are highly toxic. Alternative non-toxic doping sources are needed to reduce the safety concerns of the deposition processes. Similar to the silane case, this can be achieved by using liquid organic doping sources. Liquid organic doping sources such as trimethylphosphate (TMP), triethylphosphate (TEP), and trimethylborate (TMB), are used with the liquid organo-silicon source TEOS to deposit BPSG by LPCVD and PECVD (89, 90) to give more compatible and safer alternative processes. The chemical structure of these materials are shown in Figure 25.

In order to improve the dopant stability in humid environment and the gettering capability of the doped glass, it is desirable to control not only the dopant concentration but also the dopant bonding state in the silicate glass film.

This has lead to the investigation of organosilicon doping precursors, such as tris(trimethylsiloxy)boron and tris(trimethylsiloxy)phosphorous, in which the doping atoms, such as boron or phosphorous, are already bonded directly to the Si-O matrix. The chemical structure of these materials are shown in Figure 26. These molecules have relatively strong Si-O-B and Si-O-P back bones. The Si-O-B and Si-O-P bonds are believed to be largely retained in the deposition reactions to give doped glass films with stable boron and phosphorous bonds. The films deposited with these new doping sources have better surface morphology and the dopant concentrations are more stable in humid environment (91-93).

Water Absorption. Similar to the PSG case, water absorption in heavily doped BPSG leads to device reliability problems. For BPSG, high concentration of either B or P leads to highly hygroscopic films. Similar to PSG, BPSG films with very high P content, above 9%, will result in P leaching out of the film after the steam pot test. Even for lower P content films where P leaching is not a problem, BPSG films still can absorb a significant amount of water from the ambient. The amount of absorbed water is a complex function of both B and P concentrations.

For undensified BPSG films, moisture can penetrate the films readily at room environment. Extended exposure of undensified films to humid room environment can cause surface devitrification which becomes evident by the formation of crystals on the surface. These crystals, often called rocks, are a separate phase of the boron, phosphorus and silicon oxide as indicated by the BPO_4-SiO_2 line in the tertiary phase diagram given in Figure 24. Device wafers coated with BPSG should proceed to densification as soon as possible or be kept in a dry ambient, such as a desiccator, in order to avoid rock formation. To avoid the instability problems of undensified BPSG and to minimize the thermal budget of glass flow, a process capable of depositing BPSG films near the flow temperature of the glass has been developed to give in-situ densification (94).

Even for densified BPSG films, water absorption may still be a problem, particularly under steam pot conditions. For example, water absorption in densified BPSG films with different B and P concentrations was measured under steam pot conditions (120 C, 2 atm steam) (95). For BPSG film with a fixed amount of B but variable amount of P, the amount of absorbed water decreases with increasing P content as shown in Figure 27. This is surprising because PSG films have higher absorbed water for film with higher P concentrations. For BPSG films with fixed P concentration but variable B concentrations, the amount of absorbed water increases with B concentration as shown in Figure 28. This behavior was clarified by measuring the water concentrations near the film surface region and the depth of water penetration into the films (95). The water content at the surface region was found to increase with the B concentration. Higher surface concentration would allow more water to diffuse into the BPSG film. The water penetration depth was found to decrease with higher P concentration. Therefore, the decrease in water absorption of densified BPSG with increasing P concentration can be explained by assuming higher P doping gives a denser film after densification thereby reducing water absorption by retarding the water diffusing rate into the film.

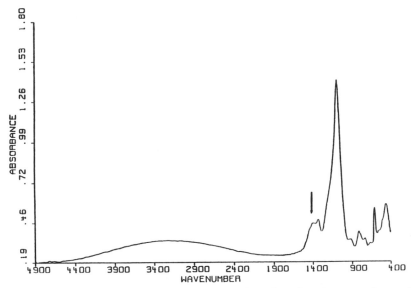

Figure 23. FTIR spectrum of BPSG showing the phosphorous absorption peak at 1330 cm^{-1} and the boron absorption peak at 1370 cm^{-1}.

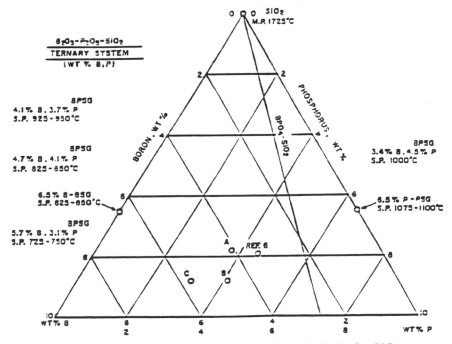

Figure 24. Ternary compositional diagram of the B_2O_3-P_2O_5-SiO_2 system with fusion temperature range in dry N_2 indicated for a few film compositions (88), reprinted with permisiion of the publisher.

TMP = trimethyl phosphate
= PO(OCH3)3

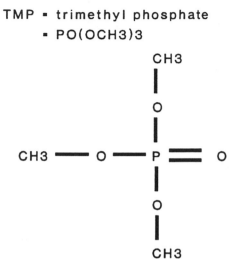

TMP = trimethyl borate
= B(OCH3)3

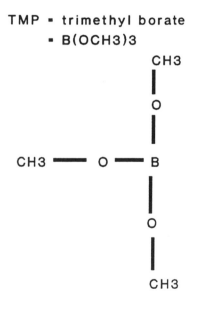

Figure 25. Chemical structure of organic liquid doping sources TMP and TMB.

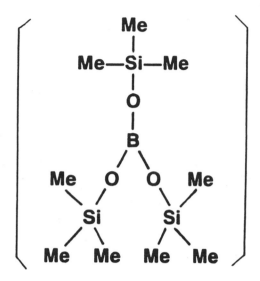

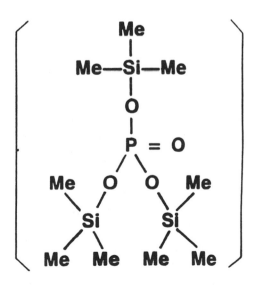

Figure 26. Chemical structure of organosilicon doping sources TTMSB and TTMSP.

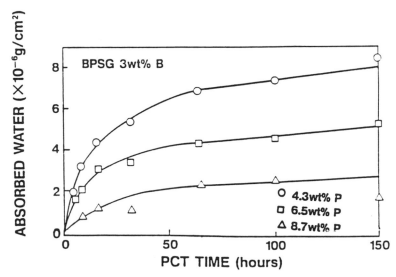

Figure 27. Water absorption of densified BPSG (at 900 C, for 15 minutes, in dry N$_2$) under steam pot conditions at 120 C, at 2 atm steam, (95)., reprinted with permisiion of the publisher.

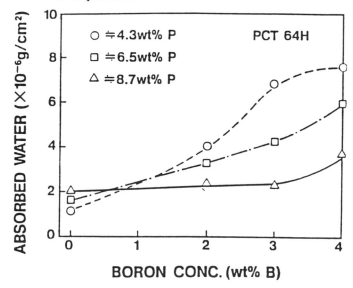

Figure 28. Water absorption in densified BPSG with fixed P but variable B concentrations (95), reprinted with permission of the publisher.

Because of the instability of highly doped BPSG films in a humid environment, very high concentrations of either B or P can not be used. Typical BPSG films are optimized around B and P concentrations of 4% to give a flow temperature of approximately 850 C.

7.0   INTERMETAL LAYER DIELECTRIC

Intermetal layer dielectric (ILD) provides the needed isolation between metallization levels. This simple task is rather difficult in practice because the ILD layer must satisfy some very stringent requirements. The most important requirement for ILD is low deposition temperatures, usually 400°C or lower, because the aluminum metallization has a low melting temperature. Although the horizontal dimensions of aluminum conductor patterns have shrunk along with the device dimensions, the thickness has not shrunk proportionally because of the need to carry the high currents required by VLSI chips. Even for sub-micron devices, the typical thickness of Al layers is still close to one micron. The Al conductor patterns, with narrow spacing and thick Al layer, create topographies with high aspect ratios (depth/width).   When a thick dielectric layer is deposited over these gaps, non-ideal step coverage often results in key-hole or void formation as illustrated in Figure 29. Once voids are formed in the ILD, they can not be removed by subsequent planarization processing steps such as etchback or polishing (38). Therefore, the second most important requirement for ILD deposition is to have good step coverage so that voids will not form even in narrow gaps. Since a high temperature flow step can not be used to improve the quality of the ILD layer, the as-deposited ILD films must be of high quality, and have low moisture content. In addition, the ILD deposition process must meet some manufacturability requirements, such as high deposition rate to keep the cost reasonable.   Also, the process must be optimized to minimize defects, such as particulate generation either from the process itself (from homogeneous gas phase reactions) or from deposited films flaking off the chamber walls.

Void Formation

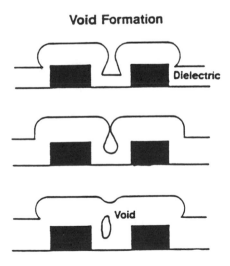

Figure 29. Schematic diagram of void formation for ILD deposited in narrow grooves.

7.1   Dielectric Materials

The most commonly used dielectric material for ILD applications is SiO_2. Doped oxides are sometimes used because they offer lower stress than undoped oxide. Silicon nitride and silicon oxynitride have also been used because of their superior step coverage and moisture barrier characteristics. Silicon nitride and silicon oxynitride, however, have much higher dielectric constants than silicon dioxide, which results in larger line capacitances and larger RC delays for signal propagation. Thus, silicon nitride is used less often as an ILD layer in high speed circuits.

Because the ILD layer must satisfy many difficult requirements, it is unlikely that a single dielectric material or deposition process will have all the desirable properties. Therefore, the ILD layer often consists of multiple layers of different dielectric materials for optimized performance. A very common approach is to use a multilayer ILD structure that includes a spin-on glass layer for step coverage and PECVD oxide layers for low moisture content (39-41).

7.2   Deposition Reactions

Various oxidation reactions of silane in APCVD, LPCVD, or PECVD reactors have been used for ILD applications because of their low deposition temperatures. The step coverage of silane oxides, however, is insufficient for fabricating sub-micron devices. Oxide films deposited with a liquid organo-silicon source, such as TEOS, have better step coverage, but oxidation of TEOS with oxygen requires a temperature too high for ILD applications. Extensive research has been carried out to find alternative organosilicon precursors that provide good step coverage at lower deposition temperatures. Some promising candidates are liquid organosilicates such as TMCTS, DES, DABS, etc. (96).

Although the search for new chemical sources with lower deposition temperatures is quite promising, it may be easier to achieve lower deposition temperature with a plasma enhanced reaction, or PECVD process. Because of their low deposition temperatures, PECVD oxide, nitride and oxynitride films have been used extensively for ILD applications. The properties of PECVD films are more complicated because they change not only with the standard CVD parameters but also with the plasma conditions. PECVD oxide films deposited from TEOS and oxygen have been widely used because they give better step coverage than films deposited with silane. The reaction is described by,

$$[19] \quad Si(OC_2H_5)_4 + O_2 \xrightarrow{\text{plasma}} SiO_2 + \text{by-roducts}$$

The film properties and step coverage characteristics of PECVD TEOS oxide are discussed in Section 7.3 and 7.4.

Another way to lower the deposition temperature of TEOS oxide is to use ozone instead of oxygen in the oxidation reaction,

[20] $Si(OC_2H_5)_4 + O_3 \longrightarrow SiO_2$ + bypoducts

This TEOS/ozone reaction has much higher deposition rates at low temperatures than the TEOS/oxygen process as shown in Figure 30 (92). TEOS/ozone oxide can be deposited in a LPCVD reactor at 400 C or lower and is often called THCVD TEOS oxide (23, 24) to differentiate it from PECVD TEOS oxide. TEOS/ozone oxide can also be deposited at or near atmospheric pressure and is often called APCVD TEOS (92, 97, 98) or SACVD (sub-atmosphere) TEOS oxide (99, 100). Other organic Si sources, such as TMCTS, OMCTS, HMDS, etc., can be used instead of TEOS in the ozone oxidation process (101). However, the oxide film properties deposited from these newer organic sources are not as well characterized as the TEOS oxide.

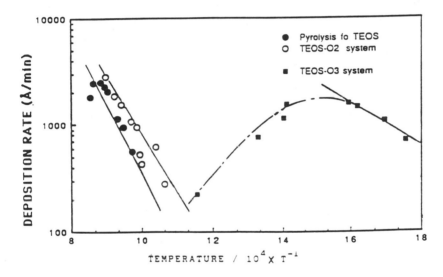

Figure 30. Deposition rate of TEOS-Ozone oxide as a function of temperature as compared to TEOS-Oxygen processes (92), reprinted with permission of the publisher.

7.3   Step Coverage

Due to the severe surface topography created by the etched metallization patterns, one of the most difficult requirements for ILD deposition to fulfill is conformal step coverage in deep crevices. Conformal deposition over severe surface topography is difficult to achieve because the shadowing effect created by surface topographies. This shadowing effect will limit the solid angle of the reactants that can reach the side walls and the bottom of a trench, resulting in thinner depositions along the side walls and at the bottom of the trench as illustrated in Figure 31. To improve the step coverage, the reactive species must migrate along the surface to eliminate the shadowing effect created by surface topographies.

PECVD Oxide from Silane and TEOS: The migration of reactive species along the substrate surface depends on the nature of the surface and the nature of the reactive species. For example, the step coverage of PECVD TEOS oxide is better than silane oxide (19, 102) as illustrated in Figure 32. Since the deposition conditions and substrate surfaces are the same in both cases, the difference in step coverage for the two films must be due to the different properties of the reactive species. In the case of silane deposition, the adsorbed reactive species, such as SiH_2, react quickly on the surface with oxygen allowing very little surface migration to take place. Under these conditions, the step coverage is rather poor. For TEOS depositions, the reactive species migrate a considerable distance along the substrate surface before they are consumed by the deposition reactions, resulting in better step coverage. Even though the exact mechanisms of the surface migration are not clear yet, computer simulation programs, such as the SAMPL program from U.C. Berkeley (103), can be used to simulate the surface contours of the deposited dielectric films by assuming the reactive species have either short or long surface migration distance. The results are as illustrated in Figure 33. The simulated profiles exhibited essentially the correct features such as the reentrant profile at a step for silane deposited oxide and positive slope for TEOS deposited oxide.

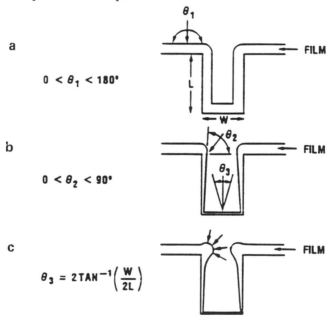

a

$0 < \theta_1 < 180°$

b

$0 < \theta_2 < 90°$

c

$\theta_3 = 2 TAN^{-1}\left(\dfrac{W}{2L}\right)$

Figure 31. Schematic diagram of film step coverage in a deep groove for a) Conformal, b) Non-conformal and directional, c) Non-conformal and non-directional, deposition processes.

Enhanced Surface Diffusion: One possible mechanism for surface migration is surface diffusion. As indicated by Equation (7), the surface diffusion rate is higher when the number of vacant surface states is large. A large number of vacant surface sites means very few sites are occupied by reactive species, resulting in low deposition rates. This approach is of limited value because low deposition rate is not desirable from a throughput point of view. Another way to increase the surface diffusion rate according to Equation (7) is to increase the reaction temperature. This also has limited merits because the low melting point of aluminum metallization prohibits high reaction temperature.

A more practical way to increase surface diffusion is to increase the ion bombardment on the surface. Higher ion bombardment will give more kinetic energy to the adsorbed reactive species to increase their mobility. More mobile reactive species should give better step coverage. This was verified by depositing oxide film in a dual frequency PECVD reactor where the high frequency component (13.6 mHz) is used to sustain the plasma and the low frequency component (300 kHz) is applied to the substrate electrode to increase the energy of the bombarding ions (104). The results are shown in Figure 34. The films deposited with low frequency ion bombardment have about 20% better step coverage.

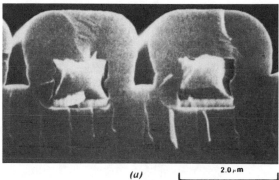

(a) 　　2.0 μm

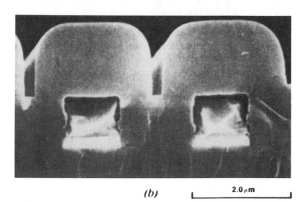

(b) 　　2.0 μm

Figure 32. Step coverage of oxide film deposited with a) Silane, b) TEOS (19), reprinted with permission of the publisher.

Sticking Coefficient: A more effective way for the reactive species to achieve long migration distances is to desorb from the substrate surface and then readsorb at another location on the surface. The desorption and readsorption process may occur many times before the reactive species are consumed by the deposition reactions. Under these conditions, the reactive species has a low sticking coefficient. Conversely, in a unity sticking coefficient process, reactive species are adsorbed and reacted at the landing site to give essentially a line of sight process when there are no gas phase collisions at low pressures as indicated by Equation (1). For example, vacuum evaporation onto a cold substrate is a process with a sticking coefficient close to unity. Figure 35 illustrates schematically three different processes with unity sticking coefficient, low sticking coefficient and surface diffusion (105). As shown in these illustrations, a process with very low sticking coefficient for filling trenches with high aspect ratios is desirable.

Many CVD deposition reactions have very low sticking coefficients, resulting in very conformal coatings. For example, polysilicon films deposited by CVD process have almost ideal step coverage. Special test structures are needed to study the desorption and readsorption process because of their long range effect. The best test structure for these studies is the over-hang structure illustrated in Figure 36. Over-hang structures with different gap width and vertical gap height were used to study the effect of long range migrations on step coverage for various CVD reactions (105). To illustrate that sticking coefficient is the dominant factor in determining the step coverage in CVD processes, various CVD film profiles were simulated with a computer program using the sticking coefficient as the only fitting parameter (106). The results are illustrated in Figure 37 (a). The sticking coefficient values that give the best fit to the experimental profiles for different CVD processes are listed in Figure 37 (b).

Flow-Like Step Coverage: Recent studies have shown that under certain deposition conditions films with better than 100% step coverage can be obtained. These films, with a flow-like surface contour, are ideal for filling narrow gaps and deep grooves. TEOS/ozone oxide films deposited at low pressures give nearly 100% conformal coatings (23), however, films deposited at atmospheric pressure (APCVD) or subatmospheric pressure (SACVD) give better than 100% step coverage with flow-like surfaces (92, 99). The overall reaction is given by Equation (20). For APCVD or SACVD depositions, the step coverage was found to be a function of ozone concentrations, changing from conformal contours to flow like contours as the ozone concentration is increased. This is illustrated in Figure 38 (107). The reason for this behavior is not well understood. It has been suggested that gas phase reactions between TEOS and ozone produces intermediate liquid-like polymer precursors with low sticking coefficient and these intermediates retain their liquid-like properties during the adsorption and migration stages until they are consumed by the SiO_2 forming reaction.

Because of their superb step coverage, the new APCVD and SACVD films are under intensive study for ILD and doped glass applications (97-100). These films have, however, a few concerns that need to be resolved such as surface dependent deposition rate (108), film cracking, and moisture absorption (93). A high temperature densification cycle can be used to minimize these problems for doped-glass applications but not for ILD applications. Because it is difficult to completely eliminate moisture in deposited oxide films, the moisture content in ILD layer is a serious concern for device reliability. Film cracking and surface sensitive deposition rate are related to water absorption because oxide films with high moisture absorption are usually porous and porous films, which crack easily, are often caused by poor surface nucleation.

Another method to obtain flow like step coverage was demonstrated with low-temperature remote-plasma deposition (109). In this work, standard silane/oxygen chemistry was carried out in a parallel plate plasma reactor. The substrates were cooled down to -110 C with liquid nitrogen. A grounded metal grid was inserted between the rf electrode and the substrate electrode to confine the plasma in the region between rf electrode and the grid. On a substrate with fine grooves, the oxide deposition was found to start from the bottom of the grooves with very little deposition on the top surface as shown in Figure 39. These characteristics were attributed to the formation of liquid-like precursors on the cooled substrate surface that would migrate to the bottom of the grooves in order to minimize surface tension energy. Submicron grooves with very high aspect ratio can be filled with this deposition technique as shown in Figure 40.

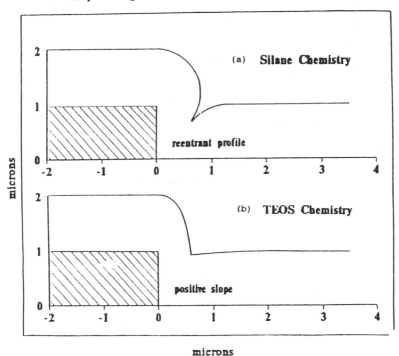

Figure 33. Step coverage of CVD oxide simulated with U.C. Berkeley's SAMPL program by assuming a) Very short surface migration distance, b) Very long surface migration distance

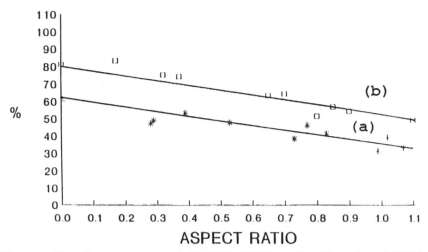

ASPECT RATIO

Figure 34. Step coverage of PECVD TEOS oxide films for a) HF ion bombardment only, b) HF and LF ion bombardment as a function of aspect ratios (104), reprinted with permission of the publisher.

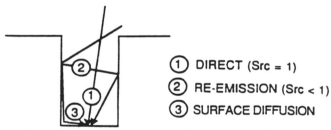

REACTANTS

① DIRECT (Src = 1)
② RE-EMISSION (Src < 1)
③ SURFACE DIFFUSION

Figure 35. Schematic diagram for deposit processes with 1) Unity sticking coefficient, 2) Low sticking coefficient, 3) Surface diffusion (105), reprinted with permission of the publisher.

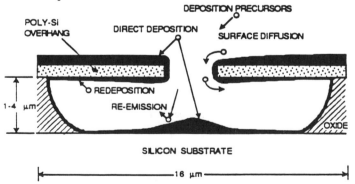

DEPOSITION PRECURSORS

POLY-Si OVERHANG   DIRECT DEPOSITION   SURFACE DIFFUSION

REDEPOSITION
1-4 μm   RE-EMISSION
OXIDE

SILICON SUBSTRATE

16 μm

Figure 36. Overhang structure used in studying the effect of desorption and readsorption on step coverage of deposited films (105), reprinted with permission of the publisher.

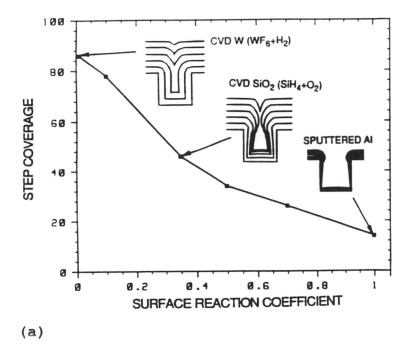

(a)

Source	Temperature C	S_{rc}
$SiH_4/O_2/PH_3$	400	0.35
SiH_4/O_2	400	0.262
DES/O_2	380	0.1
$TMCTS/O_2$	560	0.04
$TEOS/O_2$	700	0.04

(b)

Figure 37. Simulated step coverage of various CVD films a) Profiles versus sticking coefficient or surface reaction coefficient, b) Values of sticking coefficient as obtained by fitting the experimental contours (105).

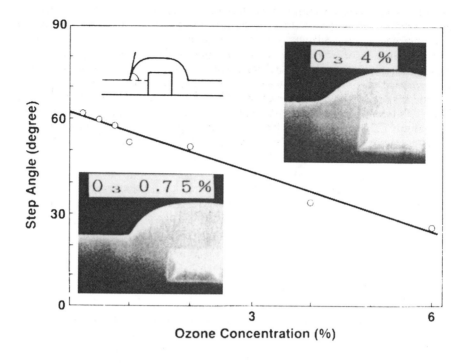

Figure 38. Step coverage of APCVD oxide deposited with TEOS-Ozone as a function of ozone concentration (107).

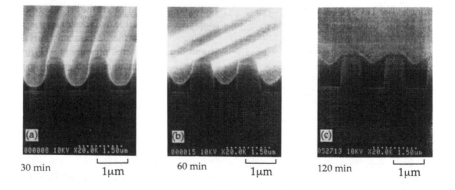

Figure 39. Flow like step coverage of silane oxide obtained by using remote PECVD and low substrate temperature (109), reprinted with permission.

0.2 µm

Figure 40. Narrow (0.2 µm) trench filled with flow like oxide deposited with low temperature remote PECVD process (109), reprinted with permission.

7.4   Water Absorption

Deposited oxide films are not as dense as thermally grown oxide, thus moisture can easily penetrate and be absorbed in the deposited films. Moisture in ILD is known to cause processing problems such as poor photoresist adhesion, film delamination, poor quality aluminum deposition, etc. (110). Moisture in ILD is also known to cause device reliability problems such as threshold voltage shift (111), transconductance degradation (112), stored charge loss (113), hot electron degradations (114), etc. Therefore, it is important to minimize the moisture content in ILD dielectric layer.

   PECVD TEOS Oxide: The water content of PECVD TEOS oxide changes with deposition conditions, ranging from too small to measure for good films to very high values for poor films (69, 115). With plasma deposition, the moisture in the deposited films can be controlled by changing the ion bombardment. It has been demonstrated that oxide films deposited with high ion bombardment usually have negligible amount of water. The amount of ion bombardment on the wafer surface can be changed by controlling the plasma power during deposition. For example, PECVD TEOS/O_2 oxide deposited in a commercial single wafer reactor (Applied Materials P-5000) contains negligible amount of water when a deposition power of 450 watts or higher is used (116). Another way to control the ion bombardment is to add a low frequency power component to the substrate electrode. The amount of water absorption can be reduced to a negligible amount by increasing the amount of LF power (117). Illustrated in Figure 41, the amount of absorbed water in the oxide film is indicated by the size of hysteresis in the stress-temperature curve as discussed in Section 5.1. The large hysteresis loop for the film deposited without LF ion bombardment is due to absorption and desorption of moisture. The hysteresis loop is very small for the film deposited with sufficient LF ion bombardment because the denser film absorbs moisture very slowly.

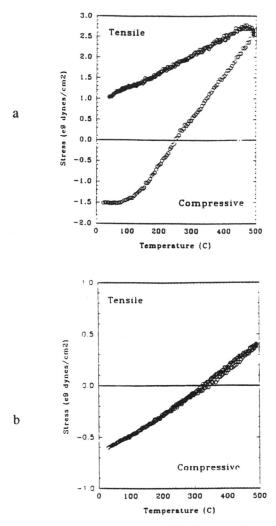

Figure 41. Stress temperature cycles for PECVD oxide films deposited with a) HF plasma only, b) Addition of 50% LF power to the substrate (117), reprinted with permission.

THCVD TEOS-Ozone Oxide: TEOS-ozone oxide deposited in a LPCVD reactor gives a very conformal, but highly hygroscopic, film capable of absorbing a great deal of water from the ambient. The water content was determined to be about 5% using FTIR measurements as shown in Figure 16. In another study, the water content was determined by measuring the hydrogen concentration of the film by using nuclear reaction analysis (23). The measured hydrogenous species concentration was about 10%. This correlates well with the MEA measurements. Due to the high water content, the dielectric constant of these wet films was abnormally high (24). The high moisture content in these films makes them unsuitable for device applications. In order to take advantage of its superb conformality without sacrificing device reliability, these films are often used in an etch-back planarization process where the bulk of the deposited THCVD TEOS oxide is etched away leaving material only in the narrow grooves as a gap filler surrounded by PECVD oxide as illustrated in Figure 42 (36).

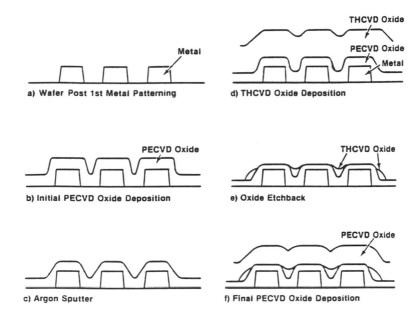

Figure 42. Schematic diagram of dep-etch ILD process using PECVD TEOS/O_2 and THCVD TEOS/O_3 oxide films (36), reprinted with permission.

APCVD and SACVD TEOS Ozone Oxide: TEOS-ozone oxide films deposited by APCVD or SACVD process are very desirable for ILD applications because of their flow like step coverage over severe surface topography. The amount of water contained in these films will determine their usefulness as ILD layer. If these films are very wet, then they can only be used as a filler in an etchback process similar to the THCVD TEOS ozone films discussed earlier. If these films are relatively dry, then they may be used as a stand alone ILD layer. The water content of APCVD and SACVD oxide films deposited under various conditions were determined with FTIR and MEA (118). The water content of TEOS ozone oxide films deposited with APCVD and SACVD processes was found to be significantly lower than films deposited with the LPCVD process. By using a low TEOS/ozone ratio, the water content in the deposited films can be reduced significantly (116). This approach may not be a very practical solution because low TEOS/ozone ratios correspond to low deposition rates. Even with optimized deposition conditions, the TEOS/ozone oxide films deposited by APCVD and SACVD still contain approximately one to two percent of water. This is still too much water to allow these films to be used as a stand alone ILD film. The TEOS-ozone oxide can be used in conjunction with other dielectric films in a multilayer structure to further minimize its moisture content and to provide protection between the wet film and the active devices as discussed in the next section.

SOG: Moisture absorption in SOG films has been studied extensively because SOG films are often capable of absorbing a large amount of water which leads to processing problems (110) and device stability problems (48). SOG, however, represents a wide variety of materials with different properties such as silicates, doped silicates, and siloxanes with different organic side groups attached. Any detailed discussion on moisture absorption of the film must take into account the specific material and its curing conditions. The discussion here will concentrate on situations that are generally applicable. For details, many original research papers are available (10).

Silicate films are formed by the condensation reaction given in Equation (9) at relatively low temperature. The resulting film is rather porous with the film density considerably lower than the thermally grown oxide. The porous silicate films are capable of absorbing a large amount of moisture and crack easily. One way to reduce the porosity and the moisture absorption is to add phosphorus doping to the film. Due to the hygroscopic nature of phosphorus oxide, the acceptable amount of P and the reduction of moisture content are rather limited. Another way to improve the situation is to use multiple coating and curing processes to facilitate the condensation and drying process. Again, the improvements are rather limited. Therefore, silicate films are limited to rather thin coatings and are usually used in an etchback process as a gap filling material to minimize the amount of material left on the wafer similar to the structure shown in Figure 42.

The siloxane films usually contain less moisture than the silicate films, and some formulations contain no detectable amount of water after curing when measured with FTIR and the more sensitive UHV thermal desorption apparatus (72). Figure 43 (a) is the desorption measurement for the SOG film after curing, no water was detected. Figure 43 (b) is the desorption measurement of the same SOG film after it was subjected to Ar sputter etching and then soaked in DI water. The processed film has a very small amount of absorbed water. This detected water is probably due to surface adsorption since the quantity is about the same magnitude as the detected Ar implanted into the film from the sputter etching process as shown in Figure 43 (c). For siloxane SOG films, the carbon content in the films can be destroyed by incompatible processing conditions, such as excessively high temperature, or an oxygen plasma, to give a porous film that can absorb a large amount of water (72) as illustrated in Figure 44. Figure 44 (a) shows the FTIR spectrum of the siloxane SOG film after curing, no water peak was detected. Figure 44 (b) shows that there is a significant water peak in the FTIR spectrum after the SOG film was subjected to an oxygen plasma resist stripping cycle. Figure 44 (c) shows that the water peak disappears immediately after a 450°C bake-out cycle. It has been demonstrated that with proper care a non-etchback process can be used with siloxane SOG without any sign of moisture out gassing problem (52).

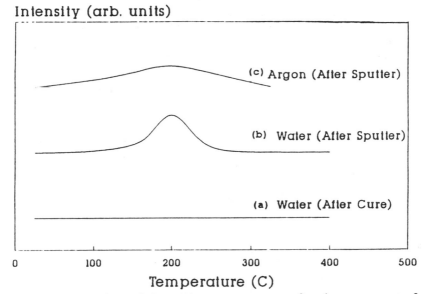

Figure 43. UHV thermal desorption measurement of moisture content of a siloxane SOG film a) Moisture cured SOG film, b) Moisture in the same SOG film after Ar sputter etch and soaking in DI water, c) Ar content in the sputter etched film (72), reprinted with permission.

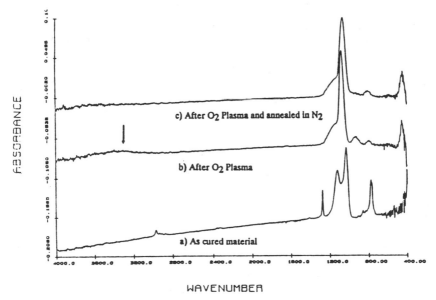

Figure 44. FTIR measurement of moisture content of a siloxane SOG film a) Cured SOG, b) Same SOG film after a plasma oxygen resist stripping cycle, c) Same SOG film in (b) immediately after a 450°C bake out cycle (72), reprinted with permission.

Methods to Reduce the Effect of Water in ILD: Since high temperature densification is not a viable option for ILD, other means are needed to minimize the water content of the deposited films and/or to reduce the effect of water content on device reliability. Some of these methods are discussed below.

Absorbing Layer: One way to reduce the effect of water in a multilayer ILD structure is to use an absorbing dielectric layer underneath the water containing ILD layer to prevent water related hydrogenous species from reaching the sensitive device regions. It has been reported that PECVD oxide with a high density of silicon dangling bonds; i.e., silicon rich oxides, can tie up hydrogenous species by forming more stable bonds (119). This is illustrated schematically in Figure 45. This technique has been applied to other types of hydrogen or water containing films such as silicon nitride films (120), SOG films (121, 122).

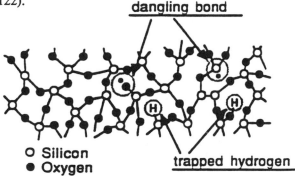

O **Silicon**
● **Oxygen**

Figure 45. Schematic diagram illustrating the capture of hydrogenous species by dangling bonds in Si rich PECVD oxide (119), reprinted with permission.

Fluorination: An interesting method to reduce the moisture content of the deposited oxide film is to incorporate some fluorine into the film. It has been shown that CVD oxide deposited with fluortriethoxysilane (which is TEOS with one of the ethoxy group replaced by fluorine) instead of TEOS as the silicon source material contains no water as measured by FTIR (123). The moisture content of SOG films can also be reduced significantly by diffusing fluortriethoxysilane into the SOG layer (50). The reduced water absorption of fluorinated oxide layer is attributed to the hydrophobic nature of the Si-F bonds contained in the film. However, the effect of F in the oxide film on the metallization and device reliability must be carefully studied before this method can be considered as an acceptable way to reduce moisture content of deposited oxide films.

Capping layer: Since the moisture content in deposited oxide films is mainly absorbed from environment, it is possible to minimize moisture absorption by capping the hydrophilic film with a more dense or more hydrophobic film. One convenient method is to use a dense PECVD oxide as a capping layer. A study was carried out to determine the water content in SACVD TEOS-ozone oxide capped by in-situ deposited PECVD oxide films. The PECVD oxide capping films used in this study are all approximately 100 nm thick but deposited with three different plasma power, 350, 450, and 550 watts. Water content of the capped films was measured with FTIR after various storage times, at room temperature, in 100% relative humidity environment. The results are shown in Figure 46. The denser PECVD capping layers deposited with higher rf-power are able to keep the SACVD films dry in a humid environment for hundreds of hours which is more than sufficient protection for normal processing delays during device fabrications. An absorbing layer underneath the ILD can also be used in conjunction with the capping process to provide increased protection to the underlying devices.

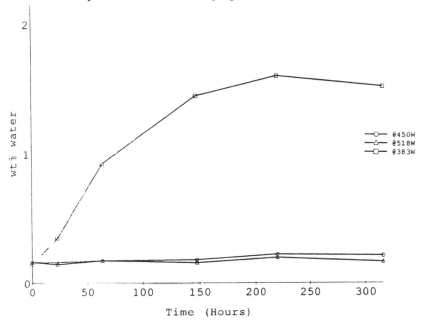

Figure 46. Moisture content of capped SACVD TEOS-Ozone oxide films as a function of storage time in 100% relative humidity environment at room temperature. The films are capped with PECVD oxide, 0.1 μm thick, deposited with different plasma power.

8.0   PASSIVATION LAYER

After patterning the final metal level, a passivation layer is deposited over the wafer surface to prevent possible mechanical and chemical damages to the underlying circuit structures during subsequent assembly and packaging operations. Commonly used inorganic dielectric materials are SiO_2, doped SiO_2, SiN and SiON. In addition, a thick layer of organic polymer, such as polyimide, is often used as a top coat to provide additional mechanical protection. As in the case of ILD, multiple layers, such as a combination of SiO_2 and SiN layers, are often used to optimize the properties of the passivation layer. For mechanical protection, the dielectric layer should be as thick as possible. However, thick films tend to crack and high stress will contribute to the failure of underlying Al conductors. Therefore, a practical limit to the passivation layer thickness is typically one to two microns. For chemical protection, SiN and SiON are preferred because they are impermeable to moisture and mobile ions. Due to the temperature limitations of the Al metallization, SiN and SiON must be deposited by PECVD process as will be described below.

8.1   PECVD SiN

The SiN film for the passivation layer is deposited at low temperatures by PECVD because of the temperature limitation of the underlying Al metallization. Silane diluted with N_2 and ammonia are used in a plasma reactor to deposit SiN according to the reaction,

$$[21]   \qquad SiH_4 + NH_3 + N_2 \xrightarrow{\text{plasma}} SiN(H) + \dots$$

The details of this reaction are not clear; recent studies indicate that the formation of aminosilanes, such as $Si(NH_2)_4$ and $Si(NH_2)_3$ are the principle precursors for the film deposition (124). The resulting film is not stoichiometric Si_3N_4 but closer to the composition of SiN with 10 to 30 % of hydrogen, thus the notation SiN(H). The composition and properties of plasma nitride, SiN, vary widely with deposition conditions. For example, SiN films deposited with SiH_4/N_2 reaction contain less hydrogen than films deposited with NH_3 but the deposition rate is lower. SiN films are used extensively for device passivation because they are inert and provides a good barrier against sodium and moisture penetration. However, with submicron VLSI devices, some reliability problems have been observed with plasma nitride passivation film due to its high hydrogen content and high stress.

High hydrogen content has been reported to degrade the device lifetime under hot electron injection conditions (125). Although the detailed degradation mechanism is still being debated, it has been suggested that the hydrogen released from the passivation layer can diffuse into the $Si-SiO_2$ interface to generate charge traps that degrade the device lifetime. In order to cause damage to the devices, the hydrogen must first be released from the PECVD nitride film and then diffuse to the device area. Therefore, the bonding state of the hydrogen in the nitride film is very important. Since the Si-H bond is weaker and more likely to break than the N-H bonds, the Si-H/N-H distribution is more important to device degradation than the total amount of hydrogen in the deposited film (34, 126).

The high stress in the nitride film can give rise to stress induced diffusion in aluminum. Since the easiest diffusion path in Al is at grain boundaries and the stress is highest at the line edges, voids are generated at inter granular boundaries, particularly at the edges of narrow lines. These voids can grow even in the absence of current flow and leads to stress migration failures in long narrow Al lines. This is discussed in more detail in Chapter 8. Films deposited at low temperature (127) or with low stress (128) are effective in reducing Al line failures. The stress migration problem generated by the SiN passivation layer could also be related to the high hydrogen content in the film. It has been shown that highly stressed SiN films contain large amounts of hydrogen (34). Furthermore, it has been shown that the hydrogen in the nitride film can chemically react with Al at elevated temperatures, such as sintering temperature, to form voids (129).

Effect of Ion Bombardment: The amount of hydrogen and the magnitude of stress in the deposited SiN film varies a great deal with the PECVD conditions. The amount of ion bombardment during film deposition has been used to control stress as well as hydrogen content in the deposited film. For example, with a dual frequency PECVD reactor, a low frequency rf (450 kHz) bias can be applied to the substrate to control ion bombardment during deposition. SiN films deposited with low frequency rf bias have less stress and reduced Si-H content without any significant changes in other film properties (104). Figure 47 shows the stress changes from tensile to compressive with increasing substrate bias.

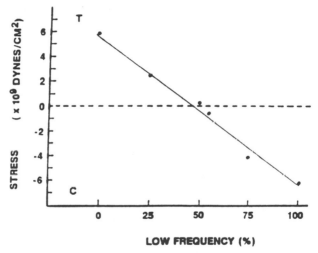

Figure 47. SiN film stress control by increasing ion bombardment using low frequency rf bias during film deposition (104), reprinted with permission.

The total hydrogen concentration as well as the hydrogen bonding configuration change with the amount of ion bombardment during film deposition. Figure 48 compares the FTIR spectra of the SiN film deposited with and without low frequency rf bias to the substrate during film deposition. Figure 48 (a), which has a distinct Si-H peak, is the spectrum for SiN film deposited without bias and Figure 48 (b), which has no detectable Si-H peak, is the spectrum for SiN film deposited with bias. Figure 48 shows that the Si-H bond in the deposited film decreases with increasing bias. This could be because the weaker Si-H are easily broken by ion bombardment and some of the released hydrogen may be captured by the nitrogen as indicated by the increase in N-H bonds. Devices passivated with low Si-H concentration SiN films have less hot electron degradation than standard SiN films with high Si-H concentration (126). Low Si-H concentration SiN films have an added benefit of being UV transparent. SiN films with low Si-H concentration are also known as UV-nitride because this property is important for certain device applications such as EPROMs which require UV to erase the stored information before programming.

Effect of Fluorination The hydrogen content in PECVD SiN films can be minimized if the source materials used in the deposition process contain no hydrogen at all. For example, SiN films deposited from SiF_4/N_2 plasma contain very little hydrogen when compared to that deposited from SiH_4/N_2 plasma (130, 131). The resulting film contains only a small amount of H but very high F concentration. The deposition rate for this reaction is rather low. Deposition rates can be improved by adding SiH_4 to the source gas. Stable films with good step coverage can be obtained for SiF_4/SiH_4 flow ratio greater than one (132). Fluorinated SiN films can also be deposited from a $SiH_4/NF_3/N_2$ plasma (i.e. NH_3 used in standard deposition process is replaced by NF_3). The deposition rates for this reaction are high, and the Si-H bond concentration decreased with increasing NF_3/SiH_4 ratio and with SiH_4/NF_3 ratio. The Si-H bond concentration decreased with increasing SiH_4/NF_3 ratio and was often below the detection limit of FTIR measurements. The fluorinated SiN film is also transparent to UV. However, the stability of the fluorinated SiN films and the effect of excessive F on the device reliability is still not yet established.

8.2   Plasma SiON

Silicon oxynitride films have some of the desirable properties of both silicon oxide and silicon nitride. When used properly, it can offer some advantages over oxide and nitride (133). For example, SiON films can be an effective barrier for mobile ions and yet have low stress and low hydrogen content. Oxynitride deposition process is a variation of the standard nitride deposition process. By adding N_2O, the reaction becomes,

$$[22] \qquad SiH_4 + NH_3 + N_2O + N_2 \xrightarrow{\text{plasma}} SiON + ...$$

The resulting SiON film is not a unique composition; it can vary from SiO_2 to SiN depending on the gas mixture. The film composition is usually characterized by measuring its refractive index, which changes from a value of 1.4 for SiO_2 to 2.1 for SiN, or its etch rate in diluted HF solution. The stress and the hydrogen content of plasma deposited SiON films are less than that of SiN films; as a result, thicker SiON films can be deposited without film cracking or peeling. Important properties for SiON passivation films are UV transmission for EPROM applications and moisture resistance for device stability.   As expected, the UV transmittance decreases and the moisture resistance increases with the index of refraction of the deposited film as illustrated in Figure 49 (134).   SiON film composition must be carefully controlled in order to achieve the desired film properties (135). With optimized deposition conditions, SiON passivation films can give good UV transmission and good moisture resistance for device applications.

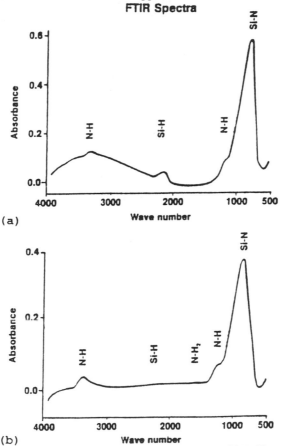

Figure 48. FTIR spectrum of H-bonds in PECVD SiN film. a) "Typical" plasma nitride with distinct Si-H peak. b) Film deposited with increasing ion bombardment has not Si-H peak (104), reprinted with permission.

Silicon oxynitride films have some of the desirable properties of both silicon oxide and silicon nitride. When used properly, it can offer some advantages over oxide and nitride (133). For example, SiON films can be an effective barrier for mobile ions and yet have low stress and low hydrogen content. Oxynitride deposition process is a variation of the standard nitride deposition process. By adding N_2O, the reaction becomes,

$$[22] \qquad SiH_4 + NH_3 + N_2O + N_2 \xrightarrow{\text{plasma}} SiON + ...$$

The resulting SiON film is not a unique composition; it can vary from SiO_2 to SiN depending on the gas mixture. The film composition is usually characterized by measuring its refractive index, which changes from a value of 1.4 for SiO_2 to 2.1 for SiN, or its etch rate in diluted HF solution. The stress and the hydrogen content of plasma deposited SiON films are less than that of SiN films; as a result, thicker SiON films can be deposited without film cracking or peeling. Important properties for SiON passivation films are UV transmission for EPROM applications and moisture resistance for device stability. As expected, the UV transmittance decreases and the moisture resistance increases with the index of refraction of the deposited film as illustrated in Figure 49 (134). SiON film composition must be carefully controlled in order to achieve the desired film properties (135). With optimized deposition conditions, SiON passivation films can give good UV transmission and good moisture resistance for device applications.

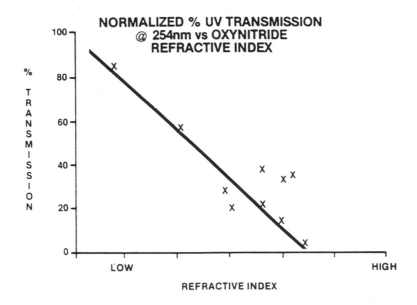

Figure 49. UV transmission at 254nm vs. oxynitride refractive index. (134), reprinted with permission.

As in the case of SiN film, SiON films are often used in conjunction with other dielectric layers, such as PSG, in order to get the desired passivation layer properties. As in the case of ILD, planarization has also been used to improved the integrity of the SiON film as well as to reduce the deleterious effect, such as stress concentration, on the underlying Al conductor patterns (136).

9.0   SUMMARY

The inorganic dielectric layers are essential parts of the multilevel metallization technology. They include doped silicate glass for under the metal applications, interlevel dielectric, and passivation layer. These dielectric films, such as SiO_2, doped SiO_2, SiN, SiON, are usually deposited by CVD processes. The fundamentals of CVD processes were briefly reviewed to provide the basic understanding between process parameters and the deposited film quality. In some applications, spin-on glass films are used to smooth over rough topography. The proper spinning and curing conditions as well as common problems for obtaining good SOG film quality are discussed.

Dielectric films must satisfy some very stringent quality and manufacturability requirements. The film properties must be carefully monitored. In addition to standard material characterization techniques, special measurement techniques, such as film stress and moisture content measurements, are needed to determine the quality of the deposited films. High stress in the deposited films not only causes failures in the deposited films, such as cracking or peeling, but also degrades the reliability of under lying conductors. Stress in blanket films is easily characterized by measuring the wafer curvature. The stress versus temperature cycles are often used to monitor the quality of the deposited films, including the moisture absorption rate. Moisture measurement techniques are described in some detail because they are important to circuit reliability. The simplest way to measure moisture is using FTIR; however, it suffers from poor sensitivity. It is often used to screen out poor quality films. The more sensitive moisture measurement methods are the moisture evolution analysis method and the UHV thermal desorption method. These methods are more complicated, often destructive, and suffer from a lack of suitable commercial equipment.

Some requirements for deposited dielectric films are common to all applications, such as uniformity, reproducibility, low defect density, and low manufacturing cost, while some are application specific. For doped silicate glasses used under the metal, the requirements are dopant concentration, dopant uniformity control, stability of the dopants in thermal cycles and humid environment. For ILD, it is step coverage and low moisture content. For passivation layers, it is mechanical integrity and moisture resistance. Moisture content is a common problem for all deposited oxide films but the relative amounts differ greatly for different deposition process. Some processes, such as the $TEOS/O_2$ oxide deposited in an LPCVD reactor, contain a great deal of moisture. They are suitable only for gap filling applications in etchback processes. Similarly, hydrogen content is a common problem for SiN films deposited at low temperatures. Again, it differs greatly for films deposited with different processes. PECVD is used extensively in depositing dielectric films because of its low deposition temperature. Increasing the ion bombardment during the deposition process generally improves film quality and reduces moisture absorption and film stresses. New deposition processes, such as APCVD $TEOS/O_3$ oxide and cryogenic deposition in a remote plasma, are being explored for their superb step coverage capability. However, the quality of the deposited films must be carefully studied. With increasing understanding of the deposition chemistry and the origin of stress and moisture in deposited films, it is hopeful the trend of rapid improvements in processes and materials will continue to satisfy the demanding requirements of multi-level interconnection in the ULSI era.

REFERENCES

1. Adams, A. C., "Dielectric and Polysilicon Film Deposition" in VLSI Technology (S. Sze, ed.), McGraw-Hill, New York, 1983.
2. Kern, W. and Ban, V. S., "Chemical Vapor Deposition of Inorganic Thin Films," in Thin Film Processes (J. L. Vossen and W. Kern, eds.), Academic Press, New York, 1978.
3. Blocher, J. M. Jr., "Chemical Vapor Deposition," in Deposition Technologies for Films and Coatings (R. F. Bunshah, ed.), Noyes Publications, New Jersey, 1982.
4. Sherman, A., "Chemical Vapor Deposition for Microelectronics," Noyes Publications, New Jersey, 1987.
5. Kamins, T., " Polycrystalline Silicon for Integrated Circuit Application," Kluwer Academic Publications, MA, 1988.
6. Hollahan, J. R. and Rosler, R. S., "Plasma Deposition of Inorganic Thin Films," in Thin Film Processes (J. L. Vossen and W. Kern, eds.), Academic Press, New York, 1978.
7. Hess, D. W. and Graves, D. B., "Plasma-Enhanced Etching and Deposition," in Microelectronic Processing (D. W. Hess and K. F. Jensen, eds), American Chemical Society, 1989.
8. Proc. of the International Conference on Chemical Vapor Deposition, The Electrochemical Society, Inc., New Jersey.
9. The International VLSI Multilevel Inter-Connection Conference (V-MIC) Proceedings, IEEE, Inc., New York.
10. "The International Conference on Chemical Vapor Deposition Proceedings," The Electrochemical Society, Inc., NJ.
11. Kern, W., and Rosler, R.S., "Advances in Deposition Process for Passivation Films," J. Vac. Sci. Tech., $\underline{14}$, p.1082, 1977.
12. Rosler, R.S., "Low pressure CVD Production Processes for Poly, Nitride and Oxide," Solid State Tech., 20(4), p.63, 1977.
13. Adams, A. C., Capio, D. C., "The Deposition of Silicon Dioxide at Reduced Pressure," J.Electrochem. Soc., $\underline{126}$, p. 1042, 1979.
14. Levin, R. M., Evens-Latterodt, K., "The Step Coverage of Undoped and Phosphorous Doped SiO2 glass films," J. Vac. Sci. Tech., $\underline{B1}$, p. 54, 1983.
15. Becker, F. S., et. al, "Low Pressure Deposition of High Quality SiO_2, Films by Pyrolysis of Tetraethylorthosilicate," JVST, $\underline{B5(6)}$, p.1553, 1987.

16. Becker, F. S., et. al, "Process and Film Characterization of Low Pressure TEOS BPSG," JVST, B4, p. 732, 1986.

17. Becker, F. S., Rohl, S., "Low Pressure Deposition of Doped SiO_2 by Pyrolysis of Tetraethylorthosilicate (TEOS)," JECS, 134 (11), p. 2923, 1987.

18. Adams, A.C., et.al. "Characterization of Plasma Deposited Silicon Dioxide," J. Electrochem. Soc., 128(7), p. 1545, 1981.

19. Chin, B. J., Van de Vem, E. P., "Plasma TEOS Process for Interlayer Dielectric Applications," Solid State Technology, 31(4), p. 119, 1988.

20. Emesh, I., et. al., "Plasma Enhanced Chemical Vapor Deposition of Silicon Dioxide Using TEOS," JECS, 136(11), p. 3404, 1989.

21. Hills, G., Harrus, A. S., Thoma, M. J., "Plasma TEOS on ILD in Two Level Metal Technology," Solid State Technology, p. 127, April 1990.

22. Thoma, et. al., "A 1.0um CMOS Two Level Metal Technology Incorporating Plasma Enhanced TEOS," V-MIC Conf. Proc. p.20, 1987.

23. Nguyen, S., et. al., "Reaction Mechanism of Plasma and Thermal Assisted CVD of TEOS Oxide," JECS, 137(7), p. 2209, 1990.

24. Nguyen A., Murarka S., "Properties of CVD TEOS Oxide: Correlation with Deposition Parameters, Annealing and Hydrogen Concentration," JVST, B8(3), p. 533, 1990.

25. Makino, T., "Composition and Structure Control by Source Gas Ratio in LPCVD SiN_x," J. Electrochem. Soc., 119(9), p.1248, 1972.

26. Habraken, F,, et.al., "Characterization of LPCVD and Thermally Grown Silicon Nitride Films," J. Appl. Phys., 53(1), p. 404, 1982.

27. Shinha, A.K., et. al.. "Reactive Plasma Deposited Si-N Films for MOS-LSI Passivation," J. Electrochem. Soc., 125(4), p.601, 1978.

28. Giunta, C. J., Chapple-Sokol, J. D., Gordon, R. G., "Kinetic Modeling of Chemical Vapor Deposition of Silicon Dioxide form Silane and Disilane and Nitrous Oxide," J. Electrochem. Soc., 137(10), p. 3237, 1990.

29. Maeda, M., Nakamura, H., "Deposition Kinetics of SiO2 Films," J. Appl. Phys., 52, p.6651, 1981.

30. Hey, H. P. W., et. al, "Ion Bombardment: A Determining Factor in Plasma CVD," Solid State Technology, p. 139, April 1990.

31. Sabin, E. and Ramiller, C., "Interaction of Plasma Oxide Deposition Variables," Electrochem. Soc. Ext. Abst., 83-1, p.225, 1983.

32. Brown, W.A., and Kamins, T.I., "An Analysis of LPCVD System Parameters for Polysilicon, Nitride and Silicon Dioxide Deposition," Solid State Tech., p.51, July, 1979.

33. Winkel, L. W., Nelson, C. W., "Improved APCVD Systems for Depositing Silica and PSG Thin Films," Solid State Technology, 24(10), p. 123, 1981.

34. Martin, R. S., Van de Ven, E. P., Lee, C. P.,"RF Bias to Control Stress and Hydrogen in PECVD Nitride," V-MIC Conference Proceedings, p. 286, 1988.

35. Marks, J., Law, K., Wang, D., "In-situ Planarization of Dielectric Surfaces Using Boron Oxide," V-MIC Proceedings, p. 89, 1989.

36.   Pennington, S., Luce, S., Hallock, D., "An Improved Interlevel Dielectric Process for Submicron Double Level Metal Products," V-MIC Conference Proceedings, p. 355, 1989.

37.   Mehta, S., Sharma, G., "A Single Pass In-situ Planarization Process Utilizing TEOS for Double Poly, Double Metal CMOS Technologies," V-MIC Proceedings, p. 80, 1989.

38.   Ting, C. H., "Dielectric Planarization for ULSI," 3rd International Symposia on ULSI Science and Technology ECS Proceedings, 91-11, p. 592, 1991.

39.   Multani, J., et. al., "Spin-on Glass Dielectric Planarization for Double Metal CMOS Technology," V-MIC Conference Proceedings, p. 474, 1986.

40.   Forester, L., et. al., "Development of a Three Layer Metal Backend Process for Application to a Submicron CMOS Logic Process," V-MIC Conference Proceedings, p. 28, 1990.

41.   Kotani, H., et. al., "An Advanced Multilevel Interconnection Technology for 0.4 μm High Performance Devices," V-MIC Conference Proceedings, p. 15, 1992.

42.   Sinha, P. K., Smythe, J. A., "Moisture and Phosphorous Sensitivity of Sacrificial SOG Dielectric Planarization Process," JECS, 138(3), p. 854, 1991.

43.   Chen, S.N., et.al.,"Spin-on Glass Characterization and Application," V-MIC Conf. Proc. p.306, 1988.

44.   Kojima, H., et.al., "Planarization Process Using Multi-Coating of SOG," V-MIC Conf. Proc., p.390, 1988.

45.   Rutherford, N.M., et.al., "A New Low Dielectric Constant Planarization Dielectric," V-MIC Conf. Proc., p.448, 1991.

46.   Homma, T., et. al., "A New Polyamide Siloxane Film for Interlayer Dielectrics in Submicron Multilevel Interconnection," V-MIC Conference Proceedings, p. 279, 1988.

47.   Oikawa, A., et. al., " Poly-silphenylene-siloxane Resin as an Interlevel Dielectric for VLSI Multilevel Interconnections," JECS, 137(10), p. 3223, 1990.

48.   Paramanik, D., et. al., "Field Inversion in CMOS Double Metal Circuits Due to Carbon Based SOGs," V-MIC Conference Proceedings, p. 454, 1989.

49.   Rucker, T., Lin, H. Y., Ting, C. H., "Chemical Modifications of Spin-on Glass for Improved Processing Performance," ECS Int. Abst., 90-1, p. 283, 1990.

50.   Homma, T., Murao, Y., "A New SOG Formation Technology Using a Room Temperature Fluoro-Alkoxy-Silane Treatment," V-MIC Conference Proceedings, p. 65, 1992.

51.   Brewer, R., Gasser, R., "Process Window Calculations for a SOG Etchback Process," V-MIC Conference Proceedings, p. 376, 1987.

52.   Ting, C. H., Lin, H. Y., Pai, P. L., Rucker, T. R., "A Non- Etchback Spin-on Glass Process for Multilevel interconnection Technology," ECS Ext. Abst. p. 366, 1988.

53. Wolter, R., Hesters, W. C. J., "Experimental Study of Metal- Metal Coated Properties Using SOG," V-MIC Conference Proceedings, p. 447, 1990.
54. Sinha, A. K., Levinstein, H. J., Smith, T. E., "Thermal Stresses and Cracking Resistance of Dielectric Films on Si Substrates," J.Appl.Phys., 49(4), p. 2423, 1978.
55. Shintani, A., et, al., "Temperature Dependence of Stresses in CVD Vitreous Films," JAP, 51(8), p. 4197, 1980.
56. Blech, I., Cohen, U., "Effects of Humidity on Stress in Thin Silicon Dioxide Films," JAP, 53(6), p. 42021982
57. Shimbo, M., Matsuo, T., "Thermal Stress in CVD PSG and SiO_2 Films on Silicon Substrates," JECS, 130(1), p. 135, 1983.
58. Klema, J., Pyle, K., Domanque, E., "Reliability Implications of Nitrogen Contamination During Deposition of Sputtered Al/Si Metal Lines," IEEE/IRPS, p. 1, 1984.
59. Curry, J., et. al., "New Failure Mechanism in Sputtered Al-Si Films," IEEE/IRPS, p. 6, 1984.
60. Yue, J. T., et. al., "Stress Induced Voids in Al Interconnections During IC Processing," IEEE/IRPS, p. 126, 1985.
61. McInerney, E. J., and Flinn P., "Diffusivity of Moisture in Thin Films," IEEE/IRPS, p. 264, 1982.
62. Bhushan, B., Murarka, S.P., "Stress in Silicon Dioxide Films Deposited Using CVD Techniques and the Effect of Annealing on These Stresses," JVST, B-8(5), p. 1068, 1990.
63. Flinn, P. A., Chiang, C., "X-ray diffraction Determination of the Effect of Various Passivations on Stress in Metal Films and Patterned Lines," JAP, 67, p. 2927, 1990.
64. Tezaki, A., et. al., "Measurement of Three Dimensional Stress and Modeling of Stress Induced Migration Induced Failures in Al Interconnects," IEEE/IRPS, p. 221, 1990.
65. Groothuis, S. K., Schroen, W., "Stress Related Failures Causing Open Metallization," IEEE/IRPS, p. 1, 1987.
66. Jones, R. E. J., "Linewidth Dependence of Stress in Al Interconnect," IEEE.IRPS, p. 9, 1987.
67. Feigl, F.J., et.al., "The Effect of Water on Oxide and Interface Charge Generation in Thermal SiO_2 Films," J. Appl. Phys., 52(9), p.5665, 1981.
68. Pliskin, W. A., "Comparison of Properties of Dielectric Films Deposited by Various Methods," JVST, 14(5), p. 1064, 1977.
69. Cox, N., et. al., " Quantitative Measurement of the Moisture Content in Thin Films," V-MIC Conference Proceedings, p. 419, 1990.
70. Kato, T., et. al., "Outgassing Spectra of CVD Insulator Films and Influence on Sputtered Al," V-MIC Conference Proceedings, p. 79, 1992.
71. Kobayakawa, M., et. al., "A Study of Outgassing from SOG Films Used for Planarization," V-MIC Conference Proceedings, p. 454, 1991.
72. Ting, C. H., et. al., "A Non-etchback SOG Process for Multilevel Interconnection Technology," ECS Ext. Abst., 88-2, p. 366, 1988.

73. Balk, P., Eldridge, J.M., IEEE Proceedings, 57, p. 1558, 1969.
74. Bowling, R. A., Larrabea, G. B., "Deposition and Reflow of PSG," JECS, 132(11), p. 141, 1985.
75. Amstrong, W. E., Tollivor, D. C., "A Scanning Electron Microscope Investigation of Glass Flow in MOS Integrated Circuit Fabrication," JECS, 121, p. 307, 1974.
76. Kern, W., et. al., "Optimized CVD of BPSG Films," RCA Review, 46, p. 117, June, 1985
77. Kern, W., Smelter, R., BPSG for Integrated Circuits," Solid State Technology, p. 171, June 1985.
78. Avigel, I., "Intermetal Dielectric and Passivation Related Properties of Plasma BPSG," Solid State Technology, p. 217, October 1983.
79. Shibata, M.,et.al., "Deposition Rate and Phosphorous Concentration of PSG Films in relation to PH_3/SiH_4+PH_3 Mole Fraction," J. Electrochem. Soc., 122(1), p.157, 1975.
80. Shibata, M., et.al., "Deposition Rate and Phosphorous Concentration of PSG Films in Relation to O_2/SiH_4+PH_3 Mole Fraction," J. Electrochem. Soc., 122(1), p.155, 1975.
81. Wong, J., "Vibrational Spectra of Vapor Deposited Binary Phosphosilicate Glass," Journal of Non-Crystalline Solids, 20, p. 83, 1979.
82. Levin, R. M., Adams A. C., "Low Pressure Deposition of PSG Films," JECS, 129(7), p. 1589, 1982.
83. Levy, R.A., et.al., "Evaluation of the Phosphorous Concentration and its Effect on Viscous Flow and Reflow of PSG," J. Electrochem. Soc., 132(6), p.1472, 1985.
84. Nagashima, N, et. al., "Water Absorption and Densification of PSG Films," JECS, 121(3), p. 434, 1974.
85. Tanikawa, E., et. al., "Chemical Vapor Deposition in an Evacuated System," Proceedings of the 4th International Conference on CVD ECS, p. 261, 1973.
86. Wong, J., "A Review of Infrared Spectroscopic Studies of Vapor Deposited Dielectric Glass Films on Silicon," Journal Electronic Mat., 5(2), p. 113, 1976.
87. Arai, E., Terunuma, Y., "Water Absorption in CVD BSG Films," JECS, 121(5), p. 676, 1974.
88. Kern, W., Schnable, G., "Chemically Vapor Deposited BPSG for Si Device Applications," RCA Review, 43, p. 423, 1982.
89. Katsumata, Y., et. al., "Properties of Low Pressure BPSG Films Deposited Using TEOS, TMP, TEB, O_2," ECS Ext. Abst., vol. 89-2, p. 309, 1989.
90. Levy, R. A., et. al., "A new LPCVD Technique of Producing BPSG Films by Injection of Miscible Liquid Precursers," JECS, 134(2), p. 430, 1987.
91. Treichel, H., et. al., "A Novel PECVD VPSG Process from TTMSB and TTMSP Glow Discharge," ECS Ext. Abst., vol. 90-1, p. 205, 1990.

92. Nishimoto, y., et. al., "Dielectric Film Deposition by Atmospheric Pressure and Low Temperature CVD Using TEOS Ozone and New Organometallic Doping Sources," V-MIC Conference Proceedings, p. 382, 1989.

93. Fujino, K., et. al., "Doped Silicon Oxide Deposition by Atmospheric Pressure and Low Temperature CVD Using TEOS and Ozone," JECS, 138(10), p. 3019, 1991.

94. Integrity CVD System, Lam Research Corporation.

95. Yoshimaru, M., et. al., "Moisture Resistance of BPSG Films," IEEE/IRPS Proceedings, p. 22, 1987.

96. Smolinsky, G., Wendling, T., "Measurements of Temperature Dependent Stress of SiO_2 Films Prepared by a Variety of CVD Methods," JECS, 132(4), p. 950, 1985.

97. Kotani, H., et. al., "Low Temperature APCVD Oxide Using TEOS-Ozone Chemistry for Multilevel Interconnections," IEDM, P. 669, 1989.

98. Ikeda, Y., et. al., "Ozone/Organic Source APCVD for Conformal Doped Oxide Films," Journal of Electronic Materials, 19(1), p. 45, 1990.

99. Lee, P., et. al., "Sub-Atmospheric CVD of TEOS/Ozone USG and BPSG," V-MIC Conference Proceedings, p. 396, 1990.

100. Yieh, E., et.al., "Low Temperature Sub-Atmospheric CVD USG/PSG for Gap Filling and Planarization of Advanced Sub-Micron Memory Devices," ECS Ext. Abst. 92-1, p. 248, 1992.

101. Nishimoto, Y., et. al., "New Low Temperature APCVD Method Using Polysiloxane and Ozone," CVD XI ECS Proceedings, PV90-12, p. 410, 1990.

102. Magnella, C. G., Ingwersen, T., Flech, E., "A Comparison of Planarization Properties of TEOS and SiH_4 PECVD Oxide," V-MIC Conference Proceedings, p. 366, 1988.

103. SAMPL Program, EECS Department, UC Berkely.

104. Van de Ven, E., Connick, I., Harrus, A., "Advantages of Dual Frequency PECVD for Deposition of ILD and Passivation Films," V-MIC Conference Proceedings, p. 194, 1990.

105. McVitte, J. P., et. al., "LPCVD Profile Simulation Using a Re-Emission Model," IEDM, p. 917, 1990.

106. Cheng, L. Y., et. al., " Sticking Coefficient as a Single Parameter to Characterize Step Coverage of SiO_2 Processes," V-MIC Conference Proceedings, p. 404, 1990.

107. Fujino, K., et. al., "Silicon Dioxide Deposition by Atmospheric Pressure and Low Temperature CVD Using TEOS and Ozone," JECS, 137(9), p. 2883, 1990.

108. Fujino, K., et. al., "Surface Modification of Base Materials for $TEOS/O_3$ APCVD," JECS, 139(6), p. 1690, 1992.

109. Shin H., et. al., "High Fluidity CVD of Silicon Oxide from SiH_4 and O_2 Plasma," Ext. Abst. SSDM, p. 201, 1991.

110. Chien, C., et. al., "Defects Study on SOG Planarization Technology," V-MIC Conference Proceedings, p. 404, 1987.

111. Noyori, M., et. al., "Secondary Slow Trapping - A New Moisture Induced Instability Phenomenon In Scaled CMOS Devices," IEEE/IRPS Proceedings, p. 113, 1982.
112. Noygori, M., Nakata, Y., "Interaction Between Water and Scaled CMOS FET's with PSG Passivation Films," JECS, 131(5), p.1109, 1984.
113. Crisenza, G., et. al., "Chargeless in EPROM Due to Ion Generation and Transport in Interlevel Dielectric," IEDM, P. 107, 1990.
114. Cottrell, P., Troutman, R., Ning, T., "Hot Electron Injection in n-Channel IGFET's," IEEE Trans. Elect. Dev., ED-26, p. 560, 1979.
115. Patrick, W. J., et. al., "Properties of PECVD O$_2$/TEOS Silicon Dioxide," 3rd Symposium on ULSI Sci. and Tech., PV91-11, p. 692, 1991.
116. Kwok, K., Appl. Mat. Inc., Unpublished Data.
117. Harrus, A. S., et. al.,"Water and Water Related Compounds in PECVD Oxides," ECS Ext. Abst., 91-2, p.332, 1991.
118. Rasti, H., Intel Corp. Unpublished Data.
119. Yoshida, S., et. al., "Improvement of Endurance to Hot Carrier Degradation by Hydrogen Blocking p-SiO," IEDM, p. 22, 1988.
120. Takahashi, J., et. al., "Water Trapping Effect of Point Defects in Interlayer Plasma CVD SiO$_2$ Films," V-MIC Conference Proceedings, p. 331, 1992.
121. Jain, V., et. al., "Improved Hot Carrier Reliability of Submicron MOS Devices by Modifying the PECVD IMO Film," V-MIC Conference Proceedings, p. 272, 1991.
122. Jain, V., et. al., "Internal Passivation for Suppression of Device Instability Induced by Backend Processes," IEEE/IRPS Proceedings, p. 11, 1992.
123. Homma, T., Murao, Y., "A Room Temperature CVD Technology for Interlayer in Deep Submicron Multilevel Interconnection," IEDM, p. 289, 1991.
124. Smith, D.L., et.al., "Mechanism of SiN$_x$H$_y$ Deposition from NH$_3$-SiH$_4$ Plasma," J. Electrochem. Soc., 137(2), p.614, 1990.
125. Sun, R. C., Clemens, J. T., Nelson, J. T., "Effect of Silicon Nitride Encapsulation on MOS Device Stability," IEEE/IRPS Proceedings, p. 244, 1980.
126. Harrus, A. S., Van de Ven, E. P., "New Passivation Schemes Needed for VLSI," Semiconductor Int., p. 124, May 1990.
127. Yamaji, T., et. al., "Suppression of Migration in Al Conductors by Lowering Deposition Temperature in Plasma CVD SiN Passivation," IEEE/IRPS, p. 84, 1991.
128. Menz, K.D., et.al., "Low Stress Oxide/Nitride Passivation" Topography and Influence on Electrical Devices," V-MIC Conf. Proc., p. 384, 1990.
129. Peek, H., Wolters, R., "Bubble and Cavity Formation in Al-Plasma SiN Structures," V-MIC Conference Proceedings, p. 165, 1986.
130. Fujita, S., et. al., "Silicon Nitride Films by Plasma CVD from SiH$_4$-N$_2$ and SiF$_4$-N$_2$-H$_2$," IEDM, p. 630, 1984.
131. Nguyen, S., et. al., "Plasma Deposition and Characterization of Fluorinated Silicon Nitride," ECS Ext. Abst., G2-1, p. 209, 1992.

132. Flamm, D. L., et. al., "A New Chemistry for Low Hydrogen PECVD Silicon Nitride," Solid State Technology, p.43, March 1987.
133. Chu, J.K., Sachdev, S., and Gargini, P., "Plasma CVD Oxynitride as a Dielectric and Passivation Film," ECS Ext. Abst. 83-2, p.510, 1983.
134. Alexander, K., Hicks, J., Soakup, T., "Moisture Resistive, UV Transmissive Passivation for Plastic Encapsulated EPROM Devices," Proceedings of IEEE/IRPS, p. 218-222, 1984.
135. Stinebaugh, W. J., et. al., "Correlation of Gm Degradation of Sub-Micrometer MOSFET's with Reflective Index and Mechanical Stress of Encapsulation Materials," IEEE Trans., ED-36(3), p. 542, 1989.
136. Gaeta, I. S., Wu, K. J., "Improved EPROM Moisture Performance Using SOG for Passivation Planarization," IEEE/IRPS, p. 122, 1989.

5

ORGANIC DIELECTRICS IN MULTILEVEL METALLIZATION OF INTEGRATED CIRCUITS

KRISHNA SESHAN And DOMINIC J. SCHEPIS
International Business Machines
East Fishkill New York
And
LAURA B. ROTHMAN
International Business Machines
Yorktown Heights, New York

1.0   GENERAL INTRODUCTION

We turn our attention now to the interlevel dielectric used in multilevel metal interconnections, and in particular, organic dielectrics which have recently become the subject of much work due to their unique properties. Advances in integration and scaling have opened new opportunities for these materials. Let us examine this further.

Future scaling of silicon integrated circuits to submicron dimensions has three consequences. First, the major component of propagation delay will transfer from the device to the interconnection wires. Second, the increased circuit integration, together with scaling, will require more interconnections, and finally, the wires will have to carry higher current densities. Therefore multilevel interconnection technology will have to consider the use of new conductors, insulators and planarized three dimensional structures to satisfy the demands of scaling and integration.(1)

One method of reducing the propagation delay is to use organic dielectrics with lower dielectric constants. Going hand in hand with this is the switch to lower resistivity materials, mainly copper-based alloys. This chapter concentrates on the properties and the processing of organic dielectric materials. In this introduction, we examine some of the reasons for the use of the "conventional" AlCu wiring with SiO_2 as the dielectric and the requirements that any new wiring scheme will have to meet.

Multilevel metal/dielectric interconnection technology (dimensional ground rules and electrical properties) strongly impacts chip size and performance and plays an important part in chip cost, yield, and reliability. Therefore, several considerations are involved in changing from one wiring/dielectric scheme to another. Historically, the AlSi or AlCu metal conductor in a SiO_2 or glass interlevel dielectric evolved first and is the industry standard. This scheme generally uses AlCu sputtering for deposition and reactive ion etching (RIE) to etch the metal. The dielectric, oxide or glass, is deposited by one of several well known techniques including sputtering, plasma enhanced chemical vapor deposition (PECVD), or spin-on glass (SOG). Via holes are then etched in the dielectric by RIE, followed by deposition of the next layer of metal. This so called "conventional etch technique", where the conductor is deposited over oxide with open vias, patterned and etched, has dominated present interconnection technology. This scheme, however, has its performance limitations (1) and becomes difficult when used in VLSI/ULSI systems as will become evident with further reading.

The choice of AlCu metallurgy evolved primarily because Al is easy to deposit and etch, compared with Cu, Ag or Au, which are other choices for low resistance conductors. It was found that Al, when carrying high current densities, is prone to electromigration. The most common metallurgy in the industry is Alx%Cu (where x is between 1 and 4%), a composition reached after extensive research, notably at Motorola by Jim Black and co-workers. It is believed that the Cu segregates to the Al grain boundaries and prevents electromigration.(2, 3) Difficulties in changing to Cu based metallurgy include the fact that Cu is not easily etched by RIE and is easily oxidized by residual moisture.

The choice of SiO_2 is also historical. The dielectric must be robust, free of pinholes, and one which can be patterned and etched with wet solutions or by plasma. It should also have a low dielectric constant, so that its capacitance is low and thus allows high speed signal propagation. SiO_2 satisfies all of these requirements. Thermally grown oxide for the recessed oxide layer, as well as sputtered or CVD silicon dioxide for successive ILD layers, are robust insulators for MLM applications. More complex structures have been suggested using dual dielectrics such as oxide-nitride repeating layers which can offer reduced defect densities. New organic dielectrics must match and even surpass these properties to be considered as a replacement for these proven insulators.

A change from this conventional process is required for several reasons.

First is the loss of planarity in the conventional process. Sequential deposition of multiple layers causes topology to develop which leads to difficulties in both reliability (step coverage) and lithography. Lithographically it is difficult to pattern a non-planar surface due to depth of focus limitations. The topology also causes shadowing during metal deposition. Some form of planarization is necessary, requiring many additional processing steps, and a further complication is the difficulty in gap free deposition of dielectric between metal lines.

A second reason to change from $AlCu/SiO_2$ is the need for higher performance, which calls for low resistance and a reduced dielectric constant. The dielectric constants of SiO_2 and Si_3N_4 are 3.9 and 7.5, respectively. These are high compared with organic dielectrics which have dielectric constants ranging from about 2.5 to 3.5. The resistance of AlCu alloys is in the range of 3.2 μOhm-cm, higher than copper-rich alloys which are less than 2.0 μOhm-cm. Solomon (4) has argued for cryogenic operation which reduces metal resistance. Perhaps a more attractive alternative involves the use of a Cu-rich alloys, with resistivity less than 2 μOhm cm (5), and an organic dielectric insulator like polyimide with a lower dielectric constant of about 2.9. Such a structure has been demonstrated by Small & Pearson (1) for an experimental 64kB SRAM chip.

A third reason for change is that density drives the requirement for a smaller line to line pitch. This unfortunately increases crosstalk between interconnections. This can be offset by the use of organic dielectrics with lower dielectric constants.

A planar Cu-polyimide wiring scheme described by Small & Pearson is shown in Figure 1. The first tungsten contact is made in PECVD oxide. A trilayer insulator structure of Si_3N_4/ polyimide /PECVD Si_3N_4 is then deposited. The organic dielectric constant is between 2.8 and 3.1. The Si_3N_4 layer acts as a Cu diffusion barrier, forms an adhesion layer between the polyimide layers, prevents Cu hillock formation, and prevents solvent absorption by the completed polyimide layer when the layer above is spun on.The metal described is 500Å Ta/Cu/Ta, where the Ta helps adhesion and prevents diffusion of Cu. Low contact resistances were reported with 90% yield on 64 kB SRAMs. Reliability was also good, demonstrated after the chips were treated for 12 thermal cycles (30 minutes at 400 C) and 100 cycles to 77 K. No effects of stress, such as change in line resistance, and diffusion effects, as measured by diode leakage measurements, were observed.

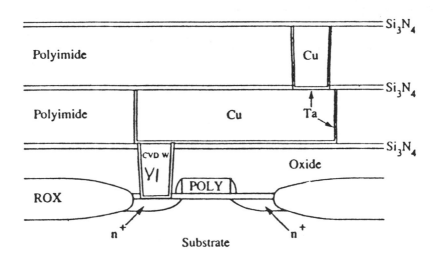

Figure 1. A planar Cu-polyimide wiring scheme, after Small et al. (1). Reprinted with permission.

The change to an organic dielectric will call for a new process sequence. The design of such a sequence will require qualification of new materials. A judicious design reduces cost and maintains or improves reliability. Cost, complexity, reliability, and performance issues pose a conundrum.(6) One of the main attractions of organic dielectrics is their low cost; in fact polyimides are used in many cost-performance offerings, as pointed out in Section 2. Photosensitive polyimides are particularly attractive, as several photoresist apply-and-remove steps may be eliminated.

This chapter talks specifically about the use of organic dielectrics in multilevel metallization. This section is meant as an overview of the emergence of organic dielectrics and as an introduction to some applications. These materials are reliable, inexpensive, defect free and can withstand high (400 C) processing temperatures.

Section 2 of this chapter presents the reader with a historical introduction which describes the realization in the late 1970s of the need for planarity and low dielectric constant, and the search for high temperature organic dielectrics. Section 3 introduces the reader to some of the chemistry of polyimides. Sections 4 and 5 describe processing steps to build a multilevel metallization scheme, and sections 6 and 7 conclude with some reliability and performance issues. A discussion of some possible future trends for this technology are presented in section 8.

This chapter is written as both a general introduction to the beginning engineer, as well as a reference for the experienced engineer. References are given to several companies actively involved in this growing field which the authors hope will increase the reader's interest in these new materials. Let us begin with a discussion of the historical evolution of organic dielectrics.

2.0   HISTORICAL PERSPECTIVE

The application of organic dielectrics to ICs began in the early 1970s when the use of polyimide for multilevel metal interconnections was reported by the Japanese.(7)   The first publication talked about a completely new planar method permitting step-free multilevel interconnections. It was the desire to achieve some degree of planarization which provided the incentive to investigate polyimides as dielectric materials.

The need for planarity arose in the 1970s due to the need for multilevel metal. Initially just two levels of metal were required, and Hitachi introduced a material called PIQ which was incorporated into routine production in 1975. The PIQ material was described as a thermally stable polyimide resin specially synthesized for use on semiconductor devices.(8) For the via hole etch process, a hydrazine solution was utilized. The desirable properties which were noted were low residual stress (compared with CVD SiO_2), high temperature stability (450 C, compared with conventional polyimides at 400 C), good breakdown voltage strength, and low dielectric constant. The reliability tests reported showed excellent results.

Within the U. S., however, the use of hydrazine as an etchant was too hazardous for use in a manufacturing environment. Alternative via hole processes were developed for polyimide films. In 1975 Yen (9) described a technique where a partially cured polyimide film could be etched with a caustic positive photoresist developer solution. In 1978, IBM introduced a new semiconductor memory technology called SAMOS which incorporated polyimide as part of the dual dielectric insulators between metal layers and as a top layer passivation coating.(10, 11) The dual dielectric was composed of silicon dioxide and polyimide layers. The via holes were etched in the polyimide using one mask level where the photoresist developer additionally served as the polyimide etchant. After the photoresist strip, a second photoresist layer was applied to etch the vias through the silicon dioxide layer. Through the use of two dielectrics and two mask steps, an extraneous hole will occur only in the unlikely event that dielectric or photoresist random defects are coincident. This provided a significant reduction in defect density. The use of the polyimide cushion on the rigid silicon dioxide layer also provided improved mechanical properties, resulting in better integrity of metal line and metal interconnections.

A cross-sectional diagram of the SAMOS multilevel metal structure is shown in Figure 2. The two levels of polyimide can be seen, with the second layer being much thicker for passivation purposes. The lines terminate on lead-tin pads which are bonded by flip-chip technology onto the package.

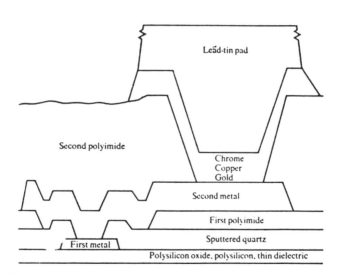

Figure 2. General view of metallization and passivation layers used in SAMOS. The structures are roughly to scale, with the exception of the much-reduced pad height after Larsen (11). Reprinted with permission.

One of the unique features of the SAMOS process was the built-in redundancy. After the second level of metallization, the wafers were tested for functionality. Chips which were non-perfect but repairable could have the failing addresses written into the second level metal by blowing the redundancy fuses. The final chip seal was provided by a second layer of polyimide, the thickness of which was sufficient to cover all metal exposed during the fuse-blowing operation. Only a polyimide layer could sufficiently cover the structure after the fuses were blown. This was one of the key aspects which kept polyimide in the process.

In the late 1970s and early 1980s the need to go beyond two levels of metal was being recognized by many.(12, 13) It was felt that the emerging techniques of VLSI would place increasing burdens on the technologies necessary to interconnect components of higher complexity and density on a single silicon surface. It was predicted that at least three levels of metal would be required with high resolution vias. The insulating layers should have reduced pinhole densities, superior dielectric properties, adhesive and temperature tolerant properties and provide no contamination to the underlying silicon devices. Most important was the need to provide gradual changes in surface slope and some self levelling or planarization to permit good step coverage in overlying high resolution conductor lines. In addition, an inexpensive, low temperature, controllable batch operation was highly desirable. (These same statements could be applied today.) The need for a planarizing material was two-fold: in order to attenuate the topography so that high resolution lithography can be done, and to provide reliability in the metal interconnections.

The first planar process, which was described by Sato (7), involved using polyimide to supposedly planarize the underlying metal pattern, followed by an etchback of the polyimide to expose the "bumps" or metal interconnections. Later it was shown (14) that due to the many geometry effects it is not possible to completely planarize a typical metal pattern with a polyimide film. Good edge coverage, however, is obtained with all polyimide films even when full planarization is not achieved. By making use of certain geometry effects, such as providing narrow spaces that will be planarized by polyimide, a process was developed for fabricating a planar multilevel metal-insulator structure.(15, 16) An example of the planarity achieved on a four level metal interconnection scheme using polyimide is shown in Figure 3 below.

There were several ways to obtain the integration density that was being projected. Other alternatives to multilevel metal were smaller ground rules or larger chips. It is apparent from the literature that many companies decided to implement the latter of these alternatives rather than explore the use of polyimide.

Possible reasons for the cautiousness with the use of polyimide were concerns about polarization and water absorption.(17) However, reliability data amassed over several years and specifically addressing polarization, water absorption, and possible corrosion consequences in non-hermetic packages have not contradicted the viability of polyimide.(18-20)

Several polyimide investigators have asked the question as to why polyimide is not more widely implemented in the semiconductor industry. Eight years after the first paper by the Japanese there had been several publications on the use of polyimide but very little about their actual implementation. Part of this may be due to the resistance to make changes - evolutionary vs. revolutionary practices within the semiconductor industry. One inhibitor to change are the skills available and the practices used in an organization. Change to new technologies is invariably resisted, particularly where the role of management has evolved into one of improving the efficiency of the existing systems rather than preparing and innovating for the "new". As a result, technologies are invariably introduced by newcomers to the industry. Despite the rapid advances in silicon chip technology in the past 20 years, VLSI has really evolved through steady process improvements. Therefore current skill levels and practices built up over the years in semiconductor fabrication discourage change to new technologies and new materials.(21)

Figure 3. SEM after four layers of metal completed. The polyimide is removed to illustrate planarity, after Rothman (15). Reprinted with permission.

The use of polyimide for packaging applications such as multichip modules started to appear in the 1980s. The people working on packaging technologies were already used to dealing with organic materials, so there was probably less of a barrier to its use compared with semiconductor technology. The multichip module technology borrows ideas from both the hybrid or printed circuit board and from the semiconductor areas.

In the late 1980s some reports of polyimide applications started to appear in the literature.(18) IBM introduced the use of polyimide in a metallized ceramic package with an insulating layer of polyimide to allow for an additional wiring plane. Figures 3 through 6 show examples of some of the packaging applications of polyimides. The films used in the packaging were generally thicker and the ground rules were much larger which made for relaxed tolerances on critical dimensions. Other articles appeared (20, 22, 23, 24, 25, 26, 27) which described the use of polyimide for high density thin film wiring on ceramic packages. The low dielectric constant of polyimide was the driving force behind its use. The excellent processing characteristics of polyimide coupled with its chemical resistance and high temperature stability were desirable.

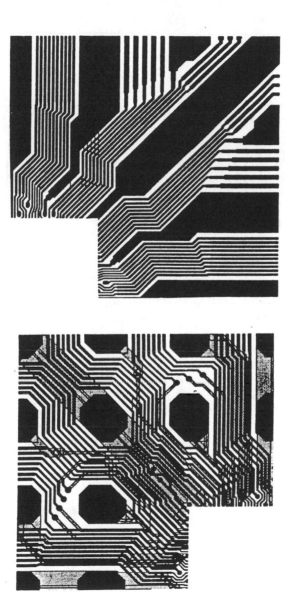

Figure 4. Photomicrograph of metal lines in polyimide on metallized ceramic after Homma and Posocco (18). Reprinted with permission.

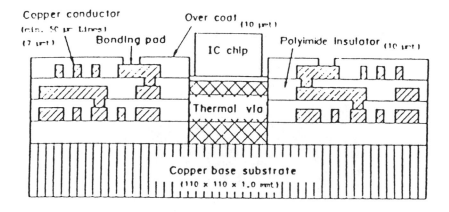

Figure 5. A schematic of the MCPM (Multilevel Copper-Polyimide-Module) structure, after Kimbara et al. (24). Reprinted with permission.

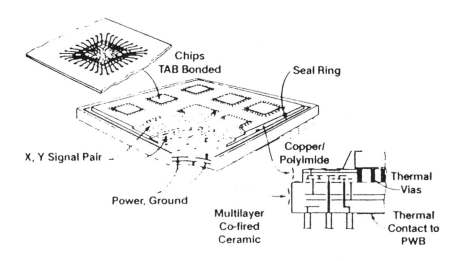

Figure 6. A multilayer thin film package structure showing the use of the copper polyimide wiring scheme, after Chao et al. (27). Reprinted with permission.

Significant advances in the development of new polyimide materials designed specifically for the semiconductor applications also occurred in the 1980s. Coefficients of thermal expansion (CTE) could be modified to match the CTE of metals or inorganic materials used in the fabrication of ICs. Resistance to moisture absorption could be achieved. Later, improved photosensitive polyimides became available. The list of advantages provided by polyimide included:

1. The use of a solution and the spin coating technique which allows: a) the possibility of cassette-to-cassette in line processing; b) planarization of underlying topography resulting in more reliable wiring; c) lower defect levels (compared to vacuum processes); d) low cost process; e) low temperature process without detrimental effects such as radiation damage or field enhanced contamination.
2. Vias are easily patterned by wet or dry etching with good slope control
3. Low dielectric constant (< 3.5)
4. Chemical resistance
5. High temperature stability
6. Versatility of chemical synthesis allows the optimization of polymer properties for specific applications.
7. Low stress allows for thicker films providing the ability to improve transmission line characteristics.

More applications of polyimide started to appear in the literature in the late 1980s. Polyimides are most popular as top passivation layers where they provide protection from moisture, corrosion, ion transport, and damage during packaging. Another type of passivation layer is the alpha particle barrier that is applied to high density memory devices to prevent soft errors. Polyimides are free of radioactive emitters and they effectively absorb alpha particles emitted by ceramics or other inorganic materials.(28, 29, 30) In addition, polyimide can be used as a stress buffer when applied at either the wafer level or at the packaging level. The use of polyimide as an interlayer dielectric has been implemented mainly on high density multilayer packages. The Japanese are leading the industry in the use of photosensitive polyimides as interlayer dielectrics in high density interconnect applications. Hitachi, Mitsubishi, Toshiba, and NTT have all announced HDI packages using photosensitive polyimides.(23, 24, 31)

The following sections will provide details on polyimides and other organic dielectrics, their processing and reliability. Hopefully the reader will find many applications for these versatile materials.

3.0 FUNDAMENTAL CHEMISTRY OF ORGANIC DIELECTRICS

Thus far in our discussions on organic dielectrics, we have discussed some of the advantages of these materials and their historical evolution. Let us now explore some of the chemical structures and physical properties that arise from these materials.

3.1   Materials Options

The class of materials that form these organic dielectrics are known as polymers. Polymers are composed of repeating units of smaller organic materials known as monomers. These monomers contain one or more reactive groups which under certain conditions react with one another to form long chains. For example, the schematic below shows a monomer A, which undergoes polymerization to form a chain of monomers.

[1]   A   -----> A-A-A-A-A-A-A-A-A

In other cases, two different monomers may react with one another to form a chain of monomers with an alternating structure as shown below:

[2]   A + B   -----> A-B-A-B-A-B-A-B-A

A polymer that is made of more than one component is called a copolymer. Most of the materials discussed in this section fall into this class. They generally form chains of very long length which can wrap around each other and add to the mechanical strength of the materials.

Most of these materials are high molecular weight polymers made from building blocks of aromatic organic monomers. The high temperature properties are generally obtained through a polymer cross-linking mechanism. This forms what is known as a class of compounds called network polymers. The materials begin as long chain polymers with reactive groups which can begin to cross-link at elevated temperatures and form 3-dimensional networks of polymer. Linear chain polymers such as Teflon do not contain these groups and will not cross-link upon further heat treatment.

The most common organic dielectrics used as multilevel metal dielectrics are materials that can withstand high processing temperatures, have low dielectric constants, and are easily integrated with semiconductor materials. A list of some of the more common organic dielectrics and some of their properties are shown in Table I below. While this is clearly not an exhaustive list of organic dielectrics, they describe some materials from several classes of polymers. For a more thorough treatment of high performance polymers, see Bureau, et al.(32)

Table I. Table showing the glass transition temperature Tg, the dielectric constant, solvent, and dielectric strength for various commonly used organic dielectrics.

Polymer	T_g C	ε_ρ	Solvent	Dielectric strength	Ref.
PMDA-ODA	450	3.6	NMP	4.0E3 V/mil	1
BTDA-ODA	-	3.5	NMP	4.0E3 V/mil	1
Teflon	<200	2.2	-	-	4
Benzo-cyclobutene	<360	2.6	Xylene	-	5
Polyquinoline	288	2.8	Cyclopent-anone	7.3E6V/cm	6
Thermid	320	3.1	NMP	-	57
PIQ	450	3.5	DMAC	1.0E6 V/cm	78
PPQ	365	2.7	Toluene	2.0E6 V/cm	6
Paralene	70	2.7	CVD	7.0E3 V/mil	2
Polyimide Siloxanes	>200	2.9	Diglyme	1.2E6 V/cm	3

Some materials which have the best dielectric properties exhibit certain undesirable properties. Teflon, for example has a dielectric constant of about 2 at room temperature. Its melting point however, is in the range of only 260-280 C.(33) This provides certain limits to the type of processing that can be done after the dielectric layer is in place. In general, metal depositions which occur after the organic dielectric is in place cannot exceed the glass transition state temperature (Tg) of the polymer. These metals are usually sputtered or evaporated on to the polymers which requires the polymers be compatible both chemically and mechanically to the deposition process.

To get a more general view of some of temperatures involved in conventional semiconductor and metallization processing, several process steps and their temperatures are plotted in Figure 7. As shown in the figure, only organic materials with reasonably high melting points are usable for today's multilevel metal interconnect schemes.

As shown in Table I, several leading candidates for organic dielectrics have emerged due to their combinations of superior processability, thermal stability, and dielectric properties. Many of these polymers contain aromatic rings or heterocyclic structure. Chains made from these compounds tend to have more thermal stability than simple aliphatic (saturated) structures. As we have discussed earlier in the historical section, polyimides have emerged as the most common family of polymers studied for multilevel interconnect technology. More recently enhancements to these materials have resulted in fluorinated polyimides (FPI) which have excellent thermal and dielectric properties. One fluorinated polyimide was evaluated at Motorola with an integrated tungsten via plug process.(34) Several new variations of Polyquinolines are also gaining attention due to their relative moisture insensitivity.(35) Dupont is also publishing data on a new Teflon material called Teflon-AF.(36) This fluorinated polymer has one of the lowest dielectric constants (epsilon approx. 1.89) of any polymer described for microelectronics applications.

For the sake of example, let us examine the most widely used class of materials, polyimides, for their chemistry and processing properties.

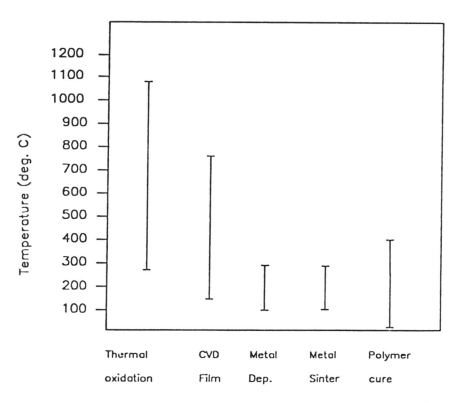

Processing temperatures for semiconductor fabrication.

Figure 7. Examples of processing temperatures for various semiconductor processing operations. No process over 400 C can follow the polymer cure.

3.2   Polyimide Structure

Polyimides are a general class of compounds which consist of a polyamic acid monomer. Upon polymerization, these compounds undergo condensation reactions to form the polyimide polymer chains. Figure 8 shows several dianhydride materials and a diamine. Pyromellitic dianhydride (PMDA) and

Benzophenone tetracarboxylic acid dianhydride (BTDA) are shown to react with the diamine oxydianiline to yield the polymers PMDA/ODA and BTDA/ODA respectively.

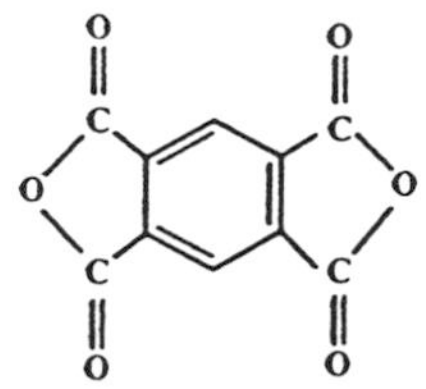

PMDA

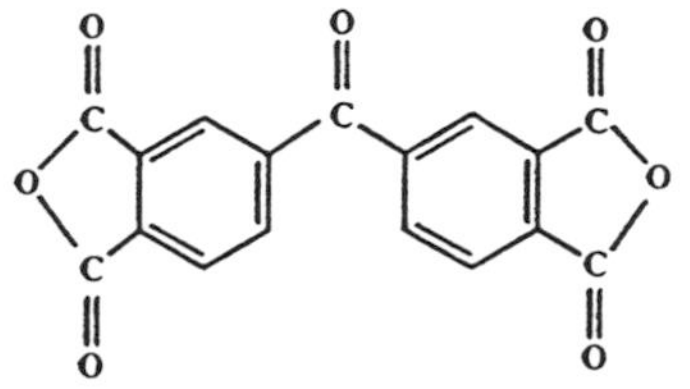

BTDA

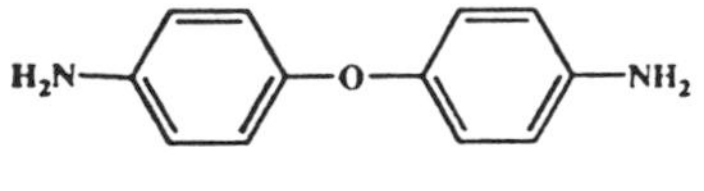

ODA

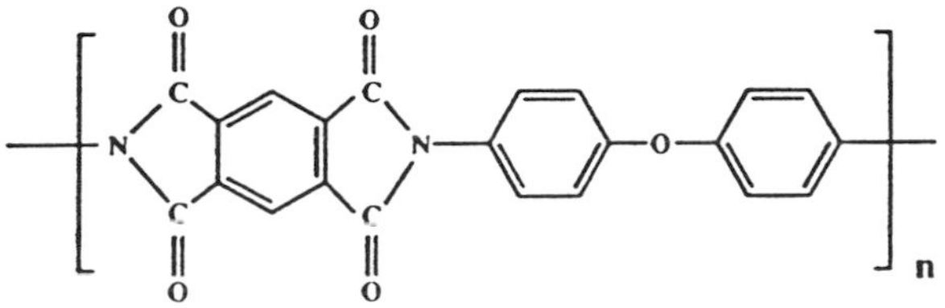

PMDA/ODA

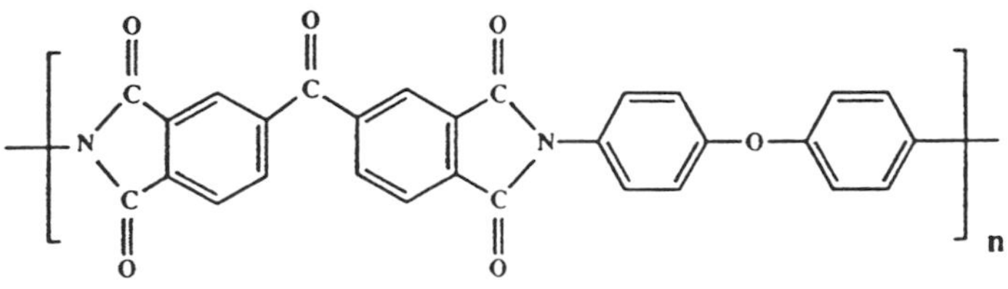

BTDA/ODA

Figure 8. The chemical structures of various monomers and polymers used in the synthesis of commercial polyimides.

The process of the formation of the polyimide polymer is a condensation reaction. For the sake of clarity let us again look at the PMDA/ODA reaction in Figure 9. Nucleophilic attack of the amine on the carbonyl group of the PMDA gives the amide intermediate. Upon further heating, the remaining acid moiety can react with another amine in solution or the amide on the adjacent leg of the molecule. Upon the loss of two molecules of water, the final condensed phase of polyimide is formed.

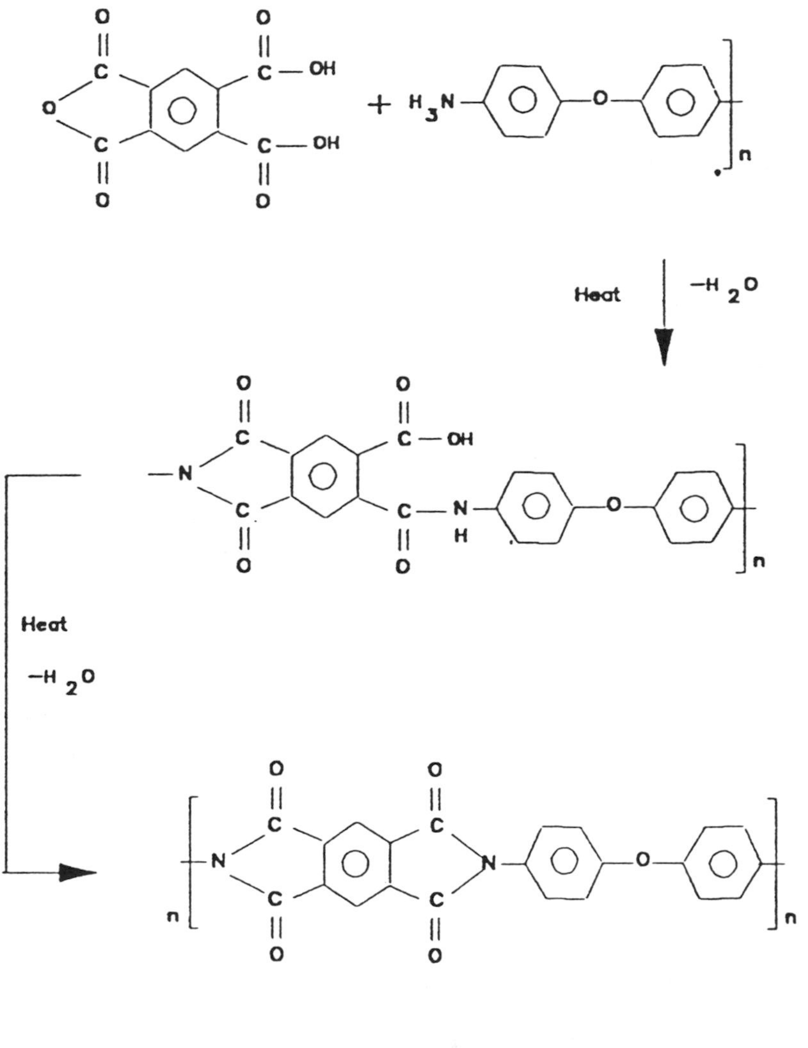

Figure 9. Diagram of the reaction sequence in the synthesis of polyimides.

Upon continued heating of many of the polyimide films, it can be seen that cross-linking between adjacent polymer films can be achieved. This cross-linking is important in imparting many of the high temperature properties of these films. This baking of the polymer to achieve the idealized film property is known as curing. Much work on curing has been done in the recent years to optimize the cure cycles with specific applications.

Figure 10 shows an idealized cure cycle for several other polyimide structures. Here again, high temperatures are required both to insure complete imidization as well as to drive off the water formed as a reaction by-product. A more detailed discussion of curing will be discussed in section 4 of this chapter, relating to specific process design.

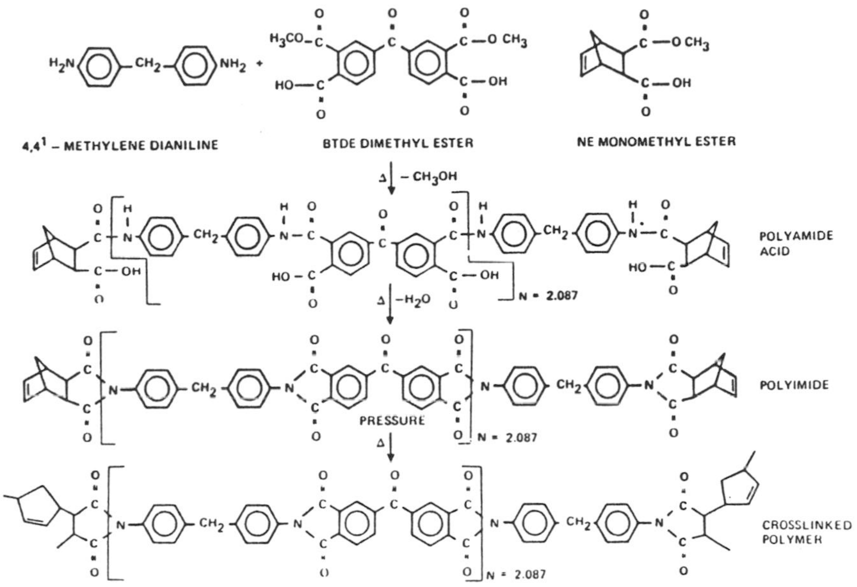

Figure 10. Shows the idealized cure sequences of PMR-15 resin.

3.3   Depositing Polyimides

The most common method of depositing polyimide films is spin casting or spraying from a polyamic acid solution. Here the choice of solvents, viscosity, and many other factors needs to be considered in order to choose the correct material for a specific application. The film is first baked at a temperature of about 120 C to evaporate off the solvent. The most commonly used solvent in many polyimide films is N-methyl, 2-pyrollidone (NMP). The cure cycle may be done in several steps at temperatures ranging from about 250-450 C for up to several hours. Here, the use of thermogravimetric analysis (TGA) is useful to determine the decomposition points for a particular polyimide film. A TGA curve for a typical polyimide is shown in Figure 11.(37) The solid line in the diagram refers to the percent weight loss as a function of temperature. The dotted line represents the derivative of the weight loss curve. The derivative defines two major transitions at about 200 C and 250 C as shown by the vertical dotted lines. This is useful in determining volatile components of the organic dielectric. To ensure proper adhesion of the polyimide layer to the substrate, adhesion promoters are generally applied. Here, as with many photoresists, organosilanes are popular materials. The silyl or siloxane moiety of the adhesion promoter forms a good bond with the available hydroxyl groups on surfaces such as silicon dioxide. The organic ligands provide a good surface for the polymers, giving rise to strong adhesion between the films. A more thorough discussion of adhesion is discussed in a later section.

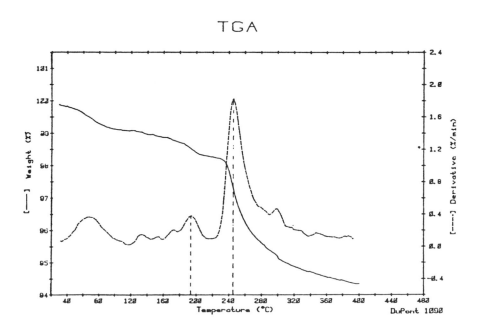

Figure 11. Shows the TGA analysis of a polyimide done in He from 40-400 C.

3.4   Moisture Absorption

One of the most critical properties in the use of polyimides in multilevel interconnect dielectrics is their ability to absorb water. The films tend to be hygroscopic after curing when allowed to come in contact with the atmosphere. Water itself is a by-product of the condensation reaction to form polyimides as mentioned earlier. Furthermore, the polymer contains many oxygen atoms which may impede the liberation of water due to hydrogen bonding. Usually, after sufficient curing, however, the films will contain very little water content.

Water in the polyimide films affects several of the physical, material, and electrical properties of the film. The amount of water absorbed in polyimide dielectrics has been shown to be directly related to the dielectric constant.(38) A graph of the moisture content vs. dielectric constant is shown in Figure 12.

Humidity affects many other properties of the polyimide films besides dielectric constants. The AC dielectric strength is decreased with increasing water content. Furthermore, the dissipation factor increases dramatically with increasing humidity. The dissipation factor, or tan delta as it is sometimes referred to, is a measure of the signal loss in the dielectric as a function of frequency. This loss factor is associated with the slower rate of change of polarization with respect to the electric field. Electronic and molecular polarization contribute to the dissipation factor in organic dielectrics.

As is shown in Table II, humidity affects the dielectric properties of polyimide films tremendously. The dielectric strength drops precipitously with increasing humidity along with a considerable rise in the dielectric constant.

Table II. Relative humidity vs. the electrical properties of Kapton.

% Relative Humidity	AC Dielectric Strength V/μm [V/mil]		Dielectric Constant	Dissipation Factor
0	339	8600	3.0	.0015
30	315	8000	3.3	.0017
50	303	7700	3.5	.0020
80	280	7100	3.7	.0027
100	268	6800	3.8	.0035

(For calculations involving absolute water content, 50% RH in our study is equal to 1.8% water in the film and 100% RH is equal to 2.8% water, the maximum adsorption possible regardless of the driving force.)

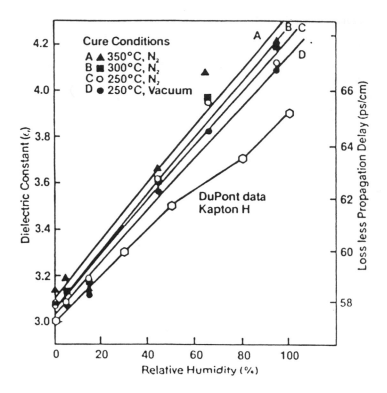

Figure 12. Shows the variation of dielectric constant with relative humidity.

3.5 Solvent Effects

The solvents used in the various polymer films can have a major effect on the properties of the materials. For example, spin casting a film from a low boiling point solvent may cause striations in the film uniformity and a high degree of mechanical stress and thinning if the solvent is boiled off too quickly. High molecular weight polymers generally require a highly polar solvent to keep them in solution over a long time period. In the case of polyamic acids, the precursors to polyimides, the solvents even affect the rate of imidization. Solvents which have too high a boiling point may also be difficult to remove completely during the cure cycle, which makes them more likely to outgas during subsequent processing. This can be a major reliability impact on multi-level metal interconnection processes, especially when a cap of nonporous insulator is coated over the polyimides. Outgassing of solvents can exert pressure from within the matrix causing distortion or breakage of metal lines. Finally, if the solvent used can hydrogen bond to oxygen atoms in the organic dielectric, it can make the solvent much more likely to remain trapped inside the polyimide, even after full cure cycles.

3.6   Oxidation

Since the chemical bonds in the polyimide tend to be quite stable, very little oxidation or weight loss of the film is observed at temperatures below 300 C. With the exception of solvents, the film will begin to show weight loss at temperatures above 400 C, especially in an air ambient. When the film is processed in an inert gas, the temperature the film can withstand increases to more than 500 C. The isothermal weight loss can be seen for various temperatures and ambients in Figure 13.

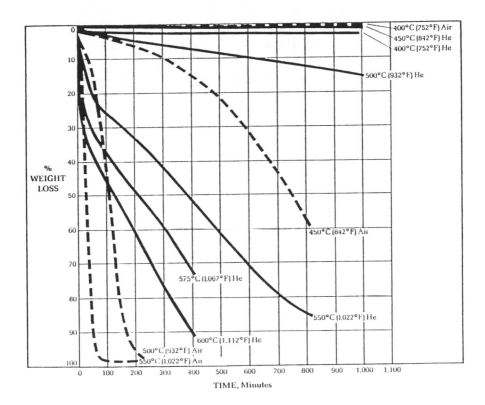

Figure 13. The isothermal weight loss vs. time in different ambients, from (33). Reprinted with permission.

3.7   Dimensional Stability

The dimensional stability of the film or shrinkage is dependent on several factors. The differences in the coefficients of thermal expansion between the film and substrate cause residual stresses to be placed on the films during manufacture. The shrinkage in a normal multilevel metallization scheme is also very critical. If the shrinkage causes the polymer to pull away from the metal lines, an air gap becomes available which can provide a channel for corrosive agents or moisture. The thermal coefficient of expansion varies from polymer to polymer but must be considered when dealing with reduced lithography tolerances. Some typical values for Kapton polyimide films are given in Table III. The stress put on the films and the wiring during curing can affect the overall reliability and integrity of the structure. These stresses may limit the overall dimensions of the multilevel metal structure. These same dimensional characteristics which affect the bulk films become important when dealing with photosensitive polyimides. This topic will be discussed in a later section.

Table III. Shows the thermal expansion coefficient of Kapton at different temperature ranges.

Temperature Range	ppm/ C
23-100 C (73-212 F)	18
100-200 C (212-392 F)	31
200-300 C (392-572 F)	48
300-400 C (572-752 F)	78
23-400 C (73-752 F	46

(Type HN film, 25 μm thermally exposed)

3.8   Metal-Polymer Interactions

A great deal of work has been done to understand the interactions between metal and polymer surfaces. Let us begin with a discussion of an ideal case where individual polymer molecules are free to interact with metal atoms. The field of organometallic chemistry is very complex and the literature is filled with cases of organometallic molecules, charge transfer complexes, and metallocenes. We do not have the time in this text to investigate all of the proposed structures; however, let us take a simple look at an ideal metal to polyimide surface.

It has been theorized by Chou (39) that the reaction of chromium metal with PMDA-ODA is a delocalized bonding between the Cr atom and the PMDA monomer. Figure 14 describes a condition whereby the d-orbitals of the chromium atom can overlap the pi-electrons on the PMDA molecule to form a stable charge transfer complex. Photoemission spectroscopy was used to obtain data which was analyzed using quantum chemical calculations of the lowest unoccupied molecular orbital for the PMDA molecule. Results show that the most energetically stable configuration comes from this d-orbital which overlaps constructively with the pi-orbital of the PMDA monomer.

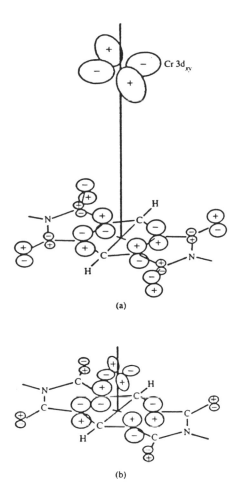

(a)

(b)

Figure 14(a). Highest occupied molecular-orbital (HOMO) diagram for the lowest unoccupied molecular orbitals (LUMO) of the PMDA monomer when chromium atom is sufficiently distant to be non-interacting. (b) Chromium over the six-member central ring of the PMDA monomer. The phases of the d(xy) levels add constructively to the pi-levels of the monomer, giving rise to a stable bonding configuration. After Ho et al. (40). Reprinted with permission.

If we extrapolate this understanding of these transition elements with unfilled d-orbitals for bonding, we can expect there to be similar complexes formed between polyimides and these metal surfaces. If we use a measurable property such as adhesion to examine the polymer-metal interface, we might expect these refractory metals to have a higher bonding energy than those of the noble metals, which have few or no unfilled d-orbitals available for these bonding interactions. Indeed, it has been observed (40) that the adhesion energy and peel strength for Ti and Cr are greater than for those of Al and Cu, which are

traditional metals used in semiconductor metallurgy schemes. The observed
trend for these metals is shown in Figure 15 below.

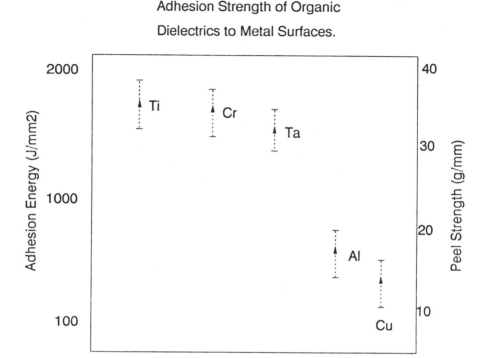

Figure 15. Shows the variation of adhesion energy and peel strength for 3d
transition metals and 4s noble metals, data compiled from various sources.

If indeed this trend is observed for most dielectric polymers, then process
design engineers need to develop a metallurgy which is consistent with all of the
necessary properties of VLSI interconnects. In order to obtain the low
resistivities such as Cu and Al and also have the enhanced adhesion of some of
the transition elements, several schemes have been suggested (41) which build
on a layered structure of a chromium liner over and under the polymer for
adhesion, with a sandwich of Al in between. This has been shown not only to
improve the adhesion and conductivity of the lines, but to improve the
electromigration lifetime as well. These added benefits have led to a host of
novel metallization schemes and alloys which can be used with both organic and
inorganic dielectrics.

While the understanding of the bonding between metals and polymers is complex, it is important to note that the the metal-polymer interface is very dependent on how the films are put down, in what order, at what temperature, and using what technique. Surely the case of coating an organic film over a metal surface produces one type of interface, while depositing a metal over an existing organic surface may produce another. It has been suggested that the deposition of a metal on a polymer film by sputtering can lead to a degradation of the polymer film surface which can be shown by spectroscopy to resemble a carbonized layer.(42) This organic-metal interface would clearly not resemble the idealized bonding structure proposed in Figure 14. Therefore, the process design engineer must carefully consider not only the metals involved, but the method of integration which will produce an interface specific to their application.

3.9   Photosensitive Organic Dielectrics

In the recent literature (43) there has been much discussion on photosensitive dielectrics. These materials have enormous potential since they can be selectively patterned, which leads to a reduction in the number of steps needed to build a multilevel metallization structure. In general, a photoactive component (PAC) is built into one of the arms of the monomers used to build the polymer. After illumination at the proper wavelength, the PAC is converted to a base-soluble material which can be removed by development in aqueous base solutions. This eliminates the need for an additional photoresist layer which is required to pattern conventional polyimides. The materials have certain limitations, however, such as control of the critical image dimensions. This is mainly caused by excess solvent in the photosensitive materials which cannot be baked out without decomposition of the PAC. There are also the mechanical stresses in the films which must be factored into the final linewidth variation.

3.10   Summary

Organic dielectrics are just beginning to show promise in applications such as insulators for multilevel interconnections. A great deal has been learned about these novel materials over the past several years. We have seen that the driving force towards moving to organic dielectrics is their superior electrical characteristics. Furthermore, these materials are lightweight, easy to apply, and available from several commercial sources.(14)

While specific organic materials have been mentioned thus far in our discussions, we in no way have scratched the surface of describing the vast field of organic dielectrics. In the case of polyimides alone, we have only mentioned the condensation of several amines with electrophilic anhydrides. In reality, there are nearly infinite combinations of such materials which can be polymerized to yield thin films of this type. Unfortunately, many are not suitable due to their instability at high temperatures or their dielectric constants.

In summary, it has been suggested by many working in the field that the ideal organic dielectric is yet to be found. Much work in future years should yield a host of new candidates, each with their own improvements in dielectric characteristics, moisture sensitivity, and thermal budget. The next decade should be an exciting time for research in these fascinating materials.

Now that we have explored some of the basic material properties and have discussed their strategic importance in semiconductor processing, let us now focus on specific applications and how these materials integrate with the substrate to form a complete metallized structure. In the next two sections, applications of organic dielectrics will be described with respect to multilevel interconnection technology. Section 4 will concentrate on isolated process steps, while section 5 considers specific process integration issues.

4.0   PROCESSING OF POLYMER FILMS

The performance advantages that organic dielectrics offer make them attractive in designing advanced semiconductor chips and packages. In order to take full advantage of these materials, a process must be designed which fully integrates the metallization steps with the polymer. Some of the difficulties of this integration come about from the inherent nature of the polymeric materials when exposed to the harsh chemical environment and elevated temperatures of semiconductor multilevel metal (MLM) processes.

Specialized equipment is used to apply polymer films, cure them, and incorporate metal wires and interconnects within them. As we shall see, certain processes must be developed specific to organic dielectrics where they differ from conventional inorganic materials.

In all processes, the polymer must be applied in a uniform and controllable fashion, with good adhesion to the substrate. Proper curing is critical, followed by process steps to deal with the problem of water absorption. The next few sections concentrate on these unit process steps and related issues.

4.1   Substrate Preparation and Polyimide Coating

Commercial polyimides come in various forms including thick films which may be applied by roll on applicators such as Riston.(33) In addition, films are applied by spin-on methods as well as spray techniques. In some cases the films may also be screen printed onto substrates. Spin coating is most often used and is done in the same fashion and using the same equipment as photolithography materials. Spin-on coatings are the method of choice since they produce films with excellent defect densities. Substrates are held to a spinning chuck with the aid of a vacuum and the polymer is applied.

A wafer is spun in the range of 3500-5000 rpm for a period of time approaching 60 seconds to give a uniform coating. After the initial coating, the wafers are baked at a moderate temperature to drive out the bulk of the solvent.

Most modern day spin-coating equipment is integrated with spin and bake stations in the same enclosed environment. The tools allow for coating of adhesion promoters and multiple layers of polymers. This coating is then cured by a multi-step anneal process.

In many cases, the polymer may not have good adhesion to the surface of the substrate unless an adhesion promoter is used prior the polymer coating. These substances are often organosilanes such as A-1100 or other similar materials. Certain polymers, however, do not require these adhesion promoters since the polymer films may exhibit good wetting to the existing surface. Since the organic dielectric may be coated on both chip and packaging substrates as well as metals and other layers of dielectric, let us examine the topic of adhesion more closely.

4.2 Polyimide Adhesion

Adhesion of the polyimide films to the underlying substrate is of paramount importance. Since the polyimide dielectrics are generally applied to a metal surface underlayer, much work has been done on the adhesion of polyimides to metals. The adhesion of polyimide to copper/chromium systems has been studied as a function of surface pretreatment.(44) It was suggested that for a series of PMDA-ODA systems, that exposure of the substrate to low energy sputtering with Ar or O_2 ions improves adhesion, while for BPDA-PDA systems, O_2 sputtering is more effective than Ar. In general, a surface pretreatment is practiced when coating most polymers, either by physical surface modification or by the application of an adhesion promoter. In some cases another polymer film can be used as a seed layer, providing the hydrophobic surface generally preferred by these films. Much work on polyimide surface modification has been performed by Shaw and Lee.(45) Some companies advertise organic polymers that themselves provide good adhesion to many surfaces so efficiently that no adhesion treatment is necessary.(35, 46)

Adhesion to ceramic surfaces has been studied by Buchwalter (47). In general adhesion promoters of the organosilane type can be used on ceramic substrates or on silicon wafers (48) with good results, while on metal surfaces the situation is more complex. Since the adhesion of the polymer is extremely surface specific, it is necessary to adjust the adhesion process to those exact materials.

Adhesion in general is usually measured with a peel test. The film is pulled away from the substrate, and the force that is applied is measured by a peel strength apparatus. The force is generally measured in g/mm. This provides a relative measurement to compare various materials. No standard test for an absolute adhesion measurement has been developed. A discussion of the use of the peel test for reliability testing appears in section 6. A schematic of a peel strength apparatus is shown below in Figure 16.(14) A force transducer is fixed on a sliding stage which is connected to the test sample by a long rod. In this particular apparatus the test sample consists of a strip of wafer of known width attached by two-sided tape to a glass slide. The wafer is then broken off leaving the film intact. The film is pulled at a consistent angle at a constant rate. This particular setup was enclosed in an environment of dry nitrogen. Several variations on this setup exist which perform the same basic functions.

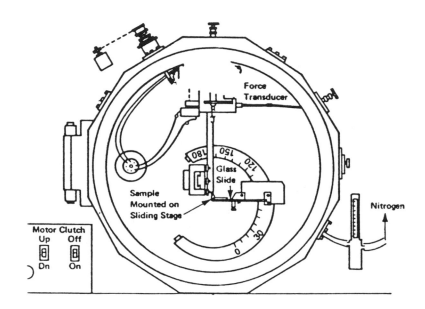

Figure 16. Schematic of a peel force test apparatus after Rothman (14). In this apparatus, the polyimide film is peeled away from the substrate. Reprinted with permission.

Many investigators have measured polymer adhesion on top of metal and ceramic substrates. Inversely, if polyimide is applied first and then subsequently coated with a metal film such as Cr, the polymer should first be RF sputtered to improve the adhesion.(49) This is thought to improve the chemical bonding between the metal and the polymer. Additionally, the adhesion of the Cr-polyimide interface is significantly degraded by exposure to high temperature and humidity environments. It is suggested that the degradation results from hydrolysis of the polyimide or the metal-oxygen bond. This conclusion is further supported by Seshan, (50) who found that the isotope O^{18} was trapped at the polymer-metal interface after allowing diffusion of water through the film. This build up of tracer material was in direct correlation to the loss of adhesion at that interface.

Finally, the adhesion can be affected by the bake cycle or "curing" of the dielectric material. Let us examine this more closely.

4.3   Curing of Polyimides

After the application of organic dielectrics, the films are baked at high temperatures in a process known as curing. The curing of films is important in order to impart the necessary properties of the organic dielectric. The cure cycle is generally done in several steps of increasing temperature, which drives out solvent and provides full imidization. The optimum cure conditions have been found experimentally and also have been modeled.(51) Commercial films of PIQ L100, for example were modeled to find the optimum conditions for internal stress, imidization, and molecular orientation.

The type and duration of curing of polyimide films after spin coating is critical to the final film characteristics. The polyamic acids and the polymer backbones must be cured at a high enough temperature to affect cross-linking without decomposing them. The cure cycle also affects planarity. If the cure is done in a vacuum at a low enough temperature, a planarizing effect is achieved.(52) Alternately, the films may be cured in an inert ambient using carefully controlled furnaces. Curing is generally done over a wide range of temperatures with the final bake approaching the T_g for the film. In many cases for polyimides, this final cure temperature is about 400 C. The curing is generally performed in a conventional furnace tube with a dry nitrogen ambient or in an integrated photolithography coating tool. Tight control on temperature, generally on the order of +/- 1 degree C, is important for reproducible film characteristics.

The length of time is also very important. The cure cycle dictates the amount of imidization which occurs and therefore the final film characteristics. Figure 17 demonstrates this for three different film thicknesses and two temperatures. As shown in the figure, the cure cycle can be extended such that, after a certain time, a constant degree of imidization is achieved where the data converges. This time can be adjusted for each polymer to ensure that a nearly complete imidization occurs. It is important to design the cure cycle toward the right of this curve where reproducible films can be realized.

4.4   Diffusion of Water

Diffusion coefficients for water in some polyimides have been measured by Chang at Motorola.(53) Diffusion of water through polyimide is fairly rapid, resulting in problems such as corrosion of metal surfaces as well as adhesion loss. The water content in the film also strongly affects the dielectric constant. The variation of the low frequency dielectric constant is directly proportional to the absorbed moisture.(54) These three negative affects of humidity make it imperative that the polyimide process takes place in a well controlled environment and that the package be hermetically sealed after the chip or package is complete. A graph of the effect of relative humidity versus dielectric constant was shown earlier in this chapter in Figure 12. Moisture absorption can be a long term reliability issue as will be discussed in a later section. Catastrophic immediate damage will occur if polyimide multilayer systems with absorbed water are rapidly heated during processing.

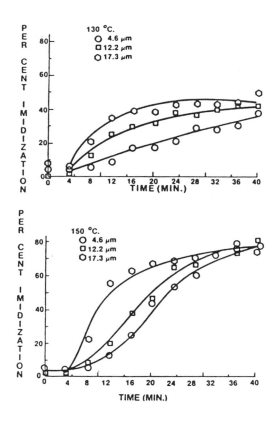

Figure 17. Shows the percent of imidization vs. time for cure at 130 and 150 deg. C.

4.5   Summary

We have examined some of the unit process steps and concerns that are part of using polymers to build multilevel metal structures. It will become apparent that incorporating these processes into a complete integrated structure is considerably more complicated and requires addressing interactions between the various layers of materials as the structure is vertically integrated.

In the next section, section 5, some of these steps of process integration will be examined in more detail and examples of organic dielectrics in multilevel metal semiconductor structures will be described.

5.0 PROCESS INTEGRATION WITH ORGANIC DIELECTRICS

There are several possible ways to build a multilevel metal interconnect structure.(15) In one case, the metal lines are first fabricated and the organic dielectric is then applied. In an alternate procedure, a dielectric is deposited and patterned and the metal is deposited into the grooves etched in the dielectric. The process will clearly vary based on the individual substrate, thermal budget, and organic material. Therefore, the individual application will drive the thermal budget and the process limitations. Thus far we have described generic techniques to build a structure with organic dielectrics. Assuming the first layer to be an SiO_2 based passivation layer, the first layer of organic material is deposited. If only a single layer is to be used, the processing would become simplified. Repeating this process to build multiple layers involves some additional considerations.

Numerous other details will affect the issues encountered during integration. In some applications, an inorganic RIE etch stop or passivation layers may be placed between organic dielectric layers. These thin films have little effect on the electrical performance but may ease the critical process integration steps. On the negative side, these thin inorganic etch stop layers are subject to cracking or delamination at later levels. The type of metallization will also drive the type of integration process that is used. For tungsten plugs, the polyimide vias may first be coated with a liner to both provide a nucleation site for the tungsten deposition as well as to buffer the polymer from the compressive stress that is generally associated with metal deposition.

Let us examine some of the procedures involved in the fabrication process.

5.1 Processes for Forming MLM Structures

A cross-section of a multilevel metal structure that could be found as part of the wiring scheme on a semiconductor substrate is shown schematically in Figure 18. It consists of several repeating alternating layers of metal and dielectric stacked upon one another to produce the final structure. In general, the layers closest to the semiconductor substrate contain the smallest wires with the upper layers used to distribute the input-output lines and provide power to various parts of the chip. Let us examine Figure 18 starting at the layer labeled M1.

This wiring level may contain many centimeters of metal, connecting contacts on a non-planar surface which is generally built upon an inorganic dielectric, usually SiO_2.

This dielectric should act as a getter of impurities and also be non-corrosive to the metal contacts below. There is also a concern about metal ion contamination sometimes present in organic materials (eg. Na, K, Ni, Fe) which could poison the semiconductor junctions if allowed to come in contact with the silicon. This is especially true of today's polysilicon electrode devices, where metals diffuse quickly through the polysilicon grain boundaries into the junctions. For these reasons, polyimides have not been used as the primary insulator at the semiconductor surface. Generally, an oxide layer is used as the first dielectric layer, usually as a thin film. The oxide layer may be pure or doped with gettering ions such as phosphorus.

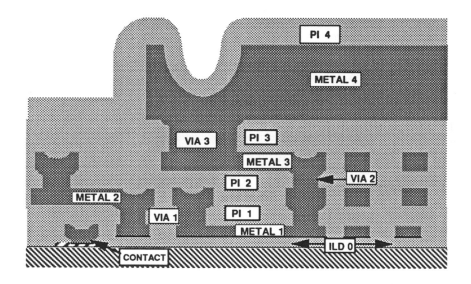

Figure 18. Shows the cross-section of a multilayer metal to polyimide structure. Notice that ILD 0 is oxide.

Upon this surface the next levels of metal are fabricated and are isolated from one another and the substrate by organic dielectrics. A simple schematic of the process steps required to build a multi-level metal interconnect technology using organic dielectrics is shown in Figure 19. It can be seen that the process consists of repeating basic building blocks of insulator and metal.

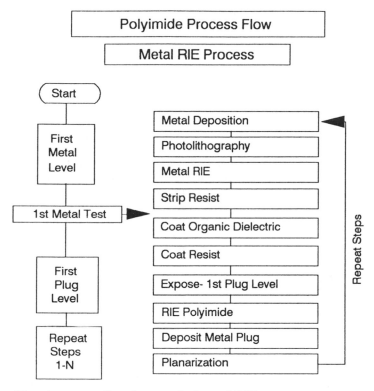

Figure 19 . A process flow for a typical metal RIE process.

General process design is built around specific steps. Each step has specific effects which must be considered. The schematic that follows gives a generic process flow for a particular type of process known as lift-off.

	Steps which repeat	Comments
1.	Apply adhesion layer	Preclean of surface required.
2.	Spin-on polyimide	May be multi-coat film procedure.
3.	Multi-step cure (100 C/200 C/300 C/400 C)	Multi-step; optimize for specificprocess. Ambient effects, N_2 during final cure. Also stress
4.	Apply hardmask layer (SiO_2)	
5.	Apply photoresist (bake)	
6.	Expose	
7.	RIE hardmask	Anisotropic RIE
8.	RIE polyimide	Anisotropic RIE with undercut.
9.	Metal Deposition	Directional
10.	Lift off	

Initially, pillars of metal are formed over predetermined areas in the M1 metal line. The pillar is formed in a contact through the first polyimide layer. As mentioned earlier, this may be performed by the opening of a via hole in the dielectric using conventional photolithographic techniques. This is usually accomplished by the curing of a blanket polyimide layer over the substrate followed by an inorganic film such as SiO_2 which is used as a RIE hardmask. The substrate is then coated with conventional photoresists, and the pattern is transferred into the hardmask layer using plasma etching in a gas such as CF_4. The gas is then changed to oxygen to etch the polyimide layer below. Control of the RIE conditions can cause anisotropic etching of the film or intentional undercuts of the hardmask layer if desired. In general, an undercut profile is desired when a lift-off metal is evaporated in a directional manner. The excess metal is then lifted off by a release layer to form a pillar of metal which is brought to the surface of the organic dielectric for contact by M2. The lift-off method has been used in recent years due to its planarity and relative ease of process. A schematic of the lift-off process is shown in Figure 20. The process has been used with both inorganic and organic films and for both semiconductor and packaging applications.

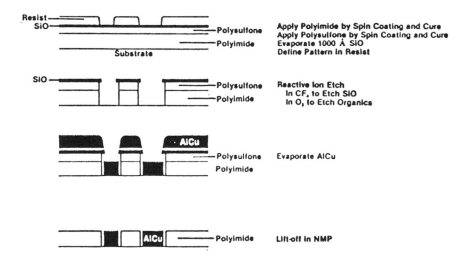

Figure 20. Process description for forming the first level of metal via a lift-off process (15). Reprinted with permission.

After the pillars have been created, the next level of metal is deposited and defined. The conducting metal lines can be defined by methods such as RIE or wet chemical etching. The metal layers will conformally cover the organic underlayer and replicate the topography already present in the structure. A structure that is as planar as possible is desirable for reliable metallization structures and lithography.

The process of patterning, etching, and deposition becomes more restricted as the number of levels increases. The metal lines sandwiched between multiple organic layers are subject to the stress and deformations imposed by the polymer material's physical characteristics.

The lift-off process is extendable down to submicron dimensions but becomes more difficult since the small lines are subject to falling over. For larger patterns, isotropic metal etching has been the method of choice due to simplicity. VLSI technologies, however, require anisotropic metal RIE or other well controlled selective deposition techniques.

5.2   Patterning of Organic Dielectrics

The polymer etching is generally performed using oxygen RIE in an anisotropic plasma etcher. At the first organic insulator level, the oxygen RIE provides a selective etch to the M1 level, since both zero level dielectric and the metal are not etched by oxygen RIE. At subsequent levels, however, the etch becomes more complicated by the fact that vias which are not fully covering metal lines below may be overetched down below the metal line. The subsequent filling of these vias may cause shorting or reliability problems at later levels. This has led some investigators to incorporate thin "etch stop" layers in between the organic dielectrics. These thin films have little effect on electrical performance but may ease the critical process integration steps. Difficulties in these etch stop layers arise from the tendency for the organic layers to outgas during subsequent processing. This leads to cracked films or stresses affecting the planarity of the structure. The adhesion of the organic layers to the thin dielectric etch stop layers may also be subject to failure later on in processing.

More recently, an alternative approach for building multilevel metallurgy has been described, making use of photosensitive polyimides. The general basic structure is formed but uses a different process sequence. One of the attractive features of this system is the relative simplicity of processing steps. Figure 21 shows a comparison of the number of steps required to form a single layer of metal isolated by organic dielectrics. A description of some of the options employing these photosensitive materials will not be discussed here but has been described by others.(43) The photosensitive materials basically combine the photolithography and etching step into one process. Commercial photosensitive materials available today, however, do have their limitations. The photosensitive organic films are cured at much lower temperatures to prevent damage to the photoactive component (PAC). They therefore contain more solvent and are at about twice their completely cured thickness. This causes aspect ratio problems and dimensional instability. If the developed image undergoes shrinkage over time, this will lead to difficulties in controlling critical dimensions required for VLSI. For this reason, the materials have thus far been used primarily in packaging and relaxed groundrule applications.(54)

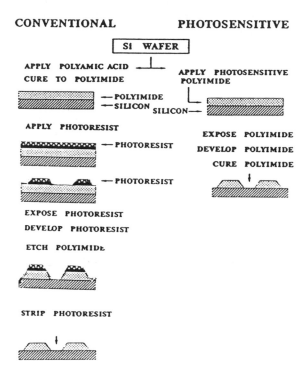

Figure 21. A comparison of the process steps using conventional and photosensitive polyimides.

5.3 Planarization

For several reasons, it is desirable that the multilevel metal insulator layers be as planar as possible. First, the film depositions of metal and insulator are most uniform when deposited as a planar thin film, as almost all deposition systems have a degree of non-conformality. This results in films on vertical surfaces being of differing thickness to films on horizontal surfaces. In the removal of metal and insulator films, planarity is even more critical. Since the removal of films is usually performed with anisotropic plasma etching, a non-planar surface results in rails or "stringers" around the highest objects which may result in short circuits.

More importantly, the step coverage of metals as they go over topography may cause reliability problems which may not show up as a time-zero fail.

Electromigration fails could be expected to form in these regions of incomplete metal thickness.

Finally, today's lithography tooling is designed for small geometries with large numerical apertures. A by-product of this engineering is that the depth of focus for these tools becomes very small, in many cases less than 1000 nm. If the non-planarity of the multilevel metal stack exceeds this topography, the lines on the next lithography level may not print uniformly across the chip.

One of the advantages of spin-on layers is that they tend to have a planarizing effect when coated over topography. They also have a very low defect density as compared with CVD films. While planarization on the wafer surface appears to be uniform at the local level, the surface is never fully planar after a single thin coat of organic dielectric. It has been shown that coating several thin layers produces a more planar coating than one coat of equal thickness.(15) It is therefore important to adjust the coating thickness and cure to coincide with the particular structure being planarized. A curve showing the degree of planarity as a function of numbers of coatings is shown in Figure 22.

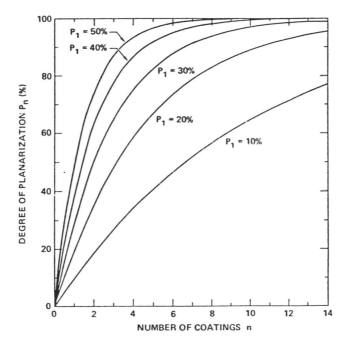

Figure 22. Diagram showing the degree of planarization vs. the number of coatings; P is the packing factor. As the structure becomes more dense, a larger number of coatings are required for planarization.

The planarity of the structure is not only affected by the number of coatings but also by the dimensions of the structures themselves. In general, when the linewidth and the spaces between the lines are kept small, the planarization will be more complete as opposed to larger lines and spaces. Furthermore, a particular substrate may have an isolated line surrounded by organic dielectric in one area and many lines close together in another area. This can lead to localized planarization of the isolated line only. In this case, it is important to concentrate on the areas which are most difficult to planarize on each particular chip layout. Figure 23 shows the effect of planarization as a function on linewidth and spaces for a particular polyimide film. In most cases, even multiple coats may not afford total planarized structures.

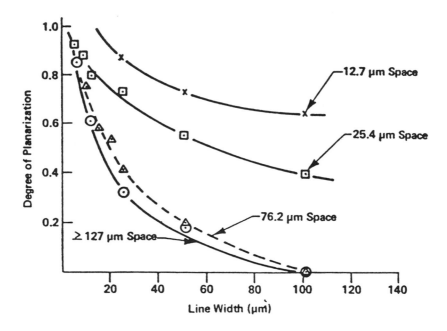

Figure 23. The degree of planarization vs. line width and line spaces.

Other factors affecting planarization relate to film characteristics, solvents, application techniques, and the film curing cycles. These need to be optimized for the individual user applications.

Thus, while local planarization is often achieved, complete global planarization is not achieved with polyimide coating. Several approaches have been used to address this. In the first approach, a photoresist is coated over the polyimide and exposed in the opposite density to the metal definition mask. The polyimide can then be removed over the metal lines by RIE etching, leaving a polyimide plug between the metal lines. This approach has been studied extensively by Chang, et al.(55) Another novel approach has been studied by

Chakravorty (56), who reports that uniform shrinkage of two polyimide layers leads to a planar structure.

As an alternate approach for producing structures with increased planarity, many engineers are investigating chemical-mechanical polishing as a method of planarization.(1, 57) This method produces planar surfaces; however, organic materials may crack or become distorted under the stresses of the polishing process. The polishing process for organic dielectrics is much more difficult due to the mechanical properties of the cured films. Some of the effects caused by metal or polishing stresses are wrinkling, cracking, adhesion loss between layers, and blistering.

5.4   Thermal Budget Considerations

The thermal budget for multilevel metallization processes has always been limited, since metals such as aluminum have relatively low melting points. In the case of organic dielectrics, processes are even more constrained since many of these materials break down at temperatures much above 400 C.

Furthermore, when several metals are used in alternating layers, metal intermixing can become a problem. Since electromigration and adhesion concerns can be improved by using multiple layers of metallurgy to form the primary conducting wire, it is important to choose the right combinations of materials and thicknesses. Several multilayered metallurgy schemes have been proposed and summarized by Mattox.(58) Some examples of thin film metallurgy are shown in Figure 24.

An Example of a Thin Film Multilayer Structure

2-layer	3-layer	4-layer
Au/Ti	Au/Pd/Ti	Au/Rh/Pt/Ti
Au/Cr	Au/Pt/Ti	Au/Ni/Cu/Ti
Au/Mo	Pd/Cu/Ti	etc
Au/Nb	Au/W/Ti	
	Au/Mo/Ti	

Figure 24. Table showing examples of various metallurgy composites used in multilevel metal wiring.

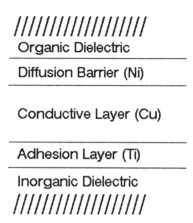

Figure 25. Schematic of the arrangement of the adhesion conduction and barrier layers commonly used in metallization schemes shown in Figure 24. (58) Reprinted with permission.

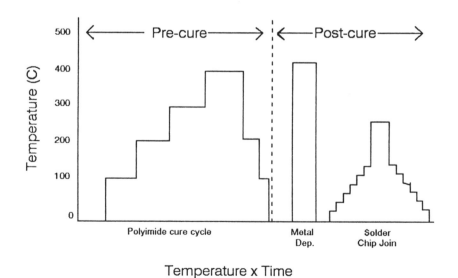

Figure 26. The Temperature-Time diagram for the thermal processes that the last layer of polyimide has to withstand.

The metals are arranged in a 3-layer fashion as shown below, in Figure 25, with the adhesion layer being the Ti or Cr. Problems related to thermal budget begin to compound as the structure is built. The first layer of metal is annealed both after deposition and after subsequent repetitions of polyimide curing and metal anneals. This further influences metal mixing. To illustrate this principle, a temperature-time curve is shown in Figure 26. It can be seen that the thermal budget issues define the material and process constraints for subsequent metal levels.

5.5 Examples of Organic Dielectrics in Semiconductor Technologies

While a great deal of studies of organic dielectrics have occurred over the years, the use of these materials in actual products has been limited. One of the earliest uses of a polyimide passivated metallurgy in semiconductor memory products was the SAMOS process, developed by IBM in the 1978 time frame.(11) Here a silicon dioxide and polyimide dual dielectric was employed at the first level of metal to reduce defect densities and also help improve reliability. The polyimide also provided improved mechanical properties. After holes were etched down to the first level metal, a second layer of metal was deposited and patterned. A second polyimide layer was then deposited across the wafer to fully passivate the structure. In this case, the film was somewhat thicker than that used at the M1 level. The polyimide was then removed over the M2 pads and in between chips. The final contact metallurgy was then evaporated through a mask to form the chips lead-tin pad connections. In the case of the SAMOS process a series of 5 metals was used. As discussed earlier, a thin chromium layer serves at a seal because of its excellent adhesion to both aluminum and polyimide. The chromium is also provides good corrosion resistance from the solder connections. A copper conductor was then used as the primary conductor along with a gold passivation layer. The lead-tin pads were then deposited. The resulting structure is illustrated schematically in Figure 2 back in Section 2.

It is interesting to note that a total of 5 metal evaporations were possible within the boundaries of the polyimide heat constraints. The thermal stability of the polyimide allowed the implementation of this advanced metallurgy into an LSI memory product. This technology was used to create a highly successful family of 18-64K bit memory chips.

In 1987, IBM also reported the use of polyimide dielectrics in a triple layer bipolar chip.(59) Again, the resultant product was found to have high reliability and planar surfaces for metallization. The process is also low cost since it does not require CVD tools for the dielectric deposition.

Some more modern examples of polyimide integration in multilevel interconnection structures have been demonstrated by Chang.(34) This integrated structure demonstrates the use of organic dielectrics to fabricate a full four levels of metal in a VLSI product. The structure is shown in Figure 27.

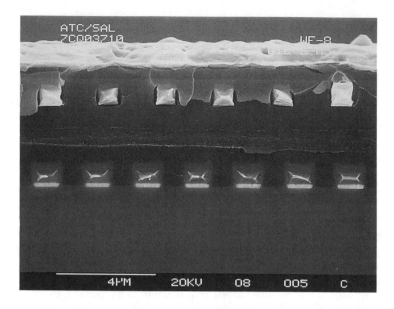

Figure 27. Cross section SEM showing 4 layers of metal and polyimide. Figure provided by Li Chang of Motorola, unpublished.

Small and Pearson (1) also describe the use of polyimides for advanced logic products to reduce wiring delay, as was shown schematically in Figure 1 of Section 1.0 of this chapter. The general flavor suggests that organic dielectrics will be studied by all companies that have an eye on the future of semiconductor manufacturing. The examples above already have demonstrated that the technology is mature enough to move from the laboratory into the manufacturing line.

Polyimides have also successfully been integrated into multi-chip packages by Honeywell.(60) These packages offer the same advantages for signal propagation as semiconductor metallization structures.

Finally, more researchers are describing techniques and applications for organic dielectrics for microelectronics. The growing interest suggests that the advantages offered by these films is finally beginning to be recognized.

5.6   Summary

During this discussion methods of fabricating multilevel metal structures have been simplified in order to bring up the various parts of building such structures. In reality, the cross-section shown back in Figure 18 is highly idealized, showing no detail of the metal or insulator stack. In general, the metal layers are actually composites of alloys, adhesion layers, and passivation layers. The insulator stack may contain other dielectrics such as silicon oxides or nitrides for plasma etch stops or passivation.

The process design to build a multilevel metal-polyimide structure involves the details of metal-metal, metal to polyimide, and polyimide-oxide interactions and their environment. Features such as pillars, plugs, contacts, and lines can be put down by evaporation or sputtering and patterned with wet or dry (plasma) etching. We have seen that there is no unique set of steps to build a multilevel metal structure. The need for adhesion layers may complicate the integration by requiring additional depositions and process steps. Coating these films over topography also presents a challenge by requiring either multiple coatings, specific cure conditions, or planarization techniques. While creative solutions for most of these integration problems have been demonstrated, their complexity may have up to now limited the use of organic dielectrics. Several examples of organic dielectrics in device structures have now been reported by a increasing number of companies and universities. It is expected that as more of these products become commercialized, the use of organic dielectrics in future products will substantially increase.

As we conclude this section on process integration, it is fitting that we now discuss the reliability of these films in actual multilevel metal applications. Section 6 that follows will discuss these issues.

6.0 RELIABILITY

The reliability of metal to polyimide multilayers falls into two broad categories: "during-build" design or reliability issues and "post-build" or "burn-in" reliability issues. This section discusses these aspects separately, although the basic mechanisms underlying the degradation of reliability are often the same.

"During-build" reliability testing and modeling is, and should be, concerned primarily with adhesion, adhesion monitoring, and the degradation of adhesion during the manufacturing cycle. Attention should be given to stress generation and propagation. Stress modeling should be used for optimizing shapes and aspect ratios of posts, pillars, and lines. Special attention should be given to the role of moisture. Both moisture and thermal aging can have important effects upon the mechanical properties of the polymer. Trapped charge and dielectric breakdown related to processing needs to be studied.

"Burn-in" or long term reliability testing is concerned with long term temperature-humidity cycles, conformation to mil-specs (military specifications), or ASTM (American Society for Testing Materials) specifications for commercial products. In special application specific ICs (ASICS), the customers may impose more stringent temperature and leakage requirements. Very often it is the electromigration of metal that limits the line lifetime. Therefore the electromigration testing of thin, narrow metal lines should be and is an area of intense study and concern. A review of the use of polyimides in the Japanese semiconductor industry is given by Makino (19). It is often difficult to find extensive detailed reports of such testing as the data is often considered proprietary.

Developing new testing methods, understanding the significance of pull and peel testing, electrical characterization of electrical nets with narrow width and pitch at high frequency, and the development of new test methods can all be considered as being within the broad scope of general reliability testing and modeling. This discussion will be limited to some of the key issues, with sections 6.1 through 6.4 addressing "during build" questions and section 6.5 dealing with long term reliability.

6.1 Adhesion and its Connection to Diffusion of Metal into Polyimide: The Interphase and Interface Stress.

Pull and peel testing and wire bond pull testing have been used extensively to study adhesion degradation by Rothman (14) and, despite the micromechanical complexities, remain as the main vehicles to quantitatively study adhesion. Generally the adhesion between metal and polymer degrades with temperature cycles during build. Cycling Ti-Cu-Ni-Au multilayers to 360C in "factory" nitrogen, with Ti as the adhesion layer, has been shown to degrade adhesion. Oxidizing ambients (like "factory nitrogen" which contains traces of oxygen) cause more degradation than reducing (forming gas) ambients.(61) There have also been extensive studies of modeling adhesion and the micromechanics of the peel test. For details see Kim et al.(49) and related papers.

Much of early reliability deals with whether the metal polyimide stack will survive the heat cycles of the manufacturing process. The main topics of concern are:

1. Adhesion of the metal to the polyimide and the different layers of polyimide to each other.
2. Adhesion degradation during manufacture.
3. Role of ambient gases while processing.
4. Thermal cycles during manufacture.
5. Stress effects that occur; for example the shrinkage of the conductor metal after evaporation.

Adhesion depends critically on the state of the metal to polyimide interphase and on the amount of diffusion of metals into the polyimide. Therefore an understanding of the diffusion of metals into polyimide forms a basis for understanding adhesion. Much work at IBM has concentrated on the understanding of the diffusion mechanism, and this is described next.

Diffusion of metals into interlayer dielectrics give rise to several related reliability concerns. These concerns are the same for both silicon dioxide and for organics. These concerns are:

1. Metal penetration from lines and change of dielectric constant.
2. Metal (Cu) diffusion into the device silicon.
3. Shorting of metal lines and pillars.
4. Adhesion degradation.
5. Thermal effects enhancing above mechanisms.

Cu and Al have been shown to diffuse into polyimide, and Gupta, Faupel, Ho, and co-workers have argued that the transition metals Cr and Ti may, in a complexed state, diffuse much more slowly in polyimide. However, since Cu diffuses readily into PI, structures which clad the Cu and protect the sidewalls have to be devised. Such structures have been described in the literature. For example, Cu interconnections with Cr-Cu-Cr at 0.8 and 0.5 μm line widths, defined using lift-off with Si templates and a trilayer resist stack, have been described by Rogers et al.(62) Reliability requires complete cladding of Cu. 0.8 μm double level metal structures with the sidewalls protected by PECVD SiON deposition over liftoff CrCuCr have been constructed and were shown to pass the reliability tests. This paper shows that when the proper structure is achieved, the structures are reliable, passing the standard tests.

In order to study the penetration of metals into polyimides, tracers have been used and diffusion coefficients have been calculated. Some of the published results are shown in Figure 28. The basic technique is to apply a coat of radioactive copper, microsection the sample using ion beams, and mass analyze the beam for the radioactive copper. This process produces plots of copper penetration as a function of time and temperature. This is then used to determine the diffusion constant.(63)

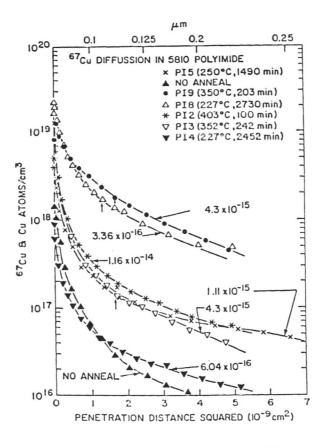

Figure 28. Shows the diffusion of isotopic copper into 5810 polyimide, after Gupta (63). Reprinted with permission.

The diffusion of copper in polyimide and in inorganic dielectrics has been compared by Gupta (see Figure 29)(63) and modeled by Faupel and co-workers.(64) It was found that Cu diffusion in polyimides is faster at the same temperature and that the diffusion of Cu in SiN:H is the least, followed by 4% PSG. Copper diffusion in PMDA-ODA is 10x as compared to BPDA-PDA. Copper solubility in polyimide is 10x compared to 4% PSG. Copper diffusion in PMDA-PDA and 4% PSG at 400 C is about 1 µm in 70 hours.

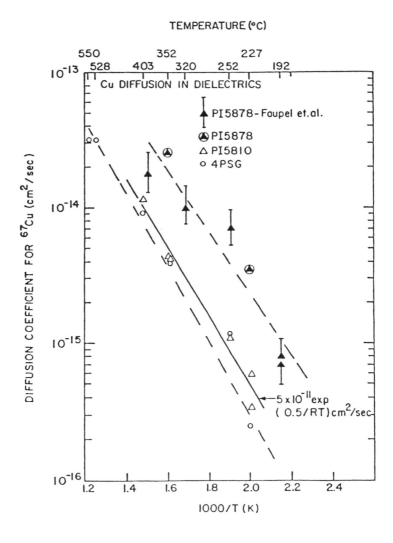

Figure 29. The diffusion of Cu into Phospho-Silicate Glass (PSG) as compared with polyimide. Gupta (63), reprinted with permission.

Monte Carlo calculations of the penetration process can be seen in Figure 30.(64, 65) These simulations provide corroborative evidence of the penetration discussed above. Notice especially the copper enriched layer below the metal. This layer plays an important part in adhesion.

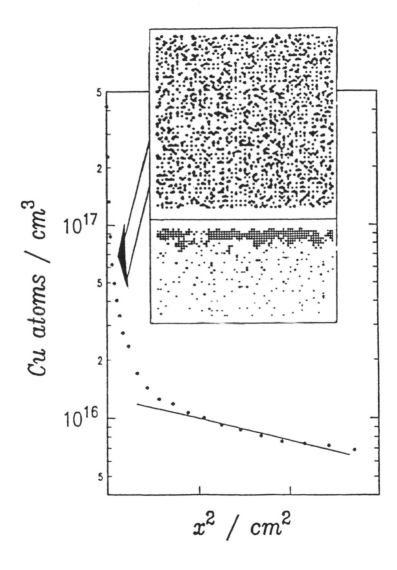

Figure 30. Monte Carlo simulation of copper diffusion into Polyimide after Silverman (65). Reprinted with permission.

Das and Morris (66) have reported self gettering behavior of ion-implanted Cu in polyimide. They attribute this to diffusant / diffusant interaction; this is further evidence in support of a cluster type mobility of Cu in polyimide. Silverman (65) has simulated metal atom cluster mobility in metal/polymer interfaces. This view of the diffused metal polyimide interface is also supported by Wool and Long (67), and this point of view has been used as a basis for further modeling by Seshan and Lacombe.(50)

The "Interphase" and its Models: The Cu diffusion experiments and the Monte Carlo simulations show that the penetration of metal into polyimide is an active, never ceasing process. This area just below the metal, we have chosen to call the "interphase", and a finite element rendering of it is shown in Figure 31. It was modeled as having two mechanical "phases" (61); it can change during the manufacturing process and during reliability test thermal cycling. This is the reason that we have chosen to call the layer between the metal and the polyimide a "transition" layer.(61)

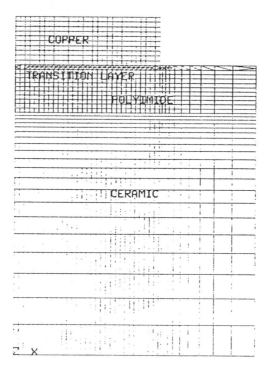

Figure 31. Shows a finite element model, based on Figure 30, to calculate the local stress effects of the copper rich "transition" layer, after Seshan & Lacombe (50). Reprinted with permission

This interphase model can be used to understand changes in properties and help explain some testing results. This is particularly useful in understanding the effect of moisture, as discussed below.

Modeling with the Metal-Polyimide "Interphase":   The results of finite element modeling of the interface, taking into account the transition layer are shown in Figure 32. It is the thickness, the mechanical properties of the "transition" region and its change with time that dominates adhesion. An understanding of this transition layer will remain a major problem in the reliability arena. The more subtle problem is the change in the "local" dielectric properties and its impact on performance.

Figure 32 shows the stresses in the interphase region when the metal is deposited at temperature and cooled, and then the mismatch stresses are imposed. The resultant changes in normal and shear stresses were calculated.(50) The conclusion of that work was that at critical thicknesses, with sufficient loading of the metal particles, delamination can occur, as is observed in early reliability testing. This affect can be avoided by reducing the thermal cycles and avoiding an oxidizing ambient.

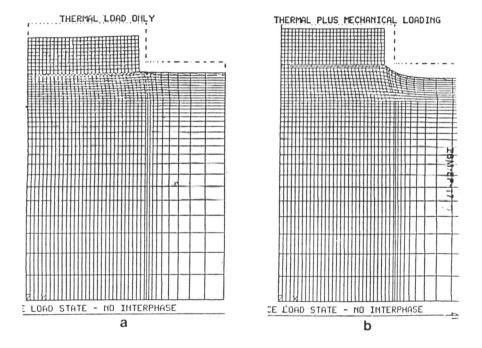

Figure 32. Shows the contraction calculated using the model in Figure 31. Figure 32(a) shows the results, only of the contraction due to the cooling of the metal. Figure 32(b) shows the results of the differential contraction of the metal and the polyimide. After Seshan & Lacombe (50). Reprinted with permission.

6.2   Effect of Moisture Ingress

Concepts from the "interphase" can be used to understand and model the adhesion when moisture is introduced. The effect of moisture ingress into polyimide/metal interfaces is an area of continuing concern. Adhesion tests of thin metal multilayer films to polyimide showed that both Cr and Ti, used as adhesion layers, degraded when exposed to an oxidizing ambient.(61) This degradation was attributed to the formation of Cr_xO_y and Ti_xO_y at the interface. It was shown that forming gas (H_2+N_2) was more benign than N_2 annealing, which is slightly oxidizing because "factory" nitrogen was found to be contaminated with oxygen. There also is a change in the failure mode.

Thus two different failure modes were observed. When tested just after metal deposition, using peel testing, high strength (30-50 gm/mm) cohesive failures were observed. In this failure mode small pieces of the polyimide were torn out with the metal. When the interface was exposed to ambient, the adhesion strength dropped to 5-10 gm/ mm, and low strength adhesive fails were seen. In this failure mode, a "new" low strength, "interphase" layer seemed to have formed. These failure modes are illustrated in Figure 33.

A thermodynamic explanation was advanced, which depended on the penetration of moisture to the metal-polymer interface region. These hypotheses are supported by the data in Figure 34. The polyimide-ingressed moisture would react with the metal clusters in the transition region. Ti would be more susceptible than Cr, because Ti has the added ability to dissolve hydrogen and expand and embrittle.(61) This seems to explain the observed experimental results that Ti adhesion layers failed faster than Cr layers. The mechanical modeling (61) also support this argument.

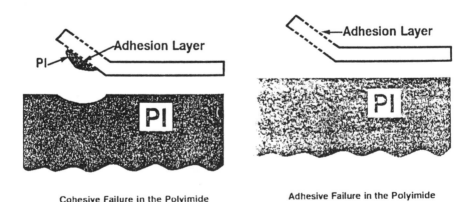

Cohesive Failure in the Polyimide   Adhesive Failure in the Polyimide

Figure 33. Shows the different failure modes observed when deposited metal is peeled off cured polyimide. Cohesive failure involves tearing out of the polyimide. Adhesive failure involves a "clean" separation between the metal and the polyimide, implying a "transition" layer as shown in Figure 31.

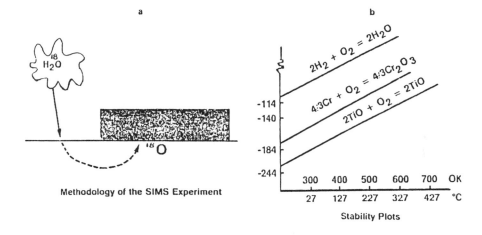

a

b

$2H_2 + O_2 = 2H_2O$

$4:3Cr + O_2 = 4:3Cr_2O_3$

$2TiO + O_2 = 2TiO$

-114
-140
-184
-244

300   400   500   600   700   OK
27   127   227   327   427   °C

$^{18}H_2O$

^{18}O

Methodology of the SIMS Experiment

Stability Plots

Figure 34(a). Shows the methodology of the SIMS experiment where isotopic water containing O^{18}, was introduced into the metal/polyimide interface. SIMS showed an accumulation of O^{18} near the interface. **(b).** Shows the stability plots relevant to Figure 35; the interface adhesion layer, usually Ti or Cr, will be able to reduce the water and form oxides of Ti or Cr, from Seshan et al. (61). Reprinted with permission.

The ingress of moisture into polyimides can be quantitatively measured. An optical technique to measure the refractive index of polyimide when exposed to water showed that, as humidity increased, the refractive index increased. This could be caused by swelling as a result of the incorporation of water, perhaps at a molecular level. The other argument is that there are micro-pores.(68) Using a laser technique, TM & TE modes were measured, these giving the refractive index for both parallel and perpendicular modes generated by coherent laser light. The micro-void model would cause the refractive index to increase. The swell model would cause the refractive index to decrease. Data suggested that both mechanisms are operative. When the amount of water is small, the swelling model applies; for larger amounts (over 1%), the void model appears to hold. This was considered important for the reliability of non-sealed low cost hybrid systems.(68)

6.3   Mechanical

Thermal aging, moisture ingress, and prolonged exposure to temperature and humidity cause the tensile strength to decrease, and the elongation to drop. The result is that films are prone to crack; these trends can be seen below. In the modeling work, data from Figure 35 was found to be useful and is included here for the purpose of reference.(33)

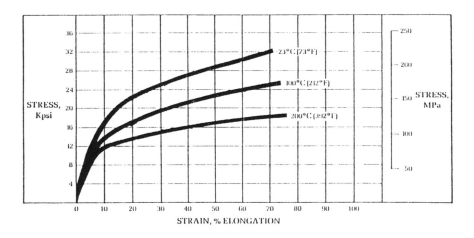

Figure 35. Shows the stress-strain curves for cured polyimide films. Data from Dupont (33).

In conjunction with a drop in strength, polyimides also creep; this is of concern when there are local stress raisers. Over time there is a possibility of dimensional changes; published data for polyimides are shown in Figure 36.(33)

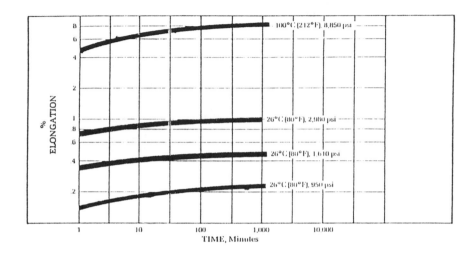

Figure 36. Tensile creep properties of cured polyimide films.(33). Reprinted wit permission.

Homma (69) reports a deformation mechanism of polyimide insulated Al alloy lines, caused by shrinkage stresses of the resin on top of the chip. This caused the polyimide and the 1st level alloy line to deform, while the upper layers were unharmed. Reliability was improved by modeling, changing the adhesion, and optimizing the film thickness.

6.4   Electrical properties

Many of the dielectric tests for oxides also apply to polyimides. Dielectric integrity tests for reliability include time zero dielectric breakdown (TZDB), time dependent dielectric breakdown (TDDB), and trapped charge measurements. Although the requirements on organic dielectrics are not as severe as on oxides, nevertheless questions about what values are to be used for extremely thin polyimide layers is becoming an issue.

The effect of temperature on AC dielectric strength, dielectric constant, dissipation factor and volume resistance are shown in Figures 37 and 38 from the properties compiled for Kapton by DuPont.(33) It must be remembered that both mechanical and electrical property changes occur.

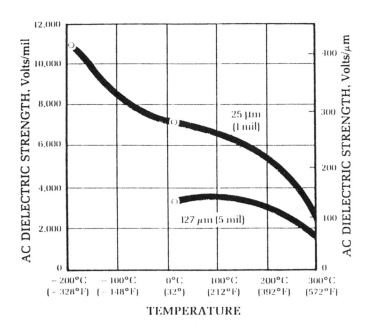

Figure 37. Shows the decrease of dielectric strength with temperature. The defects shown in Figure 39 will enhance these effects.

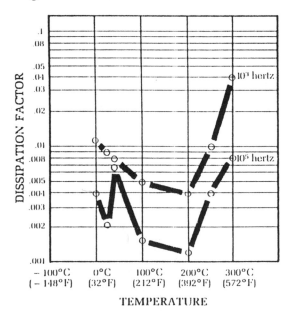

Figure 38. Shows the dissipation factor vs. temperature. Notice that the dissipation factor increases with temperature. This will adversely affect the performance of metal wiring in polyimide.

Figure 39 shows some of the origins of dielectric integrity and reliability problems.

Dielectric Integrity and Reliability

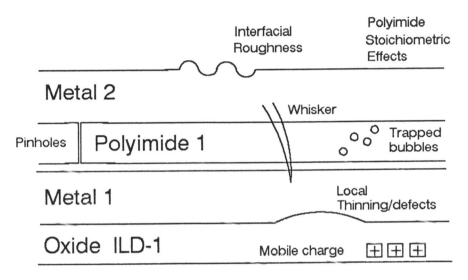

Figure 39. Shows the electrical defects that could reside between the metal/polyimide/ceramic multilayer. As the metal line pitch decreases, or higher operating temperatures are used, these become more important for reliability. Diagram compiled from different sources including Bakoglu (71). Reprinted with permission.

Some of the issues in dielectric integrity and reliability are shown in Table IV. These properties are measured by the usual CV/IV measurements used to test oxides. The breakdown of the dielectric and the effect of the trapped charge are important, depending on whether the application is in power or in performance.

Table IV. Table showing the various dielectric properties that must be monitored, that may impact dielectric reliability.

DIELECTRIC INTEGRITY AND RELIABILITY BY CV/IV
Measurement of Dielectric Quality
* **Time zero dielectric breakdown (TZDB)**   Voltage -to-breakdown (V_{BD})   Field -to-Breakdown (F_{BD})   Self Healing behavior (V_{SB},K)
* **Time dependent dielectric breakdown (TDDB)**   Time-to-breakdown (t_{BD})   Charge-to-breakdown (Q_{BD})   Electric field acceleration (G, β)
* **Oxide trapped charge**   Oxide trapped charge centroid ($\overline{X}$)   Oxide cross section (σ)   Oxide trapped charge distribution ($n_{ot}(x)$)
* **Statistical data analysis**   Weibull plots   Process control charts

6.5 Long Term Reliability

Long term reliability of multilayer metals in polyimide includes tests for:
1. Heat resistance
2. Planarization and eventual viscous flow
3. Withstanding wire bonding, die attach
4. Withstanding the pressure cooker test (PCT)
5. Withstanding temperature/humidity (T/H) test
6. Effect of incomplete imidization
7. Effect of sodium on device performance
8. Protection of devices from alpha particles
9. Propagation of soft error rates (SER)
10. Effect of via chain resistance and its change
11. Effect of electromigration
12. Effects dealing with the absorption of water
13. Effects of Si_3N_4 passivation

There have been several papers dealing with various aspects of these issues, and in all of them polyimides compare very favorably with TEOS, PSG, and other conventional dielectrics.

The most extensive tests were carried out by A.W. Lin (20) of AT&T, who compared nine commercial polyimides for multilevel interconnection applications.   In his tests, DuPont PI 2555 performed the best.   This was compared against other formulations which included PI2540, PI2545, PI2550, PI2560, PI2562, PI 2566, and PI2590. Photosensitive PIs were also included in the test; these included Hitachi's PIQ, PAL 1000 and photosensitive PI, Rhone-Poulenc Nolimid 32, Lat 10*, 50*, Kermid 601, Ciba Geigy, Upjohn Polyimide 2080, Epoxy Tech Epo Tech 390 3M Photosensitive PI, Siemens Photosensitive PI, and EM chemical HTR-2 photosensitive PI.

Lin subjected these materials to the (THB) Temperature Humidity Bias test 85C, 85% RH, 180V DC bias with 3 mil (75 μm) spacing. The following is a summary of the findings. Thermoplastic preimidized materials showed poor chemical resistance. Compared with RTV, considered the best encapsulant for IC's, with leakage of 10E-9 amps at the THB test, all the polyimides showed increased leakage with time, with currents in the 10E-9 to 10E-7 amps. The data for DuPont PI255 is compared with RTV in Figure 40.

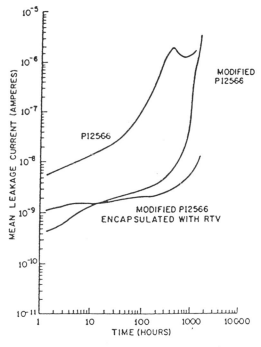

THB TEST RESULTS OF TiPdAu TRIPLE TRACKS
COATED WITH PI2566 OR MODIFIED PI2566

Figure 40. Encapsulated leakage current graph

When the best polyimide is encapsulated with RTV, the leakage current is reduced to lower values at 100 hours as shown below. These tests showed that contaminants have a severe effect on increasing the leakage current. Mobile ions including sodium are possible suspects.

Degradation of metallized via chains with polyimide has been studied by Homa (18), who developed a special test structure which allowed testing of wiring crossovers, which were subject to T/H testing and measured by four point probe for opens and shorts. They report that via resistance showed little or no change and the voltage bias test with 10 volts over a 675x72 crossover area showed no change in insulation resistance. Their conclusion from testing 236,000 vias with 10 V bias with 85/85 stress test showed no via and interlevel failures. It appears from this paper that if particular attention is paid to cleanliness and process detail, highly reliable structures can be obtained.

Although the uptake of moisture by the polyimide is always a concern, there have been a number of studies in the literature showing that it is possible to passivate the polyimide surface and achieve good reliability. Hefner et al. (70) compare PIQ-13 and Si_3N_4 moisture uptake. Taking conventional oxide dielectrics as standard, the pressure cooker test (PCT) showed that a passivating Si_3N_4 coating over the polyimide gave comparable moisture uptake to conventional oxide dielectrics. Low leakage currents of 1 pA were observed, and it was concluded that the Si_3N_4 can also act as an etch stop for subsequent layers. Other long term reliability tests have been reported in the literature Makino (19) and Lin.(20)

6.6 Summary

It can be seen that the reliability measurements for this technology are very important anytime a new material is introduced into the commercial market. Organic dielectrics have thus far shown great potential for the building of high reliability structures. Let us now move on to a discussion of the performance advantages of organic dielectrics in multilevel metal interconnects.

7.0 PERFORMANCE ADVANTAGES OF ORGANIC DIELECTRICS

A discussion of organic dielectrics would not be complete without the discussion of their performance advantages over conventional dielectrics. This section will make some simple comparisons with more conventional dielectrics and conclude by discussing what the authors consider to be the ultimate limits in the use of polymers.

7.1 Performance Comparisons

Organic dielectrics offer properties which make them desirable over many other inorganic dielectrics. One of the most important property is the dielectric constant, which has a direct effect on the transmission of signals in the wiring levels.   Simply stated, the lower the dielectric constant, the higher the performance that can be attained.   Let us discuss this matter in a little more detail and compare some of the effects of lowering the dielectric constant in a metallized structure.

The importance of having low dielectric constants in chip and package wiring, in order to enhance performance, can be shown by a simple calculation of interconnection wiring capacitance.   This is a simplification of a more complete treatment by Bakoglu (71) and Glasser.(72)

As chip dimensions increase, wiring capacitance of on-chip wires approaches the driver gate capacitance. At least at this point, wiring capacitance dominates the circuit delay.   If chip sizes are increased beyond this critical size, larger global nets will not perform as fast as smaller nets. The same is true at the board level, where chip-to-chip capacitance is an order of magnitude higher than on-chip capacitances.

The chip level wiring capacitance can be modeled as shown in Figure 41, where a driver is connected to a receiver via an interconnection of length l_{int}. The capacitance between the wire and ground includes the signal layer dielectric constant, here assumed to be oxide,with a dielectric constant of about 4 pF/cm. The driver in this example is a CMOS driver with a rise time of 500-2000 psec. The assumption is that the rise time t_r is much larger than the time of flight, which is assumed to be about 2 psec. Under these conditions a simple lumped circuit is valid. Faster circuits will be treated later.

Figure 41(a) shows a simple equivalent circuit to illustrate the importance of capacitance in determining wiring delay. A more complete model would include distributed capacitances, inductances, reflections, etc. The gate delay increases with the on-resistance of the driver, the capacitance of the wire, C_{int}, and the capacitance of the receiver gate, C_{gate}. When C_{int} is larger than C_{gate}, wiring capacitance dominates the delay.

Figure 41(b) shows a realistic model for an actual bipolar IC driver coupled via wires to an external circuit and then to a CMOS IC.

which is given, to first order, as:

$$[3] \qquad T_{50\%} = R_{tr}\left(C_{int} + C_{gate}\right)$$

where C_{int} is the interconnection capacitance, C_{gate} is the gate capacitance, and R_{tr} is the driver resistance. The fifty percent delay $T_{50\%}$ is defined as the delay from the time the input potential reached midway between V_{dd} and ground to the time the output reached the same point. For the purpose of this argument we take a simple form of C_{int} as:

$$[4] \qquad C_{int} = \frac{\varepsilon_{ox} W_{int} l_{int}}{t_{ox}}$$

with W_{int} as the interconnect width and t_{ox} the ILD thickness. Values of C_{int} turn out to be about 2 pF/cm.(73)

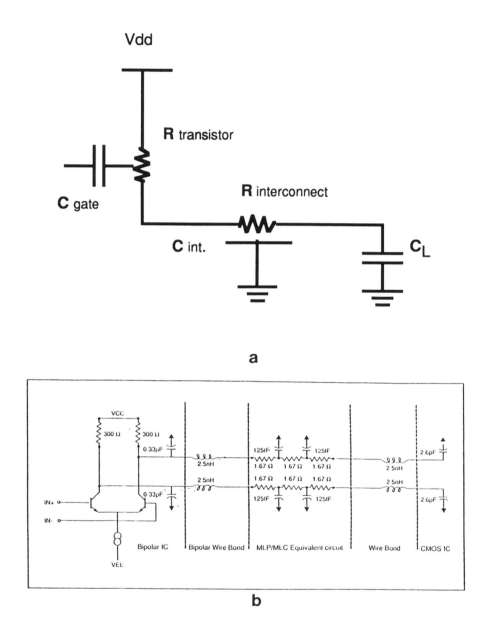

Figure 41. (a) Shows the simple model of a driver and an receiver and the equivalent electrical circuit. **(b)** Shows a more complex and realistic driver-wire-MLM-to receiver in a typical IC. Under certain conditions the simple model in Figure 41(a) can predict the behavior of 41(b).

The equivalent circuit shown in Figure 41(a) is used to determine the $T_{50\%}$. Using these two formulae we can reach some important conclusions about wiring delay. Suppose chip dimensions are increased. At some point the interconnection capacitance will equal the gate capacitance. The wire length at which this point is reached is given by:

[5] $$C_{gate} = 2.0 \frac{pF}{cm} l_{int}$$

For a C_{gate} value of the order of 0.6 pF (typical number for a 2 μm channel 10μ x 20μ device), one calculates the critical length as 0.3 mm.

This leads to several conclusions:

1. First, the capacitance of a 0.3 mm long wire equals the input capacitance of a large CMOS inverter.

2. Second, as the inverter is made smaller, this critical length will decrease. Therefore, neglecting other factors, as integration increases, larger chip sizes become increasingly slow.

3. Third, the critical paths of large logic chips are usually dominated by wiring rather than device capacitance, in state-of-the art device technology.

4. When one is dealing with GaAs and fast logic technologies, the wiring designer has to force a compromise between several wiring layers, the interlevel dielectric constant, and wire size and shape.

There are several factors that go into optimizing the wiring delay. The choice of the dielectric is a significant one. It helps to choose a dielectric with as low a dielectric constant as possible. The second is the shape of the conductor, especially its width over height (W/H) ratio and the insulator thickness. The decrease of wiring capacitance with these factors is shown in Figure 42, which shows the decrease of wire capacitance with geometry and with the dielectric constant fixed at 2 and 4, modified after Edelstein, et al. (74) and Bakoglu and Meindl.(71) Wire capacitance decreases as W is decreased with respect to H. There must be a balance, however, between width and height to optimize the resistive and capacitive components. This relationship continues until W approaches the dielectric thickness, where the capacitance levels off at about 1 pF/cm.

Certainly other factors are involved in optimizing the wiring delay such as the resistance of the conductor, temperature, and other factors, which for brevity will not be discussed here.

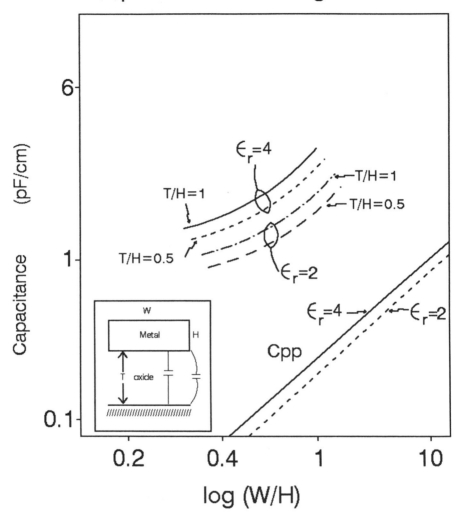

Figure 42. A calculation of the capacitance of a single runner, when it is placed in oxide and in polyimide; the graphs illustrate the effect of the decrease in the dielectric constant. Modified from a diagram from Bakoglu & Meindl (71).

With faster technologies, where much faster rise times are involved, like 20-100 psecs, transmission line analysis has to be used. Here the line inductances have to be considered, and the analysis becomes more complex. Nevertheless the simple argument about using lower dielectric constant insulators still holds. In order to make a comparison, Table V showing on-chip rise times is useful.

Table V: The rise time of different semiconductor technologies, after Bakoglu (71)

Technology	On Chip Rise Time	Off Chip Rise Times
CMOS	500 - 2000	2000 - 4000
Bipolar	50 - 200	200 - 400
GaAs	20 - 100	100 - 250

Let us define the critical length l_{crit} to be the propagation velocity V_{prop} times the rise time t_r.. It then can be shown that under the condition that the size of the package is larger than l_{crit}, time of flight arguments will still hold. This will predict that there is a direct advantage to the use of lower dielectric constant materials.

Used in the right regime, the propagation speed versus dielectric constant is given in Figure 43 below. Clearly the propagation velocity increases as one reduces the dielectric constant.

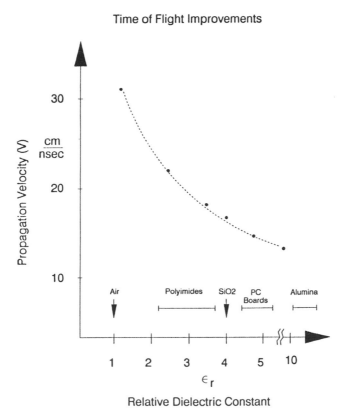

Time of Flight Improvements

Figure 43. Shows the time of flight improvement as a function of the dielectric constant.

Figure 44 shows l_{crit} vs. rise time t_r for various technologies and one arbitrary value of propagation velocity. The numbers are derived from Bakoglu.(71) This graph shows the regimes where time of flight arguments can be used. Given a t_r of 500 psecs, it will be seen that time of flight arguments can be used up to 3 cm. However for a GaAs driver, with a 10 psec rise time, l crit is only 6 mm. By using an ILD with a lower dielectric constant, one can shift the l_{crit} line to higher values.

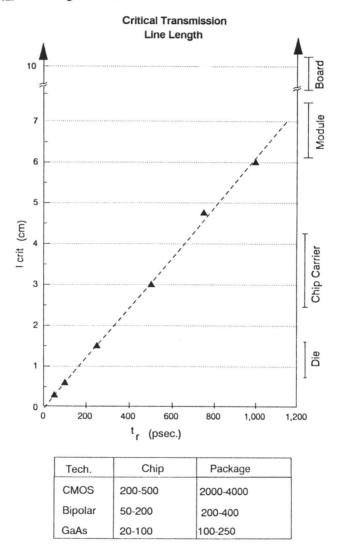

Tech.	Chip	Package
CMOS	200-500	2000-4000
Bipolar	50-200	200-400
GaAs	20-100	100-250

Figure 44. A diagram showing that for a certain rise time, the domain where the dielectric constant reduction achieves performance improvements. For a chip or a module with rise time over 1000 ps, it will be seen that dielectric constant and size reduction achieves performance improvements.

7.2   Performance Conclusions

In summary, the electrical interconnect designer is driven by several factors:
1. Line resistance should be minimized. This is the motivation for using copper-based interconnections.
2. Dielectric constant should also be minimized. This is the main motivation to implement organic dielectrics.
3. Wiring dimensions, shapes, and spacing should be optimized to minimize cross-talk.
4. Line length should be minimized and reflection matching considered. This is achieved by a judicious layout.

These factors often impose contradictory requirements; however, by a judicious application of these principles, a fully integrated multilevel metal structure can be evolved.

7.3   Factors in the Ultimate Limits to Performance

The ultimate performance of wiring structures for integrated circuits may depend on the scaling and conduction of nanometer dimension conductors.

Van Roggen and Meijer (75) have studied the conduction mechanisms in purified polyethylene single crystals. They have reported some very interesting non-linear conduction in thin 100 Angstrom, single crystals as shown in Figure 45 from their 1962 & 1988 papers. More recently this group has used a spring loaded, etched wire contact and has reported the same observation in 2 to 3 layered structures. In some of the follow up papers the authors have suggested the use of such characteristics in FET type molecular electronic devices as described in Figure 46. The interested reader is referred to the papers of Meijer and co-workers.

The important implication from the point of this chapter is as follows. As signal line pitch continues to scale to sub-micron dimensions, the anisotropy of organic polymer chain type structures will begin to manifest itself. The first property to come into question will be the dielectric constant; the fact that Meijer and co-workers have shown molecular conduction means that the "macroscopic" definition of the dielectric constant is itself in question. This in turn poses two challenges.

The first challenge is to define an "average" small dimension dielectric constant. This will affect and limit the performance of the interconnection scheme in a fundamental way; specifically it will result in a net increase in the dielectric constant, thus increasing the RC constant. This will work against the use of organic polymers as we have them now, unless new polymers are developed.

The second challenge is to measure the dielectric constant at such small dimensions. This will force the use of optical techniques to determine the refractive index and then derive the dielectric constant from the classical relation to the refractive index. However it will be evident to the reader that this is making a "classical" compromise. We believe the papers of Meijer et al. are important from these points of view towards the future.

It is anticipated that organic dielectrics will gain more popularity as these materials and processes improve. Our expectations of future trends are addressed in the following section.

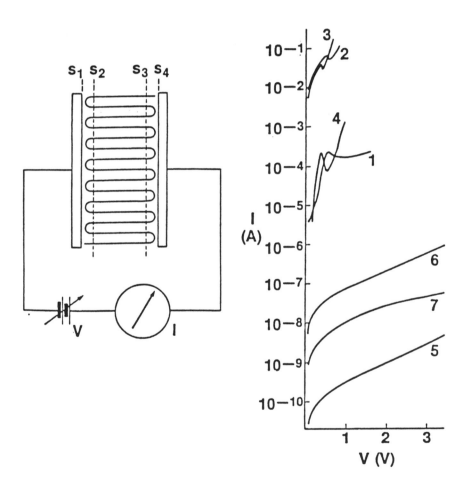

Figure 45. Experimental setup and electrical parameters of conduction in purified single crystal polyethylene.(75) Reprinted with permission.

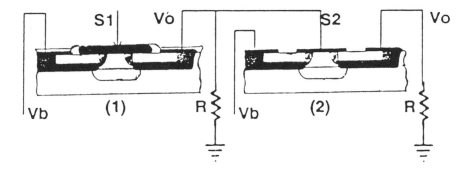

Figure 46. Two FET devices as conceived by Meijer for use in molecular electronics.

8.0   FUTURE TRENDS

Now that we have examined the present state of organic dielectrics in multilevel metallization applications, we can attempt to make some predictions about future trends.

From the literature, it is clear that from a cost, performance, reliability, and low defect rate, organic dielectrics are of great interest.

The use of polyimides as we mentioned is driven by performance requirements, due to its low dielectric constant, or for cost-performance applications. We also stated that implementing more levels of metal in advanced metallization schemes requires some degree of planarization. Polishing layers with materials of very dissimilar mechanical properties is still a fundamental problem, but not specific to polyimides.

It is also shown in the literature that the utilization of polyimides has occurred to a larger extant in Japan than in the U.S., specifically in the cost-performance low end products.(19)   From this position, there are not fundamental properties that limit the increased use of polyimides in both chip and package.

Finally, it can be said that the main reason that organic dielectrics are absent from many products is historical. Many device and process designers do not yet feel comfortable with these non-traditional materials. This is probably due to the fact that electrical engineers and metallurgists are determining the processes. If we want to see more usage of organic dielectrics in the future, a shift towards hiring polymer chemists and chemical engineers will be required. The use of polymers will also occur if there is no inorganic material which can meet the design requirements. This may happen if a low dielectric constant (<3) becomes mandatory.

Further developments in the treatment and the understanding of polyimide surface properties are also anticipated. Hiroyuki et al.(76) have shown that polyimides become highly hydrophilic upon irradiation by deep UV or UV lasers in air. This laser exposure is efficient, and these authors have shown the possibility of direct image metallization. The metal does not nucleate on the laser exposed regions, while unexposed surfaces showed uniform metal deposition. Such developments could stimulate the use of electro- and electroless deposition processes for the patterning of fine metal lines.

Activity in the CVD deposition of polyimides directly from the vapor phase is being investigated by Kowalczyk and others.(77) While this is an advanced application, it is not likely to be used in cost sensitive packages.

Considerable work has concentrated on characterizing polyimides and other organic dielectrics. Many of the polymers in use today are tradeoffs between materials with the best electrical properties and acceptable mechanical properties. In some cases this combination is difficult to achieve. At the present time polymers that are variations of Teflon are being examined seriously due to its low dielectric constant (<2.5). Although Teflon does not have the required thermal and mechanical properties, we expect to see new organic materials which do meet these requirements.

It can be said with some confidence that the optimum dielectric for microelectronics applications has not yet been found. Polymers for the future will require decreasing dielectric constants while also providing improved moisture resistance, thermal and mechanical properties, and cost.

In conjunction with these materials, new classes of materials which are hybrids between organic and inorganic dielectrics are just beginning to be investigated. The art of sol-gel technology for glass formation is being explored where one of the silicon ligands is replaced by an organic group. This new class of "glasses" demonstrate new properties of low dielectric strength and high temperature stability. The drive towards low dielectric constants has initiated a study on porous materials that would have lower dielectric constants due to the incorporation of tiny pockets of air. These organic sol-gel compounds contain large pores where a network of material is formed containing many air spaces in the structure. It is expected that these materials could be very useful if the mechanical properties are such that it survives semiconductor multilevel metal processing.

In the areas of process development, it is expected that photosensitive dielectrics will become more prevalent. With much simpler processing capability, these materials will be ideal for both chip and packaging applications where their simplicity and cost can be exploited. The limitations of these polymers today are generally in the form of critical dimension control due to solvents in the polymer which are not baked out during the restricted bake cycles. While this is a limitation in today's materials, rapid advances in photoresist technology has resolved many of these problems and it is expected that photosensitive organic dielectrics will follow close behind.

As dimensions of chips becomes smaller, device designers will be pushed to the limits of semiconductor physics to produce further gains in performance. It is expected that improvements in the multilevel metal process and materials offer the greatest potential for performance improvements. Given this reasoning and the proper paradigm shifts, it is expected that organic dielectrics should become more widespread in order to meet the growing demands in the electronics industry.

REFERENCES

1. M.B. Small, D.J. Pearson," On-chip wiring for VLSI: Status and Directions," IBM J. Res. Dev. 34(6), 858 (1990).
2. J.R. Black " Electromigration - A Brief Survey and Some Recent Results," IEEE Trans. Electron. Dev. 16, 338 (1969).
3. J.R. Lloyd and P.M. Smith, " Electromigration Lifetime of Al/Cu Thin Film Conductors," J. Vac. Sci. and Technol. 1(12), 455 (1983).
4. P. M. Solomon,"The Need for Low Resistance Interconnections in Future High Speed Systems," SPIE 947, 104 (1988).
5   V. Ramakrishna, A.S. Oberai, P. A. Farrar, D.W. Kemmer, P.A. Totta, N.G. Koopman, and M.B. Small, "Future Requirements for High Speed VLSI Interconnections," Proc. 4th IEEE VLSI Multilevel Interconnection Conference, pp. 27-32 (1987).
6. L.S. White, R. Blumenthal and H. McAdams," The Conundrum of ULSI Interconnections: Manufacturability, Reliability, and Performance," in: Tungsten and Other Advanced Metals for ULSI Applications in 1990, (G.C. Smith and R. Blumenthal, eds.), pp. 3-17 (1991).
7. K. Sato, S. Harada, A. Saiki, T. Kimura, T. Okubo, and K. Mukai, "A Novel Planar Multilevel Interconnection Technology Utilizing Polyimide," IEEE Trans. Parts, Hybrids, Packag. PHP-9 (3), 15 (1973).
8. A. Saiki, S. Harada, T. Okubo, K. Mulai, and T.A. Kimura, "New Transistor with Two Level Metal Electrodes," J. Electrochem. Soc.124,1619 (1977).
9. J.C. Yen, Proc. Electrochem. Soc. 1975 Spring Meeting, (75-1) Abstr. 170 (1975).
10. H.C. Cook, P.A. Farrar, R.A. Uttecht, and J.P. Wilson, "Structure for Improving the Passivation of Semiconductor Chips," IBM Tech. Disclosure Bull. 16, 728 (1973).
11. R. A. Larsen, "A Silicon and Aluminum Dynamic Memory Technology," IBM J. Res. Dev. 24 (3), 268 (1980).

12. T.O Herndon and R.L. Burke, "Inter-Metal Polyimide Insulation for VLSI," Kodak 1979 Interface Symposium, Oct. (1979).
13. A. Wilson, "Polyimide Insulators for Multilevel Interconnections," Thin Solid Films 83, 145 (1981).
14. L.B. Rothman "Properties of Thin Polyimide Films," J. Electrochem. Soc. 127(10), 2216 (1980).
15. L.B. Rothman, "Process for Forming Passivated Metal Interconnection System with a Planar Surface," J. Electrochem. Soc.130(5), 1131 (1983).
16. L.B. Zielinski, U. S. Patent No. 3,985,597 (1975).
17. S. Mastroianni "Multilevel Metallization Device Structures and Process Options," Solid State Tech. p.155, May (1984).
18. T. R. Homa, and A. J. Posocco, "Reliability of Metallized Ceramic / Polyimide Substrates," IEEE Trans. Comp. Hybrids and Mfg. Tech. CHMT-9 (4), 396 (1986).
19. D. Makino, "Application of Polyimide Resin to Semiconductor Devices in Japan," IEEE Elec. Insulation Magazine 4(2), 15 (1988).
20. A. W. Lin, " Evaluation of Polyimides as Dielectric Materials for Multichip Packages with Multilevel Interconnection Structure," Proc. Electron. Components Conf.,IEEE pub. 89-CH2775-5, pp.148-154 (1989).
21. M.G. Sage, "The Future for Electronic Packaging and Interconnection," Proc. Tech. Program of National Electronic Pkg. and Prod. Conf., NEPCON West p. 611. (1987).
22. T. Lane, F. Belcourt, and R. Jensen, " Electrical Characteristics of Copper / Polyimide Thin Film Multilayer Interconnects," IEEE Trans. Comp. Hybrids and Mfg. Tech. CHMT-12(4), 577 (1987).
23. H. Takasago, K. Adachi, and M. Takada, "A Copper/Polyimide Metal-Base Packaging Technology," J. Elec. Mat.18(2), 319 (1989).
24. K. Kimbara, A. Dohya and T. Watari, " Polyimide-Ceramic Substrate for Supercomputer Packaging," Mat. Res. Soc. Symp. Proc. 167, 33 (1990).
25. D. Volfson and S. Senturia, "A Process for Multilevel Copper-Polyimide High Density Interconnect Structures, Flex Circuits and TAB Tape," Circuit World 16(3), 32 (1990).
26. T. Tessier, G. Adema and I. Turlik," Polymer Dielectric Options for Thin Film Packaging Applications," IEEE, p. 127 (1989).
27. C. Chao, K. Scholz, J. Leibovitz, M. Cobarruviaz and C. Chang, "Multi-layer Thin Film Substrates for Multi-chip Packaging," IEEE Trans. Comp. Hybrids and Mfg. Tech. 12(2), 180 (1989).
28. "Overcoats Protect RAMS from Alphas," Electronics Review, p. 41, Sept. 11 (1980).
29. Polyimide Prevents Alpha Errors," Electronic Design, Sept. 1 (1980).
30. K. Kitade, "The Soft Error Preventive Effect of Polyimide Resin Coatings," Japan Semi. Tech. News, p. 42, Feb. (1982).
31. M. Pottinger " Second Generation Photosensitive Polyimide Systems," Solid State Tech. p. S1, Dec. (1989).

32. J.M. Bureau, F. Bernard and D. Broussoux, "High Performance Polymers for Packaging and Interconnections in Microelectronics," Revue Technique Thomson-CSF, 20-21(4), 689 (1989).

33. Technical Bulletin, E. I. Dupont de Nemours & Co. "Kapton, Summary of Properties," p.4-23 (1988).

34. Li-Hsin Chang, D. Weston and J. Sellers, "Chemical Vapor Deposition Tungsten Via Plug Process Development with Polyimide Interlevel Dielectric in a Multilevel Metal System," J. Vac. Sci. Technol. B 10(5), 2277 (1992).

35. Technical Bulletin, PQA-4015 Polyquinoline Dielectric Coating, Allied Signal Corp., 1090 So. Milpitas Blvd. Milpitas, CA 95035 (1992).

36. J. Harwood, "Teflon-AF - A New Polymer for Electronics," 8th IEMT Int. Electron. Mfg. Technol. Conf., Publ IEEE Services, Piscataway, NJ, cat. 90CH2833-2, pp. 503-507 (1990).

37. M. S. Chace, "An Apparatus for Evolved Gas Analysis: Linking a Thermogravimetric Analyzer to an Ion Trap Detector," Publication of Finnigan MAT Corp., Application Report No. 218.

38. D. Hawks, (private communication).

39. N. J. Chou, J. Kim, S.P. Kowalczyk, Y.H. Kim and T. Oh, "Adhesion, Reaction, and Stability of Metal/Polyimide Interfaces," Proc. Mat. Res. Soc., AEPM Sym., p. 137 (1990).

40. P.S. Ho, B.D. Silverman, R.A. Haight, R.C. White, P.N. Sanda, and A.R. Ross, "Delocalized Bonding at the Metal-Polymer Interface," IBM J. Res. Dev. 32(5), 658 (1988).

41. R.R. Tummala and E.J. Rymaszewski, Eds. Microelectronics Packaging Handbook, (Van Nostrand, Reinhold, 1989).

42. D. Akihiro, W. Toshihiko and N. Hideki, " Packaging Technology for the NEC 3/SX X Supercomputer," in Proc. 40th Electronics Components and Technology Conf. 1, pp. 525-33 (1990).

43. R.G. Frieser, S.P. Ashburn, F.M. Tranjan, T.D. DuBois and S.M. Bobbio, "A Review of Simplified Photolithographic Techniques for Image Transfer in Planarized Very Large Scale Integrated Circuits Technology," J. Vac. Sci. Technol. B 8(4), 643 (1990).

44. D. L. Pappas, J.J. Cuomo and K. G. Sachdev, "Studies of Adhesion of Metal films to Polyimide," J. Vac. Sci. Technol. A 9(5), 2704 (1991).

45. K. W. Lee, S. P. Kowalczyk and J. M. Shaw, "Polyimide Surface Chemistry," Polymer Reprints, ACS 31(1), 712 (1990).

46. K. W. Paik, H. S. Cole, and N. H. Hendricks, " Studies of PQ-100 Polyquinoline Films for Multichip Module (MCM) Applications, "Proc. Mat. Research Soc. Sym. San Francisco, CA., (1992).

47. L.P Buchwalter, "Adhesion of Polyimides to Metals and Metal Oxides," J. Adhesion Sci. and Technol. 1(4), 341 (1987).

48. N. Majid, S. Dabral and J.F. McDonald," The Paralene-Aluminum Multilayer Interconnection System for Wafer Scale Integration and Wafer Scale Hybrid Packaging," J. Electronic Materials 18(2), 301 (1989).

49.   D.G. Kim, T.S. Oh, S. Molis,S.P. Kowalczyk and S. Kim, "Effect of RF Sputtering on T/H susceptibility of Cr/Polyimide Adhesion," Electronics Packaging Materials Science V., Symposium, pp.65-70, (1990).

50.   K. Seshan and R. H. Lacombe, "The Metal-Polyimide Interface Under Degradative Ambients," in: Metallized Plastics 2, ECS Symp. K. S. Mittal ed., (Plenum Press, New York, 1990).

51.   M. A. Van Andel and W. F. M. Gootzen, "Polyimides for Passivation of VLSI Circuits," Electronics Packaging Materials Science IV Symposium, pp.183-193 (1989).

52.   K.L. Mittal, Editor, Polyimide: Synthesis, Characterization, and Applications, 1-2, (Plenum Press, New York, 1983).

53.   Li-Hsin Chang and H. Tomkins, " Method for Measuring Diffusion of Moisture in Polyimide," Appl. Phys. Lett. 59(18), 2278 (1991).

54.   K. K. Chakravarty, J. M. Cech, C.P. Chien, L.S. Lathrop, M.H. Tanielian and P. L. Young, J. Electrochem. Soc. 137(3), 961 (1990).

55.   Li-Hsin Chang and R. Goodner, "Improved Planarization Techniques Applied to a Low Dielectric Constant Polyimide Used in Multilevel Metal ICs," Proc. SPIE Int. Soc. Opt. Eng. 1596, 34 (1991).

56.   K. K. Chakravorty, M. H. Tanielian, " Planarization in Multiple Polyimide Layers," Appl. Phys. Lett. 60(14), 1670 (1992).

57.   S. Roehl, L. Camilletti, W. Cote, D. Cote, E. Eckstein, K.H. Froehner, P.I. Lee, D. Restaino, G. Roeska, V. Vynorius, S. Wolff and B. Vollmer " High Density Damascene Wiring and Borderless Contact for 64 M DRAM," Proc. 9th IEEE VLSI Multilevel Interconnection Conf. pp. 22-24(1992).

58.   D. M. Mattox, "Electronic Thin Films," J. Electrochem. Soc.10, 1 (1983).

59.   P. Burggraaf, "Polyimides in Microelectronics," Semiconductor International, pp. 58-62, March (1988).

60.   R.J. Jensen, J.P. Cummings, and H. Vora, "Copper/Polyimide Materials System for High Performance Packaging," IEEE Trans. Comp. Hybrids and Mfg. Tech. CHMT-7(4), 348 (1984).

61.   K. Seshan, S.N.S. Reddy, and S.K. Ray, " Adhesion of Thin Metal Films to Polyimide," IEEE Trans. Comp. Hybrids and Mfg. Tech. CHMT (1987).

62.   Rogers, B., Bothra, S., Kellam and M. Ray "Issues in a Submicron Copper Interconnection System Using Liftoff Patterning," Proc. 8th Internat. IEEE VLSI Multilevel Metal Interconnection Conf., pp.137-143 (1991).

63.   D. Gupta, "Diffusion in Thin Films," in Encyclopedia of Applied Physics, 5, pp.75-86 (VCH Publishers, New York, 1993).

64.   F. Faupel, D. Gupta, B. D. Silberman and P. S. Ho, " Monte Carlo Simulation of Interface Penetration," Appl. Phys. Lett. 55(4), 1234 (1989).

65.   B.D. Silverman, "Single Particle Diffusion Into a Disordered Matrix," Macromolecules 24 (9), 2467 (1991).

66. J.H. Das and J.E. Morris, "Diffusion Simulations of Gettering of Ion Implanted Copper in Polyimides," Proc. of the 1990 IEEE Tech. Conf., pp. 259-265, IEEE cat. 90TH0313-7 (1990).

67. R.P. Wool and J.M. Long, "Structure of Diffused Polymer-Metal Interfaces," Proc. American Chem. Soc., Wash. DC Meeting, Polymer Preprints, Div. of Polymer Chem. ACS 31 (2), 558 (1990).

68. D.A. Horsma, "Measurement of Refractive Index and Birefringence of Polyimide Thin Film Dielectrics as a Function of RH and Temperature Before Aging," Proc. 7th Electronic Materials Congress, ASM Intl., Materials Park, Ohio, Sept. (1992).

69. Y. Homma, N. Sakuma, T. Nishida and I. Yoshida, Proc. 8th Intl. IEEE VLSI Multilevel Interconnection Conf., pp. 249-255 (1991).

70. A. Hefner, R. Isernhagen, M. Lentmaier and E. Waschler, Proc. 5th Intl. IEEE VLSI Multilevel Interconnection Conf., pp. 476-483 (1988).

71. H.B. Bakoglu, "Circuits, Interconnections and Packaging for Packaging for VLSI," Chpt. 1-4, (Addison-Wesley, 1990); J.D. Meindl, "Optimal Interconnection Circuits for VLSI," Trans. Elec. Dev. ED-32(5), 903 (1985).

72. L.A. Glasser and D.W. Dobberpuhl, "The Design and Analysis of VLSI Circuits", (Addison-Wesley Pub. Co., 1985).

73. L.W. Schaper and D.I. Amey, "Improved Electrical Performance for Future MOS Packaging," IEEE Trans. on Comp. Hybrids, and Mfg. Tech. CHMT-6, 282 (1983).

74. D.C. Edelstein, "3-D Capacitance modeling of Advanced Multilevel Interconnection Technology," Proc. of the SPIE 1389, p. 352, (1991).

75. A. van Roggen and P.H.E. Meijer, "The Effect of Electrode-Polymer Layers on Polymer Conduction.Part 2:Device Summary" in Molecular Electronic Devices," Ed. R.E. Carter and H.W. Siatkowski, pp. 427-437 Pub. (Elseiver Science Pub. N. Holland 1988).

76. H. Hiraoka and S. Lazare, "Surface Modification of Kapton and Cured Polyimide Films by ArF Excimer Laser: Application to Imagewise Wetting and Metallization," Applied Surface Science 46 (1-4), 264 (1985).

77. S.P. Kowalczyk, C. D. Dimitrakopoulos, and S. E. Molis, "Growth of Polyimide Films by Chemical Vapor Deposition and Their Characterization," Mat. Res. Soc. Symp. Proc. 227, pp. 55-59 (1991).

78. R.F.W. Pease and O-K. Kwon, "Physical Limits to the Useful Packaging Density of Electronic Systems," IBM Res. Dev. 32(5), 12 (1988).

6

PLANARIZATION TECHNIQUES

JEFF OLSEN
Applied Materials
Santa Clara, California
And
FARHAD MOGHADAM
Intel Corporation
Santa Clara, California

In this chapter we discuss planarization of the dielectric layers in multilevel interconnect systems. We begin with a discussion of the various motivations for pursuing planarized topography. This leads to a discussion of the nomenclature and concepts of planarization technology. The remaining sections of the chapter present detailed descriptions and discussions of the mainstream techniques used to achieve dielectric planarization.

The first step in nearly every planarization technique is to coat the underlying topographic features with a layer of dielectric material. This becomes "planarized" during the steps which follow. Until recently, this first step has been trivially simple. Now, however, in selecting a process for this task the engineer is confronted with small, difficult to fill gaps between adjacent interconnect structures, and filling these gaps is an increasingly challenging element of planarization technology. Various techniques which address these gap filling challenges are discussed, and these are treated as gap filling "modules". A gap filling module, when integrated with one or more additional modules which bring about complementary planarization effects (topography smoothing and step height reduction), is an integral element of a complete planarization process.

"Global planarization" has become an important topic recently. The basis for this is discussed in Section 2. Three planarization techniques have evolved to address global planarization. Two of these are discussed in section 7 on chemical-mechanical polishing techniques. The third -- the "planarization-block mask" or "N-mask" technique -- is discussed in Section 4 on resist-etchback techniques.

1.0   WHY PLANARIZE?

Most methods used for planarization are complex undertakings. None of them produce truly planar surface topography. Why do we go to all the trouble? What is the payback we hope to achieve? At each dielectric level of the device, the motivations and paybacks are different. These are briefly pointed out in the next few paragraphs, before moving on to Section 2 in which they are discussed in greater detail.

At the pre-metal dielectric level (ILD0), planarization gives rise to improved step coverage of the overlying metallization layer. This in turn improves the patternability, line resistance, and reliability of those interconnects.

The benefits of *pre*-metal planarization continue to appear even later in the fabrication process as well. Each of the subsequent interconnect layers, once it is patterned, produces new topography which superimposes onto whatever topography has propagated from the underlying interconnect structure. The greater the attenuation of pre-metal topography by planarization of the premetal dielectric, the smaller its contribution to this stackup effect. Minimizing this stackup greatly diminishes the difficulty of planarization processes occurring later in the fabrication sequence, for example between metal 2 and metal 3.

Planarization of the *inter*-metal dielectric layers (ILD-1, ILD-2, ...) yields exactly the same benefits as those described above.

Planarization of the final passivation dielectrics is done to address a different set of issues. Many devices rely on a layer of PECVD nitride or oxynitride for hermetic protection. These films are most vulnerable to penetration by contaminants where they cover the vertical sidewalls of the underlying interconnects. Using planarization techniques, these vertical surfaces can be transformed into sloping ones prior to the deposition of the nitride or oxynitride, eliminating the root cause of this vulnerability, and improving the reliability of the packaged device.

In addition, planarization of the passivation dielectrics can prevent polyimide and epoxy-resins from entering small gaps between adjacent interconnect lines during plastic packaging processes. This protects the interconnects from deformation due to shrinkage stresses which arise when the resins cure. In addition, it can reduce parasitic coupling between adjacent lines since the polyimide, which can have a relatively high dielectric constant after exposure to ambient water vapor, is excluded from the gaps between closely spaced lines.

2.0   CONCEPTS

2.1   Nomenclature

In the semiconductor fabrication industry, the term "planarization" is broadly applied to a variety of processes. In general, these are processes which produce or modify a dielectric film with a specific focus on its surface topography, and with the following objective: to make the surface topography of the dielectric less severe, for the benefit of some subsequent process step. "Less severe" means:

(a)   An absence of small gaps, or more precisely, an absence of gaps having large aspect ratio.

(b)   Gradually sloping transitions, rather than abrupt vertical
        steps, from the top to the bottom of a topographic feature.
(c)   Minimal elevation difference between the tops of the highest
        features and the bottoms of the lowest features.

Achieving this third characteristic, however, is not always regarded as a positive contribution to the fabrication process. While it gives rise to improved deposition and patternability of the overlying metallization, it increases the difficulty of etching vias to underlying features. This is discussed in section 2.5.

A "planarization technique" is a process or sequence of processes which diminishes the severity of the dielectric topography by contributing to at least one of the three characteristics described above. Many of the techniques contribute to more than one of these characteristics.

Before describing these techniques in detail, we will examine the key issues related to these three elements of planarization: gap filling, topography smoothing, and step height reduction.

2.2   Gap Filling

Gaps are produced as a consequence of the interconnect patterning process (Figure 1), and they exist at each interconnect level, providing electrical isolation between adjacent interconnect lines.

a

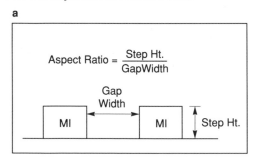

b

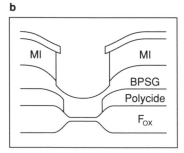

Figure 1.   Gap between adjacent interconnect lines (**a**). Baselayer (BPSG) loss due metal-1 overetch, overhanging antireflective coating, and the stackup of underlying steps (**b**) increase the difficulty of gap filling.

The ease, or difficulty, of filling a gap with a dielectric material is strongly affected by the dimensions, the aspect ratio, and the geometry of the gap. The gap structure depicted in Figure 1(a) is quite straightforward; however the gaps found on real integrated circuits are generally more involved, as shown in Figure 1(b). This illustrates how the *nominal* parameters of an *individual* interconnect layer can understate the actual difficulty of a gap filling task.

Voids. When using a CVD process as a gap filling tool, or as an element of a gap filling sequence, an important challenge is to prevent the formation of "voids" in the gaps. A void is a consequence of the fact that a film deposited in a gap grows laterally from each sidewall as well as vertically from the top and bottom surfaces (Figure 2). The gap becomes filled when the vertical surface of the film growing from one sidewall meets the vertical surface of the film growing from the opposite sidewall. If these surfaces meet near their tops before they meet at their bottoms, this closes the gap while leaving it unfilled at the bottom. Additional deposition into the unfilled bottom portion is of course impossible, since there is no longer an opening at the top through which reactive species can enter. The unfilled volume is called a void.

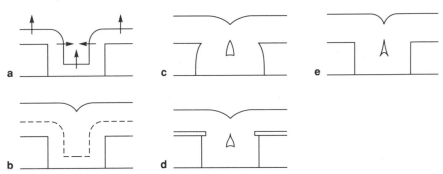

Figure 2. Movements of the growth fronts as a film is deposited into a gap **(a)**, and the resulting film profile after the gap is filled **(b)**. Voids produced due to: **(c)** re-entrant metal profile; **(d)** overhanging anti-reflective coating; or **(e)** non-conformal deposition.

Three situations are commonly encountered which can cause this premature closure at the tops of the gaps. These are illustrated in Figure 2(c,d,e). This illustrates the strong link between the performance of a metal etch process and the effectiveness of the gap filling process. It also illustrates the motivation behind the drive for conformal deposition of CVD dielectric films.

The width of the smallest gap in an interconnect pattern imposes an important constraint on the design of the gap fill process: it dictates the maximum amount of film which can be deposited without closing and forming a void in the smallest gaps.

What are the consequences of building IC devices with voids in the dielectrics? One hazard arises when etching back the dielectric -- a commonly used step in many planarization techniques. Figure 3 illustrates the trench produced when a void is re-opened during an etchback process. This can cause difficulties during subsequent processes which deposit films over this harsh feature (1).

Very little is known about the consequences of *unopened* voids in IC devices. Nevertheless, there is a widely held intuition that devices are better off without voids than with them. One school of thought is that voids are acceptable so long as they are "buried voids" i.e. if the tops of the voids are no higher than the tops of the underlying steps, such that an etchback process would be incapable of reaching and opening the voids. In actual practice it is difficult to assure that *all* voids are buried voids. Voids formed in the smallest gaps may indeed be buried, but those formed in larger gaps will be at higher elevations.

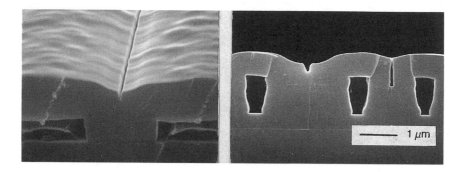

Figure 3. Trenches in an interlevel dielectric, caused by voids which become exposed during a planarization etchback.

2.3   Topography Smoothing

Let us first distinguish between *topography smoothing* and *step height reduction*. Generally, planarization processes produce topography smoothing as well as step height reduction effects; however the two effects are different, and the motivations for achieving one are different than the motivations for achieving the other (discussed below).

Topography smoothing techniques transform abrupt vertical steps into gradually sloping ones. In doing so, the height of the step may or may not be reduced. Figure 4(a) illustrates topography smoothing with and without concurrent step height reduction.

Range. Planarization "range", or "leveling length", can be used to quantify topography smoothing capability. This is defined and illustrated in Figure 4. In Figure 5, several common planarization techniques are ranked in terms of range (2).

Step height reduction occurs when the planarization range is greater than half the distance between adjacent steps (Figure 6). Techniques characterized by a planarization range of less than 5~10 μm are often called "local" or "short range" planarization processes; techniques having ranges of 10~100 μm are "long range" planarization processes. "Global" planarization can be achieved when the range is long enough to planarize the largest distances between adjacent steps, which may be on the order of millimeters. If these largest gaps can be planarized, then all smaller gaps will be planarized as well, and the dielectric surface across the entire die will be at uniform elevation.

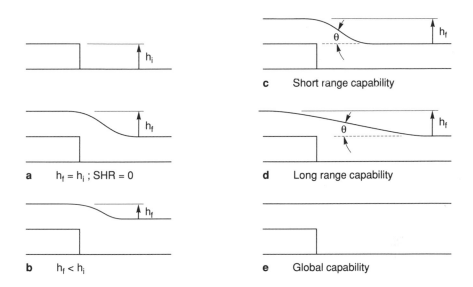

Figure 4.   Topography smoothing without (**a**) and with (**b**) step height reduction. Planarization "Range" (**c,d,e**) is the horizontal distance from an isolated feature at which the final step height, h_f, is equal to the initial step height, h_i. The greater the planarization range, the smaller the sidewall angle, θ.

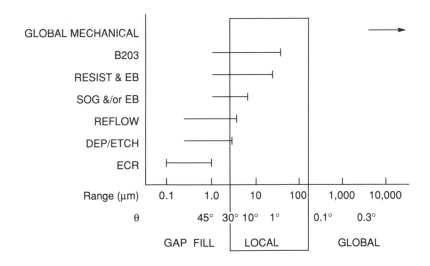

Figure 5.   Various planarization techniques ranked in terms of Range capability, R. From ref. 2; reprinted by permission of the publisher: the Electrochemical Society, Inc.

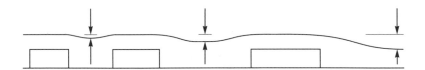

Figure 6.   Step Height Reduction occurs only within the range of planarization. When a gap width is more than twice the range, the full height of the underlying step appears at the surface of the planarizing layer.

Motivations.   What determines the range of planarization needed for a given multilevel interconnect system? Notice that the greater the range, the more gradual the slope of the dielectric surface from the top to the bottom of a feature (Figure 4(c,d,e)) or, stated geometrically, the smaller the "sidewall angle" or "tangent angle", θ. Gradually sloping transitions are preferable to abrupt vertical transitions for the following reasons.

Step Coverage of Overlying Films.   Oftentimes, the step coverage of a deposited film is strongly dependent on the underlying topography. For example, an aluminum sputter deposition process may produce step coverage of ~100% when the underlying sidewall angle is 45 degrees, but only ~20% (or less) when the underlying sidewall angle is 90 degrees (i.e. vertical). Interconnect lines patterned from this aluminum film are therefore only a fraction of their nominal thickness wherever they cross underlying topographic features. The greater the sidewall angle of a feature, the smaller this fraction, and the smaller the local thickness of the line. Local loss of thickness, of course, equates to local loss of current carrying cross section, which increases the vulnerability of the line to electromigration. Furthermore, capacitive and resistive losses are increased if an interconnect traverses steps due to the increased line length and reduced cross sectional area. Finally, grain growth is affected where aluminum is sputtered onto steps, and the interconnects are thus vulnerable to accelerated electromigration and stress migration failures (Figure 7) (3).

Another example of an improvement to an overlying film arises in the case of final passivation layers. Many devices, particularly those to be packaged in plastic, rely on a layer of PECVD nitride or oxynitride for hermetic protection. These films are most vulnerable to penetration by contaminants where they cover the sidewalls of the underlying interconnects. Two factors contribute to this vulnerability:

1. The film deposited onto vertical sidewalls is exposed to an ion bombardment flux which is predominantly oblique relative to the film surface, whereas film deposited onto the tops of steps and the bottoms of gaps -- surfaces which are horizontal -- is exposed to an ion bombardment flux which is predominantly perpendicular to the film surface. This gives rise to a somewhat less dense film locally on the sidewalls, which results in locally diminished hermeticity (Figure 8(a)).

2. At the bottom corner of any vertical step onto which a film has been deposited, an infinitesimal discontinuity exists between the film

segment grown horizontally from the sidewall and the segment grown vertically from the base surface of the step, or gap (Figure 8(b)). The result is an interface or boundary along which contaminants can penetrate the nitride layer.

The root cause of both of these effects disappears if the PECVD nitride is not confronted with vertical surfaces as it is deposited. Hence, "planarized" passivation layers exhibit improved hermeticity (4).

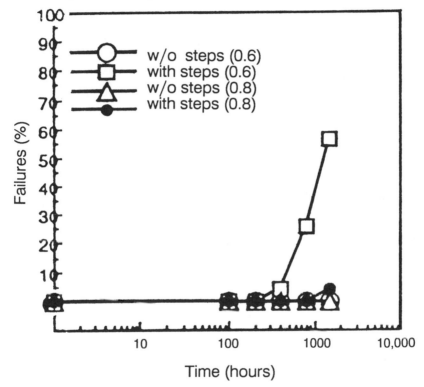

Figure 7. Comparison of failure rates of interconnect lines constructed on top of planar and non-planar topography. The metal is Al-1%Si, with 0.5 μm PECVD nitride passivation coating. The failures are stress-migration failures during "storage tests" at 175°C. From ref. 3; reprinted by permission of the publisher: SEMI.

Imaging of Overlying Resist. The surface topography of a dielectric film is replicated in the metallization layer which is subsequently deposited. Steps in this layer give rise to abrupt resist thickness variations (Figure 9), which may exceed the depth of focus of the exposure tool. Exposure is therefore not uniform, causing localized linewidth variations and loss of critical dimension (CD) control (5).

A second source of linewidth variation is the scattered reflection from sloped surfaces of the metal (Figure 9). This gives rise to local exposure non-uniformities and linewidth variations in the vicinity of an underlying step. This is known as "notching" (3,5). Smoothing the topography beneath the metal layer diminishes the notching effects.

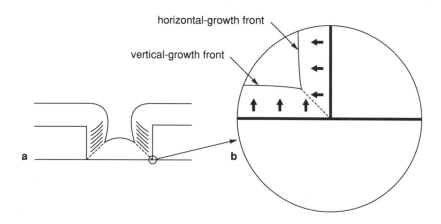

Figure 8. Sites at which passivation films are vulnerable to contaminant penetration. Where PECVD films are deposited onto vertical surfaces such as **(a)**, ion bombardment is oblique and the film is less dense than that deposited onto horizontal surfaces. Boundaries are created at the bases of steps **(b)**, where vertical and horizontal growth fronts join as the film is deposited.

Removal of Overlying Films.   The effectiveness of anisotropic etching is greatest when the film to be etched lies on a flat surface. Difficulties arise where a metal film covers an abrupt step. Here, the vertical distance from the top surface of the film to the bottom surface is much greater than the nominal thickness of the film, as shown in Figure 10. Consequently, the metal etch process requires an extensive overetch in order to clear these thick accumulations of metal along the sidewalls. Extensive overetching is generally undesirable, particularly in etchback processes used to produce plugs, where even slight overetching is deleterious to the plugs. On the other hand, insufficient overetching can leave residual "filaments" or "stringers" of metal at these sidewall areas, which can cause electrical shorts from one circuit element to another.

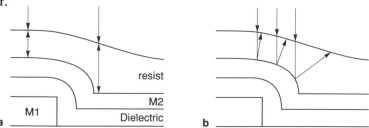

Figure 9.   Sources of local exposure non-uniformities, which lead to linewidth variations in the patterned metal. Path length differences due to resist thickness differences **(a)**; reflective scattering **(b)**.

The more gradual the slope of the underlying step, the less severe this accumulation effect (Figure 10), and the shorter the required overetch.

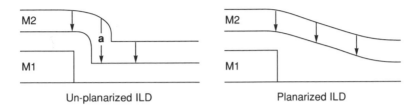

Un-planarized ILD   Planarized ILD

Figure 10. When the interlevel dielectric is not planarized, extensive metal overetch is required to clear metal from sites such as **(a)**.

Global Planarization ... *Why* ? From these discussions it is clear that a planarization technique having a range of 10 μm, for example, results in topography which is more gradual than one having a range of 3 μm; and this in turn leads to benefits which appear during subsequent deposition and patterning steps. What benefit, though, is there to be gained by achieving "global" planarization, in which the planarization range is on the order of tens of thousands of μm? The paragraphs which follow discuss the issue of *step height reduction*; from that discussion we will see the motivation for the drive to achieve global planarization.

2.4 Step Height Reduction

In a previous section, we pointed out that a planarization technique which achieves *topography smoothing* may or may not produce *step height reductions* as well. Likewise, the techniques which accomplish step height reductions may do so over short distances or over long distances ... depending on which technique is used. The point is, again, that topography smoothing and step height reductions are different effects.

The term "step height reduction" is not a precisely correct one, however we will continue to use it in this text since it is in widespread use in our industry. It is worth taking a moment, though, to clarify its meaning. When one says that a technique has accomplished a step height reduction, this does not of course mean that the height of the step itself has been reduced. It means that the height which exists at the Metal-1 level, for example, exists to a lesser extent at the top surface of the ILD-1. Step height reductions do not *reduce* the height of existing features, just as an electrical transmission line does not reduce the amplitude of a signal presented to it. A step height reduction process, like a transmission line, *attenuates* the amplitude of a feature (or input signal) rather than propagating it at full strength.

To quantify step height reduction (SHR), the final step height is usually expressed as a fraction or a percentage of the initial step height:

[1] $SHR = (1 - (h_f / h_i)) \times 100$
        where h_f is the final step height:
          after planarization.
        h_i is the initial step height:
          before planarization.

The terms "full planarization" and "partial planarization" are commonly used; these indicate the magnitude of the step height reduction. "Full planarization" means 100% SHR, i.e. the topography of underlying steps is absent at the surface of the dielectric layer. "Partial planarization" indicates that underlying topography has at least partially propagated through the dielectric layer and appears at its surface.

Range and Step Height Reduction.   A planarization technique which produces step height reductions usually does so to varying degrees, depending on the characteristics of the underlying topography. In the simplest case, consider a conformal CVD process which has produced a film thick enough to fill a small gap (Figure 11(a)). At that gap, the topography which exists at the metal-1 level is no longer present at the top surface of the dielectric. Nearly one hundred percent step height reduction has been achieved. However, the thickness of the film which filled those small gaps is insufficient to fill larger gaps. In these, the step height existing at the metal-1 level exists undiminished at the surface of the dielectric as well ... zero percent step height reduction.

Planarization techniques which make use of spin-on liquid planarizing materials generally give rise to planarization characteristics depicted in Figure 11(b). The smallest gaps are filled by the CVD film, and larger gaps are filled by the spun on material, thus the planarization range is greater than in the example above. There are some gaps, however, which are too wide to be completely filled by the spun on material, and in these gaps the planarization is not complete, i.e. the SHR is less than 100%. And of course in the largest gaps there is 0% SHR. (This simple example ignores the effects of line width and pattern density on planarization; these will be discussed in detail in a subsequent section).

From these examples it is obvious that the performance of a planarization technique cannot be characterized in terms of step height reduction alone; it is also necessary to describe the range: the distance over which step height reductions can be attained. It is meaningless, for example, to simply state that a technique is "capable of producing full planarization", or "... sixty percent planarization".

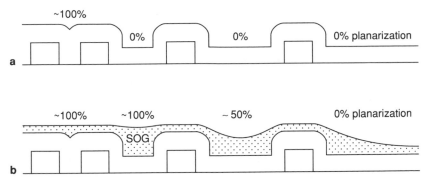

Figure 11. The Step Height Reduction factor (SHR) is a function of the planarization method (its Range) *as well as* the dimensions of the gaps being planarized.

There are two main motivations for seeking step height reductions.

Motivations. The most obvious is that each topography-producing process (such as the patterning of each interconnect layer) superimposes its features onto whatever topography has managed to make its way to the surface of the underlying dielectric layer (Figure 1(b)). Unless the propagation of topography from one layer to the next is suppressed, this superposition of layer after layer can give rise to a cumulative topography which is so severe that subsequent deposition, etch, and lithography processes fail.

The second motivation has arisen due to the desire to extend the use of optical lithography to half-micron and smaller technologies. High numerical aperture (NA) exposure tools are used to address the resolution demands of these technologies, which means that the high resolution is achieved at the expense of depth of focus. In the case of i-line (365 nm) steppers, for example, half-micron feature resolution capability reduces total depth of focus to approximately 1.2 μm ; and 0.35 μm resolution capability reduces total depth of focus even further. This depth of focus budget is consumed by substrate non-flatness, focus errors, and other factors in addition to circuit topography, thus the circuit topography must be limited to a rather small fraction of the total budget (2,3,5,6).

Since the surface to be exposed must be confined within the focal limits of the exposure tool from one edge of the exposure field to the other, there can be no steps in the surface even in the largest gaps or in sparsely populated peripheral areas. The range of the planarization technique must therefore be at least half the size of the largest gap (see Figure 6), which can be on the order of tens of millimeters (7). A planarization technique capable of this is said to be capable of "global" planarization.

2.5   The Cost Of Achieving Long Range Or Global Planarization

Along with the benefits described above, step height reduction processes produce one important consequence which is not a benefit. Figure 12 depicts an interlayer dielectric on which full, global planarization has been achieved. No topographic steps exist on its surface, despite the presence of steps in the underlying layers.

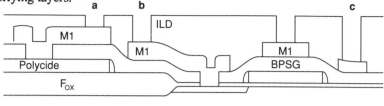

Figure 12. Differential via depths at **(a)**, **(b)**, and **(c)** caused by global planarization of the dielectric over metal-1.

Unless forbidden by design rules, vias to metal-1 may be placed anywhere including locations (a), (b), and (c). Large differences will exist in the depths of these three vias ("differential vias"), which can lead to the following difficulties. First, by the time the via etch process clears the via at site (c), it will have overetched those at sites (a) and (b). This deteriorates dimensional control and profile control of these vias, and can sputter metal from the exposed interconnect onto adjacent oxide and resist surfaces. Second, at site (a), where the dielectric is thinnest, a certain minimum thickness is required in order to minimize capacitive coupling between this interconnect layer and the next. If this lower thickness limit is 5000 Å, for example, the dielectric thickness at site (c) will be 12,000 Å. This deep, high aspect ratio via can be difficult to fill with metal plugs. A blanket tungsten CVD process, followed by a blanket etchback can oftentimes produce plugs capable of filling these differential vias; however aluminum deposited by PVD is generally incapable of this. Selective tungsten plug technology will also have difficulty with this situation: by the time plugs have filled the vias at (a), they will still be short by several thousand angstroms at (c).

This cost or penalty associated with long range or global planarization must be confronted at *one* level of the device, but thereafter it does not re-occur. In this example, it is confronted at the ILD-1 level, through which the via-1 openings will be difficult to etch and fill. Once this is done, however, all metal-2 structures will reside at a uniform elevation, and the root cause of this difficulty will no longer exist.

2.6   Integration Of Planarization Techniques

In the sections which follow, various planarization techniques will be discussed. Some are primarily *gap filling* tools, which may or may not contribute substantially to *step height reductions* as well. Others are capable of long range step height reductions, for example, but may contribute nothing toward gap filling. Techniques which achieve one objective without contributing significantly to the others must usually be supplemented with one or more additional schemes which addresses the balance of the overall planarization task. An example of this is the chemical-mechanical polishing (CMP) technique. It can produce very long range planarization, but must be preceded by a dielectric deposition process capable of filling the gaps and coating the underlying interconnect structures. Some of the schemes to be discussed contribute substantially to both gap filling and planarization. For example spin-on-glass is often used to fill or partially fill the gaps which remain after an initial coating of CVD oxide; at the same time it contributes to the smoothing of underlying topographic features.

Now that the general issues and driving forces associated with dielectric planarization have been introduced and discussed, each of the remaining sections in this chapter focuses on specific planarization techniques, discussing the step-by-step execution of the technique, the effects of the technique on device topography, key technological and manufacturing issues, and novel adaptations which improve the basic capability or expand the applications of each technique.

3.0 THERMAL FLOW of BOROPHOSPHOSILICATE GLASS (BPSG FILMS

This is one of the most straightforward planarization techniques. Although its application is limited to planarization of refractory metal interconnects such as polycides (i.e. it cannot be used after aluminum-based interconnects are in place), it has gained widespread acceptance in that role and has been a workhorse technology since the early 1980's.

There are two essential elements of the technique: (a) chemical vapor deposition of a BPSG film onto the polycide interconnects; and (b) a subsequent annealing step which causes the BPSG to soften and flow, resulting in a smooth surface topography. The effects are illustrated in Figure 13. The flowed film is left on the device as the permanent poly-to-metal-1 dielectric (ILD-0). We will discuss in detail the key issues related to the BPSG deposition and to the annealing or flowing of the film. In addition, several embellishments to this basic two-step scheme will be discussed, since these are often used in actual practice to achieve some ancillary objectives.

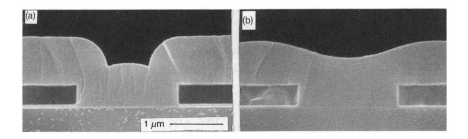

Figure 13.   BPSG as-deposited (**a**), and after flow-anneal (**b**).

3.1   The BPSG Film

A variety of CVD tools and processes have been used to deposit BPSG films for this application (8-19). A complete discussion of these is beyond the scope of this chapter; however we will discuss those parameters which influence the planarizing behavior of the film during the flow-anneal. These are: (a) the composition of the film, specifically the concentration of boron and phosphorus; and (b) the step coverage and thickness of the deposited film.

Glass Transition and Flow.   Glassy materials do not undergo distinct transitions between solid and liquid phases. Melting behavior is usually characterized by a "glass transition temperature"; viscous flow and planarization behavior of BPSG are usually characterized by a "flow temperature". Glass transition temperature is the lower of the two. It is the temperature at which the rate of molecular rearrangement and structural transformation increases sharply, such that internal stresses in the glass are relieved within a matter of minutes. At this temperature the viscosity of the glass is still high enough ($\sim 10^{13}$ Poise) that viscous flow is negligible; however above this temperature the glass behaves more like a viscous liquid than a solid material.

The flow temperature of a glass is higher than its glass transition temperature. At the flow temperature, viscosity is low enough ($10^7 \sim 10^8$ Poise) that a film will flow over or into topographic features in response to the driving force of its own surface tension. Determining the onset of flow is rather subjective; consequently a precise flow temperature is difficult to establish. Correlations between flow temperatures and glass transition temperatures of various BPSG and PSG films have revealed that the flow temperature of such a glass is generally 270~300 Celsius degrees higher than its glass transition temperature (20).

Composition and Stability.   BPSG is a ternary glass system, and its composition strongly affects its melting and flow behavior (Figure 14) (8,9,15,16,18,20,21). Introduction of B_2O_3 and P_2O_5 components to the silica network gives rise to discontinuities and local distortions of the structure. The greater the concentration of these components, the lower the temperature at which glass transition and viscous flow occur (Figure 15). Composition is usually expressed in terms of weight percentages (wt%, or w/o) of boron and phosphorus. Generally, an increase in boron concentration of 1 wt% reduces the transition temperature (and the flow temperature) of a BPSG to a greater extent than does a 1 wt% increase in phosphorus concentration. This is largely because one percent boron by weight corresponds to nearly three *mole percent* B_2O_3; however one percent phosphorus by weight corresponds to about one *mole percent* P_2O_5.

When designing a BPSG planarization process, there is usually a thermal budget which limits the maximum temperature and/or time available for the flow-anneal. The budget exists in order to preserve shallow junction profiles and to avoid degradation of silicide materials. The upper limits are typically in the range of 800~900ºC, and ten to thirty minutes. The thermal budget imposes a lower bound on the concentrations of boron and phosphorus in the BPSG. For example, when the flow-anneal is constrained to a maximum temperature of

900°C, the BPSG film typically must contain at least 3~4 wt% boron and 3~4 wt% phosphorus. At 900°C, the viscosity of this glass will be low enough to allow the film to flow into a shape having a smooth surface topography within thirty minutes.

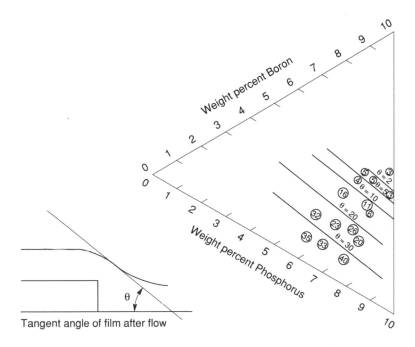

Figure 14.   The SiO_2-rich corner of the SiO_2-B_2O_3-P_2O_5 ternary phase diagram, showing compositions which give constant tangent angle, θ, after flow-anneal. The circles locate each composition and contain the value of θ measured at that composition. From ref. 16; reprinted by permission of the publisher: the Electrochemical Society, Inc..

The more restrictive the thermal budget, the higher the concentrations of boron and/or phosphorus that are needed to achieve sufficient film flow. However, there are upper bounds on the allowable concentrations of boron and phosphorus as well. Glasses containing high concentrations of phosphorus can be highly hygroscopic (22) and may, in the presence of water, participate in chemical and electrochemical reactions capable of corroding aluminum interconnects (23-25). Therefore practical phosphorus concentrations are typically limited to 8~10 wt%. (Before BPSG became widely used in this application, heavily doped PSG (7~8 wt%) films were deposited and flowed (26). This upper limit on phosphorus concentration precluded the use of flow-anneal temperatures below ~1000°C; this was the motivation for the transition to the "new" material -- BPSG -- in the mid-1980's.)

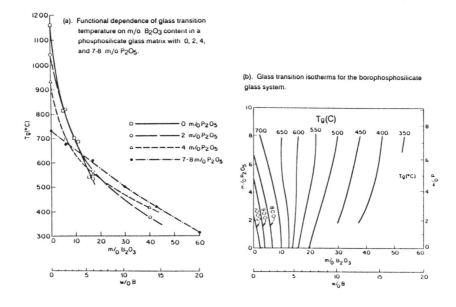

Figure 15. Effects of boron and phosphorus concentrations on glass transition temperatures of BPSG films. From ref. 20; reprinted by permission of the publisher: the Electrochemical Society, Inc..

The maximum amount of boron which can be tolerated seems to depend on the amount of phosphorus present in the glass. The greater the concentration of phosphorus, the lower the allowable concentration of boron. The upper limit on the sum of these two components is generally believed to be in the range of 18~24 mole percent, depending on the methods (PECVD, LPCVD, APCVD) and conditions (pressure, temperature, RF power, chemistry, ...) of deposition. Glasses at or above this upper limit can react rapidly with ambient water vapor, leading to segregation of crystalline precipitates of boric acid at the surfaces of the films (9,11,13,15). In addition, a variety of microstructural defects can form *during* certain flow-anneal processes. Amorphous defects have been observed within the bulk of BPSG films (16) and are thought to be caused by phase separations of phosphorus rich regions from the BPSG matrix during the flow-anneal. Crystalline defects have been observed on the surfaces of BPSG films (27-29), and these are thought to be due to phase separation of BPO_4 crystallites from the amorphous BPSG matrix, which precipitate on the surface of the film.

These instabilities are a principal source of yield "crashes" and other difficulties associated with this planarization technique. Visual inspections of the films can generally detect these defects. Some -- such as boric acid crystallites -- can be removed by a simple water rinse; others -- particularly BPO_4 crystallites -- are difficult or impossible to remove. When *not* detected visually, BPSG defects generally lead to electrical shorts in the overlying interconnect layer.

Generally, the less time that elapses between the deposition of the BPSG and the subsequent flow-anneal, the less vulnerable the films are to the formation of these defects. For this reason, most fabs have "sit time" rules which ensure that the BPSG films do not age indefinitely before undergoing the flow-anneal. After the flow-anneal, the films generally are stable.

The behavior described above can be understood by bearing in mind the extremely hygroscopic nature of B_2O_3 and P_2O_5 and the fact that these components make up a substantial fraction of a BPSG film. The stability of such a film depends upon two factors: the ease with which water is able to penetrate the glass structure and reach these hygroscopic components; and the reactivity of the components once a water molecule does arrive (22). Both factors depend on the following structural and compositional characteristics of the glass network.

Component Concentration. CVD glasses of pure SiO_2 can be highly resistant to water penetration. At small concentrations of P_2O_5 and/or B_2O_3, these components are adequately protected from water vapor by the glass network which is predominantly SiO_2. As the concentrations of these components are increased, however, they make up an increasingly large fraction of the glass structure, which obviously decreases its resistance to water penetration, making it less stable than one having lower concentrations of P_2O_5 and B_2O_3.

Component Protection. The ability of a silicate glass structure to protect P_2O_5 and B_2O_3 components from water penetration depends on the porosity of that glass structure (30,31). The porosity of a glass structure is determined to a large extent by the CVD method used to deposit it (32-34). Processes which deposit films at high temperatures (600~800°C) or in the presence of ion bombardment are capable of producing dense glass films which are more resistant to water penetration than films deposited at lower temperatures and in the absence of ion bombardment.

Component Structure. Reactivity of boron oxide and phosphorus oxide components is influenced by their local bonding environments. Water molecules which manage to penetrate the glass and reach the boron and phosphorus oxides can be expected to react readily with strained bonds and network terminations such as non-bridging oxygen atoms and hydroxyl groups. Highly reactive sites such as these are likely to be present in a BPSG film which is a solid solution or mixture of B_2O_3, P_2O_5, and SiO_2 components. However, a film in which boron, phosphorus, and silicon exist in a true borophosphosilicate glass structure is likely to be relatively free of these reactive sites, and therefore more stable. Most CVD processes probably produce BPSG glass structures which are intermediate between the two extremes mentioned here, and this can account for differences in stability among different BPSG films having identical chemical compositions but produced using different CVD methods (15,19).

Stabilization Due to Annealing.   Temperatures in excess of 700~750°C cause substantial structural changes to CVD oxides deposited at lower temperatures. Following such an anneal, the glass is "densified" (35,36) and relatively free of pores, discontinuities, and bond strain. This accounts for the stability of BPSG films after undergoing high temperature flow-anneal processes (9,11).   Furthermore, annealing tends to move any interstitial phosphorus atoms to substitutional sites (37), where they are likely to be most stable.

Film Thickness and Step Coverage.   Figure 16 depicts typical CVD BPSG films in their as-deposited condition (i.e. before the flow-anneal). Once a film becomes molten during the flow-anneal, its flow is driven by internal pressure gradients created by curvature of the film surface and its surface tension.   Resistance to flow is greatest at sites where a film is thin (38). Therefore, the thickness as well as the step coverage profile of the BPSG affect the degree of flow which can be achieved at a given film composition, in a given flow-anneal process.   When the film thickness is large relative to the height of the step (Figure 16(a)), the molten glass can flow easily in the vicinity of the step edge.   If the film is thin relative to the step height (Figure 16(b)), or is locally thin due to poor step coverage (Figure 16(c)), the step is then a more substantial impediment to the flow of the glass.

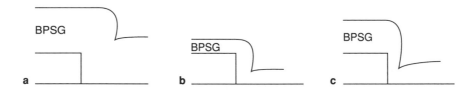

Figure 16.   As-deposited film thicknesses which are large relative to the underlying step height (**a**), small relative to the step height (**b**), and locally thin at the edge of the step (**c**). These characteristics affect the ability of a film to flow in the vicinity of a step.

3.2   The Flow-Anneal Process

Several parameters of the flow-anneal process affect the flow and planarization of BPSG films.

Temperature and Time.   The higher the temperature, the lower the viscosity of the molten glass, and the more readily it will flow. The longer the flow-anneal process, the more *completely* the glass can flow. However, there are constraints which prevent the use of arbitrarily high temperatures or long anneals.   A thermal budget usually exists, and this imposes specific limits on the temperature and/or time available for the flow-anneal.   The maximum temperature must be low enough to avoid degradation of silicides (39-41).   It is also constrained, along with the flow-anneal time, by a maximum allowable D*t product (D is the diffusion coefficient of a specific dopant in silicon, and is an increasing function of temperature; t is the time during which diffusion can occur, i.e. the length of the flow-anneal process step), which exists to control diffusion of shallow junction profiles. The upper limits are typically in the range of 800~900°C, and ten to thirty minutes.

Rapid thermal annealing (RTA) processes are capable of producing glass flow using very short flow-anneal processes (42). By subjecting a BPSG film to extremely high temperatures, its viscosity becomes so low that it is able to flow extensively within a matter of seconds. This can help achieve flow-anneal processes compatible with restrictive D*t budgets. However the temperatures needed to accomplish this are generally in the 1000~1100°C range; much higher than those used in conventional furnace flow-anneals. This makes RTA flow techniques incompatible with fabrication processes where silicide formation precedes the flow-anneal process.

Annealing Ambient.   Flow of phosphorus and boron doped silicate glasses occurs to a greater extent in a steam ambient than in an oxygen ambient, and to a greater extent in an oxygen ambient than in a nitrogen ambient (26,36). This is thought to be a consequence of interactions between the glass network and the water or oxygen molecules. Water molecules can hydrolyze silicon-oxygen bonds in the network, leaving hydroxyl structures in their place. Oxygen can have a similar effect, whenever a lone pair-containing oxygen atom substitutes for a bridging oxygen in the network. In both cases, the result is a displacement of bridging structural elements with non-bridging structural elements, creating discontinuities and local strains which reduce the melting point and viscosity of the structure (25,36,43). Thus, in steam or oxygen ambients, glass flow can be more extensive, or can be accomplished at lower temperatures, than in nitrogen ambients.

Glass flow can also be enhanced in high pressure (e.g. 10 atm.) steam ambients (44,45).

The use of steam or oxygen ambients, however, is not permissible in many fabrication processes. At the temperatures used for flow-anneal processes, either of these ambients is capable of oxidizing underlying polycides and source/drain areas. In some device designs, a layer of silicon nitride is present beneath the BPSG film. Since silicon nitride can prevent penetration of steam and oxygen, oxidation of underlying materials does not occur. This type of barrier, therefore, can allow the use of these flow-enhancing ambients (17). Without a nitride barrier beneath the BPSG, it is generally not possible to make use of steam or oxygen flow-anneal processes. In these cases N_2 is commonly used.

3.3   Planarization Characteristics

Like any planarization technique, BPSG deposition and flow can be characterized in terms of gap filling, topography smoothing, and step height reduction. The basic deposition-and-flow sequence produces a film which fills gaps with non-sacrificial material; thus it has some value as a gap filling tool. It also produces obvious topography smoothing, and can cause step height reductions of up to 100% over short ranges (Figure 17).

Its gap filling capability depends mainly upon the gap filling capability of the deposition process. The deposition-flow sequence is more capable, however, than a deposition-only process, due to the flow-anneal step and its ability to redistribute BPSG into gaps which were not filled during the deposition step. Additional gap filling capability can be achieved using techniques such as multiple-deposition-flow cycles described below.

Topography smoothing is primarily a function of the thickness of the deposited film (Figure 18) and the viscosity reduction produced by the flow-anneal. The lower the film's resistance to flow, the smoother the final topography. If it is not possible to achieve low viscosity during the flow anneal, due to thermal budget limitations or constraints on boron and phosphorus doping levels, then the length of the flow-anneal process is important with respect to topography smoothing. Step height reductions are greatest where the underlying topography is made up of narrow, tight pitch structures, where the film can readily flow from the tops of narrow lines into adjacent gaps. The smaller these gaps the more they become filled by the flowing glass, and the greater the local step height reduction. The larger the gap and/or the wider the line, the less complete the step height reduction.

The range of step height reduction depends primarily on the thickness of the deposited BPSG film. The thick-deposition-etchback techniques described below can provide substantial step height reductions over ranges of a few μm (using a blanket etchback) or over many tens of microns (using chemical-mechanical polishing).

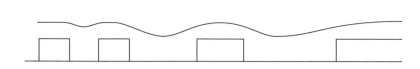

Figure 17. Response of flowed glass to the widths of underlying lines and gaps.

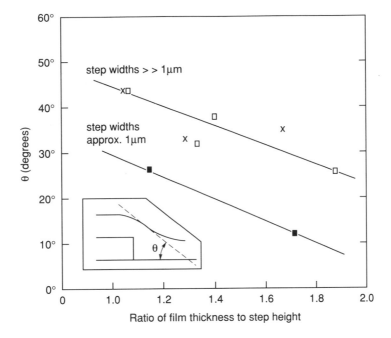

Figure 18. Effects of film thickness, step height, and line width on planarity of BPSG films after flow-anneal.

3.4 Variations Of The Basic BPSG Deposition - Flow Sequence

Flow and Reflow. The preceding section discusses the flow-anneal process, whose function is to melt a deposited film such that its topography becomes planarized. In some fabrication processes, an additional annealing process is employed as well, and its objective is different than that of the flow-anneal. This is called a "reflow" anneal. It is a second anneal, done after the flow-anneal and after contact openings have been etched through the glass (Figure 19). The reflow re-melts the BPSG, at which time the contact openings flow into a smooth rounded shape in response to the local surface tension forces. Ideally, the resulting profile will resemble that in Figure 19(b), enabling good step coverage and filling when the next-level metallization materials are deposited. Profiles such as this can be achieved when the lateral spacing between adjacent openings is large relative to the thickness of the BPSG. However, when these lateral dimensions are small the glass flows into a small-radius circular shape (46), resulting in contact profiles such as that in Figure 19(c). Also, since the silicon surface at the bottom of the opened contact is exposed to the reflow ambient, it is vulnerable to several types of damage. For example, autodoping of boron or phosphorus from the BPSG film can counterdope this silicon; and oxide growth is possible (47) if the reflow ambient is an oxidizing one such as steam or oxygen. Although reflow of the contact openings is no longer common practice, the term "reflow" remains in widespread use and is often used incorrectly when referring to the flow-anneal (pre-contact etch) process.

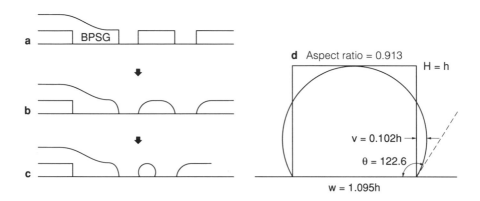

Figure 19. BPSG film after flow-anneal and contact etch **(a)**. After *reflow*, **(b)** and **(c)**. In **(c)** the sidewalls of the contact openings have assumed overhanging profiles during the reflow due to the small separation between neighboring openings. This is due to the large aspect ratio of the "slab" of BPSG between adjacent contact openings; during the reflow anneal, surface tension drives the high-aspect ratio BPSG slab toward the "bead" shape shown in **(d)**. Figure (d) is from ref. 46; reprinted by permission of the publisher: the Electrochemical Society, Inc..

Flow and Etchback of Thick Films. Films which are thick -- relative to the height of underlying steps -- encounter less resistance when flowing, compared with films which are thin. In the basic BPSG-flow planarization scheme, the film is deposited and flowed, and is then ready for contact patterning. The thickness of this film must therefore not exceed 5000~8000 Å, in order to assure that contact openings etched through it will have reasonably small aspect ratios (for the benefit of subsequent metal deposition processes). If the BPSG film, when deposited, is in this range of 5000~8000 Å, its thickness is roughly the same as the heights of the underlying steps. Therefore, these steps impose substantial resistance to the glass flow during the flow-anneal. Resistance can be reduced, and planarization improved, by depositing a thicker BPSG film. After the flow-anneal, the excess film thickness must be removed to obtain a practical film thickness. The simplest method is to use a timed blanket etchback of the BPSG; this transfers the well planarized surface topography of the thick film to the thinner film (48). An alternative method makes use of chemical-mechanical polishing to remove the excess BPSG. By pairing the BPSG glass flow with this additional planarization process, extremely long-range planarization of the final film surface can be achieved. The penalty associated with either of these thick-deposition-etchback methods is the non-uniformity of the final film thickness. For example, in the case where a BPSG CVD process capable of ±3% (3 sigma) thickness uniformity is used to deposit a 2 μm thick film; and the blanket etchback removes 1.4 μm from the flowed film at ±3% (3 sigma) etch rate uniformity, the final 6000 Å film would have actual thicknesses ranging from 5300 Å to 6700 Å.

Multiple Deposition-Flow Cycles. When the underlying topography contains closely spaced lines, it is possible for the BPSG film to pinch off and form voids when depositing in the smallest gaps. Whether or not this occurs depends on the conformality of the deposited BPSG film, the aspect ratio of the gaps, the width of the gaps, and the thickness of the BPSG. As a rough rule of thumb, void formation is possible when the thickness of the deposited film is greater than the widths of the smallest gaps. One method used to prevent void formation is the multiple-deposition-flow technique. The first deposition is relatively thin, to prevent pinch off and void formation in the small gaps. This film is then flowed, to produce a smooth topography onto which a second deposition of BPSG can be made without creating voids. This second film is then flowed to produce additional smoothing.

3.5 New Approaches To Glass Flow Planarization

With each new device technology, the effectiveness of the BPSG deposition-flow technique decreases. This is due to the scaling down of device feature sizes, specifically the interconnect pitches and the junction depths of active features. Reduced interconnect pitches translate to smaller and increasingly difficult to fill gaps between interconnect lines. Reduced junction depths translate to more restrictive thermal budgets, which limit the time and temperature available for the flow-anneal process.

Two approaches are being pursued to improve the gap-filling capability of glass-flow processes.

Filling Difficult Gaps. One approach makes use of the exceptional gap filling behavior of O_3-TEOS oxide films deposited by "high" pressure (500~760 Torr) CVD processes (see Section 8 of this chapter). Films such as this, when doped with boron and phosphorus, are capable of filling small, high aspect ratio gaps which cannot be filled using traditional BPSG CVD processes and films. This provides a straightforward means of filling high aspect ratio gaps, after which the films can undergo flow-annealing to achieve planarization (49-51). With this capability, it is possible to eliminate the need for multiple deposition-flow cycles necessary when depositing less conformal films.

Another approach involves deposition of heavily doped BPSG films at temperatures at or above their flow temperature. The high doping levels (typically 5 wt % boron with 5~6 wt % phosphorus) and high deposition temperature (typically 850°C) make it possible for the films to flow as they are deposited. This gives rise to good gap filling capability. In the past, films with high dopant concentrations such as these have often been found to be unstable in ambient air. In this one-step deposition-flow method, however, some densification of the films occurs before they exit the CVD tool and are exposed to air, due to the 850°C process temperature. This may produce films which are better able to resist ambient humidity, helping to offset the other consequences associated with the high doping levels (52,53).

Reduced D*t Flow. To address the increasingly restrictive thermal budgets imposed on fabrication processes, two new CVD glass-flow approaches are being pursued. One of these is the one-step method described in the preceding paragraph. By using high concentrations of boron and phosphorus, glass flow can be achieved at temperatures somewhat lower than those historically used, for example at 850°C. If the 850°C deposition temperature produces sufficient densification of the films, this may permit the practical use of these heavily doped films.

The other method is conceptually different from any of the glass flow processes discussed above. It involves chemical vapor deposition and simultaneous flow of a glass film, but the film in this case is boron oxide (B_2O_3): a low viscosity (~10^6 Poise at 450°C), very low melting point (~450°C) glass which enables deposition and flow at temperatures in the 350~450°C range. The planarization technique is illustrated schematically in Figure 20. The role of the B_2O_3 is that of a sacrificial planarizing agent, analogous to that of the resist in a conventional *resist-etchback* process. Its smooth, flowed surface topography is transferred to an underlying non-sacrificial CVD SiO_2 film by a non-selective blanket etch process which removes B_2O_3 at the same rate as the underlying film. B_2O_3 is highly unstable in ambient air, hydrolyzing rapidly to form metaboric acid crystals; thus, the crucial element of the technique is the need to perform this etchback and remove all B_2O_3 without permitting exposure to air until it is completely removed from the device (54,55). The equipment architecture must enable wafer transfer under vacuum from the B_2O_3 CVD chamber to the etchback chamber. Because of the low temperature at which this glass flow occurs, it is possible to use this technique for *inter*-metal dielectric planarization as well, where thermal budgets are far lower than in even the most restrictive *pre*-metal dielectric (ILD-0) planarization application.

Although BPSG deposition-and-flow techniques have been widely used for pre-*metal dielectric planarization, they are not suitable for planarization of dielectrics above aluminum-based interconnects, due to the high temperatures needed for the flow-anneal processes. The preceding paragraph discusses a variation of the glass-flow planarization technique which is compatible with* post-*metal as well as pre-metal processing. In each of the sections which follows, other techniques are discussed which also are suitable for post-aluminum as well as pre-aluminum dielectric planarization.*

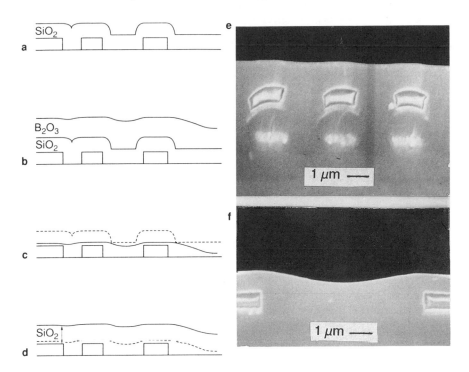

Figure 20. Low-temperature glass flow planarization. Initial layer of CVD SiO_2 (**a**). Deposition and simultaneous flow of B_2O_3 (**b**). Non-selective blanket etchback to remove all B_2O_3 before exposure to air (**c**). Second layer of CVD SiO_2 (**d**). Examples of the completed interlevel dielectric films (**e,f**). No B_2O_3 remains in the devices.

4.0   PLANARIZATION WITH SACRIFICIAL PHOTORESIST

The Resist-Etchback (REB) planarization technique was the first to make use of a sacrificial spun-on material as a planarizing agent (26,56-61). The technique is conceptually similar to the spin-on-glass etchback planarization techniques which are discussed in Section 5. Both the resist-etchback and the SOG-etchback techniques have become widely used in multilevel metal planarization processes: the resist etchback techniques being the more effective of the two when long range planarization is the objective.

The REB technique involves these essential elements. An initial layer of CVD oxide (usually SiO_2 or PSG) is deposited onto the interconnects which are to be planarized.   Next, a layer of photoresist is spun onto the wafer, "submerging" all topographic features. Its smooth surface topography is then replicated into the underlying CVD oxide layer by means of a non-selective blanket etchback process. Finally, a second layer of CVD oxide is deposited onto what remains of the first, in order to achieve the target thickness for the completed, planarized dielectric.

At first glance this appears to be straightforward. However, a number of complicating factors appear when implementing this on real device structures using real processes and processing equipment. These will be described shortly, but we will start by illustrating the step-by-step effects of this process sequence when applied to an idealized device structure using idealized process behavior.

4.1   Planarization Results: Idealized Case

Consider a case which is idealized in three respects:

1. The gaps between adjacent interconnect lines are easily filled by the initial CVD oxide film; or alternatively, the gaps are large enough that the CVD film does not fill them. Whichever the case, no voids are produced during this CVD process.

2. The interconnect level to be planarized lies on an underlying topography which is completely flat. The figures which illustrate this example (Figure 21(a)) depict a polycide interconnect level lying on topography which has been globally planarized by the front end processing. The BPSG or ILD-0 has been globally planarized as well.

3. The etchback process is able to etch the resist and the underlying CVD dielectric at exactly the same rate, and is able to do so through the entire duration of the etch process. In other words the etch is non-selective (or the etch selectivity is 1:1), and is non-varying.

CVD-1 and Resist Spin.   The first step deposits a CVD oxide film called the "CVD-1" layer. This must be thick enough that its lowest lying surface is well above the top surfaces of the underlying steps (Figure 21(a)). Next, the resist is spun on, and this is followed by a bake at 150~200°C. The surface topography now has the characteristics shown in Figure 21(b). Three types of topographical features must be considered: large solid features such as test pads; tight-pitch arrays made up of narrow lines and small gaps; and narrow isolated features. Any real device will, of course, contain other types of features as well. But the three types illustrated here give rise to the three extreme planarization results; whatever other features exist will produce a planarization result which is intermediate between the results produced at these three key features.

Notice the thickness of the resist at each of these features. This is the key to understanding the planarization results. In general, the local resist thickness at any isolated feature is a function of the linewidth of that feature: the smaller the linewidth the smaller the resist thickness on top of it. In an array, local resist thickness on top of the features is a function of three parameters: the overall size of the array, the widths of the lines, and the widths of the gaps.

Local resist thickness increases with the overall array size and with the linewidths; it decreases with increasing gap size (60,62). (In general, the effects of underlying topography on local resist thickness can be modeled to a first approximation by considering that the volume of an underlying step displaces an equal volume of resist, which re-accumulates at the resist surface above the feature, increasing the local thickness. Thus, the greater the volume occupied by the underlying structures, the greater the local resist thickness. The greater the *nominal* resist thickness, the greater the lateral dispersion of the step -- that is -- the greater the horizontal distance over which the resist surface "falls" from the top to the bottom of the step. (Figure 22).)

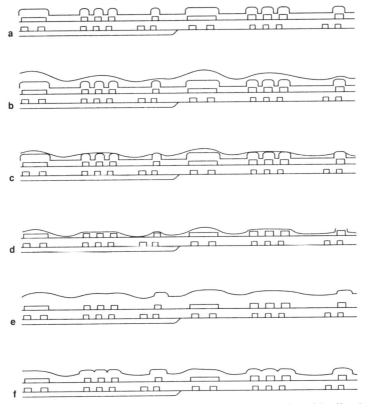

Figure 21. Surface profile evolution at various steps of an idealized resist-etchback planarization process applied to metal-1 interconnects lying on idealized (flat) topography. The first step of the etchback stops when the array areas clear (**c**). With perfect 1:1 selectivity and no microloading, the profile in (**d**) is achieved after the second step of the etchback. 100% step height reduction is achieved within the arrays. Residual steps still exist at larger and smaller features, and the SiO_2 is thicker on tops of the larger features. This will lead to differential via depths. Uniform via depths would have been possible if the first step of the etchback process had continued until the *largest* features cleared. However, this would have sacrificed planarization within the arrays and at smaller features (**f**).

This example illustrates typical resist thicknesses. At the large feature, the resist thickness on top is approximately the nominal resist thickness. At the narrow, isolated line the local resist thickness is about fifty percent of the nominal resist thickness. And at the array, local thickness is about eighty percent of the nominal thickness.

Etchback. The etchback process typically contains two steps. The first step simply removes the resist while preserving its smooth surface topography. This step is stopped just before exposing the tops of the oxide-covered steps in the array areas (Figure 21(c)). At this point, the isolated lines are already exposed, while the large features remain covered by a substantial amount of resist.

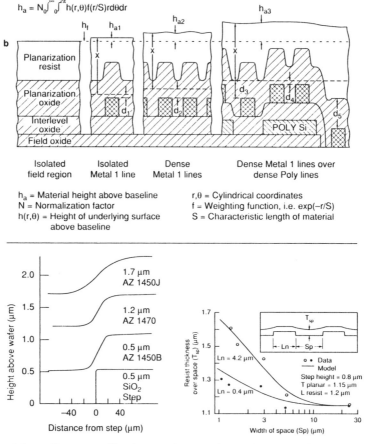

Figure 22. Factors affecting resist coating of underlying topography. "Average Height model" (a) and its graphic representation (b); from ref. 60. The effect of resist thickness on step coverage (c), from ref. 60. Local resist thicknesses in arrays of varying line and gap widths (d): experimental results compared with the "Average Height" model (from ref. 62). Figures reprinted by permission of the publisher: the Electrochemical Society, Inc..

The second step now begins. The objective of this step is to preserve the smooth surface topography of the resist while replicating it into the underlying CVD oxide. This is accomplished, in principle, by establishing the RIE process conditions such that the resist and the oxide etch at equal rates (63). This step proceeds until most or all of the resist has been cleared from the surfaces of the oxide. The surface topography of the oxide is shown in Figure 21(d). It is the same as that of the spun resist, except at the small isolated features. Here, a portion of the underlying step height appears at the surface of the oxide. This is a consequence of the differential resist thicknesses, which led to non-simultaneous exposure of the various features during the first step of the etchback, as mentioned above. A second consequence of this is seen at the large features. Here the resist was thickest (the "resist mounding" effect), it cleared last, and consequently the remaining oxide thickness is greater on these features than on the smaller features.

CVD-2. Following an ashing process to remove any residual resist, the final step is a second CVD process (CVD-2) which increases the total oxide thickness to the desired value. Since the preceding steps produced a more-or-less planar film surface, this final CVD process does not change the topography (Figure 21(e)).

Topography. In this idealized case the completed intermetal dielectric film now has these characteristics:

(a) A generally smooth, fluent surface topography with long, gentle transitions from the tops to the bottoms of the topographic features. This long planarization range is achieved because the nominal resist thicknesses (1.5~2.5 μm) usually used for these processes are large relative to the underlying step heights. As a result the resist is able to flow freely over the features, and it has sufficient capacity to disperse the amplitude of the underlying steps.

(b) Complete step height reductions in the arrays but not at smaller isolated features. At these features, some fraction of the original step height remains, depending on the size and isolation of the feature.

(c) Uniform dielectric thickness on top of the array and on any smaller features where the resist thickness was less than that on the array. On features larger than the array, however, the dielectric is thicker. As a consequence, vias etched at these sites will be deeper than those at other sites.

Tradeoffs. Characteristics (b) and (c) are generally undesirable in an interlevel dielectric. By modifying one or both endpoints of the etchback process, it is possible to eliminate one but not both of these characteristics (Figure 21(f)). There is a fundamental tradeoff between these two parameters: small feature planarization may be optimized only at the expense of differential via depths, and vice versa.

4.2 Planarization Results: Real World Cases

In the previous case, we specified three ideal conditions and were thereby able to avoid three situations which deteriorate REB planarization results.

1. No voids were produced during the CVD-1 process.

2. The interconnects being planarized lay on a completely flat underlying topography.

3. Furthermore, they remained equal through the entire second step of the etchback process.

In actual practice, these ideal conditions are not likely to be achieved.

Voids. Voids can form during the CVD-1 deposition process if that layer is thick enough to close the smallest gaps between adjacent interconnect lines (however this is highly dependent on several factors including the conformality of the CVD film, the thickness and profile of the interconnect lines, etc.). These can become exposed during the etchback process, resulting in trenches similar to those in Figure 3.

Underlying Topography. Figure 23(a) illustrates a typical situation in which the front end isolation structures as well as the polycide interconnects create topography which is not completely eliminated by the BPSG flow process. As a result, some metal-1 structures are at higher elevations than others. The resist coating above the low-elevation metal-1 structures is thicker than that above the high-elevation structures (Figure 23(b)), since the planarization range of the spun resist is greater than that of the flowed BPSG. What are the consequences? During the etchback, resist clears from the tops of high elevation features before clearing from similarly sized low elevation features (Figure 23(c)). As a result, the following steps will produce a completed ILD having either larger via depth differentials than in the preceding "ideal" case, or one having larger residual steps (Figure 23(d,e)).

Selectivity Loss. Although etchback processes can be engineered to produce equivalent resist and SiO_2 etch rates under some nominal conditions, a number of factors can cause these rates to become unmatched. First, resist etch rates are generally sensitive to any polymer deposits present on internal reactor surfaces; they are sensitive to the curing cycle of the resist; and they are highly sensitive to oxygen flow during the etch process. In a manufacturing environment, these parameters can fluctuate from run-to-run or day-to-day, causing variations in resist-to-SiO_2 etch rate ratios and therefore to planarization results.

Second, resist etch rates are also sensitive to the relative amounts of SiO_2 and resist surface area exposed during the etch process. This is a fundamental problem. This ratio changes sharply when the etchback first uncovers the areas of the CVD-1 layer on the tops of the features, causing the resist etch rate to increase (64,65). As the etchback proceeds, the ratio -- and hence the resist etch rate -- continues to increase, due to the increasing exposure of the CVD-1 material and concurrent removal of resist (Figure 24). Not only does this ratio vary from the beginning to the end of a given etchback, it also varies from one circuit design to another, due to variations in density of interconnect patterns, and in the distributions of interconnects between high elevation sites and low elevation sites. What are the consequences of unequal resist-to-oxide etch rates? These are illustrated in Figure 25. Planarity degrades whenever the rates are different, regardless of which material is etching faster or slower.

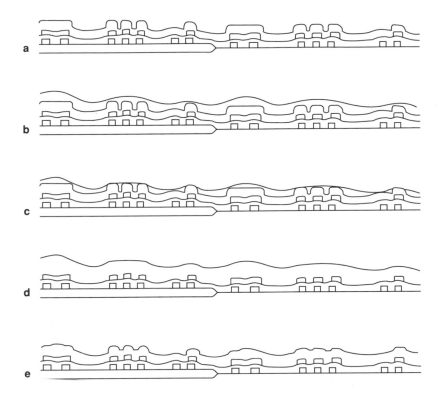

Figure 23. Idealized resist-etchback planarization process applied to metal-1 interconnects lying at *various elevations* due to underlying topography which is not fully planarized **(a)**. Resist-profile before etchback **(b)**. End of the first step of the etchback **(c)**, with CVD-1 oxide exposed at high elevation arrays but with low elevation arrays still submerged. *If* the second step of the etchback begins here, array areas and small features will be planarized but via depth differentials will be large **(d)**. Alternately, the first etchback step may be extended until additional resist clears before proceeding to the second step. Via depth differentials can be reduced but steps will appear in the final topography, particularly at high elevation arrays and small features **(e)**. This tradeoff is more acute in this example than in Figure 21, due to the larger differences in local resist thicknesses caused by underlying topography. Degradation due to microloading has not been included here.

Non-ideal Planarization. In a non-idealized situation, then, planarization results will be somewhat poorer than those described above in the *Idealized Case*. Step height reduction will not be 100% -- even in the tight pitch array areas -- due to the dependence of etch rate ratio on the changing surface areas of exposed resist and SiO_2. The tradeoff between step height reductions and via depth differentials will be more acute, due to elevation differences within the same interconnect level. Furthermore, variations in planarization results will occur: from lot-to-lot due to different circuit designs and interconnect patterns; from run-to-run due to changing polymer accumulation on reactor walls; and from time-to-time due to drift of O_2 mass flow controllers and to imperfect curing of resist.

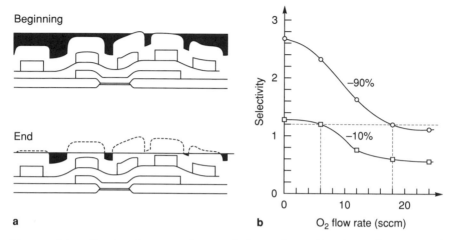

a   b O_2 flow rate (sccm)

Figure 24. During the second step of the resist etchback, surface area of exposed oxide increases from a small fraction near the beginning of the step to large fraction near the end (**a**). As this happens, selectivity deteriorates (decreases) as illustrated in (**b**), where exposed resist area is ~90% near the beginning of the step and ~10% near the end. Selectivity (oxide etch rate divided by resist etch rate) drops as the area ratio decreases. (Figure (b) is from reprinted ref. 65 by permission of the publisher; copyright 1988 IEEE.)

These, along with particle contamination due to polymer exfoliation from etch chamber walls, are the major sources of manufacturing yield loss associated with the resist-etchback planarization technique. Although some of these difficulties can be diminished by clever engineering of the etchback process (65) and by careful control of manufacturing logistics and process endpoints, the technique is neither as straightforward nor as effective as one might first expect.

4.3   Enhancements Of The Basic Resist-Etchback Technique

A number of variations of this technique are possible. These are briefly described here.

Alternative Planarization Materials. In the preceding discussion, the spun on planarizing material is referred to as "resist"; however in many cases materials other than photoresist have been used. One of these materials is a novolac-resin which lacks the photoactive compounds used in photoresists and therefore flows more freely when baked at 200°C (64,66), producing greater step height reductions of large features. Another material, a polyimide, is used as the spun-on sacrificial layer in reference 67, where the planarization technique was designed to produce partial, short range planarization. Other alternative materials are engineered specifically to flow when baked and are capable of producing very long range planarization: step height reductions of 80~90% over distances of 100 μm (5, 68-70).

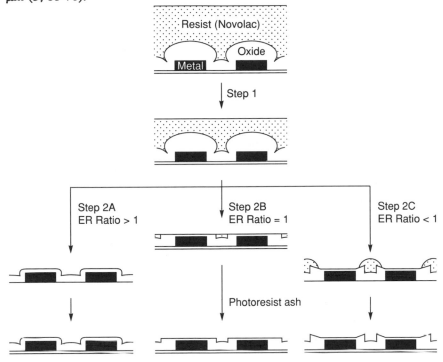

Figure 25. Consequences of unequal resist and oxide etch rates. Etch Rate Ratio ("ER RATIO") is defined here as the resist etch rate divided by the oxide etch rate. From ref. 64; copyright 1987 IEEE; reprinted by permission of the publisher.

Integration with Vias and Vertical Interconnections. The resist-etchback technique can be an integral part of certain plug and via formation techniques. For example, it is possible to pattern the planarization resist with the via pattern prior to the planarization etchback. The etchback, then, performs two functions: it transfers the topography of the resist into the underlying CVD-1 film -- as described above -- while simultaneously transferring the via openings to the dielectric as well (71). This eliminates the need for a separate resist spin/cure cycle and via etch process later on, reducing the number of steps in the overall fabrication process. This approach demands a very thick (1.5~2 μm) CVD-1 layer, since it is not possible to deposit a second layer to achieve the desired dielectric thickness after the planarization etchback (due to the via openings). Furthermore, this technique is only applicable when the interconnect level to be planarized lies on a fully planarized underlayer. This is because the etchback process tends to facet the via openings in the resist, which leads to tapered via profiles. Tapered vias landing on high-elevation interconnects will erode laterally -- deteriorating CD control -- as they are overetched while resist clears from the low-elevation surfaces.

Other via schemes make use of vertical "pillars" for connections between different wiring levels. Construction of the pillars typically precedes the dielectric deposition and planarization processes, so the pillars increase the vertical topography and hence the difficulty of the planarization task. Resist-etchback processes have been used in these applications (72-74), modified such that the non-selective etchback step stops once the tops of the pillars are exposed.

Integration with Advanced Gap-Filling Techniques. Resist-etchback techniques used in ULSI processes must overcome the increasingly difficult challenge of filling the gaps between closely spaced interconnect lines without producing voids. To do this, the techniques used to produce the CVD-1 layer are evolving from simple conventional CVD processes toward multi-step sequences employing tools such as O_3-TEOS CVD (Section 8 of this chapter), deposition-etch iterations (75,76), CVD-with-SOG (77); etc. These multi-step gap filling modules generally have greater gap filling capability than conventional CVD process. Once the gap-filling task is accomplished, the topography smoothing and step height reduction tasks are then addressed by the resist spin, etchback, and associated steps as described above. Further discussions of the gap filling tools and techniques can be found in sections 6 and 8 of this chapter.

4.4   Global Planarization Using Resist-Etchback Methods

An important enhancement of the basic resist-etchback technique is the so-called "Planarization Block Mask" or "N-Mask" technique (62,76,78-81). It makes use of an extra masking step, which is used to create dummy topographic features, or "planarization blocks", in areas where functional features do not exist. This essentially eliminates large spaces, leaving only small gaps: gaps between closely spaced interconnect features, between interconnects and adjacent planarization blocks, and between adjacent planarization blocks (Figure 26).

Once this is done, a conventional resist spin, bake, and planarization etchback sequence is used to planarize the steps. Conventional resist-etchback techniques are capable of accomplishing close to 100% step height reductions of small features so long as the gaps between them do not exceed 10~20 μm; since the planarization blocks remove any gaps larger than this from the entire pattern, the resist-etchback effectively planarizes the entire pattern. This is global planarization.

The key engineering tasks are related to the design of the N-mask (62,82). The design is nominally the negative of the interconnect mask; however, compensation must be made for the thickness of the CVD oxide on the sides of underlying interconnects as well as for lithography misalignment tolerances. In addition, blocks are not placed between closely spaced interconnect features where the initial CVD oxide fills the gap, or where it leaves a gap smaller than the lithography resolution limit plus alignment tolerances. In practice this means that gaps of 3 μm or so may remain after the planarization blocks are in place. Gap variations in the N-layer can lead to local thickness variations when the second resist layer is spun on, and these can translate to residual non-planarities of ±1000 Å or so after completion of the resist-etchback (Figure 27).

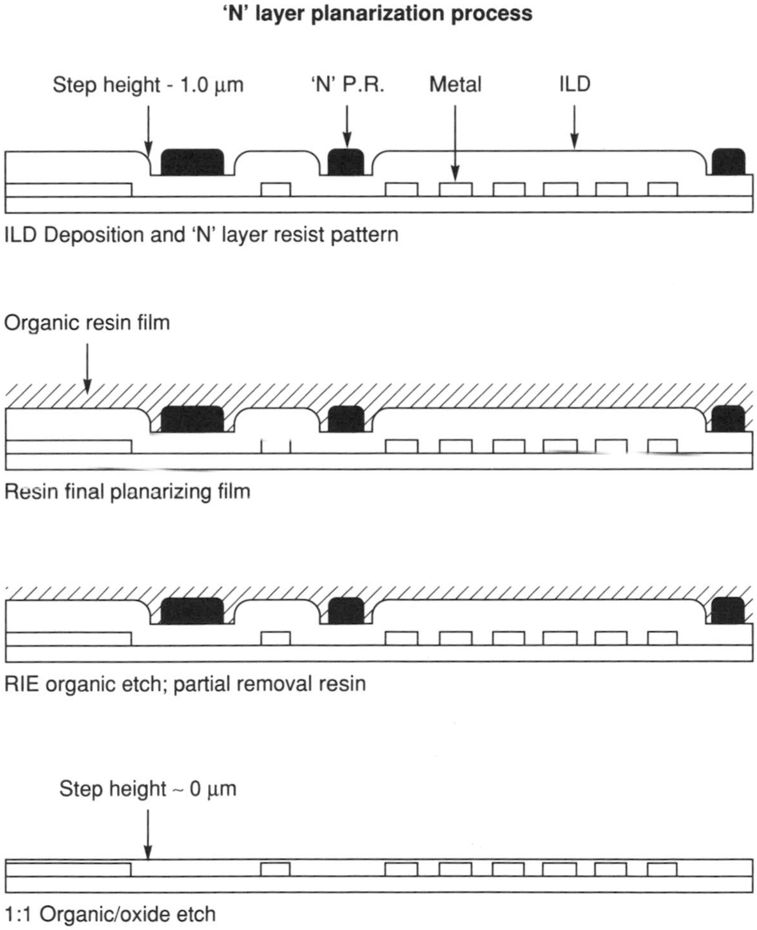

'N' layer planarization process

Step height - 1.0 μm   'N' P.R.   Metal   ILD

ILD Deposition and 'N' layer resist pattern

Organic resin film

Resin final planarizing film

RIE organic etch; partial removal resin

Step height ~ 0 μm

1:1 Organic/oxide etch

Figure 26. Schematic illustration of the "Planarization Block Mask" (PBM) or "N-mask" planarization technique. From ref. 81; copyright 1990 IEEE; reprinted by permission of the publisher.

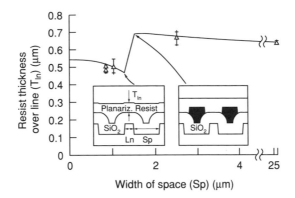

Effect of PBM on resist thickness over 0.5 μm
linewidth array of variable space.

Figure 27.   Sources of resist thickness variations in PBM or "N-mask"
planarization technique. From ref. 62; reprinted by permission of the publisher:
the Electrochemical Society, Inc..

*Spin-on-glass planarization techniques -- which are conceptually very
similar to the resist-etchback techniques -- are the topic of the next section.*

5.0   PLANARIZATION USING SPIN-ON-GLASS

Spin-on-glass (SOG), as the name implies, is a material which can be
applied to a wafer by spin coating while in a liquid state. Because of this, it
flows over the underlying topography; filling or partially filling gaps, and
forming smooth ramps or fillets at the edges of steps. The liquid material is
then converted to a silicon oxide glass film by a sequence of baking and curing
processes. Planarization techniques which use spin-on-glasses have achieved
widespread acceptance for intermetal dielectric applications in double-level metal
devices (83-93). These techniques generally produce *short range* planarization,
however, which oftentimes is inadequate for three- and four-level metal devices.
     Like all multi-step planarization techniques, development of spin-on-glass
planarization processes requires careful selection and optimization of the process
sequence. The unstable nature of spin-on-glass materials demands that careful
attention be paid to other issues as well, specifically:
     (a) Selection of the type of SOG, and optimization of the baking and
     curing processes. These strongly effect the material properties and
     stability of the resultant glass films. They also have some affect on
     the degree of planarization achieved.
     (b) The types of processing which will occur after the spin-on-glass
     film is applied, baked, and cured; for example via etching, photoresist

stripping, etc. Certain types of processing can attack and degrade certain types of SOG films, leading to yield degradations and/or device reliability problems.

The emphasis of this discussion is on the performance of various spin-on-glass planarization processes, and on the process integration implications of each. Properties and characteristics of the SOG materials and films themselves are discussed in a separate chapter ("Inorganic Dielectrics"); therefore these are not discussed extensively in this chapter.

5.1   Elements Of SOG Planarization Techniques

All SOG-based planarization techniques are multi-step sequences. Many variations exist, but most make use of the following elements in one way or another: oxide CVD processes; SOG processes (spin-coat, bake, and cure); and SOG etchback processes. Before describing the various techniques, we will first discuss these common modules or building blocks from which nearly all SOG planarization techniques are constructed.

The Oxide CVD Processes.   Most SOG-based intermetal dielectrics make use of two layers of CVD oxide. One layer is deposited directly onto the underlying interconnect structures, prior to the SOG. This is called the "CVD-1" layer. "CVD-2" is deposited after the SOG film has been produced. If the planarization sequence uses an SOG etchback, this precedes the deposition of the CVD-2 film. What are the functions of these CVD films?

CVD-1.   In general, SOG films adhere more tenaciously to CVD oxide films than to metals. This is due to the nature of the interface bonds, and to the relative thermal expansion characteristics of the materials. Furthermore, CVD oxide films deposited directly onto underlying metal can contribute to hillock suppression. SOG films -- being relatively soft -- are less effective at hillock suppression. Finally, the CVD-1 layer prevents exposure of the underlying interconnects during the overetch portion of SOG etchback processes.

CVD-2.   This layer provides good adhesion between the interlevel dielectric and the overlying metallization layer. In addition, it can help protect underlying SOG material from the subsequent metal etch and resist strip processes; and from exposure to ambient water vapor, which is aggressively absorbed by SOG under certain conditions.

The Spin-on-Glass Processes.   Spin-on-glasses are applied to wafers as liquid mixtures, then are transformed to solid glass films by carefully engineered baking and curing steps. During the spin-coating step, the SOG solution is a liquid mixture which is driven across the wafer surface by centrifugal forces. Once these cease, capillary forces become dominant, driving the film flow over steps, into gaps, and generally tending to produce a smooth fluent surface. At the same time, however, this planarizing viscous flow is impeded by the film "setting up" as the solvents rapidly evaporate.

Baking and Curing.   The bake step -- typically in the 125~200°C range -- drives most of the remaining solvents from the mixture, and initiates hydrolysis of its silica constituents. Curing, at 425°C typically, initiates condensation polymerization (cross linking). This produces a solid film composed of Si-O backbones with some number of organic and silanol sidechains. The solvent loss, hydrolysis, and cross linking reactions result in substantial volume reductions, which in turn give rise to large tensile stresses ("shrinkage stresses") in the films. Consequently, spin-on-glass films generally have low resistance to cracking.

The sidechain groups mentioned above largely determine the key material properties of a spin-on-glass (94-100). SOG mixtures are designed to produce films which are either *silicate* ("inorganic") -glasses, or *polysiloxane* ("organic") -glasses. Each type has certain positive and negative attributes, which will be discussed shortly.

Silicate SOG's. These are typically produced from solutions of TEOS dissolved in organic solvent and water. TEOS reacts with the water, producing intermediate silanol compounds ranging from R_3-Si(OH) to Si(OH)$_4$. Baking and curing evaporates the solvents and initiates condensation polymerization of adjacent silanol groups, which eventually produces a rigid silicate network. Most of the network discontinuities are residual silanol groups, giving the films a hygroscopic nature.

Some of the formulations contain some small fraction of P_2O_5 compounds as well, and are designed to produce phosphosilicate glass structures, which are somewhat more resistant to cracking compared with pure silicates.

Polysiloxane SOG's. These are produced from solutions of methyl-siloxane polymers in alcohol and ketone solvents. Curing initiates condensation reactions producing silicate chains, with some sites retaining silanol groups and others retaining organic sidechains from the precursor polymer. Some fraction of the organic groups are thermodynamically stable enough to remain in the film through the 425°C curing process. In the final glass structure, organic groups substitute for bridging oxygen atoms in the SiO$_2$ network, introducing local distortions and network discontinuities. This network is less rigid than a pure silicate one. In addition, the siloxane films are relatively resistant to water absorption, since most network discontinuities terminate with organic groups and thus are locally hydrophobic. Proper baking and curing is a crucial element of the siloxane SOG technology (88,95,101). It must completely drive out the solvent and promote comprehensive cross linking and condensation without denuding the organic groups from the glass.

Planarization Characteristics. Figure 28 illustrates the surface topography after SOG coating. The topography being planarized here is far simpler than that which confronts planarization processes in real-world applications, however the basic functioning of the SOG techniques is accurately portrayed. Notice that *full planarization* (100% Step Height Reduction) has been achieved in the small gap. In the open field areas adjacent to these lines, the full step height reappears within a few µm of the structure. The technique has achieved gap filling as well as short range planarization: results which are typical of mainstream SOG-based planarization techniques.

The Etchback Process. Spin-on-glass techniques can be divided into two broad categories: *Etchback* techniques, and *Non-etchback* techniques. In this section we simply introduce the etchback process and point out its key features. In the sections which follow we will discuss the pro's and con's of the etchback and non-etchback approaches.

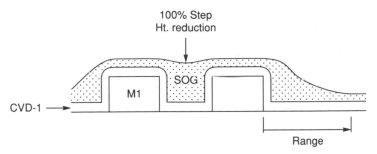

Figure 28.   Example of SOG planarization of simple topography. At this point, the SOG may be partially removed by an etchback step, or it may be coated with a second CVD oxide layer.

The SOG etchback process is a blanket (maskless) etch which, ideally, removes SOG at the same rate as the underlying CVD-1 oxide. In doing so, it transfers the smooth surface profile of the SOG (Figure 28) to the underlying CVD-1 layer (analogous to the etchback process used in Resist-Etchback planarization techniques). Usually, the process is designed to stop once it has cleared all SOG from the top surfaces of the underlying interconnect structures (85,102-104).   SOG is allowed to remain in the gaps between these interconnects, and it becomes a permanent part of the finished device. (This is different from the resist-etchback processes, in which the planarizing resist must be removed entirely from all areas of the device). The purpose of this partial SOG etchback process is to ensure that vias from any subsequent interconnect level to the underlying level do not penetrate or expose SOG material. The motivation for imposing this rule will be discussed shortly.
In sections 5.4 through 5.6, we will illustrate the complications caused by real-world device topography and by real-world process performance.

5.2   Non-Etchback Techniques

The most straightforward SOG planarization techniques are the non-etchback techniques. These are sometimes called "non-sacrificial" or "leave-on" techniques as well. Their key characteristic is that the entire SOG layer is left in place and becomes a permanent layer in the device (86-93).
A typical non-etchback sequence begins with the deposition of the CVD-1 layer. If the smallest gaps between the interconnect lines are 6000 Å wide, for example, the thickness of CVD-1 would typically not exceed about 3000 Å. If this film is too thick, it will reduce the size of the small gaps to the point where they can no longer be filled by the subsequent SOG layer.
Next the SOG is spun on. The thicker the layer, the more effective the planarization, but the greater the tendency of the SOG to crack when cured. Typically, the SOG is applied in two or three separate thin coats. Each coat is followed by a baking step, and sometimes by a curing step as well, prior to the application of the next coat. Using this iterative approach, it is possible to

produce multi-layer films having better cracking resistance than individual films of the same thickness produced by a single coat-bake-cure cycle (88,95,100). Multi-coat polysiloxane SOG films are typically built up to thicknesses of 3000~7000 Å; while silicate-PSG SOG's are typically built up to 1500~2500 Å. Thicknesses much greater than these can accumulate at certain topographic features, however (Figure 29), and it is in these areas where SOG films are generally most vulnerable to cracking. Because of this, one cannot judge a film's resistance to cracking simply from its intrinsic properties and its nominal thickness; the underlying topography must be considered as well.

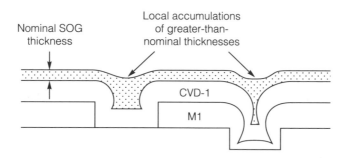

Figure 29. Local accumulations of SOG induced by underlying topographic features.

Local planarization -- or gap filling and smoothing -- has now been achieved. The degree of planarization depends on the SOG solution properties (viscosity, solvent volatility, ...), the application methodology (thicknesses and number of coats), and the characteristics of the underlying topography (step heights, gap widths, feature widths, ...). In general, small steps, small gaps, and narrow features give rise to the greatest planarization. Figure 30 illustrates some of these trends (77,88,95,105-109).

The final step is the deposition of the second CVD SiO_2 layer, CVD-2. Its thickness is selected based on the desired final thickness of the overall interlevel dielectric. One constraint arises if a so-called "wet-dry" process is to be used to etch vias through the dielectric. In these cases, CVD-2 must be sufficiently thick to prevent the "wet" or isotropic portion of the etch from reaching the underlying SOG. This type of etch process generally etches SOG's faster than CVD oxides; this of course would undercut the CVD-2 layer and destroy the desired via shape.

Forbidden Gaps. The thicknesses of all three layers -- CVD-1, SOG(s), and CVD-2 -- must be engineered as a system to avoid the occurrence of forbidden gaps. Film thicknesses selected to produce optimum results in the smallest gaps can surprisingly lead to disastrous results in gaps which are slightly larger. This is illustrated in Figure 31. The narrow trench produced at the forbidden gap gives rise to poor local step coverage (Figure 32) and patternability of the subsequently deposited metal layers (106,107). ("Forbidden gaps" are discussed further in section 5.3).

Poisoned Vias.   In order for a non-etchback technique to be successful, via poisoning effects must be overcome. Via poisoning is a consequence of the SOG layer being temporarily exposed on the sidewalls of vias (Figure 33(a)) (110). Most process flows involve one or more wet processing or cleaning steps after via etching and before the next metal layer is deposited. Furthermore, the vias are exposed to air -- which is a source of water vapor -- while the wafers are in queues between process steps in the fab. Since SOG films are relatively porous glass structures, they aggressively absorb water when given these opportunities. The water absorbed locally at the via sidewalls tends to desorb from these same sites while under vacuum in the metal deposition system, interfering with the deposition of metal in the via holes. Via resistance is degraded due to localized metal thinning and/or oxidation of the underlying metal where it is exposed to the outgassing in the via holes (87,91,99,110-116). These are called poisoned vias or "black vias", due to their dark appearance when inspected in the fab.

Avoiding Poisoned Vias.   Three strategies have evolved which avoid or diminish vulnerability to via poisoning.

1.   Silicate-type Spin-On-Glasses.   One approach is to use silicate-type SOG's rather than polysiloxane types. With these, vulnerability to via poisoning can be diminished but not eliminated. Silicate SOGs are relatively free of the organic components present in polysiloxane SOG's. These organic components are responsible for much of the via poisoning behavior in polysiloxane SOG's. These groups generally incorporate into the glass as terminations and terminal branches of the silicate backbone. As such, they give the glass a relatively high resistance to water. However, they are vulnerable to attack by oxygen plasmas (98,111,117-119), forming volatile CO_x compounds and leaving a discontinuous, porous glass network.

When is the SOG on the via walls exposed to such conditions? An oxygen plasma is commonly used for dry resist stripping processes. If used to remove the via mask photoresist, it can attack and decompose the exposed SOG in the vias. Once the SOG has been denuded of organic groups, by whatever means, the glass structure which remains is highly hygroscopic, and in this condition the exposed SOG will aggressively seize whatever water is available (Figure 34).

Silicate-type SOG's, on the other hand, are relatively free of organic groups and therefore do not decompose in oxygen-plasma conditions, leaving them less vulnerable to wet processing or water vapor (Figure 34). These types of SOG's are the most commonly used for non-etchback applications, specifically because they are less prone to via poisoning behavior. Unfortunately, the absence of organic groups in silicate-type SOG's is what gives rise to the greatest disadvantages of these materials. They produce rigid glass structures which are less resistant to cracking than polysiloxane-type glasses. This limits the thicknesses at which these films can be used, which in turn limits the degree of planarization which can be achieved. Furthermore, even when very thin layers are used, they are vulnerable to cracking where they accumulate in small topographic gaps, as discussed above. So, when silicate SOG's are used as solutions to via poisoning, one trades off the better planarizing properties and cracking resistance obtainable from polysiloxane SOG's.

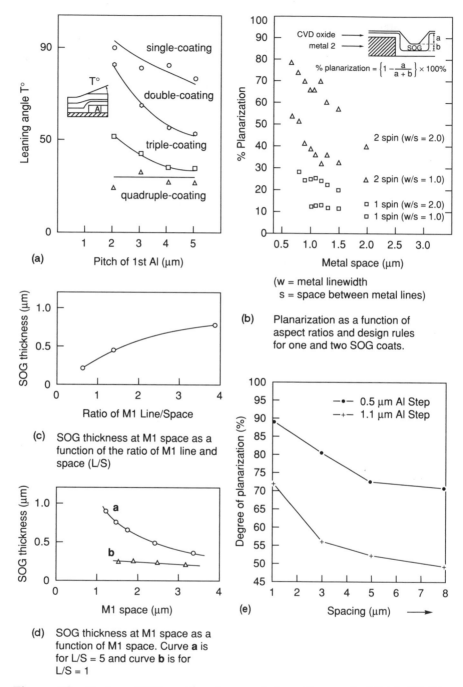

Figure 30. Various SOG coating characteristics. From refs. 95 **(a)**, 77 **(b)**, 107 **(c,d)**, and 88 **(e)**. Copyright IEEE; reprinted by permission of the publisher.

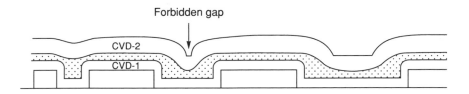

Figure 31. "Forbidden gap" in an interlevel dielectric.

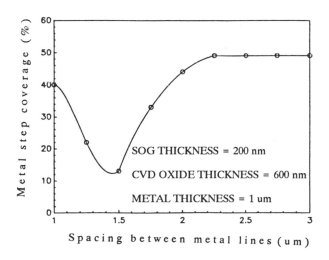

Figure 32. Degradation of metal-2 step coverage over forbidden gaps in the underlying ILD. The height of the M1 steps is 1μm, and the thicknesses of the ILD and M2 layers are as shown. From ref. 106; copyright 1987 IEEE; reprinted by permission of the publisher.

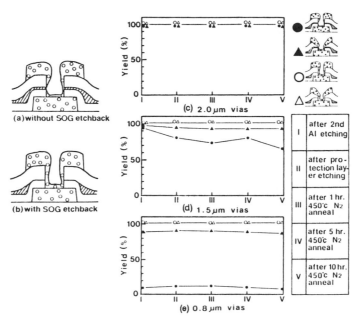

Figure 33. Vias through interlevel dielectrics produced using non-etchback (a) and etchback (b) SOG schemes. Effects of via poisoning on via string (4000 via) test structures (c,d,e). From ref. 110; reprinted by permission of the publisher: SPIE.

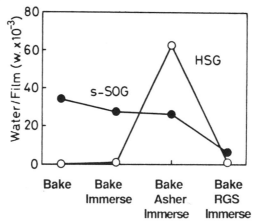

Figure 34. Effects of oxygen plasma ashing on the moisture resistance of silicate SOG (s-SOG) and polysiloxane SOG (HSG). Water content is measured by mass spectroscopy after each of the four steps shown: bake (200°C, 450°C), bake + immerse (10 min. in water), bake + asher + immerse, and bake + RGS + immerse. Water contents are normalized to the film weights. ("RGS" is a "Reactive Glass Stabilization" process, which modifies the surface of the SOG film and improves its resistance to water penetration). From ref. 119; reprinted by permission of the publisher: the Electrochemical Society, Inc..

2. Timely Baking.   If polysiloxane-type SOG's are used, via poisoning can be controlled or eliminated if precautions are taken to bake and dehydrate the films immediately preceding the deposition of metal into the via holes (101,113,120). This is most effective if done in-situ with the metal deposition process. The baking temperature must be high enough to desorb most or all of the water from the SOG at the via walls. It must also be at least as high as the temperature at which the metal will be deposited. Otherwise, the temperature of the SOG will increase as it progresses from the baking step to the deposition step, initiating additional desorption of water from the vias as the metal is depositing.

3. SOG Etchback.   Finally, a third solution to via poisoning is to remove SOG from any areas through which vias will be etched. This is done with a SOG etchback process. Techniques which make use of this are called *etchback* techniques; they are discussed in the next section.

5.3   Etchback Techniques

The root cause of via poisoning can be removed from an interlevel dielectric if SOG is removed from all areas through which vias will be subsequently etched (Figure 33). When this is done, it allows the freedom to make use of polysiloxane SOG's, and can diminish or eliminate the need for in-situ baking prior to subsequent metal deposition. There are tradeoffs, however. These will be discussed in detail later in this section. First the basic process sequence of an etchback technique will be described.

Process Sequence.   Etchback techniques begin, like non-etchback techniques, with a CVD process which deposits a layer of SiO_2 directly onto the underlying interconnects. This is the CVD-1 layer, and it serves generally the same functions as its counterpart in the non-etchback SOG techniques (sect. 5.1). In addition, it is the layer in which the subsequent SOG etchback process must stop, in order to avoid exposing the underlying interconnects. This requirement imposes a lower bound on the design thickness of this layer -- a constraint not imposed by non-etchback techniques. This is an important consequence which will be analyzed in sections 5.5 and 5.6.

The next step is the SOG spin, followed by baking and curing. As in the non-etchback sequences, this spin-bake-cure cycle may be repeated two or three times to build up a relatively thick layer having better cracking resistance than a similarly thick layer produced from a single spin-bake-cure cycle. As mentioned earlier, polysiloxane SOG films are typically built up to thicknesses of 3000~7000 Å; while silicate-PSG SOG's are typically built up to 1500~2500 Å. SOG can accumulate in small gaps to thicknesses much greater than these nominal values (Figure 29), and it is at these sites where cracking failures are most apt to occur. Once the SOG layer is completed, local planarization -- or gap filling and smoothing -- is achieved.   Figure 30 illustrates some of the important parameters affecting planarization.

The etchback process is now performed. Its objective is to remove those areas of the SOG layer which lie directly above any interconnect feature. This will ensure that the subsequent via etch through this dielectric to those features will not penetrate any SOG. The via walls will therefore be free of exposed SOG surfaces, and not vulnerable to exposure to oxygen plasmas, wet cleaning, ambient water vapor, and so on. To accomplish this SOG removal without destroying the local planarization achieved in the previous step, the etch process must remove the CVD-1 oxide at the same rate as the SOG. In this sense the SOG etchback step is analogous to its counterpart in the Resist-Etchback planarization techniques. Unlike the resist etchback processes, it is not necessary for the spun-on material to be removed from the *entire* surface of the CVD-1 layer. SOG must be removed from all horizontal metal surfaces eligible for via placement, but need not be removed from the gaps between interconnect lines.

Finally, the second CVD dielectric layer -- CVD-2 -- is deposited. Its function is to add to the residual thicknesses of the CVD-1 and SOG layers to produce the correct final design value for the interlevel dielectric. It also coats and protects areas of SOG exposed but not entirely removed during the preceding etchback step. This prevents water or air from reaching the SOG during subsequent processing, it provides a stable surface to which the subsequent metallization layer can adhere well, and it provides protection to the SOG during the overetch and resist strip steps used when etching that metal.

Forbidden Gaps. Non-etchback techniques can be plagued by "forbidden gaps": gaps in which the cumulative thicknesses of the CVD and SOG layers produce narrow, high-aspect crevices (Figs. 31 and 32). Etchback techniques are more resistant to this effect. Two aspects of the etchback techniques can contribute to the solution. First, since etchback techniques generally allow the use of polysiloxane-type SOG's -- which are often precluded from non-etchback techniques -- relatively thick coatings of SOG can be used. These thicker coatings give smoother and more complete filling of intermediate-size gaps -- where forbidden gap effects occur -- compared to thinner coatings (Fig 35(a)). Second, by using multiple coating-and-etchback cycles, the planarization process can be made less vulnerable to forbidden gaps (Figure 35(b)).

5.4   Planarization Results: Idealized Cases

In this section and the next, the planarization results of spin-on-glass processes are discussed in detail. First, a simple case is considered. Although the behavior of each process step in the sequence is assumed to be ideal, this case will reveal some planarization effects which are not obvious when considering the *nominal* case illustrated earlier in section 5.1 and in Figure 28. Following this, we will discuss additional complications that occur in real-world cases where the process steps do not behave ideally, and where the underlying topography imposes some difficulties as well.

In the idealized cases in this section, the following simplifications are made. First, we will ignore the loss of planarization due to shrinkage of the spun on glass during baking and curing. Second, the smallest gaps are large enough to be easily filled by the CVD-1 oxide layer and the SOG. Third, the

SOG etchback process has an etch rate ratio of exactly 1:1 (SOG etch rate-to-CVD oxide etch rate), and this ratio is constant throughout the etchback process. The underlying topography is realistic in that it contains a range of feature sizes and pattern densities, however in this idealized case: (a) all metal-1 features lie at the same elevation -- the pre-metal dielectric has completely planarized all underlying topography; and (b) "forbidden gap" features have been excluded.

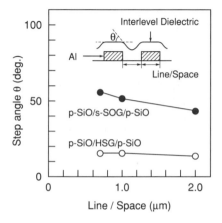

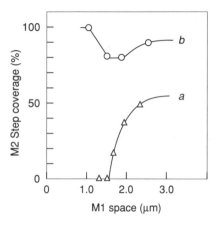

a   Planarization capability of three-layered interlevel dielectrics over 0.7 μm step height.

b   M2 step coverage as a function of M1 space. Curve a is for single SOG planarization and curve b is for two cycles SOG planarization.

Figure 35.   Topography within a gap improves as the thickness of the SOG layer is increased (**a**). The upper curve in (**a**) represents an interlevel dielectric using 2000Å of (silicate) SOG; the lower curve represents a 6000Å (siloxane) layer (from ref. 119). Overlying metal step coverage is poor in a forbidden gap (**b:** curve (a)). The forbidden gap can be eliminated with an additional SOG coat/etchback cycle (from ref. 107). Figures reprinted by permission of the publishers: the Electrochemical Society, Inc. (**a**); and IEEE (**b**).

Non-etchback Results.   Figure 36(a) and (b) illustrate the results of a non-etchback SOG planarization process. This is "short range" or "local" planarization. The Planarization Range is 3~5 μm or less, i.e. gaps larger than 6~10 μm are not planarized (0% Step Height Reduction). However, the dielectric surface makes smooth, low angle transitions from the tops to the bottoms of these larger gaps, such that the subsequent metal deposition can produce good step coverage, and can be etched without leaving "stringers" or "filaments" (Section 2.3).

The thickness of the SOG coating is different at sites *(i)*, *(ii)*, *(iii)*, and *(iv)*, due to the differences in characteristic sizes of those features. This makes the total ILD thickness different at each site, which will eventually give rise to differential via depths (Figure 36(b)). The SOG thicknesses in the gaps is different from site to site as well. The small gaps in the dense array area contain greater SOG thicknesses than larger gaps.

Etchback Results. The etchback sequence begins with essentially the same two elements as the non-etchback process: deposition of the CVD-1 oxide layer; followed by one or more SOG spin, bake, and cure cycles (Figure 36(a)). The etchback case may require a thicker CVD-1 layer, since a portion of it will be consumed by the etchback step. Also, etchback schemes usually make use of polysiloxane SOG's, which oftentimes cannot be used in *non*-etchback schemes. Therefore, the SOG can be thicker in the etchback case, producing a somewhat longer planarization range. The etchback step, in this ideal case, simply transfers the surface topography of the SOG into the underlying CVD-1 layer by etching both at the same rate. The etch step finishes once the SOG has been removed from all horizontal surfaces above the underlying interconnects (Figure 36(c)). SOG is allowed to remain in gaps between or adjacent to the interconnects, since vias will not be placed in these locations. Finally, the CVD-2 layer is deposited (Figure 36(d)). This step generally does not modify the surface topography, although if forbidden gap geometries exist, it is at this point that the forbidden gap itself would be produced. We have excluded that situation, though, from this idealized case.

The intermetal dielectric is now completed, and its topography has generally the same characteristics as those described above for the ideal non-etchback results. The important differences: no SOG remains where vias will be placed; and the smoothing or planarization may be longer range (if thick layers of polysiloxane-type SOG are used). Differential via depths occur in this etchback scheme, just as they do in the non-etchback scheme.

5.5   Planarization Results: Real-World Cases

The planarization characteristics described in the previous section carry forward into real-world situations, but in real world situations those results are degraded by non-ideal behavior of the individual process steps, and by complications caused by non-planar underlying topography. These are discussed in this section.

A real-world topographic situation is illustrated in Figure 37. In addition to the variety of feature sizes, line widths, and gap widths present in the *idealized case*, there now are difficulties imposed by the underlying topography which is only partially planarized (typical of pre-metal dielectrics smoothed by BPSG-flow techniques). In this situation, the interconnects lying on top of the partially planarized dielectric reside at various elevations. Figure 37 illustrates the evolution of the intermetal dielectric topography throughout a non-etchback and an etchback SOG planarization process. This illustrates some of the consequences of non-planarized underlying topography.

What are the non-ideal behaviors that occur in real-world processes? The key ones are: (a) shrinkage of SOG during bake and cure steps; and (b) etch rate changes in the SOG etchback processes.

Certain consequences of these non-ideal situations are common to both non-etchback and etchback -type planarization schemes. Others appear only in the etchback schemes.

Non-etchback Schemes.

SOG Shrinkage.   Substantial volume reductions occur when SOG films are baked and cured. This is mainly due to solvent loss during the lower temperature bake, and to condensation reactions as well as structural densification during the higher temperature cure. The amount of shrinkage is highly dependent on the type of SOG and on the particulars of the baking and curing steps; but generally the total shrinkage is in the range of 15~40%. When a small gap is filled or partially filled with SOG, the accumulation of SOG in the gap is what produces the local planarization (109). However, when the SOG shrinks, say, 25%, the absolute thickness loss is greater where the SOG is thick (in the gaps) than where it is thin (on the tops of steps). Twenty-five percent shrinkage causes a loss of 2000 Å, for example, in a small gap but only 500 Å at the tops of the steps, creating a loss of planarity as shown in Figure 38.

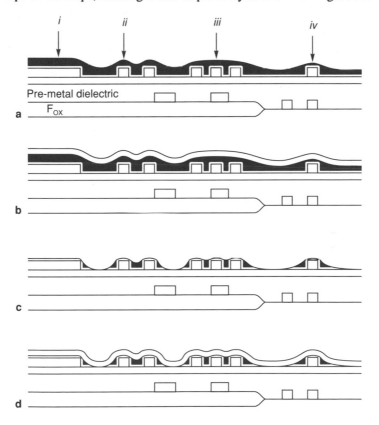

Figure 36.   Idealized SOG non-etchback planarization processes applied to idealized topography (i.e. metal-1 lies on a fully planarized pre-metal dielectric layer). After deposition of CVD-1 layer and SOG coating (**a**). Completed interlevel dielectric ... non-etchback scheme (**b**). Etchback scheme, at the end of the SOG etchback step (**c**). Etchback scheme, after deposition of the CVD-2 oxide layer. Note the local SOG thickness differences at sites *i, ii, iii,* and *iv*. These give rise to differential via depths at these sites.

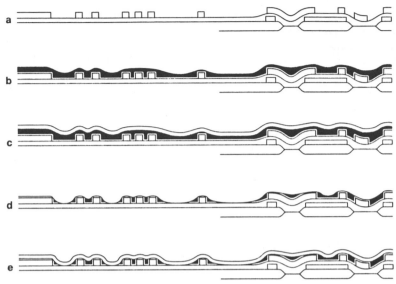

Figure 37. SOG planarization techniques applied to real-world device topography. Metal-1 lines before (**a**) and after (**b**) the CVD-1 deposition and SOG coating steps. The completed interlevel dielectric produced by a non-etchback SOG technique (**c**). Intermediate and final results from an etchback SOG technique: after the etchback step (**d**), and after the CVD-2 oxide deposition step (**e**).

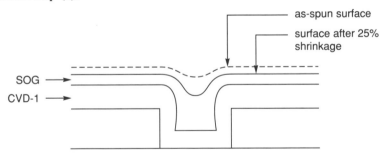

Figure 38. Loss of planarity due to SOG shrinkage.

SOG Accumulations. Local SOG thickness depends on various dimensions and characteristics of nearby underlying features (Figure 30). It is thickest where features are large, and where gaps are narrow. It is thinnest where features are narrow and gaps are wide. This gives rise to differential via depths in the completed, planarized dielectric, even in the idealized case described earlier (Figure 36). In real world cases where underlying topography is not necessarily flat, features from one interconnect level may be stacked onto unplanarized features from a previous level (Figure 39), creating gaps which are deeper than those in the idealized case. The deep gaps occurring in real world device structures can accumulate greater thicknesses of SOG than the shallower, ideal-case gaps, and this has two consequences. First, in very narrow gaps (Figure 39*(i)*) vulnerability to cracking is increased, as this larger volume of material shrinks during subsequent baking and curing steps (77). Second, in wider gaps (Figure 39*(ii)*), SOG can "pool up" in the low lying areas. After the CVD-2 oxide layer is deposited, vias placed at sites such as *(ii)* will be deeper than those placed at sites such as *(iii)*. These via depth differentials add to those caused by feature size variations.

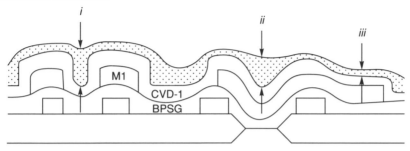

Figure 39.   Effects due to underlying topography. SOG accumulates to a thickness roughly equal to the depth of a gap such as at *(i)*, and is vulnerable to cracking. After the CVD-2 oxide layer is deposited, vias placed at sites such as *(ii)* will be deeper than those at *(iii)*.

Etchback Schemes.   The consequences of SOG shrinkage and of accumulation between stacked features -- which occur in non-etchback schemes as described in the preceding paragraphs -- occur as well in the etchback schemes. In addition, other effects occur which are unique to the etchback schemes.

Etch Rate Variations.   Although etchback processes can be engineered to produce equivalent SOG and CVD oxide etch rates under some nominal conditions, a number of factors can cause these rates to become unmatched. This situation is very similar to that with the resist-etchback planarization techniques. First, SOG etch rates are generally sensitive to the presence or absence of polymer deposits on internal reactor surfaces; they are sensitive to variations in the baking and curing cycle(s) of the SOG; and they are strongly sensitive to oxygen flow during the etch process. In a manufacturing environment, these parameters can fluctuate from run-to-run or day-to-day, causing fluctuations in SOG-to-CVD-1 etch rate ratios and therefore to planarization results.

Second, SOG etch rates are also sensitive to the relative surface areas of CVD oxide and SOG exposed during the etch process (Figure 40) (121,122). This ratio changes sharply when the etchback first uncovers the areas of the CVD-1 layer on the tops of features, causing the SOG etch rate to increase. As the etchback process proceeds, the ratio -- and hence the SOG etch rate -- continues to increase, due to the increasing exposure of the CVD-1 layer and concurrent removal of SOG (see Figure 24 for an illustration of the analogous effect which occurs during resist-etchback processes). Not only does this ratio vary from the beginning to the end of a given etchback process, it also varies from one circuit design to another, due to variations in density of interconnect patterns, and in the distributions of interconnects between high elevation sites and low elevation sites. The consequences of these unequal SOG-to-CVD-1 etch rates are illustrated in Figure 41.

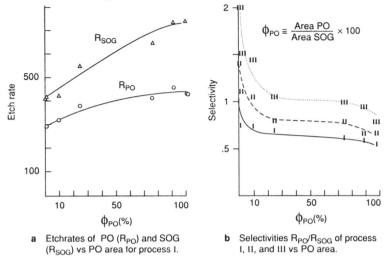

a   Etchrates of PO (R_{PO}) and SOG (R_{SOG}) vs PO area for process I.

b   Selectivities R_{PO}/R_{SOG} of process I, II, and III vs PO area.

Figure 40.   Dependence of SOG etch rate (R_{sog}) and PECVD oxide etch rate (R_{po}) on exposed areas of the two materials (**a**). ϕ_{po} is the ratio of PECVD oxide surface area to SOG surface area. Selectivity loss due to this effect (**b**), for three different etchback processes I, II, and III. Selectivity, S, is the ratio of the PECVD oxide etch rate (R_{po}) to the SOG etch rate (R_{sog}). When this falls below 1.0, planarization degrades during the etchback. From ref. 121; copyright 1988 IEEE; reprinted by permission of the publisher.

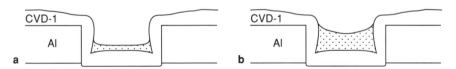

Figure 41.   Loss of planarity during SOG etchback step, due to acceleration of SOG etch rate as the CVD-1 oxide layer becomes exposed. The etchback process in (**a**) has a greater sensitivity to this micro-loading effect than that in (**b**).

In real-world situations, then, planarization results are somewhat poorer than those described earlier in the ideal case. Planarization will not be 100% -- even in the tight pitch array areas -- due to the dependence of etch rate ratio on the changing surface areas of exposed SOG and CVD SiO_2.

Furthermore, *variations* of planarization results occur: from lot-to-lot due to different circuit designs and interconnect patterns; from run-to-run due to changing polymer accumulation on reactor walls; and from time-to-time due to drift of O_2 mass flow controllers and to imperfect baking and/or curing of SOG. These variations are a main source of manufacturing yield losses associated with the SOG etchback planarization techniques. Incomplete removal of the SOG layer from via sites can result in high via resistances due to via poisoning effects. Excessive removal of the SOG layer can degrade planarity, causing thinning and/or shorting in the overlying metal layers. Although some of these difficulties can be diminished by careful optimization of the etchback process and by careful control of manufacturing logistics, the etchback-type schemes are highly vulnerable to a wide variety of external perturbations.

Coping with Features at Various Elevations. When all of the features to be planarized reside at the same elevation on the device, as they do in the *idealized case* described earlier, the etchback process is finished once it clears the SOG from all horizontal surfaces of the CVD-1 layer on the tops of the interconnects. The SOG is thicker on the tops of large solid features such as wide lines or pads ("mounding"), compared to its thickness on the tops of narrow isolated features. Therefore the CVD-1 layer is exposed first at the tops of the narrow isolated features, and is exposed to continued etching while the SOG clears from the tops of the larger features.

However, when these interconnect features reside at different elevations, due to incomplete planarization of the previous interlevel dielectric, the etchback process is confronted with a more demanding task. In Figure 42, the greatest accumulation of SOG needing to be cleared occurs at site *(i)*. In the worst case, the SOG thickness here can be equal to the sum of the nominal SOG film thickness, the polycide thickness, and one-half of the field oxide thickness (94,106,123). While the etchback process is clearing the SOG from sites like this, it is removing the CVD-1 layer from the tops of narrow isolated features such as *(ii)*, where the SOG cleared first. If the CVD-1 layer is not thick enough to endure this overetching, the underlying interconnect will become exposed. This is generally undesirable, since planarization will be degraded as the etchback proceeds beyond this point. Furthermore, the exposed interconnect surface can be a source of sputtered metal contamination during the etchback process; and, if it is aluminum, the interconnect will be vulnerable to hillock growth if exposed during subsequent SOG curing or CVD-2 deposition steps.

This long SOG etchback dictates a lower bound on the thickness of the CVD-1 layer: it must be thick enough to endure the etchback. This causes a difficult situation in cases where small gaps exist (site *(iii)*). It will be discussed in the next section: "Process Window".

Additional Complications. Although we have just illustrated various difficulties imposed by non-ideal geometries occurring in real-world device structures, in this section we still have not discussed two additional geometric features. These are *forbidden* gaps, and *"small"* gaps: ones small enough to cause pinching off and void formation during the deposition of the CVD-1 layer. Forbidden gap features are discussed in preceding sections 5.2 and 5.3). Although not discussed again in this section, they of course are a complicating factor in real world cases.

Small, difficult to fill gaps are also not discussed in this section. We have assumed that the smallest gaps are large enough to be easily filled by the CVD-1 oxide and the SOG. The next section discusses the process window of SOG processes, and it is in that discussion that the issue of small gaps becomes important.

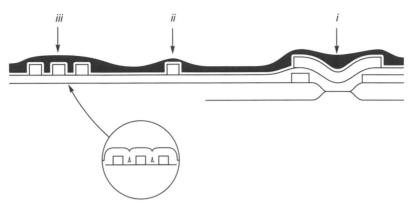

Figure 42. The SOG etchback process must be long enough to clear the SOG at sites such as (*i*). During this etchback, narrow isolated features such as (*ii*) will clear first, and the thickness of the CVD-1 layer must be sufficient to prevent exposure of the underlying metal while the etch continues. The thickness of the CVD-1 layer cannot be arbitrarily increased, however, due to the possibility of void formation in the smallest gaps (*iii*, and inset).

5.6   Process Window

The etchback technique imposes a lower bound on the thickness of the CVD-1 layer. Because of this, the technique is increasingly difficult to apply successfully as gap dimensions shrink from one device generation to the next.

Figure 42 illustrates why this is so. Recall that the SOG etchback process must remove an amount of SOG sufficient to clear the top surfaces of interconnects lying at the bottoms of gaps such as that at location *(i)*. While doing so, the etchback removes oxide from the CVD-1 layer at sites such as *(ii)*. In order to prevent exposure of the interconnect at *(ii)*, the thickness at of the CVD-1 layer must be at least equal to the accumulated thickness of SOG at *(i)*.

Unfortunately there is an *upper* bound on the thickness of the CVD-1 layer as well. The upper bound is imposed by the widths of the smallest gaps in the interconnect level being planarized. The CVD-1 layer must be thin enough to allow SOG solution to enter the gaps which remain (Figure 42, site *(iii)*). If it is too thick, it will produce either narrow crevices which can resist penetration by the SOG, or it will completely fill the gaps itself. Completely filled gaps are undesirable *if* the CVD-1 layer produces voids (Figure 42-inset), because the etchback can re-open them, producing trenches in the dielectric similar to those in Figure 3.

With each new device technology generation, the operating window between the lower and upper limits on the CVD-1 layer thickness shrinks, and the need for effective gap filling techniques becomes more acute. Using the calculation methodology proposed by Pai et. al. in Ref. 106, for example, one can see that the operating window is impossibly small when engineering an etchback SOG scheme to the device geometries typically used in "half-micron technologies", unless it is possible to produce the CVD-1 layer without forming voids in the small gaps. This is one of the motivations driving the development of new, high performance gap filling processes such as ECR CVD and O_3-TEOS CVD processes.

5.7   Tradeoffs: Etchback vs. Non-Etchback Techniques

Which type of SOG planarization technique is best: etchback or non-etchback ? There is no consensus on this. Several tradeoffs exist, and these must be balanced to determine which is best for any particular case. The major tradeoffs are discussed here.

Simplicity and Yield.   In practice, the etchback process requires very careful monitoring from run-to-run, because of its ability to sharply degrade the results of the entire planarization process if not performing optimally. Optimal performance and consistent yield are elusive due to a number of external factors which strongly affect it (Section 5.5).   On the other hand, non-etchback techniques generally require thorough dehydration of the SOG in the vias prior to metal deposition, and this involves an additional process step as well as careful control of fab logistics to minimize delay time between dehydration and metal deposition. Yield loss due to high via resistance can result if something goes wrong.

Intermittent yield losses due to SOG cracking can occur using either etchback or non-etchback techniques. These can be caused by minor variations in the heating or cooling rates of the films during the cure cycle, which can occur due to variations in lot sizes or in loading and unloading procedures. In addition, the viscosity of SOG solutions tends to increase during prolonged storage, or when ambient temperatures fluctuate, which results in greater-than-anticipated film thicknesses when they are spun on. If this goes undetected, the etchback process can fail to completely clear SOG from the via sites, which can lead to via poisoning and cause high via resistances. In non-etchback schemes, the consequences can be excessive via depths, which give rise to poor coverage by sputtered metals.

Planarization Results.   It is generally possible to achieve better planarization when using polysiloxane-type SOG's, and the etchback techniques are generally more compatible with this type of SOG. On the other hand, a certain amount of planarity is lost during the etchback process whenever the etch rates of the SOG and the CVD-1 film become unequal. This can happen for a variety of reasons (Section 5.5).

Process Window.   The etchback technique imposes a lower bound on the thickness of the CVD-1 layer. This shrinks the window within which the technique works well, and mandates the use of difficult control limits on each process step in the sequence. Furthermore, this lower bound makes the etchback technique increasingly difficult to engineer as interconnect dimensions shrink from one device generation to the next.

Learning Curve.   Although the step-by-step descriptions of the etchback and non-etchback techniques seem relatively straightforward, the key to successfully implementing either type lies in the mastery of many of the subtler aspects. Much of this detailed know-how is acquired only by trial and error; it is rarely discussed in the open literature. Temperatures and ramp rates of the bake and cure cycles; "sit time" limits between certain process steps; optimization of etchback processes to compensate for etch rate shifts during the etchback, and from one run to the next; ... once these and all the other nuances have been successfully reckoned with during the development and implementation of a particular technique in a particular fab, this accumulated knowledge provides an immense motivation to retain that technique even in the face of potential advantages offered by alternative techniques.

5.8 Additional Applications of and Enhancements to the Basic SOG Techniques

The principal application for spin-on-glasses has been for inter-metal dielectric planarization. However, they have also been used for planarization of pre-metal dielectrics (between polycide and metal-1), and in planarized passivation schemes.

Pre-metal Planarization.   When used for pre-metal dielectric planarization, the SOG provides an alternative to the conventional BPSG deposition and furnace-flow techniques (92,93,103,104,122,124). Because the underlying interconnects are refractory materials such as polysilicon and tungsten silicide, curing of the SOG can be done at temperatures much higher than those permitted in inter-metal planarization applications. This produces a dense, stable SOG film which is relatively free of residual organic contamination and is more resistant to water penetration than those produced using lower temperature (e.g. 425°C) cure processes. Typically, the phosphorus doped silicate type SOG's are used for these applications.

Passivation Planarization.   Planarization of the final passivation dielectrics is becoming increasingly important, as discussed earlier in Section 2.3. The objective of this is to provide a smooth surface, free of narrow crevices and vertical or re-entrant steps, onto which the nitride or oxynitride film can be deposited. This can improve the hermeticity of the passivation layer (4), reduce parasitic capacitance between adjacent interconnect lines, provide protection from epoxy-resin shrinkage stresses during plastic packaging processes, and improve electromigration resistance of the final-level interconnect lines (3). Spin-on-glass is one of the tools used to achieve this planarization.

SOG as a Gap Filling Tool.   Because SOG is able to flow into relatively small gaps (the exact capability varies from one type of SOG to another), and can be left permanently in devices, it can be used to help accomplish the task of gap filling, which is an increasingly difficult element of nearly all planarization processes. For example, in a three-level metal process, planarization of the second metal level can be substantially more difficult than planarization of the preceding interconnect levels. A major source of this difficulty is the deep, high aspect ratio gaps between adjacent metal-2 lines. These arise because metal-2 is usually relatively thick, and because its topography superimposes onto whatever residual non-planarity has propagated from underlying interconnect and isolation structures. If these gaps cannot be filled by an oxide CVD process, one solution is to replace that process with a three-step sequence of a thin CVD oxide, SOG, and a second CVD oxide. This is discussed in Reference 77 and is illustrated in Figure 43. The SOG layer helps the second CVD oxide layer to fill the gaps without pinching off and forming voids. This allows that layer to be relatively thick, which is the prerequisite for the resist-etchback planarization steps which follow.

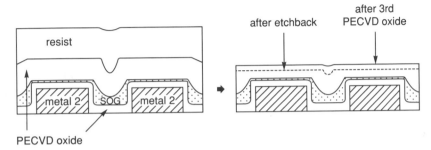

Figure 43.   SOG used to help fill high aspect ratio gaps between metal-2 lines. Once the gaps are filled, a resist-etchback process is used to achieve long range planarization.   From ref. 77; copyright 1990 IEEE; reprinted by permission of the publisher.

Preventing SOG Accumulations in Narrow Crevices.   SOG which accumulates in deep, narrow crevices between interconnect lines is highly vulnerable to cracking while curing, and during subsequent temperature cycles. Resistance to cracking can be improved if these accumulations can be prevented. If the CVD-1 layer is capable of filling the smallest gaps without pinching off and forming voids, this eliminates the possibility of SOG accumulating in these gaps. Of course, even if the small gaps are completely filled by the CVD-1 layer, larger gaps will be only partially filled, and in these larger gaps it is still possible for SOG to accumulate. However, if the CVD-1 layer produces a rounded or positively tapered profile in these partially filled gaps, rather than a narrow vertical or re-entrant profile, the SOG which accumulates has a reduced tendency to crack (107). This is one of the motivations behind the drive to develop CVD processes with excellent gap filling capability. With such a CVD process, the SOG is relieved of the need to fill difficult trenches ... this is done instead by the CVD-1 dielectric. The SOG layer then contributes smoothing and short range planarization ... tasks for which it is well suited.

Integration of SOG with Advanced Gap Filling Techniques. If the CVD-1 layer can fill the small gaps without pinching off and forming voids, and can produce smooth, positively rounded or tapered profiles in the larger gaps, this removes two of the difficulties associated with SOG planarization. First, it can lead to improved cracking resistance, as discussed in the preceding paragraph. Second, it removes the upper bound on the thickness of the CVD-1 layer (Section 5.6), expanding the window in which etchback-type SOG processes can be used. Two recently developed CVD technologies address this. One of them makes use of O_3-TEOS chemistry at "high" pressures

(500~760 Torr). The other makes use of CVD and physical sputtering processes occurring simultaneously and at very high rates in an electron-cyclotron resonance (ECR) discharge. These advanced gap filling technologies are discussed in detail in Sections 6 and 8 of this chapter.

Advanced SOG Materials. Advancements in the SOG materials themselves are also helping to improve the performance of the techniques. These activities are directed at improving the ability of SOG solutions to flow into very small crevices, improving the planarization range or relaxation distance of the solutions, reducing shrinkage and shrinkage stresses during curing processes, and reducing the temperatures needed to achieve thorough curing.

In the photoresist-etchback and the spin-on-glass-etchback planarization techniques, films are applied and then etched back in such a way that the final surface topography is smoothed and planarized. The planarization effects are due to the behavior of the spun on planarizing material ... photoresist or SOG. Another family of planarization techniques also takes advantage of deposition and etchback processes, but does not rely on a spun on material. These are called "dep-etch" techniques, and are discussed in the section which follows.

6.0   DEPOSITION-ETCHBACK TECHNIQUES

Many variations of deposition-etchback *(dep-etch)* schemes are in use. Some are tailored to address difficult gap filling challenges; others are tailored to produce local or short range planarization; and some do both. Before discussing each dep-etch scheme in detail, some perspective can be gained by classifying them in terms of their key traits. Since we are interested in these techniques as elements of planarization processes, the most useful classification is to distinguish between techniques that produce "faceting", those that produce sidewall "spacer" structures, those which "radius" steps, and those which combine two or more of these effects. Those capable of forming either facets or spacers can be useful as gap filling tools, while those which "radius" (increase the radii of) steps can be used to achieve short range planarization. Table I organizes the techniques into these categories. The paragraphs which follow introduce the concepts of facet formation, spacer formation, and radius growth. Once this is done, each of the dep-etch techniques is discussed in detail.

6.1   What Can Be Accomplished By Dep-Etch ?

Faceting. This effect can be achieved by etch processes designed to produce physical sputtering. Physical sputtering is the ejection of atoms from a film's surface due to ion bombardment. The mechanism is physical in nature, not chemical. The incident ions simply transfer energy to the film by collisions as they impact and penetrate; if an atom at or near the surface acquires sufficient energy to overcome the local binding forces, it will eject (sputter) from the film.

Momentum transfer from the incident ion to the film is most favorable when the angle of incidence is 45 degrees. Consequently the sputtering yield and rate are greatest at that incidence angle, and exhibit strong angle dependency (Figure 44) (125,126). In Figure 45 the ion flux is normal to the substrate, however the ions impinge at *various* angles relative to the surface of the step. The etching rate at the corners is greater than that on the horizontal steps, which in turn is greater than that on the vertical sidewalls. As the sputter etching proceeds, the profile of the film evolves as shown in Figure 45(c,d,e). Some fraction of the material ejected from the top corners of a gap re-deposits onto the adjacent sidewalls (127-132), causing the gap to gradually shrink. So the effect of physical sputter etching is to facet the profiles of gaps and steps, while contributing to the filling of small gaps as well. Argon gas, when ionized, has been found to be highly effective at sputtering SiO_2, and is widely used in these physical sputter etching processes. Oxygen produces a somewhat lower sputtering yield (133) and is occasionally used as well.

Many of the planarization schemes discussed later in this section make use of these effects. In some cases, this faceting process is applied to an existing film as shown above; in others it occurs while the film is being deposited. In both cases the fundamental effects are the same.

Spacer Formation.   Figure 46(a) illustrates a gap in which oxide spacers have been produced on the sidewalls. The profile of the gap, which was originally re-entrant with overhanging sidewalls, is transformed into a positively curved one by the formation of the spacers. It can now be reliably filled with spin-on-glass, or by a conformal CVD film (Figure 46(b)). This transformation, achieved using a dep-etch technique, is a tool used as a means to an end, as will be discussed shortly.

Radius Growth.   When deposited onto a step, the profile radius of a CVD oxide film increases as the thickness of the film increases (Figure 47(a)). If the film is non-conformal, the profile deteriorates as the film thickness (and radius) increases (134). However, if the films are conformal, the profile improves with increasing film thickness (134-138) -- the abrupt rectangular profiles of the steps transform to ones with smooth, rounded, and positively sloped transitions from the tops to the bottoms. Figure 47(b,c,d) illustrates how this radius-growth effect gives rise to short range planarization, or "local smoothing". Small gaps are fully planarized while larger gaps and isolated steps are radiused and smoothed, producing low angle transitions from the tops to the bottoms of the steps. In contrast with the *faceting* and *spacer formation* effects discussed above, *radius-growth* produces its planarizing effect in the deposition step without assistance from any etching effects. Usually an etch step does follow the thick deposition, but its only role is to reduce the thickness of the film -- which may be three to four times the height of the steps -- to a value which can be tolerated by subsequent process steps (such as via etching and filling). In doing so, it transfers the thick-film profile to the thinner film. This radiusing behavior, and the scheme used to produce it, is a technique used in some of the planarization schemes described later in this section.

Table I: Deposition-Etchback Techniques

	TYPE	DEPOSITION METHOD RF Bias	Etch-back Type	Deposition-Etchback Mode	Purpose of the Technique.
FACET FORMING:					
1. Dep-Etch Dep	CVD	No	sputter	sequential	Gap Fill.
2. Iterative Dep-Etch	CVD	No	sputter	sequential	Gap Fill.
3. Bias sputtered oxide	PVD	Yes	sputter	simultaneous	Gap Fill.
4. Bias-CVD	CVD	Yes	sputter	simultaneous	Gap Fill.
5. Bias-ECR CVD	CVD	Yes	sputter	simultaneous	Gap Fill.
SPACER FORMING:					
6. Dep-Etch Dep	CVD	No	R.I.E.	sequential	Gap Fill.
7. Iterative Dep-Etch	CVD	No	R.I.E.	sequential	Gap Fill.
RADIUS-GROWTH:					
8. Dep-Etch	CVD	No	R.I.E.	sequential	Planarization.

COMBINATIONS of *Gap-Filling* Dep-Etch with *Planarizing* Dep-Etch:

9. No. 3, 4, or 5 above *plus* long planariz'n etch.	PVD or CVD: (see above)	Yes	sputter	(see text)	Gap Fill and Planarization.
10. Dep-Etch-Dep-Etch.	CVD	No	(see text)	sequential	Gap Fill and Planarization.

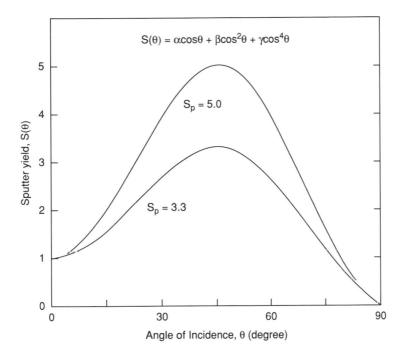

Figure 44. Sputter yield as a function of ion incidence angle. From ref. 126; reprinted by permission of the publisher: the Electrochemical Society, Inc..

Combining the Effects of Dep-Etch Schemes. The dep-etch techniques discussed in the paragraphs which follow are used to achieve specific planarization objectives. Each of these techniques makes use of one, and sometimes two or even three, of the basic effects described above: faceting; spacer formation; and radius-growth.

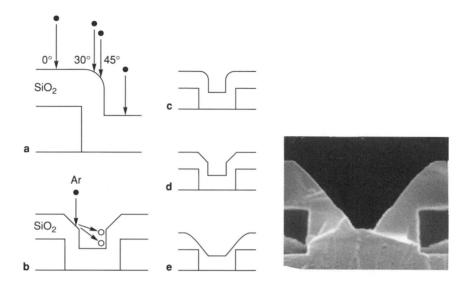

Figure 45.   Incidence angles of ions arriving normal to the substrate (**a**). Re-emission and re-deposition of previously deposited material (**b**). Evolution of the surface profile of the deposited film due to sputter etching (**c~e**).

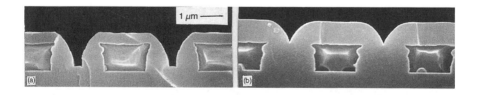

Figure 46.   A spacer-forming dep-etch sequence used to compensate for a retrograde metal etch profile (**a**). A subsequent CVD oxide deposition fills the gaps (**b**).

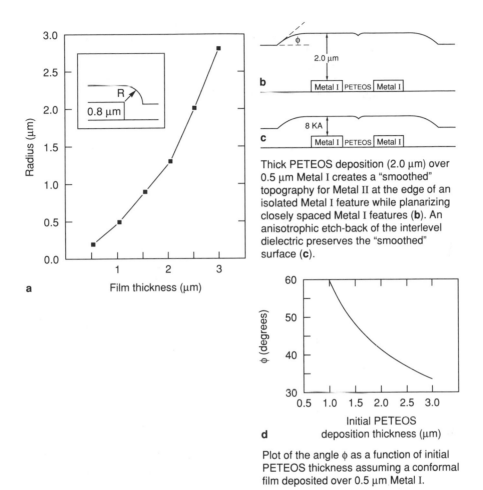

Thick PETEOS deposition (2.0 μm) over 0.5 μm Metal I creates a "smoothed" topography for Metal II at the edge of an isolated Metal I feature while planarizing closely spaced Metal I features (b). An anisotrophic etch-back of the interlevel dielectric preserves the "smoothed" surface (c).

Plot of the angle φ as a function of initial PETEOS thickness assuming a conformal film deposited over 0.5 μm Metal I.

Figure 47. Radius growth due to thick deposition of conformal PE-TEOS SiO$_2$ (a). Deposition of a thick, conformal film (b) followed by a blanket RIE process (c) gives rise to local planarization. (Figures (b~d) are reprinted from ref. 137, by permission of the publisher; copyright 1987 IEEE.

6.2   Facet-Forming Dep-Etch Techniques

The faceting effects produced by physical sputter etching can be exploited to help accomplish difficult gap filling tasks. They can also be used to achieve full planarization of certain types of features, and in fact much of the literature has emphasized this use of the techniques. However, in practical applications, it is the gap filling power of these techniques that makes them useful.

Faceting effects can be applied to a previously deposited oxide film by simply following the deposition process with a sputter etch process. This is the basis of the *sequential dep-etch* techniques (Table I). Faceting effects can also be applied *while* a film is being deposited: the deposition and the sputter etching occurring simultaneously. This happens in the "bias sputtered oxide", "bias-CVD", and "bias-ECR-CVD" processes. Whether the deposition and sputter etching are sequential or simultaneous, the first order effects on the surface profile are the same. These are illustrated in Figure 48.

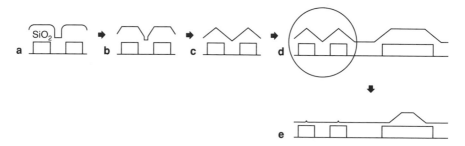

Figure 48.   Surface profile evolution as sputter etching proceeds from (**a**) through (**e**).

In Figure 48(a), an initial layer of oxide coats the interconnect lines. In 48(b), the first effects of tapering are seen: a substantial amount of the film has been removed from the top edges of the steps; and some small fraction of that material has been re-deposited onto the vertical portions of the adjacent sidewalls. The tapered surface is approaching a 45 degree angle, and the width of the gap is shrinking. After extended sputter etching (or extended deposition and simultaneous sputter etching), the profile in (c) is reached. The pyramidal shape occurs because the tapered part of the profile effectively moves laterally as the etchback proceeds. If the interconnect lines are narrow, the tapered surfaces from each side of the line travel toward each other, eventually meeting at the middle. In (d) we see what has occurred elsewhere on the device, where gaps between lines are larger and where the lines themselves are wider. All gaps have been slightly decreased in width, but this is not sufficient to fill the larger gaps; therefore no step height reduction is achieved. The edges of all steps, regardless of linewidth, are now tapered at angles of about 45 degrees.

At this point, there are two options. One is to make use of an "extended planarization etch", in which physical sputtering is carefully balanced with the deposition such that the very narrow lines eventually become completely planarized (125,139,140). This gives rise to the surface topography illustrated in Figure 48(e). The wider features are somewhat narrower than in Figure 48(d) as a result of the extended planarization etch, however their heights have not been diminished. This extended planarization etch is described in greater detail in the discussion of bias-sputtered oxide below.

The second option is to end the dep-etch scheme at point (d), and proceed to a separate planarization process capable of smoothing or eliminating the remaining steps over the large features (141-143). The dep-etch scheme has accomplished what it is good at -- which is gap filling -- and this is oftentimes the most difficult task of the entire planarization effort. This option is usually more practical than the first, since the *extended planarization etch* is only capable of achieving planarization of narrow features.

The remainder of this section 6.2 describes several techniques which make use of physical sputter etching to produce faceting behavior. Although the techniques themselves are quite different from one another, bear in mind that they all produce essentially these results. The main differences -- in terms of planarization results -- are in the abilities to fill small, high aspect ratio gaps.

Sequential Dep-Etch Techniques. These are generally more straightforward compared to the simultaneous dep-etch processes. Furthermore, because the deposition and the sputter etching processes are done sequentially, the deposition can be done in a reactor or chamber designed specifically for such a process, while the etchback can be done in a separate tool or chamber designed specifically for it. It is not necessary to compromise either process for the benefit of the other.

Dep-Etch-Dep Sequence. One sequential scheme is the so-called "Dep-Etch-Dep" sequence. It is used in cases where the gap filling capability of the CVD oxide is insufficient to fill the smallest gaps without pinching off and forming voids. Its objective is to produce the surface topography shown in Figure 48(d). To do this, an initial layer of CVD oxide is deposited, stopping before the small gaps are filled. A physical sputter etch process is used to taper the top edges of the gaps, transforming the vertical trenches into ones with "v"-shaped openings at the tops. At the same time, the re-deposited material slightly reduces the widths of the trenches remaining at the bottoms of the gaps. The longer this etchback, the more tapered the trenches become (or, the greater the "acceptance angle" of the trenches), which makes them easier to fill during the oxide CVD process which follows. However, the etchback step must end before the tapered surfaces of the oxide, which are "moving" laterally, reach and expose the top corners of the underlying interconnect lines. Finally, a second deposition is done, this time onto the tapered surface profile of the first oxide layer. If the gaps have been sufficiently tapered, this second CVD layer can fill them. The thickness of the second CVD layer can be selected based on what is needed for the planarization process which will follow (76,141-149). For example, if a resist-etchback process is to be used, this faceted oxide film would serve as the CVD-1 layer (Section 4.3), and would typically be in the range of 1.5~1.8 µm thick.

The effectiveness of this scheme depends on the depth of the gap and, to a lesser extent, on the aspect ratio of the gap (150). When a gap is relatively shallow, the sputter etch step produces a profile with a large acceptance angle such as that in Figure 49(a). This can be easily filled by the second CVD oxide film. However if the gap is deep, the profile shown in 49(b) is obtained, and this is more difficult to fill. Longer etching of the profile in 49(b) would eventually lead to a profile similar to 49(a), however by that time the corners of the underlying interconnects would be uncovered. The gap width is a factor as well: the narrower the gap, the thinner the first oxide layer must be (to avoid pinching off the gap); and the thinner the first oxide layer, the greater the constraint on the amount that can be etched back (from the angled surfaces). Figure 50 gives an example of the gap filling capability of a specific sequential dep-etch scheme.

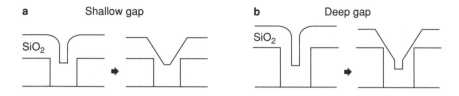

Figure 49. SiO_2 profiles after sputter etching: the effect of gap depth.

(Notice that another *type* of dep-etch-dep scheme is described later in this section. Although both schemes bear the same name, they use different types of etch processes, and therefore produce different results.)

Iterative Dep-Etch Cycles.   The tougher the profile of the gap, the greater the difficulty of filling it using a dep-etch-dep scheme. Specifically, small gaps which are more than ~1 μm deep, gaps which are re-entrant, and gaps with overhanging structures such as anti-reflective coatings, are not easily filled. However, a second iteration of the deposition and etching cycle can sometimes overcome this difficulty. The sequence would be: dep-etch-dep-etch-dep. The second deposition, like the first, is stopped before the gaps are closed; and the second etch, like the first, tapers the profiles, increasing the acceptance angles for the subsequently deposited layer. As the number of iterations increases, the gap filling ability increases (144,150-152). Of course the overall process cycle time increases as well, and in actual practice it is uncommon for more than two iterations (i.e. more than dep-etch-dep-etch-dep) to be used.

Simultaneous Dep-Etch.   The simultaneous dep-etch processes make use of the same faceting behavior used in the sequential schemes discussed above, but with the deposition and sputter etching processes occurring simultaneously. A simultaneous dep-etch process can be thought of as an iterative dep-etch sequence in which each deposition and each etchback is infinitesimally short, and the number of iterations is very large. Because of this, the gap filling capability of a well engineered simultaneous process usually exceeds that of its sequential counterpart. The disadvantage of the simultaneous approach is that the deposition and etch processes are forced to operate in the same process chamber and under identical process conditions. Therefore, one process or the other must be "force fit" into a chamber or an operating regime to which it is not well suited. The well known examples of simultaneous dep-etch processes are bias-sputtered-oxide (or bias-sputtered-quartz), bias-CVD, and bias-ECR-CVD.

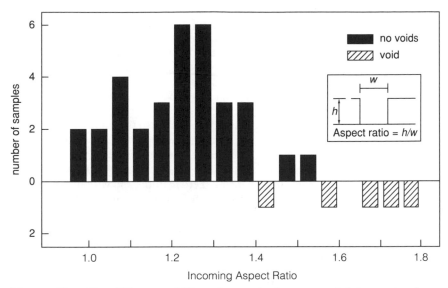

Figure 50. Gap filling capability of a particular sequential dep-etch scheme used in an intermetal dielectric application. The gaps are 8000Å deep. (Intel-Sematech EIP Report, 1991; courtesy of Intel Corp.)

Mechanism.   Figure 44 shows the dependence of sputter yield on the angle of incidence. Qualitatively, this same curve also expresses the etch *rate* versus angle of incidence. Most deposition processes also exhibit deposition rate variations as a function of the local surface angle, and an example of such a relationship is superimposed with the etch rate curves in Figure 51. This superposition of the two behaviors is what occurs in a simultaneous dep-etch process. At any point on the step's surface, the *net* deposition rate is equal to the difference between the local deposition rate and the local etch rate. Since the net deposition rate is lowest in the vicinity of the surfaces angled at 45 degrees, an angled surface develops.

In *simultaneous* dep-etch processes, various results can be achieved, depending on the relative rates of the two processes (125,130,139,140,153). This is shown in Figure 52. Moving from left to right across the figure, the rate of the sputter etching increases relative to the rate of deposition. In case (a), the sputter etching rate is nil, meaning that the net deposition rate -- as a function of the incidence angle theta -- is simply that of the deposition process. This is the condition occurring in CVD processes having little or no ion bombardment, or in a bias-sputtered-oxide deposition process with no substrate bias.

In (b), some faceting occurs but its contribution is small relative to that of the deposition process. The surface does not acquire a distinctive faceted shape, nevertheless the small, angle-dependent reduction of the net deposition rate reduces the tendency of the deposited film to pinch off as the gap closes. In (c), the deposition rate still exceeds the etch rate over the entire surface of the step, regardless of incidence angle, however the rates are much closer to each other.

Very little net deposition occurs on the angled surfaces, so the film quickly acquires a tapered profile. As its thickness increases, the angled surfaces converge, eventually producing a pyramidal profile over the narrowest lines. Once this happens, no further accumulation is possible on the top of this peaked structure, but accumulation continues on all other surfaces, gradually burying the peaked features. Wider features also acquire faceted shapes, however they take longer to reach a peaked shape, and the burying process does not begin until this peaked shape is attained.

In case (d) the surface profile evolution is similar to that in (c). However a special condition exists in (d): the deposition rate on the *angled* surfaces exactly matches the etch rate on them. As a result, once the steps become faceted, no deposition occurs on the angle surfaces. As in (c), the angle surfaces gradually migrate toward each other, producing peaked features on which no further accumulation is possible. This condition is reached sooner in (d), though, since the film thickness does not increase on the angle surfaces. And, as in (c), accumulation continues on other non-angled surfaces, gradually burying the peaked features.

In (e), another special condition exists: the deposition rate and etch rate are exactly matched on the *horizontal* surfaces. Thus, the film thickness on horizontal surfaces remains constant as the process proceeds, while a net removal of oxide occurs from the angled surfaces. The profile evolves as shown in the figure, provided that a pre-existing film is in place. This condition gives rise to the planarization mechanism described by Mogami, et. al. in reference 139.

In cases (f) and (g), the local sputtering rate exceeds the local deposition rate at all positions on the surface, causing lateral migration of angled surfaces, erosion and disappearance of peaked features, *and* loss of thickness from horizontal surfaces as well. Case (g) is the special condition in which no deposition mechanism is active. This, of course, is simply the sputter etch process used in the sequential dep-etch schemes.

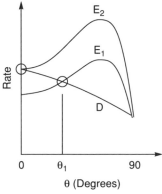

Figure 51. Angular dependence of local deposition rate (D) onto a step, superimposed with the corresponding sputter etching rates (E_1 and E_2) (from ref. 139). *Net* deposition rate at any point on the step is equal to the difference between the local deposition and etching rates. Depending on the relative rates of the two processes, various topographies can evolve. Reprinted by permission of the publisher: American Institute of Physics.

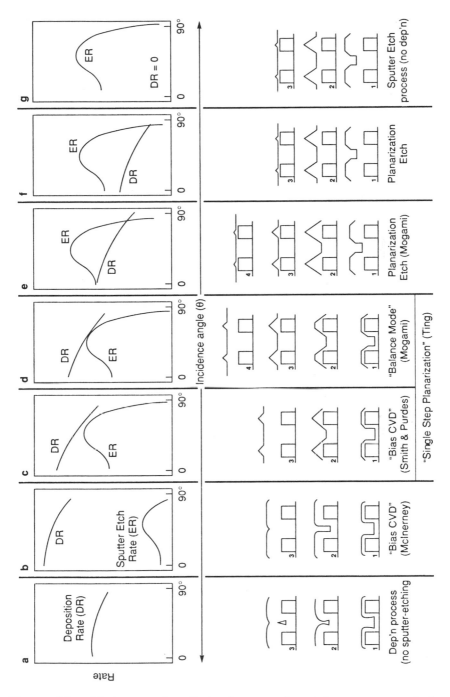

Figure 52. Various modes of simultaneous dep-etch processes, and the surface profiles which evolve.

Gap Filling.   Void free filling of gaps having extremely difficult aspect ratios has been reported using simultaneous dep-etch techniques.   Redeposition of sputtered material into the lower portions of gaps contributes to the gap filling ability (130,131,143). In addition, with deposition and faceting occurring simultaneously, gaps can be "held open" as the deposition into them proceeds; whereas when deposition and faceting occur sequentially, gap widths are reduced by the deposited film before the faceting mechanism is able to make its contribution. This diminishes what can be accomplished by the sputter etching process.

Substrate Bias.   In bias-sputtered-oxide, bias-CVD, and bias-ECR-CVD processes, rf power is applied to the pedestal which holds the wafer. This can be used to control the potential of the substrate relative to that of the discharge: a key parameter governing the behavior of the process. Exactly how this occurs varies from case to case, depending on several factors such as the reactor configuration, the discharge pressure, and the rf frequency. For this discussion it is sufficient to say that, in general, the greater the rf power applied to the substrate pedestal, the more negative its potential becomes relative to that of the discharge, and the greater the acceleration of ions as they "fall" from the discharge potential to the substrate potential.

We will use the term "rf bias" to refer to the rf *power* applied to the pedestal, although strictly speaking this is not itself a bias but rather a tool used to *create* a bias. The true bias is the potential difference, or voltage drop, from the discharge to the substrate.

Why is the substrate biased? Because the ion bombardment resulting from the bias is what gives rise to the physical sputtering mechanism. This contributes to the gap filling mechanism (Figure 53), and improves the quality of the films where they are deposited onto sidewalls of steps (154-158). In addition, ion bombardment generally helps to densify the dielectric films being deposited (156,159).

Although energetic ion bombardment is the fundamental mechanism responsible for the faceting behavior which is the foundation of these processes, it is also the source of most of the difficulties associated with them. Because much of the energy of the bombarding ions is transferred to the wafer as heat, the wafer can reach temperatures in the 350~500 degree C range. Thus the sputter etching rate, which dictates the throughput of these processes, can be improved only at the expense of temperature control, and vice versa. This is an important issue in *inter*-metal dielectric applications, since the aluminum-based interconnects as well as many types of contact barrier structures can be degraded by exposure to temperatures in this range. In addition, several types of device-level anomalies have been linked with these types of processes (Figure 54) (129,141,152,160). This is a somewhat less critical issue in bipolar technologies, and indeed it is the bipolar applications where these types of techniques have primarily been used.

Bias-Sputtered Oxide Process.   In this process, deposition of the SiO_2 is by physical vapor deposition (PVD), which distinguishes it from all other dep-etch techniques discussed in this section (Table I). The SiO_2 is sputtered from a target made of high purity silica, after which it deposits onto the substrates. This is done at pressures in the 1~10 millitorr range. A separate rf bias is applied to the substrates, and this along with an argon ambient gives rise to physical sputter etching of the substrate. The faceting results in a tapered surface profile, enhanced filling of small gaps, suppression of seam propagation at the bottoms of steps, and densification of the deposited oxide film. Bias-sputtered-oxide deposition is primarily used as a gap filling process; it produces a surface profile made up of faceted structures wherever there are wide underlying interconnect structures. The primary drawbacks associated with this technique are the following. First, large amounts of sputtered oxide accumulate on the walls of the chamber. Yield can deteriorate or crash as these find their way onto the wafers. Lengthy cleaning procedures are required at frequent intervals to control this. Second, the time required to process a small batch of wafers is typically on the order of several hours and throughput is on the order of 3~4 wafers per hour. Finally, bias-sputtered-oxide processes have been linked with various types of device damage. During the sputter deposition process, the devices are exposed to irradiation by high energy secondary electrons originating at the target, and certain types of damage have been found to correlate with this electron energy (129,141).

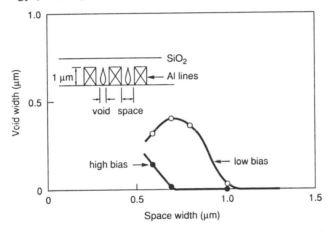

Figure 53.   Gap filling and void formation with high and low substrate bias settings in a bias-sputtered-oxide deposition process. From ref. 130; reprinted by permission of the publisher: SEMI.

Bias-CVD Process.   This technique makes use of simultaneous PECVD and physical sputter etching processes (151-153,161). The rates of the two processes are set such that the film surface topography evolves with angled surfaces at the edges of underlying steps. This process regime is characterized by (c) of Figure 52. A different implementation of this technique makes use of a process regime in which the sputter etching rate is a small fraction of the deposition rate, as depicted in (b) of Figure 52. Under these conditions, the faceting is sufficient to prevent "breadloafing" or "cusping"; which would otherwise occur when depositing the oxide. However, the faceting is not sufficient to produce a tapered profile; the profile is simply conformal with the underlying step. Tapering and gap filling is achieved by multiple iterations of a deposition-etch sequence, in which this bias-CVD process is followed by a partial etchback using a physical sputter etch process (with no deposition component). Like the bias-sputtered-oxide process, this technique has value as a gap filling tool but requires that a supplemental planarization process be used if long range planarization is sought.

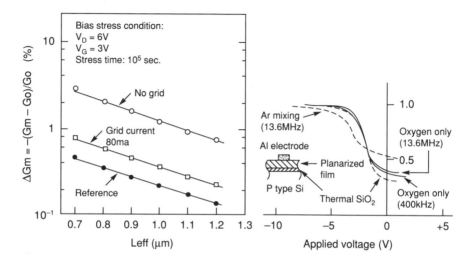

a G_m degradation by bias stress test for LDD n-ch MOS transistors coated with the bias sputtered SiO_2 films, with or without grid electrode.

b C-V characteristic curves of the planarized films formed by combination with oxygen sputtering and Ar mixed sputtering.

Figure 54. Examples of device degradation linked to deposition processes in which simultaneous oxide deposition and sputter etching are used to achieve faceting. From refs. 141 (**a**), and 160 (**b**). copyright IEEE; reprinted by permission of the publisher.

Bias-ECR-CVD Process.   This, like the bias-CVD process, is a simultaneous execution of PECVD and physical sputtering processes, in which the rates of each are adjusted such that the profile of the deposited film evolves in the desired way. The novel aspect of this technique is the means used to produce the discharge (162) (discussed below). This discharge enables one to achieve ion densities two-to-three orders of magnitude greater than what is possible with the capacitive discharges used in bias-sputtered-oxide and bias-CVD processes. Because of this, very high sputter etching rates can be achieved. A relatively high rate deposition process can thus be superimposed with this, while maintaining the proper relationship between the deposition and etching rates. This relationship, of course, determines the evolution of the profile of the deposited film, just as it does in the other simultaneous dep-etch techniques.

The ECR Discharge.   Magnet coils arranged around the periphery of the plasma chamber (Figure 55) impose a static axial magnetic field within the plasma chamber.   In this field, electron motion is constrained to circular or helical paths, with angular speed dictated by the magnetic field strength.

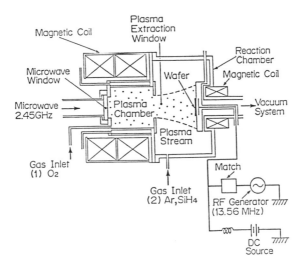

Figure 55.   Bias-ECR CVD tool.

Microwave power is introduced to the plasma chamber, perpendicular to the magnetic field flux lines. Resonance coupling is attained when the angular frequency of the electrons equals the incident microwave frequency. Under these conditions, an electron always remains in phase with the electrical field, and its energy continually increases. This ECR condition enables the plasma to effectively absorb the microwave energy at low pressures in the 10^{-5} to 10^{-2} torr range. The resulting secondary collisions produce plasmas having ionization densities in the range of 10^{10} to 10^{12} cm^{-3}.

The operating microwave frequency is 2.45 GHz: an authorized industrial frequency supported by commercially available equipment. At this frequency, the ECR condition requires 875 Gauss (G), which is established at the port orifice by adjusting current to the magnet coils.

In SiO_2 deposition processes, the microwave power level determines oxygen ion current density, which in turn determines the film deposition rate (Figure 56(a)) (158,163).

The Ion Beam. The magnetic flux density decreases gradually along the axis of the electromagnetic coil, from the ECR region toward the substrate holder, as the magnetic force lines diverge to the wall of the reaction chamber. High energy electrons, in helical motion excited by ECR, are accelerated by interaction with the magnetic field gradient, moving along the magnetic force lines. Therefore a static electric field is self-generated along the plasma stream between the plasma chamber and the wafer pedestal, with the negative potential directed toward the wafer pedestal and increasing with distance from the port orifice, corresponding to the gradient of the magnetic field intensity.

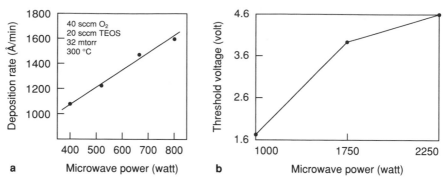

Figure 56. Effects of microwave power in bias-ECR CVD processes, from refs. 158 **(a)** and 156 **(b)**. (The two studies were carried out in different reactor types using different process chemistries, thus direct *comparison* the power levels is not meaningful). Reprinted by permission of the publishers: the Electrochemical Society, Inc. (a), and IEEE (b).

Energetic ions originating from the ECR region follow the diverging magnetic field lines produced by the coil. The energy of the ion bombardment is on the order of 20 eV at the substrate pedestal, supplementing the effect of the plasma sheath at its surface.

When only the magnetic field of the coils around the discharge tube is applied, the lines of magnetic force are divergent, so the plasma stream is divergent as well, with little diffusion of electrons or ions across the magnetic force lines. Therefore the plasma density just above the substrate is non-uniform -- even if the discharge is uniform in the plasma chamber -- resulting in non-uniform process results as well as oblique incidence of ion bombardment.

For this reason, additional magnets are used to help shape the magnetic field. One approach makes use of two magnetic coils (163,164) located coaxially below the wafer pedestal. The inner coil generates a "cusp" magnetic field which opposes the main field, and the outer coil generates a "mirror" field which augments the main field. In combination with the divergent field by the ECR coil, this creates a low magnetic field region just above the substrate, surrounded by a relatively high magnetic field region. In the low magnetic field region the plasma can diffuse isotropically, and ions impinge normally onto the substrate due to the sheath electric field. The high magnetic field region prevents the plasma from diffusing out across it.

Another method of shaping the ion beam (160) makes use of magnet coils not only around the discharge tube but also around the reaction chamber. When the additional magnetic field is correctly applied, a cylindrical plasma can be formed with the direction of the net magnetic force lines nearly perpendicular to the substrate.

RF bias, applied to the substrate pedestal, provides the means to achieve high ion bombardment energies, which are necessary for faceting, re-emission, and re-deposition to occur.

Gap Filling. Narrow, high aspect ratio gaps can be filled at high rates using bias-ECR-CVD when high bias levels are applied to the wafer pedestal.

SiO$_2$ deposition processes using precursors such as TEOS, which gives rise to relatively conformal deposition, require less bias than processes using SiH$_4$, which deposit films that tend to cusp and pinch off (157,158). Bias-ECR-CVD processes are typically carried out at pressures in the 1 mTorr or less range, so scattering probabilities are low. With proper shaping of the magnetic flux lines, bombarding ions as well as reactive species approach the device topography vertically, and this may contribute to gap filling performance as well (130,165).

Issues.   When applying bias-ECR-CVD to semiconductor fabrication processes, several issues arise. Wafer temperatures can reach levels of 400~500°C, due to the large amount of energy transferred from the ion beam to the wafer to achieve high rate sputter etching. In addition to energetic physical bombardment from the ion beam, UV irradiation of the wafer occurs as well, due to the microwave discharge; device threshold voltage shifts and gate oxide degradations have been encountered when applying bias-ECR-CVD processes to MOS devices (Figure 56(b)) (156,160,166).

Like the bias-sputtered-oxide and bias-CVD processes, bias-ECR-CVD processes have value as gap filling techniques but require that supplemental planarization processes be used if long range planarization is sought (167).

The key manufacturing and yield issues are not likely to be recognized until bias-ECR-CVD processes become widely implemented in high volume production fabs. One potential issue is the accumulation of oxide deposits on the walls of the reaction chamber. The configuration of the chambers makes it difficult to arrange the biases to accomplish plasma cleaning of these surfaces, therefore disassembly and manual cleaning is necessary, and some ongoing or intermittent impact on yields may be anticipated. A second potential issue is the fundamental difficulty of achieving uniform film deposition from the divergent ion beam, especially on 200 mm and larger wafers.

6.3   Spacer-Forming Dep-Etch Techniques

Oxide spacers produced on the sidewalls of steps can help ease the task of filling gaps with re-entrant or overhanging profiles, as shown in Figure 46. The gap profile produced by the spacers can be reliably filled with spin-on-glass, or by a conformal CVD film. Spacers are produced using a dep-etch technique which does not rely on physical sputtering or faceting in the etchback process. (Some small degree of physical sputtering does occur, but it is not the dominant component of the etch mechanism, as it is in the faceting-type processes discussed above.)

Dep-Etch-Dep Sequence (for Gap Filling).   The first step is a CVD process which coats the interconnects with a layer of oxide. The thicker this layer, the wider the spacers will be after the etchback; however, it must be thin enough to avoid pinching off the smallest gaps (Figure 57(a)). In the second step, most of the initial oxide layer is removed by a blanket reactive ion etchback process, leaving just enough to avoid uncovering the interconnects. Because of the anisotropy of the RIE process, filament-like stringers (called "spacers") are left on the sides of the steps, having positively contoured profiles

(Figure 57(b)) (136,168-171). The gap width is reduced, which tends to increase the difficulty of filling it; but because its acceptance angle is now somewhat larger, particularly at its top where the next deposited layer would tend to pinch off, the *net* effect in many cases is to make the gap easier to fill. The optimum spacer profile is achieved when the RIE process includes a minor physical sputtering component. In this case the lower walls of the spacers can have a slight positive taper (171), making the gap easier to fill. This is done in the third step: an oxide CVD process (Figure 57(c)).

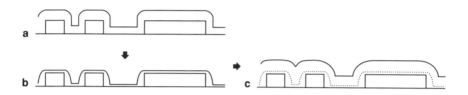

Figure 57. Deposition of CVD oxide (**a**) followed by blanket RIE to produce spacers (**b**) on the interconnects. A second CVD oxide deposition (**c**) to achieve the desired ILD thickness.

At this point a separate planarization process such as resist-etchback or chemical-mechanical polishing can be used to increase the range of planarization. The role of the dep-etch-dep sequence is to address the gap filling portion of the planarization process, so that a sufficiently thick oxide film can be delivered to the planarization process which follows. For example, resist-etchback, SOG-etchback, and chemical-mechanical polishing processes all require that thick, void-free oxide films be in place before their etchback steps can begin. The dep-etch-dep sequence produces the CVD-1 layer for these techniques.

Iterative Dep-Etch Cycles.   The narrower and deeper the gap, the greater the difficulty of filling it using a spacer-forming dep-etch-dep scheme. A second iteration of the deposition and etching cycle can sometimes overcome this difficulty. The sequence would be: dep-etch-dep-etch-dep. As the number of iterations increases, the gap filling ability increases, but so does the process cycle time, and in actual practice it is uncommon for more than two iterations (i.e. more than dep-etch-dep-etch-dep) to be used.

In which cases are *spacer*-forming dep-etch schemes more effective than *facet*-forming (sputtering based) dep-etch schemes? The spacer-forming schemes are generally better able to fill gaps which have re-entrant or overhanging sidewalls (Figure 46) (172). This is due to the way the surface profiles evolve during the etchback process. Also, during the spacer-forming etchback process, the oxide removal mechanism is dominated by reactive processes rather than by physical momentum transfer, therefore it imposes less heating and physical damage to the wafer compared with the faceting-type etchback processes.

6.4   Radius-Growth Dep-Etch Techniques

When *radius-growth* behavior was first mentioned near the beginning of this section on dep-etch techniques, we showed how this effect can be used to achieve short range planarization, or local smoothing. In contrast with the faceting and spacer forming effects discussed above, radius-growth produces its planarizing effect mainly in the deposition step, with only a minor contribution from etching effects. The primary role of the etchback step is simply to reduce the thickness of the film. This radius-growth behavior, and the scheme used to produce it, is a simple, straightforward planarization technique (134-138,170,173).

Dep-Etch Sequence for Planarization. The sequence begins with the deposition of a thick oxide film using a CVD process capable of conformal step coverage. PE-TEOS SiO_2 CVD processes are commonly used for this. When deposited onto a step, the profile of a conformal CVD film improves as its thickness increases (Figure 47). The abrupt rectangular profiles of the steps transform to ones with smooth, rounded transitions from the tops to the bottoms. This radiusing of the steps gives rise to local smoothing, or short range planarization (Figure 47). Small gaps are fully planarized while larger gaps and isolated steps are rounded and smoothed, producing low angle transitions from the tops to the bottoms of the steps. Generally, the thickness of the film must be roughly three to four times the height of the underlying steps in order to achieve a sufficiently rounded surface topography.

Usually an etchback step follows the thick deposition. Its role is to reduce the thickness of the film to a value which can be tolerated by subsequent process steps such as via etching and filling. In doing this, the etchback process contributes very little to the planarization; it simply transfers the thick-film profile to the thinner film. Some smoothing does occur at sharp inflections of the profile (Figure 58), but the main planarizing effects are a consequence of the radius increase due to the thick deposition.

Because the etchback step stops after *partially* etching the first CVD layer, it does not produce a distinct endpoint signal. Achieving the correct final interlevel dielectric thickness therefore requires good ongoing monitoring and control of the etchback process. Day-to-day variations in the ILD thickness will result in variations of via depths, which can cause difficulties with those via fill technologies which are sensitive to via depth.

An important characteristic of the etch process is its effect on the "seams" produced during the deposition process. Seams form wherever the two film surfaces growing horizontally from sidewalls of a gap meet each other (Figure 59). Although the gap is filled, a discontinuity or boundary exists at the seam. The discontinuity does not propagate upward, because once the gap is filled, subsequent deposition occurs onto a horizontal surface rather than onto two vertical surfaces. When exposed to a chemical etching environment, the seam itself is rapidly penetrated, after which the adjacent bulk material is immediately exposed to the etchant, which then attacks isotropically. (Chemical etching is often done to delineate or decorate cross sectioned samples prior to SEM analysis. The chemical etchant has access to these seams from the cleaved cross section surface, even though it *cannot* penetrate from the horizontal surface of the film. SEM micrographs then show vacancies where the seam and the immediately-adjacent oxide had been (Figure 59). These are referred to by many names, for example: "pseudo voids", "low density regions", "soft spots", etc.).

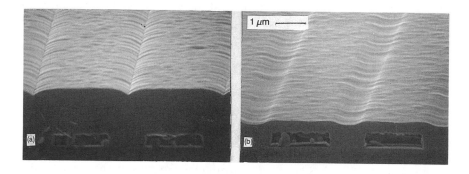

Figure 58.   Thick layer of PE-TEOS SiO_2 after deposition (**a**) and after blanket RIE (**b**).

In an actual etchback process, only the *horizontal* surfaces of the interlevel dielectric films are generally exposed to the etching species. Nevertheless, as the films etch the seams can be uncovered and exposed. Fortunately the seams are not vulnerable to *physical* etching mechanisms such as oxide RIE processes; they etch at the same rate as the bulk oxide material. Therefore, accelerated etching of the seams can be avoided so long as the etch mechanism is predominantly a physical one.

The radius-growth techniques are not good choices for certain planarization applications. The radius increase becomes pronounced only when the thickness of the deposited film is substantially greater than the height of the underlying step. If the steps are greater than roughly 8000 Å, very thick films (3~4 μm) are required, which in turn require long etchback processes. This lowers the throughput of the overall sequence, as well as increasing the non-uniformity of the final film. The local smoothing produced by radius-growth does not give rise to long range planarization. In some applications, this local smoothing is adequate; in others it is not. When local smoothing *is* the objective, however, this technique is a very simple and straightforward one, free of difficulties associated with spun-on liquids, etch rate matching between spun-on and CVD films, local loading and etch rate variations during the etchback, differential via depths, and so on.

Of course, radius-growth techniques work only if thick films can be deposited without producing voids as they fill the gaps. While this may be possible if the aspect ratios of the gaps are relatively small (roughly 0.8:1 or less), it is not so straightforward when the gaps are more difficult. For these situations, the *radius-growth* dep-etch sequence can be added to the end of a *gap-filling* dep-etch process, resulting in a dep-etch-dep-etch sequence capable of small gap filling as well as local smoothing. Section 6.5 elaborates on this technique.

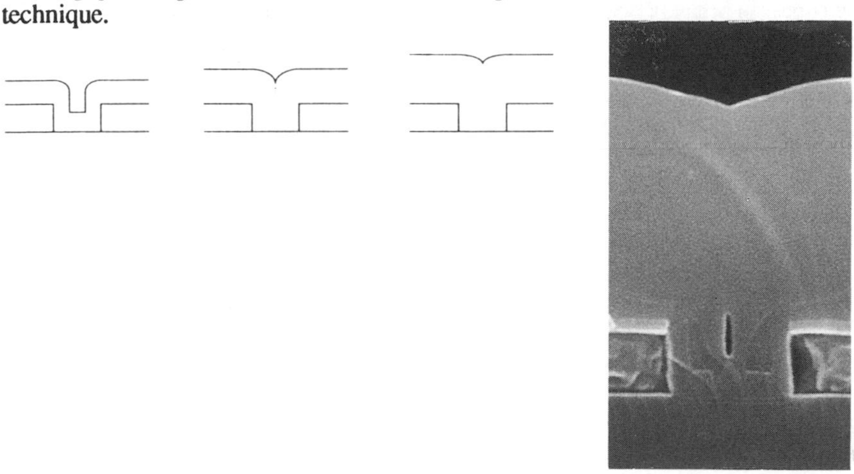

Figure 59. Boundary or "seam" created where vertical sidewalls meet when filling a gap with a deposited film. This will appear as a "pseudo-void" or "porous region" if the cross section is decorated in HF solution.

6.5 Dep-Etch For Gap-Filling Plus Dep-Etch For Planarization

Most of the dep-etch techniques described above are best suited to the gap filling module of a planarization process. Two of them, however, are capable of accomplishing short range planarization or topography smoothing. By splicing together a dep-etch *gap filling* technique and a dep-etch *smoothing* or *planarization* technique, one can accomplish the entire dielectric planarization task. In some cases it is possible to consolidate the entire sequence of processes into a single processing tool (174,175).

Extended Planarization Etch.   Three of the gap filling techniques make use of simultaneous dep-etch processes: bias-sputtered oxide, bias-CVD, and bias-ECR-CVD. In these cases, it is possible to reduce or eliminate the deposition mode of the process, such that only the physical sputter etching behavior persists. This sputter etching process can be used to remove the facet shaped features which remain after the gap filling process has been completed. As the etch proceeds, the angled surfaces of facet shaped features gradually migrate laterally as the etch proceeds. Eventually, the entire feature is consumed by this process, leaving a completely planarized surface over the feature (125,139,140). The wider the feature, however, the longer the time required for this lateral erosion process to consume the entire feature. This is called an "extended planarization etch".

This planarization etch process can be executed in one of two ways. If there is no deposition whatsoever, the lateral migration of the angled surfaces will be accompanied by a much slower vertical migration (thinning) of the horizontal surfaces. This corresponds to condition (g) in Figure 52. If the deposition process is not eliminated altogether, but instead is balanced with the etching process, it is possible to offset this vertical etching and eliminate the thickness loss from the horizontal surfaces. This condition still results in net loss of material from angled surfaces (conditions (e) and (f) of Figure 52), so the faceted features gradually are eroded and the narrow features planarized.

Since device topography is made up of interconnect features of various widths, and since only the narrower ones are planarized by an extended planarization etch process, the final surface topography is as shown in Figure 48(e).

Overlapping Dep-Etch Sequences.   When one of the gap-filling dep-etch-dep sequence is appended with a radius-growth dep-etch sequence, the result is a dep-etch-dep-etch sequence capable of filling difficult gaps as well as producing local smoothing (138,145,147,148,172,174,175).   The second deposition step of the dep-etch-dep combines its function with the first deposition step of the dep-etch sequence, therefore they can be combined or overlapped into a single, thick, deposition process.

Also, by adding the gap-filling dep-etch sequence at the beginning of the radius-growth sequence, an additional benefit is obtained. Seam formation is eliminated or else is localized to the lowest portions of the gaps, where the seams are well protected from exposure to chemical etching or other subsequent wet processing steps. This is because, when the thick deposition process is preceded by a dep-etch sequence, the thick deposition is confronted with *tapered* steps or oxide spacers in the gaps, rather than with vertical surfaces. As a gaps close, the laterally growing oxide surfaces do not join head-on; instead they join at positive angles -- one segment at a time -- facilitating fusion rather than abutment.

Most of the dep-etch techniques discussed in this section contribute primarily to the gap-filling element of dielectric planarization, and oftentimes must be supplemented with some type of longer-range step height reduction process. When global *planarization is sought, one of these options is chemical-mechanical polishing ... the topic of the section which follows.*

7.0 PLANARIZATION BY CHEMICAL-MECHANICAL POLISHING

Mechanical polishing, assisted by chemical action, can be used to bring about the selective removal of high elevation features, resulting in a planarized topography. This is called "chemical-mechanical polishing", or "CMP."

Two categories of CMP planarization techniques are used for multi-level metallization applications. In the *dielectric planarization* category, interconnects are first coated with a thick dielectric layer. Chemical-mechanical polishing is then used to reduce topographic steps while removing most of the dielectric film. In the *recessed metal* category, it is the metal film which is removed by polishing. The polishing step acts as a surface clearing etchback. This technique is used in a multilevel metallization architecture called "Damascene", in which a novel method is used to pattern and fabricate interconnects and vias (149,176).

The high numerical aperture (NA) stepper lenses used with optical lithography in sub-half micron multi-level metal technologies have limited focal depth ... typically only a few thousand angstroms. All surfaces to be imaged must fall within this depth of focus across the entire exposure field. Spun-on planarizing liquids such as SOG or resist, when applied at thicknesses which are practical, are not capable of planarizing features separated by more than 20~100 μm. Chemical-mechanical polishing is a technique capable of planarization of distances up to several millimeters (Figure 60), providing a means of achieving *global* planarization (7,177).

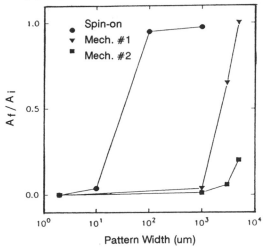

Figure 60. Planarization range of chemical-mechanical polishing process, compared with that of a spun on polymer. A_f/A_i is the ratio of the final step height to the initial step height. When A_f/A_i ~0, the step is fully planarized. From ref. 7; copyright 1990 IEEE; reprinted by permission of the publisher.

7.1   Planarization Techniques Using CMP

Dielectric Planarization.   Figure 61 illustrates one of the two categories of CMP-based planarization techniques: *dielectric planarization* CMP. This is conceptually similar to the other planarization techniques discussed in this chapter, in that its task is to remove from the SiO_2 surface the topographical features which have propagated from the underlying interconnect structures. In the case of CMP, this is possible due to the higher removal rate of the high elevation features relative to those at lower elevations. In a variation of this process, pillars or studs are built onto the interconnect lines. These become the via connections to the next metal level. The interlevel dielectric is deposited onto these line/pillar structures; then the CMP step removes and planarizes it, stopping when the tops of the pillars are cleared (149,178,179).

Recessed Metal. Reference 176 introduces the *Damascene* recessed metal architectures in which metal CMP processes are used. The desired interconnect pattern is etched into a planar SiO_2 layer, leaving grooves where the lines are to be. Next, a thin adhesion layer is deposited, followed by a thicker conformal metal layer such as tungsten. This fills the grooves and blankets the horizontal surfaces of the SiO_2. A CMP process is then used, which removes the exposed metal from the horizontal surfaces (the "overburden") but not the metal recessed in the grooves. At this point the interconnect lines are in place and they are co-planar with the surface of the dielectric. A *"Dual* Damascene" architecture is possible as well (Figure 62) (176), in which the via holes are patterned and etched along with the grooves for the interconnect lines. The conformal metal is deposited into the grooves and via holes, forming not only the interconnect lines but plug connections to the underlying interconnect level as well.

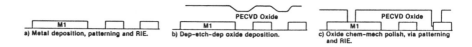

a) Metal deposition, patterning and RIE.   b) Dep-etch-dep oxide deposition.   c) Oxide chem-mech polish, via patterning and RIE.

Figure 61. Schematic illustrations of the dielectric planarization CMP process flow. From ref. 149; copyright 1991 IEEE; reprinted by permission of the publisher.

Engineering and process integration issues are discussed in detail below, after introducing some of the fundamental characteristics of the polishing processes. In the case of dielectric planarization in particular, empirical characterization of planarization performance is extremely cumbersome, so a good understanding of the fundamentals is very helpful.

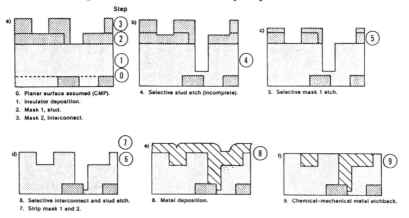

0. Planar surface assumed (CMP).
1. Insulator deposition.
2. Mask 1, stud.
3. Mask 2, interconnect.

4. Selective stud etch (incomplete).

5. Selective mask 1 etch.

6. Selective interconnect and stud etch.
7. Strip mask 1 and 2.

8. Metal deposition.

9. Chemical-mechanical metal etchback.

Figure 62. Schematic illustration of the "Dual Damascene" recessed metal process flow. Dielectric CMP is used to produce the initial planar surface; metal CMP is used to remove the metal overburden. From ref. 176; copyright 1991 IEEE; reprinted by permission of the publisher.

7.2   Polishing Mechanism

Chemical-mechanical polishing methods have only recently been introduced and adapted to multi-level metallization processing applications. Therefore, the CMP equipment now in use is just beginning to evolve from that which has traditionally been used for the polishing of bare silicon substrate surfaces. The key tooling elements are shown schematically in Figure 63. Briefly, wafers are held by rotating wafer chucks or heads, such that the wafer surface is exposed to the opposing surface of a polishing pad, which is rotated as well. A controlled amount of pressure or down-force can be applied, and a slurry fed at a controlled rate flows between and wets the polishing pad and the wafer surface. The nominal polishing rate is strongly influenced by the rotational velocities of the wafer chuck and polishing pad, and by the down-force or pressure applied to the wafer.

Polishing Pads.   The polishing pad, saturated with slurry, is the component which makes physical contact with the film surface and carries out the polishing mechanics. Pads are typically made of polyurethane impregnated, chemically resistant Dacron felts, or other proprietary compounds engineered to produce the desired compliance, porosity, wear characteristics, and so on.

Pad properties affect not only the removal rate and surface finish of a film, but also the planarity parameters such as planarization rate, planarization range, and so on. Generally, softer pad materials give rise to higher removal rates and are less prone to scratch the films being polished. Harder pads, however, remove topography more efficiently, and have other planarizing characteristics which are favorable (2,7,180,181). (These are discussed in section 7.3.)

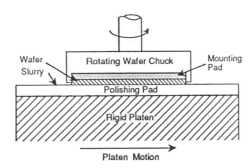

Figure 63.   Chemical-mechanical polishing tool. The wafer is held face down by the chuck. From ref. 187; reprinted by permission of the publisher: SEMI.

Slurries.   These transport the abrasive agents and chemical agents to and from the wafer surface. The abrasives are typically particles having roughly the same hardness as the film being polished. They are typically 10~1000 Å in size, and are suspended in an aqueous solution, making up roughly 1~30% of its weight. Acids or bases are added to the solution, depending on the material to be polished (specific examples are discussed below).

The material removal mechanism is somewhat different in SiO_2 CMP processes than in metal CMP processes, however both types involve material removal by abrasive action, in the presence of concurrent chemical processes. It is mainly the nature of the chemical processes which differs from one CMP process to another.

SiO_2 Polishing.   Silica particles are commonly used as abrasives in SiO_2 polishing slurries. These are suspended in aqueous solutions to which a base such as potassium hydroxide is added (180). By controlling pH in the range of 10~11, agglomeration of the abrasive particles is minimized. The mechanism by which SiO_2 is removed is not fully understood, however it is clear that water plays an active role in the process; abrasives suspended in non-aqueous solutions exhibit substantially lower polishing rates than those suspended in water (182,183). A widely accepted model of the process involves two mechanisms. First, the surface layers of the glass are modified by dissociative chemisorption, in which water molecules attack strained Si-O-Si links, cleaving them and leaving $Si(OH)_x$ complexes as reaction products. The glass's resistance to this attack is reduced by the mechanical pressure imposed by the polishing fixture, and probably by the potassium hydroxide as well, which itself is capable of attacking SiO_2 glasses. Next, these so-called "soft hydroxides" are removed by the abrasive mechanical action of the colloidal silica (184). Mechanical damage of the surface is prevented by the fact that the colloidal silica particles in the slurry are no harder than the oxide being removed.

Metal Polishing.   Polishing of Al-Cu is done with slurries containing Al_2O_3 abrasives in acidic aqueous solution. In the absence of acid, the Al-Cu and SiO_2 are removed at similar rates. Acid additives tend to increase the removal rates, however nitric acid (HNO_3) surprisingly lowers the SiO_2 removal rate while increasing that of Al-Cu (178). This selectivity to SiO_2 adds a measure of control to surface-clearing metal CMP processes such as those used to produce plugs.

Tungsten, deposited by CVD, is the material used in Damascene processes. Slurries containing either silica or alumina abrasives have been used, along with a metal etchant and a metal passivating agent in acidic aqueous solution. The CMP mechanism involves continuous competition between etching reactions which remove tungsten, and passivation reactions which oxidize the surface and impede etching. In the presence of mechanical abrasion, the passivation film is disrupted, allowing the etching to proceed unimpeded. Since high elevation features are subjected to the greatest abrasive action, these features are etched faster than low elevation features. A weak base additive (ethylenediamine) is a key contributor to surface passivation and is necessary for protection of the tops of recessed interconnects during the overetch portion of the polishing process (185,186).

7.3   Step Removal By Polishing

Up to this point, the discussion has been about removal of material by chemical-mechanical polishing, but nothing has been said about the removal of topographic features from non-planar surfaces. While this may at first seem to be a trivially straightforward consequence of a polishing process, the dynamics of the transformation from a non-planar surface topography to a planar one are considerably more complex than one might imagine.

Selectivity.   The surfaces of high elevation features ("up areas") are eroded faster than those at lower elevations ("down areas"). This high-to-low selectivity is the fundamental effect which brings about the gradual disappearance of steps during a polishing process. Much of the difficulty in characterizing and modelling CMP planarization is due to the fact that selectivity is strongly dependent on so many factors. For example, selectivity is a function of feature dimensions (and therefore is not constant from one pattern to another); it is a function of the structure and surface condition of the polishing pad (and therefore is not constant from one pad to another, nor over the life of a pad); and it is a function of the height of the step being removed (and therefore it changes continuously throughout a polishing process as the steps erode).

High-to-low selectivity is thought to arise due to the compliant nature of polishing pads. Figure 64 depicts the localized compression occurring within a pad as a wafer's topographic features are driven across it. The pad applies pressure to the substrate through the slurry, with the greatest pressures occurring where the pad is most compressed. More pressure is therefore applied to raised surfaces and leading edges than to low surfaces and trailing edges, resulting in local differences in material removal rates.

As the steps erode, the down areas are exposed to increasing pressure, and the removal rate from these surfaces increases, approaching that of the up areas. As this happens, the rate of step height reduction decreases toward zero. In the latter stages of the polishing process, selectivity is so low that the small residual steps, for all practical purposes, cannot be fully removed by continued polishing.

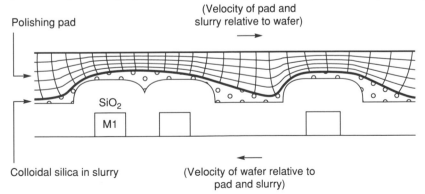

Figure 64.   Schematic representation of polishing pad compression and localized pressures applied to topographic features. Pressures are greatest at the top leading edges of steps, and are lowest in the wakes of the steps.

To a first approximation, this height reduction can be modeled as a logarithmic decay; that is, the step height reduction rate at any instant is proportional to the remaining step height at that instant (7,179,187). A *Planarization Rate Constant* (7,187) can be used to characterize this decay, although this model is not completely accurate as we will show later.

"Down Area" Polishing. Since material is removed from down areas as the up areas are eroded (i.e. since selectivity is not infinitely high), and since the rate of removal from down areas increases (selectivity decreases) as the steps erode, the rate of step height reduction decays as the polishing process proceeds, as mentioned above. This is illustrated in Figure 65, where the instantaneous step height as well as the nominal amount of oxide removed are normalized relative to the initial step heights. In addition, material is removed faster from wide down areas than from narrow down areas (Figure 66). On a device structure, this results in slower step height decay (slower planarization) of steps adjacent to large field areas and faster planarization of steps adjacent to smaller field areas (Figure 65). If a down area is large enough, the removal rate from this area is so similar to that from the up areas that virtually no step height reduction is possible. This would be the case, for example, in Figure 66 for any down area wider than about 3 mm. One of the factors affecting the design of a CMP planarization process, then, is the length over which planarization must be achieved. This can be on the order of several millimeters, where devices contain large up areas such as test pads or dense arrays adjacent to sparsely populated peripheral areas and scribe lines.

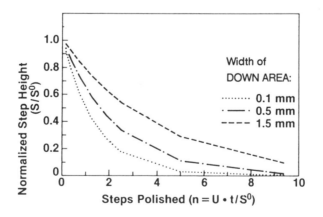

Figure 65. Logarithmic reduction of step height, S, for large steps adjacent to "down areas" of various sizes (0.1, 0.5, and 1.5 mm). Initial step heights, S^o, are 0.6μm. The parameter n indicates the nominal thickness of oxide removed by polishing for t minutes; normalized to the initial step height, S^o. From ref. 181; copyright 1991 IEEE; reprinted by permission of the publisher.

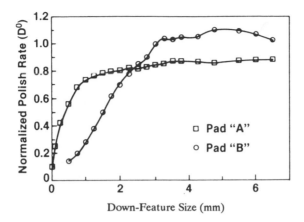

Figure 66.   Polishing rates of the "down" surfaces (normalized to the polishing rate of large "up" surfaces) as a function of the size of the down area. Where the normalized rate, D^o, is ~1, down surfaces and up surfaces are polished at similar rates so planarization does not occur. Pad "A" is more compliant than pad "B". From ref. 181; copyright 1991 IEEE; reprinted by permission of the publisher.

The polishing time required to planarize a structure, or to remove some given fraction of its initial step height, is dependent on the width of the down area adjacent to it. More specifically, the time to planarize depends on the width of that area as well as the ability of the polishing pad to rigidly span that distance. The stiffer the pad, the more rigid the span and the slower the material removal from the bottoms, a situation which gives good high-to-low selectivity and efficient planarization. The more compliant the pad the less capable it is of bridging large distances, and the faster its removal of material from the down areas between up features. This gives poorer selectivity, and the time to planarize is longer. This can be seen in Figure 67(a), where the planarization rate constant, "p", is used as a measure of the rate at which a specific feature is planarized (7,187) under a specific set of conditions. The higher the value of "p", the more efficiently planarization occurs.

[2] $ln\ S(t)/S^o\ =\ -p * D(t)$
where:
$S(t)$ = step height at any time, t, during the polishing process (μm)
S^o = initial step height (μm)
p = planarization rate constant (μm^{-1})
$D(t)$ = amount of dielectric removed from a large flat area after polishing for t minutes.
The surface condition of a pad, which changes continuously with use, strongly affects the efficiency of planarization as well. Figure 67(b) illustrates the effects of pad aging and wear on planarization of various features.

These factors, as well as certain parameters of the polishing tool (down force, rotational velocities of wafer holders and polishing tables, etc.), collectively determine the levelling distance of a pad. When polishing features separated by distances greater than a pad's levelling length, there is no high-to-low selectivity, and planarization does not occur.

The greater a pad's levelling length, the longer the range of the planarization process. Ideally, the levelling length is large enough to bridge and planarize the widest down areas present in an interconnect pattern or between adjacent die. On the other hand, the greater the levelling length, the greater the response of the pad to wafer scale non-planarities such as bow, thickness variations, etc. Thus an optimal levelling length is one which is large relative to device scale features yet small relative to wafer scale features. Data such as that in Figure 66 can be used to ascertain the characteristic levelling length of a particular pad, under a particular set of machine conditions.

"Up Area" Polishing.   The local material removal rates from the tops of large up features are not necessarily the same as the *nominal* polishing rate, i.e. the removal rate from a blanket film on a flat test wafer. Furthermore, they are not necessarily the same from one up area to another, nor are they constant through the duration of a polishing process. Here we will discuss some of the factors which influence local polishing rates from up areas.

If an up feature is continuous and is large enough (relative to the levelling length of the pad), the removal rate from the up area on its top (denoted "U") is very similar to that from a blanket film on a flat test wafer. A feature such as an array is not continuous; roughly forty percent of its overall area might be up area (the tops of the interconnect lines), and the balance is down area (the gaps). The relative amount of up area can be characterized by a *pattern factor* (181), which in this example is 40%. The stiffer a pad is, the longer its levelling length, and the larger a feature must be for the pad to respond to it as a "large" feature. If a pad has a levelling length of 2 mm, for example, and the largest array has a diameter of 1 mm and is adjacent to large open field areas, the pad will respond to an *effective* pattern factor determined by the up area within the array relative to the levelling area of the pad.

The pattern factor of a feature directly affects the polishing rate of material from its up areas. Specifically, the local removal rate from the up areas of a feature is approximately equal to the nominal removal rate divided by the pattern factor. This in turn affects the planarization of that feature, as shown in Figure 68.

The logarithmic decay model which we have used up to now is based on the assumption that the polishing rate from a feature's top surface(s) *remains constant* throughout the polishing process, while the polishing rates of the down areas increases as the steps erode. In fact, however, the polishing rate from a given up area is not constant; it increases during the course of the polishing process. This results in a divergence from the logarithmic decay behavior, as shown in Figure 69(a). The rate increase occurs once the step profiles become rounded by the polishing action (Figure 69(b)) (179,188). Rounding is thought to be a consequence of local pad pressure enhancement at the leading edges of the steps (Figure 64).

When designing or characterizing CMP planarization processes, it is necessary to identify those device structures which are the most difficult to planarize. These are likely to be the largest, densest interconnect structures adjacent to wide unpatterned field areas, scribe lines, and or wafer edges. While planarization of these structures is occurring slowly, it proceeds more rapidly at smaller arrays, smaller gaps, small isolated features, and so on (Figure 70). It is also necessary to establish the criteria for allowable residual step heights, that is, at which point can a feature be considered "planarized". This is necessary because, in the latter stages of the polishing process material is removed nearly as fast from down areas as from up areas, and step heights no longer decrease at appreciable rates.

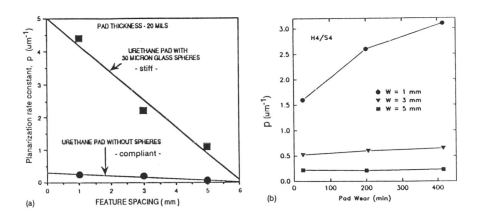

Figure 67.   Effects of pad characteristics and age on planarization performance.   The "planarization rate constant", p, is an indicator of the efficiency of step erosion, which is greatest where adjacent "down" areas are small (i.e. small "FEATURE SPACING"), and which is highly dependent on pad characteristics and pad age. From refs. 187 **(a)** and 7 **(b)**. Reprinted by permission of the publishers: SEMI **(a)**, and IEEE **(b)**.

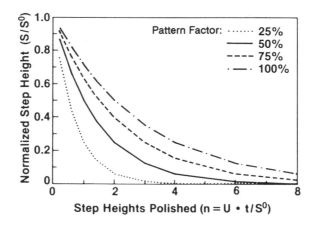

Figure 68. Erosion of various interconnect arrays having different "pattern factors". The parameter n indicates the nominal thickness of oxide removed by polishing for t minutes; normalized to the initial step height, S^0. From ref. 181; copyright 1991 IEEE; reprinted by permission of the publisher.

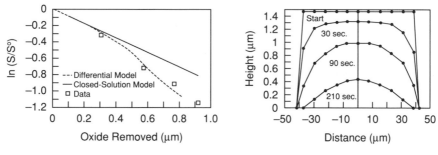

a Step height vs. oxide removed for modelled and experimental results.

b Data from profilometer traces taken after various polishing times on an 80 μm wide square feature (Suba IV pad).

Figure 69. (a) Step height erosion models assuming *constant* removal rate from "up" areas ("Closed-solution Model"), and assuming accelerated removal from "up" areas due to feature rounding ("Differential Model"); from ref. 181. Experimental data plotted for comparison, from ref. 7. (b) Profile evolution of a rectangular step as it is polished (from ref. 179). This rounding effect gives rise to the acceleration of oxide removal rate from "up" areas during a polishing process. Figures are reprinted by permission of the publishers: IEEE (a), and the Electrochemical Society, Inc. (b).

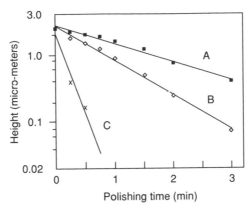

Figure 70. Erosion rates of (A) a large, isolated step ($150\mu m^2$), (C) a small isolated step ($8.5\mu m^2$), and (B) a small step ($15\mu m^2$) at the center of a large array of similar steps. From ref. 179; reprinted by permission of the publisher: the Electrochemical Society, Inc..

7.4   Technology And Manufacturing Issues

Although CMP processes are clearly capable of delivering planarization over very long distances, they are relatively new and unprecedented entrants in most back end technologies. Much of the work in this area is still directed at understanding and characterizing some of the fundamental aspects of the processes. In addition, several of the manufacturing issues are relatively unfamiliar ones, due to the unique nature of these processes.

Because the polishing mechanism is capable of planarizing features separated by great distances, it follows that the mechanism can be sensitive to such features as well. In other words, the planarization results achieved at any feature are affected not only by the characteristics of *that* feature but also by *neighboring* features, even if those features are several millimeters away. Some die are adjacent to test die; others are adjacent to wafer edges ... the planarization of these die can be different than the planarization of *identical* die located *away* from these features (7).

Non-uniformities of CMP processes are substantial. In the dielectric CMP applications, large amounts of SiO_2 are typically removed by CMP. When a CMP process having non-uniformity of 10%, three sigma, for example, is used to polish back a 2 μm thick SiO_2 film to a thickness of 7000 Å, this step alone may introduce 1300 Å of non-uniformity to the final film thickness. In addition, results drift from wafer to wafer, due to the dynamic nature of the polishing pads. Polishing rates as well as selectivity and range are strongly dependent on the condition of the pad surface, and this condition continuously changes as the number of wafers processed increases. Pad materials wear, and the porous surfaces become clogged with abrasive constituents of the slurry with increasing exposure to the polishing environment. Substantial process drift can occur over a span of a few wafers (2,7,179,187).

These factors result in considerable uncertainty as to the amount of material being actually removed from a given wafer. To guard against overpolishing, a typical dielectric CMP manufacturing approach is to intentionally underpolish the SiO_2 layer, then inspect and re-polish it, using an adjusted polishing time. This cycle is repeated until the thickness observed at inspection is within spec limits.

Slurries tend to dry rapidly on a wafer once it leaves the polishing table. To prevent this, wafers are typically placed quickly in megasonic wet tanks to remove the slurry before it dries. In a fab, logistical difficulties which delay this can allow the slurry to dry up, resulting in high counts of large particles on the wafers.

New pads generally require "break in" periods, during which they polish dummy wafers for the sole purpose of stabilizing the pad's condition and performance. This conditioning step can consume roughly one third of a pad's useful life, and of course the polishing tool is unavailable for productive use while this is in progress.

Damascene Issues.   The Damascene technique requires that the deposited metal be capable of filling narrow trenches without forming voids. In the Dual Damascene technique, deep, high aspect vias must be filled as well. CVD tungsten has been used for this application due to its relatively conformal step coverage, however its resistivity is roughly three times that of Al-Cu alloys. This imposes a speed penalty on the device, or requires a wiring strategy which works around this difficulty (176).

Another design rule is imposed on the widths of the metal lines. Because they are produced by lateral growth of deposited film from the sidewalls of the grooves, these grooves -- and therefore the interconnect lines -- must all have identical widths. Wide test pads cannot be produced by this technique.

Finally, various defects can be produced during the overetch portion of the surface clearing polishing processes. The relatively soft metal is vulnerable to smearing and scratching while exposed to the abrasive elements and shear forces which drive the CMP process. Certain slurry mixtures are unstable (185), tending to coalesce and form large abrasive particles which give rise to scratch defects in the polished surfaces. This makes their use in manufacturing a tricky undertaking. Accelerated consumption of the dielectric and metal *within* an array can occur if the dielectric surface area is small relative to that of the embedded metal lines, even under CMP conditions which otherwise produce very high metal-to-dielectric etch-rate ratios. The elevation of these array areas thus becomes lower than that of field areas or of arrays which are less dense (Figure 71). This is known as "thinning" (176), or "insulator erosion" (189), and it necessitates design rules to avoid high ratios of exposed metal-to-insulator. Further loss of planarity occurs due to "dishing" (189); *within* large dense arrays the polish rate is higher at the center than at the edges, and that of the tungsten lines is greater than that of the dielectric (Figure 71).

In tungsten CMP, chemical attack of the exposed metal can occur due to subtle changes in the slurry chemistry (185) which disrupt the balance between surface passivation and chemical etching.

Yield.   Until the CMP planarization techniques become widely implemented in high volume production fabs, one can only speculate as to their vulnerabilities to yield degradations or crashes. Based on the issues mentioned here, though, one can begin to anticipate what these difficulties might be.

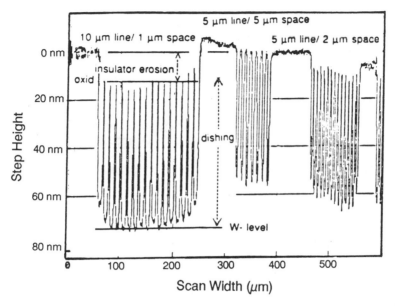

Figure 71. Insulator erosion (or "thinning") and "dishing" which arise during the overetch of tungsten polishing processes used for Damascene interconnect technologies. Both effects are influenced by the pattern factor of the interconnect array. This profilometer trace illustrates the residual steps in and between three different arrays. From ref. 189; copyright 1992 IEEE; reprinted by permission of the publisher.

Dielectric CMP techniques can be implemented only after a thick dielectric layer is in place. Depositing such a layer without producing voids in the small gaps is an increasingly difficult challenge. One of the newer tools for doing this -- oxide CVD using O_3+TEOS -- is discussed in the section which follows.

8.0   GAP FILLING USING CVD OZONE-TEOS OXIDES

Planarization techniques such as resist-etchback, dielectric CMP, and some of the SOG schemes require that a thick dielectric layer be in place as a prerequisite to the step-height removing steps. This layer is usually deposited by a CVD process, and is called the CVD-1 layer. When device feature sizes were large, nearly any CVD process was capable of depositing a suitable CVD-1 layer. However, the task is becoming increasingly difficult as feature sizes shrink, due to the tendency of CVD oxide films to form voids when deposited into small gaps. In many cases this gap filling task represents the most difficult obstacle in the entire planarization sequence. Several of the *dep-etch* schemes (section 6) have evolved specifically to address this gap filling challenge. Another technology that can be applied to the gap filling task is the ozone-TEOS (O_3-TEOS) CVD process; a process capable of depositing SiO_2 films having exceptional step coverage characteristics and gap filling capabilities.

8.1 Conformal And Hyper-Conformal Step Coverage

Figure 72 illustrates various step coverage characteristics: conformal, non-conformal, and hyper-conformal. Most oxide CVD processes are characterized by somewhat non-conformal step coverage, and this leads to undesirable consequences in most planarization schemes (see for example sections 2.2, 5.6, and Figure 3). O_3-TEOS SiO_2 films can be deposited with conformal or hyper-conformal step coverage; the film profile evolves without cusping or breadloafing, and it does not pinch off and produce voids as the gaps close. Hyper-conformal (*beyond*-conformal) step coverage is a characteristic in which the film initially conforms to the shape of the underlying step but, as its thickness increases, its profile evolves in such a way that the rectangular shapes of the underlying gaps transform to smooth, rounded depressions. When step coverage is hyper-conformal, the films are capable of filling gaps having extremely difficult profiles (Figure 72(d)).

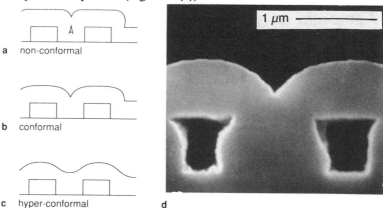

a   non-conformal

b   conformal

c   hyper-conformal   d

Figure 72. Conformality characteristics of CVD oxide films: non-conformal (**a**), conformal (**b**), and hyper-conformal (**c**). Films having hyper-conformal step coverage are capable of filling gaps having difficult profiles such as that in (**d**).

The first O_3-TEOS oxide films used in mainstream device fabrication applications were deposited using CVD processes carried out at pressures in the 60~80 torr range (138,145,147,148,174,175,190,191). These processes deposited films conformally (Figure 73(a)). Within a few years, a second generation of O_3-TEOS CVD processes was developed; these were carried out at "high" pressures (500~760 torr). In this regime, hyper-conformal step coverage can be achieved (Figure 73(b)). Furthermore, in this process regime, it is possible to produce oxide films having substantially better material properties than those deposited in the "low" pressure regime. Specifically, the density, resistance to water penetration, and resistance to cracking are improved (49,50,192).

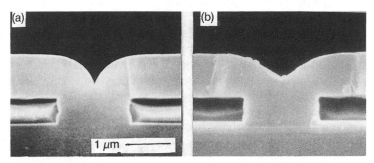

Figure 73. Gap filling characteristics of *conformal* O_3-TEOS oxide deposition at 60 torr (**a**), and *hyper-conformal* O_3-TEOS oxide deposition at 600 torr (**b**).

8.2   Applications

Inter-metal Dielectrics. O_3-TEOS oxide films are rarely used as stand-alone intermetal dielectrics; they are usually used along with PECVD oxide films in integrated, multi-layer dielectric structures. The O_3-TEOS oxide is deposited only until the necessary gap filling and/or topography smoothing has been achieved. The remainder of the dielectric structure is generally composed of PECVD oxide films.

The first layer of a multi-layer intermetal dielectric is usually PECVD oxide. O_3-TEOS CVD processes are strongly affected by underlying materials (192-196); one function of this initial PECVD oxide layer is to coat the various underlying dielectric and interconnect materials with one single material before the O_3-TEOS CVD process begins.

Because of the high process pressures used for O_3-TEOS CVD, it is not possible to implement these as plasma-enhanced CVD processes. Since the SiO_2 is not subjected to ion bombardment during its deposition process, its stress is tensile, and the SiO_2 glass structure is somewhat less dense than that of SiO_2 films produced by PECVD processes. In an inter-metal dielectric, the layer of dense, compressive PECVD oxide beneath the O_3-TEOS oxide film provides protection from thermal expansion of underlying metal layers, improving cracking resistance of the O_3-TEOS oxide. A PECVD oxide film on top of the

O_3-TEOS oxide layer can provide similar protection from overlying metal interconnects, and also protects the O_3-TEOS oxide from ambient water.

Sacrificial Schemes.   In some applications, much of the O_3-TEOS oxide layer is etched back from the device to remove it from sites at which vias may be placed. This is done for similar reasons that SOG films are etched back: to avoid the possibility of water penetration into O_3-TEOS oxide when it is exposed on the sides of vias following via etching. One of these etchback approaches is illustrated schematically in Figure 74. This is an iterative dep-etch gap filling process sequence, in which the O_3-TEOS oxide is deposited in the third step to fill the small, faceted gap produced by the first dep-etch cycle. A subsequent blanket RIE step removes the O_3-TEOS oxide from the tops of the metal lines, and a final PECVD process encapsulates the underlying oxide films and increases the thickness of the intermetal dielectric to the desired value (51). At this point, the gaps are filled, the interconnects are coated with SiO_2, and the wafer is ready for further planarization. Any of several planarization schemes may be chosen (resist-etchback, SOG-etchback, chemical-mechanical polishing, etc.) to bring about whatever step height reductions are sought (138,197).

Non-Sacrificial Schemes.   In other schemes, most or all of the O_3-TEOS oxide layer is left on the device. Of course, vias to the underlying interconnects will penetrate this layer, so it is necessary to take measures to desorb water from the vias shortly before the next layer of metallization is deposited.

One of these non-sacrificial approaches is illustrated in Figure 75; the O_3-TEOS oxide layer is used as one of the CVD dielectrics in an etchback SOG scheme. Specifically, it is used to produce a CVD-1 layer which is free of voids. This capability removes a thickness constraint on the CVD-1 layer (section 5.6), so that it can be sufficiently thick to withstand long exposure to local overetching during the SOG etchback. This enables the use of *etchback* SOG schemes for planarization of advanced devices with small gaps (193,198). A second function of the O_3-TEOS oxide layer is to prevent accumulations of SOG in narrow, high aspect ratio gaps where SOG's are vulnerable to cracking (100,107).

When used in SOG planarization schemes such as this, the O_3-TEOS oxide layer fulfills the gap filling function, relieving the SOG of that burden. The SOG is only used to provide local topography smoothing; a task for which it is well suited.

Another scheme makes use of the flow-like surface profile of the O_3-TEOS SiO_2 film to achieve local smoothing *without* the use of spin-on-glass. With a sufficiently thick layer of O_3-TEOS oxide, a smooth, rounded surface profile is produced. A blanket etchback can then be used to reduce the thickness of this layer, and a final PECVD oxide layer can be deposited to protect the O_3-TEOS oxide and produce the targeted interlevel dielectric thickness (Figure 76) (192,199). This approach is very similar to the dep-etch schemes described in section 6.4 which make use of the *radius-growth* principle. Due to the hyper-conformal characteristics of O_3-TEOS SiO_2 films, these need not be as thick as films which are simply conformal. Nevertheless, the taller the metal step, the greater the oxide thickness that is required; so this technique is most viable for applications where the steps to be planarized are in the 5000~8000 Å range.

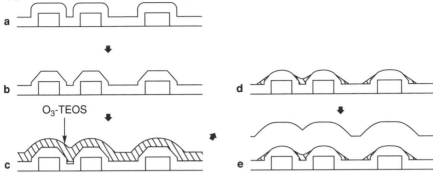

Figure 74. Gap filling scheme using sacrificial O_3-TEOS oxide, along with PE-TEOS oxide. Initial layer of PE-TEOS oxide (**a**), followed by faceting sputter etchback (**b**). O_3-TEOS CVD layer fills the small gaps (**c**). Blanket RIE removes O_3-TEOS oxide from sites where vias may be placed (**d**). Final deposition of PE-TEOS oxide builds the total thickness (**e**) such that a subsequent long range or global planarization process can be applied.

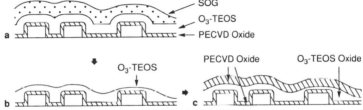

Figure 75. O_3-TEOS oxide used as the CVD-1 layer in an etchback SOG scheme. The O_3-TEOS oxide layer prevents accumulation of SOG in the narrowest gaps (**a**), and provides a thick underlayer to contain local overetching during the SOG etchback (**b**). A second layer of PECVD oxide encloses the O_3-TEOS oxide which remains (**c**).

Pre-metal Dielectric Applications. O_3-TEOS CVD processes were mentioned in the discussion of BPSG flow planarization techniques earlier in this chapter (section 3.5). When small gaps are present in a pre-metal interconnect pattern, conformal or hyper-conformal deposition of the dielectric films may be necessary to achieve void-free gap filling. O_3-TEOS CVD processes can be used for these applications. In one approach, a thin layer of undoped O_3-TEOS SiO_2 is deposited prior to the BPSG layer, filling the smallest gaps and transforming the underlying topography to one which is slightly smoother. The subsequent BPSG film is then confronted with a less challenging topography. Alternatively, boron and phosphorus doped O_3-TEOS oxide films (BPSG) can be used as the flowglasses themselves (49-51,196). In either approach, the O_3-TEOS oxide films undergo high temperature annealing during the flow-anneal process. This results in a considerable amount of densification of these films, which substantially improves their resistance to ambient water vapor and to wet processing (200).

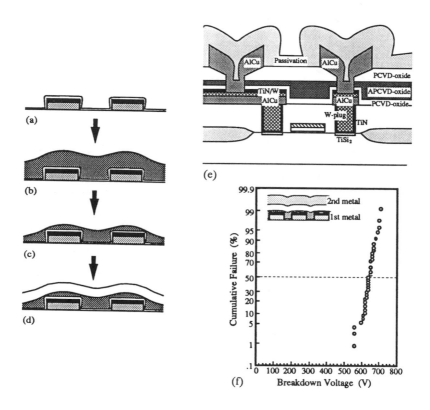

Figure 76. O_3-TEOS SiO_2 integrated with PECVD SiO_2 layers in an interlevel dielectric. Fabrication sequence begins with 1000Å of PECVD oxide **(a)**. 10,000Å of O_3-TEOS SiO_2 is deposited next **(b)**, followed by *partial* etchback **(c)**. Final layer is 5000Å of PECVD SiO_2 **(d)**. Complete double-metal structure is shown in **(e)**. Electrical breakdown strength of the composite interlevel dielectric (8000Å) is shown in **(f)**. From ref. 199; copyright 1991 IEEE; reprinted by permission of the publisher.

If an alternative to flow-anneal planarization is used, phosphorus-doped O_3-TEOS oxides may be used in place of BPSG as the pre-metal dielectrics (201). The phosphorus enables the film to getter mobile ion contaminants. Boron is unnecessary since the film need not flow. Since the underlying materials are refractory materials, one can make use of moderate temperature annealing processes (600~800°C) to densify the O_3-TEOS PSG film.

Passivation Applications.   Section 2.3 discusses some of the motivations for planarization of the final passivation layers. O_3-TEOS oxide is one of the tools used to bring about a smooth surface topography, such that the subsequent deposited PECVD nitride film is not forced to cover vertical or reentrant sidewalls, or acute corners at the bases of steps (193). These are the sites at which the nitride layer would be most vulnerable to contaminant penetration.

8.3   Issues

When integrating O_3-TEOS oxide films into a fabrication sequence, one must bear in mind the hygroscopic nature of these glasses. If the process architecture and manufacturing logistics are not established with this in mind, it can be a source of manufacturing difficulties and intermittent yield problems. Although O_3-TEOS oxide films are dry when deposited, they absorb water whenever given the chance. One source of water, of course, is ambient air (Figure 77); so in inter-metal dielectric films the O_3-TEOS oxide layer is vulnerable while the wafers sit in the fab after the via etching process. While this situation is analogous to that in a non-etchback siloxane SOG scheme, in which a layer of SOG is exposed on the via sidewalls after via etching, a key difference is the resistance of the two types of materials to dry resist stripping processes after the via etch. In polysiloxane SOG's, the organic components in the glass are essential to its resistance to water penetration. Once these are denuded from the SOG by a dry resist strip process, what remains is a highly fragmented and porous silicate structure into which water readily diffuses and reacts (98,111,113-115,117-119). This is not the case with O_3-TEOS SiO_2 films, which contain virtually no organic components and are not attacked by dry resist strip processes. Hence, dry resist strip processing is permissible in planarization schemes using O_3-TEOS oxide films.

Water absorbed by O_3-TEOS oxide is desorbed when the films are heated to 350~450°C, and this mechanism provides a means to rid the films of water before critical processing steps. For example, water can be desorbed from via walls by baking the wafers under vacuum in the metal deposition tool prior to the via filling process. When the bake and the metal deposition are done in-situ, one can ensure that the metal is deposited onto thoroughly dried oxides.

O_3-TEOS oxides used in *pre*-metal flow glass applications become densified during the flow-anneal. Once this is done the films are no longer prone to absorb water.

The key dielectric planarization processes and techniques have now been introduced and discussed. Whereas the preceding sections 3.0 ~ 8.3 have reviewed the mainstream applications of these techniques, the final section of this chapter will briefly speculate as to the role of these and other techniques in the foreseeable future.

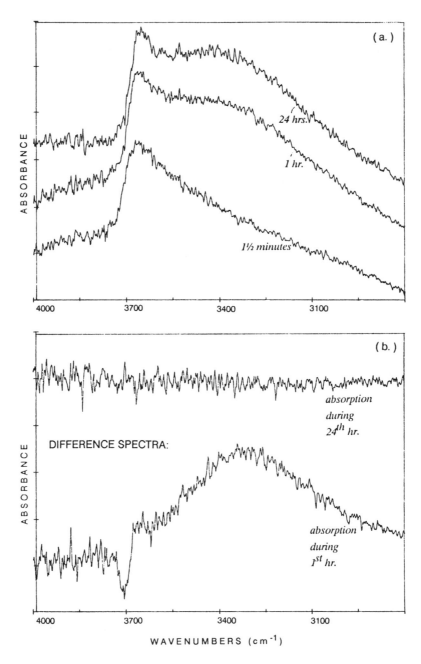

Figure 77. Infrared absorbance spectra of O_3-TEOS SiO_2 films aging in air. The films are dry initially , but rapidly absorb ambient water vapor during the first hour of exposure **(a)**. Absorption continues at a slower rate during the next 24 hours, and is nearly indetectable after 24 hours **(a,b)**.

9.0   OUTLOOK

There is little doubt that the effort and resources directed at *global* planarization technologies will be considerable. This is driven by the economic incentives to retain optical lithography methods into the sub-half micron era -- an ambition made increasingly difficult by two trends which will certainly not reverse: (1) that toward decreasing feature sizes, and (2) that toward increased numbers of interconnect levels. Two technologies -- dielectric chemical-mechanical polishing, and resist-etchback with planarization block masks -- are attracting most of the development resources and are likely to be the frontrunner technologies into at least the mid-1990's.

Gap filling, which is a prerequisite to the long range or global step-height-reduction techniques, has already become a highly challenging module in most planarization technologies. This task will become increasingly difficult as interconnect pitches shrink while their thicknesses do not. Bias-ECR-CVD and O_3-TEOS CVD processes are new tools being aggressively applied to this challenge. Meanwhile, incremental improvements to the established sequential dep-etch techniques have been found to extend the usefulness of that approach beyond what were once considered their practical limits. *Directional* or *anisotropic* CVD has been proposed as an alternative to gap *filling* ... however at this time that technology is not being driven by industrywide efforts and has not progressed beyond initial feasibility experiments.

Deposition and flow of BPSG is still the most widely used technique for planarization of pre-metal dielectric (ILD-0) films. However, temperatures and times allowed for the flow-anneal processes are ratcheting downward as D*t budgets shrink. Within these constraints viscous flow is sharply diminished, and flowed BPSG -- at least at the concentration levels currently considered acceptable -- is unlikely to provide planarization beyond very short range smoothing. This causes difficulties with respect to via depths if subsequent dielectric layers are globally planarized, as discussed earlier. It is likely that some of the techniques presently used for *inter*-metal dielectric planarization -- techniques capable of filling difficult gaps and of long range or global planarization -- will find increased use at the pre-metal dielectric level as well.

In general, techniques capable of local or short range planarization are likely to be limited to use primarily for planarization of the dielectrics between, or after, the *upper* interconnect levels. This is because of the via-related difficulties mentioned in the previous paragraph. SOG techniques, for example, may be suitable for smoothing of the dielectrics between M2 and M3 of a three-level-metal structure, or after M3 but beneath the PECVD nitride passivation layer; with long-range or global planarization techniques used for the ILD-0 and ILD-1.

Recessed metal interconnects, and the techniques used to produce them, can reduce the need for many of the dielectric planarization processes. One of these techniques, *Damascene*, is discussed in Section 7. This Damascene technique makes use of CVD tungsten for the interconnects; however an important impediment confronting recessed metal technologies is the lack of a means to fill the dielectric recesses with *low*-resistivity metals. Aluminum CVD, copper

CVD, and electroless copper deposition are three novel processes which may help overcome this hurdle, but none have progressed beyond the early stages of development.

REFERENCES

1.   T. Abraham, M. Ayukawa, "Process Sensitive Test Patterns for Evaluating Multilevel Interconnect Structures," *Proc. IEEE VLSI Multilevel Interconnection Conf.*, (1988), p.221.

2.   K. Monnig, R. Tolles, A. Maury, R. Legget, S. Sivaram, "Overview of Planarization by Mechanical Polishing," *Ext. Abstr. Electrochemical Society Meeting*, 91-1 (1991) p.586.

3.   A. Isobe, "Planarization (From the Viewpoints of Fine Patterning and Reliability)," *Technical Proceedings, Semicon Japan*, (SEMI) 1991, p.271.

4.   I. Gaete, K.J. Wu, "Improved EPROM Moisture Performance Using Spin-On Glass (SOG) for Passivation Planarization," *IEEE Proc. Reliability Phys.* (1989).

5.   M.W. Horn, "Antireflection Layers and Planarization for Microlithography" *Solid State Technology*, Nov. 1991, p.57.

6.   S.R. Wilson, J.L. Freeman,Jr., C.J. Tracy, "A Four-Metal Layer, High Performance Interconnect System for Bipolar and BiCMOS Circuits," *Solid State Technology*, Nov. 1991, p.67.

7.   P. Renteln, M.E. Thomas, J.M. Pierce, "Characterization of Mechanical Planarization Process," *Proc. IEEE VLSI Multilevel Interconnection Conf.*, (1990), p.57.

8.   C. Ramiller, L. Yau, "Borophosphosilicate Glass for Low Temperature Reflow," *Technical Proceedings, Semicon West*, (SEMI) 1982, p.29.

9.   W. Kern, G.L. Schnable, "Chemically Vapor-Deposited Borophosphosilicate Glasses for Silicon Device Applications," *RCA Review* 43:423 (1982).

10.   C. Dornfest, "The Effect of Reducing Deposition Temperature in an Atmospheric Pressure BPSG Process," *Ext. Abstr. Electrochemical Society Meeting*, 85-1 (1985) p.347.

11.   W. Kern, W.A. Kurylo, C.J. Tino, "Optimized Chemical Vapor Deposition of Borophosphosilicate Glass Films," *RCA Review*, 46:117 (1985).

12.   K.H. Hurley, L.D. Bartholomew, D.T. Bordonaro, "BPSG Films Deposited by APCVD," *Semiconductor International*, Oct 1987, p.91.

13.   A.J. Learn, W. Baerg, "Growth of Borosilicate and Borophosphosilicate Films at Low Pressure and Temperature," *Thin Solid Films*, 130:103 (1985).

14.   T. Foster, G. Hoeye, J. Goldman, "A Low Pressure BPSG Deposition Process," *J. Electrochem. Soc.* 132:505 (1985).

15. F.S. Becker, D. Pawlik, H. Schafer, G. Staudigl, "Process and Film Characterization of Low Pressure TEOS BPSG," *J. Vac. Sci. Technol.*B, 4:732 (1986).

16. D. Williams, E.A. Dein, "LPCVD of Borophosphosilicate Glass from Organic Reactants," *J. Electrochem. Soc.* 134:657 (1987).

17. P.B. Johnson, P. Sethna, "Using BPSG as an Interlayer Dielectric," *Semiconductor International*, Oct. 1987, p.80.

18. J.E. Tong, K. Schertenleib, R.A. Carpio, "Process and Film Characterization of PECVD BPSG Films for VLSI Applications," *Solid State Technology*, Jan. 1984, p.161.

19. K. Law, J. Wong, C. Leung, J. Olsen, D. Wang, "Plasma-Enhanced Deposition of BPSG Using TEOS and Silane Sources," *Solid State Technology*, Apr. 1989, p.60.

20. K. Nassau, R.A. Levy, D.L. Chadwick, "Modified Phosphosilicate Glasses for VLSI Applications," *J. Electrochem. Soc.* 132:409 (1985).

21. S. Chittipeddi, A.N. Velaga, A.K. Nanda, W.T. Cochran, R.N. Graver, "The Flow Characteristics of BPSG (TEOS Based) Glass," *Proc. IEEE VLSI Multilevel Interconnection Conf.*, (1991), p.68.

22. Nagasima, H. Suzuki, K. Tanaka, S. Nishida, "Interaction Between PSG Films and Water," *J. Electrochem. Soc.* 121:434 (1974).

23. W.M. Paulson, R.W. Kirk, "The Effects of Phosphorus-Doped Passivation Glass on the Corrosion of Aluminum," IEEE *Proc. Reliability Phys.* (1974) p.172.

24. R.H. Dawson, G.L. Schnable, "Passivating Composite for a Semiconductor Device Comprising a Silicon Nitride (Si_3N_4) Layer and Phosphosilicate Glass (PSG) Layer," U.S. Patent 4,273,805; June 16, 1981.

25. R.A. Bowling, G.B. Larrabee, "Deposition and Reflow of Phosphosilicate Glass," *J. Electrochem. Soc.* 132:141 (1985).

26. A.C. Adams, C.D. Capio, "Planarization of Phosphorus Doped Silicon Dioxide," *J. Electrochem. Soc.* 128:423 (1981).

27. M. Susa, Y. Hiroshima, K. Senda, T. Takamura, "Borophosphosilicate Glass Flow in a PH_3-O_2 Ambient," *J. Electrochem. Soc.* 133:1517 (1986).

28. I. Banerjee, B. Tracy, P. Davies, B. McDonald, "Use of Advanced Analytical Techniques for VLSI Failure Analysis," IEEE *Proc. Reliability Phys.*, (1990) p.61.

29. G.L. Schnable, A.W. Fisher, J.M. Shaw, "Devitrification in BPSG Films Used in VLSI," *J. Electrochem. Soc.* 137:3973 (1990).

30. W.A. Pliskin, H.S. Lehman, "Structural Evaluation of Silicon Oxide Films," *J. Electrochem. Soc.* 112:1013 (1965).

31. W.A. Pliskin, "Use of Infrared Spectroscopy for the Characterization of Dielectric films on Silicon," *Semiconductor Silicon*, p. 506 (1973).

32. A. Takamatsu, M. Shibata, H. Sakai, T. Yoshimi, "Plasma Activated Deposition and Properties of Phosphosilicate Glass Films," *J. Electrochem. Soc.* 131:1865 (1984).

33.   Y. Shioya, M. Maeda, "Comparison of Phosphosilicate Glass Films Deposited by Three Different Chemical Vapor Deposition Methods," *J. Electrochem. Soc.* 133:1943 (1986).

34.   H.P.W. Hey, B.G. Sluijk, D.G. Hemmes, "Ion Bombardment: A Determining Factor in Plasma CVD," *Solid State Technology,* Apr. 1990, p.139.

35.   W. Kern, "Densification of Vapor Deposited Phosphosilicate Glass Films," *RCA Review*, 37:55 (1976).

36.   R.M. Levin, "Water Absorption and Densification of Phosphosilicate Glass Films," *J. Electrochem. Soc.* 129:1765 (1982).

37.   O.K.T. Wu, A.N. Saxena, "Effect of Annealing on Chemical State of Phosphorus in SiO_2 Films," *J. Electrochem. Soc.* 132:932 (1985).

38.   D.E. Bornside, R.A. Brown, S. Mittal, F.T. Geyling, "Global Planarization of Spun-on Thin Films by Reflow," *Appl. Phys. Letters* 58:1181 (1991).

39.   C.Y. Ting, F.M. d'Heurle, S.S. Iyer, P.M. Fryer, "High Temperature Process Limitation on $TiSi_2$," *J. Electrochem. Soc.* 133:2621 (1986).

40.   R.K. Shukla, J.S. Multani, "Thermal Stability of Titanium Silicide Thin Films," *Proc. IEEE VLSI Multilevel Interconnection Conf.,* (1987), p.470.

41.   C.M. Osburn, "Formation of Silicided, Ultra Shallow Junctions Using Low Thermal Budget Processing," *J. Electronic. Mater.,* 19:67 (1990).

42.   J.S. Mercier, "Rapid Flow of Doped Glasses for VLSI Fabrication," *Solid State Technology,* July 1987, p.85.

43.   I. Barsony, H. Anzai, J. Nishizawa, "Structural Impact of Lamp Annealing in Device Back-End Processing," *J. Electrochem. Soc.* 133:156 (1986).

44.   R.R. Razouk, L.N. Lie, *Ext. Abstr. Electrochemical Society Meeting,* 82-1 (1982) p.138.

45.   S.P. Tay, J.P. Ellul, "High Pressure Technology for Silicon IC Fabrication," *Semiconductor International,* Sept, 1986, p.122.

46.   R.A. Levy, K. Nassau, "Reflow Mechanisms of Contact Vias in VLSI Processing," *J. Electrochem. Soc.* 133:1417 (1986).

47.   H. Ozaki, S. Mayumi, S. Ueda, M. Inoue, "Contact Resistance Behavior in Borophosphosilicate Glass," *Proc. IEEE VLSI Multilevel Interconnection Conf.,* (1987), p.323.

48.   C.A Fieber, E.P. Martin, H.Z. Chew, G.W. Hills, N. Selamoglu, S.A. Lytle, "Superior Metal Step Coverage and Dielectric Quality in a Simple Two-Level Metal 1.0 μm CMOS Technology," *Proc. IEEE VLSI Multilevel Interconnection Conf.,* (1989), p.55.

49.   Y. Nishimoto, N. Tokumasu, K. Fujino, K. Maeda, "Dielectric Film Deposition by Atmospheric Pressure and Low Temperature CVD Using TEOS, Ozone and New Organometallic Doping Sources," *Proc. IEEE VLSI Multilevel Interconnection Conf.,* (1989), p.382.

50.   P. Lee, M. Galiano, P. Keswick, J. Wong, B. Shin, D.N.K. Wang, "Sub-Atmospheric Chemical Vapor Deposition (SACVD) of TEOS-

Ozone USG and BPSG," *Proc. IEEE VLSI Multilevel Interconnection Conf.*, (1990), p.396.

51.   J.G. Lee, S.H. Choi, T.C. Ahn, C.G. Hong, P. Lee, K. Law, M. Galiano, P. Keswick, B. Shin, "SACVD: A New Approach for 16 MBit Dielectrics," *Semiconductor International*, May 1992, p.116.

52.   D.W. Freeman, M.A. Logan, L.F. Wright, J.R. Monkowski, "Planarized BPSG Deposited From Organometallic Sources," *Ext. Abstr. Electrochemical Society Meeting*, 88-2 (1988) p.337.

53.   W. Kern, J. Hartman, "Simultaneous Deposition and Fusion-Flow Planarization of BPSG," *Technical Proceedings*, Semicon Korea, (SEMI) 1990, p.19.

54.   J. Marks, K. Law, D. Wang, "In-situ Planarization of Dielectric Surfaces Using Boron Oxide," *Proc. IEEE VLSI Multilevel Interconnection Conf.*, (1989), p.89.

55.   S. Pennington, D. Hallock, "A Low-Temperature, In-situ Deposition and Planarizing PSG Process for Filling High-Aspect-Ratio Topography," *Proc. IEEE VLSI Multilevel Interconnection Conf.*, (1990), p.71.

56.   L.K. White, "Planarization Properties of Resist and Polyimide Coatings," *J. Electrochem. Soc.* 130:1543 (1983).

57.   H. Fritzche, V. Grewal, W. Henkel, "An Improved Etch-back Process for Multilevel Metallization and Its Reliability Results for CMOS Devices," *Proc. IEEE VLSI Multilevel Interconnection Conf.*, (1986), p.45.

58.   L. Koyama, M. Thomas, "New Double Planarization Process for Multilevel Metallization Using Oxide ILD," *Proc. IEEE VLSI Multilevel Interconnection Conf.*, (1985), p.45.

59.   W. Geiger, A. Sharma, "An Optimized Planarization Process for a Multilayer Interconnect," *Proc. IEEE VLSI Multilevel Interconnection Conf.*, (1986), p.128.

60.   R.H. Wilson, P.A. Piacente, "Effect of Circuit Structures on Planarization Resist Thickness," *J. Electrochem. Soc.* 133:981 (1986).

61.   A. Shepela, B. Soller, "A Model for RIE Dielectric Etchback Planarization," *J. Electrochem. Soc.* 134:714 (1987).

62.   T.H. Daubenspeck, J.K. DeBrosse, C.W. Koburger, M. Armacost, J.R. Abernathey, "Planarization of ULSI Topography Over Variable Pattern Densities," *J. Electrochem. Soc.* 138:506 (1991).

63.   G.E. Gimpelson, C.L. Russo, "Plasma Planarization with a Non-Planar Sacrificial Layer," *Proc. IEEE VLSI Multilevel Interconnection Conf.*, (1984), p.37.

64.   B. Vasquez, R. Goodner, "Planarized Oxide with Sacrificial Photoresist: Etch Rate Sensitivity to Pattern Density," *Proc. IEEE VLSI Multilevel Interconnection Conf.*, (1987), p.394.

65.   L. deBruin, J.M.F.G. vanLaarhoven, "Advanced Multiple-Step Resist Etchback Planarization," *Proc. IEEE VLSI Multilevel Interconnection Conf.*, (1988), p.404.

66.   C. Jang, S.R. Chen, T.F. Klemme, J. Lerma, H.M. Naguib, "A Planarization Process for Multilevel Metallization Using AZ-Protective Coating as a Sacrificial Layer," *Proc. IEEE VLSI Multilevel Interconnection Conf.*, (1987), p.357.

67. V. Grewal, A. Gschwandtner, G. Higelin, "A Novel Metallization Technique for Advanced CMOS and Bipolar Integrated Circuits," *Proc. IEEE VLSI Multilevel Interconnection Conf.*, (1986), p.107.
68. P.L. Pai, C.H. Ting, "Effect of Via Filling on the Via Resistance and Surface Topography," *Ext. Abstr. Electrochemical Society Meeting*, 88-2 (1988) p.364.
69. C.H.Ting, P-L Pai, Z. Sobczack, "An Improved Etchback Planarization Process Using A Super Planarizing Spin-On Sacrificial Layer," *Proc. IEEE VLSI Multilevel Interconnection Conf.*, (1989), p.491.
70. V. Comello, "Planarizing Leading Edge Devices," *Semiconductor International*, Nov 1990, p.60.
71. G.W. Hills, H.P.W. Hey, "Via Planarization: Application to a 0.9μm Two Level Metal Technology," *Ext. Abstr. Electrochemical Society Meeting*, 88-1 (1988) p.186.
72. M.T. Welch, C. Garcia, "Pillar Interconnections for VLSI Technology," *Proc. IEEE VLSI Multilevel Interconnection Conf.*, (1986), p.450.
73. T.A. Bartush, "A Four Level Wiring Process for Semiconductor Chips," *Proc. IEEE VLSI Multilevel Interconnection Conf.*, (1987), p.41.
74. M.T. Welch, R.E. McMann, M.L. Torreno, "Scalable VLSI Metallization," *Proc. IEEE VLSI Multilevel Interconnection Conf.*, (1987), p.51.
75. C.A. Bollinger, D. Grube, S.A. Lytle, E.P. Martin, J.A. Shimer, H.R. Siddiqui, "An Advanced Four Level Interconnect Enhancement Module for 0.9 Micron CMOS," *Proc. IEEE VLSI Multilevel Interconnection Conf.*, (1990), p.21.
76. A. Nagy, J. Helbert, "Planarized Inorganic Interlevel Dielectric for Multilevel Metallization -- Part 1," *Solid State Technology*, Jan. 1991, p.53.
77. N. Parekh, A. Butler, W. Doedel, W. Heesters, L. Forester, "Feasibility of a Novel Modular Aproach for Planarization of A Submicron Triple-Level Metal CMOS Process," *Proc. IEEE VLSI Multilevel Interconnection Conf.*, (1990), p.453.
78. B.C. Feng, "Planarization of Integrated Circuit Surfaces Through Selective Photoresist Masking," U.S. Patent 3,976,524, 1976.
79. D.J. Sheldon, C.W. Gruenschlaeger, L. Kammerdiner, B. Henis, P. Kelleher, J.D. Hayden, "Application of a Two-Layer Planarization Process to VLSI Intermetal Dielectric and Trench Isolation Processes," IEEE *Transactions on Semiconductor Manufacturing*, 1(4):140 (1988).
80. A. Schiltz, M. Pons, "Two-Layer Planarization Process" *J. Electrochem. Soc.* 133:178 (1986).
81. S.R. Wilson, J.L. Freeman, C.J. Tracy, "A Four Metal Layer, High Performance Interconnect System for Bipolar and BiCMOS Circuits," *Proc. IEEE VLSI Multilevel Interconnection Conf.*, (1990), p.42.
82. E.W. Scheckler, D.E. Lyons, A.R. Neureuther, W.G. Oldham, "Process Simulation and Experiment for RC Parasitics in Multilevel Metallization," *Proc. IEEE VLSI Multilevel Interconnection Conf.*, (1989), p.299.

83. P. Elkins, K. Reinhardt, R. Tang, "A Planarization Process for Double Metal CMOS Using Spin-on-Glass as a Sacrificial Layer," *Proc. IEEE VLSI Multilevel Interconnection Conf.*, (1986), p.100.

84. J.K. Chu, J.S. Multani, S.K. Mittal, J.T. Orton, R. Jecmen, "Spin-on-Glass Dielectric Planarization for Double Metal CMOS Technology," *Proc. IEEE VLSI Multilevel Interconnection Conf.*, (1986), p.474.

85. L.B. Vines, S.K. Gupta, "Interlevel Dielectric Planarization with Spin-on-Glass Films," *Proc. IEEE VLSI Multilevel Interconnection Conf.*, (1986), p.506.

86. C.H. Ting, H.Y Lin, "Planarization Proces Using Spin-On-Glass," *Proc. IEEE VLSI Multilevel Interconnection Conf.*, (1987), p.61.

87. C. Chiang, N.V. Lam, J.K. Chu, N. Cox, D. Fraser, J. Bozarth, B. Mumford, "Defects Study on Spin On Glass Planarization Technology," *Proc. IEEE VLSI Multilevel Interconnection Conf.*, (1987), p.404.

88. D.L.W. Yen, G.K. Rao, "Process Integration with Spin-on-Glass Sandwich as an Intermetal Dielectric Layer for 1.2μm CMOS DLM Process," *Proc. IEEE VLSI Multilevel Interconnection Conf.*, (1988), p.85.

89. S.N. Chen, Y.C. Chao, J.J. Lin, Y.H. Tsai, F.C. Tseng, "Spin-On Glasses: Characterization and Application," *Proc. IEEE VLSI Multilevel Interconnection Conf.*, (1988), p.306.

90. J. Nulty, G. Spadini, D. Pramanik, "A Highly Reliable Metallization System for a Double Metal 1.5 μm CMOS Process," *Proc. IEEE VLSI Multilevel Interconnection Conf.*, (1988), p.453.

91. H.W.M. Chung, S.K. Gupta, T.A. Baldwin, "Fabrication of CMOS Circuits Using Non-Etchback SOG Processing for Dielectric Planarization," *Proc. IEEE VLSI Multilevel Interconnection Conf.*, (1989), p.373.

92. P.L. Pai, "Planarization for 0.5 μm CMOS/BiCMOS Technology," *Proc. IEEE VLSI Multilevel Interconnection Conf.*, (1990), p.450.

93. H. Chung, S. Wong, S. Lim, "Non-Etchback Silicate Spin-on Glass for Advanced BiCMOS Technology," *Proc. IEEE VLSI Multilevel Interconnection Conf.*, (1991), p.376.

94. S. Morimoto, S.Q. Grant, "Manufacturable and Reliable Spin-on-Glass Planarization Process for 1μm CMOS Double Layer Metal Technology," *Proc. IEEE VLSI Multilevel Interconnection Conf.*, (1988), p.411.

95. H. Kojima, T. Iwamori, Y. Sakata, T. Yamashita, Y. Yatsuda, "Planarization Process Using a Multi-Coating of Spin-on-Glass," *Proc. IEEE VLSI Multilevel Interconnection Conf.*, (1988), p.390.

96. C. Chiang, D.B. Fraser, "Understanding of Spin-on-Glass (SOG) Properties from Their Molecular Structure," *Proc. IEEE VLSI Multilevel Interconnection Conf.*, (1989), p.397.

97. M.P. Woo, J.L. Cain, C.O. Lee, "Characterization of Spin-on-Glass Using Fourier Transform Infrared Spectroscopy," *J. Electrochem. Soc.* 137:196 (1990).

98. F. Gualandris, L. Masini, A. Borghesi, "A Comparitive Evaluation of Spin-on-Glass Cure by FTIR Technique," *J. Electr. Mat.*, 20:299 (1991).

99.   J.D. Romero, M. Khan, H. Fatemi, J. Turlo, "Outgassing Behavior of Spin-on-Glass," *J. Mat. Res.* 6:1996 (1991).
100. H. Suzuki, "Planarization Technology Using Spin-Coating with Organic Materials," *Technical Proceedings,* Semicon Japan, (SEMI) 1991 p.287.
101. S. Tsou, C.L. Chen, S. Chen, H.K. Chou, J.J. Hsu, F. Chen, G.W. Liang, "The Backend Process Integration of an Advanced Double Metal Technology for Sub-μm High Speed CMOS and BiCMOS SRAM," *Proc. IEEE VLSI Multilevel Interconnection Conf.,* (1991), p.34.
102. W. Geiger, A. Sharma, "An Optimized Planarization Process for a Multi Layer Interconnect," *Proc. IEEE VLSI Multilevel Interconnection Conf.,* (1986), p.128.
103. M.D. Bui, T.A. Streif, K.E. Schoenberg, R.H. Dorrance, P.P. Procter, "Contact Glass Planarization Using a Double Etch Back Technique and Spin-on-Glass Sacrificial Layer," *Proc. IEEE VLSI Multilevel Interconnection Conf.,* (1987), p.385.
104. G. DeGraaf, A.L. Butler, R. Penning De Vries, "An Investigation of Advanced First Dielectric Planarisation Techniques in Combination with a Tungsten Plug Metallisation Process," *Proc. IEEE VLSI Multilevel Interconnection Conf.,* (1988), p.357.
105. L.K. White, "Approximating Spun-On, Thin Film Planarization Properties on Complex Topography," *J. Electrochem. Soc.* 132:168 (1985).
106. P.L. Pai, W.G. Oldham, C.H. Ting, "Process Considerations for Using Spin-On Glass as a Planarizing Dielectric Film," *Proc. IEEE VLSI Multilevel Interconnection Conf.,* (1987), p.364.
107. M. Kawai, K. Matsuda, K. Miki, K. Sakiyama, "Interlayer Dielectric Planarization with TEOS-CVD and SOG," *Proc. IEEE VLSI Multilevel Interconnection Conf.,* (1988), p.419.
108. L.K. White, "Modelling Spin-On Film Planarization Properties," *Proc. IEEE VLSI Multilevel Interconnection Conf.,* (1989), p.487.
109. D.E. Bornside, "Mechanism for the Local Planarization of Microscopically Rough Surfaces by Drying thin Films of Spin-Coated Polymer/Solvent Solutions," *J. Electrochem. Soc.* 137:2589 (1990).
110. T. Tokunaga, N. Owada, "Effects of Multichamber Processing on Reliability of Submicron Vias," *Multichamber and In-situ Processing of Electronic Materials,* Proc. SPIE 1188:61 (1990).
111. C.H. Ting, H.Y. Lin, P.L. Pai, T. Rucker, "A Non-Etchback SOG Process for Multilevel Interconnect Technology," *Ext. Abstr. Electrochemical Society Meeting,* 88-2 (1988) p.366.
112. H.G. Tompkins, C. Tracy, "Desorption from Spin-on-Glass," *J. Electrochem. Soc.* 136:2331 (1989).
113. R.A.M. Wolters, W.C.J. Heesters, "Experimental Study of Metal-Metal Contact Properties Using Spin-on Glass," *Proc. IEEE VLSI Multilevel Interconnection Conf.,* (1990), p.447.
114. H.G. Tompkins, C. Tracy, "Tightly Bound H_2O in Spin-on-Glass," *J.Vac.Sci.Technol.*B, 8:558 (1990).

115. M. Kobayakawa, A. Arimatsu, F. Yokoyama, N. Hirashita, T. Ajioka, "A Study of Outgassing from Spin-on-Glass Films Used for Planarization," *Proc. IEEE VLSI Multilevel Interconnection Conf.,* (1991), p.454.

116. N. Hirashita, M. Kobayakawa, A. Arimatsu, F. Yokoyama, T. Ajioka, "Thermal Desorption Studies of Phosphorus-Doped Spin-on-Glass Films," *J. Electrochem. Soc.* 139:794 (1992).

117. A.D. Butherus, T.W. Hou, C.J. Mogab, H. Schonhorn, "O_2 Plasma-Converted Spin-on-glass for Planarization," *J. Vac. Sci.Technol.*B, 3:1352 (1985).

118. B.G. Bagley, W.E. Quinn, C.J. Mogab, M.J. Vasile, "The Effect of Reactor Configuration on the Oxygen Plasma Conversion of an Organosilicon to SiO_2," *Mat. Letters,* 4:154 (1986).

119. S. Ito, Y. Homma, E. Sasaki, S. Uchimura, H. Morishima, "Application of Surface Reformed Thick Spin-on-Glass to MOS Device Planarization," *J. Electrochem. Soc.* 137:1212 (1990).

120. L. Forester, W. Doedel, K. Osinski, W. Heesters, "Development of a Three Layer Metal Backend Process for Application to a Submicron CMOS Logic Process," *Proc. IEEE VLSI Multilevel Interconnection Conf.,* (1990), p.28.

121. C. Hausamann, P. Mokrisch, "The Dependence of Oxide and Spin-on-Glass Etchrates on Their Area Ratio," *Proc. IEEE VLSI Multilevel Interconnection Conf.,* (1988), p.293.

122. F. Whitwer, D. Milligan, "A Submicron Double Level Metal Process for High Density Memory Applications," *Proc. IEEE VLSI Multilevel Interconnection Conf.,* (1990), p.49.

123. R.M. Blewer, R.A. Gasser, "Process Window Calculations for an SOG Etchback Process," *Proc. IEEE VLSI Multilevel Interconnection Conf.,* (1987), p.376.

124. L. Forester, A.L. Butler, G. Schets, "SOG Planarization for Polysilicon and First Metal Interconnect in a One Micron CMOS Process," *Proc. IEEE VLSI Multilevel Interconnection Conf.,* (1989), p.72.

125. C.Y Ting, V.J. Vivalda, H.G. Schaefer, "Study of Planarized Sputter-Deposited SiO_2," *J. Vac. Sci. Technol.* 15:1105 (1978).

126. H. Kotani, H. Yakushiji, H. Harada, K. Tsukamoto, T. Nishioka, "Sputter Etching Planarization for Multilevel Metallization," *J. Electrochem. Soc.* 130:645 (1986).

127. C.L. Standley, R.E. Jones, L.I. Maissel, "Sputtered SiO_2 Deposited Over a Step," *Thin Solid Films,* 5:355 (1970).

128. T.N. Kennedy, "Step Coverage of Sputtered Insulators," *J. Vac. Sci. Technol.* 13:1135 (1976).

129. J.S. Logan, J.H. Keller, R.G. Simmons, "The RF Glow Discharge Sputtering Model," *J.Vac.Sci.Technol.,* 14:92 (1977).

130. H. Kitahara, "Planarization of SiO_2 Insulating Inter Layers," *Technical Proceedings,* Semicon Japan, (SEMI) 1987, p.162.

131. J.S. Logan, J. Costable, F. Jones, J.E. Lucy, "RF Sputter Deposition of SiO_2 Films at High Rate," *J.Vac.Sci.Technol.A,* 5:1897, 1987.

132. K-K. Lin, T-L. Tung, F.A. Leon, J. Garcia-Colevatti, E. Andideh, R.J. Patterson, K. Mack, F. Moghadam, "Optimizing Argon Sputter-Etch and Redeposition in ILD Gap Fill Using Computer Simulation," *Proc. IEEE VLSI Multilevel Interconnection Conf.*, (1992), p.462.

133. R.E. Jones, H.F. Winters, L.I. Maissel, "Effect of Oxygen on the Sputtering Rate of SiO_2," *J.Vac.Sci.Technol.*, 5:84 (1968).

134. R.M Levin, K. Evans-Lutterodt, "The Step Coverage of Undoped and Phosphorus-Doped SiO_2 Glass Films," *J. Vac. Sci. Technol.*B 1:54 (1983).

135. J.S. Mercier, H.M. Naguib, V.Q. Ho, H. Nentwich, "Dry Etch-Back of Overthick PSG Films for Step Covereage Improvement," *J. Electrochem. Soc.* 132:1219 (1985).

136. H.M. Naguib, C. Jang, T.F. Klemme, K. Wong, A. Rangappan, W.W. Yao, R.T. Fulks, "The Evaluation of Planarization Techniques for Double Level Metallization in 1.2µm CMOS Technology," *Proc. IEEE VLSI Multilevel Interconnection Conf.*, (1987) p.93.

137. M.J. Thoma, W.T. Cochran, A.S. Harrus, H.P.W. Hey, G.W. Hills, C.W. Lawrence, J.L. Yeh, "A 1.0µm CMOS Two Level Metal Technology Incorporating Plasma Enhanced TEOS," *Proc. IEEE VLSI Multilevel Interconnection Conf.*, (1987), p.20.

138. S.L. Pennington, S.E. Luce, D.P. Hallock, "An Improved Interlevel Dielectric Process for Submicron Double-Level Metal Products," *Proc. IEEE VLSI Multilevel Interconnection Conf.*, (1989), p.355.

139. T. Mogami, M. Morimoto, H. Okabayashi, E. Nagasawa, "SiO_2 Planarization by 2-Step RF Bias Sputtering," *J. Vac. Sci. Technol.*B, 3:857 (1985).

140. B. Singh, O. Mesker, D. Devlin, "Deposition of Planarized Layers by Biased Sputtered Quartz," *J. Vac. Sci. Technol.*B. 5:567 (1987).

141. Y. Hazuki, T. Moriya, "A Damage Free Perfect Planarization Method Using Bias Sputtered SiO_2," *Proc. IEEE VLSI Multilevel Interconnection Conf.*, (1986), p.121.

142. Abraham, "Reactive Facet Tapering of Plasma Oxide for Multilevel Interconnect Applications," *Proc. IEEE VLSI Multilevel Interconnection Conf.*, (1987), p.115.

143. M. Abe, Y. Mase, T. Katsura, O. Hirata, T. Yamamoto, S. Koguchi, "High Performance Multilevel Interconnection System with Stacked Interlayer Dielectrics by Plasma CVD and Bias Sputtering," *Proc. IEEE VLSI Multilevel Interconnection Conf.*, (1989), p.406.

144. C. Kaanta, W. Cote, J. Cronin, K. Holland, P.I. Lee, T. Wright, "Submicron Wiring Technology with Tungsten and Planarization," *Proc. IEEE VLSI Multilevel Interconnection Conf.*, (1988), p.21.

145. K. Law, J. Wong, M. Chang, J. Olsen, D. Wang, "Chemical Vapor Deposition -- A Critical Technology," *Microelectronics Manufacturing and Testing*, Sept., 1988.

146. D. Moy, M. Schadt, C-K. Hu, F. Kaufman, A.K. Ray, N. Mazzeo, E. Baran, D.J. Pearson, "A Two-Level Metal Fully Planarized Interconnect

Structure Implemented on a 64kb CMOS SRAM," *Proc. IEEE VLSI Multilevel Interconnection Conf.*, (1989), p.26.

147. J.M Perchard, H.E. Smith, R. O'Conner, J. Olsen, K. Law, "Characterization of a Multiple-step In-situ PECVD TEOS Planarization Scheme for Submicron Manufacturing," *Multichamber and In-situ Processing of Electronic Materials,* Proc. SPIE 1188:75 (1989).

148. S. Mehta, G. Sharma, "A Single-Pass, In-Situ Planarization Process Utilizing TEOS for Double-Poly, Double-Metal CMOS Technologies," *Proc. IEEE VLSI Multilevel Interconnection Conf.,* (1989), p.80.

149. R.R. Uttecht, R.M. Geffken, "A Four-Level Metal Fully Planarized Interconnect Technology for Dense High Performance Logic and SRAM Applications," *Proc. IEEE VLSI Multilevel Interconnection Conf.,* (1991), p.20.

150. G.C. Schwartz and P. Johns, "Gap-Fill with PECVD SiO_2 Using Deposition/Sputter Etch Cycles," *J. Electrochem. Soc.,* 139:927 (1992).

151. E.J. McInerney, "An In-Situ Planarized PECVD Silicon Dioxide Interlayer Dielectric," *Proc. IEEE VLSI Multilevel Interconnection Conf.,* (1986), p.467.

152. E.J. McInerney, S.C. Avanzino, "Planarized SiO_2 Interlayer Dielectric with Bias-CVD," *IEEE Trans. Electron Devices,* ED-34:615 (1987).

153. G.C. Smith, A.J. Purdes, "Sidewall Tapered Oxide by PECVD," *J. Electrochem. Soc.* 132:2721 (1985).

154. J.S. Logan, F.S. Maddocks, P.D. Davidse, "Metal Edge Coverage and Control of Charge Accumulation in RF Sputtered Insulators," *IBM Res. Dev.,* 14:182 (1970).

155. S.V. Nguyen, K. Albaugh, "The Characterization of Electron Cyclotron Resonance Plasma Deposited Silicon Nitride and Silicon Oxide Films," *J. Electrochem. Soc.* 136:2835 (1989).

156. C. Chiang, D.B. Fraser, "Electron Cyclotron Resonance (ECR) Deposited SiO_2 Films for Interlayer Dielectric Application," *Proc. IEEE VLSI Multilevel Interconnection Conf.,* (1990), p.381.

157. C.S. Pai, J.F. Miner, P.D. Foo, "Biased ECR CVD Oxide Deposition Using TEOS and TMCTS," *Proc. IEEE VLSI Multilevel Interconnection Conf.,* (1991), p.442.

158. C.S. Pai, J.F. Miner, P.D. Foo, "Electron Cyclotron Resonance Microwave Discharge for Oxide Deposition Using TEOS," *J. Electrochem. Soc.,* 139:850 (1992).

159. J.L. Vossen, "Control of Film Properties by RF Sputtering Techniques," *J. Vac. Sci. Technol.* 8:512 (1971).

160. T. Fukuda, M. Ohue, T. Kikuchi, K. Suzuki, T. Sonobe, N. Momma, "Planarized SiO_2 Formation by a New Microwave Plasma System," *IEDM Technical Digest,* International Electron Devices Meeting, IEEE (1989), p.665.

161. D.T.C. Huo, M.F. Yan, C.P. Chang, P.D. Foo, "Planarization of SiO_2 Films Using Reactive Ion Beam in Plasma Enhanced Chemical Vapor Deposition," *J. Appl. Phys.* 69:6637 (1991).

162. S. Matsuo, M. Kiuchi, "Low Temperature Chemical Vapor Deposition Method Utilizing an Electron Cyclotron Resonance Plasma," *Japan. J.Appl.Phys.*, 22:L210 (1983).

163. T. Ebata, D. Denison, M. Logan, "Improved ECR Oxide Deposition," *Technical Proceedings*, Semicon Japan (SEMI) 1991, p.385.

164. S. Nakamura, S. Nakayama, "ECR Plasma Deposition Under a Controlled Magnetic Field," *Ext. Abstr. Electrochemical Society Meeting*, 88-2 (1988) p.439.

165. M.W. Horn, S.W. Pang, M. Rothschild, G.A. Ditmer, "Planarizing a-C:H and SiO_2 Films Prepared by Bias ECR Plasma Deposition," *Proc. IEEE VLSI Multilevel Interconnection Conf.*, (1989), p.65.

166. P.H. Singer, "ECR: Is the Magic Gone?," *Semiconductor International*, July 1991, p.46.

167. R. Chebi, S. Mittal, "A Manufacturable ILD Gap Fill Process with Biased ECR CVD," *Proc. IEEE VLSI Multilevel Interconnection Conf.*, (1991), p.61.

168. S.J.H. Brader, J. Rogers, S.C. Quinlan, "A Planarizing Spacer Technique for Double Level Metallisation," *Proc. IEEE VLSI Multilevel Interconnection Conf.*, (1986), p.93.

169. Y. Chasset, P. Launay, J.L. Liotard, "Statistical Analysis of Planarized Double Metal Structures," *Proc. IEEE VLSI Multilevel Interconnection Conf.*, (1986), p.114.

170. B. Lee, A. Pierfedrici, E.C. Douglas, "Dielectric Planarization Techniques for Narrow Pitch Multilevel Interconnects," *Proc. IEEE VLSI Multilevel Interconnection Conf.*, (1987), p.85.

171. G.W. Hills, A.S. Harrus, M.J. Thomah, "Plasma-Assisted Deposition and Device Technology: Interlevel Dielectric Considerations," *Dry Processing for Submicrometer Lithography*, Proc. SPIE 1185: (1989).

172. R.E. Oakley, S.J. White, N.P. Armstrong, T.L. Rees, C. Mallardeau, M. Roche, C. Paillet, D. Thomas, A. Hefner, R. Isernhagen, E. Waschler, M. Lentmaier, H.A. Williams, "ESPRIT Project: Narrow Pitch Four Level Metallisation Development," *Proc. IEEE VLSI Multilevel Interconnection Conf.*, (1988), p.238.

173. G.W. Hills, M.J. Thomah, M.L. Chen, W.T. cochran, A.S. Harrus, C.W. Lawrence, C.W. Leung, H.P.W. Hey, "A High Performance Submicron CMOS Two-Level Metal Technology Incorporating a Plasma CVD TEOS Interlevel Dielectric," *Proc. IEEE VLSI Multilevel Interconnection Conf.*, (1988), p.35.

174. D.N.K. Wang, S. Somekh, D. Maydan, "Advanced CVD Technology," *Ext. Abstr. Electrochemical Society Meeting*, 87-1 (1987) p.299.

175. D.N.K. Wang, K. Law, J. Wong, "A Single-System Approach to Intermetal Dielectrics," *Technical Proceedings*, Semicon Japan, (SEMI) 1987.

176. C.W. Kaanta, S.G. Bombardier, W.J. Cote, W.R. Hill, G.J. Kerszykowski, H.J. Landis, D.J. Poindexter, C.W. Pollard, G.H. Ross, J.G. Ryan, S. Wolff, J.E. Cronin, "Dual Damascene: A ULSI Wiring

Technology," *Proc. IEEE VLSI Multilevel Interconnection Conf.*, (1991), p.144.

177. M.E. Thomas, S. Sekigahama, P. Renteln, J.M. Pierce, "The Mechanical Planarization of Interlevel Dielectrics for Multilevel Interconnect Applications," *Proc. IEEE VLSI Multilevel Interconnection Conf.*, (1990), p.438.

178. K.D. Beyer, W.L. Guthrie, S.R. Makarewicz, E. Mendel, W.J. Patrick, K.A. Perry, W.A. Pliskin, J. Riseman, P.M. Schaible, C.L. Standly, "Chem-Mech Polishing Method for Producing Coplanar Metal-Insulator Films on a Substrate," U.S. Patent 4,944,836 (1990).

179. W.J. Patrick, W.L. Guthrie, C.L. Standley, P.M. Schiable, "Application of Chemical-Mechanical Polishing to the Fabrication of VLSI Circuit Interconnections," *J. Electrochem. Soc.* 138:1778 (1991).

180. W.J. Cote, M.A. Leach, "Wafer Flood Polishing," U.S. Patent 4,910,155 (1990).

181. P.A. Burke, "Semi-Empirical Modelling of SiO_2 Chemical-Mechanical Polishing Planarization," *Proc. IEEE VLSI Multilevel Interconnection Conf.*, (1991), p.379.

182. D. Cornish, L. Watt, *Br. Sci. Instr. Res. Assoc. Report*, R295, (1963).

183. T. Izumatani, *Treatise on Mater. Sci. and Technol.*, vol. 17, Acad. Press, New York (1979).

184. L.M. Cook, "Chemical Processes in Glass Polishing," *Jour. Noncrystalline Solids*, 120:152 (1990).

185. F.B. Kaufman, D.B. Thompson, R.E. Broadie, M.A. Jaso, W.L. Guthrie, D.J. Pearson, M.B. Small, "Chemical-Mechanical Polishing for Fabricating Patterned W Metal Features as Chip Interconnects," *J. Electrochem. Soc.* 138:3460 (1991).

186. C. Yu, A. Laulusa, M. Grief, T.T. Doan, "Chemical-Mechanical Polishing of CVD W and Sputtered Al Thin Films for Microelectronics," *Proc. IEEE VLSI Multilevel Interconnection Conf.*, (1992), p.156.

187. M.E. Thomas, P. Renteln, S. Sekigahama, J.M. Pierce, "Mechanical Planarization Process Characterization," *Technical Proceedings*, Semicon Japan, (SEMI) 1991, p.295.

188. B. Davari, C.W. Koburger, R. Schulz, J.D. Warnock, T. Furukawa, M. Jost, Y. Taur, W.G. Schwittek, J.K. DeBrosse, M.L. Kerbaugh, J.L. Mauer, "A New Planarization Technique, Using a Combination of RIE and Chemical Mechanical Polish (CMP)," *IEDM Technical Digest*, International Electron Devices Meeting, IEEE (1989), p.61.

189. S. Roehl, L. Camiletti, W. Cote, D. Cote, E. Eckstein, K.H. Froehner, P.I. Lee, D. Restaino, G. Roeska, V. Vynorius, S. Wolff, B. Vollmer, "High Density Damascene Wiring And Borderless Contacts for 64 M DRAM," *Proc. IEEE VLSI Multilevel Interconnection Conf.*, (1992), p.22.

190. S.V. Nguyen, D. Dobuzinski, D. Harmon, R. Gleason, S. Fridmann, "Reaction Mechanisms of Plasma- and Thermal-Assisted Chemical Vapor Deposition of TEOS Oxide Films," *J.Electrochem.Soc.*, 137:2209 (1990).

191. J.N. Cox, J.Z. Ren, J.M. Van Horn, K.W. Kwok, "Water Trapping and Detrapping in Thin Film Dielectrics: Temperature Dependence and Water-Trap Dynamics," *Ext. Abstr. Electrochemical Society Meeting*, 92-1 (1992) p.250.

192. H. Kotani, M. Matsuura, A. Fujii, H. Genjou, S. Nagao, "Low-Temperature APCVD Oxide Using TEOS-Ozone Chemistry for Multilevel Interconnections," *IEDM Technical Digest*, International Electron Devices Meeting, IEEE (1989), p.669.

193. B. Ahlburn, R. Nowak, M. Galiano, J. Olsen, "Advanced Dielectric Techniques for the Fabrication of 16 Megabit-Generation Devices," *Proc. ULSI Science and Technology*, 91-11, Electrochem. Soc. (1991) p.617.

194. K. Fujino, Y. Nishimoto, N. Tokumasu, K. Maeda, "Dependence of Deposition Rate on Base Materials in TEOS/O_3 AP CVD," *Proc. IEEE VLSI Multilevel Interconnection Conf.*, (1990), p.187.

195. K. Fujino, Y. Nishimoto, N. Tokumasu, K. Maeda, "Dependence of Deposition Characteristics on Base Materials in TEOS and Ozone CVD at Atmospheric Pressure," *J. Electrochem. Soc.* 138:550 (1991).

196. K. Fujino, Y. Nishimoto, N. Tokumasu, K. Maeda, "Reaction Mechanism of TEOS and O_3 Atmospheric Pressure CVD," *Proc. IEEE VLSI Multilevel Interconnection Conf.*, (1991), p.445.

197. Y.C. Shih, C.S. Pai, K.G. Steiner, W.G. Wilkins, "O_3-TEOS Integrated Process for Interlevel Dielectric Applications," *Proc. IEEE VLSI Multilevel Interconnection Conf.*, (1992), p.109.

198. T. Doi, Y. Mori, M. Kawai, K. Taniguchi, K. Uda, K. Sakiyama, "Planarization Technology of Inter-metal Dielectrics Layer for 0.35 μm Devices," *Proc. IEEE VLSI Multilevel Interconnection Conf.*, (1992), p.163.

199. Y. Takata, A. Ishii, M. Matsuura, A. Ohsaki, M. Iwasaki, J. Miyazaki, N. Fujiwara, J. Komori, T. Katayama, S. Nakao, H. Kotani, "A Highly Reliable Multilevel Interconnection Process for 0.6μm CMOS Devices, *Proc. IEEE VLSI Multilevel Interconnection Conf.*, (1991), p.13.

200. K.G. Donohoe, P. Lee, unpublished results.

201. E. Yieh, K. Kwok, B.C. Nguyen, H. Nobel, C. Basa, D. Cote, B. Neureither, "Low-Temperature Sub-Atmospheric CVD USG/PSG for Gap Filling and Planarization of Advanced Submicron Memory Devices," *Ext. Abstr. Electrochemical Society Meeting*, 92-1 (1992) p.248.

7

LITHOGRAPHY AND ETCH ISSUES FOR A MULTILEVEL METALLIZATION SYSTEM

GREGORY W. GRYNKEWICH And JOHN N. HELBERT
Motorola, Inc.
Advanced Custom Technologies
Mesa, Arizona

1.0   INTRODUCTION

The requirements of VLSI/ULSI multilevel metallization (MLM) systems have led to numerous challenges in the photolithographic and etch areas. These challenges arise from two interrelated sources: the continuing trend to smaller geometries and the materials that must be used to build a successful MLM. Smaller geometries often create a need for new materials, and the ability to process those materials can limit usable geometries. Photolithography and etch technologies with the capability of producing submicron MLM structures are an integral part of the successful fabrication of the high density, high performance, and highly reliable interconnect technologies of the nineties and beyond.

The purpose of this chapter is to familiarize the reader with photo and etch issues that are particular to MLM technology. The chapter is not intended to be a tutorial on photo and etch. Instead, we intend to review the pitfalls and issues that are present in MLM processing, and to present solutions and proposed solutions. We assume that the reader is new to MLM, but is already familiar with basic concepts and practices in photolithography and etch.

The chapter will be organized into two broad sections: Section 2 will address photolithography and Section 3 will concern etch. Within each section, the discussion will follow the natural order of building a MLM interconnect structure. Each section will include discussions on manufacturing issues and finish with comments on future trends.

2.0 INTRODUCTION TO PATTERN TRANSFER TECHNOLOGY

Photolithography technology, the combination of the exposure tool and the image transfer process, is vital to integrated circuit fabrication, (1) or more generally, semiconductor device manufacturing. Nearly every primary device fabrication step requires a process compatible masking layer, which is capable of providing a desired circuit level pattern. This indirect patterning is required because either the layer is not directly patternable technologically, or it cannot be accomplished economically. Resist-image transfer layers, as their name implies, "resist" individual layer processing steps to enable electronic devices to be fabricated vertically layer by layer on a thin silicon crystal wafer.(1) For devices with multilayer metal (MLM) interconnects, these individual layers are insulating dielectrics or metallic interconnect layers. The ability to produce high quality MLM layers depends upon both the lithographic tool imaging capability, quantified by the Modulation Transfer Function (MTF) (2) of the tool optics which in part depends upon the tool objective lens numerical aperture (NA), and the resist process quality or CMTF.(3,4) The electrical yield for MLM device layers depends upon these quantities for device element packing density and the ability of the lithographic tool to align layer to layer; the combination represents the total overlay capability of the system which must be better than the design rules for the circuit.

In addition to providing device manufacturability from circuit element definition, (1) manufacturing yield, and circuit density points of view, the lithographic process is also capable of influencing device performance. The resist lithographic resolution and critical dimension (CD) control, for example, can directly influence device reliability and electrical performance. Historically, the resist CD requirements have reduced approximately 20-30% every two years, (5) thus pushing some older lithographic tools to their limits and making them obsolete except for the fabrication of older more mature devices.

The capability of the lithographic process is determined to a large degree by the wavelength of the electromagnetic energy source used to carry out the selective patterning image exposure process. Typically, visible light is used of wavelengths ranging from 300-420 nm for most photo stepper equipment. Most of the stepper and process results for this chapter will be for I-line, which is at a visible spectrum wavelength of 365 nm. The light is imaged on the wafer through chromium metal patterned transparent quartz masks using refractive optics.(5) Electromagnetic energy sources of wavelengths <300 nm can be provided by deep UV (DUV) producing systems such as high pressure mercury arc lamps or laser systems.(6) Further reductions in wavelength are achieved by employing focused electron beam sources, like those found in scanning electron microscopy (i.e., 10-30 keV), or those found in soft X-ray systems (2-20 Å wavelength). Most importantly, the wavelength in most cases determines what type of resist can be employed, because the energy of the lithographic tool must be coupled to the resist to insure conversion of electromagnetic energy to radiation chemical energy occurs.

Although the lithographic properties of resists can determine circuit density and performance, the resist must first of all be device MLM layer process compatible, or it is of academic importance only. Unfortunately, the literature abounds with resist systems of great lithographic capability, but they cannot be employed in the commercial fabrication of semiconductor devices because they are not capable of withstanding or "resisting" certain required processes. In the tool and process applications sections of this chapter, actual process compatible processes and tools will be discussed. In essence, the perspective is a practical user's view, where actual device fabrication experience exists and the process and manufacturing issues are real.

The preceding paragraphs point out that the resist and alignment tool make up a total lithographic system; both can influence the final result, but since the tool aerial image, (6) or the light contrast across the mask edge, is fixed at the wafer plane by the tool manufacturer, the process engineer is left with only the resist process optimization as a primary variable of influence for circuit pattern density. The ability of the stepper to provide stacked primary layer patterns depends upon the tool alignment precision, which can be layer dependent, how well the tool is maintained, and the quality of the tool's alignment system. These tool performance variables will be discussed further later in the chapter.

2.1 Resist Principles

A manufacturable submicron MLM device process requires a photoresist that integrates well with the other fabrication processes, and it must also have excellent lithographic properties as well, especially for ≤ 0.5 µm advanced CMOS or BiCMOS VLSI device fabrication. The photoresist used in the fabrication of these devices is vital to the quality of the final VLSI product. The photoresist must retain its image feature size when subjected to different temperatures and etching processes. Dry etching process temperatures sometimes exceed the image flow temperatures of some current photoresists, especially in harsh MLM backend processing. Therefore, thermal stability and dry etch compatibility are vital photoresist evaluation parameters for MLM process applications.

In addition, the photoresist must be lithographically capable of producing quality submicron device features, especially submicron via cuts and metal spaces. The contrast, exposure latitude, depth of focus, and linearity of a photoresist must all be adequate to meet the design rule needs of the device being produced. This section provides results of an evaluation of the different lithographic and process compatibility properties of two model 4th generation positive photoresists, resist A and resist B, to demonstrate example photoresist process capabilities for the fabrication of next-generation submicron devices with MLM backends. The results are provided merely as example data for two resists; therefore, no resist recommendation is intended nor implied.

Positive Photoresists: These resist materials (4,5,7,8) are the workhorses of modern integrated circuit (IC) manufacturing technology. All new very large scale IC (VLSIC) fabrication lines employ high resolution positive toned material, while the older lines with more mature products still rely heavily on negative toned resists. Positive toned resists develop away to create recessed relief images in the exposed areas with safely-disposing dilute aqueous base developer solutions. When employed, they can be used at all device levels by simply changing the density of the reticle.

Positive photoresists are composed or formulated from several components: polymeric resins of molecular weight of the order of 1-10K grams/mole, photoactive molecular organic additives (PAC) or non-photoactive dissolution inhibitors, leveling agents (SLA), optional dyes to reduce substrate reflectivity effects, sensitizers, surfactants for developer wetting, and organic spinning solvents. The resin molecular weights are intentionally chosen to be low to insure solubility in the polar basic developers. The photoactive species also acts as a dissolution inhibitor, that is, it prevents development in the unirradiated regions of the film needed to resist (i.e., mask) further processes (see section 3). The leveling agents prevent undulations on the resist surface by plasticizing the resin or by providing a resist solution with lower surface tension to improve wetting at wafer spin.

PAC Influence: Photoactive compounds, or sensitizers, are usually naphthoquinone diazides (i.e., PACs) like those pictured in Figure 1.(9) The diazide (DAQ) moiety of this molecule absorbs in the visible region of the spectrum; but most importantly, it undergoes a photochemically-induced radiation chemical reaction, the photoelimination of the azo nitrogen, that results in a solubility change in the dissolution inhibitor photoproduct (References 4, 5, 7, 8 and references therein). It is this energy conversion process from electromagnetic energy to chemical reaction product which results in the observed resist behavior.

PHOTOACTIVE DIAZOQUINONE (DAC) COMPONENTS (PAC)

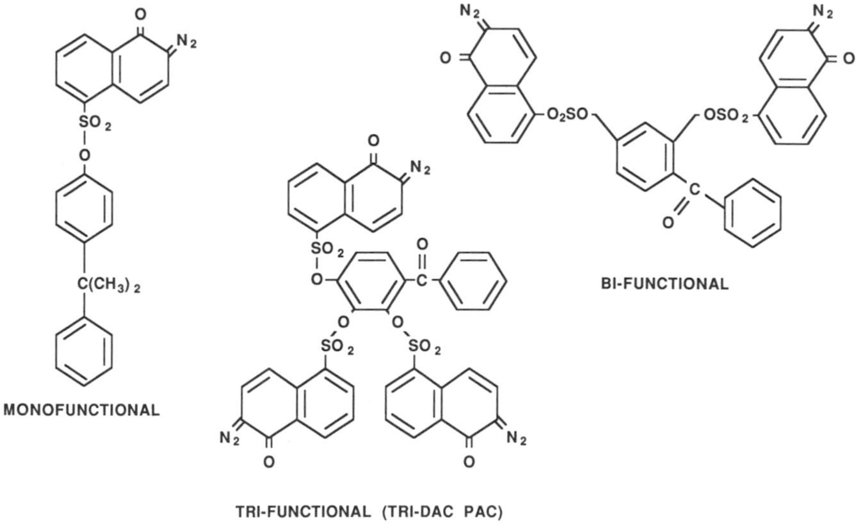

Figure 1.   Structural formulae for photoactive diazoquinone (DAC) components of positive photoresist. Note that q=3 for the tri-functional PAC.

It turns out that this conversion process is fairly efficient as determined by basic quantum efficiency measurements for some PACs. This quantity, defined as the ratio of the number of molecules reacting to photoproduct to the number of photons absorbed, $\emptyset$, can be as large as 10^6. Values larger than 1 are usually associated with a free radical chain reaction mechanism, while most photoresist photochemical reactions have values ranging from a few hundredths to a few tenths. The quantum efficiency for acetone, a model carbonyl-containing compound (i.e., C=O containing PAC) like those of Figure 1, was measured to be 0.17.(9) The quantum efficiencies for the PACs of Figure 1 were determined to all be about 0.3 at the typical optical exposure wavelengths.(9) Actually, these values are quite high when compared to other energy conversion processes, thus, these photoprocesses are very energy efficient, roughly 30%. Even greater efficiency, 50%, has been observed for some resists by other researchers.(10)

In the acetone example above, the light is being absorbed by the specific carbonyl chromophore group, which in turn leads to the chemical reaction. The light, which is merely absorbed in the resin or the substrate and not at the specific chromophore, does not provide contributions to $\emptyset$. In other words, only the bleachable absorption of the resist over the exposure spectrum is important in the lithography (see Figure 2).

AZ 5214
1.45μM COATING, UNEXPOSED & 5 SEC EXPOSURE

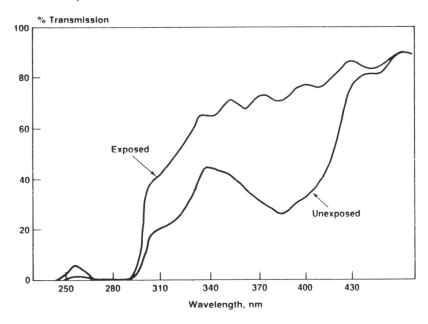

Figure 2. Bleaching curves for AZ 5214 mid-UV resist. The bleached portion of the spectrum is the difference between the exposed and unexposed spectra of the figure. (Courtesy of AZ Photoresists.)

Absorption of light in the resist is given by the Beer-Lambert law: $I/I_o = 10^{-Ecl}$,(11) where E is the molar extinction coefficient, c the chromophore concentration, and l the resist film thickness. Arden, et al.(12) have shown high E can lead to poor resist image edge walls and larger CD variation, and should be judiciously chosen in designing the positive photoresist. E is a linear function of A plus B, the resist absorption parameters to be discussed later in more detail, with both A and B being corrected for concentration.

It is pretty clear that the photochemical quantities of interest to photoresist design are Ø and E. Both quantities can be measured empirically, as outlined in Ref. 11. Resist sensitivity is influenced by Ø, but E is merely a measure of the film absorption and may not represent absorption which leads to useful radiation chemical change in the resist as a result of a photochemical reaction. For example, conventional photoresists have large E at wavelengths less than 300 nm, but are very poor resists at those wavelengths due to the high absorption of the novolac resin alone, regardless of the Ø value of the PAC involved. Obviously, E must not approach 1, or the system is useless at those wavelengths but must have some value intermediate (i.e., 0.3-0.5) so the "skin absorption effect" can be avoided. This ensures the resist image will be cleared to the substrate, and that the resist image edge wall will not be severely degraded (i.e., undercut) from normal due to the high resist absorptivity.(5, 12)

The composition of the PAC can influence both the spectral response and the contrast or resolution of the resist. Daniels and coworkers (13) have also shown the importance of PolyDAQ substitution of the PAC upon resist contrast or effective aerial image(i.e., CMTF improvement) of the total resist system. The resist can be designed to provide image resolution better than the resolution limited tool performance, a result which is becoming more prevalent; that is, photolithography has gone from aligner limited with low contrast resists to resist performance, or contrast limited.

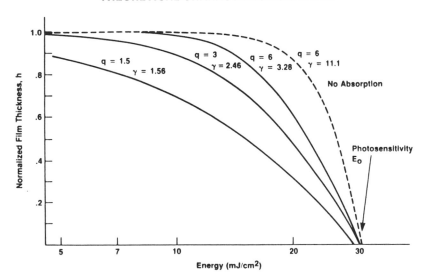

Figure 3. Theoretical characteristic curves for positive photoresist assuming a polyphotolysis mechanism. (Courtesy of Shipley Co. and Reference 13.)

Successful or high contrast positive resist design, requires a very nonlinear response between exposed and unexposed resist. For any degree of polyphotolysis, q, the general dissolution rate can be given by:

[1] $$R = r_o(1 - e^{-Ec})^q$$

where, r_o is the fully photolyzed dissolution rate.(13) The potential influence of polyphotolysis upon resist contrast is demonstrated in Figure 3, where it is seen that as q increases, the resist contrast increases. Of course, these theoretical limits are rarely obtained due to the complexity of the total system; but over the last three years, significant gains in resist contrast have been achieved by several commercial positive resist manufacturers (e.g., Shipley , JSR, OCG, and others), and average q values approaching 4 and above have been achieved.

2.2   Resist Selection Performance Criteria

Resist Parameter Screening:
Sensitivity and Contrast: A resist is characterized functionally by six basic parameters. The first three are related to the exposure absorption characteristics, the Dill A, B, and C parameters, (14) while the second three are related to the empirical dissolution characteristics of the resist, E1, E2, and E3 in the specific developer system.(15) All six parameters are easily measured empirically, as outlined by Refs. 14 and 15, and can be employed in modeling packages to theoretically calculate image profiles and other important quantities.

A and B are related to E, the resist absorbtivity by AM +B, where M is the local PAC concentration. A and C are related, and are really determined by the photoactive component quantum efficiency, $\varnothing$, (16) and track each other in value up and down as well as determine the resist sensitivity. A is also increased by resist photoactive component(PAC) loading, (16) usually 15-25 % by weight of the resist formulation compositionally. B is determined by the transparency of the PAC ballast molecule, the resist formulation resin transparency, and whether absorbing dyes are present or not in the resist formulation.

Table I contains some example resist parameter data. Older resists are represented in the Table by the OFPR 800 data. They are characterized by lower contrast or resolution, poorer CD linearity (i.e., larger values), and lower A and greater B parameters. Notice that the contrast, linearity, and Depth of Focus (DOF at 0.5 μm CD) values of the table are all well correlated.

Choosing a resist for metal layer 1 and MLM layer applications beyond requires a careful judicious selection, which is based upon the six parameters above. The choice also depends upon what the lithographic requirement is; i.e., the resolution required or contrast, the topography of the substrate, the reflectivity of the substrate or wafer, and of course, the the processing cost. Older IC fab lines utilize less expensive resists with typically lower contrast or lower PAC q values, smaller A and larger B parameters, while new fab lines requiring higher metal or via layer resolution will employ resists with greater values of contrast and q, and lower B and greater A parameters. Resists with nonlinear resist dissolution characteristics as demonstrated in Refs. 13 and 17 are required for advanced high resolution fabrication facilities. Nonlinear characteristics are primarily driven by the esterification degree of the PAC (i.e. q values>1) (see Figure 1). The optimum Dill A value has been established for newer advanced resists at between 0.6-0.8 as demonstrated in Reference 17. B values for newer undyed resists range between 0.02 and 0.06 μm^{-1}, while C values range typically between 0.01 and 0.03 cm^2/mJ. Via resists suitable for advanced applications usually have a low B value, while for older devices with >1.5 μm dimensional resist requirements lower A and greater B values are more typical. Resist contrast is inversely related to the sum of the A and B parameters, (18) both corrected for local concentration; and as we will see in later sections, there will be dyed resist systems with higher B values (see Table I) that are still not capable of suppressing substrate reflectivity effects upon linewidth CD control, and more complex processing must be employed (see multi-layer process section).

Table I: Resist Performance Parameters

Resist System	A μm^{-1}	B μm^{-1}	Contrast	Linearity μm	CD DOF @ 0.5 μm μm	Swing Curve Ampl'de mJ/cm^2	Swing Curve Slope mJ/cm^2/kÅ
OFPR 800 (Ito)	0.50	0.11	~1.3	0.6μ	NM	NM	NM
System 9	0.44	<0.06	3.4	~0.5	≈0.5	60	16.2
Shipley 500	0.85	0.050	3.8	0.4	2.0	31.	6.4
Shipley 518L†	≥0.85†	>0.05†	<SPR500†	NM	NM	25	7-9
JSR 500	0.75	0.069	7.0	0.40	2.0	40	8-9
OCG 897i	0.73	0.07	4.3	0.45	2.0	38	9
PFI-D11A (undyed)*	0.92	0.50	NM	NM	1.20	NM	NM
MC-231A*	0.76	0.22	NM	NM	1.65	NM	NM
MC-232A*	0.76	0.34	NM	NM	1.50	NM	NM
MC-233A*	0.76	0.46	NM	NM	1.50	NM	NM
MC-234A*	0.76	0.58	NM	NM	1.35	NM	NM

NM - No measurment and/or equipment was not a wafer stepper.

† Dyed Shipley 500, B is > B for SPR 500.

* Sumitomo dyed resist data courtesy of M. Hanabata.

Process Swing Curve Evaluations for Operating Points due to Reflective Interference and Bulk Effects: The basic data needed to establish resist thickness requirements for example lithographic testing are found in Figures 4-7.(19) This data must be generated because of the need to know how the resist system responds to reflective interference between light reflected from the test substrate and forward light within the film.

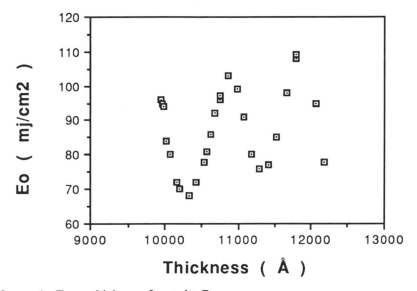

Figure 4. Eo vs thickness for resist B.

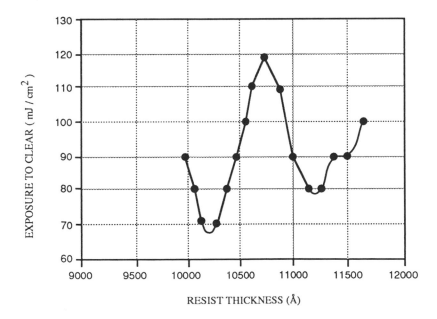

Figure 5. Eo vs resist thickness for resist A.

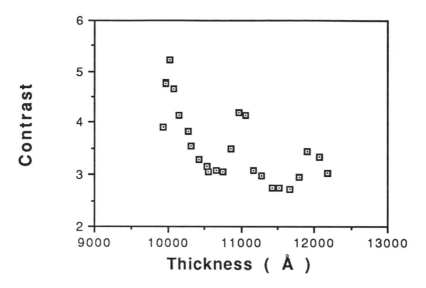

Figure 6. Contrast vs resist thickness for resist B.

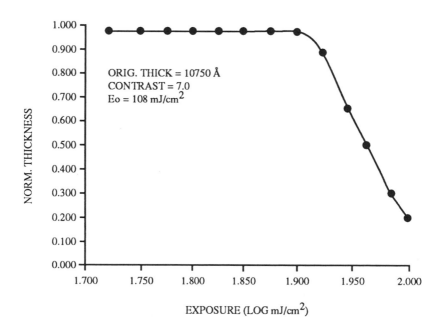

ORIG. THICK = 10750 Å
CONTRAST = 7.0
Eo = 108 mJ/cm^2

EXPOSURE (LOG mJ/cm^2)

Figure 7. Normalized open frame exposed image thickness vs log exposure for resist A. The contrast value was obtained by a least squares fit of the values and the thickness of resist was lithographically equivalent to that at the second maximum from Figure 6.

To generate the data of Figures 4-7, resist dispense spin speeds are varied from 4500 rpm to 6400 rpm at 100 rpm increments on silicon test wafers (i.e., no thin films present). The preexposure resist thicknesses are measured at each spin speed on a Prometrix FT-600. Each wafer with a different resist thickness is then exposed on a Canon 2000i wafer stepper with a 10X10 exposure matrix starting at an exposure dose of 50 mJ/cm^2 with an increment of 2 mJ/cm^2. This enables the determination of the exposure to clear value(or the first exposed square where the resist is completely developed), Eo, to be determined for each thickness after development. Actual high resolution MLM images are usually generated at exposure values $\geq$2Eo for reference.

Figure 4 shows the graph of Eo vs. resist thickness. Incidentally, resist image critical dimension, or CD, plots vs. exposure, have the same exact cyclical shape. The 10.9 kÅ thickness on the graph was chosen as an optimum example operating point for the resist A evaluation, because it was at a swing curve maximum. This choice was made due to a number of considerations; first, resist image tops are more square at the maximum, the resistance to under developed images (i.e., scumming image spaces) created by thickness variation is greatest at a Eo max, process contrast values are largest for operation at swing curve maxima (Figure 4); plus, at least a micron of resist is usually needed to provide etch masking. Processes requiring thicker resists for masking purposes dictate operating at maxima at greater thickness values along the horizontal axis of the swing curve. Resist A curves are found in Figures 5 and 7. Notice the operating maxima for both example resists are very similar indicating similar resist refractive indices.

To establish the contrast data for resist B of Figure 6, dispense spin speeds and exposures were varied as for the Eo studies above. The post exposure and develop resist image thickness was measured for each open frame exposure field of the exposure array for each resist thickness. From this, the contrast for each resist thickness was calculated from a least squares fitting algorithm. In most cases, the phase of Eo and contrast are the same; (19) therefore, contrast is usually also at a max when Eo (or image dimension) are at a max. This provides better image dimensional control, another reason to operate at a resist thickness max.(19)

The swing curve of Figure 4 has three peaks (~ every 1.0 kÅ) with operating thickness potential. The peak at 10.9 kÅ is optimum because of the reasons above and because the photoresist thickness must be at a value higher than 1.0 μm due to etching process constraints; metal thicknesses are typically $\geq$0.7 μm at metal levels 1-3. In addition, the larger thickness maxima operating points have significantly lower contrast values or poorer resist image dimension control capability, which also affects device layer to layer total overlay; therefore, they must be avoided unless RIE needs dictate otherwise. The swing curve for resist A (Figure 5) shows a similiar result (10.75 kÅ) for operating thickness. A comparison of the max to min ranges of both resist swing graphs shows that for resist A, thickness has a slightly greater effect, or that resist A is slightly more transparent than resist B. Comparing the slopes of the swing curves, generated by either the min to min or max to max connecting lines, provides a measure of the resistance to image dimensional changes over device topography (i.e., the bulk effect); the data shows that resist B is slightly more tolerant to topography. The differences observed here for the latter parameters are, however, fairly small and may even be close to measurement uncertainty limits.

Figure 6 is the contrast resist thickness response curve obtained for resist B photoresist. At the optimum operating thickness (10.9 kÅ), resist B has a contrast of 4.3. The data of Figure 7 establishes the contrast, at the operating thickness, of resist A to be 7.0, a significantly higher value than that for resist B, thus, resist A is the higher contrast/resolution resist of the two.

Depth of Focus (DOF) and Exposure Latitude: To obtain resist DOF data, resist B and resist A coated wafers were exposed with a test reticle on a Canon 2000i (0.52 NA) stepper using a 13 X 13 matrix array. The test reticle had CDs ranging from 0.3 µm to 2.0 µm. Machine focus was varied by column with values ranging from -1.8 to 1.8 µm with a delta focus of 0.3 µm per column. The exposure dose was varied by row with values ranging from 110 to 230 mJ/cm^2 with a delta exposue of 10 mJ/cm^2 per row. The 0.5 µm CD was measured in each die using a Hitachi SEM after development. From this type of data, the best focus and exposure dose to CD size can be determined for each photoresist.

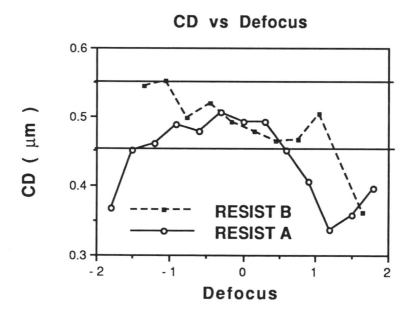

Figure 8. Resist image CD vs defocus for resist B and resist A resist systems under equivalent exposure conditions.

Figure 8 contains the resist image critical dimension vs. defocus data for each photoresist at the same best relative exposure, or the isofocal explosure condition (190 mJ/cm^2 for resist B and 210 mJ/cm^2 for resist A). The depth of focus values observed were both approximately 2.1 µm for a line CD specification of 0.5 µm +/- 10%. Both photoresists have good depth of focus which ensures CDs are printed within the given specification, even when process variations occur (i.e., changes in wafer flatness, topography, etc.). It is not atypical for example, for local total indicated range (LTIR) flatness values (20) for individual wafer die at the wafer edges to be of the order of 0.5-2 microns. Thus, processes with large DOF are very important. DOF is designed into a resist by increasing the PAC q value through increased PAC esterification and by resin molecular weight tailoring.(21)

CD Linearity: Figure 9 shows the actual Hitachi SEM image size vs. nominal size for the resist B. Figures 9 and 10 allow the resist example systems to be compared under equivalent processing conditions for CD linearity. Linearity is a fundamental parameter because it measures the ability of the resist to delineate the smallest feature possible within the standard ±10 % CD criterion. The test reticle has all line sizes in the unbiased condition. It is important that there be no bias when printing different sized features simultaneously. Resist A can print from 2.0 µm down to 0.4 µm CD sizes with no bias, while resist B can print only down to 0.45 µm CD sizes without bias. This observation is consistent with the higher observed resist contrast performance for the resist A system, since it is well known that high contrast resists also have greater CD linearity. CD linearity is basically determined by the degree of PAC esterification or contrast and the effect of that upon the resist development nonlinerity. For advanced designs such as 64M DRAM and beyond, linearity less than 0.35 µm will be required.

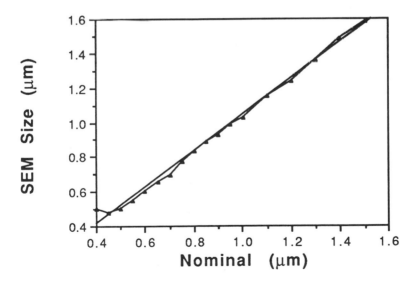

Figure 9. Linearity curve for resist B.

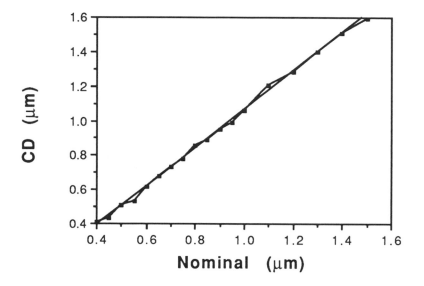

Figure 10. Linearity curve for Resist A.

Resist Image Edgewall: This parameter defines the criteria for the depth of focus determination given above. The useable edge wall angle definition varies, but is typically from 80-85 degrees. Figure 11 contains some typical resist image SEM cross sections illustrating >85 degree edge walls over a fairly large range of focus for Shipley SPR 500 resist. The resist image edge wall angle also plays an important role in the RIE etch process bias; i.e., the more sloped the resist image the greater the RIE process bias, even under anisotropic etch conditions. When large values of the B parameter or both A and B parameters are large, resist image edge walls degrade to values less than 80 degrees, and resolution suffers as well.

Resist CD Latitude: Figure 12 shows the CD vs. Exposure Dose data for each photoresist at the best relative focus offset condition (-0.45 μm for resist B and -0.4 for resist A). The conditions for this comparison represent equivalence, i.e., both offset values represent a CD data set centered on the offset value. The exposure latitudes were both approximately 18% over a CD specification of 0.5 μm +/- 10%. These values indicate that both example resists are capable of printing CD's within specification when process variations (e.g., resist thickness, lamp illumination problems, substrate reflectivity, etc.) occur within +/-18%.

SPR 500/CANON 0.5μ IMAGES
THROUGH DEFOCUS

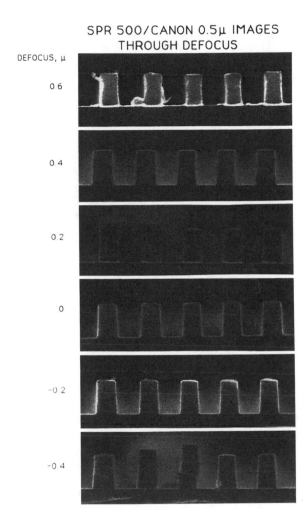

Figure 11. Resist image edgewalls for SPR 500 features printed on a Canon 2000i stepper through defocus.

Process Compatability Evaluations:

Etch Resistance: It is an obvious advantage for a photoresist to have a high level of device process reactive-ion etch (RIE) selectivity or etch resistance. This means the photoresist is resistant to etching during the etching processes (see Section 3.0). If the resist erodes or flows during the etch process, the device cannot be made with reproducible overlay or high yield.

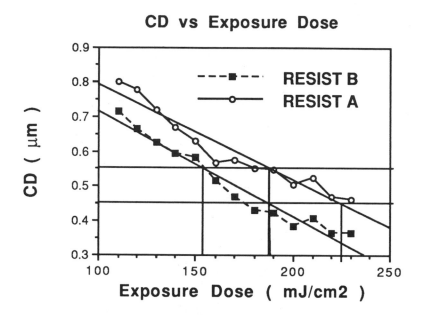

Figure 12. Exposure latitutde curves for resist B and resist A.

Resist B and A example photoresist coated wafers were measured for thickness on a Prometrix SM-200, both with and without Fusion Deep Ultraviolet Radiation (DUV) image stabilization. An oxide test wafer was etched with the resist B and resist A coated wafers in an Applied Materials (AME) 8110 RIE to determine selectivity. Here, selectivity is simply the ratio between the oxide film etch rate and the resist etch rate. An Al test wafer was also etched with one wafer of each photoresist to determine the Al RIE selectivity values. The Al and oxide(TEOS plasma enhanced deposited glass) thickness steps were measured on an Alpha-Step profilometer. The thicknesses of the photoresist coated wafers were measured on a Prometrix SM-200 to determine the etch induced resist loss during the etch process.

Table II contains the selectivity results of each resist for the TEOS interlevel dielectric and metal etch processes, which can be employed to fabricate BiCMOS MLM devices. The selectivity values for resist A (4.30 and 2.00) are marginally higher than the values for resist B (3.95 and 1.86), meaning the resist A is slightly more resistant to the metal and ILD etch chemistries than resist B.

Thermal Image Flowing: The thermal stabilitiy of the images for a photoresist process is an important factor because some backend metallization etch processes are very long and significantly high temperatures can be achieved, especially under high overetch conditions which are prevalent. DUV curing of the photoresist image after development stabilizes the pattern to this type of image flow through both radiation and thermal crosslinking, but unfortunately some production areas do not have this capability or cannot afford to provide this extra process. As a result, this type of fundamental testing is important.

Table II: Resist etch selectivities vs AlCu metal and via TEOS glass materials

Resist Systems	Selectivity(material/resist)	
	TEOS	Metal
Resist A	4.3	2.0
Resist B	3.9	1.9
Resist C	6.0	2.8
Resist D	5.3	2.2

The standard vendor processes were followed when preparing wafers for thermal image flow testing. To determine the effects of DUV, four control wafers for each photoresist did not have the DUV treatment. One wafer (DUV) from each photoresist was baked on a hot plate for 180 seconds at 110°C. Two wafers (DUV and No DUV) from each photoresist were baked on a hot plate for 180 seconds at 120°C, 130°C, 140°C, and 150°C.(22) The wafers were then SEM cross-sectioned for 0.5 μm line and 10.0 μm pad CDs. Both sizes were required because larger images always flow at lower temperatures, (22) and because at least two image size domains should be evaluated for flow.

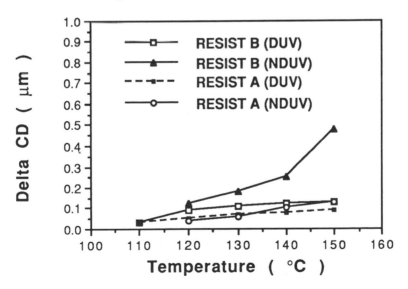

Figure 13. Delta CD vs image flow temperature for 0.5 μm CDs of resist B and A resists.

The results of this work shown in Figure 13 demonstrate that the resist B (No DUV) photoresist images begin to flow starting at a hotplate bake temperature of 120°C. Resist A, on the other hand, is a much more thermally stable system (without DUV) and exhibits good image stability at hotplate temperatures up to 150°C. Of course, resist images of both resists exhibited high temperature stability after DUV curing.

2.3 General MLM Resist Processing

Resist Adhesion Requirements: Resist image adhesion processes for most MLM applications are standard vapor phase hexamethyl disilazane (HMDS) processes, preferably at reduced pressures in stand alone reaction chambers or high throughput wafer track system modules.(23) Since most dielectric layers exposed to lithography on the backend are oxide substrates which are handled well by standard HMDS processing, very few problems are encountered with this processing. Wafer rework also is fairly well understood and routine.(23)

Metal image adhesion, on the other hand, can be a problem, especially when rework is encountered or when wet etching is practiced as for older fab lines. Freshly prepared Al, AlCu, and W wafers, however, do not usually present an adhesion problem for photoresist images. In fact, test image failures or lifting are usually impossible to induce. Furthermore, no evidence of a chemical reaction between the HMDS and the Al films has been obtained from ESCA studies, so this treatment is probably ineffective on metal substrates except for the concurrent wafer dehydration which also occurs in the process. Problems arise when Al metal is ashed or contaminated in any other way. Then, the only way to save the wafers is to sputter or etch the film surfaces to provide a near virgin type surface, hopefully with minimal metal thickness loss and further contamination.

Resist Application: This process is well documented (24, 25) and is very routine except where substrate topography approaches $\geq 2X$ thickness of the resist patterning layer. Methods to overcome these rare coating problems, when they occur, are found in Ref. 26. With standard via and metal MLM processing, resist applications are very routine, and thickness variation can be controlled to ≤ 100 Å, 3 sigma all families of variation, assuming that good wafer planarity is maintained.

Most resist coatings are very conformal because the viscosity of the film is rising quickly during coating, thus freezing in the topography to be planarizied due to local surface tension gradients.(27) Achieving greater planarization requires increasing the solvent fraction or increasing the residual solvent in the film, (27) which does not seem very practical currently.

Prebake/Exposure/Development Processing: Prebake temperatures and exposure levels are determined through careful process optimization responce surface experiments (19) carried out usually by the resist manufacturer and from Bossung curve generation (Ref. 2, and see CD vs. exposure data for metal and via layers in section 2.7), respectively. If the user must determine these parameters, then Refs. 2 and 28 describe the methodologies. The ARC process optimization of Section 2.4 is a very representive model process optimization methodology sequence for any type of photo process.

Currently, nearly all high end processes; i.e., positive resist processes at 1 μm and below, are run with 0.26 N TMAH (2.38 % tetramethyl ammonium hydroxide water solution) developer without surfactant wetting agents. Most resists used at this level of processing contain surfactants, therefore, resist wetting at the develop step with the surfactant-free developer is not a problem. Some manufacturers do use surfactant containing developers, supposedly to prevent puddle development induced undeveloped spots defects, but these defects can be prevented through developer prewetting, developer arm program optimization, and exhaust level adjustments.

Resist Postbake: Most positive resist processes in MLM processing use today require a post expose and predevelopment bake (PEB) process to improve process contrast and to prevent standing wave interference effects upon the resist image side walls.(19) Adhesion may also be improved, but the most improvement comes in the form of improved resist image CD control, which occurs due to the contrast improvement of the PEB process.

SPC Methods of Process Control: Standard SPC methods as outlined in Ref. 2 are necessary for MLM lithography or for any type of lithography, for that matter. There are many statistical tools available to control a process so the product is made with high quality. Quality is the conformance to target performance levels and statistical process control uses methods to reduce process variations so continuous improvement is possible.

Multivariate Studies: Multivariate studies determine the process variation without manipulation of any factors. Data are collected so that several families of variation can be analyzed to determine which family contributes the most variation to the process. The effect of each of these families on the total variation can be determined. One example of a multivariate study is to collect data for critical dimension control using product material. The data can be collected on two wafers out of each lot processed, measuring five sites on a wafer (top, center, bottom, left, and right). The families of variation are then site-to-site, wafer-to-wafer, and lot-to-lot. When we speak of all families of variation, we mean the rms of all the families or the total overall variation.

Nested Variance. The total process variability can be partitioned into major sources of variability using the concept of nested variances:

[2] $$s_{total}^2 = s_l^2 + s_w^2 + s_s^2$$

In this example, the data being collected are critical dimension measurements, x_{ijk}, for the ith lot, the jth wafer, and the kth site. The grand average is:

[3] $$\beta = \sum_{i=1}^{L} \sum_{j=1}^{W} \sum_{k=1}^{S} x_{ijk} / (SWL).$$

where L is the number of lots inspected, W is the number of wafers inspected in the lot and S is the number of inspected sites per wafer. The average of lot i (for i=1,...,L) is:

$$[4] \qquad \beta_i = \sum_{j=1}^{W} \sum_{k=1}^{S} x_{ijk} / (SW).$$

The average of wafer j within lot i (for j=1,...,W and i=1,...,L) is:

$$[5] \qquad \beta_{ij} = \sum_{k=1}^{S} x_{ijk} / s.$$

The total variation of lot means is:

$$[6] \qquad s_l^2 = \sum_{i=1}^{L} (\beta_i - \beta)^2 / (L-1).$$

The pooled variation of wafer means within the lots is:

$$[7] \qquad s_w^2 = \sum_{i=1}^{L} \sum_{j=1}^{W} (\beta_{ij} - \beta i)^2 / (L(W-1)).$$

The pooled within-wafer variability is:

$$[8] \qquad s_s^2 = \sum_{i=1}^{L} \sum_{j=1}^{W} \sum_{k=1}^{S} (\beta_{ijk} - \beta_{ij})^2 / (SL(W-1)).$$

The variance components for each family of variation are calculated from their estimates after accounting for nested variation:

$$[9] \qquad s_S^2 = S_S^2,$$

$$[10] \qquad s_S^2 = S_W^2 - S_S^2 / S,$$

[11] $s_l^2 = S_l^2 - S_w^2 / W.$

Control Charts: There are two principal types of control charts. The first is a variables control chart, for monitoring outputs such as ß, R, and s (for the mean, range, and standard deviation estimate, respectively).

One type of variables control chart for normally distributed outputs has three zones, a green zone, a yellow zone, and a red zone.(2) About the target center, the green zone is ±1.5 sigma. The yellow zone is outside the green zone, about the target center, to ±3 sigma. Outside of the yellow zone is the red zone. The rules for using this control chart are to make no adjustments if the outputs are plotted randomly in the green zone. Adjustments are made to the process if six consecutive points are in the green zone on the same side of the centerline, two consecutive points are in the same yellow zone, or a single point is ever in the red zone.

C_p and C_{pk}: Two process capability indices are C_p and C_{pk}. These indices are used to judge the capability of the process to meet specifications after the process is under statistical control. Standard deviation and mean values are used in the calculations and it is assumed the distributions are normal.

C_p describes the process variation with respect to the existing specification window.

[12] $$C_p = \frac{(USL - LSL)}{6\sigma}$$

where USL and LSL are the upper and lower specification limits, respectively, and σ is the total process variation.

C_{pk} describes how well the process is centered relative to the specification window.

[13] $$C_{pk} = \min(USL - \beta, \beta - LSL) / 3\sigma$$

where ß is the mean value of the process distribution.

Ideally, C_{pk} should be as large as C_p. A typical minimum standard of manufacturability is $C_{pk} \geq 1.33$, but a C_{pk} goal of 2.0 must be achieved for product production through continuous improvement methods. If CP_k and CP are not equal, then this measured data is not centered at the process target value. Process capability indices are valuable tools to ensure designs are manufacturable.

MLM Resist Process SPC: For resist application, the thickness control is dictated by the swing curve as shown in section 2.2. From the Figure 4 example of that section, if resist thickness varies over ±500 Å the full range of fluctuation in CD is possible. To minimize this contributor to the CD control budget, the resist thickness is usually controlled to ±50-100Å as measured by a Prometrix thin film analyzer. The Cp and Cpk values typical for a developmental device fabrication pilot line are shown in Figure 14. Notice the specification is ±100Å and that for advanced pilot lines Cp and Cpk values ranging between 1.0 and 1.5 are very typical. The one process shown in Figure 14 with a value below those values is for an older resist process where lower values are acceptable; i.e., that layer is not a critical layer and doesn't control the process variability for the actual primary process involved. The other data, however, is for resist layers used at device critical layers; therefore, tight thickness control is appropriate for them and is achieved.

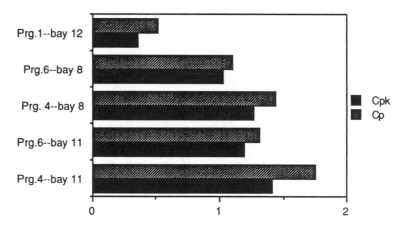

Figure 14. Resist Cp and Cpk values for several example pilot line coat processes.

In addition to the daily resist thickness qualifications above, the resist application wafer tracks and all of the possible combinations of equipment are also qualified daily for funtional Eo as defined in Section 2.2. This additional Eo functional test provides a redundant check for thickness, plus, it provides a functional photospeed check, which is not guaranteed just by a resist thickness check alone. The Eo matrix for all lithographic tools and their combinations is a half fractional factorial design, so that if any tool is out of specification the "out of spec condition" is detected and corrected in real time before the day's production. Example data is found in Figure 15. Examples of out of control conditions detected by this system are resist out of date for shelf life, thickness drifts out of spec, batch to batch resist photospeed or sensitivity shifts, developer CO_2 absorption and strength changes, and errors ocurring in developer tank fills with the wrong developer or the wrong resist has been placed on the wrong track dispense pump.

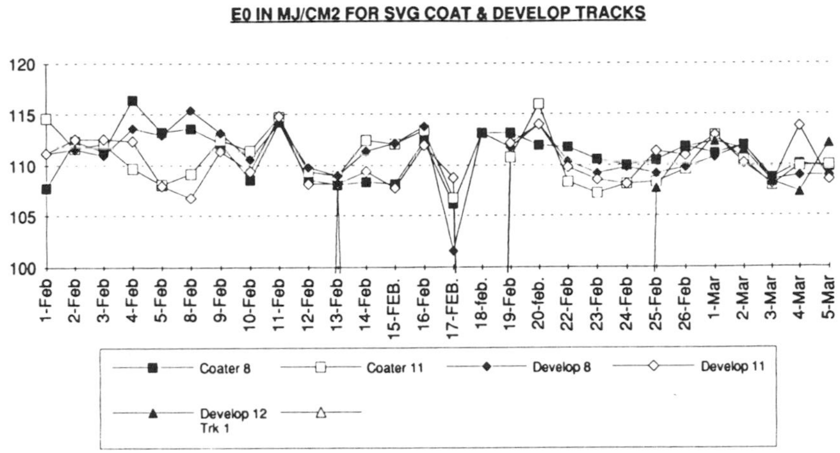

E0 IN MJ/CM2 FOR SVG COAT & DEVELOP TRACKS

Figure 15. Daily Photolithography Matrix results for SVG coat and develop track examples.

2.4 Resist Processes For Reflective and Topographical MLM Situations

Dyed and Thinned Single Layer Resist (SLR) Processes: The two most obvious things to do to improve resist imaging performance on the most difficult substrates (i.e., those requiring greatest resolution and with reflective topography as for MLM metal and via levels), is to reduce the resist thickness and add dyes to it, respectively. Unfortunately, resist thinning is most feasible with multilevel lithography processes, (29) where a thin top resist layer is allowable. Resist thinning in itself accomplishes nothing towards reflective image notching relief, the main observed problem for reflective topographical MLM situations. Furthermore, resist thinning presents a severe problem to step coverage and metal etching because of poor selectivity, and is in fact usually prohibitive. All of these negatives aside, resist thinning has been shown by IBM researchers (30) to improve linewidth control by 15% and focus control by 35%, when and if it is feasible to do it. Therefore, the advantages of resist thinning are really only achievable through bi and tri-layer lithographic processes, further reinforcing the restrictive applicability of resist thinning.

The more practical solution to reflective notching problems, those observed on reflective surfaces due to either feature topographical or metal grain boundary relief structures, is provided by resist dyeing. Most device fabrication areas will select this option over multi-layer Portable Conformable Mask, PCM, (29) processes due to greater simplicity and lower costs. The resulting dyed and usually thicker (~2 microns) material is still a single level resist process, but without the added complexity of multi-layer processes. Dyeing resists requires a price be paid in lower resist contrast (31) and greater exposure time, (32) but dyeing does provide greater process latitude (25) and reflective notching can be effectively eliminated (33) or at least minimized. Sandia workers (34) have provided a dyed system for H and G line steppers which has a very small exposure penalty, just 15 mJ/cm^2. Bolson et al. (33) have demonstrated similar results, and an approximately 60% gain in exposure latitude was achieved or a reduction in K from the Raleigh resolution equation from 1.1 to 0.6 was effectively obtained. On the negative side, adding dye to the resist formulation may lead to a larger standing wave foot, (35) lower depth of focus (Table I), and reduced image edge wall angles consistent with the observed bulk contrast reductions reported by Pampalone. (31) Most dramatically, Brown and Arnold (35) have observed a 3-fold increase in CD exposure latitude, a result which explains the wide acceptance of this technique for metal layer lithography by older mature production fab lines, lines where metal CDs are larger and advanced planarization techniques beyond POT processing are not common (see planarization section).

Multilayer Resist Processing-An Optimized Organic ARC Process for I-Line Lithographic Fabrication of 0.5 µ Devices: When dyed resists, which have been widely used to prevent notching reflections in older product line fabs at >1.5 µm metal linewidths, fail to provide the necessary reflective notching relief, multilayer processes must be employed.

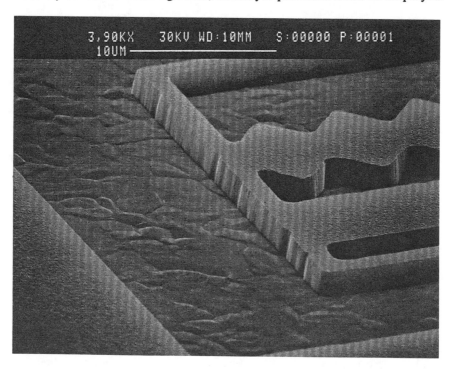

Figure 16. SEM micrograph of notched resist images on large grained aluminum film.

Dyed resist useage is limited by reduced contrast, resolution loss due to greater film thicknesses, and poorer general effectivity.(35) An example of the failure of dyed resist processing to prevent metal grain boundary reflective notching is found in Figure 16; the severe reflective notches are easily seen along the resist pattern edge. The notching effect caused by the substrate reflectivity of this example can be reduced by employing inorganic or organic antireflective coatings (ARCs) in a bilayer structure, i.e., the ARC layer is below the resist patterning layer on the wafer surface. Bilayer ARCs have been widely used in optical lithography to prevent or minimize problems of linewidth variation caused by substrate topographical reflective notching, i.e., for resist metal line images running over vertically raised steps on the substrate. (35-42) Antireflective coatings also help reduce standing wave and thin film interference or swing curve effects, (40) a result similar to the effect of increasing the resist B parameter by resist dyeing, and IBM has futher shown they can also be top ARC coatings as well.(43)

Inorganic layers, sputtered thin films, such as a-Si, V, TiW (35,36), TaSi and TiN, (37) and TiON, (38) have been tested and employed as ARC layers. Economic factors, however, favor the use of organic ARCs based upon equipment costs.(35) The organic ARC referred to primarily in this chapter is ARC-XLT, manufactured by Brewer Science.(39) Bilayer ARCs from this manufacturer have been and continue to be widely used, even in volume MOS production lines.

ARC Processing Example: Organic ARC bake conditions need to be carefully optimized and controlled.(35,40) The bake latitude for the ARC process is found to be extremely critical to the lithographic results.(39,40) This section outlines a strategy for ARC process optimization, which led to improved lithographic results, as verified by metal electrical snake structure probe data. Most significantly, through ARC process optimization, including the develop process, it is found that the ARC application resolution range can be extended to much smaller geometries than the 0.9-1.0 μm limits previously reported; (35,36,39,40) these references attributed that limitation to development undercut of the ARC layer, created by the isotropic nature of it. Furthermore, the greater bake process latitude reported in Ref. 39 has been confirmed.

ARC Process Optimization Strategy: The optimization strategy entails the use of the following two response variables which help by simplifying the type of experiments to be performed: a) ARC thickness remaining after development: Experiments that use this variable as a response are very simple to perform, in that they do not require the use of photoresist or an exposure tool. The rate of ARC development is probably the most important factor in determining lithographic performance for a fixed ARC/photoresist system. This response variable is used to explore and obtain a broad process window, while the final conditions were determined by further experimentation within this window using the response mentioned in b, b) Minimum CD remaining without lift off: ARC undercutting has been a limiting factor in determining the upper limit of resolution with regards to lithographic performance. A small circular pillar-like pattern is best for this testing because this structure is very susceptable to undercut lifting. This response variable plays a key role in ARC process optimization. An example of a pillar functional test working specification for 0.5 μm MLM processing would be 0.6

±0.1 µm. Furthermore, this simple functional test provides an SPC ARC control test for routine daily qualifications.

The following sequence of experiments were all integral parts of the optimization strategy:

1) Identification of bake and develop methods: The following variable combinations will be discussed here: a) Convection bake / Immersion develop, b) Contact bake on a hot plate / Single spray puddle develop.

2) Determination of bake uniformity: An acceptable level of ARC thickness uniformity following bake has to be achieved and verified by ellipsometry.

3) Screening design experiment to determine significant variables: This experiment was a 2^{7-4} fractional factorial (44) using ARC thickness after development as a response. Wafers were coated with ARC only. Resist softbake /PEB were taken into account in the process simulation. The development time used was 5 - 10% of the actual develop time for resist coated wafers. Variables used in this design were:

a) Dehydration bake, b) Delay between ARC coat / Dehydration bake, c) Delay between Convection bake / ARC coat, d) Bake temperature, e) Delay between resist coat / ARC bake, f) Resist softbake, and g) Post exposure bake. These variables may vary depending on the type of bake used. Variables such as proximity bake, wafer gap from hot plate may be included as desired.

4) Determination of development rate of ARC in developer: ARC thickness remaining after develop was determined for a series of development times. From the screening experiment, bake temperature and time were determined to be the only significant variables. Hence, a two factor Central Composite inscribed (CCI) design was (45) run using these two factors and ARC thickness, post develop, as the response variable. This experiment led to an empirical result with a broad enough process window for further experimentation to take place. The final ARC thickness at the end of the photoresist/ARC develop process has to be zero. However, the ARC thickness should not approach zero early on in the develop process because this will lead to severe undercutting, hence, the reason for process optimization.

5) Minimum feature size remaining without lift off on metal wafers: The two factor Central Composite Inscribed design above was run using a range of ARC bake temperatures and times. Minimum CD without pattern liftoff was determined. Wafers were checked for possible ARC scumming. ARC bake conditions were selected for lowest possible CD remaining without the presence of ARC scum, which occurs readily at high bake temperatures (>180 C) due to excessive ARC curing.

6) Process Verification using SEM X-Section/electrical Probe: An electrical probe structure in the form of a standard snake pattern was used to verify process conditions. CDs may also be determined electrically using linewidth split cross bridge structures. Cross sections using the SEM reveal final ARC/resist profiles. SEM measurement waveforms may also be used to qualitatively determine profiles for the presence of an ARC foot.

Optimized Process Comparisons: Results of the Screening design experiment for Convection bake/Immersion develop suggested that the only significant variables were bake temperature and bake time from RS1 empirical modeling.

Plots of ARC thickness remaining were obtained for different temperature/time combinations using a 2-factor CCI experiment. The statistical analysis confirmed bake temperature, bake time squared, and the product of bake temperature and bake time to be significant factors for this response.

A follow up 2-factor factorial experiment was conducted using these conditions and using metal snake probe yield as a response. Neither temperature nor time was determined to be statistically significant for probe yield, however, the product of temperature and time was still significant.

The optimum process conditions were determined to be 163 C bake for 41 min. This is a 1.3 μm line/ 0.7 μm space pattern, and ARC undercutting is clearly visible as reported in Refs. 35, 36, 39, and 40. Table III clearly illustrates improvement in CD control with the optimized ARC process compared to the unoptimized process. The ARC process optimized for immersion develop can also be used successfully in conjunction with track development. However, the presence of an ARC foot can be a problem if the same bake conditions are used. The foot problems are most likely related to problems with Convection baking which include 1) temperature variation in the oven as a function of position, 2) heat loss upon opening of oven, and 3) operator errors leading to overbake. The presence of the ARC foot combined with the problems mentioned above made it necessary to switch to track baking of the ARC, which provides better temperature and time control, and it increases wafer process throughput. Plus, wafer track processing is more manufacturable.

Table III: Mean CD and total CD variation for resist processes with and without ARC

Process	Mean CD μm	CD Variation 3-sigma
SPR 500/unoptimized ARC	1.10	0.35
SPR 500/optimized ARC	1.04	0.09
SPR 500(dyed)/optimized ARC	1.08	0.13
SPR 500/no ARC	severe notching	>0.4
SPR 500(dyed)/no ARC	notching	>0.4

By employing optimized wafer track contact baking and development, 0.6 line/space patterns have been successfully achieved as verified by actual etched metal probe results and the SEM cross section of Figure 16. Metal lines from the process ranged between 0.5 and 0.6 μm due to process bias, but these features were far below the limits established by previous studies.(35, 36, 39,

40) Furthermore, the footed features which plagued the convection baking process have been eliminated. Metal snake yields using a 1.0 line/space pattern were determined to be 94.8 %. 90% snake yields using a 1.3 μm line/0.7 μm space were also obtained.

In conclusion, the process optimization of an organic ARC has been accomplished and submicron feature capability demonstrated, as verified by electrical probe and SEM methods (see Figure 17). The resulting optimized ARC example process is currently being employed to manufacture advanced BiCMOS and CMOS devices with submicron features. Note, that the same optimization methodology can be applied to any lithographic process such as the photoresist process, and this is usually done by the resist manufacturer to define a recommended baseline process prior to customer receipt.

Figure 17. SEM cross section micrograph of immersion developed resist/ARC process on AlCu at 0.6 micron space resolution.

2.5 Stepper Performance Criteria

Evaluation   Results   of   Two   State-of-the-Art I-line Lithography Tools: Two state-of-the-art I-line wafer steppers have been critically evaluated for performance at 0.65 μm and 0.5 μm, respectively, using results from standard tool performance metric tests and live BiCMOS product.(46) These example steppers are chosen to provide data for state-of-the-art tools from a US and a Japanese supplier. Their test performance results are not directly comparable because they are not of the same generation of tools, but the results of their testing does provide useful general tool characterization guildelines and also shows machine improvement trends between tool generations. Direct on-wafer production data is also provided. The major performance components investigated were lens parameters, alignment capability and stage precision, and overall production performance on typical critical MLM layers.

Lens Performance. Stepper A and B both have recent generation 5:1 reduction lenses. The stepper A lens has a 0.45 numerical aperture (NA) objective, capable of printing a 21 mm circular diameter field (max. square is 14X14 mm), while the stepper B lens has a 0.52 NA and a 28 mm (max. square is 20X20 mm) diameter field. Assuming a fixed resist CMTF, resolution and depth of focus are functions of numerical aperture such that line width resolution improves with increasing NA, and depth of focus (DOF) decreases with increasing NA for reasonable operating settings. With present generation resist technology, the larger NA lens of stepper B is not significantly affected by lost DOF. Stepper B performed with an overall depth of focus greater than 2.0 µm for 0.50 µm features over the full 20X20 mm square field and to a ± 10 % linewidth control criteria, while the stepper A exhibited 1.2 µm DOF for 0.65 µm features over a 14X14 mm square and the same cutoff criteria. An improved resist system was used to qualify stepper B, however, and simulations using Prolith 2 (image transfer modeling software) (47) predicted that the expected DOF of the stepper A at a 0.65 µm feature size using the same resist as for stepper B would be approx. 2.0 µm also, if empirically measured. The added usable depth of field from the higher NA stepper B is most likely due to either improved optical coherence or its ability to do a best fit focal plane tilt adjustment for the exposure die (~2X difference between machines). Stepper A does not have this function and must rely more upon the relative flatness of the wafer, which generally is about 0.50 µm LTIR within a die for the wafers used for these evaluations. All linewidth measurements taken were for resist on bare silicon on an Hitachi 7000 SEM.

Lens distortion, after removing all correctable aberrations, was measured for both steppers across the area of their respective maximum square fields. The specification for stepper A was 100 nm for the maximum error and 80 nm for stepper B. Stepper B exhibited a maximum X component error of 57 nm and a maximum Y error of 41 nm. Stepper A performed at 81 nm and 96 nm for its respective maximum X and Y components of error, therefore, the stepper B lens has better or lower intrinsic distortion and over a larger exposure field. Of course, larger distortion values than these will lead to poorer overlay results layer to layer if multiple randomly matched tools are employed or if their lenses are not well matched for field curvature.

Stage Performance: The two example steppers in this study accomplish their alignment through a grid model based on a few sample mapped die, and then blind step the wafer through exposure. The necessity for stage stepping precision and accuracy are essential components to accomplishing the overlay required for most modern device structures. The measurement of stage precision is accomplished by stepping and exposing die such that there is an interlocking of alignment verniers in both the X and Y stepping directions, wherever two fields butt against each other. Stage precision is calculated by taking the standard deviation of the X and Y alignment errors where the fields butt together. The stage precision for stepper A was measured to be 90 nm for X and 91 nm for the Y direction, while the stepper B stage performed at 21 nm and 24 nm for X and Y, respectively. In addition, the stepper B stage is clearly more stable with time than that for stepper A, as verified by Figures 18 and 19. The difference in stage precision performance between these tools accounts for a large portion of the tool overlay error disparity to be discussed below.

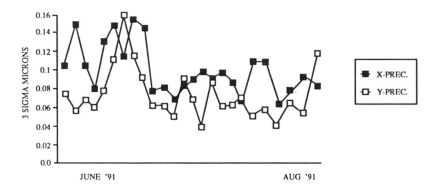

Figure 18. Stage stability data for a stepper A i-line stepper.

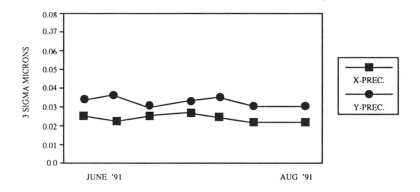

Figure 19. Stage stability data for a stepper B.

GCA Dark Field Alignment Target Requirements: There is a substantial amount of empirical evidence suggesting that MLM dark field alignment results can be influenced heavily by process variation effects upon wafer alignment targets. For this reason, target optimization is almost always essential for most dark field alignment steppers (eg, GCA, Ultratech, and Nikon machines), and when it is combined with advanced target analysis software algorithms, BiCMOS device alignment improvements are realized.

A 5X GCA reduction stepper system has been fitted with a developmental software algorithm, which has the ability to distinguish between "flyer" or outlier alignment signal data and normally distributed data. "Flyer" software utilizes residual analysis from the least squares grid model, as opposed to simply fitting the grid model and exposing the wafer. Prior to the installation of this software, alignment variability on a difficult BiCMOS metal to via masking sequence was as much as 0.45 µm, three sigma, due to the presence of poor alignment target signals in random die on the mapped wafer. After installation of the "Flyer" software and careful empirical alignment target optimization, these outlier signals were discriminately eliminated with resulting observed alignment errors less than the 0.25 µm three sigma machine specification. This improved overlay capability leads directly to reduced rework, product cycle time, and when combined with µDFAS target optimization can lead to increased stepper longevity for one more advanced generation, or design rule shrink of devices.

Dark Field Alignment Background: Dark field (DF) alignment systems, first reported by Wilczynski, (48) are common to several current I-line wafer aligners. They provide layer to layer overlay by their ability to acquire the previous layer local alignment grid, which allows the stepper to accurately step and print the next layer. This grid is detected by illuminating the previous layer alignment targets with either actinic or non-actinic light, capturing the reflected light off of the wafer target topography in a dark field detection cone, and then modeling the laser stage metered positions of these detected local signals.(49)

Unfortunately, the detection quality of the target signals depends upon the resist thickness over the target mark, (50) the thin film reflectivity of the mark structure, the detection system signal/noise ratio, the signal asymmetry, the pitch and size of the target pattern array, the polarity of the target mark (i.e., raised or recessed), and the type of the mark (eg, dashed squares or other geometrical shapes, (50) segmented rectangles, or simple continuous lines). Multiline Pseudorandom alignment mark structures have been shown to improve DF signal intensity over that for conventional solid single line targets.(50, 51) For example, Wanta et al. (51) found a definite mark type effect on process latitude and alignment consistency using a dashed zig vs. standard line marks for a 1.2 µm CMOS trench process printed using the Ultratech dark field aligning system.

Lambson (52) has shown that MLM backend alignment performance is highly mark size and target polarity or phase (i.e., mesa vs. recessed cut marks) dependent for both vias and metal targets. Casteel and Shamma (53) found that optimized DF targets improved registration accuracy dramatically, and that 1) target polarity and sidewall slope, 2) target height, 3) resist transparency, and 4) the amount of resist asymmetric thickness modulation around the targets were all important to DF alignment quality. Finally, Gaboury and Wilson (54) found that target phase is the most statistically significant DF alignment target variable, and after target optimization their rework rate decreased by a factor of 3.5X, hence, the value of DF target optimization cannot be overemphasized.

Wafer Mapping: Most wafer steppers utilize a mapping type alignment method, as opposed to die by die alignment, in order to increase the photolithography throughput of a given production area (2) without sacrificing wafer distortion determination accuracy. Mapping is executed in the same manner as a normal die exposure pass, only without actinic exposure. Mapping produces the wafer target data needed to provide compensation corrections for linear stage grid errors that occur due to mechanical stage drift and the various process induced wafer shape changes. User selected independent variable alignment sites are chosen within the stepper job file, and then these target locations are acquired by the stepper DF alignment system prior to actual device field stepping and exposure. A file of target location error data is created. From the mapping data, six key linear grid errors are fit for each wafer.(55) Once a fit is obtained for the wafer errors, the stepper then steps all the die on the wafer, exposes them utilizing the correctable linear compensation errors from the modeling, but with no regard for the "goodness of fit" of the grid model.

By observing the local alignment signals of several wafers and as displayed on the DFAS monitor of the 5X stepper, one finds that for many processes there are alignment targets that have been damaged or altered in the device fabrication process itself. Example problems are the over/under etching of targets, resist build up over the targets, and contamination on the back of the wafer or on the chuck during the previous level exposure, etc. When mapping includes these types of damaged alignment target signal (s), their contribution to the wafer distortion mapping output can provide enough leverage to the software generated least squares grid model, to cause severe misalignment of the printed new layer. In fact, the old software does not take "goodness of fit" in account at all, hence, large actual die alignment errors can occur if targets have degraded detection wise.

General Target Sizing Results: As for the references cited in the background section, (49-54) careful target optimization is found to be critical to successful MLM device fabrication utilizing DF alignment systems. For the 5X steppers employed, it was very important to find the target type, size at fixed pitch, and polarity for each substrate or the machine MLM alignment

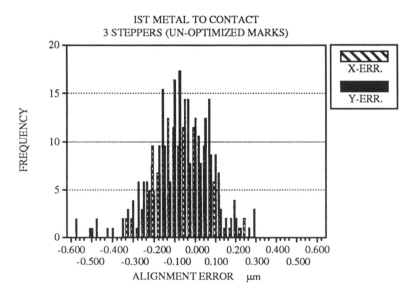

Figure 20. First metal to contact alignment distribution with unoptimized wafer targets at contact.

specifications could not be met on live product. By employing methods very similar to those in reference 54, results even better than specification are achieved on many device layers. Representative data, with and without target optimization, for another difficult metal level are shown in Figures 20 and 21. Clearly, the misalignment distribution has been reduced by nearly a factor of 2X, and the device alignment specification for this critical layer has been met, where before target optimization, it was not. It is very significant to note here that all the data following is also for optimized targets as in Figure 20. For target layers other than metal layers, segmented targets were generally most successful for target GCA MLM optimization experiments. However, as in the references cited earlier, (50-53) target polarity or phase is found to be very important in addition to the target shape. Although target optimization is important to all device layers, the importance of it is most critical to device backend or MLM layers where larger misalignment values are typically observed.

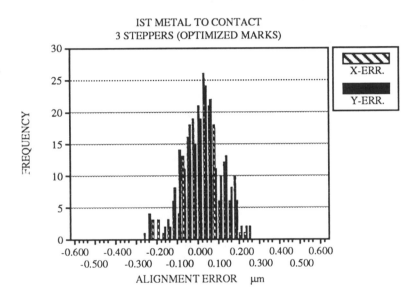

Figure 21. First metal to contact alignment distribution with optimized wafer targets at contact to be compared to Figure 20.

Flyer Software Discription. A beta-site version of an advanced "Flyer" mapping software package has been evaluated and the results are reported in this section.(56) The software was designed to analyze the grid modeling residual errors (i.e., large non-linear grid errors) from wafer mapping, and then compensate or eliminate the process induced errors or "Flyers" obtained for those local alignment targets possessing enough leverage to result in misalignment of the wafer. The purpose of this software is to model the wafer grid errors using all selected data sites, and then based on the amount of residual error, discard data which is extremely non-linear to obtain a new modeling fit with lower residual error and within the distributional design criterion. By reducing the amount of misregistration caused by a poor signal (as opposed to a true misplaced die), alignment accuracy is improved while variability is minimized within the capability of the current tool's configurational capability (i.e., stage and alignment systems). This can translate directly into a cost savings as these steppers can then be used to exceed the manufacturer's alignment specification, thus, increasing their device fabrication longevity.

The Flyer software initially collects the X and Y mapping corrections. The software analyzes the data to determine the standard deviation (1 sigma) and then compares the data to two times that value. If any X or Y mapping corrections from the initial data is >2 sigma, it is discarded from the mapping data set.

The new data set now consists of X and Y corrections that are ≤2 sigma. This data is then modeled in the usual manner. The entire procedure is transparent to the user and is accomplished by the SmartSet computer automatically.

Statistical Analysis Methodology: Classical multivariate analysis of variance (57,58) based on misregistration vector responses was used in the analysis of alignment data. The data was collected by wafer mapping, and then the misregistration measured between the exposed metal and etched via MLM layers.(57) Of course, analysis with and without the GCA "Flyer" software enabled were accomplished to show software effectiveness.

Multivariate/Univariant Analysis of Variance:
- comparison of the alignment vector of means with respect to trial level
- comparison of individual alignment means with respect to trial level

[14]   MANOVA Model: $(dx,dy) = \mu + T_i + e_{ij}$ $i = 1,2,3$
$$j = 1, 2,...,40 = (n_i)$$

Where the following assumptions are made:
(dx,dy) is the vector of alignment errors measured,
μ = overall level mean vector, e_{ij} are independent and Multivariate normally distributed random error terms, and
T_i = the i th treatment effect vector (Trial level) under the constraint $\Sigma_i n_i$ ($n_i \times T_i) = 0$.

[15]   Anova Model: $dx_{ij} = \mu + T_i + e_{ij}$
where, $i = 1,2,3$
and $j = 1,...,40$

μ is an overall level mean (constant component to all observations), T_i is the effect of the ith trial under the summation constraint, and e_{ij} are the random error terms, which are independently and normally distributed.

Experimental Results for the Model Study: Live product wafer testing of the "beta-site Flyer" software was accomplished on a "hot AlCu" second metal layer to via alignment, which traditionally caused the most widespread error in overlay for MLM applications. An IVS optical metrology tool was employed to monitor layer to layer overlay performance.

To create wafers with known large to moderate intentional errors, an experiment was designed such that both stepper and process induced errors were strategically placed at specific die locations on the via layer wafers by a special job file. Four die were dropped in and exposed on a second exposure pass after most of the standard die array had already been printed. The specific errors which were induced by these drop in die are wafer dependent, such that each wafer received only one type of error, but for four different die by means of intentionally misadjusting stepper software and configuration parameters. The errors introduced were 20 ppm wafer scale, 0.50 μm translation in X and Y, 12 ppm grid orthogonality, and a -2.3 μm out of focus condition. These wafers then received a via etch, and then were processed through a second level of metalization prior to 2nd metal alignment. Figure 22 indicates the wafer sites with the errors listed, while the rest of the wafer die were printed with standard exposure and grid parameters. The trials were performed as follows:

Trial 1:   5 second metal level wafers were mapped using only 12 "good" die with optimized alignment mark structures and sizes. The beta-site test software was disabled for this test. (Figure 22)

Trial 2:   The wafers from trial 1 were reworked and then mapped using 16 alignment targets, 4 of which had process induced errors incorporated in order to create poor, but recognizable alignment targets (Figure 23 middle and bottom left hand signal traces). The remaining ten alignment locations were the same as trial 1. Flyer software was not utilized on this trial.

Trial 3:   The wafers from trial 2 were reworked and aligned with the same alignment map as trial 2, but the beta-site flyer software was enabled.

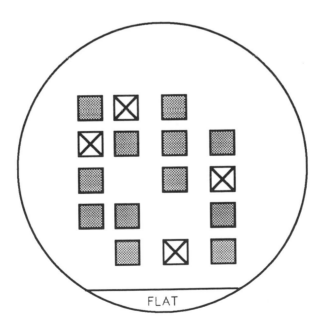

Figure 22. Target mapping locations with drop-in flyer die (X). Trial 2 and 3 use all sites, trial 1 uses only the shaded non-flyer locations.

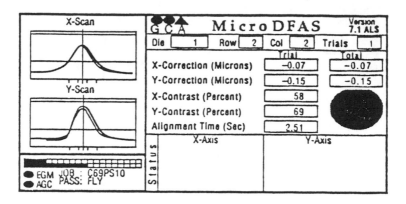

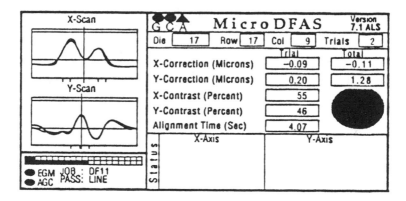

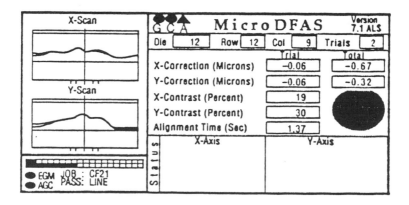

Figure 23. MicroDFAS example traces illustrating a good signal (top) and poor signal traces.

The mapping alignment signal variation induced by the intentional process and software changes outlined in trials 1-3 above, caused target signal degradation as manifested by asymmetric signals and target signatures with large signal to noise variation. This is contrasted to those of "normal die" which are generally symmetrical and singly peaked (see top trace of Figure 23). Some of the signals were affected to such a degree that the alignment system actually read an error of up to 1 μm because the "Flyer" software was disabled. The original μDFAS software will actually use this corrupted information to create a wafer grid file with very large modeling residuals and resulting wafer misalignment (Figure 24).

Wafer Number	Wafer Error	Trial 1[*]		Trial 2[**]		Trial 3[**]	
		Residuals		Residuals		Residuals	
		X	Y	X	Y	X	Y
1	Scale (X/Y) 20PPM	164	210	406	997	133	131
5	Baseline Shift (X/Y) 0.5 μm	116	160	1816	2319	41	56
8	Orthogonality 12 PPM	167	113	453	991	14	79
9	Focus Shift ~2.3 μm	140	166	1066	1575	109	196

[*] Standard mapping pass without any poor alignment signals with the beta site Flyer software disabled.
[**] Mapping process includes the 4 process induced outlier die per wafer.
Beta-site Flyer software is disabled for Trial 2.
Beta-site Flyer software is enabled for Trial 3.

Figure 24. Grid residuals created by the intentional errors for trials 1-3.

Pairwise comparisons were performed on the trials 1-3. Tukey's studentized range test for the X alignment data and the Y alignment data indicated that with a minimum significant difference of 50 nm, the results from trial 2 (flyer software disabled, Flyer die included in map) were significantly different than those of trial one and three for both data axes.(59) The trial 1 results, which signify standard mapping with no Flyer data points or software compensation for residual analysis, were in accord with the trial 3 results. Trial 3 results encompassed poor quality signals in the map, but with subsequent model fitting and refitting based on residual error analysis (Table IV).

The data of Table IV clearly indicates that the beta-site software tested was able to decrease the intentional misalignment contributions from the bad mapping targets by approximately 146 nm in X and 251 nm in the Y direction(average values relative to trial 1 and trial 3). Even more significantly, the alignment variability (3 sigma) of the wafers sampled was also reduced by a factor of 2 (Figures 25, 26 and 27). Obviously, effects of this magnitude can influence device layer overlay and yield significantly.

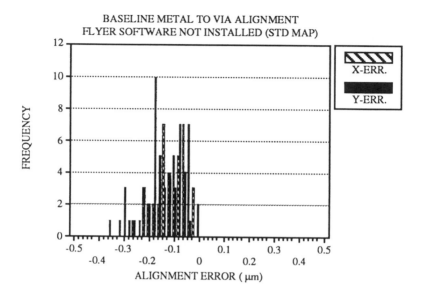

Figure 25. Alignment distributions before "Flyer" software installation.

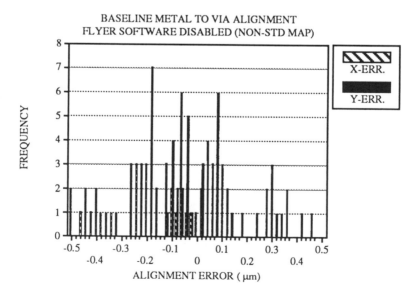

Figure 26. Alignment distributions with "Flyer" software installed, but disabled.

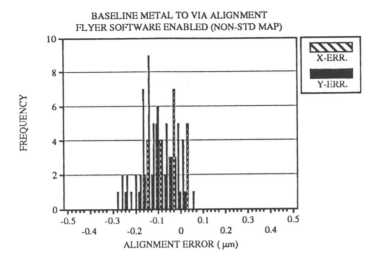

Figure 27. Alignment distributions with "Flyer" software installed and enabled.

Table IV: MLM Flyer Software Analysis Results

	Simultaneous Lower Confidence Limit	Difference Between Means	Simultaneous Upper Confidence Limit
Trial Comparison:			
Dependent Variable = XDATA:			
Trial 1 - Trial 2	0.0728	0.1281	0.1835 ***
Trial 1 - Trial 3	-0.0917	-0.0363	0.0197
Trial 2 - Trial 3	-0.2198	-0.1644	-0.1090***
Dependent Variable = YDATA:			
Trial 1 - Trial 2	-0.3297	-0.2728	-0.2159***
Trial 1 - Trial 3	-0.1001	-0.0432	0.1001
Trial 2 - Trial 3	0.1726	0.2295	0.2864***

Data are in μm.
Alpha = 0.05 Confidence = 95 %
Comparisons Significant at the 50 nm level are noted by *** :

The modeled grid residuals when fit by a standard linear regression model indicate a strong positive correlation with misalignment data from IVS measurements of printed wafers. In Figure 28, X and Y alignment standard deviation data for trials 1,2 and 3 are plotted vs. the X and Y grid modeling residuals (± 3 sigma nm residuals). The high observed R2 values of 0.958 and 0.999 for X and Y alignment, respectively, suggest the higher the residual error of the GCA Smart Set software output, the greater the degree of misalignment and overall overlay variation on product (Figure 28).

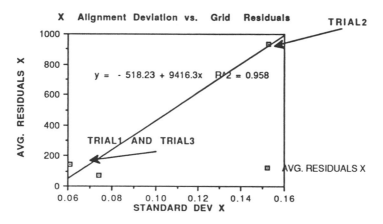

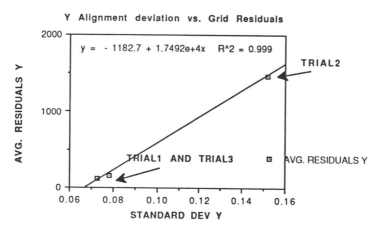

Figure 28. X and Y alignment errors vs grid residuals for trials 1-3.

"Flyer" software for GCA DFAS wafer distortion mapping can be effective in decreasing the effects of poor wafer targets upon die misalignment. The specific beta- site software employed produced superior alignment data for product wafers by safeguarding against the inclusion of mapping data from poor alignment targets. The system performed particularly well for the difficult second metal to via alignment sequence, and probably would do well for the other difficult aligning back-end alignment sequences as well. Furthermore, the beta-site software worked well for recent production lots which had particularly poor alignment signals at the edges of the wafer due to poor etching of the previous layer. The 3-sigma alignment results without flyer software were 0.44 µm for X with observed grid residuals at 600 nm, but the same wafers mapped and aligned using the "Flyer" algorithm were measured at 0.118 µm (3- sigma) in X with grid residuals 209 nm. There does not appear to be any significant difference in alignment performance when the wafer map has no die with poor target signals, with or without the enhanced mapping software enabled. The major role of the mapping software is to compensate for problem layers which often have some outlier data points in the grid model, which causes overall poor alignment performance on production wafers.

Alignment Capability: The MLM alignment capability for both example tools was tested using a technique where a pattern is printed from consecutive layers to provide a pair of interlocking alignment box verniers, which can be read by an optical metrology IVS system. The optical metrology system used in these evaluations measures the relative error of the first to the second layer and has an accuracy of approximately ± 7.5 nm. The metric used to compare overlay performance was the mean plus three times the standard deviation (3-sigma) of each of the X and Y errors measured throughout the respective maximum stepper square die.

Using resist patterns on unprocessed virgin Silicon wafers, stepper B using the standard alignment illumination bandwidth, yielded data with an X error of 62 nm and a y error of 70 nm across the full 20X20 mm square field. Stepper A performed with an X error of 196 nm and a Y error of 150 nm for the 15X15 mm field. The major difference between tool modeling systems is the stepper A system models user selected die locations through a single iteration, open loop, while that for stepper B examines the residuals from the initial iteration and refits the data as required to accomplish a better fit. The refit is very useful on backend material, where the targets of a die or two might have been destroyed in the process resulting in a poorer relative alignment information set. Using this data set (i.e.., no Flyer Software enabled), leveraged with discordant data points, in most cases leads to a wafer model with poorer alignment results than that for the stepper B compensated case.

Using production alignment performance data for the metal 1 and via 1 critical MLM layers of an advanced Bipolar device on stepper A and an advanced BiCMOS device on the stepper B as metrics, a technical comparison of these two steppers on real product wafers is accomplished for a pilot production environment. The standard deviation for each layer's alignment (X and Y) is found in Table V (note, that process variation is included in these values, but also the process variability is approximately equal for both devices and should only cause an equal translation without confounding the metric).

In summary, stepper B is a volume production worthy tool capable of producing device patterns with 0.50 µm device rules, while stepper A is also production worthy, but for a less critical device with 0.65 µm device rules. This is not too surprising because the numerical apertures of the two tools are significantly different, i.e., 0.52 vs. 0.45, respectively. The major difference between these stepper examples is primarily a target acquisition issue as shown above. Stepper A DF targets must be optimized, at the expense of significant engineering time, for every primary device layer in order to achieve the alignment capability listed above for production material, while the stepper B Bright Field stepper target detection system was compatible with all device structure levels and no target optimization was required. Furthermore, the difference between tool alignment results, beyond that accounted for on the basis of stage precision, must be attributed to the superior target acquisition and recognition system of stepper B.

Table V: MLM ALIGNMENT (3-sigma) Data

	Stepper A*		Stepper B**		
	X***	Y	X	Y	Sample Size
METAL 1	243	258	165	132	98 / 147
VIA 1	309	285	186	153	111 / 59

* Optimized DF targets were employed with the dark field alignment system.
** Stepper B employs bright field alignment with improved target mark design.
*** Data in nm units.

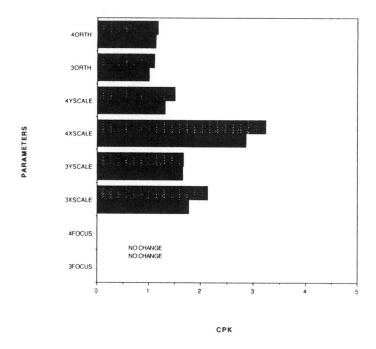

Figure 29. SPC data for intrafield stepper tool parameters.

Stepper SPC Methods of Process Control: The photo equipment of any fab facility must be maintained under SPC methods and controlled as for

resist coating and other applications in section 2.4.(2) Examples of stepper SPC
reporting are included in Figure 29. The example fab here is the same as for the
resist SPC example which runs a half fractional factorial every day to monitor
equipment in every combination to verify machine qualification for that day of
production. A system like this one is also employed by IBM.(60) This system,
called the daily matrix, has detected problems early before costly rework for every
system monitored. Example control charts for this system which depends upon
observed Eo values are located in Figures 30.

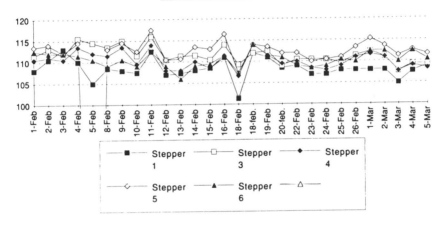

Figure 30. Daily Photolithography Matrix Eo functional test data used to test
illumination levels for a bank of 5X i-line steppers.

Other daily checks for the steppers include simple wafer functional visual
checks for lens contamination and stepper wafer chuck contamination. Figures
31 and 32 show the effects of wafer chuck particle fall-on and lens fall-on
contamination, respectively. Figure 31 is a snake pattern with 2 micron pitch,
while in Figure 32 the open frame of the lens is employed for the entire
exposure print field. If these problems are observed, the equipment is taken
down and the chuck cleaned and vendor service is called in to remove the dirt
from the objective lens, respectively.

Reticle Illumination Effects: How the reticle, containing the device
layer information, is illuminated can have a large effect upon the stepper
patterning performance. The NA of the condensor illumination optics can
determine the resolution of the stepper through the coherence factor, (2) the ratio
of the condensor NA to the objective lens NA. The Depth of Focus for a
particular size feature close to the resolution limit of the tool can also be
improved dramatically by illuminating the reticle with a condensor system
employing non-standard apertures which affect oblique illumination.(61) The
improvement is achieved by restricting the number of diffracted orders employed
to compose the image at the wafer plane, i.e., only one diffracted order and the
zero order ray are needed and are employed. As a result, a smaller area of the
optics are needed and the DOF is improved.

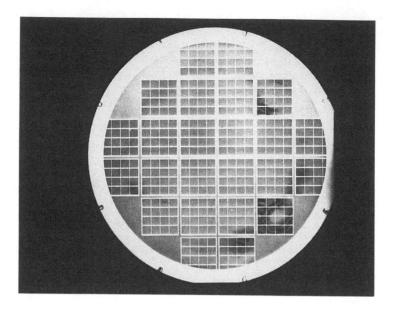

Figure 31. Optical micrograph of a wafer with an out of focus condition for a 2 μm pitch snake pattern resulting from stepper wafer chuck contamination under the out of focus die.

Figure 32. Optical micrograph of exposed and developed open frame exposures showing a defect on the objective lens of the stepper. Note the defect appears in every partially developed frame.

A result identical to the one above can also be achieved by employing phase shifting layers in the fabrication of the reticle itself.(62) This method, however, requires sophisticated reticle processing, inspection equipment that is immature at this date, and advanced computer algorithms must be developed for circuit layout. Nevertheless, research continues on this method of resolution and DOF improvement, but tools with the oblique illuminator apertures are being delivered in the 1st quarter of 1993 and will probably be employed in production before the phase shifting reticle technique. Actually, a recent publication has demonstrated <0.25 μm lithography for I-line steppers by employing both oblique illumination and attenuated phase shift technology, thus, the lifetime of optical lithography may be extended at the expense of X-ray and other technologies.(63)

2.6   Metrology

MLM Issues: To prevent shorting of high density interconnects due to photolithographic bridging, the photolithographic metrology must have the ability to measure 10% of the nominal CD and/or the nominal alignment tolerance ideally. The task then is to establish this type of gauge capability for both the CD and alignment measurement tools used to monitor MLM fabrication verification data. For example, the SPC capability of an Hitachi 7000 to measure 0.5 μm features in resist is between 20-30 nm, and the IVS ACV4 capability is between 20 and 33 nm, depending upon the MLM layer being measured. Clearly, these capabilities are larger than the 10% ideal target, but these numbers are probably adequate until newer metrology equipment can be developed or measurement methods or algorithms improved. Notice that at larger design rules, i.e., 0.8 μm CD/0.25 μm alignment, these capabilities are very adequate.

MLM Metrology Calibration: CDs are calibrated either electrically (see Figure 33) or by SEM cross sections (i.e., the CD at the image base is referenced). The offset in Figure 33, the difference between the electrical CD data and the Hitachi SEM data, is the sum of the RIE etch bias and the Hitachi bias. For metal layers, split cross bridge electical structures are routinely used to calibrate either ADI or ACI CDs since they are absolute measurements. For via type structures, via resistance measurement for long via chains may be employed for calibration or electrical test structures recently developed by Lin can be employed.(64) More typically at via, SEM cross section calibrations are done. Here, metrology SEM measurements are typically compared to analytical SEM cross section measurements at 100000X for the same features. Of course, the pitch for these patterns is well known to facilitate the calibration. Example MLM SEM cross sections for via and metal CDs are found in Figures 34 and 35.

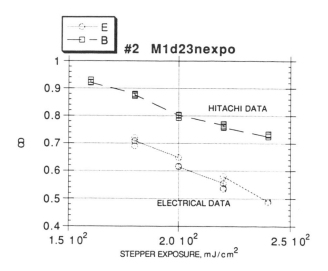

Figure 33. CD vs stepper exposure for 0.6 micron metal 1 patterns illustrating CD calibration and the difference between SEM data and electrical probe data.

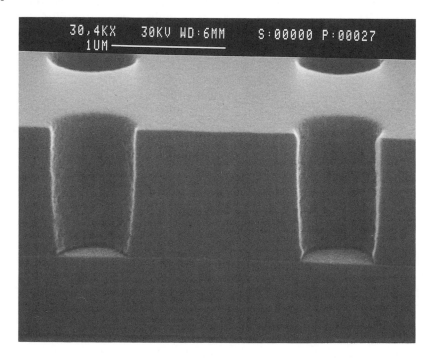

Figure 34. SEM cross section for a 0.6 micron via printed with a Canon 2000i stepper.

Figure 35. SEM cross section for a 0.6 micron metal 1 line on AlCu at 1.2 micron pitch printed with a Canon 2000i stepper.

For alignment, electrical structures are also prevalent like those employed by Prometrix commercial electrical measurement systems, but these sytems are hardly ever used on actual circuit reticles due to space issues, or if they are included, they are only found in PC cells. More typically, IVS or KLA optical alignment systems are employed to check alignment of product wafers during fabrication and those tools are SPC controlled vs. secondary standard wafers in a continuous mode as for all the other lithographic tools discussed previously.

2.7   MLM Photolithographic Applications

General MLM CD/Alignment Requirements: Total device layer to layer overlay is a complex situation, but to first order requires that the reticles stack well, the stepper alignment be good (i.e., 3-4X smaller than the CD), and that the CDs for both layers are within specification. Even though the latter terms are usually the smallest terms, they can be large for poor processes and lead to poor circuit yield. Typically, CD control is specified at ±10% of the design rule or better. As of 1992, this requirement translates to approximately ±0.1 μm, 3-sigma CD contol for critical layers.

Substrate Reflectivity Issues: Variations in wafer substrate film stacks can have a significant effect upon the resist critical dimension (CD) and exposure level for layers patterned to fabricate advanced four level metal BiCMOS MLM devices. In the fabrication of these VLSI devices, patterning is frequently performed on film stacks of varying thickness and optical properties.

PROLITH 2 (47) can and has been used to simulate lithography behavior on actual device film stacks, and the results compare favorably to data collected from actual MLM product wafers. These simulations can be used to accurately predict the exposure changes needed to compensate for changes in film thickness and film stack upon CD. In most cases studied, less than 3% deviation between the experimental and simulated results is observed. Thin film reflectivity is also observed to have a strong influence on CD variation.   The significant improvements in CD variation have been generally correlated with reductions and/or optimizations in substrate reflectivity. Electrical probe CD data for backend metal layers has also been evaluated for thin film notching behavior, and the CD variation is minimized by applying the results from PROLITH reflectivity analysis.

The constructive and destructive interference of light reflected from each film stack layer gives rise to reflectivity minima and maxima which can have an adverse effect on CD control.(65) Frontend and backend CDs, printed under fixed processing conditions, can vary up from ±5-25% or a large part of the usual CD variation budget due to film thickness and reflective notching effects. On metal layers, grain boundaries cause large local reflectivity variations which can cause notching (see Figure 16).(66)

Brunner (67) has developed a simple analytical expression, based upon the Fabry-Perot Etalon optical model for thin film interference, which accurately predicts the effect of thin film interference upon CD behavior. This expression can be used to account for reflectivity effects upon resist CDs, and examples of reflectivity interference related peak to peak amplitude minimization for both resist top surface and substrate antireflective coatings have been demonstrated.(67) Furthermore, detailed example results for the latter case are provided here.

In this section, PROLITH 2 was used to simulate reflectivity and CD behavior on actual device film stacks and the results compared to data collected from product wafers. Results are described from a study of thin film lithographic effects in the fabrication of advanced BiCMOS devices employing Shipley System 9 (900 series resist in tables) or SPR 500, or JSR 500 high performance, resists. The TiW inorganic ARC films were sputter deposited and MRC films.

Results for Metal and Via MLM Layers: The film stack shown in Figure 36 is representative of the layers encountered in BiCMOS and Bipolar devices at metal and via photo steps. The only difference between metal and via layers is the oxide thickness, 1.5 kÅ versus 7.5 kÅ. In this case, the aluminum-copper (AlCu) is sufficiently thick to behave as bulk Al which is opaque to the exposing wavelength. Hence, any layer lying beneath the Al layer can be ignored for the purposes of the reflectivity calculation. The PROLITH 2 simulations in Figure 37 show how the substrate reflectivity varies with TiW thickness in the absence of an oxide overlayer. It can be seen that TiW behaves as an inorganic antireflection coating, and can reduce the reflectivity (at 365 nm) of the underlying AlCu from 65% to less than 30%.

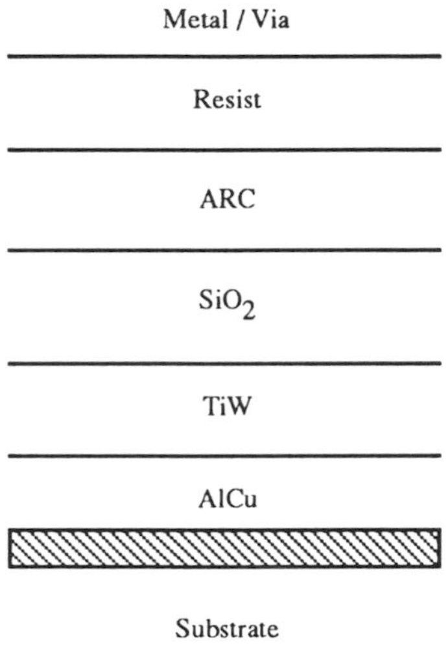

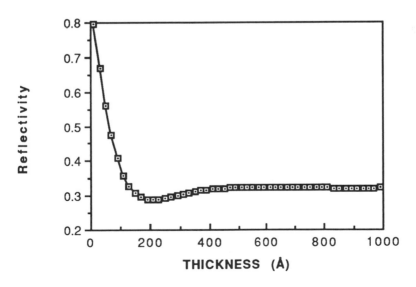

Figure 36. Film stack of BICMOS and Bipolar devices at metal and via photo steps.

Figure 37. Dependence of reflectivity on TiW thickness. Substrate is 6.5 kÅ AlCu on silicon.

When an oxide and resist layer are present, the reflectivity behaves as shown in Figure 38. Interference effects are responsible for the modulation in reflectivity with oxide and TiW layer thicknesses. At a fixed TiW thickness, variations in the oxide thickness can cause a 13% change in R (Note: The reflectivity changes from min to max when the oxide thickness changes by 1.25 kÅ). It is also easily noticed from Figure 38 that ARC is a much more efficient reflectivity suppressor that TiW (i.e., see triangle data at bottom). The simulations in Figure 38 (right hand photo) show that over this reflectivity range a ±0.05 μm change in CD can occur. When the ARC is included, this CD change becomes more tolerable at ±0.03 μm and the impact of the interference effects are effectively minimized.

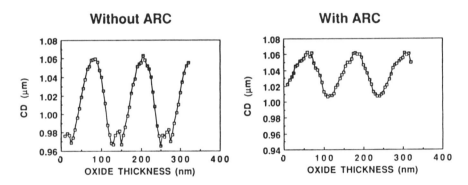

Figure 38. Effect of oxide thickness on CD variation film stack is resist/oxide/TiW/6.5 Å AlCu.

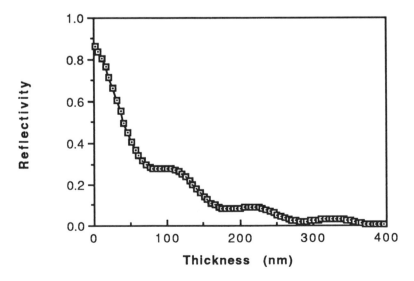

Figure 39. Dependence of reflectivity on ARC-XLT thickness. Substrate is 6.5 kÅ AlCu on silicon.

The theoretical reflectivity curve in Figure 39 shows that ARC-XLT can reduce the AlCu reflectivity to below 10%. When used in conjunction with TiW, the overall reflectivity and modulation is reduced significantly as shown in the right hand curve in Figure 38. Figure 40 shows there is good agreement between theoretical and experimental results.

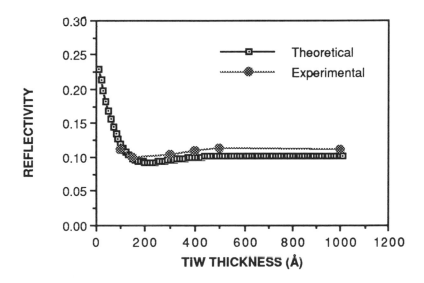

Figure 40. Comparison of experimental vs. theoretical reflectivity; film stack is resist/ARC/oxide/TiW/A/Cu.

The data shown in Table VI demonstrates the effect of reflectivity on via CD control. In via patterning experiments, vias printed with only resist exhibited high CD variation (~0.25 μm 3-sigma) due to substrate reflectivity and reflective notching. Actually, larger CD variation (>0.30 μm) has been observed on live product depending upon the lot. In an extensive multivari experiment with notching and non-notching lots, almost the entire via CD variation was attributed to the presence of grain boundaries in the underlying metal. The use of a higher performance Shipley dyed resist (510L) did not improve the CD variation. When an optimized ARC was used in conjunction with the resist, CD variation dropped by 35% for the Shipley 915 dyed resist process and notching was minimized. By using 510L which is a higher contrast dyed resist, 0.11 μm 3-sigma variation was obtained. The significant improvement in CD variation observed with ARC and 510L/ARC has been correlated with the reduction and optimization of substrate reflectivity.

Table VI: Mean CD and Total CD Variation for Via MLM Patterning

Resist Systems	CD Mean	Std. Dev. (3-sigma)
915(no ARC)*	1.09 µm	0.24 µm**
510L(no ARC)***	0.97	0.26
915/ARC	0.94	0.15
510L/ARC	0.93	0.11

* Shipley dyed resist for I-line.
** Significant amount of notching appears without ARC. Values can be as high as 0.30 µm.
*** Shipley positive resist generation after 900 series. L means dyed material.

The results of linewidth experiments on metal substrates are show in Table VII. Wafers patterned only with 915 resist showed waviness and notching which contributed to poor linewidth control (0.22 µm 3-sigma). Similar to the via results described above, the addition of ARC reduced CD variation by 40% and minimized the effect of notching. Our CD control target of ±0.1 µm was exceeded using 511 + ARC.

Table VII: Mean CD and Total CD Variation for Metal 1 MLM Patterning

Resist Systems	CD Mean	Std. Dev. (3-sigma)
915(no ARC)*	1.31 µm	0.22 µm
915/ARC	1.33	0.12
511/ARC**	1.28	0.09

* Shipley dyed resist for I-line.
** Shipley undyed positive resist generation after 900 series.

Thin film reflectivity has been demonstrated to be a major issue in MLM lithography and has been shown to have a strong influence on device CDs, level exposures, and CD variation. 0.09 and 0.11 µm 3-sigma CD control can be achieved on difficult metal and via layers, respectively. The significant improvement, i.e., reduction in CD variation, has been correlated with a reduction and optimization of substrate reflectivity, consistent with the work of Brunner.(67) In patterning the highly reflective layers encountered in BiCMOS backends, the use of a high contrast resist in conjunction with ARC-XLT can significantly improve CD variation by reducing the effect of substrate reflectivity. This bilayer process effectively suppresses the effects of reflective notching and interference swing contributions to CD variation below that allowed by device design CD tolerance levels. An undyed resist/ARC process is currently employed in patterning metal layers where ±0.10 μm or better CD control is required for device yield.

Process Biasing: Metal features are usually biased larger than their design to allow for size errosion during fairly harsh etch processes (see Section 2.2). Older resists used to fabricate mature devices require at least +0.2 μm bias (i.e., here, total bias for both image sides is used) be added to achieve the required resist image CD before etch processing. For some very old processes, biases as large as 0.5 μm have even been employed, but these were for fairly isotropic etch processes and much lower contrast resists.

Today's resist processes are higher contrast materials and photo biases are now zero or near zero, i.e., the chromium image from the reticle is printed divided by the reduction ratio of the tool. Metal etch biases may still be ~0.1 μm, but they too are being reduced through improved processing and by implementing next-generation single wafer etchers (eg, Applied Materials 5000 etchers).

Via images were typically biased to be undersized so that greater CD control could be achieved through greater exposures. Recently, however, processes have been improved so substantially that truely zero bias processes are now achievable for both photo and etch biasing. Typical biases for older via/contact processes were of the order -0.2-.3 μm, but it is fairly apparent that for next-generation dense cicuits near-zero photo and etch biases are required.

Metal Alignment: Several sets of typical alignment data for metal 1 has been reviewed in earlier sections of the chapter, i.e., in section 2.5. In general, however, layer to layer alignment and the design requirements at metal levels 1-3 are very similar, with metal 4 (M4) alignment and the M4 CD specifications being rather loose in tolerance. Alignment at these M1-M3 critical metal layers depends upon target and target aquisition quality, stepper stage precision and repeatablity, and the target discrimination software as disscussed in earlier sections. When compared to pre-metallization layer levels or device front end layers, metal alignment values are usually one to a few tenths greater, and consequently, some stepper vendors have begun to specify a smaller or better alignment capability at pre-metal layers than those for the MLM layers.

DOF Budget and Focus Offsets: To successfully print MLM features, the wafer field must stay within a certain focus range or patterns may be missing in parts of the die or print field. The resist image DOF as shown in Figures 41 and 42 must be wider than the DOF budget determined by summing wafer, lens, and reticle/pellicle budget contributions.(62) Table VIII contains an example budget for a 0.5 μm metal 2 or via 1 processes. Since the data of Figure 41 is greater than this budget, the metal 1 patterns should print accross the field with fidelity with an estimated 0.7 μm surplus tolerance. Note, that the budget may be somewhat smaller or better for via layers because they are all

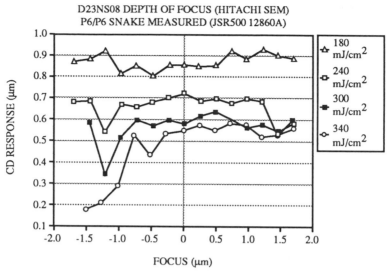

Figure 41. Metal 1 CD vs defocus or the JSR 500/ARC process.

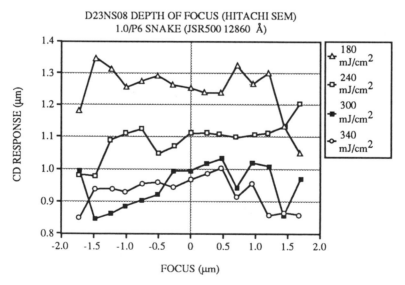

Figure 42. Metal 1 CD vs defocus or the JSR 500/ARC process with -0.6 micron focus offset.

printed to a single metal plane determined by the metal thickness and they are small square images that do not extend over great lengths as metal line interconnects do. The metal 3 budget may be slightly worse because 2 layers of planarization errors must be budgeted for instead of just the one contributor encountered at metal 2. Similarly, there may be a frontend topographical contributions that deminish the budget at metal 1, but overall the budget estimates should all be less than the resist process DOF. This has been verified on MLM wafer line monitors, i.e., >95% metal snake yields have been routinely achieved with the example DOF budget estimates.

Table VIII: DOF Budget for 5X Canon 2000i with N-layer planarization at Metal 2 or Via 1

Contributor	3-sigma Contribution
Lens Field Curvature	0.4 µm
Focus Repeatability	0.1
Field Leveling Repeatability	0.15
Reticle Flatness, etc.	0.1[a]
Wafer LTIR	0.5
Residual Planarity topography[b]	0.3[c]
Total DOF =	1.55 µm[d]

[a] B.J. Lin Ref 62.
[b] Doubles at Metal 3.
[c] Contribution may be closer to 0.1 µm since the vias all land on metal of the same thickness.
[d] At M1 approx. the same DOF, but at M3 the DOF is ≈ 1.8 µm

Figures 41 and 42 are referenced to demonstrate another principle governing MLM requirements, that is, focus offsets may and usually are required. They are required primarily because the effective thickness of the resist must be corrected for the refractive index of the resist (~1.5); as a result, negative offsets for most layers range fom -0.3 µm to -0.6 µm. The offset centers the defocus CD data at machine zero so that wafer flatness or other errors, which can be as large as 0.5 µm and as great as several microns at edge die of severely distorted wafers, do not create a yo-yo effect where the wafer CDs move up and down the steeply varying CD vs. defocus curve on the negative side of defocus. From Figure 41, it is apparent that without an defocus offset, probably half of the optical fields will be precariously near the CD cliff edge (left hand side of Figure 41) with little process tolerance or margin available .

Planarization Processes: (also see Chapter 6)

Types of processes: The DOF budgets for many new devices are very tight, especially those at DUV R&D facilities where the DOF of the resist/tools are currently ~1 µm TIR. Additionally, wafer flatness on an exposure field basis, termed local total indicated range (LTIR), can be much greater than the 0.5

μm usually budgeted for wafer flatness and can have values as high as several microns. Without planarizing steps between metal layers to achieve some degree of global planarization, the topography alone can be as great as 2-3 μms. Obviously, employing advanced 4 or 5 level MLM processing dictates some sort of planarizatiion processing from 1st level ILD to 4th or 5th level metal or the stepper images needed to be printed at 3rd via or third metal will probably be out of focus and not print with fidelity.

The three processes in current development (68) are Chemical Mechanical Polishing (CMP), and two different 2-layer etchback techniques, I-layer or negative blocked 2-layer (69) and N-layer, termed positive resist 2-layer blocked.(70) The last technique will be discussed here because it is probably the most mature process from a defectivity point of view and experience with it is extensive.(71, 72) The first two processes provide better global planarity, but require equipment development and negative resist development, respectively. IBM researchers have also reported using both N-layer and CMP together to achieve excellent global and local layer planariazation.(73)

Two-Layer Photo Planarization Process: The conventional POT process, (71) i.e., where the dielectric layer is planarized by the application of one sacrificial layer followed by an etchback of that layer, is always considered first as a leading candidate for the planarization of any device MLM backend due to its simplicity. However, it is hampered by the fact that conventional, commercially available organic materials and spin-on glasses can only planarize underlying features up to a finite distance from the edge of the feature, and neither group of materials is capable of more than step rounding or smoothing in reality. This distance is approximately 10-20 μm, and is known as the planarization length. Thus, for large, >20 μm isolated features separated by more than 20 μm such as bonding pads, etc, the degree of planarization achieved with conventional materials spun to workable thicknesses (< ≈2 μm) decreases dramatically, to the point where there will be vast numbers of features on typical circuitry left completely unplanarized. For these types of structures, the only thing gained using conventional POT processing is simple rounding of the edges, which does improve things by helping maintain metal line continuity over steps and by eliminating shorting bars or picket fences during metal etch at those locations.

The Two Layer Photo (TLP) process has been described in the literature (74-76) as a means, through the use of an extra masking step per ILD layer, to overcome the shortcomings of conventional POT and provide nearly perfect dielectric planarization independent of underlying topology. Figure 43 a-f describes the method. After the given metal layer is etched, the initial TEOS layer is deposited. In the standard POT process, the sacrificial organic would be spun on at this point, giving a large variation in the vertical altitude of the organic surface due to the dependence of the planarizing ability of the coated layer on the dimensions of the underlying topography.(77-79) This problem becomes more severe as feature sizes continue to shrink. With the TLP process, the problem of global planarization is lessened substantially by the use of the additional masking step. A photoresist layer is spun onto the structure (Figure 43b), and is patterned with a mask that is roughly an inverse image of the underlying metal pattern (Figure 43c). This leaves resist in the field areas between the metal lines, effectively plugging these areas and planarizing the structure. In order to achieve planarity, the thickness of this resist layer (known as an "N-layer") must be equal to that of the step height, which is the sum of the metal thickness and the degree of overetch into the underlying glass layer during the metal etch step.

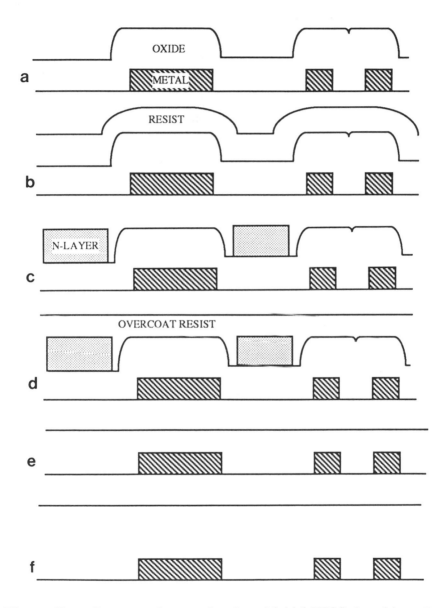

Figure 43. a. Structure after metal etch and initial TEOS deposition. **b.** After N-layer resist spin. **c.** After N-layer patterning. **d.** After overcoat and resist spin. **e.** After etchback planarization. **f.** After TEOS redeposition.

The N-layer is not an exact reverse image of the underlying metal because of several reasons. First, the TEOS deposited onto the sides of the etched metal steps effectively widens the features by an amount equal to 60% of the deposited glass thickness, independent of feature size. Second, there is a limit to how small the N-layer "plugs" can be made. Since patterning resist layers planarize small gaps fairly well without TLP and adhesion and overcoat compatibility concerns exist for small N-layer images, n-layer plugs are usually larger than the lower limit of 1- 2.0 μm. Finally, alignment tolerance comes into play. This tolerance is aligner specific, but is typically set to be ≤0.25 μm, meaning that there is a nominal 0.25 μm or smaller gap between the edge of the glass and the edge of the N-layer plug. This is necessary because if the tolerance is made too tight, normal misalignments could result in full-thickness N-layer structures ending up on top of features, to the detriment of surface planarity. Thus, areas on the circuit with gaps between glass sidewalls smaller than 1 + (0.25x2) = 1.5 μm will not get plugged. For any thickness (ILD$_t$) used for the initial TEOS deposition for first ILD, the areas with a gap between neighboring metal features of size N$_{min}$ of less than

[16]   Max unplugged Dim. = N$_{min}$+2 x (ILD$_t$ x sidewall coverage)

will not be plugged. For future generations of back end product, the adhesion and alignment tolerances will be tightened, resulting in a larger percentage of the circuit getting N-layer plugs and n-layer plugs as small as 0.5 μm may be easily obtained.

After the N-layer feature has been patterned, the wafers are coated with another organic layer which is similar to the sacrificial organic used in the POT process. A typical thickness of this organic layer is about 8500 Å for all three ILD's, but thinner layers of the order of 0.3 μm are more attractive for etchback time reduction reasons and films this thin are still capable of the misalignment gap filling required. A thinner resist overcoat is attractive out of concern for the overall nonuniformity of the glass thickness after etchback planarization. The thicker the resist/TEOS stack, the larger the magnitude of thickness variation after the etch, assuming the percent nonuniformity remains constant as a function of stack thickness. Its only purpose is to fill in the gaps that the N-layer has not plugged due to the photolithographic alignment considerations (Figure 43d). The overcoat layer does not have to be photo sensitive and should not be for cost reduction reasons.

Great care must be taken in the treatment of the wafers after completion of the N-layer patterning before the overcoat layer is spun. The N-layer resist must be either baked to high temperatures (70) or deep UV cured to prevent dissolution of the bottom N-layer resist images. If this N-layer image insolubility process does not occur, planarization of the surface will not occur due to dissolution of that layer at the overcoat step. Deep UV curing has been reported (75) to be a necessary component of the TLP process, and is used extensively. A high temperature bake will accomplish the same purpose, but temperatures high enough to prevent image dissolution (>160° C) will also cause greater shrinkage of the N-layer resist to leave larger gaps. Because this shrinkage occurs to the same degree in all directions, a gap of several microns can open up between the edge of the N-layer plug and the edge of the glass. For this reason lower temperature baking followed by a deep UV cure is the recommended process. Others have used a layer of sputtered quartz to separate the two resist layers, but the deep UV cure is a simpler technique, without the added deposition and has been demonstrated to work well.(71, 72)

N LEVEL FACTORIAL ANALYSIS

	NO DUV		DUV	
	120°C	160°C	120°C	160°C
5214	NO PATTERNS 15500	SOME PATTERNS 28600	PATTERNS 27600	PATTERNS 28000
910	NO PATTERNS 12900	SOME PATTERNS 15100	PATTERNS 26200	PATTERNS 25700

ONLY SIGNIFICANT VARIABLE IS RESIST

Figure 44. N-layer 3-factor factorial analysis results to determine the minimum full thickness process required to prevent N-layer dissolution at the overcoat process.

The example N-layer process depicted is the result of three basic experiments. First, the temperature where N-layer image shinkage was zero or as small as possible to prevent pattern pull away and film stress needed to be determined. Next, the early literature was unclear as to whether DUV curing, high T curing, or both were necessary or what order was essential for N-layer double coating. By doing the experiments of Figures 44 and 45, it was determined that DUV was required and that it should occur after baking. Futhermore, baking before DUV is necessary to prevent "eyelash" defects from forming around the N-layer image edges as shown in the Figure 46. "Eyelash" images will transfer to the underlying TEOS layer during etchback and there is a considerable amount of deleterious topography associated with them, so they must be avoided. Any photoresist can be used for the N-layer process; in fact, at least four different resists have been successfully tested. Images for all of these resists are rendered insoluble to the planarization overcoat before etchback by the 120°C baking followed by the several hundred mJ/cm^2 DUV cure/exposure.

FACTORIAL ANALYSIS

N LEVEL TREATMENT

		DUV FIRST		DUV LAST	
		120°C	140°C	120°C	140°C
OVERCOAT BAKE TEMP	120°C	E, E	R, R		
	140°C	R, R	R, R	E, E	

E = EYELASHES

R = RETICULATION

(ALSO HAVE EYELASHES)

Figure 45. 3-factor factorial results of the experiment designed to determine the processing conditions that prevent the "eyelash" problem.

Early on, there was concern about the limitation of the N-layer resist's ability to fill the narrowest crevice left in the glass after initial TEOS deposition. This problem never materialized, however, and the resist overcoat easily fills gaps of approximately 1500 Å without leaving voids. Any gaps that the resist does not fill should be plugged by the TEOS re-deposition. After the overcoat layer is spun on, it is also important not to overbake the wafers, as excessive temperatures cause layer "bubbling," which causes a phenomenon to occur near the boundary between N-layer plugs and glass edges reminiscent of localized resist reticulation. This problem becomes more severe as the metal thickness increases, which can cause local N-layer build up. The cause of this problem is believed to be due to excessive volatile product build up in the film that cannot escape quick enough as for thinner film regions. This is a common DUV curing problem as resist thicknesses are increased.

The resist/TEOS sandwich is next etched back to within some targeted distance above the top of the metal using a chemistry that etches resist and TEOS at roughly the same rate (figure 43e), and then another TEOS deposition brings the final ILD thickness up to the specified target (figure 43f).

A great deal of discussion has transpired regarding the need for the complex processing associated with the TLP N-layer technique. Until other techniques such as CMP are established in volume production as cost effective, TLP proccess will be employed extensively, at least in R&D fabs doing advanced MLM processing.

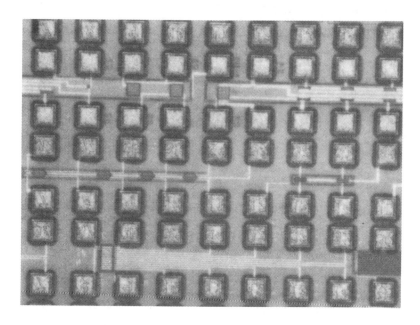

Figure 46. Optical micrograph illustrating the "eyelash" problem of improperly processed N-layer images around MLM bonding pads.

Alternate Processes to TLP Etchback: There has also been an ongoing effort to investigate the claims of several photoresist vendors of the perfection of materials giving far better planarization lengths than today's resists. At this writing, impressive degrees of planarization (in excess of 80%) have been achieved with some of these materials such as Futterex PC2-1500.(68) Problems remain with purity and stability, however. It is hoped that use of these materials in the simpler POT-like process will be a viable alternative to N-layer processing in the near future.

Of course, this type of single layer process would require an organic material capable of planarizing substantially large pads, 100 microns square and isolated. A careful survey of the literature turned up several ideas, some involving multiple layers and various processing sequences. These were all tested, the results of which are found in Table IX. The results of the Table clearly indicate a lack of planarization for large feature geometries and a large discrepancy between the degree of planarization between center and edge of the test wafers. Even though this center to edge phenomena is not well understood, the results are also not acceptable, because the maximum planarization value observed was only 27%.

Table IX: Planarization measurements for literature systems before 1988

Organic Systems	Planarization* at Wafer Center, %	Planarization* at Wafer Edge, %
SINGLE LAYER SYSTEMS		
OCG WX-214(Special Spin)**	1.8	24.2
OCG WX-214(Std. spin)	~0	18.5
Polystyrene	0	8.5
TLS(Polysiloxane)	3.1	23.6
FUTTEREX (early version)	~0	19.1
BILAYER SYSTEMS		
OCG WX-214/OFPR-800***	~0	25.4
OCG WX-214/OFPR-800	~0	21.7
Polystyrene/OFPR-800(180 C)	~0	26.8
OFPR-800/Polyvinyl alcohol	~0	26.8
OFPR-800/Polyacrylic acid	1.5	19.3

* Calculated from P=100X (Pad Step Ht.- Meas. Step Ht. at Coat)/Pad Step Ht.
** Special spin has a very short spin of a few seconds and the wafer drys with the chuck stationary not spinning as for Std. WX-214 is a standard photoresist.
*** Dynachem standard photoresist.

Stillwagen (80) was also aware of these limitations, and was able to formulate a series of polymeric compounds which were capable of planarization due to highly viscous flow and long spin times. Unfortunately, these compounds are not practicle for use to the electronics industry, but they did demonstrate satisfactory planarization could be achieved and provide an existance proof of that. Since that work, several polymer systems have been developed by OCG and Futterex, which are capable of thermal flow after coating followed by thermally induced crosslinking reactions to harden them. This creates planarity over 1 micron topography of upto 70-90% for Hunt and Futterex test materials, respectively. The later data occurred for 200 C cures, and should be adequate for initial POT process experiments without N-layer lithographic steps. Even though these advances are impressive, it appears that CMP research will win out over the long term and these POT like processes may fade away.

Via Layer Processing: Via layer images of an MLM process are square small contact geometries and allow the layers of metal to be connected at specific sites. So, for every layer of metal there is a via level which follows it, including even the last layer of metal which must be contacted through passivation dielectric in order to place the chip into a package. The via images are printed usually on a TEOS oxide or any type of insulating dielectric layer, then transferred into the dielectric to the lower metal surface using RIE as in Section 3.

Via Layer CD control: The issues for via CD control are very similar to those for metal layers, i.e., reflectivity, focus, and exposure. As for metal layers, image focus can be very important.

In Figures 47 and 48, the importance of putting in a focus offset to center the CD vs. the range of defocus data is demonstrated. Without an offset, the process is peformed precipitously close to the right hand data of Figure 47 which is going out of CD control quickly at less than 1 μm of positive defocus (see 370 and 420 mJ/cm^2 data). The CD spec is 0.5±0.1 μm and notice there is an interaction of exposure and defocus at large minus and positive defocus values. By working at the machine defocus of zero from Figure 47, if any thing happens across the wafer or the print field, the data will go out of specification quickly. In comparison the data of Figure 48 is better centered at zero machine focus after putting in the -0.7 μm offset and now any changes that occur upto ±1.4 μm in local field planarity will have only a minimal effect on the via CD. Notice, that this empirically determined offset value is very close to the theoretical value you would calculate to compensate DOF for resist thickness, (62) the resist thickness 1.28 kÅ divided by the refractive index of 1.6 or -0.8 μm. The data of both Figures also allows the etch bias of this example process to be estimated to be about +0.05 μm, i.e., the via gets a little larger after etch by using the fixed 420 mJ/cm^2 data from both plots. As via CDs decrease with time, processes with no photo or etch bias will be required.

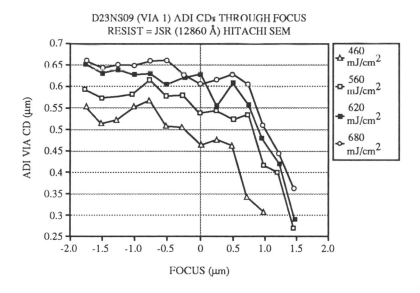

Figure 47. Via 1 CD vs defocus or the JSR 500 process.

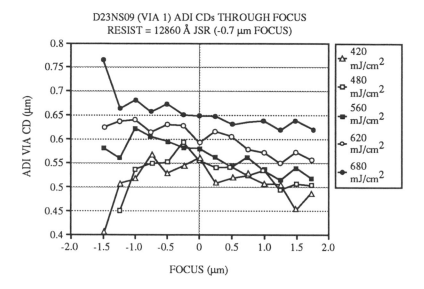

Figure 48. Via 1 CD vs defocus or the JSR 500 process with -0.7 micron focus offset.

Via CD variation can also be affected by metal reflectivity in the same way metal CDs are affected, because they too are printed to the same type of metal grainy surface (see ARC section). If refelective notching is a problem at the preceeding metal layer, it most likely will also be a problem at the via layer as well since the metal has not changed. In fact, studies on 1 μm vias cut to metal requiring ARC demonstarted greater CD control when the ARC was also used at the via layer. The assymetric notched via images with no ARC are found in Figure 49 with the heavily notched via CD line aids, also printed with the via. After employing ARC at via, the reflective notch induced CD variability is reduced from between 27-37%, from separate studies.

For smaller vias at 0.5 μm, CD variability can be as large as 0.13 μm (all families of variation for 10 lots of real wafer data, 5 wafers/lot, 5 sites/wafer 3-sigma), without ARC with the largest family of variation being wafer via site to site variation. This is really not too surprising because reflective notching does vary via to via as is seen in Figure 49. By employing an ARC at via, that variation can be reduced to within the ±0.1 μm specification level. Vias printed with only resist exhibited "reflective notching dominated" CD 3-sigma variation of 0.10 μm greater, or double the specification, than that observed where an optimized ARC process was employed under the resist to minimize substrate reflectivity.

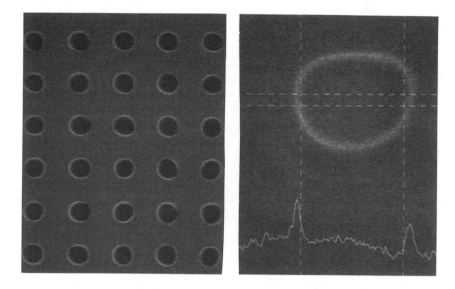

Figure 49. SEM micrographs of notched vias illustrating asymetric vias that occur without an ARC undercoating.

Via Layer Alignment: Via alignment results for MLM vias aligned on the GCA example tool of the stepper section are mean +3 sigma of 0.23-0.26 μm (Figure 50). When this level of misalignment exists, the SEM cross sections of stacked vias look like that shown if Figure 51. Since that tool is specified to have 0.25 μm capability, these results are in line. But to achieve these results, a careful wafer target study as outlined in the wafer target section had to be carried out.

Alignment results for the Canon Stepper example are found in the histogram of Figure 52. Here the specification is mean± 3 sigma=0.21 μm, and the data of the figure is all within the dashed vertical line specification edges. The targets used to generate this live wafer lot data, however, did not have to be optimized due to the greater process compatibility of the Bright Field Canon alignment system. The Canon data is better, because of the improved target aquistion system and because the Canon stage precision is >2X more precise.

2.8 Summary and Future Predictions

Resist technology for the future will probably focus upon the area of chemically amplified resists (81) and shorter wavelength DUV ARCs which will be dry developed and not wet developed as the example ARCs of this chapter. Surface imaging (82) cannot be ruled out either, due to the great depth of focus potential for these systems.

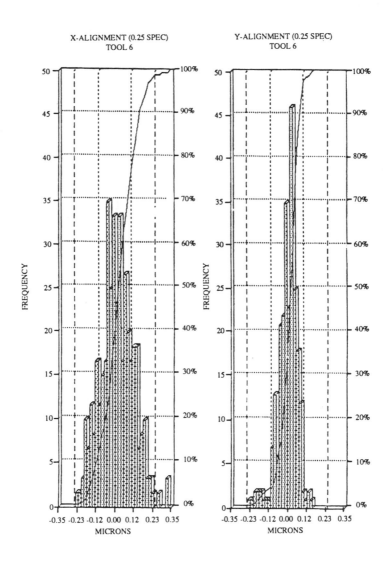

Figure 50. X & Y misalignment data for via 1 printed on randomly matched GCA Autosteppers.

Wafer stepper improvements appear to be centered around wavelength reduction for resolution reduction, stage precision data level reduction for improved alignment and overlay, alignment system upgrades to phase sensitive systems for improved overlay, and oblique illumination subsystems, offering again the most important thing to MLM, improved DOF capability. Oblique illumination will require greater light intensity from illuminators or more sensitive resists due to the inherent intensity losses of those systems.

The MLM photo processes and stepper tool examples of this section were basically for 0.8-0.7 Biploar and 0.5 µm BiCMOS design rule technologies. With futher design rule shrinkage to 0.35 µm and below, zero bias processes and alignment tools with greater capablilities will be required. For example, the authors of ref. 63 have demonstrated 0.25 µm capability with a stepper with oblique illumination combined with attenuated phase shift reticles, which is one of the easier phase shift techniques to employ. Of course, advances in phase shift technolgy processing and inspection must also occur before these technology goals can be achieved. Advances in DUV stepper technology and phase shift reticle technology are also pushing the frontiers of lithography and may eventually achieve 0.18 µm production lithography by the year 2000.

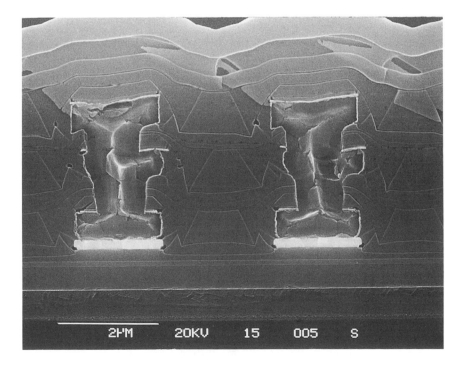

Figure 51. SEM cross section of stacked vias printed with the same order of magnitude misalignment as that shown in Figure 50.

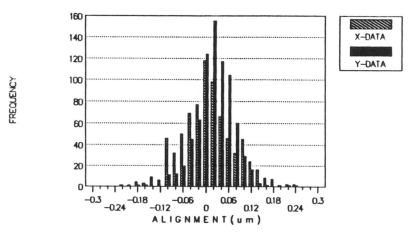

Figure 52. X & Y misalignment data for via 1 printed on randomly matched Canon 2000i steppers.

3.0   ETCH

This purpose of this section is to introduce the reader to issues concerning etching materials commonly used to manufacture the advanced MLM structures found in devices today. This section is slanted towards processing issues. Knowledge of basic concepts in plasma etching and reactive ion etching (RIE) is assumed; readers interested in reviewing plasma physics or the principles of RIE will find excellent coverage of these topics in other general references (83 - 88). The intent of this section is to provide the reader with a basic understanding of etch issues that are of particular concern to the development of a successful MLM structure. Although the section is not intended to be a tutorial on etch, some explanations of basic principles will be required and included. Many of the concepts and problems that are discussed are generally applicable and therefore useful for processing single metal layer structures.

The section will be organized in the natural order of constructing a MLM system. As such, contact etching will be discussed first. Contact etching was deemed sufficiently different from via etching so as to warrant its own section. Contact etch will be followed in order by metal etch, planarization etch, via etch, and passivation (or pad) etch. The discussion on planarization etch will be limited, since Chapter 6 is devoted to planarization techniques. These sections will be followed by a brief discussion of future trends and needs.

3.1   Contact Etch

General Considerations: The purpose of a contact etch is to open holes in the first interlayer dielectric (ILD0) so that metal can be deposited in those holes and make contact to underlying single crystal silicon and polysilicon. Although via etch shares a similar purpose with contact etch, contact etch warrants a separate section because the underlying silicon makes selectivity a much greater issue. In addition, it may be required that contacts of greatly differing depths be opened in a single etch (Figure 53). As an added complication, the silicon underlying the contacts will often be capped by a thin (several hundred Ångstroms) metal silicide (89) that must be preserved. This is a particular challenge since the shallow contact will be opened long before the contact to single crystal substrate; the silicide in the shallow contact will be exposed to the plasma while the deep contact continues to be etched, increasing the demands for high selectivity.

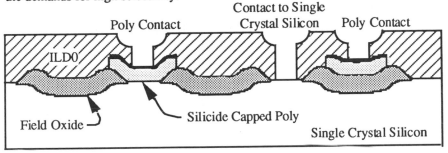

Figure 53. Contacts to polysilicon and single crystal silicon. Note the different depths which must be etched. As drawn, these contacts have wineglass profiles from a slope etch.

Another important problem in contact (and via) etching is associated with the aspect ratio (AR) of the contacts. The AR is the ratio of the height to the width of the opening. When the AR becomes large, a phenomenon known as etch lag can occur. Etch lag refers to the fact that very small, deep openings tend to etch more slowly that larger ones. So if a design requires both 1.2 and 0.6 μ diameter contacts to be opened simultaneously, the 0.6 μ contact may require a much longer etch time. The industry is now at a point where etch lag has become an important problem, and it is discussed in detail below.

High aspect ratio contacts and vias are also harder to fill with metal, especially with aluminum based metallization systems. The use of barrier metals in contacts (and liners in vias) generally increases the AR and makes filling even more difficult. As contact diameters have become smaller, this problem has become worse since ILD thicknesses are remaining relatively constant. Isotropic slope etching and tapering contacts and vias during the etch make them easier to fill. Slope and taper etching are discussed below.

Etch Chemistries: The material being etched in a contact etch is almost universally some form of silicon dioxide (BPSG, PSG, PETEOS, etc.). Therefore, fluorine based chemistries are used. Unfortunately, fluorine based chemistries are also used to etch silicon, polysilicon, and silicides, the very materials upon which the etch must stop. This difficulty can be overcome by adjusting the carbon to fluorine ratio of the etch chemistry. Increasing the C/F ratio increases the rate of polymer formation on surfaces in the plasma chamber. The C/F ratio can be increased either by adding fluorine scavengers such as hydrogen, or by adding fluorocarbon gases that are relatively low in fluorine, such as C_2F_6. Highly polymerizing conditions increase the selectivity of oxide to silicon and silicides. Increased selectivity arises from the fact that the thickness of polymer deposited on SiO_2 remains small, but under the same conditions, polymer tends to build up on silicon. The film remains thin on the oxide because ion stimulated reactions between the film and the oxide convert the polymer into gaseous products such as CO, CO_2, and COF_2.(90) The oxygen in these by-products comes from the silicon dioxide. On the other hand, there is no oxygen for silicon or silicides to contribute, so these reactions can not occur. Instead, the polymer remains in place on silicon and silicides, where it effectively blocks the etch. Highly polymerizing conditions also tend to minimize the concentration of fluorine atoms, which attack silicon rapidly. This again contributes to increased selectivity. Therefore, the best conditions for achieving the needed selectivity are those that promote polymerization.

It should be noted that photoresist erosion also contributes to polymer formation. In fact, when small (submicron) contacts are being etched, it is easier to control the polymer formation through resist erosion than through gas phase deposition reactions. So for small contacts and vias, adjusting power and pressure to control the resist erosion rate may be more beneficial than changes in the gas flows.

Polymerizing conditions and relatively low fluorine radical concentrations are key to achieving the selectivity needed in a contact etch. The rate of polymerization is limited, however. Excessive polymerization can cause the etch to slow to an unacceptable rate, or even stop completely. It can also contribute to problems with etch lag. The polymer deposition rate also can also have an effect on the cross sectional shape of contacts (Figures 54 and 55). Very high deposition rates can cause the profile to be sloped and the bottom to be undersized. Too low of a deposition rate will cause the profile to be bowed. So, selectivity must be balanced by etch rate and profile considerations. These opposing trends can be balanced by using a mixture of gases. For example, if pure CF_4 is used, the etch will not be selective enough. But if some hydrogen is added to scavenge fluorine and promote polymerization, (91 - 94) excellent selectivity can be achieved. The trade off is that the etch rate is slowed; both the SiO_2 and Si etch rates diminish, but the Si etch rate decreases more rapidly than that of the SiO_2. Other gases that have been used for the same purpose include CHF$_3$, (95 - 97) C_2F_4, (98) C_2H_2, (99) and C_2H_4.(92, 100) Also investigated have been C_2F_6 and C_3F_8, both alone and in combination with hydrogen.(91, 92) All of these gases and gas mixtures produce more polymer than pure CF_4. Because of its availability and ease of handling, CHF$_3$ is most commonly used, although mixtures of other fluorocarbons with hydrogen are not uncommon. When contacts of different depths are etched, very high selectivity is required, and therefore highly polymerizing conditions are needed. Polymerizing conditions also promote selectivity to silicides. By balancing the etch conditions, selectivity to silicon in the range of 4 - 10:1 and to silicides ranging from 7 - 15:1 can be achieved. Of course, the exact selectivity achievable depends on the etch tool being used. Schemes that use etch stops have also been developed (101, 102) which relaxes the requirement for etch selectivity.

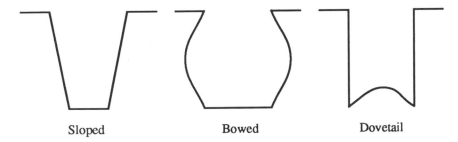

| Sloped | Bowed | Dovetail |

Figure 54. Contact profiles that deviate from anisotropic. Sloped profiles, when unintended, consume too much area. Bowed and dovetail profiles are difficult to fill with metal. Sloped profiles indicate too high of a polymerization rate. Bowed profiles result from too little polymerization. Dovetail profiles are caused by etching at excessively low pressure.

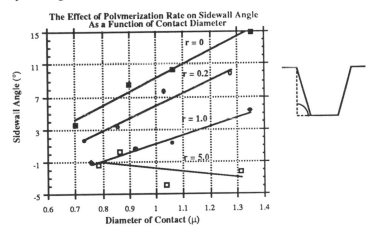

Figure 55. The effect of polymer deposition rates on contact sidewall angles. The deposition rates were varied by adjusting the CF_4/CHF_3 ratio (r). Higher values of r result in less polymer deposition. Note that the amount of sloping also depends on the contact diameter. In the case of r = 5, the profiles were bowed. (S. Ramaswami, unpublished results, 1992)

Note that selectivity requirements are further strained by the fact that it is often impossible to endpoint a contact etch. Because the open area (the area of oxide to be etched) in a contact pattern is very small (a few percent at best), standard means of endpointing are of little use. Changes in optical emission spectra are too small to detect with most systems, and laser interferometry can not be used because there is no area large enough to direct the laser. In principle, purposely opening up large areas around the perimeter of the wafer (outside of the patterned area) can alleviate both of those problems. However, that approach only works when large contacts are being etched, because etch lag causes the large open areas (used for endpoint detection) to etch at a different rate than the contacts themselves. The greater amount of open area will also tend to slow etch rates and reduce uniformity. Although it is possible to compensate for the loss of etch rate, recovering the uniformity may be impossible (especially if resist erosion is being relied upon for polymer formation). Because of these problems, most contact etches are stopped after an empirically determined amount of time has passed.

Tapered Contacts: As mentioned above, high aspect ratio contacts can be difficult to fill with metal. One way to reduce the effective aspect ratio of contacts is to purposely taper their profiles. In this approach, the etch is carried out in such a way that the diameter of the contact increases continuously from the bottom to the top of the opening. Tapered profiles can be either stepped or continuous (Figure 56). In either case, the tapered profile is generated by a controlled erosion of the photoresist.

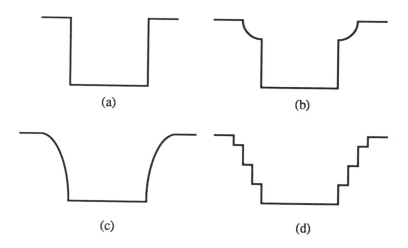

Figure 56. Different possible contact profiles. (a) Perfectly anisotropic, (b) Result of a combination slope and straight etch, (c) Tapered, and (d) Stepped or tiered.

Stepped or tiered profiles are created by stepwise etch programs. The etch is designed to alternately erode photoresist and etch oxide. For example, a short, high pressure, low bias oxygen plasma can be used to isotropically remove some of the photoresist, followed by a low pressure, high bias CHF_3/CF_4 plasma to etch some of the oxide. A tiered contact profile is produced by repeating this cycle a number of times. When running any stepwise etch such as this, it is important to be aware of the fact that abruptly turning the RF on and off can contribute to gate oxide degradation. Therefore, when implementing a stepwise etch, it is best to ramp the RF on *and* off.

Continuously tapered profiles are generated in a single step. Here, it is usually necessary to start with resist profiles that are already rounded. The rounding can be achieved by thermal flow of the resist prior to etch, or the resist can be intentionally faceted by an RIE step. The key to a successful etch in this instance is to etch the resist and the oxide at about the same rate. This can be attained by adding oxygen to the contact etch chemistry or by increasing the ion bombardment, which will increase the resist erosion rate. By etching the rounded resist and the oxide at a similar rate, the sloped profile of the resist will be transferred to the oxide. Wafer temperature also affects etch rates and profiles.(103 - 105) Higher temperatures (130° C) give straighter profiles because of higher polymer deposition rates at lower temperatures. Lower temperatures also increase selectivity to silicon because the silicon etch rate slows with decreasing temperature.

Although it is conceptually easy to generate tiered or tapered contacts, it is more difficult in practice. Both processes are sensitive to photoresist thickness, which affects the resist erosion characteristics. The relative etch rates of resist and oxide must also remain constant. If the process drifts to favor one over the other, contact profiles will deviate significantly from what is desired. Factors which can alter the relative etch rates include moisture in the etch chamber and chamber cleanliness. (Specifically, fluorocarbon polymer contamination in the chamber can cause large increases in the oxide etch rate. This is discussed in more detail in the Planarization section.) Another difficulty with these processes is maintaining selectivity to the underlying substrate. Adding oxygen or increasing ion bombardment to increase the resist erosion rate will also decrease the selectivity to silicon. Finally, CD control using these processes is complicated by the resist erosion. Because the resist is being intentionally removed in the lateral direction, contact diameters grow. Unless the process is tightly controlled, CDs can become unacceptably large, especially in areas of high pattern density where contacts are closely spaced.

Isotropic Slope Etch: Another way of reducing the aspect ratio of a contact is to carry out an isotropic etch before the anisotropic etch. The isotropic etch, commonly called a slope or round etch, is used to etch only part of the ILD thickness. The anisotropic etch, or straight etch, is used to clear the ILD and open the contacts to silicon. The combination of these two etches gives these contacts a characteristic wineglass shape (Figure 56). By definition, an isotropic slope etch will etch laterally as well as down. Because of that, the top portion of the contact will have a very low AR. In fact, the sloped part of the contact can be neglected when considering the metal filling characteristics. Only the straight portion of the contact need be used to determine the AR. By using an appropriate slope etch, contacts can be readily filled that would otherwise be voided.

Slope etches can be done either wet or dry. Wet slope etches are done using solutions of HF or buffered oxide etchant (BOE). Solutions such as 10:1 or 50:1 HF are in common use. In general, the more concentrated the etchant, the more difficult it is to control the etch. Wet slope etches also require good resist adhesion. If adhesion is poor, etchant will be drawn along the interface by capillary action and etch the oxide in a greater area than intended. That will result in abnormal profiles and oversized CDs.

Dry slope etches are usually done in discharges which produce high concentrations of fluorine radicals. Fluorocarbons mixed with oxygen or gases such as NF_3 and SF_6 are often used. High pressures and flow rates help to ensure isotropic profiles. For a given chemistry, oxide etch rates are mainly dependent upon wafer temperature and the pressure since dry slope etches are chemical in nature. In single wafer tools, wafer size and flow rates have little effect on the etch rate. Because of the high flow rates, loading is also minimal in single wafer tools: the difference in oxide etch rate between blanket oxide and wafers patterned with a few percent open area has been found to be less than 10 %.(106)

Under the conditions of a dry slope etch, photoresist becomes heavily fluorinated to a depth of a few thousand Ångstroms.(106, 107) This skin etches 5 - 10 times more rapidly than unfluorinated resist, and is therefore an important factor in controlling CD loss. This fluorinated skin will erode quickly in both the isotropic and anisotropic portions of the etch. Reduced power will help control the extent of fluorination if there is excessive CD loss due to this process.

Slope etching increases the critical dimension of contacts at the top of ILD0. So, just as in the case of tapered contacts, CD control is crucial. Because the slope etch is a separate step, it is easier to control than taper etching. However, the slope etch approach has the disadvantage of creating sharp cusps at the juncture between the slope and straight etches (where the stem meets the bowl of the wineglass). Metal step coverage over these cusps can suffer. To reduce the sharpness of the cusp, a step similar to a taper etch is sometimes added to the anisotropic etch. The resist is intentionally eroded laterally so that some of the horizontal area of the cusp is exposed. During the ensuing straight etch the cusp is smoothed and flattened by the faceting effects of the ion bombardment.

These approaches to decreasing the aspect ratio of contacts can all be made to work well, and are in widespread use today. However, trends toward smaller contact dimensions and tighter packing densities are rendering these techniques obsolete. With very small, densely packed contacts, the increased CDs caused by sloping or tapering are not acceptable. So in more advanced devices, only anisotropic contact etching can be used. In this case, the burden of filling the contacts falls upon the metal deposition. This is the driving force behind such technologies as tungsten plug. The lack of aspect ratio reduction relaxes that requirement of the etch. However, the very small diameter contacts which must be etched increase the demands on etch performance because of etch lag.

Etch Lag: Etch lag refers to the tendency of small openings to etch more slowly than larger ones. More correctly, high aspect ratio features etch more slowly than their low AR counterparts.(108) For example, if 0.6 and 1.2 μ diameter contacts of the same depth are etched simultaneously, the 1.2 μ contact will open before the 0.6 μ one if etch lag is present. This phenomenon is quite general; etch lag has been observed in a wide variety of etch tools and process chemistries.(109, 110) It is important to note that AR is a dynamic variable. The initial AR is that of the patterned resist. Because the selectivity to resist of a contact etch is greater than one (typical selectivity to resist in a contact etch ranges from 2 - 5:1), the AR will increase as the etch progresses, which the makes lag more severe as well. Besides the dependence on AR, the severity of etch lag is affected by the extent of polymerization in the etch chemistry, the pressure, and the local density of the pattern being etched. Increasing any of those factors will make etch lag worse. Many different mechanisms have been proposed to account for etch lag. For example, increased AR could result in reduced mass transport, inhibiting the ability of reactants to diffuse into, and products to diffuse out of, small spaces. Physical or geometric shadowing could reduce ion bombardment at the bottom of a high AR feature, reducing the ion stimulated reactions that drive the etch. Excessive polymerization (or lack of polymer removal) would exacerbate ion shadowing effects. Furthermore, ion deflections by surface charges and by collisions in the sheath could slow the relative etch rates of high AR features, because a higher percentage of ions would be deflected into the sidewalls than in a low AR feature. This effect would also reduce the ion bombardment at the bottom of the contact, thereby reducing the etch rate. Although all of these mechanisms may contribute to etch lag, empirical evidence seems to indicate that collisions in the sheath are the most important.

A plasma sheath is a region depleted of electrons between the plasma and an adjacent surface. (The sheath forms because, due to their different masses, electrons and ions pass from the plasma to a nearby surface at different rates. Most of the potential difference between the plasma and the surface is contained in the sheath.) As ions traverse the sheath, they are scattered by collisions with neutral species. Collisions in the sheath are a function of both the sheath width, d, and the mean free path, L. The critical parameter is the ratio d/L because the probability of a collision within a distance ∂ is:

[17] $P(\partial) = 1 - \exp(-\partial/L)$

So if the sheath width is equal to the mean free path, then the probability of an ion having a collision in the sheath is 0.63. Doubling the mean free path while holding the sheath width constant reduces the probability of a collision to 0.39. Therefore, to avoid collisions in the sheath, the mean free path should be much greater than the sheath width.

Lowering the pressure will increase both the L and d. However, L will increase faster than d, so overall, collisions in the sheath are reduced as the pressure is lowered. This is consistent with observations that the incidence angle of argon ions changes from about 18° off normal to 10° upon reducing the pressure from 200 to 30 mT.(110) Other work found that the probability of a CF_4 molecule experiencing a collision in the sheath was about 0.6 at 5 mT (d ~ L under the conditions studied).(109) At pressures above 10 mT, the collision probability was essentially unity. At higher pressures, ions underwent multiple collisions while traversing the sheath.

From the above discussion, it is obvious that one way to reduce etch lag is to lower the pressure. Lower pressure does reduce ion scattering and the associated etch lag. However, notice that to avoid essentially all collisions, it is necessary to operate at extremely low pressures (< 10 mT). Not only is this pressure range beyond the capability of many current etch tools, but there are also penalties associated with operating in the low pressure regime. As might be expected, etch rates slow dramatically due to the deficiency of active etchant species. In addition, low pressures reduce selectivity and increase DC self bias voltages which can lead to increased RIE damage. Also, very low pressures can cause the profiles at the bottom of etched features to distort into the so-called dovetail shape (Figure 54).

Some of the problems associated with low pressure etching can be mitigated by changing the etch chemistry. By increasing the polymerization rate of the system (e.g., replacing CF_4 with CHF_3), improvement in profiles can be obtained, although at the expense of etch rate. A more elegant process change is to use electronegative gases. Electronegative gases are those which produce high concentrations of electronegative species, such as fluorine radicals, in a discharge. By using an electronegative gas, such as SF_6, the sheath width can be greatly reduced. That effectively decreases d/L and therefore reduces the sheath collisions and etch lag. (Because sheath width is roughly proportional to DC bias, the effects of using an electronegative gas can be approximated by monitoring the DC bias.)

Damage and Residues: Once the substrate has become exposed in a contact etch, continued exposure to the plasma (for example, during the overetch) can result in damage to the contact. In addition, residues can be left in the contact consisting of plasma derived polymers, ILD material, and sputtered substrate. Both the damage and the residues can degrade the electrical performance of the contact. Cleaning the residues out of contacts is done using the same methods that are used to clean vias. That subject is discussed in Section 3.4.

One of the most common etch induced problems is dopant deactivation. Although both n- and p-type dopants are deactivated to some extent, (112) the problem is much more severe with p-type dopants. In particular, boron can be deactivated by hydrogen derived from etch gases such as CHF_3.(112 - 114) The deactivation is caused by hydrogen atoms complexing with the boron atoms, which prevents them from acting as acceptors. Annealing the wafers after etch has a positive, but somewhat variable, effect. Annealing breaks the B-H bonds and causes the hydrogen to out diffuse. Interestingly, low temperature anneals (100 - 200° C) show an initial increase in sheet resistance due to hydrogen redistribution.(113) Continued annealing at low temperature decreases the sheet resistance as the hydrogen diffuses out of the silicon. Besides the effects of hydrogen, thin areas of lattice damage, caused by ion bombardment during the etch, also appear to contribute to degradation of electrical performance.(114) All of these considerations mean that it is important to minimize total etch times so that any loss of electrical performance is minimized.

Summary: Like all plasma etches, contact etch must balance several opposing and interacting forces to achieve the desired result. A successful contact etch must have high selectivity to substrate, generate good profiles, and produce no unannealable damage; all achieved with acceptable etch rates, good CD control, and good uniformity. Achieving good selectivity is accomplished by having a significant rate of polymerization. Good profiles are generated by fine tuning the polymerization rate so that it is not too fast or too slow. Damage is minimized by ensuring that the length of the etch is not excessive. Although the rules outlined in this section are qualitative (by necessity --

quantitative rules are specific to individual etch tools), it is hoped that they will serve as a guide toward the goal of producing a good contact etch.

3.2 Metal Etch

Overview: Metal etch presents some of the most difficult challenges in etch in the entire MLM module. There are several reasons for this: (1) Etch chemistries are very aggressive to photoresist, which makes selectivity a concern and maintaining CDs difficult, (2) The presence of alloying agents such as copper can result in residues being left after etch, (3) The presence of barrier metals or caps such as TiW and TiN requires that dissimilar materials must be etched without undercutting or leaving a "foot," and (4) Post etch corrosion is always a major concern, especially in copper containing alloys. None of these problems is unique to an MLM system, however, the presence of several levels of metal means that there are that many more opportunities for one or more of these problems to occur.

Aluminum and aluminum alloys are by far and away the most commonly used metals in MLM schemes. The alloys usually consist of aluminum with relatively small percentages of copper and/or silicon added to aid in the prevention of electromigration and junction spiking. Other metals such as tungsten and molybdenum (115) have been used (see Chapter 3). Because of the relatively high resistance of these refractories compared to aluminum, they are often used only for the first level of metal. Other materials that are often encountered in metal etching are Ti, TiW and Ti/TiN barriers, TiW or TiN caps for stress relief or to act as anti-reflection coatings (ARCs), and oxide or other hardmasks to assist in maintaining CDs. A good metal etch, therefore, may be required to etch through several different materials, and result in an anisotropic etch with no residues, no corrosion, minimal undercut or foot, good CD control, and minimal loss of underlying substrate.

Because of their popularity, the bulk of this section will concentrate on aluminum and aluminum alloys. In fact, most of the discussion will concern aluminum alloys, because pure aluminum is not used in advanced MLM interconnect systems. Other metals will be briefly discussed near the end. Additional information on metal etching can be found in several good review articles.(88, 116 - 119) Please note that for convenience, we will generally use the term "aluminum" to refer to both aluminum and aluminum alloys.

Etch Chemistries: An aluminum etch consists of several steps. First, a breakthrough, or initialization, step is needed to remove the native oxide that forms when the metal is exposed to air. This native Al_2O_3 is not attacked by chlorine or HCl plasmas.(120) Although the oxide is only about 50 Å thick, it is tenacious and a highly physical, sputter etch is required to remove it. Therefore, discharges which produce relatively heavy ions are used under conditions of low pressure and high DC bias. Mixtures such as BCl_3/Cl_2 and $Ar/BCl_3/Cl_2$ are in common use. However, many other gases can be used for this step, e.g., CCl_4 or pure argon. Although it is believed that ion bombardment or sputtering is the primary mechanism for removal of the native aluminum oxide, it has been suggested that chemical reduction by BCl_x or CCl_x species may play a role.(117)

The next step, usually called the main etch, is designed to etch away the bulk of the aluminum. Chlorine based etches are almost universally used. In fact, Cl_2 and BCl_3 are almost always found in the main etch chemistry. Chlorine is the primary etchant and BCl_3 acts as a source of heavy ions for bombardment. The BCl_3 also contributes to the atomic chlorine concentration in the plasma by way of dissociation reactions. Silicon tetrachloride ($SiCl_4$) is sometimes used instead of BCl_3.

At normal temperatures, the etch by-product of chlorine based etches is Al_2Cl_6, (120, 121) which is not very volatile. Because of this, metal etchers are operated at elevated temperatures, commonly around 50 - 90° C. Not only must the cathode (where the wafer sits) be heated, but the entire chamber and sometimes the vacuum plumbing are heated to prevent a build up of Al_2Cl_6 in the system. At temperatures below about 125° C, aluminum bromide is somewhat more volatile than the chloride. Bromine based etches have therefore been examined. Numerous combinations of Br_2, BBr_3, and HBr, combined with Ar, Cl_2, and BCl_3 have all been investigated.(122 - 126) These etches tend to have excellent selectivity to photoresist (as much as 15:1 has been reported (125)). However, difficulties with profile control and slow etch rates have forestalled the widespread use of bromine based etches.

Numerous other gases are used to adjust the chemistry of the main etch. The addition of carbon containing gases promotes polymer formation which is essential to CD and profile control (see below). This is generally referred to as "adding carbon to the system," although it is not the carbon alone that forms the polymers. The gases most commonly used for this purpose include CCl_4, $CHCl_3$, CF_4, CHF_3, and CH_4. In general, hydrogen containing gases will produce more polymer than a corresponding hydrogen free gas. For example, under a given set of conditions, more polymer will usually be produced from CHF_3 than from the same flow of CF_4. Nitrogen may also be added to assist in resist erosion; in that case the photoresist is used to contribute carbon to the system. Oxygen may also be added to further adjust the amount of carbon available for polymer formation. The role of oxygen is to reduce the amount of polymer present, rather than add to it.

In some etch tools, carrier gases such as argon or helium are used. These are needed in higher pressure systems to achieve uniform and stable plasmas. Argon may also be added to provide a source of heavy ions, which contribute to the directionality of the etch by way of ion bombardment.

After the aluminum is etched, the photoresist must be stripped before the wafers are exposed to air so that corrosion problems are avoided. Oxygen based plasmas are used for this step. Sometimes, species such as N_2O are added to the plasma to increase the etch rate. Ammonia has also been added to decrease sensitivity to post etch corrosion.

Finally, fluorine-chlorine exchange reactions are often carried out, again to reduce the potential for post etch corrosion. In these reactions, residual chlorine bound to the aluminum after the main etch is replaced by fluorine. This is covered in more detail in the section on corrosion, below.   The most common gases used for this purpose are CF_4 and CHF_3.

Critical Dimensions: Metal critical dimensions (CDs) are especially crucial in MLM systems.   Metal levels in MLM modules will be connected to subsequent levels by way of vias.   So in addition to the negative effects on current carrying capability, metal CD loss in an MLM module will consume overlay tolerance for the alignment of a via level to the previous metal level.   If the CD loss becomes too large, slight misalignment can cause the via to miss part or all of the metal.   This would at best result in an unreliable via, and at worst would result in a totally failed, open via.   As metal lines get smaller and smaller, less and less CD loss is tolerable.   Not only are the lines getting smaller, but the spaces between the lines are shrinking as well.   Although relatively rare, CD growth can also occur.   With very small spaces, CD growth can easily cause bridging of metal lines and the shorting that goes along with it. An increase in CDs is usually attributable to prograde profiles, so the reader is referred to the section on profiles for information concerning the control of overly large CDs.   This section will address the much more common problem of CD loss.   Because it is often impossible to achieve zero CD loss ($0.1\ \mu$ loss is not uncommon), reticles are sometimes sized so that metal features are slightly larger than the final design requires.   This so-called biasing of the reticles affords some relief to the metal etch.   However, as minimum spaces decrease, limitations in the ability of steppers to print the small spaces present on biased reticles are beginning to limit the applicability of this technique.

Atomic and molecular chlorine will etch aluminum in the absence of a plasma.(117, 127)   This plasma free etching is chemical in nature, and as such is completely isotropic.   Because of this, a scheme is needed to force the etch to be anisotropic.   Anisotropy is achieved by the combination of two effects: ion bombardment and sidewall passivation.   The passivation is often achieved by adding carbon to the plasma, in the form of the gases such as $CHCl_3$ and CF_4, or by causing the photoresist to erode at a controlled but acceptable rate.   In a plasma, the chemically unsaturated carbon species derived from these sources polymerize and deposit on both vertical and horizontal surfaces on the wafer. Etch conditions are selected such that positive ions bombard the wafer surface from a direction perpendicular to horizontal surfaces.   Due to the normal incidence of the ion bombardment, polymer deposited on the horizontal surface is removed, while the polymer formed on vertical surfaces remains in place and protects the sidewalls from lateral attack by the active etchants.

Although addition of carbon containing gases is popular, there is a trend in the industry to eliminate their use.   The main reason for this trend is a manufacturing issue; addition of these gases causes etch chambers to become dirty faster (polymer deposits throughout the chamber, not just on the wafer), and therefore incur more down time for cleans.   To maintain CDs and profiles without the addition of carbon, conditions are selected so that the photoresist erosion rate is enhanced.   Under these conditions, the photoresist is the sole source of carbon for sidewall passivation. Because of the proximity of the resist to the wafer surface, polymers derived from it are more likely to deposit on the wafer than are polymers derived from added gases.   Therefore, little or no added carbon is required.   Because less carbon is added, and because most of the deposition is on the wafer and not on the chamber walls, the etch chamber remains relatively clean. A difficulty with this approach is that the resist erosion rate must not be excessive. Obviously, enough resist must remain throughout the etch so that the integrity of the pattern is not compromised. Despite this difficulty, successful etches using this approach have been realized. These etches normally use BCl_3 and Cl_2, sometimes combined with N_2 or Ar, as the only gases in the main etch. Power and pressure can be adjusted to fine tune the resist etch rate. Increasing power or decreasing pressure will result in a higher resist erosion rate and greater anisotropy, but there is a risk of CD loss from an excessive erosion rate.

Another approach to CD control involves the use of hardmasks or non-erodable masks. Hardmasks consist of thin layers of a material, usually SiO_2, deposited directly on top of the metal. The resist for the metal etch is also used to define the pattern in the hardmask. Although not widely employed, the presence of a hardmask can be of enormous benefit if resist breakdown occurs. Resist breakdown refers to partial or total loss of photoresist over a feature during the course of the etch. As the resist erodes, it tends to pull back from the edges of the features being defined. If a hardmask is not present, the underlying metal is exposed and etched, resulting in ragged metal and reduced linewidth. Hardmasks are especially useful when the metal to be etched is very thick.

Although hardmasks can be of great assistance in maintaining metal CDs, there are several penalties associated with their use. First, it complicates the process by introducing another layer that must be both deposited and etched. When deciding on the need for a hardmask, it is important to consider that the hardmask etch must be done with the metal photo pattern in place. Loss of resist will occur during the hardmask etch, so overall selectivity (of the entire metal etch process) will be reduced compared to a process without a hardmask. Long overetches (on the order of 100 %) are often used on the hardmask to ensure that the hardmask itself is clear and does not end up blocking the etch. These long overetches add to the overall selectivity loss. Furthermore, the hardmask etch may leave veils of hardmask material which need to be removed (see Veils and Veil Removal section), and the long overetch will cause the metal containing component of the veils to be worse. Finally, it must be decided where to etch the hardmask. For example, if the hardmask is made of oxide, the etch can be done in an oxide etch tool or in the metal etcher itself. When done in an oxide etch tool, an additional processing step is added, which is undesirable. And if the metal being etched is a copper containing alloy, etch blocking can occur (see Corrosion section). On the other hand, if the hardmask is etched in a metal etcher, then the tool is being required to do something that it was not designed for. It may be difficult or even impossible to develop an acceptable hardmask etch in a metal etch system. Furthermore, the latter approach requires oxide etch gases to be introduced into a metal etch tool, and potential memory effects may lead to deleterious effects on subsequent processing steps.

Despite all of the above considerations, hardmasks are being successfully used in MLM processes today.(128) By carefully choosing etch conditions, all of the above problems have been solved. In fact, in some MLM systems, where the final metal level is several microns thick, the use of a hardmask is essential to CD control since the resist thickness is limited, and some resist breakdown is certain to occur.

Profiles: Although it is desirable for metal etch profiles to be near 90°, many deviations from the ideal are possible. Profiles may turn out to be retrograde, where the width of the top of a feature is larger than the bottom, or prograde, where the top is narrower than the bottom (Figure 57). More subtle phenomena can also occur. For example, isolated features can be prograde while dense (tightly spaced) features on the same die have good profiles. Alternatively, isolated features may have excellent profiles, while dense features are bowed. All of these problems can and do occur, and the following are some general methods of correcting these profile problems.

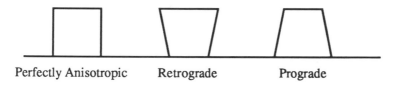

Perfectly Anisotropic   Retrograde   Prograde

Figure 57. Some possible profiles of etched metal lines.

If all features are either retrograde or prograde, the solution is relatively simple. If the profiles are retrograde, the cause is generally ascribed to loss of sidewall passivation. A retrograde profile can come about as follows: as the etch proceeds deeper into the metal, the surface of the metal being etched, and therefore the sidewall at that point, becomes further and further removed from the photoresist. If the resist is providing most of the sidewall protection and its erosion rate is too slow, then it is unable to provide adequate protection. In such a case, the metal nearest the resist (i.e., near the top of the metal) may be sufficiently protected by the resist derived polymers, but further down, the polymer from the resist is depleted to where it can no longer adequately protect the sidewall. So, as the etch goes deeper into the metal, the protection degrades and the metal is undercut. The profile that results is retrograde. The solution to this problem is to increase the amount of carbon in the etch chemistry. This can be done by adding carbon containing gases or by increasing the resist erosion rate. If carbon containing gases are being used, increasing the pressure can also effectively increase the carbon content. Additionally, lowering the total flow rate will increase residence times and may allow sufficient polymer formation to take place. In any case, the increased polymer formation on the sidewalls of the metal protect it from lateral attack during the etch. Note that only a small addition of carbon should be necessary. If gross amounts of carbon were needed, then the profiles would not have turned out retrograde. Instead, a large CD loss would have occurred from undercut at both the top and bottom of the metal feature.

Prograde profiles, on the other hand, are usually caused by either excessive polymer deposition or by insufficient ion bombardment. Either of these, or a combination of the two, will allow polymers to build up on surfaces being etched. If the deposition rate is not enough to block the etch completely, it can still prevent etching on slightly sloped sidewalls. If this occurs, prograde profiles will result. The solution will be found by decreasing the polymer deposition, increasing the ion bombardment, or some combination of both. The polymer deposition can obviously be decreased by lessening the flows of any carbon containing gases or by increasing the selectivity to photoresist. Increased selectivity to resist can be achieved by raising pressure or lowering power. Also, reducing the flows of gases such as nitrogen will lower the resist erosion rate. In addition, raising the substrate temperature will tend to reduce polymer deposition by decreasing the overall adsorption of polymer precursors (actually by increasing the desorption rate). If carbon containing gases are used, lowering the pressure will tend to lessen polymer formation, and will also increase the ion bombardment energies (less energy loss due to collisions in the sheath). Both of these are beneficial. Increased ion bombardment can also be achieved by raising the RF power.

Retrograde and prograde profiles are problems that must be overcome, but are readily understandable in the context of polymer deposition and sidewall passivation. More interesting problems arise when the profiles are pattern dependent. The most commonly observed effects of pattern dependent metal etching are: [1] isolated line profiles are good, but tightly spaced lines have bowed profiles, and [2] tightly spaced features have good profiles but isolated lines are prograde. Both of these are caused by the way ions are deflected from the perpendicular around and in these particular features. Conditions of low pressure and high power tend to give rise to [1], while high pressure and low power tend to result in [2].

In the first case, where isolated features are good but dense features exhibit bowed profiles, too much ion scattering is taking place in the tight spaces. This scattering results in off-normal ions colliding with sidewalls in the small spaces, which breaks down the sidewall protection and allows the profiles to bow. The scattering in this case arises from reflections off of faceted photoresist (Figure 58). The solution is to increase the pressure and lower the power. Both of these have the effect of reducing the ion energy, so that collisions with sidewalls do not remove enough polymeric passivation to allow lateral etch. Addition of carbon to the system can also be beneficial by increasing the thickness of the sidewall protection, and therefore reducing the likelihood that it will break down.

The second case, where isolated features are prograde but dense features have good profiles, can result from high recombination rates of active etchants and high passivation rates. This is similar to the situation when all features are prograde, but in this case, ions deflections in tight spaces tend to straighten the profile and prevent it from becoming prograde. The solution here involves lowering the pressure and raising the power. Lower pressure reduces both the recombination and passivation rates. Increased power removes more polymer as it forms (by increasing the ion bombardment energies) and therefore reduces the overall passivation rate. Increasing the chlorine flow and raising the cathode temperature will also help reduce the extent of these differential profiles. The increased chlorine flow serves to replace etchant lost to recombination, and raised cathode temperature reduces the extent of passivant formation.

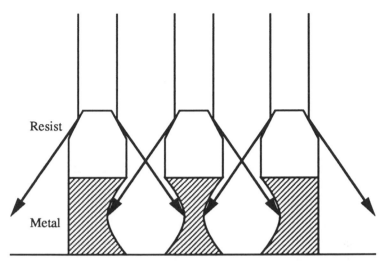

Figure 58. Generation of bowed profiles from ions scattering off of faceted photoresist. Note that metal profiles are straight on the edges of the array, where no adjacent feature is present to cause scattering.

Selectivity to photoresist: As with any etch, selectivity to both photoresist and underlying substrate is important. Low selectivity to photoresist can lead to small CDs and ragged, notched, or even missing metal features. If a process has marginal selectivity, then small, isolated metal features will degrade first. These features will be ragged and have small CDs, while denser features will look good. They degrade first simply because there is less resist per unit area (relative to denser features) to provide carbon for polymer formation and the resultant sidewall protection. Note that good selectivity to photoresist in metal etching is only in the range of 2:1 - 3:1.

Selectivity to photoresist is especially difficult in metal etching. The chlorine based etch environment is very aggressive towards resist. However, it is not the chlorine chemistry itself which is so aggressive. Note that polysilicon can be etched in chlorine based chemistries without extensive attack of the resist. Rather, it is the etch by-products that contribute to the breakdown of the resist. Monomeric aluminum chloride ($AlCl_3$) is a Lewis acid and is commonly used in organic chemistry to catalyze Freidel-Crafts alkylation and acylation reactions.(129) It is likely that phenyl groups present in the novolac resins of positive resists react in this fashion and contribute to the breakdown of the resist.(117) The elevated etch temperatures needed to volatilize the Al_2Cl_6 by-product also increase the resist etch rate.

There are a number of strategies which can be used to insure sufficient resist integrity. First of all, the photoresist can be subjected to deep ultraviolet radiation and a high temperature bake. This so-called "deep uv" treatment is done after the resist has been patterned. The initial exposure to deep uv serves to create a tough, highly crosslinked skin on the resist. The presence of the skin makes it possible to hardbake the resist at higher than normal temperatures without flow or reticulation. During the bake, the resist further crosslinks, which renders it more durable to the hostile conditions found in metal etching plasmas. On the etch side, sources of carbon can be added to the plasma chemistry to increase the selectivity to resist. Commonly used for this purpose are CF_4, CHF_3, and $CHCl_3$. As discussed above, addition of these gases has the added benefit of contributing to sidewall passivation and, therefore, CD control. Minimizing the length of the breakthrough step will also help preserve the resist. Breakthrough steps are, by design, highly physical in nature and therefore very low in selectivity. Similarly, minimizing the length of etch and overetch steps will help preserve the resist. It is therefore best to endpoint etch steps wherever possible, and use the smallest percentage of overetch that will ensure that the wafer is completely clear. Endpointing aluminum etches can be readily done using optical emission spectroscopy.(130) This is done by monitoring an emission line of an aluminum containing species and observing the intensity drop as the wafers clear. A commonly used emission for this purpose is the 396 nm line generated by the decay of excited aluminum atoms. Finally, selectivity to resist generally increases in a given etch system as the pressure is increased and the power is decreased. These parameters must be varied with caution, however, since anisotropy will tend to suffer as the pressure is raised and the power is reduced.

Selectivity to substrate: If selectivity to the underlying substrate is low, the aspect ratio of small spaces can increase significantly. This can give rise to difficulty with the subsequent filling of these spaces with the next level of ILD. Excessive substrate loss will also increase step heights after ILD deposition and make it more difficult to planarize. Typical oxide losses from metal etching are around 1 - 2 kÅ. Organic ILDs etch faster; losses around 3 kÅ are not uncommon. In an MLM system, these step height increases are often additive, and so the effect of ILD loss increases from level to level. Also, if the ILD gets too thin, it is possible that device performance will be degraded due to increased RC delay and crosstalk between levels. For all of these reasons, it is critical to maximize the selectivity of metal to underlying substrate.

The most common ILD is some form of silicon dioxide. The most common oxides used are PETEOS deposited oxide, PSG, and BPSG. Loss of these oxides in metal etching can come both from sputtering and from chemical etching if any fluorine was added during the etch (for example, to assist in maintaining CDs and profiles). The sputter loss of ILD is aggravated if a high bias overetch is used to clear residues that may be left from alloying agents commonly present in the aluminum. This is especially true when copper is added to the metal, since copper etch by-products are not volatile and are usually sputtered away.

Because the ILD begins to erode as soon as the metal starts to clear, minimizing etch and overetch times helps minimize the loss of ILD. By endpointing the etch, the minimum etch time can be achieved. Additionally, minimizing the physical component of the etch will tend to reduce the loss of oxide, not only because the sputtering is reduced, but also because oxide requires physical bombardment to achieve an appreciable chemically enhanced etch rate. Both of these approaches must be tempered by the fact that the etch must be residue free. To remove residues, some sputtering (for copper residues) or fluorine plasma treatment (for silicon residues) is needed. So some loss of oxide is inevitable.

Significant oxide can also be lost if fluorine containing gases such as CHF_3 were used during the main etch. These gases assist in profile and CD control, but deposition of fluorocarbon polymers will occur if they are used. This polymer will not restrict itself to the wafer surface, but will be deposited all over the inside of the plasma chamber. It is possible for subsequent etch steps, such as an in situ resist strip, to volatilize the polymer and release significant concentrations of fluorine radicals, which will attack the oxide. By cutting back on the amount of fluorocarbons in the etch, the oxide lost through this mechanism in a given etch run can be minimized. It is also helpful to reduce the build up of fluorocarbon polymers from run to run. If fluorocarbon polymers are allowed to accumulate, the ILD loss will become more severe as more wafers are processed. Here, adding small amounts of oxygen to the plasma during the steps that use fluorocarbons is beneficial. Including a small amount of oxygen in those steps can greatly reduce the accumulation of fluorocarbon polymers while not undermining their useful effects. An alternative is to periodically clean the chamber by running an oxygen plasma on dummy wafers. However, this latter approach incurs the penalty of a loss of throughput.

A special case of selectivity to substrate occurs when polyimide or another organic ILD is used. Selectivity to organic ILDs is comparable to that of resist, so it is only on the order of 2 - 3:1. This means that relatively large losses of ILD (around 3 kÅ) will occur during the metal etch. There are also many commonly used procedures that are no longer applicable because the ILD will be severely attacked. For example, complete in situ stripping of the resist can not be done without disastrous results. However, a short oxygen plasma to remove the highly chlorinated skin is possible, and it does contribute significantly to corrosion reduction. Similarly, fluorine/chlorine exchange reactions must be used with caution so that unacceptable amounts of the ILD are not removed.

Metal Stacks: Metal one is often different from other layers due to the presence of a barrier metal. The barrier metal is present to prevent junction spiking, which occurs due to the alloying of aluminum and silicon. Later levels of metal may contain similar underlayers which are used, for example, to line vias to assist in filling them with metal. Problems that can occur with such structures are undercutting of the barrier or liner, undercutting the aluminum, and leaving a "foot" of underlayer material (Figure 59). Metallization schemes that use plug technology for contact filling have no barrier metal exposed during metal one etch, so these concerns are eliminated.

Commonly used barriers or liners include TiW, TiN and Ti/TiN. All of these can be etched in fluorine and chlorine containing chemistries. Chlorine and BCl_3 mixed with fluorocarbons or SF_6 are commonly used. Although most of the general principles outlined above apply to underlayer etch, some additional pitfalls can occur because of the need to etch two dissimilar materials. Note that it is necessary to etch both the metal and the underlayer in situ. That is, it is not practical to etch the aluminum in one tool and the underlayer in another, because the aluminum will usually corrode if it is exposed to air after metal etch with the resist in place (see Corrosion section). Repatterning the wafers with photoresist between the metal and underlayer etches is not practical because of the alignment limitations of steppers. There is also a risk of causing corrosion in some metal systems, especially those containing copper, when the metal is

exposed to photoresist developers. Because of these difficulties, the burden falls
on the etch to produce good profiles from a single process.

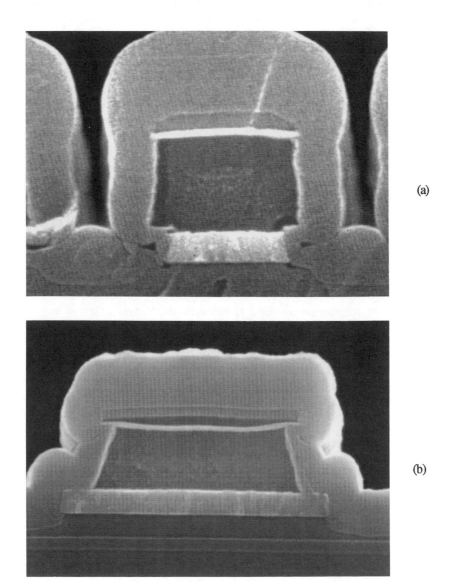

(a)

(b)

Figure 59. Potential profile problems with metal underlayers. (a) Undercut,
(b) A "foot."

A common problem is maintaining anisotropic profiles through both the
aluminum and underlayer etches. The problem can manifest itself as either an
undercut of the underlayer or as the presence of a foot of underlayer material after

etch. Undercut can arise during either the underlayer etch itself, or can be generated in post etch steps. If it occurs during the underlayer etch, adding sources of carbon can help. However, if the underlayer is TiW and fluorocarbon gases are added, then the problem may be made more severe. This is because undercut of TiW can be caused by high concentrations of fluorine radicals.(131, 132) In that case, it may be preferable to reduce the addition of fluorocarbon, and, if necessary, increase the resist erosion rate slightly to aid in sidewall passivation. When TiW is used as the underlayer, increasing the chlorine content of the plasma, by adding either Cl_2 or BCl_3, may help. Chlorine helps passivate the tungsten in the underlayer sidewalls by forming involatile tungsten chlorides. Undercut of TiN/Ti can be caused by an excess of chlorine in its etch chemistry. The undercut can be reduced by lowering the chlorine concentration, adding a source of carbon, or increasing the resist erosion rate.

Undercut can also occur during post underlayer etch steps, for example, during a resist stripping step.(131) The mechanism involves revolatilization of fluorocarbon polymers that were deposited during previous etch steps. Reduction of the amount of fluorocarbons being used or limiting the amount of their deposition by introducing small flows of oxygen into the system can be of tremendous benefit. Care must also be taken when fluorine/chlorine exchange reactions are carried out to assist in corrosion prevention. The fluorine present in these plasmas can again cause undercut. Often, limiting the time of a fluorine/chlorine exchange step can eliminate undercut while providing sufficient corrosion protection.

Another common problem is leaving a foot of underlayer metal. The presence of a foot will tighten the space between the base of metal features which can increase crosstalk between metal lines. The small space at the base of the metal may be hard to fill with ILD. It may even bridge small spaces and cause electrical shorts. If a foot is present in small spaces, while it is absent in features next to large open areas, it is probably due to a form of etch lag. This type of etch lag phenomenon is becoming more important as the size of spaces is being reduced (see the Contact section for further discussion of etch lag). This problem is attacked in a similar fashion to that of feature dependent profiles (see above). To remove a foot that is only present in small spaces, it can be beneficial to raise the pressure of the underlayer etch. This will only have an effect if the foot is formed by a depletion mechanism, that is, one that would result in residues being left in small spaces. Increasing the pressure increases the concentration of active etchants, and tends to clear out these tight spaces.

Another mechanism giving rise to a foot has to do with the etch chemistry. As mentioned above, the etch chemistry for underlayer etch commonly contains fluorine and chlorine. With TiW, excessive chlorine in the underlayer etch can inhibit the removal of tungsten, since tungsten chlorides are not volatile. If enough tungsten chlorides are formed such that the ion bombardment can no longer remove them from horizontal surfaces, then a foot can result. The obvious solution here is to reduce the amount of chlorine in the underlayer etch and/or overetch by reducing the Cl_2 or BCl_3 flow. Too much of a reduction, however, may cause the TiW to undercut.

When dealing with this sort of etch problem, it is important to do a stepwise evaluation of the etch. An excellent way to do this is to take cross section SEMs at several points in the etch. Good places to take the SEMs are after the aluminum etch, after the aluminum overetch, after the underlayer etch, and after the underlayer overetch. In general, a series of such SEMs will show exactly where the problem is coming from. For example, a foot could be forming because the aluminum etch is giving sloped profiles, and have nothing to do with the underlayer etch itself.

Corrosion: Whenever aluminum is etched, there is a potential for corrosion. Figure 60 shows some possible results of corrosion. This propensity to corrode is due to the electrochemical potential of aluminum; without its protective native oxide coating, aluminum would dissolve in water. Aluminum is also amphoteric, so it will be attacked by both acids and bases. Chlorine based etch chemistries only make matters worse, since chloride catalyzes the corrosion reaction. A simplified corrosion reaction scheme is presented in Equations 18 to 20. Note the catalytic nature: the HCl formed in Eq. 20 will continue to react with the aluminum through Eq. 19. Also notice that the hydrolysis of chlorine (Eq.18) is not necessary for corrosion to occur. If Al-Cl bonds are present, they will be attacked by moisture (Eq. 20) and start the corrosion process going (so aluminum chloride plays an important role in both corrosion and photoresist degradation, and must be carefully managed (133)). Finally, the corrosion by-product is represented here as a stoichiometric aluminum oxyhydroxide. In reality, mixtures of oxides, hydroxides, and oxyhydroxides are more likely to form. Because of its catalytic nature, small amounts of chlorine on a wafer can cause catastrophic corrosion once the wafer is exposed to the moisture in air. The addition of copper to the alloy only causes the metal system to become more sensitive to corrosion, because of the differing electrochemical potentials of the two metals. When aluminum and copper come in contact, an electrochemical corrosion cell, essentially a battery, is set up. Because aluminum is anodic relative to copper, the aluminum tends to chemically oxidize, and therefore corrode. Similar effects occur when aluminum is in contact with barrier metals.(134) Despite all of these obstacles, aluminum and its alloys are routinely processed today and corrosion is controlled.

[18] $$Cl_2 + H_2O \Rightarrow HOCl + HCl$$

[19] $$2Al + 6HCl \Rightarrow Al_2Cl_6 + 3H_2$$

[20] $$Al_2Cl_6 + 4H_2O \Rightarrow 2AlO(OH) + 6HCl$$

The corrosion reaction requires the presence of moisture to proceed. Therefore, the first line of defense in controlling corrosion is controlling humidity. In fab areas where aluminum is etched, low humidity is essential, especially if copper containing alloys are being used. Maintaining the relative humidity below 35 % (at 20° C) is generally sufficient, but for high copper alloys (2 to 4 % copper), it may be desirable to control the humidity to 25 % (at 20° C) or below. Controlling the humidity to such a low level is very expensive, however, and may require isolation of the etch bay from the rest of the fab. It is also somewhat misleading to refer to *relative* humidity as what is necessary to control, because the *absolute* humidity is what is actually important. Unfortunately, absolute humidity is difficult to measure. However, the relative humidity and the temperature can be used to determine the absolute humidity, and this in turn can be used to decide whether or not the moisture content of the air is sufficiently low.

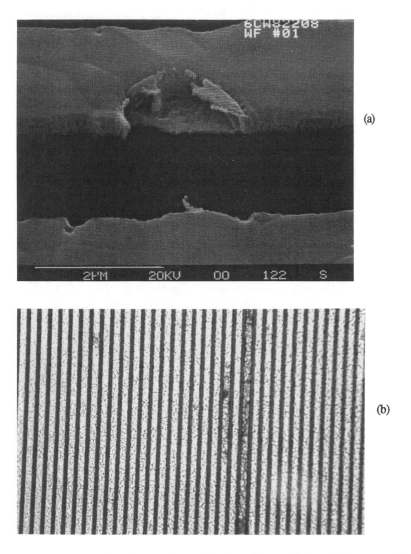

(a)

(b)

Figure 60. Two results of corrosion. (a) Loss of metal linewidth caused by post etch corrosion, (b) Optical micrograph of corroded metal lines. Corrosion can cause *many* more problems than these.

Besides control of humidity, there are many things that can be done with the metal etch process to render the wafers less sensitive to corrosion. One common practice is to remove as much chlorine as possible from the wafers before they are exposed to the air. There are a number of ways to reduce the chlorine concentration. First, it is crucial to remove some, and preferably all, of the photoresist before the wafers are exposed to air. In fact, if no resist is removed, the wafers will often corrode in the time it takes to get them to a microscope for inspection. This rapid corrosion occurs because, during the course of a metal etch, chlorine from the plasma is incorporated into the resist, mostly in a thin skin near the surface of the resist. Once exposed to ambient moisture, the chlorine rapidly hydrolyzes to HCl (Eq. 18) and attacks the aluminum. If the resist is stripped in the metal etcher, the chlorine is removed along with it. Stripping the resist goes a long way toward reducing the sensitivity of wafers toward corrosion. However, note that if the metal being etched is on a polyimide substrate, then it is not possible to strip the resist completely since the polyimide would also be removed. In that case, a partial resist strip is still very effective in reducing the tendency towards corrosion. For polyimide substrates, the strip should be designed to remove several hundred Ångstroms of resist, which will remove the bulk of the chlorine rich skin.

Stripping the resist makes such a big difference that most modern metal etch tools incorporate stripping capabilities into their design. However, stripping the resist does not remove all of the chlorine from the wafer surface. There is still enough chlorine remaining bound to the aluminum that corrosion can, and will, still occur; it will just take longer to become apparent. There are several solutions to this problem. One is to immerse the wafers in deionized water immediately upon their removal from the etch tool.(135) The water rinse serves to wash the chlorine away from the wafer where it can do no harm. The water will also help the aluminum regrow its protective oxide coating. This approach, resist strip followed by immersion in water, works very well with pure aluminum and aluminum-silicon alloys. However, if copper is added to the alloy, electrochemical corrosion will still occur. In fact, water immersion often worsens corrosion in Al/Cu alloys. A slightly more sophisticated approach has been taken by some etch tool manufacturers, who have incorporated a nitrogen purged, water rinse station as an integral part of the etch system. The water rinse can be done at either ambient or elevated temperatures, after which the wafer is spin dried. The wafer can be further dried with high temperature nitrogen after this water rinse. This type of system has been used effectively with copper containing alloys.

Although resist stripping and water rinsing are effective in controlling corrosion, they are not necessarily sufficient for all metallization systems and etch tools. As mentioned above, water rinses may actually cause corrosion rather than prevent it. Furthermore, stripping photoresist does not remove all of the chlorine on the wafer. Some chlorine remains, bound directly to the aluminum. Additional practices which have been proven effective are fluorine/chlorine exchange reactions and polymeric encapsulation. The idea behind F/Cl exchange reactions is to remove the chlorine bound to the aluminum. Usually, this is accomplished by simply exposing the etched and stripped wafers to a fluorine containing plasma. Gases such as CF_4 and SF_6 can be used for this purpose. The highly reactive fluorine radicals readily displace chlorine that is bound to aluminum. Thus, Al-Cl bonds are replaced by Al-F bonds. Whereas Al-Cl bonds will react with ambient moisture and start the corrosion process going, Al-F bonds are very stable and do not react. Furthermore, fluorine will not catalyze any corrosion reactions. This process is commonly used with much success in the industry.

Polymeric encapsulation is also used to control corrosion. In this method, a thin, hydrophobic, fluorocarbon polymer is intentionally deposited on the wafer surface to act as a barrier to moisture when the wafer is removed from the etch tool. This film is deposited as the last step in the etch process, and is only an interim measure; it must be removed. The best gas for this process is CHF_3. The idea is to protect the wafers from moisture until another treatment can be applied. Although the removal of this film would seem to add an additional processing step, in reality it may not, because metal etched wafers are often ashed in a separate tool after metal etch (despite the fact that the resist has already been removed), in an attempt to grow back the protective native aluminum oxide. The plasma ash will, of course, remove the polymer that was deposited.

Some other approaches to controlling corrosion include the regrowth of the protective native oxide mentioned above and volatilization of chlorine from the wafer surface. Regrowth of the native oxide may be done in a furnace under oxygen (135) or in an oxygen plasma. Furnace methods usually expose wafers to elevated temperatures (300 to 450° C) for extended periods of time (30 to 60 minutes). This not only regrows aluminum oxide, but also tends to drive off any remaining chlorine. Plasma grown aluminum oxides are known to be of low quality, nevertheless, experience has shown that exposure of etched wafers to such a plasma is an effective anti-corrosion measure. Another treatment is to expose the wafers to vacuum at elevated temperatures in an attempt to volatilize any chlorine on the wafer, and thus reduce their tendency to corrode. Usually, temperatures around 200 - 250° C suffice. With either method, it is important to be aware of the temperature if copper containing alloys are used. Temperatures around 325° C will cause Ø-phase Al_2Cu to form, which can give rise to another set of corrosion problems (see below). It is best to use temperatures below about 250° or above 400° C with Al/Cu. At the lower temperatures, copper diffusion does not take place to any appreciable extent. The higher temperatures are sufficient to keep high copper dissolved and uniformly distributed. If higher temperatures are used, it is best to cool the wafers quickly so that Ø-phase does not have a chance to form.

Native oxide regrowth and chlorine volatilization can be combined by exposing the wafers to an oxygen plasma at elevated temperature. In order to use this method successfully, it is necessary to apply this treatment before the wafers have been exposed to air. This fact, in turn, implies that the etch tool must be capable of varying the wafer temperature over the entire range of temperatures needed for the etch, 50 - 90° C during the etch and 150 - 250° C during the passivation. The only practical way to implement this today is with a multiple head etch tool, or cluster tool, with separate chambers configured to accommodate the different steps.

As can be seen from the above discussion, there are a large number of weapons capable of attacking corrosion. It is rare that any single one of these methods is sufficient in eliminating corrosion. Generally, some combination of several of these techniques is required. The particular techniques that are needed depend upon such things as the alloy that is being etched and the tool that is available to do the etch. In addition, some of these approaches are either not applicable or are very limited if an underlayer is present, or if an organic substrate is being used. For example, F/Cl exchange reactions must be judiciously applied when a underlayer is present, because the underlayer may undercut severely. Native oxide regrowth is not possible with polyimide ILDs. However, by careful application of these methods, very robust processes can be developed and corrosion can be kept under control.

Before leaving the subject of corrosion, there are additional concerns with copper containing alloys that should be addressed. These are forms of corrosion that, although they do not come from the etch itself, can affect the results of the etch. In copper containing alloys, there is always a possibility that the Ø-phase of the alloy will form. Theta phase Al/Cu is stoichiometric Al_2Cu, and is copper rich relative to the bulk of the metal, even in the higher copper alloys (up to 4 % Cu) used in microelectronics manufacturing. The copper rich Ø-phase forms by taking copper from the surrounding alloy, so it is generally located in the center of a copper poor region. The copper poor region is anodic relative to the Ø-phase, and therefore a small electrochemical corrosion cell is set up. All that is necessary to get corrosion from this system is water. Although moisture in the air will generally not cause any of the so-called "Ø-phase corrosion" to occur, immersion in water or water containing solutions will. Pure water is not particularly aggressive toward Ø-phase (although given sufficient time, water will attack metal with Ø-phase present), but many wet processing solutions commonly used to process wafers are. Examples are developers and post etch veil removal solutions.

When Ø-phase corrosion occurs, the metal will have small (about 1 µ diameter) "fish eyes" on the surface, which can be observed with an optical microscope. In the center of the fish eye will be a nodule of Ø-phase (Figure 61). Surrounding the Ø-phase, and making up the bulk of the fish eye, will be corrosion by-products comprised of aluminum oxide and oxyhydroxides. These corrosion by-products are sufficiently thick to cause etch blocking. So if Ø-phase corrosion occurs during resist development or rework, circular residues will remain on the wafer after etch. The corrosion by-products are also insulating, so if the Ø-phase corrosion occurs after etch (for example during veil removal), and if a via is subsequently opened on that spot, a failed via can occur.

The best solution to this problem is to minimize the exposure of wafers to water. It also is beneficial to use a protective layer on top of the metal. If such a layer is used, then the top of the metal is protected during photolithographic and post etch processing. A thin TiN capping layer, which is often present as an anti-reflective coating, is sufficient to prevent these problems. Thin TiW caps, used for stress relief, and, of course, hardmasks will also reduce concerns about Ø-phase corrosion. Note that after etch, metal sidewalls are still exposed and are sometimes attacked. Fortunately, sidewalls seem to be less sensitive to this attack. It is possible that material sputtered onto sidewalls during the etch affords some protection.

Veils and Veil Removal: Metal etch has a strong physical component needed to achieve anisotropy, and because of that, material accumulates on the sidewalls of the resist in the course of an etch. This material will generally contain sputtered metal and ILD, and can also include components

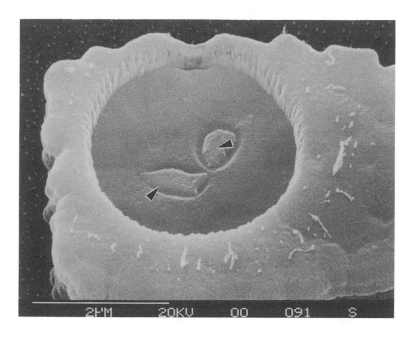

Figure 61. A scanning electron micrograph of a "fish eye." Note the nodules of Ø-phase in the center (arrows).

of any hardmask, cap, or sublayer that is part of the metallization system. Once the photoresist is stripped, the inorganic components of this accumulated material will remain. This remaining material is referred to as "veils," rabbit ears", or "picket fences" (Figure 62). The veils may fall inward onto the metal lines, stand up relatively straight, or droop outward. If they are not removed, they will disrupt the subsequent ILD layer. For example, with a conformal oxide deposition as the next ILD, if the veils droop outward, they will prevent the oxide from filling smaller spaces between metal features. If the veils stand up, and polyimide is spun on as the ILD, puddling of the polyimide will occur resulting in unacceptable thickness variations in the ILD. In fact, no matter which way the veils fall, and no matter what material is used for the ILD, the veils will cause problems. Therefore, they must be removed.

The exact veil composition will depend on such factors as the composition of the metal alloy, the composition of a hardmask (if present), the composition of any capping materials or sublayers that are in use, and the composition of the underlying ILD. Metallic components of the veils will tend to be at least partially oxidized, because the photoresist will have been stripped in an oxygen based plasma. So veils are a complex mixture of materials all of which must be removed, preferably in a single step, without detrimental effects to the metal features. Wet chemical treatments are used to perform this task.

The most common veil removal processes are based on alkaline solutions or solutions based upon buffered oxide etchant (BOE). Alkaline-based processes often use wet resist stripping solutions to remove the veils. The BOE based processes are often formulated in polyhydric alcohols (glycols) to reduce the attack of acid on the metal.

Figure 62. An SEM of veils after metal etch.

Etching Metals Other than Aluminum: Aluminum based alloys are not the only metals that are used in MLM systems. Other metals, most notably tungsten, are also in use. Multilevel metal schemes have been developed that use refractory metals for all levels of metallization. However, it is much more common to find metals such as tungsten employed only at first metal because of their relatively high resistance compared to aluminum. This section will briefly discuss issues associated with etching these materials. The emphasis will be on tungsten etching, because it is the most commonly used refractory metal in MLM processing today.

Using tungsten as first metal has some processing advantages. Because it is easy to deposit tungsten using CVD techniques, excellent conformality can be achieved (Figure 63). This fact means that tungsten can more readily fill small contacts than can aluminum. From an etch perspective, a major advantage of tungsten is that it does not corrode. However, etching tungsten does have its complications. Although no underlayer metal is required, tungsten does not stick well to oxide. So a glue layer must be used and therefore must be etched. The most common glue layers are TiW and Ti/TiN. The presence of a glue layer brings into play the same potential for undercutting or leaving a foot that barriers or liners have in aluminum etching.

Tungsten etch processes have been developed using both fluorine and chlorine based etches. Chlorine based etches (136 - 138) suffer from very low selectivity to resist, and are not in use much today. An extremely wide variety of fluorine based etches have been developed. For the most part, these etches use SF_6, NF_3, or CF_4 as the primary etch gas. However, if they are used alone, these gases all result in severe undercut of the mask. So it is common practice to combine these gases with others so that an acceptable etch results. For example, additions of CHF_3 (139) or CH_4 (140) will reduce lateral attack of the tungsten by passivating sidewalls with deposited polymers. Chlorine can also be added to maintain profiles. Tungsten chlorides are not very volatile, so any that form on sidewalls help to reduce the extent of lateral etching. Because the addition of chlorine causes loss of selectivity to resist, bromine has been added in the form of HBr or CH_3Br.(140) Bromine addition helps maintain profiles, but is less aggressive to resist than chlorine. Additional sidewall protection can be realized by adding CCl_4, which contributes chlorine and polymers, or N_2, which enhances polymer formation by way of increased resist erosion. Because ion bombardment is an important factor in the desorption of the main etch by-product WF_6, (139, 141) heavy ion precursors such as Ar and BCl_3 are also used. Consequently, typical tungsten etch chemistries consist of mixtures of gases, optimized to give the best compromise between etch rate, selectivity, and profile.

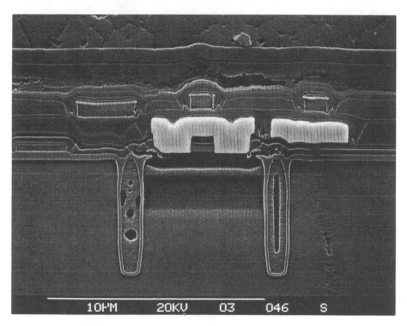

Figure 63. An SEM cross section of an MLM system using tungsten as first metal. Note the conformal filling of the contact. Also note that the anisotropic profile of the tungsten also extends through the TiW glue layer. Metals 2 and 3 are Al/Cu.

Once the tungsten is etched, the glue layer must be etched without undercut or leaving a foot. Processes similar to those used when etching underlayer metals under aluminum work well. Mixtures of fluorine and chlorine generally give the best results. Most tungsten etch chemistries will also attack the glue layer, however, under most tungsten etch conditions, glue layers will undercut once they begin to clear. For TiW glue layers, chemistries such as CHF_3/Cl_2 are effective. TiN can be etched in chlorine based chemistries such as Cl_2/HBr.

TiN has also been reported to be an effective etch stop when tungsten is etched in a $SF_6/Cl_2/HBr$ mixture.

Another metal which finds limited use in MLM is molybdenum. Molybdenum is usually etched in chlorine based chemistry. Carbon tetrachloride combined with oxygen has been found to be effective, (142, 143) as have $SF_6/Cl_2/O_2$ (142) and Cl_2/O_2 mixtures (144). In chlorine/oxygen chemistries, the by-products of the etch are molybdenum oxychlorides, $MoOCl_3$ and $MoOCl_4$, both of which are volatile. Molybdenum on TiW has been etched using SF_6/Cl_2 in a one step process.(142)

Summary: Metal etches are generally chlorine based. However, chlorine gas alone will not etch metal anisotropically. To achieve anisotropy, additional gases must be added. Most commonly used are fluoro- and chlorocarbons to increase sidewall passivation and heavy ion precursors to enhance ion bombardment. Fluoro- and chlorocarbons can also help maintain CDs. Inert gases may also be added to increase resist erosion which promotes sidewall passivation. Selectivity to resist in metal etching is comparatively poor, but adding carbon containing gases, reducing the power, and increasing the pressure can raise selectivity to an acceptable level. If underlayers are present, they must be etched in situ without undercut or leaving a foot. Care must be taken to minimize the tendency towards corrosion, not only immediately after the etch when the wafers are exposed to air, but also during any post etch processing such as veil removal. In situ resist stripping, fluorine-chlorine exchange reactions, polymeric encapsulation, high temperature bakes, and native oxide regrowth are all proven anti-corrosion techniques.

3.3   Planarization Etch

Overview: After etched metal features are coated with an ILD, there are irregularities present on the surface. If these irregularities, or steps, are not smoothed out, several problems occur. There will be depth of focus issues in the photo area which may lead to poor pattern quality. Metal coverage over these steps will also be poor, which can lead to reliability problems or even broken metal lines. After metal etch, "stringers" (thin strips of metal left along the edges of steps due to the excess metal thickness over those steps) will typically be left behind. Stringers can cause catastrophic shorting to occur. These problems are illustrated in Figure 64. The way to avoid these problems is to planarize the ILD surface before the next level of metal is deposited.

There are many different planarization strategies, and Chapter 6 covers them in detail. In this section, we are only concerned with processes that involve etching, the so-called etchback processes. Etchback processes all use the same generic approach (Figure 65). After the ILD has been deposited, a sacrificial layer is introduced onto the wafer surface in such a way that its surface is essentially flat. An etch is then carried out which is carefully designed to etch the sacrificial layer and the ILD at the same rate (often called the non-selective or 1:1 step). The etch is run long enough to remove all or most of the sacrificial layer. By doing this, the smooth topography of the sacrificial layer is imparted to the underlying ILD. The two most common sacrificial layers are photoresist and spin on glass (SOG).

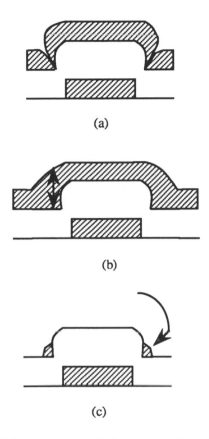

(a)

(b)

(c)

Figure 64. Some of the problems which can arise from lack of planarization. (a) Poor metal step coverage. (b and c) Metal stringers. The stringers result from the thicker metal initially over the edge of step in ILD, indicated by the arrow in (b).

The critical step in etchback methods is the 1:1 step. It is important to ensure that the 1:1 selectivity of the sacrificial layer to the ILD is not only achieved, but is also maintained throughout the course of the etch. If 1:1 selectivity is not maintained, gross deviations from planarity will occur (Figure 66). Note that by convention, etchback selectivity is cited as the ratio of the etch rates of sacrificial layer to ILD. Therefore, reference to selectivity greater than unity implies that the sacrificial layer is etching faster than the oxide. Because of its importance, most of this section concerns exactly how 1:1 selectivity is attained.

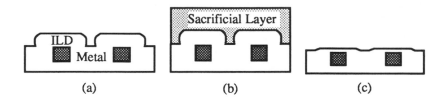

Figure 65. Etchback planarization scheme using sacrificial layer. (a) Initial structure, (b) After sacrificial layer has been applied, (c) After 1:1 etch.

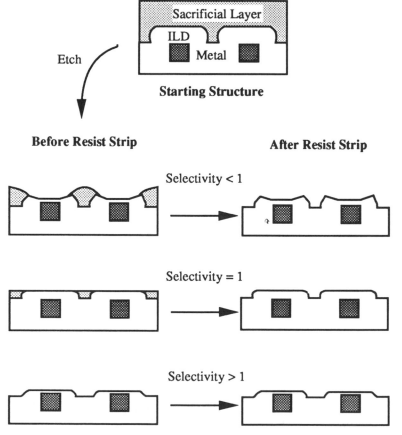

Figure 66. Results of etchback planarization at different selectivities. (a) Resist:ILD < 1:1. Note the shape of the ILD near the edges of the steps, (b) Resist:ILD ~ 1:1. Note that the step height here is less than that of either (a) or (c), (c) Resist:ILD > 1:1. Note the thickness of the ILD over the metal, and that the top of the ILD between the metal lines is below the top of the metal.

Bulk Resist Removal: The most common sacrificial layer is photoresist, which is typically spun on to a thickness of around 1 μ. When resist is used, the etchback is a two step process. The first step is a simple blanket organic etch, and the second step is the 1:1 etch. The first step is used to remove the bulk of the photoresist. Although this step is not strictly necessary (that is, a 1:1 etch step from beginning to end would suffice), it is included for throughput considerations -- the etch rate of the resist in this step is much faster than during the 1:1 etch. The second step, the 1:1 etch, is considerably more challenging.

The bulk resist removal step is readily done using an oxygen plasma. The amount of resist to be removed by this step depends upon the resist thickness, the height of the steps to be planarized, and the dimension and density of the features being planarized (Figure 67). Large metal features, such as bonding pads, will have full resist thickness on them after resist coat. Dense metal features will also have full resist thickness. This is because conformal ILD depositions will merge as small spaces between metal features fill. The ILD over isolated metal features will have less resist thickness over them, because as the resist is spun on, it will tend to go on flat over these features. So the resist thickness over a small isolated feature will be essentially the nominal resist thickness minus the step height. The thinnest resist will be found over metal features that are both small and far enough apart that the ILD does not completely fill the gaps between the metal lines. This situation will occur whenever the spacing between metal features is about 20 % greater than the ILD thickness, because conformal ILD grows laterally at about 60 % of the vertical rate. In this case, the resist both spins on flat and sinks into the dips in the ILD which occur between the underlying metal lines. These variances necessitate a compromise in the resist removal step. This step should go long enough to remove the bulk of the resist without exposing too much of the ILD in areas where the resist is thinnest.

Because the degree of planarity can be greatly affected by the amount of resist removed in resist etchback schemes, it is best to stop the resist removal step by endpointing it. Endpointing can be done using either optical emission or laser interferometric techniques. In the former, the intensity of an appropriate emission line is monitored for changes. For example, the 483 nm CO line can observed and, as oxide becomes exposed, its intensity will drop because there is less resist exposed to the plasma. By stopping the etch when the intensity has reached an empirically determined value, consistent amounts of resist can be removed from run to run. Alternatively, or simultaneously, the oxygen emission line at 777 nm can be monitored for an increase in strength.

In laser interferometry, a laser beam is reflected off of the sample into a detector. Because some of the beam reflects off of the top surface of the resist, while some of it penetrates to the underlying substrate, an interference pattern results. Because the resist thickness varies as the etch proceeds, the interference pattern changes with time as the two reflections move from constructive to destructive interference and back again. Therefore, monitoring the intensity of the self-interfering beam as a function of time results in a sinusoidal pattern. Each cycle corresponds to the removal of a fixed thickness of resist. The thickness removed is a function of the laser wavelength. Obviously, the etch can be stopped after a certain number of cycles have passed, which ensures that the same amount of resist has been removed in each run. However, it is important to realize that laser interferometry only measures the resist removed from the one point where the laser strikes the wafer. So if the uniformity of the etch is not good, planarity across the wafer surface may suffer. More importantly, when laser endpointing is used in a batch tool, all of the wafers are endpointed based upon one point on one wafer. Across batch uniformity is critical in such a case. Because optical emission samples the (relatively) homogeneous plasma, it is less than susceptible laser techniques to excursions in uniformity. Note that laser techniques can also be applied to endpoint the 1:1 etch.

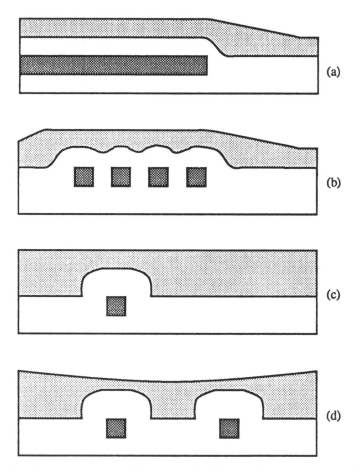

Figure 67. Resist coverage of features of differing pitch. (a) Large feature, such as a bonding pad. (b) Dense metal features. Note that the ILD has merged between the small spaces, and that the surface of the ILD is essentially planar. (c) Small isolated feature. Note that the resist is thinner over the top of the ILD covering this feature.   (d) Small features separated by greater than 1.2 X the ILD thickness. Note how the resist sags due to the volume to be filled.

Achieving 1:1 Selectivity: Most ILDs which require an etchback are composed of some form of silicon dioxide. Many different compositions are used, such as PETEOS deposited oxide, PSG, BPSG, etc. However, the general approach to achieving 1:1 selectivity is the same in all cases. Because oxide will be etched, it is common practice to use fluorocarbons as the etch gases. Etches are frequently based upon CF_4, CHF_3, and C_2F_6, or mixtures of them. However, pure fluorocarbon discharges will etch oxide more rapidly that photoresist. Therefore, just enough oxygen is added to the system to increase the resist etch rate to that of the resist. The exact ratio of fluorocarbon gas to oxygen will depend upon the particular oxide being etched, the choice of fluorocarbon, the conditions of the etch (power, pressure, etc.), and the etch tool being used (in fact, there are often small selectivity differences between two supposedly identical tools, which may need to be compensated for by adjusting flows slightly). For a given set of conditions, the oxygen flow is the easiest and most important variable to adjust when searching for a set of 1:1 conditions.

By varying the etch parameters, it is a relatively simple matter to find an appropriate oxygen flow that will give rise to 1:1 selectivity, at least at the start of the etchback. Unfortunately, during the course of the etch, the selectivity tends to drift away from the ideal 1:1 because the resist etch rate tends to accelerate.(145, 146) The reason for this is a form of microloading. At the beginning of the 1:1 step, the wafer surface is nearly completely coated with photoresist. As the etch progresses, more and more oxide is exposed at the expense of photoresist. As the area of exposed photoresist decreases, its etch rate increases while the oxide etch rate remains essentially unchanged. On going from 0 to 90 % exposed oxide, a doubling of the resist etch rate was observed while the oxide etch rate decreased only 10 %.(145) The longer the etch goes, the less resist is present, and the faster its etch rate becomes. So the rate at which the selectivity changes becomes worse as the etch progresses. Note that the accelerated resist etch rate results in reduced planarization -- oxide steps are greater than expected (Figure 66).

The reason why the resist etch rate is such a strong function of the amount of exposed oxide has to do with the concentration of oxygen in the plasma. As the amount of resist exposed to the plasma decreases, there is less of it present to consume oxygen, so the relative oxygen concentration of increases. Therefore, there is more active etchant per unit of resist, and the resist etch rate naturally increases. The excess oxygen present will also increase the amount of fluorine radicals present in the plasma, and fluorine consumes resist rapidly. Finally, as the oxide is etched, it contributes its oxygen to the plasma (the O in SiO_2). As more oxide is exposed and etched, more oxygen is contributed, and there is a corresponding rise in the resist etch rate. This explains why the increase of the resist etch rate associated with percent resist coverage is much reduced with aluminum (145) or silicon (146) substrates (compared to oxide).

To further complicate matters, the severity of the selectivity change is a function of the pattern density of the previous metal level.(146) If the metal level is very sparse, then these effects will be minimized because the photoresist will clear off of the wafer everywhere at nearly the same time. As the previous metal's pattern density increases, the drift away from average 1:1 selectivity increases since the percentage of resist coverage decreases as the etchback proceeds (more steps in the oxide give rise to more exposed oxide before all of the resist clears). In principle, if the metal density were to become very large, the severity of this problem would again start to diminish. In fact, it would go to zero in the fanciful extreme of 100 % metal pattern density, i.e., blanket metal. In practice, metal pattern densities high enough to diminish this problem are only found locally in current MLM structures.

One simple way to deal with the problem of drifting selectivity is to set up the etch so that the average oxide etch rate is slightly greater than the resist etch rate. This will result in decreasing the excessive oxide step heights that would otherwise occur. This adjustment can be achieved by decreasing the O_2 flow slightly. With a reduced oxygen flow, the resist etch rate will be slowed which will serve to offset the increased photoresist etch rate occurring later in the etch. A selectivity of 0.8:1 (resist:oxide) is often sufficient to realize improved local planarity. However, this method has the drawback that it does not have a built in capability to handle selectivity changes due to mask density differences. If the mask densities being processed are sufficiently different to cause a problem, different etches must be created for the different mask sets being run. In a fab running many different part types, this can rapidly become cumbersome.

A second method, although not in widespread use, is much preferable to the above technique. This approach uses some sort of feedback loop to dynamically adjust the selectivity while the etch is in progress. For example, by monitoring the 483 nm CO emission line, a multiple step process was developed which incrementally reduced the oxygen flow during the etch.(145) Such a process adjusts the etch chemistry to offset the microloading effects, and superior planarity can be achieved. This type of scheme also has the advantage of automatically compensating for pattern density differences. The disadvantage of this strategy is that it requires some method of determining how much oxygen is needed, and some way to change the oxygen concentration while the etch is in progress. Conceptually, emission spectroscopy can be used to monitor either by-products, such as the CO used above, or etchant concentrations such as that of oxygen (with, e.g., the 777 nm line). In practice, it is difficult to determine these concentrations accurately enough to attain a stable and repeatable process.

Finally, it is also important to realize that the selectivity will be altered by the cleanliness of the etch chamber. Compared to a clean chamber, a tool that has been contaminated with fluorocarbon polymers will be low in oxygen and high in fluorine radical concentration for a given planarization etch chemistry. This will increase the oxide etch rate relative to the resist etch rate. As more wafers are planarized, the chamber will become cleaner, and the resist:oxide selectivity will drift toward higher values. Contact and via processes tend to be high in polymer deposition. Because of this problem, most wafer fabrication areas dedicate separate tools to via etch and planarization. If it is not possible to dedicate tools, it is critical to clean the etch chamber with an oxygen plasma prior to a planarization etch if any via processing has been done since the last planarization.

Although the above discussion has referred to photoresist as the sacrificial layer, similar arguments apply when SOG is used instead. Spin on glass as a sacrificial layer is more common at ILD0 than at subsequent levels because the high temperature that are often used to cure SOG are not compatible with metallized wafers. However, there are some advantages to using SOG. It is not necessary to remove all of the SOG during the process since it is a glass and can withstand subsequent processing such as ILD or metal depositions. (It is critical, however, to remove any SOG that is located where a via will be placed, or there is a risk of "poisoning" the via from moisture outgassing from the SOG.) Because SOG is more like oxide than is photoresist, it would seem that obtaining a 1:1 selectivity is somewhat easier. However, some forms of SOG are derived from siloxane polymers and, even after curing, contain residual hydrocarbons. They are also typically less dense than other oxides used in semiconductor manufacturing. These factors cause the etch rate of SOG to be faster than other forms of oxide, and sensitive to such factors as the oxygen concentration in the plasma.

Planarization etches with sacrificial SOG once again are usually based on fluorocarbons, and so the use of CF_4, CHF_3, and C_2F_6 are common. Gases such as oxygen or helium are added to adjust the selectivity to the required 1:1. Just as in the case of resist etchback processes, the selectivity of SOG to oxide will depend on the composition of the substrate oxide, the choice of fluorocarbon, the etch conditions, and the etch tool. Microloading effects are also present. Also analogous to resist processes, altering the etch gas ratios is the easiest way to achieve the desired selectivity. For example, in a mixture of C_2F_6 and O_2, decreasing the O_2 flow decreases the SOG etch rate with little effect on oxide etch rate due to the hydrocarbon content of the SOG.(147) Note that with SOG, the thickness spun on is considerably less than that of resist in resist etchback schemes. The SOG thickness eliminates the need for the rapid initial (sacrificial layer only) etch used in resist etchback processes, which can improve process uniformity.

The SOG etch rate is also affected by parameters that are not applicable to resist processes. It has been shown that cure temperature has an effect on the SOG etch rate.(148) The etch rate also depends on whether the cure was carried out in vacuum or air. In general, conditions that render the SOG more like silicon dioxide and less like a siloxane polymer will reduce the SOG etch rate. Therefore, higher temperature cures, which densify the SOG better, and vacuum bakes, which remove a higher percentage of the solvent, both serve to reduce the etch rate and bring it closer to that of a true oxide.

Summary: Planarization etchbacks use a process with 1:1 selectivity of a sacrificial layer to ILD to transfer the inherent planarity of the sacrificial layer into the ILD. Either photoresist or spin on glass can be used as sacrificial layers. The 1:1 selectivity is achieved by adjusting the etch chemistry, which usually means adjusting the oxygen flow. Microloading effects can reduce planarity by causing the etch selectivity to move away from the desired 1:1. The drift is caused by the sacrificial layer's etch rate accelerating as it is removed. These microloading effects not only vary across a wafer with a given mask, but also vary with pattern density differences from mask to mask. Microloading effects can be compensated for by either adjusting the nominal selectivity away from 1:1, or by adjusting the chemistry dynamically as the etch progresses.

3.4 Via Etch

General Considerations: Via etch is very similar to contact etch. For oxide ILDs, many of the principals outlined in Section 3.1 on contact etch also hold true for via etching. The discussions in Section 3.1 concerning etch chemistries, tapered profiles, slope etch, and etch lag are just as valid for etching vias in oxide ILDs as they are for etching contacts. Therefore, we will not repeat them here. It should also be noted that, just as in the case of contact etching, it is very difficult to reliably endpoint a via etch. The reason for this is the same in both cases: the very small percentage of open area makes obtaining a strong enough signal for endpointing nearly impossible. Because of this, most via etches are stopped after a predetermined amount of time has passed.

There are some significant differences between via and contact etching, however. One obvious difference is that via etches stop on metal, whereas contact etches stop on silicon, silicides, and poly. This fact makes it easier to achieve selectivity to substrate in a via etch, especially with aluminum based metallizations. If the substrate is molybdenum or tungsten, then selectivity will be slightly less compared to aluminum. However, the required selectivity is still easy to achieve. With aluminum metallization, potential corrosion of the metal after via etch (especially during post via cleans) becomes a concern. There is, of course, no such concern after a contact etch.   Because via etches expose metal levels, post etch cleans must also be different from their contact counterparts. Via etching becomes radically different from contact etching when an organic ILD such as polyimide is used instead of oxide. Via etches may also be required to remove capping materials, such as antireflective coatings, present on top of the metal. Finally, there is an issue of interfaces present at the bottom of vias after the next level of metal has been deposited. Via etches can contribute to this undesirable interface, and need to be designed to minimize or eliminate its presence.

Organic ILDs: When polyimide or other organic materials are used as an interlayer dielectric, the via etch must be considerably different than when the ILD is composed of oxide. First of all, the etch chemistry must be changed. Instead of fluorine based plasmas, organic ILDs are etched with oxygen plasmas. Low pressure and high bias conditions are used to achieve anisotropy. However, oxygen plasmas are good for etching most organics, so the selectivity to resist is nil. Due to the lack of selectivity, hardmasks are often used when etching vias in organic ILDs. Oxide and various metals (e.g., Al, Mo, and Ti) have been used as non-erodable masks in polyimide etching. Oxide hardmasks are generally preferred because they are non-conductors. The absence of selectivity also means that the resist thickness used to pattern organic ILDs must be greater than the ILD itself (especially with no hardmask). Additionally, since organic ILDs planarize when they are applied, the thickest part of the ILD must be considered when the resist thickness is chosen.

Sloped or tapered profiles can also be generated in organic ILDs. To do this, a hardmask must be used, again because of the 1:1 selectivity with photoresist. Typically, a high pressure, low bias oxygen plasma is used first to generate the slope in the via profile. This is followed by a low pressure, high bias step to generate the anisotropic portion of the via. Because the high pressure step is largely isotropic, this approach can be used with or without a hardmask. If no hardmask is present, then resist erosion techniques can also be used, just as in the case of inorganic ILDs. Either flowing the resist prior to etch or purposely faceting the resist during the etch can be used to generate tapered profiles.(149)

If a hardmask is used, then it may be necessary to remove it. Oxide hardmasks can be removed in fluorocarbon plasmas or in wet dips. Such things as CF_4 plasmas or dips in 50:1 HF solutions work well. Conducting hardmasks must be removed.

Post Etch Cleans: Residues, often called veils, are generally left inside of vias after etch. These veils interfere with the subsequent metal filling of the vias, and poorly filled or voided vias result. Therefore, the veils must be removed after the vias are opened. Veil removal can be done by wet chemical methods or by a combination of wet and dry processes. The exact method that is appropriate will depend upon the composition of the veils, which in turn depends upon the materials being used in the MLM system.

Veils will be composed of complex mixtures of metal and ILD material, which are sputtered onto sidewalls during the etch, and plasma deposited polymers generated from the via etch chemistry. As such, they typically contain oxides of Si and Al, as well as fluorocarbon polymers. In addition, since aluminum becomes rapidly fluorinated upon exposure to a fluorine containing plasma, (150, 151) aluminum fluorides may be present on the sidewalls of the via and on the surface of the exposed metal substrate. Veils may also contain material derived from metal caps if they are present, so oxides of Ti and W may also be included. These materials are not deposited in the via simultaneously. Rather, they tend to be laid down one after another as the materials from which they are derived are sequentially exposed. That fact can make the veils surprisingly difficult to remove.

Veils must be removed without attacking either the ILD or the underlying metal. The veils must also be removed without causing the metal to corrode. Most veil removal processes are either wet chemical treatments, or combinations of wet and dry treatments. Wet chemicals used to remove veils are generally alkaline. Many commercially available photoresist strippers, often based on organic amines, can be used effectively to remove the inorganic component of veils. Solutions of tetramethylammonium hydroxide (TMAH) or other organic ammonium compounds are also effective. Resist strippers may also remove the organic portion of veils, however, there are times when they are not effective, and dry processes are needed. Photoresist ash cycles are usually adequate to remove the polymeric component of veils. If a combination wet/dry process is required, it is important to note that the order of the treatments affect the outcome. For example, wet/dry may work effectively, while reversing the sequence may not. This unexpected behavior has to do with the cross sectional composition of the veil. If organics were deposited first and followed by the inorganic components, then the inorganic material is exposed to the beginning of the veil removal process. Wet chemicals can remove these inorganics, but ashing will not. So, if an ash is done first, it will have little effect on the veil. The organic component will be protected from the ash by the intervening inorganic material. A subsequent wet process may remove the inorganics, but leave the organic components in the via. On the other hand, by running a wet/dry sequence, the inorganics will be removed by the wet treatment, and the organics will be left exposed for easy dry removal.

Caps and Interfaces: Thin caps of TiN or TiW are sometimes deposited on top of metals to act as antireflective coatings or to improve interconnect reliability. Because of interface issues with the underlying or next metal layer, the via etch is often required to remove them in order to improve via resistance. Fortunately, these thin coatings etch in fluorine based plasmas, and they will usually etch rapidly in the via etch chemistry itself. However, they can be more difficult to etch if they have alloyed with the underlying metal. Adding a short step designed to etch the cap will solve this problem. This step should be less polymerizing than the via etch itself, be higher in fluorine radical concentration, and use conditions of low pressure and high bias. Therefore, plasmas based on CF_4/O_2 or SF_6/Ar work well.

3.5   Pad Etch

The purpose of pad or passivation etch is to remove the passivating dielectric over bonding pads so that electrical connections can be made to them. Most passivation schemes use plasma deposited (or enhanced) nitride (PEN) or phosphosilicate glass (PSG), or layered combinations of the two. The combination of these two materials is intended to hermetically seal the die and provide them with a barrier to mobile ions. Both PEN and PSG are etched in fluorine based chemistries. It is possible to etch both layers in oxide-like chemistries. However, two step etches often give better results. For example, etching PEN in roughly equivalent flows of CHF_3 and O_2 affords some selectivity to the underlying oxide. Then for the PSG, increasing the CHF_3 to O_2 ratio provides a faster oxide etch rate which increases throughput.

Because bonding pads are large, CDs are not crucial. Because CD requirements are so lax, pad etch is often thought of as being trivial. However, most experienced etch engineers have had problems with pad etches at one time or another. For example, it is possible to undercut the PSG. If the wafers are to undergo a gold bump process before being packaged, then the process is almost certain to fail. This is because there will be a physical discontinuity in the metallic coatings which are sputtered onto the wafers prior to electroplating the gold. The discontinuity results in an open circuit in the electroplating bath, so the gold does not plate. Note that if wire bonding is used instead of gold bumping, slight undercut of the PSG will not be a concern. However, undercut of the PSG can cause reliability problems if it opens pathways for mobile ions to migrate into underlying layers. Pad etches also will generally result in the formation of veils which must be removed. Just as with contacts and vias, veils in pads can cause electrical discontinuities. Veil removal procedures are identical to those used after via etches (see Section 3.4). Finally, it is possible to alter the surface of the metal after a pad etch by, for example, forming aluminum fluorides on the surface (150, 151) or by roughening the surface sufficiently to cause a relatively thick layer of (native) aluminum oxide to form. In such a case, high resistance gold bumps or unreliable wire bonds can result.

By putting a little time and effort into the development of a pad etch, these problems can be readily avoided. Undercut can be eliminated by adjusting the polymer deposition rate of the etch. Using a two step etch makes this easier to do. Problems with altered aluminum surfaces often result from the fact that, in the interest of throughput, high powers are used to achieve fast etch rates. High powers result in high bias and high ion bombardment energies. Excessive ion bombardment tends to degrade the exposed metal surface. Reducing the power, while it may increase throughput slightly, will not only reduce the extent of degradation, but will also tend to reduce the extent of veil formation.

3.6   Summary

Plasma etching has developed over the years from a simple means of ashing photoresist to a technique of major importance in the semiconductor industry. Processes have been developed to etch such diverse materials as insulators, semiconductors, and metals. To develop useful plasma etches, it is important to understand not only the interactions of the large number of parameters that control the properties of the plasma, but also how those plasma characteristics affect the materials (both physically and electrically) exposed to the etch. To do this requires knowledge of many diversified fields, such as chemistry, physics, and materials science. It is hoped that the qualitative descriptions presented in this section have given the reader a basic understanding these interactions from a processing point of view, as well as a general idea of what to do when faced with specific processing problems.

There is no doubt that the future holds many challenges for etch. The semiconductor industry is driven by the need for high performance. It is this need that drives its technology. The two major ramifications that arise from this need are the unending trend toward smaller geometries and the development and application of new materials. Both of these factors will demand even greater performance from etch in the future.

The trend to smaller geometries means that it will be necessary to etch very small features. Presently, MLM modules with metal lines, spaces, contacts, and vias all being sized around 0.6 μ are nearing production. By the turn of the century, these sizes will be cut in half. As feature sizes shrink, the importance of such parameters as CDs, profiles, and selectivities becomes magnified. Metallization schemes are currently being developed that contain six levels of metals and organic ILDs. Meanwhile, increasing wafer sizes make it more difficult to achieve the uniformities necessary to produce good product over all of the wafer area. Many of today's etch tools will not be capable of producing acceptable features in the future. However, many new technologies are being actively developed by tool manufacturers which should be able to meet these needs.

The etcher of the future will probably make use of a high density plasma (HDP) source to generate active etchants. High density plasma sources make it possible to operate at very low pressures while minimizing ion energies and therefore etch induced damage. Among the approaches to HDP are electron cyclotron resonance (ECR), transformer or inductively coupled plasmas (TCP or ICP), and RF whistler or helicon sources. All of these approaches have much greater ionization efficiencies and lower ion energies than current tools such as magnetically enhanced reactive ion etchers (MERIE) (Table X). The lower pressures used in these technologies make it possible to etch small features, obtain anisotropic profiles, and minimize etch lag. The high density plasma sources generate sufficient numbers of active etchants to achieve good etch rates at the low pressures used. And the low ion energies minimize plasma induced damage and give these systems the potential for very high selectivity to resist. Unfortunately, it is not yet clear whether an independently controlled bias will need to be applied to the wafer to attain anisotropy, in which case the benefits of low ion energies are negated. It has also proven difficult to attain good uniformity in these systems, especially with larger wafer sizes.

Table X. Characteristics of Future Etch Technologies

Technology	Ionization Efficiency	Ion Energies (eV)
MERIE	10^{-4} - 10^{-3}	50 - 200
ECR	10^{-2}	10 - 50
ICP	10^{-2}	10 - 30
Helicon	10^{-1}	0 - 10

Other types of hardware innovations in store for the future revolve around controlling etch systems more tightly. For example, electrostatic chucks are the likely next step beyond today's helium backside cooling in controlling wafer temperatures more uniformly. Continued improvements in such things as mass flow controllers and pressure control will also improve the capability of etch tools.

The other set of challenges to etch comes from the application of new materials which are needed to meet the requirements of device performance. The driving force towards new materials in MLM schemes is RC delay, circuit delay caused by the resistance of conductors and the parasitic capacitance of insulators. To minimize RC delay, more highly conductive metals and low dielectric constant insulators are required. This need has led to investigation of copper metallization and organic ILDs. Copper is very difficult to etch by traditional methods. It can be etched at very high temperatures in chlorine based plasmas.(152, 153) However, major issues with corrosion and even simple oxidation of copper remain unresolved. The best approach for copper metallization developed so far is to etch grooves in the ILD where metal lines will be, blanket deposit the copper, and use chemical mechanical polishing to remove the excess. This scheme avoids copper etching entirely. It also eliminates the need for planarization etching, however, it increases the demands on oxide etching.

Whatever the future may hold, there is no doubt that many interesting problems and equally innovative solutions will be forthcoming in the area of etch.

Acknowledgements:

The excellent wafer support over the years of the ACT technicians is acknowledged by both of us (especially D. Rodgers by GWG). The excellent photo engineering technical support of K. Way, S. Malhotra, C. Smith, and W. Waldo is most appreciated by JNH. GWG would like to thank J. Pearse, T. Roche, R. Goodner, and S. Ramaswami for helpful discussions.

REFERENCES

1.   D.A. McGillis, "Lithography,"*VLSI Technology*, S.M. Sze, ed., 267 (1983).

2.   W. Waldo, Chapter 4 in Handbook of VLSI Microlithography , W. Glendinning and J. Helbert, Eds., Noyes Publications, Park Ridge , NJ (1991).

3.   S. Wolf and R.N. Tauber, "Silicon Processing for VLSI Era," Silicon Processing, 413, Lattice Press, Sunset Beach, CA. (1986); J.S. Peterson and A.E. Kozlowski, "Optical Performance and Process Characterizations of Several High Contast Metal-ion-free Developer Processes," SPIE 469, 46 (1984).

4.   L.F. Thompson, C.G. Willson, and M.J. Bowden, "Introduction to Microlithography," Wash. D.C.: ACS Symposium Seriess, 219 (1983).

5.   M.J. Bowden in: Materials for Microlithography, L.F. Thompson, C.G. Willson, and J.M.J. Frechet, Eds., ACS Symposium Series, 266, 39 Wash. D.C. (1984).

6.   M.C. King, "Principles of Optical Lithography," Chapter in VLSI Electronics-Microstructures Science, N.G. Einsproch, Ed., New York: Academic (1981).

7.   L.F. Thompson and R.E. Kerwin, "Polymer Resist Systems for Photo- and Electron Lithography," Annual Review of Materials Science, 6, 267 (1976).

8.   M.J. Bowden and L.F. Thompson, " Resist Material for Fine Line Lithography," Solid State Technology, May: 72 (1979).

9.   W. DeForest, "Photoresist Materials and Processes," New York: McGraw-Hill (1975); M. Kaplan and D. Meyerhofer, RCA Review, 40, 167 (1979).

10.   M.P.C. Watts, "A High Sensitivity Two Layer Resist Process for Use in High Resolution Optical Lithography (for VLSI)," SPIE, 469, 2 (1984).

11.   J.G. Calvert and J.N. Pitts, "Photochemistry," New York: Wiley (1967).

12. W. Arden, H. Keller, and L. Mader, "Optical Projection Lithography in the Submicron Range," Solid State Technology, July: 143-150 (1983).

13. P. Trefonas and B.K. Daniels, "New Principle for Image Enhancement in Single Layer Positive Photoresists," SPIE, 771, 194 (1987).

14. F.H. Dill, J.A. Tuttle, and A.R. Neureuther, "Modelling Positive Photoresist," Proceedings of Kodak Interface, 24, (1974).

15. W.P. Hornberger, P.S. Huge, J.M. Shaw, F.H. Dill, "The Characterization of Positive Photoresists," Proceedings of Kodak Interface, 44, (1974); M. Esterkamp, W. Wong, H. Damar, A.R. Neureuther, C.H. Ting, and W.G. Oldham, "Resist Characterization: Procedures, Parameters, and Profiles," SPIE 334, 182 (1982).

16. B.K. Daniels, P. Trefonas III, and J.C. Woodbrey, "Advanced Characterization of Positive Photoresists," Solid State Technology, September, 105-109 (1988).

17. T. Kitaori, S. Fukunaga, H. Koyanagi, S. Umeda, and K. Nagasawa, "A Study of Photosensitizer for I-line Lithography," SPIE 1672, 242 (1992).

18. C.A. Mack, "Development of Positive Photoresists," Journal of Electrochemical Society, 134, 148 (1987).

19. W. Waldo and J. Helbert, "Lithographic Process Development for High Numerical Aperture I-Line Steppers," SPIE 1088, 153 (1989); K. Way, C. Smith, S. Malhotra, and J.N. Helbert, Motorola ACT Technical Report No. 190, (1993).

20. J.A. Underhill, D.L. Lunding, and M.L. Kerbaugh, "Wafer Fatness As A Contributor to Defocus and to Submicron Image Tolerances in Step-and-Repeat Photolithography," J. Vac. Sci. Technol. B 5(1), 299 (1987).

21. M. Hanabata, "Material Design of Photo Sensitive Agent 0.35 μm Single Layer Within Target," Nikkei Microdevices, April, 51 (1992); M. Hanabata, F. Oi, and A. Furuta, "Novolak Design for High Resolution Positive Photoresists (IV) Tandem Type Novolak Resin for High Performance Positive Photoresists," SPIE 1466, 132 (1991).

22. D. DeMuynck, S. Malhotra, and J. Helbert, "Next Generation i-Line Photoresist Evaluation for 0.6μm Photolithography," Proceedings of 1991 Motorola SPS Technical Enrichment Matrix, 1, B.9 (1991).

23. J. Helbert, N. Saha, N. and P. Mobley, "Adhesion Aspects of Resist Materials," Opportunities and Research Needs in Adheison Science and Technology , G.G. Fuller and K.L. Mittal, Eds., 3-17, Hitex Publications, June (1988).

24. D. Meyerhofer, "Characteristics of Resist Films Produced by Spinning," J. Appl. Phys. 49, 3993 (1978).

25. J. Helbert, Chapter 2 in Handbook of VLSI Microlithography , W. Glendinning and J. Helbert, Eds., Noyes Publications, Park Ridge , N.J., (1991).

26. J.N. Helbert and N. Saha, "Application of Silanes for Promoting Resist Patterning Layer Adhesion in Semiconductor Manufacturing," in Silanes and Other Coupling Agents, K.L. Mittal, Ed., 439 (1992)

27. P. Sukanek, "A Model for Spin Coating with Topography," J. Electrochem. Soc. 136, 3019 (1989).

28.   J.W. Bossung, "Projection Printing Characterization," Proc. Soc. Photo-Optical Instrum. Eng., 100, 80 (1977).

29.   B.J. Lin, "Multi-Layer Resist Systems," Chapter 6 in "Introduction to Microlithography," L.F. Thompson, C.G. Willson, and M.J. Bowden, Eds., Wash. D.C.: ACS Symposium Seriess, 219 (1983).

30.   J.A. Bruce, J.L. Burn, D.L. Sunding, and T.N. Lee, "Characterization of Linewidth Variation for Single and Multiple Layer Resist Systems," in: Proceedings of Kodack Microelectronics Seminar-Interface (1986).

31.   T.R. Pampalone, F.A. Kuyan, "Improving Linewidth Control over Reflective Surfaces Using Heavily Dyed Resists," in Proceedings of the 1985 Kodak Microelectronics Seminar Interface, (1985).

32.   C.A. Mack, "Dispelling the Myths About Dyed Photoresists," Solid State Technology, 31, 125 (1988).

33.   M. Bolsen, G. Buhr, H.J. Merrem, and K. Van Werden, "One Micron Lithography Using a Dyed Resist on Highly Reflective Topography," Solid State Technology, 29, 83 (1986).

34.   C.L. Renschler, R.E. Stienfeld, and J.L. Rodriquez, "Curcumin As a Positive Resist Dye Optimized for g- and h-line Exposure," JECS, June, 1586 (1987).

35.   R. D. Coyne and T. Brewer, "The Use of Anti-Reflection Coatings for Photoresist Linewidth . Control," Proceedings of the Kodak Microelectronics Interface, 83. 40 (1983).

36.   Y.-C. Lin, A.J. Purdes, S.A. Saller and W.R. Hunter, "Linewidth Control Using Anti-Reflective Coating," IEDM International Electron Device Meeting, San Francisco CA, 399 (1982).

37.   C. Nölscher, L. Mader, M. Schneegans, "High Contrast Single-Layer Resist and Antireflection Layers--An Alternative to Multilayer," SPIE, 1086. 242 (1989).

38.   T.R. Pampalone, M. Camacho, B. Lee, and E.C. Douglas, "Improved Photoresist Patterning Over Reflective Topographies Using Titanium Oxynitride Antireflection Coatings," J. Electrochem. Soc., 136, 1181 (1989).

39.   B. Martin, A. N. Odell, J.E. Lamb III, "Improved Bake Latitude Organic Antireflective Coatings for High Resolution Metallization Lithography," SPIE, 1086, 543 (1989).

40.   Y-C. Lin, V. Marriott, K. Orvek, and G. Fuller, "Some Aspects of Anti-Reflective Coating for Optical Lithography," SPIE, 469. 30 (1984).

41.   S. Kaplan, "Linewidth Control Over Topography Using Spin-On Ar Coating," Proceedings of the KTI Microelectronics Seminar, 307 (1990).

42.   K. Harrison and C. Takemoto, "The Use of Anti-Reflection Coatings for Photoresist Linewidth Control," in Proceedings of the Kodak Microelectronics Seminar, Interface 83, (1983).

43.   S. Miura and C. Lyons, "Reduction of Linewidth Variation Over Reflective Topography," SPIE, 1674, 147 (1992).

44.   G.E.P. Box, W.B. Hunter, and J. Hunter, "Statistics for Experimenters," Wiley, New York (1978).

45. G.R. Bryce, and D. Collette, "Statistically Designed Experiments For Process Optimization," Microelectronics Manufacturing and Testing, Oct., 25 (1984).

46. C. Smith, J. Helbert, and B. Gadsen, "Target and Mapping Optimization for Improved Alignment Results for Advanced BICMOS Device Fabrication," Proceedings of 1992 Motorola SPS Technical Enrichment Matrix, 2, 51.1, (1992).

47. C.A. Mack, "Prolith: A Comprehension Optical Lithography Model," Optical Microlith,IV. Proc..SPIE, 538, 207 (1985).

48. J. S. Wilczynski, "Optical Step and Repeat Camera with Dark Field Automatic Alignment," J. Vac. Technol., 16 (6), Nov./Dec. (1979).

49. D. R. Beaulieu, et al., "Dark Field Technology - A Practical Approach to Local Alignment," SPIE, Optical Microlithography VI, 772, 142, (1987).

50. H. Ohtsuka, et al., "Simple Technique for Double Structured Alignment Marks for Field-By-Field Alignment," Proc. Kodak Interface, p. 1 (1985).

51. M.S. Wanta, et al., "Characterizing New Darkfield Alignment Target Designs," Proc. Kodak Interface, p. 169 (1987).

52. C. Lambson and A. Awtrey, "Alignment mark optimization for a multi-layer metal process," Proc. of the KTI Microlithography Seminar Interface '91, 38 (1991).

53. E.D. Casteel, et al., "Effect of Alignment Mark Design and Processing Factors on Overlay Accuracy," SPIE, Optical/Laser Microlithography II, 1088, 231 (1989).

54. P. Gaboury, et al., "Optimizing Metal Level Alignment Targets to Minimize the Impact of Process Variation on Target Acquisition," Proc. KTI, 115 (1991).

55. GCA Technical Publishing , "Autostep 200 System Operation Manual." Document # 06942461 , GCA Corp. ,(1989).

56. C. Smith, J . Helbert, and B. Gadsen, "Target and Mapping Software Optimization to Achieve Improved Dark Field Alignment Results for Advanced BICMOS Device Fabrication," Microlithography World in press (1993).

57. R.A. Johnson and D.W. Wichern , "Applied Multivariate Statistical Analysis," 171, Prentice Hall , Englewood Cliffs, N. J., (1988).

58. SAS/STAT User's Guide Vol. 2 Version 6.0 , 978, SAS Institute Inc., Cary , N.C., (1990).

59. J. Neter, W. Wasserman and M.H. Kutner, "Applied Linear Statistical Models," 527, Richard D. Irwin Inc., Homewood , Illinois, USA., (1985).

60. A. Martin, et al., "Elimination of Send Ahead Wafers in an IC Fabrication Line," SPIE, 1673, 640 (1992).

61. N. Shiraishi and S. Hirukawa, "New Imaging Technique for 64-M Dram," SPIE, 1674, 741 (1992); K. Tounai, H. Tanabe, H. Nozue, and K. Kasama, "Resolution Improvement with Annular Illumination," SPIE, 1674, 753 (1992).

62. B.J. Lin, "Quarter- and Sub-Quarter-Micron Optical Lithography," Patterning Science and Technology II, W. Greene and G.J. Hefferton, Eds., The Electrochemical Society, Inc. Penning, N.J., 3 (1992).

63. E. Tamechika, et al., "Investigation of Single Sideband Optical Lithography Using Oblique Incidence Illumination," J. Vac. Soc, Technol. B10, 3027 (1992).

64. B.J. Lin, J.A. Underhill, D. Sunding, and B. Peck, "Electrical Measurement of Submicrometer Contact Holes," SPIE, 921, 164 (1988).

65. J.D. Cuthbert, "Optical Projection Printing," Solid State Technology, 59, Aug. (1977).

66. M. Bolsen, G. Buhr, H.J. Merrem, and K. van Werden, "One Micron Lithography Using a Dyed Resist on Highly Reflective Topography," Solid State Technology, 83. Feb. (1986).

67. T. Brunner, " Optimization of Optical Properties of Resist Processes," SPIE, 1466, 297 (1991).

68. P. Singer, "Searching for Perfect Planarity," Semiconductor International, March, 43 (1992).

70. A. Schiltz and M. Pans, "Two-Layer Planarization Process," J. Electrochem Soc., 133, 178 (1986).

71. A. Nagy and J. Helbert, "Planarized Inorganic Interlevel Dielectric for Multilevel Metallization - Part I," Solid State Technology, Jan, 53 (1991).

72. A. Nagy and J. Helbert, "Planarized Inorganic Interlevel Dielectric for Multilevel Metallization - Part II," Solid State Technology, March, 77 (1991).

73. V. Comello, "Wafer Processing News," Semiconductor International, March, 28 (1990).

74. A. Szxena and D. Praminik, "Manufacturing Issues and Emerging Trends in VLSI Multilevel Metallizations," Proc. 1986 V-MIC Conf. p. 9, IEEE, New York.

75. M.J. Thoma, et al., "A 1.0 u m CMOS Two Level Metal Technology Incorporating Plasma Enhanced TEOS,"Proc. 1987 V-MIC Conference, 20, IEEE, New York.

76. D.N.K. Wang, S. Somekh, and D. Maydan, "Advanced CVD Technology," Proc. 1st Int'l Symp. on ULSI Science and Technology, 712 (1987).

77. L.E. Stillwagon, R.G. Larson, and G.N. Taylor, "Spin Coating and Planarization," SPIE, 771, 186 (1987).

78. L.E. Stillwagon and R.G. Larson, "Fundamentals of Topographic Substrate Leveling," J. Appl. Phys., 63, 5251 (1988).

79. R.H. Wilson and P.A. Piacente, "Effect of Circuit Structure on Planarization Resist Thickness," Proc. 1984 V-MIC Conf. p. 30. IEEE, New York.

80. L.E. Stillwagon, "Planarization of Substrate Topography by Spin-Coated Films: A Review," Solid State Technology, June, 67 (1987).

81. A. Lamola, C.R. Szmanda, J.W. Thackeray, "Chemically Amplified Resists," Solid State Technology, August, 53 (1991).

82. C. Garza, E. Solowiez, and M. Boehm, " Equipment, Materials and Process Interactions in a Surface-Imaging Process: Part II," SPIE, 1672, 403 (1992); M.A. Hanratty and M.C. Tipton, "Deep UV Lithography for Prototype 64 Megabit DRAM Fabrication," SPIE, 1674, 894 (1992).

83. D. M. Manos and D. L. Flamm, Eds., "Plasma Etching, An Introduction," Academic Press, San Diego, CA (1988).

84. B. Chapman, "Glow Discharge Processes," John Wiley and Sons, New York, NY, 1980.

85. J. W. Coburn, "Plasma Etching and Reactive Ion Etching," AVS Monograph Series, American Institute of Physics, New York, NY (1982).

86. C. M. Melliar-Smith and C. J. Mogab, "Plasma-Assisted Etching Techniques for Pattern Delineation" in "Thin Film Processes," J. L. Vossen and W. Kern, Eds., Academic Press, New York, NY, 1978.

87. J. A. Mucha and D. W. Hess, "Plasma Etching" in "Introduction to Microlithography. Theory, Materials, and Processing" L. F. Thompson, C. G. Willson, and M. J. Bowden, Eds., ACS Symposium Series no. 219 (1983) p. 215.

88. D. L. Flamm and V. M. Donnelly, "The Design of Plasma Etchants," *Plas. Chem. Plas. Proc.*, 1, 317 (1981).

89. S. Gupta, J-S. Song and V. Ramachandran, "Materials for Contacts, Barriers and Interconnects," *Semicond. Internat.*, 80, October, 1989.

90. V. M. Donnelly, D. L. Flamm, W. C. Dautremont-Smith, and D. J. Werder,"Anisotropic Etching of SiO_2 in Low Frequency CF_4/O_2 and NF_3/Ar Plasmas," *J. Appl. Phys.*, 55, 242 (1984).

91. R. A. H. Heinecke,"Control of Relative Etch Rates of SiO_2 and Si in Plasma Etching," *Solid State Electron.*, 18, 1146 (1975).

92. R. A. H. Heinecke,"Plasma Reactor Design for the Selective Etching of SiO_2 on Si," *Solid State Electron.*, 19, 1039 (1976).

93. R. d'Agostino, F. Cramarossa, V. Colaprico, and R. d'Ettole, "Mechanisms of Etching and Polymerization in Radiofrequency Discharges of CF_4-H_2, CF_4-C_2F_4, C_2F_6-H_2, C_3F_8-H_2," *J. Appl. Phys.*, 54 (1983).

94. L. M. Ephrath, "Selective Etching of Silicon Dioxide Using Reactive Ion Etching with CF_4-H_2," *J. Electrochem. Soc.*, 126, 1419 (1979).

95. R. A. H. Heinecke, "Plasma Etching of Films at High Rates," *Solid State Technol.*, 21, 104 (1978).

96. H. W. Lehmann and R. Widmer, "Profile Control by Reactive Sputter Etching," *J. Vac. Sci. Technol.*, 15, 319 (1978).

97. H. Toyoda, H. Komiya, and H. Itakura, "Etching Characteristics of SiO_2 in CHF_3 Gas Plasma," *J. Electronic Mater.*, 9, 569 (1980).

98. J. W. Coburn and E. Kay, "Some Chemical Aspects of the Fluorocarbon Plasma Etching of Silicon and Its Compounds," *IBM J. Res. and Develop.*, 23, 33 (1979).

99. E. A. Truesdale, G. Smolinsky, and T. M. Mayer, "The Effect of Added Acetylene on the RF Discharge Chemistry of C_2F_6. A Mechanistic Model for Fluorocarbon Plasmas," *J. Appl. Phys.*, 51, 2909 (1980).

100. S. Matsuo, "Etching Characteristics of Various Materials by Plasma Reactive Ion Etching," *Japan J. Appl. Phys.*, 17, 235 (1978).

101. F. White, W. Hill, S. Eslinger, E. Payne, W. Cote, B. Chen, and K. Johnson, "Damascene Stud Local Interconnect in CMOS Technology," *IEEE IEDM*, 301 (1992).

102. K. Ueno, K. Ohto, K. Tsunenari, K. Kajiyana, K. Kikuta, and T. Kikkawa, "A Quarter Micron Planarized Interconnection Technology with Self-Aligned Plug," *IEEE IEDM*, 305 (1992).

103. D. Jillie, P. Freiberger, T. Blaisdell, and J. Multani, "The Use of Orthogonal Design to Optimize a Sloped Contact Etch Process in a Single Wafer Etcher," *Proc. Electrochem. Soc.*, 87-6, 108 (1986).

104. T. Toyosato, T. Tamaki, and T. Tsukada, "High Selectivity SiO_2 Etching and Taper Angle Control by Wafer Temperature Control RIE," *Proc. Electrochem. Soc.*, 90-14, 716 (1990).

105. T. Ohiwa, K. Horioka, T. Arikado, I. Hasegawa, and H. Okano, "SiO_2 Tapered Etching Employing Magnetron Discharge of Fluorocarbon Gas," *Japan J. Appl. Phys.*, 31, 405 (1992).

106. J. Ding, K. Y. Fung, G. W. Hills, and X. C. Mu, "Characterization of an Isotropic Anisotropic Contact Etch," *Proc. Electrochem. Soc.*, 90-14, 437 (1990).

107. G. B. Powell, D. Drage, and W. G. M. van den Hoek, "Plasma Etch of Several Oxides at Elevated Temperatures," *Proc. Electrochem. Soc.*, 90-14, 452 (1990).

108. D. Chin, S. H. Dhong, and G. L. Long, "Structural Effects on a Submicron Trench Process," *J. Electrochem. Soc.*, 132, 1705 (1985).

109. H. C. Jones, R. Bennett, and J. Singh, "Size Dependent Etching of Small Shapes," *Proc. Electrochem. Soc.*, 90-14, 45 (1990).

110. R. A. Gottscho, C. W. Jurgensen, and D. J. Vitkavage, "RIE Lag and Micro-loading: What's the Difference," SRC Topical Conf. on Plas. Etching, 1992.

111. B. E. Thompson and H. H. Sawin, "Monte Corlo Simulation of Ion Transport through RF Glow-Discharge Sheaths," *J. Appl. Phys.*, 63, 2241 (1988).

112. J. M. Heddleson, M. W. Horn, S. J. Fonash, and D. C. Nguyen, "Effects of Dry Etching on the Electrical Properties of Silicon," *J. Vac. Sci. Technol.*, 6, 280 (1988).

113. J. Pearse and M. Grimaldi, unpublished results.

114. J. M. Heddleson, M. W. Horn, and S. J. Fonash, "Evolution of Damage, Dopant Deactivation, and Hydrogen-Related Effects in Dry Etch Silicon as a Function of Annealing History," *J. Electrochem. Soc.*, 137, 1892 (1990).

115. R. J. Saia and B. Gorowitz, "The Reactive Ion Etching of Molybdenum and Bilayer Metallization Systems Containing Molybdenum," *J. Electrochem. Soc.*, 135, 2795 (1988).

116. G. C. Schwartz, "Reactive Plasma-Assisted Etching of Aluminum and Aluminum Alloys," *Proc. Electrochem. Soc.*, 85-1, 26 (1985).

117. D. W. Hess, "Plasma Etch Chemistry of Aluminum and Aluminum Alloy Films," *Plasma Chem. Plasma Proc.*, 2, 141 (1982).
118. B. Gorowitz, R. J. Saia, and E. W. Balch, "Methods of Metal Patterning and Etching" in "VLSI Electronics: VLSI Metallization," v. 15, p. 159, N. G. Einspruch, S. S. Cohen, and G. Sh. Gildenblat, Eds., Academic Press, Orlando, FL (1987).
119. P. E. Riley, S. S. Peng, and L. Fang, "Plasma Etching Aluminum for ULSI Circuits," *Solid State Technol.*, 36, 47 (1993).
120. D. L. Smith and P. G. Saviano, "Plasma Beam Studies of Si and Al Etching Mechanisms," *J. Vac. Sci. Technol.*, 21, 768 (1982).
121. H. F. Winters, "Etch Products from the Reaction on Cl_2 with Al(100) and Cu(100) and XeF_2 with W(111) and Nb," *J. Vac. Sci. Technol.*, 3, 9 (1985).
122. K. Fujino and T. Oku, "Dry Etching of Al Alloy Films Using HBr Mixed Gases," *J. Electrochem. Soc.*, 139, 2585 (1992).
123. M. Yoneda, H. Sawai, N. Fujiwara, K. Nishioka, and H. Abe, "Highly Selective AlSiCu Etching Using BBr_3 Mixed-Gas Plasma," *Japan. J. Appl. Phys.*, 29, 2644 (1990).
124. A. L. Keaton and D. W. Hess, "Aluminum Etching in Boron Tribromide Plasmas," *J. Vac. Sci. Technol.*, A3, 962 (1985).
125. H. B. Bell, H. M. Anderson, and R. W. Light, "Reactive Ion Etching of Aluminum/Silicon in BBr_3/Cl_2 and BCl_3/Cl_2 Mixtures," *J. Electrochem. Soc.*, 135, 1184 (1988).
126. P. M. Schaible, W. C. Metzger, and J. P. Anderson, "Reactive Ion Etching of Aluminum and Aluminum Alloys in an RF Plasma Containing Halogen Species," *J. Vac. Sci. Technol.*, 15, 334 (1978).
127. D. L. Smith and R. H. Bruce, "Si and Al Etching and Product Detection in a Plasma Beam under Ultrahigh Vacuum," *J. Electrochem. Soc.*, 129, 2045 (1982).
128. S.R. Wilson, J. L. Freeman, Jr., and C. J. Tracy, "A Four-Metal Layer, High Performance Interconnect System for Bipolar and BiCMOS Circuits," *Solid State Technol.*, 34, 67 (1991).
129. R. T. Morrison and R. N. Boyd, "Organic Chemistry," Second Edition, Allyn and Bacon, Boston, MA, 1966.
130. D. M. Manos and H. F. Dylla, "Diagnostics of Plasmas for Material Processing" in "Plasma Etching, An Introduction," D. M. Manos and D. L. Flamm, Eds., Academic Press, San Diego, CA (1988) p. 312.
131. S. J. Irving and R. J. Trahan, "Sequential Plasma Etching of a Multilayer Metallization," *Proc. Electrochem. Soc.*, 85-6, 161 (1985).
132. N. Takenaka, D. Kimura, K, Nishizawa, and K. Sakiyama, "Reactive Ion Etching of an Al-Si/TiW Multilayer" in "Tungsten and Other Refractory Metals for VLSI Applications II," E. K. Broadbent, Ed., Materials Research Society, Pittsburgh, PA (1987) p. 371.
133. J. E. Spencer, "Mananement of $AlCl_3$ in Plasma Etching of Aluminum and Its Alloys," *Solid State Technol.*, 27, 203 (1984).

134. P. C. Chang and K. K. Chao, "Prevention of Corrosion During the Reactive Ion Etch of Aluminum on TiW," *Proc. Electrochem. Soc.*, 85-1, 12 (1985).

135. W-Y. Lee, J. M. Eldridge, and G. C. Schwartz, "Reactive Ion Etching Induced Corrosion of Al and Al-Cu Films," *J. Appl. Phys.*, 52, 2994 (1981).

136. D. S. Fischl and D. W. Hess, "Plasma-Enhanced Etching of Tungsten and Tungsten Silicide in Chlorine-Containing Discharges," *J. Electrochem. Soc.*, 134, 2265 (1987).

137. C. Kaanta, W. Cote, J. Cronin, K. Holland, P-I. Lee, and T. Wright, "Submicron Wiring Technology with Tungsten and Planarization," *Proc. IEEE VLSI Multilevel Interconnect Conf.*, 21 (1988).

138. C. Kaanta, W. Cote, J. Cronin, K. Holland, P-I. Lee, and T. Wright, "Submicron Wiring Technology with Tungsten and Planarization," *IEEE IEDM*, 209 (1987).

139. C-H. Chen, L. C. Watson, and D. W. Schlosser, "Plasma Etching of CVD Tungsten Films for VLSI Applications" in "Tungsten and Other Refractory Metals for VLSI Applications II," E. K. Broadbent, Ed., Materials Research Society, Pittsburgh, PA (1987) p. 357.

140. B. Jucha and C. Davis, "Reactive Ion Etching of Thick CVD Tungsten Films," *Proc. IEEE VLSI Multilevel Interconnect Conf.*, 165 (1988).

141. G. Turban, J. F. Coulon, and N. Mutsukura, "A Mechanistic Study of SF_6 Reactive Ion Etching of Tungsten," *Thin Solid Films*, 176, 289 (1989).

142. R. J. Saia and B. Gorowitz, "Reactive Ion Etching of Refractory Metals for Gate and Interconnect Applications" in "Tungsten and Other Refractory Metals for VLSI Applications II," E. K. Broadbent, Ed., Materials Research Society, Pittsburgh, PA (1987) p. 349.

143. Y. Kurogi and K. Kamimura, "Molybdenum Etching Using CCl_4/O_2 Mixture Gas," *Japan. J. Appl. Phys.*, 21, 168 (1982).

144. T. Watanabe, "Method of Reactive Ion Etching Molybdenum and Molybdenum Silicide," U. S. Patent 4,478,678 (1984).

145. L. de Bruin and J. M. F. G. van Laarhoven, "Advanced Multiple-Step Resist Etchback Planarization," *Proc. IEEE VLSI Multilevel Interconnect Conf.*, 404 (1988).

146. B. Vasquez and R. Goodner, "Planarized Oxide with Sacrificial Photoresist: Etch Rate Sensitivity to Pattern Density," *Proc. IEEE VLSI Multilevel Interconnect Conf.*, 394 (1987).

147. M. D. Bui, T. A. Streif, K. E. Schoenberg, R. H. Dorrance, and P. P. Proctor, "Contact Glass Planarization Using a Double Etch Back Technique and Spin On Glass Sacrificial Layer," *Proc. IEEE VLSI Multilevel Interconnect Conf.*, 385 (1987).

148. L. B. Vines and S. K. Gupta, "Interlevel Dielectric Planarization with Spin-On Glass," *Proc. IEEE VLSI Multilevel Interconnect Conf.*, 506 (1986).

149. C. H. ting, S. Yeh, and K. L. Liauw, "Sloped Via in Polyimide by Reactive Ion Etching," *Proc. IEEE VLSI Multilevel Interconnect Conf.*, 106 (1984).

150. T. Ohiwa, T. Arikado, K. Horioka, I. Hasegawa, T. Matsushita, K. Shimomura, and H. Okano, "Influence of Al Surface Modification on Selectivity in Via-Hole Etching Employing CHF_3 Plasma," *Japan. J. Appl. Phys.*, 31, 3731 (1992).

151. T. H. Fedynyshyn, G. W. Grynkewich, B. A. Chen, and T. P. Ma, "The Effect of Metal Masks on the Plasma Etch Rate of Silicon," *J. Electrochem. Soc.*, 136, 1799 (1989).

152. B. J. Howard and Ch. Steinbruchel, "Reactive Ion Etching of Copper in $SiCl_4$-based Plasmas," *Appl. Phys. Lett.*, 59, 914 (1991).

153. G. C. Schwartz and P. M. Schaible, "Reactive Ion Etching of Copper Films," *J. Electrochem. Soc.*, 130, 1777 (1983).

8

ELECTRO- AND STRESS- MIGRATION IN MLM INTERCONNECT STRUCTURES

MICHAEL L. DREYER
Motorola, Inc.
Mesa, Arizona

And

PAUL S. HO
University of Texas
Austin, Texas

1.0   INTRODUCTION

For the past 20 years the semiconductor industry has seen an exponential decrease in the size of critical features on integrated circuits and a corresponding increase in device packing density (1). The impact of down scaling integrated circuit geometries on the functional characteristics of the interconnect system was the subject of a number of studies. In metal oxide semiconductor (MOS) based technologies linear scaling results in a linear increase in interconnect current density (2). For bipolar technologies the scaling rules for interconnect current density follow a power law dependence on scaling factor, with the power factor approaching 2 depending on the transistor parameter scaled (3, 4). If we extrapolate the interconnect current density requirements to the year 2000, based on an exponential decrease in feature size, scaling rules predict sub-half micron linewidths capable of supporting current densities approaching 10^6 A/cm^2. This presents a significant challenge to the development of advanced interconnect technologies because of the impact of increased current density on electromigration reliability for MLM interconnect structures of small cross sectional area.

In the early 1960s electromigration was identified as a major cause of failure in integrated circuits (5, 6). Because of the technological importance of semiconductors this spurred an active effort in thin film electromigration studies. The goal of a majority of these studies was to understand those aspects of electromigration which led to interconnect failure. The information gained from these experiments has resulted in failure rates which have decreased exponentially over the last 20 years (1). This improvement in reliability has reduced the failure rate to a level so low that direct measurement becomes impractical. Therefore, to meet the failure rate criteria necessary for advanced interconnect systems a new methodology for assessing interconnect reliability is necessary. Historically, reliability assessment was performed near the end of the development cycle where the results of these experiments would have little impact on materials and processing development of the interconnects. This "back door" approach to reliability has become obsolete because it cannot meet the cycle time requirements of today's semiconductor market. To overcome this difficulty the concept of "building in" reliability was proposed as a means of ensuring that intrinsic wearout occurs only after product life has past, thus minimizing the need for reliability qualifications at the end of the development cycle (7). Implicit in this concept is a thorough understanding of the processes that cause intrinsic wearout due to electromigration.

There has been considerable progress in understanding the atomistic (driving force) and phenomenological (geometry, material, and process related (8)) characteristics of electromigration. The latter has been the focus of considerable attention particularly for conductors with linewidths in the range from 0.5 μm to 3.0 μm. Over this range, the interconnect geometry plays an important role in controlling the mass transport characteristics through its effect on the microstructure. The advent of layered thin film conductors have increased its technological importance for advanced interconnects systems, and introduces materials related effects. Recently, several studies have focused on the effects of process related imperfections such as defects, or notches on electromigration in narrow lines.

Of equal importance to the reliability of narrow interconnects is failure due to void formation induced by thermal stresses. The stress is generated by the mismatch in thermal expansion between the metal conductors and the dielectric layer. Upon cooling during thermal processing, a large tensile stress can be induced in the metal conductor, which will relax over time during subsequent operation or in storage. Mass transport driven by stress relaxation will result in flux divergence and void formation at grain boundary junctions and/or metal/dielectric interfaces. Although the driving force arises from the thermal mismatch and dielectric confinement instead of the electric current, this phenomenon is similar to electromigration in several aspects, including the roles of flux divergence and microstructural defects in controlling damage formation. In the last few years, this phenomenon and the implications for reliability have been extensively investigated and good progress in understanding is being made. Readers interested in this topic are referred to recent reviews and conference proceedings in (8-12).

The purpose of this chapter is to review the fundamentals of electro- and stress-migration and the process of damage formation leading to failure of interconnects. Recent progress in electromigration studies in multi-level interconnects will also be summarized.

2.0   THEORY OF ELECTROMIGRATION

An ideal metal can be thought of as an ordered array of positively charged ions surrounded by an electron 'gas'. Simplistically, the crystal lattice is bound by the interaction between the positively charged ions and the negatively charged electron gas. The electrostatic attraction of the ions to the gas is balanced by the repulsion between the positively charged ions. In an ideal crystal lattice the atoms are free to 'move' within their potential wells (Figure 1a). Assuming there are no temperature or chemical gradients and with no applied electric field, no net diffusion will occur. When an electric field is applied, this gives rise to an electrical current consisting of a very large number of conduction electrons. The electrons move in the direction opposite to that of the electric field vector, and is referred to as the 'electron wind'. For an ion to migrate it must be excited to its saddle point (the 'top' of its potential energy well). At this configuration, the electrons collide with the activated ion (Figure 1b) and transfer a portion of the electron's momentum to the ion imparting an 'electron wind force', F_p, in the direction of the electron wind. Opposing the wind force, there is an electrostatic or 'direct' force between the ion and applied electric field, F_E. The magnitude of the direct force is proportional to the bare valence, Z, of the atom, and its value depends on the extent that the ionic charge is screened by the conduction electron during the jumping process.

The phenomenological equations describing the electron and ion flux can be formulated according to irreversible thermodynamics as

[1] $$J_i = -L_{ii}\nabla(\mu_i + q\Phi) + L_{ie}\nabla(\mu_e + e\Phi)$$

[2] $$J_e = -L_{ei}\nabla(\mu_i + q\Phi) + L_{ee}\nabla(\mu_e + e\Phi)$$

where J_i and J_e represent the ion and electron flux, respectively. The terms L_{ii} and L_{ee} are the generalized conductivities for ions and electrons, respectively. L_{ei} and L_{ie} represent the coupling between the electron and ion flux. The chemical potentials for ions and electrons are given by μ_i and μ_e, respectively, and q represents the bare ionic charge. The electric potential, Φ, is the gradient of the applied electric field, $\Phi = -\nabla \cdot E$.

If we assume there are no chemical potential gradients, i.e. $\nabla\mu = 0$, then the ion flux term reduces to the equation

[3] $$\vec{J}_i = -L_{ii}q^*\vec{E}$$

where q* is the effective charge of the ion. Because of the interaction between the ions and electrons the effective charge is defined as

$$[4] \qquad q^* = q - \frac{eL_{ie}}{L_{ii}}$$

The difference between the bare (or intrinsic) charge and the effective charge, q*-q, is the result of the electron-ion interaction. The nature of this interaction has been the focus of most of the theoretical work on the nature of the electromigration driving force.

Experimental studies of the driving force are generally geared toward measuring the effective valence, Z^*, which is related to the effective charge by $q^*=eZ^*$. Thus the effective valence is given by the equation

$$[5] \qquad Z^* = Z - \frac{L_{ie}}{L_{ii}}$$

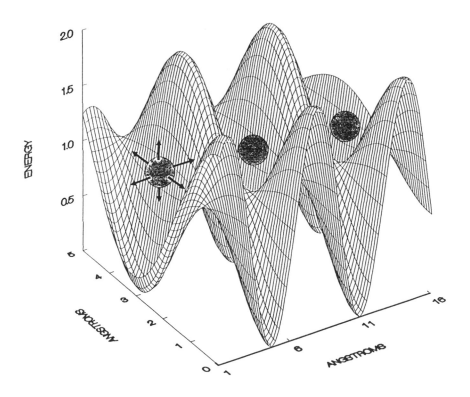

Figure 1a: Conceptualized potential well structure for an ideal metal with no external electric field applied. The arrows indicate potential directions of motion available to the atom. Energy and distance axes are arbitrary scales.

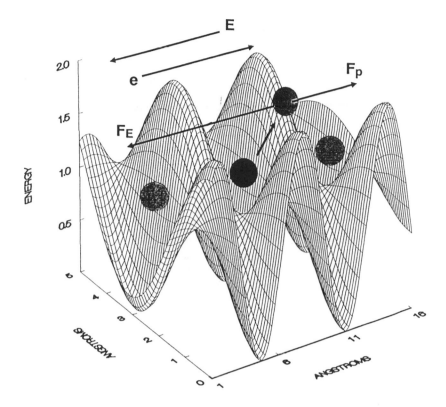

Figure 1b: Conceptualized potential well structure for an ideal metal lattice with an applied electric field, E. The atom in the center of the lattice has been activated to its saddle point and is acted on by the direct (electrostatic) force, F_E, in the direction of the electric field and the 'wind' force, F_p, resulting from momentum transfer from the electron current, e. Energy and distance axes are arbitrary scales.

Experimental results show that for metals characterized by electron conduction Z^* is generally negative and differs from the bare valence, Z. Eq. (3) is used to estimate Z^* by measuring J_i and L_{ii} (which is proportional to the diffusion coefficient), assuming there are no concentration gradients.

The effective valence, Z^*, is the quantity of interest in theoretical studies of electromigration. In the presence of an electric field an ion is subjected to the driving force, F, by the equation

[6] $\vec{F} = Z^* e \vec{E}$

A number of theoretical approaches have been used to model the dynamic scattering between the electron and ion, and subsequently, more realistic structures such as vacancies and defect complexes. Fiks (13), and also

Huntington and Grone (14) provided a semiclassical ballistic model for the driving force. Bosvieux and Friedel (15) applied quantum mechanics to calculate the force on an ion and there have been subsequent refinements to the theory. While the latter is beyond the scope of this chapter a review of the ballistic model provides insight into the nature of the electron-ion interaction. All of the theoretical development assume that the driving force is the sum of the direct and wind forces. Therefore, it is convenient to express Eq. (6) in terms of these components,

[7] $$\vec{F} = e\vec{E}(Z + Z_w)$$

where Z_w represents the valence resulting from electron-ion coupling.

According to Fiks (13) the ion is subject to an electrostatic (direct) force, $\vec{F}_d = e\vec{E}Z$, where Z is the bare valence. The electron wind force is calculated by considering the momentum transfer from the electrons colliding with the "saddle point" (jumping) ion when the electrons are ballistically scattered off the ion. The total momentum transferred per unit time due to these collisions is expressed in terms of the effective valence, where the second term of Eq. (5) is replaced with

[8] $$\frac{L_{ie}}{L_{ii}} = Z_w = n_e \lambda_e \sigma_e$$

where n_e is the electron density, λ_e is the mean free path between collisions, and σ_e is the scattering cross section for electron-ion collisions. From Eq. (5) and Eq. (8) the effective valence can be expressed as

[9] $$Z^* = Z - n_e \lambda_e \sigma_e$$

In Eq. (9), the driving force is derived by a semi-classical treatment of the wind force. If we consider a microscopic cylinder of length λ and scattering cross section, σ, with electron density, n, then the total number of electron-ion collisions is given by the product $\lambda \sigma n$. This treatment of the wind force is a reasonable approximation for metals with simple electronic structures of nearly free electrons, such as Al. For such metals, the effective charge is negative and has a value significantly different from the bare valence indicating that the wind force dominates the direct force (30).

Huntington and Grone (14) considered the initial electronic state and momentum of the electron as well as the change in the electronic state immediately after being subjected to the collisions. Subsequent modifications to the theory emphasize that the force exerted on the ion is not constant during the jump but must be integrated over the path of the jump. This is important for different diffusion mechanisms. For example, the ion-electron interaction would not be affected by the jumping of interstitial atoms since the scattering cross section remains unchanged during the jumping process, while the jumping of substitutional atoms requires activation to a saddle point, increasing its scattering cross section. In this case the effective valence is defined by incorporating a specific resistivity parameter ,

$$[10] \qquad Z_w = -Z\left(\frac{\rho_d}{N_d}\right) \cdot \left(\frac{N_l}{\rho_l}\right) \cdot \frac{|m^*|}{m^*}$$

The product $(\rho_d/N_d)(N_l/\rho_l)$ is the ratio of specific resistivities of mobile defects to that of the lattice atoms. The subscripts d and l represent the mobile defect and lattice terms, respectively. This product is a measure of the electron-ion scattering during the jump and the magnitude of ρ_d/N_d depends on the jumping mechanism. The quantity $|m^*|/m^*$ is a term which indicates a change in the direction of the driving force with the sign of the charge carriers.

The above theories do not take into account the screening of the jumping ion by conduction electrons, which is a quantum-mechanical effect. Therefore, it is unclear how the momentum transfer will be distributed among the ions in a complex defect or the surrounding crystal lattice. Despite these shortcomings the model is essentially correct; the wind force results from the transfer of the electron's momentum to the ion and dominates the direct force. A more complete review of the development and subsequent refinements in the theory of the driving force can be found in (16-30).

3.0   ELECTROMIGRATION IN THIN FILMS

The most common material for conductors in silicon based integrated circuits is an aluminum based dilute alloy. There are exceptions such as gold or tungsten, and many MLM interconnects are actually layered structures, i.e. a refractory metal layer is incorporated on top of, beneath, or within the aluminum based conductor. The more commonly used alloys are binary, Al-Si or Al-Cu or the ternary alloy Al-Si-Cu.

Aluminum has a face centered cubic (f.c.c.) crystalline structure. The dominant mechanism for diffusion in f.c.c. metals occurs via the vacancy mechanism. In all crystals some of the lattice sites are unoccupied or 'vacant'. If an atom adjacent to one of these vacancies jumps into the site, the atom is said to have diffused by means of a vacancy mechanism. The movement of large numbers of atoms or vacancies resulting from a diffusion process is described in terms of the atomic (or vacancy) flux, i. e. the number of atoms (or vacancies) per unit cross sectional area diffusing per second. If we assume that an ideal conductor is composed of a single crystal grain, then the atomic flux through the lattice is described by the formula

$$[11] \qquad \vec{J}_l = \frac{N_l D_l}{kT} Z_l^* e\vec{E}$$

where N_l is the atomic density in the lattice, D_l is the lattice diffusivity, k is Boltzmann' s constant, T is temperature, eZ_l^* is the lattice effective charge, and E is the applied electric field. The lattice diffusivity is given by the equation $D_l = D_{ol} \exp(-\Delta H_l/kT)$ where D_{ol} is the lattice diffusion coefficient and ΔH_l is the energy required to activate the atom to its saddle position ($\Delta H_l = 1.2$-1.4 eV for Al).

In reality, a thin film conductor is typically composed of many crystal grains. For conductors in which the linewidth (W) is much greater than the film's median grain size (D_{50}) electromigration occurs almost completely via grain boundary diffusion. Unlike the previous discussion where we assumed a uniform lattice structure, polycrystalline films are composed of a distribution of grain sizes, usually with a preferred orientation. The grain boundaries are modeled as arrays of dislocations that provide many vacant sites. In these regions the energy required to form a vacancy or move an atom into a vacancy position is lower than in the lattice. As a result, grain boundaries are called high diffusivity paths. The atomic flux for an ideally textured polycrystalline film is given by the equation

$$[12] \qquad \bar{J}_b = \frac{\delta}{d} \cdot \frac{N_b D_b}{kT} \cdot Z_b^* e\bar{E}$$

where N_b is the atomic density in the vicinity of the grain boundary, D_b is the grain boundary diffusivity, k is Boltzmann's constant, T is temperature, eZ_b^* is the grain boundary effective charge, and E is the applied electric field. The grain boundary diffusivity is given by the equation $D_b = D_{ogb} \exp(-\Delta H_{gb}/kT)$ where D_{ogb} is the grain boundary diffusion coefficient and ΔH_{gb} is the grain boundary diffusion activation energy ($\Delta H_{gb} = 0.4$ eV for Al). Eq. (12) represents an ideal case for grains of equal size and texture, as well as possessing grain boundaries with identical properties.

The temperature range over which the diffusional components are measured will have a first order effect on the relative contributions of grain boundary and lattice diffusion. The grain size and distribution, relative to linewidth is also expected to have a first order effect. At temperatures typical of integrated circuit operation grain boundary diffusion dominates the diffusional flux. This can be seen by comparing the relative contributions of the diffusional components, by taking the ratio of Eq. (11) and (12) for W >> d;

$$[13] \qquad \frac{J_l}{J_b} = \frac{d}{\delta} \cdot \frac{N_l}{N_b} \cdot \frac{D_l}{D_b} \cdot \frac{Z_l^*}{Z_b^*}$$

If we assume the film is composed of pure Al and that the ratio of the effective charges is less than an order of magnitude (31) then for a film with d=1.0 μm, $N_l/N_b \approx 1$, $d/\delta \approx 10^3$, $D_l/D_b \approx 10^{-7}$ and $J_l/J_b = 10^{-4}$ for temperatures at 50% of the film's melting point. A number of experimental results, using techniques involving radioisotopes as diffusion tracers or autoradiography also support grain boundary diffusion as the dominant diffusion mechanism at relatively low temperatures. Therefore, for thin film conductors with linewidths much greater than the median grain size (W >> D_{50}) and at temperatures typically used to characterize electromigration, the atomic flux is dominated by the grain boundary diffusion component.

3.1   The Kinetics of Damage Formation: Flux Divergence

In the study of electromigration the kinetics of failure has received substantial attention because of the semiconductor industry's reliance on metal thin films as interconnects in VLSI circuits. Generally an electromigration failure will occur in the form of an open or short circuit. An open circuit failure occurs when a sufficient number of atoms are locally depleted from a cross sectional area of the conductor, forming an electrical discontinuity across its width dimension (Figure 2). This type of failure can also be induced without the electron wind force, when the film is subjected to a mechanical stress and is discussed in section 8. A short circuit failure occurs when atoms depleted from one or more regions of the film accumulate locally forming whiskers or extrusions (32-34) and hillocks. Hillocks rise vertically from the film surface and can fracture the overlying dielectric layer causing an interlayer short with a coincident film (Figure 3). Hillock formation can occur when the film is in compression, or at a site of flux divergence, where atoms are accumulating. In many instances the layers forming the hillock resulting from accumulation are visible. Whiskers or extrusions can occur when the conductor is under compressive stress, with or without electromigration, and cause a similar mode of failure. Whiskers are typically long and thin with angular kinks (Figure 4) whereas extrusions tend to be larger less well defined outcroppings (Figure 5).

Short circuit failures due to whiskers or extrusions occur via the same mode as hillocks. Moreover, failure may also occur at a weak point in the seam between two dielectric layers, causing the seam to split locally allowing the metal to extrude into the seam (Figure 6). There are other areas of interest in MLM reliability such as corrosion and contact reactions which are beyond the scope of this chapter. The interested reader is urged to review the discussions in (35-37) for further information on these topics.

Figure 2: Electromigration induced open circuit failure in a 10 μm wide conductor. The top scanning electron micrograph shows numerous voids (dark areas in center conductor) and hillocks (bright areas in center conductor). The bottom micrograph is a backscattered electron image of the damage. The failure can be seen on the left side of the center conductor and is not obvious from the scanning electron micrograph. The failure was located using voltage contrast and indicates the importance of proper failure analysis techniques in reliability studies.

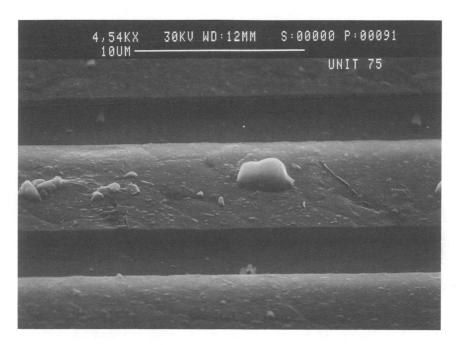

Figure 3:   Scanning electron micrograph of an electromigration induced hillock.

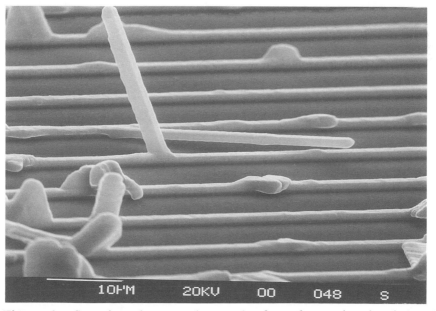

Figure 4:   Scanning electron micrograph of an electromigration induced whisker in a pure aluminum film.

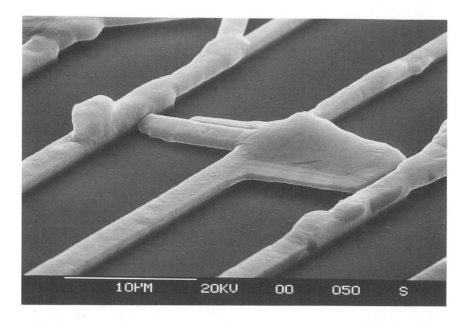

Figure 5: Scanning electron micrograph of an electromigration induced extrusion in a pure aluminum film.

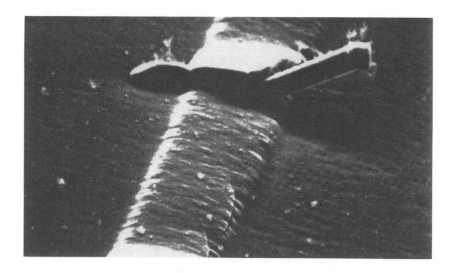

Figure 6: Scanning electron micrograph of an extrusion through a passivation seam.

In a homogeneous equiaxed grain film, with no chemical or temperature gradients the local variation in the atomic/vacancy concentration, N_b, in the vicinity of a grain boundary is given by the continuity equation

[14]
$$\frac{dN_b}{dt} = -\nabla \cdot J + \frac{N_b - N_b^0}{\tau}$$

where N_b^0 is the equilibrium specie concentration and τ is the average life of the specie at the boundary before recombining with a vacancy/atom. For an open circuit failure to occur, vacancies must coalesce to form a void, in which case N_b in Eq. (14) can be considered the vacancy concentration and τ the average life of the vacancy in the vicinity of the boundary. The local vacancy concentration given by Eq. (14) is therefore proportional to both the vacancy flux divergence and the average lifetime of a vacancy at the boundary. Under steady state condition $dN_v/dt=0$, and the local variation in the vacancy concentration is given by

[15]
$$N_b - N_b^0 = \tau \nabla \cdot J_b$$

Thus, in a local region of the film the deviation in the specie concentration from equilibrium is proportional to the flux divergence. If the film is ideal the flux divergence vanishes and the vacancy concentration does not vary from its equilibrium concentration. In terms of void formation this implies vacancies cannot coalesce to form an electromigration induced open without divergent sites in the film. The flux divergence resulting from the grain boundary flux is given by

[16]
$$\nabla \cdot J_b = \frac{\delta D_b}{d} \left[-\nabla^2 N_b + \frac{N_b}{kT} Z_b^* eE \left(\frac{\nabla N_b}{N_b} + \frac{\nabla Z_b^*}{Z_b^*} - \frac{\nabla T}{T} \right) \right]$$
$$+ \left(\frac{\nabla D_b}{D_b} - \frac{\nabla d}{d} \right) J_b$$

This includes the divergence resulting from a gradient in the atomic/vacancy concentration. Thus, if there is a gradient of any of the parameters in Eq. (16) a divergence in the flux occurs. In a real polycrystalline film a spatial variation in grain size will also produce a divergence, as indicated by the last term in Eq. (16). Once a site of flux divergence is established, continued growth of the void requires a continuous vacancy flux divergence. In general, for metal conductors with a finite cross sectional area, void growth will cause current crowding which will increase the local film temperature and accelerate void growth.

There are several sources of flux divergence arising from the microstructure of metal thin films. The configuration of grains comprising an interconnect can produce flux divergent sites. The grain diameters of a polycrystalline thin film generally follow a lognormal distribution. This distribution can lead to a divergence in the grain size when the metal film is patterned to form the interconnect, even when the distribution is single mode (one set of parameters, D_{50} and σ, describe the distribution). When the flux from a region of small diameter grains containing a large number of grain boundaries flows toward an adjacent region of larger grains with few boundaries a hillock forms at the interface between the two regions. The hillock, and therefore the flux divergence, is due to the large number of atoms flowing out of the small grain region to the interface, exceeding the number of atoms leaving the large grain region. Void formation can occur by a similar mechanism. When the flux flows from the large grain region to the small grain region there are fewer atoms entering the interface from the large grain region than are leaving the small grain region. As a result, a void forms at the interface between the two regions. Experimentally, there have been several reports of films with highly uniform grain size distributions exhibiting greater electromigration resistance compared to those with large variations in grain size because of the reduced number of such divergent sites (39, 40).

A common location for flux divergence to occur is the triple point, the region where three grain boundaries intersect. The flux divergence is caused by either a geometric inhomogeneity or by variation in the transport properties of the intersecting boundaries. The inhomogeneity can be caused by a large angle between the grain boundary and the electron current vector and can be quantified. Variations in the transport properties of grain boundaries are more difficult to quantify because it involves an in depth understanding of the grain boundary structure. In a pure metal (e.g. aluminum) variations in the atomic structure or defect concentration in the vicinity of the grain boundary can affect the interaction between the boundary and migrating ion. Additionally, films containing a large number of grains with random orientations are expected to have grain boundary transport properties that vary over a wider range compared to films with uniform orientations. For dilute alloys the segregation and adsorption of solute atoms and defects can also affect this interaction resulting in a diffusivity that may vary over several orders of magnitude between grain boundaries.

The design and location of MLM structures in a VLSI/ULSI circuit may also give rise to a flux divergence. Large pad areas connected to fine line structures can result in temperature and current density gradients, particularly when operated at test conditions associated with electromigration experiments. At high current densities the line may produce significant Joule heating, raising its temperature well above that of the adjacent pad. If the gradient is sufficiently large open circuit failures will occur at the cathode and hillocks will form at the anode. Also, interconnect structures located in the proximity of active transistors can produce temperature gradients along a portion of the interconnect length due to Joule heating at the pn junction. Temperature gradients can result from Joule heating at relatively low current densities in the uppermost metal levels of MLM systems due to the increased interlayer dielectric thickness. Also, vias, pillars, plugs, and contacts with diameters on the order of 1 μm or less can produce extremely large temperature and/or current density gradients.

In an advanced metallization system planarized conductors are connected electrically to other metal levels through W-plug vias or contacts to Si using a barrier layer to impede interdiffusion of Al and Si. Common barrier layer materials such as TiN, TiW or W are essentially impervious to electromigration relative to Al based conductors. As a result, the interface between the conductor and the contact or via represents a site of flux divergence. When an electron flux induces mass accumulation on one end of a stripe connecting two W-plug vias and depletion on the other, a mechanical stress is generated due to accumulation of atoms at the anode end and depletion at the cathode. The stress is compressive and results in a back flow of atoms in the direction opposite to the electron wind. The flux due to the mechanical stress is described by the equation

$$[17] \qquad J_\sigma = N_b \frac{\Omega}{kT} \nabla\sigma \frac{\delta}{d} D$$

where Ω is the atomic volume and $\nabla\sigma = \sigma/L$ is the gradient of the stress in a conductor of length L due to mass accumulation. For a given condition of temperature and current, if the length of the stripe is greater than some threshold length, then the electromigration flux is larger than the back flow and electromigration will occur. If the length of the stripe is less than the critical length then no electromigration will occur since the mechanical stress is large enough to oppose the electromigration driving force.

4.0 MEASUREMENT OF TEMPERATURE AND CURRENT DENSITY IN MLM STRUCTURES

To measure electromigration parameters accurately requires a precise knowledge of film temperature and current density. The film temperature is a strong function of the sample geometry and adhesion to the underlying substrate. For a conductor of length L and thickness ℓ the idealized one dimensional heat flow equation for thermal transport from the film to the substrate is given by

$$[18] \qquad K \frac{d^2T}{dx^2} - j\mu \frac{dT}{dx} + j^2\rho - \frac{H}{\ell}(T - T_0) = 0$$

where j is current density, K is thermal conductivity of the thin film, μ is the Thomson coefficient, ρ is the resistivity, T_0 is the temperature on the substrate side of the film-substrate interface and T is the temperature of the film (35, 41, 42). The term $H(T-T_0)/\ell$ is the heat loss through the conductor/substrate interface and is assumed to be proportional to ρj^2. Typically the conductor is separated from the silicon substrate by an oxide of thickness ℓ_{ox}. In this case the term H/ℓ can be replaced by $(K \cdot K_{ox})/(\ell \cdot \ell_{ox})$, where K_{ox} and ℓ_{ox} are the thermal conductivity and thickness of the oxide, respectively (41). This is only valid if the silicon substrate is assumed to be at ambient temperature. For current densities well below the fuse current and neglecting the Thomson effect, the solution to Eq. (18) gives the temperature profile along the film length. Taking the origin as the center of the conductor and the ambient temperature at $x=\pm \ell/2$ as T_o the solution to Eq. (18) yields the increase in the film temperature over the ambient due to Joule heating and is given by the equation

$$T - T_0 = \frac{\rho j^2}{\frac{H}{\ell} - \rho_0 \alpha j^2} *$$

[19]
$$\left[1 - \frac{Cosh\left(\frac{1}{2}\sqrt{Kx^2\left(\frac{H}{\ell} - \rho_0 \alpha j^2\right)}\right)}{Cosh\left(\frac{1}{2}\sqrt{KL^2\left(\frac{H}{\ell} - \rho_0 \alpha j^2\right)}\right)} \right]$$

where H/ℓ can be replaced with $(K \cdot K_{ox})/(\ell \cdot \ell_{ox})$ as discussed previously. The temperature coefficient of resistivity, α, describes the temperature dependence of resistivity according to the equation

[20] $$\rho(T) = \rho_0\left(1 + \alpha\left(T - T_0\right)\right)$$

where ρ_o is the resistivity at T_o. Under the conditions stated the heat transfer is large, $H/\ell >> \rho_0 \alpha j^2$, so that the temperature is constant along the length except at the ends of the film where a connection is usually made to a large pad. The temperature profile for an Al conductor was calculated using the appropriate parameters from Table 1 for an unpassivated Al conductor deposited on SiO_2 with T_0=200°C, L=1000µm, ℓ=0.65 µm and ℓ_{ox}=10 µm.

The profile shown in Figure 7 was calculated using Eq. (19) for current densities ranging from J=10^6 A cm^{-2} to J=7.5X10^6 A cm^{-2}. Figure 7 shows that to a first order approximation the film temperature is essentially constant along the length of the conductor except at its ends. When passivated, Eq. (19) provides a slight over estimate of film temperature since it does not account for heat flow through the top surface and side walls of the conductor.

The temperature distribution given by Eq. (19) provides an estimate of the average film temperature. Localized hot spots will occur in regions of the film where there is poor adhesion to the substrate or where voids develop. This is particularly troublesome in measurements of mass distribution using TEM or SEM since the film is unsupported in the region of interest. In life test experiments some inaccuracy is introduced in the measured failure time because the localized film temperature depends on the extent of damage formation. Damage in the form of voids or cracks produces current crowding which results in an additional Joule heating component.

Table 1: Thermal conductivities of common dielectric and conductor materials for semiconductors from ref. (43, 44).

Thermal Conductivities (K) of Common Semiconductor Materials at 27°C	
Material	K (W cm^{-1} °C^{-1})
Si (Single Crystal)	1.1-1.6
SiO$_2$ (Amorphous)	0.014
SiN	0.277
Al	2.05
Au	2.94
W	1.64
Cu	3.90

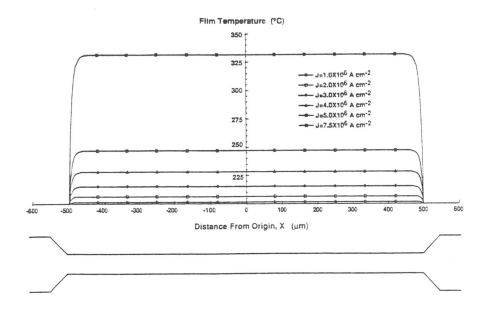

Figure 7: Temperature distribution along a planar conductor. The ambient temperature is assumed to be 200°C.

Experimentally, the film temperature can be extracted from resistance measurements made during the experiment if the temperature coefficient of resistivity (TCR) is known *a priori*. The TCR can be measured before an experiment using a non-Joule heating current to determine the conductor resistance at several ambient temperatures. By combining terms and multiplying both sides of Eq. (20) by L/A, where A is the film cross sectional area, the temperature coefficient of resistance is given as

[21] $R(T) = R_0 + \alpha R_0 T$

where R_o is the resistance of the film at T_0 (°C) and α is the TCR (°C^{-1}). For convenience T_0 in Eq. (21) is usually taken as T_0=0°C. A linear regression in the form of Eq. (21) to the experimental data allows the average film temperature to be estimated from the conductor resistance when a joule heating current is applied. Typically for Al based films the TCR is in the range 0.003 °C^{-1} ≤ TCR ≤ 0.004 °C^{-1}.

Electromigration damage in the form of voids or cracks can result in current crowding around the damage site. Current densities near voids and cracks have been calculated using computer simulations. The magnitude of these current densities may be as large as a factor of 100 greater than the average current density and is a sensitive function of the geometry of the damage. In the region of a crack or void the increased current density can increase the magnitude of the local atomic flux, accelerating the damage formation process well above the rate established by the average current density. The increased local current density will also result in localized Joule heating, as mentioned previously, further accelerating the electromigration rate.

In multilevel metallization systems electrical contact is made to adjacent metal levels through vias (see Chapter 1 for information on via structures and fabrication). The via is an opening in the dielectric material between 2 or more metal levels which allow electrical connections to be made. For via sizes (square geometry) of 0.8 μm or greater the metallurgy of the via is usually the same as the conductors. When the via size is less than 0.8 μm tungsten is generally used to 'fill' the via because it can fill high aspect ratio vias (aspect ratio is the via depth divided by via size). Calculation of the current density and temperature in via structures is more complicated than in planar conductors because of increased geometric complexity. Several computer simulation studies have investigated the current density and temperature distributions in via structures (45-48). Computer simulation of the current density can be performed using finite element analysis with commercially available programs such as ANSYS or FIDAP, by solving Laplace' s equation for the potential distribution subject to the appropriate boundary conditions. The current density can be calculated from the potential distribution and is proportional to $j = -\rho \nabla \cdot \phi$ where ϕ is the potential and ρ is the resistivity of the metal. The temperature distribution due to current crowding in the via is calculated by solving Poisson's equation $-k\nabla^2 T = \rho j^2$, where ρj^2 is the joule heating source. In a study by Kwok (45, 46) the current density was found to crowd at the entry and exit points of the via, such that the magnitude of the current density at these points was greater than the average value. Current crowding at the entry and exit point of the via created temperature peaks at these locations although their magnitude was small relative to the overall temperature distribution. Decreasing the via size while forcing the same current to flow through the via results in larger peak current densities and temperatures at the via entry and exit points. When the current enters the leading edge of the via it spreads out within the via material. The extent to which the current redistributes depends on the resistivity of the via material. In addition to current crowding, the magnitude of the peak current density at the conductor-via interface is a function of the resistivity of the interface region. For conventional Al based vias this does not produce significant Joule heating since the interface resistivity is small, however, W-plug vias typically use a Ti-W or TiN layer to promote adhesion between the W and dielectric material. The resistivity at the interface between the W and Al can therefore be significant, producing large peak current densities.

5.0 ELECTROMIGRATION MEASUREMENT TECHNIQUES

The techniques used to study thin film electromigration generally have two distinct purposes. The first is to extract information about the fundamental processes which govern electromigration. This usually requires measurements of the diffusivity, D, and effective valence, Z^*, which require specialized measurements of the ion flux or compositional changes. Measurements of the ion flux give results in terms of $D_l Z_l^*$ or $\delta D_b Z_b^*$ (see Eq. 11 and Eq. 12) so that a diffusivity measurement must be made to determine Z^*. Measurements of D in thin films can be obtained using a number of techniques such as radiotracer profiling, surface accumulation techniques using AES or ESCA, or lateral spreading measurements. For thin film electromigration the latter technique requires a specialized test structure and can measure compositional effects resulting from solute addition. The most well known experiment exploiting this technique, the cross stripe experiment, has the added advantage of providing an independent measure of Z^* and D so that a second measurement is unnecessary. An alternative to these techniques which does not require a diffusivity measurement is drift velocity technique. However, for this measurement a separate determination of film stress is required and is discussed in Section 7.1. Values for effective valence and diffusivity derived from some of these techniques are reproduced from (31) and shown in Table 2 for metallurgies commonly used in semiconductor technology .

In general, measurements of atomic flux are carried out at temperatures in the range of 50% of the melting point of aluminum and current densities are usually less than 10^6 A/cm^2. Most of the data for thin films has been obtained using structures in which the linewidth is generally much greater than the median grain size. Under these conditions it is assumed that $J_b/J_l \gg 1$ as mentioned previously. However, over the same range of test conditions and for structures where the median grain size is much greater than the linewidth there is little data. For interconnect geometries smaller than the median grain size the mass transport characteristics are still not well understood. Techniques for direct measurement of mass transport are discussed in Section 5.1.

Table 2: Electromigration effective valence and diffusivities for selected metallurgies. Sources of the data in this table are given in ref. (31).

Diffusing species	Host metal	$T, \degree K$	Z_b^*	Z_l^*	βD_b cm^2 s^{-1}	D_l cm^2 s^{-1}
Cu	Al	528	−16.8	−7	1.8×10^{-7}	2.7×10^{-14}
Cu	Al+Cu	448	−20[a]	−7[a]	2.0×10^{-8}	8.1×10^{-17}
Mg	Al+Mg	504	−40	−39	6.0×10^{-12}	1.1×10^{13}
Ni	Al+Ni	448	−3	?	1.5×10^{-12}	?
Al	Al	500	−10	−17	1.5×10^{-9}	3.0×10^{-14}
Al	Al+Cu	448	−30[a]	−17[b]	2.7×10^{-11}	5.3×10^{-16b}
Al	Al+Mg	504	−30[a]	−17[b]	4.7×10^{-11}	2.1×10^{-14b}
Al	Al+Ni	448	−30[a]	−17[b]	5.6×10^{-11}	5.3×10^{-16b}
Ag	Au	523	−9.6	−7.4	1.7×10^{-10}	1.1×10^{-18}
Au	Au	573	−10	−9.2	1.3×10^{-10}	9.2×10^{-18}
Au	Au+Ta	569	—	—	3.2×10^{-11}	10^{-18b}
Sb	Ag	750	−230	−100	2.5×10^{-7}	1.2×10^{-12}

[a]Values not actually measured but assumed in calculating βD_b.
[b]Values calculated according to the pure metal data.

The second purpose of electromigration experiments is to measure the statistics of electromigration induced failure times so that the reliability of metallization systems can be established. The results from this type of experiment are more difficult to interpret in terms of fundamental mechanisms because the failure times are indirect measurements of the atomic flux. Careful design of test structures and selection of test conditions can still provide useful information on activation energies, which are close to those reported for diffusional processes in thin films. Because of the technological importance of thin film conductors to the semiconductor industry this category of experiments has received wide attention. The statistics and general characteristics of electromigration in MLM structures are discussed in sections 6.0 and 7.0.

5.1 Direct Measurements of Mass Transport

Generally, measurements of D and Z* are obtained using direct measurements of mass transport requiring highly specialized equipment/techniques such as TEM or SEM and extensive sample preparation. The use of autoradiography and microtoming techniques have also provided useful information. Historically, these types of measurements have not been used to estimate conductor reliability but rather to investigate the more fundamental mechanisms of electromigration. With the advent of W-plug technology at least one of these methods, drift velocity measurements, are being used to study the reliability of W-Al interfaces in multi-level metallization systems (section 7). A number of newer measurement techniques have been proposed including resistometric techniques (49, 50) and 1/f noise to characterize electromigration. Neither technique provides any information on the driving force or diffusivity, but both provide measurements of activation energy that are similar to those reported for grain boundary diffusion. In addition the origin of 1/f noise is believed to be mesoscopic in nature (51) and may provide a means for studying the interaction between the driving force and the grain boundary motion of atoms. This section introduces several of the more common methods used for direct measurement of mass transport.

Mass Depletion and Accumulation: The goal of these measurements is to provide a direct measure of mass accumulation or depletion as a means for characterizing electromigration. These techniques use a specialized test structure in which a portion of the film is unsupported and the applied current is used to adjust the film temperature (52, 53). Since the film does not make intimate contact with the substrate over its entire length heat is forced to flow laterally along the length of the conductor toward the anode and cathode ends. Precise measurements of film temperature are difficult and a generalized parabolic temperature distribution is assumed so that the peak temperature can be estimated. In this technique a current is impressed on the sample while on the specimen stage of a transmission electron microscope (TEM). The variation in the temperature along the sample's length dimension creates variations in the flux according to Eq.s (11) and (12). A temperature gradient will also result in Soret (thermal) diffusion independent of the applied current stress, however, this is assumed to be negligible compared to the electromigration flux. The temperature gradient causes a flux divergence near the anode and cathode ends of the stripe. The signs of the flux divergence are opposite on either side of the temperature maximum. On the cathode side of the film voids nucleate and grow and hillocks form on the anode side. It is assumed that the flux through the film cross section located at the site of peak temperature is proportional to the rate of void growth. Thus, measuring the total void volume as a function of temperature and test conditions can provide a measure of the activation energy for electromigration. In addition, TEM measurements provide information on the morphology of void formation and growth and accurate spatial information.

Mass Distribution: This technique measures the spatial variation of mass along an unsupported film resulting from the passage of a direct current. The samples are prepared in a manner similar to the previous discussion, however instead of using TEM, a modified scanning electron microscope (SEM) is used (54). As in the previous technique, the peak temperature is determined by the magnitude of the stressing current. To record the mass variation due to electromigration, the primary electron beam scans the suspended film along the plane normal to the direction of current flow in the film. An electron collector located below the film integrates the transmitted electron beam. The attenuation of the beam is proportional to the thickness of the film. If the attenuation of the beam is measured as a function of time and the flux through the scanned region is constant then variations in the mass profile should be proportional to Eq.

(16). Thus, the drift velocity and activation energy can be determined from measurements of the variation in the mass profile relative to an unstressed film.

Another mass distribution technique utilizes the electron microprobe to determine the variations in the mass profile due to electromigration (55). Samples are not unsupported as in the other techniques and are similar to those used in conventional life test measurements. Current stress is removed from the sample and the sample is placed on the specimen stage of an SEM. The electron beam is swept perpendicular to the length of the film to observe the K_α x-rays generated. This measurement process is repeated along the length of the line to determine the spatial variation in the number of K_α lines. The increase or decrease in the number of K_α counts is proportional to the total mass transport, assuming that mass is proportional to counts. This technique is particularly useful for measuring impurity electromigration.

Lateral Spreading Measurements- Cross Stripe Method: This measurement technique is particularly well suited for measuring the effects of solute additions on electromigration, although it can be applied to the study of self-electromigration if radioisotopes are used. It was originally conceived as a means to study the effects of Cu additions on electromigration in Al films. The sample configuration is the thin film analog of the classic Kirkendall diffusion couple (56). The sample geometry shown in Figure 8 is used with region 'A' composed only of 'A' type atoms and region AB representing an alloy composed of type A and type B atoms. The two edges separating these regions correspond to the Kirkendall interfaces. For a given temperature and applied current there is a flux at the interface due to the concentration gradient and electron wind force. At the left interface the flux due to the concentration gradient is in the direction opposing the flux due to electromigration while at the right interface the flux from the concentration gradient and electron wind force superimpose in the same direction. Since the interfaces are independent, measurements of the diffusion and electromigration driving forces in each region provide a means for independently determining the diffusivity and effective charge.

An analysis of the composition profiles at the interfaces, with grain boundaries acting as fast diffusion paths provides the slope of the logarithm of the concentration as a function of distance.

If no current is applied, the diffusivity can be measured directly since Z^* is set to zero. Measurements of the diffusivity can be carried out using autoradiography (57), high resolution autoradiography (58, 59) or digital scanner techniques (58) or microtoming (sectioning) techniques (60, 61). The primary drawback to these techniques is lack of spatial resolution. As a result there is a lack of experimental data for today's submicron geometry conductors.

6.0 Life Test Method

The life test is the most common method for evaluating electromigration in thin film conductors. The primary goal of the life test is to measure the

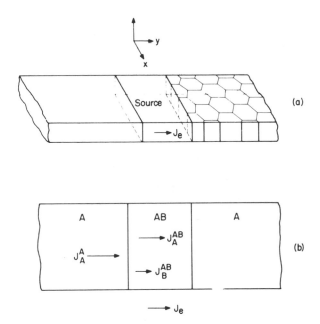

Figure 8: Cross stripe geometry used for electromigration measurement. The top figure shows the idealized grain structure used for analysis of the results. The lower figure shows the fluxes due to the electron current J_e (from ref. (31)).

statistical distribution of electromigration failure times, and their current density and temperature dependence so that electromigration design rules can be established and validated for an interconnect system. The failure time of a conductor can be thought of as an indirect measure of the atomic flux. However, the catastrophic electromigration damage which occurs at the end of a life test makes it difficult to accurately estimate the rapidly changing film temperature and current density as the conductor cross section diminishes. Therefore, life tests are not useful for measuring effective charge or diffusivity. Measurements of electromigration activation energy have values which are similar to those reported for the techniques in section 5.0 for both lattice and grain boundary diffusion mechanisms. *In situ* observations of electromigration damage formation and failure analysis are consistent with these reports, suggesting that the life test method can accurately measure the activation energy for these mechanisms if careful experimental procedures for temperature measurement are followed. Accurate measurement of the activation energy is required because life test data is typically extrapolated from test conditions to circuit operating conditions.

In a typical life test experiment a constant current and elevated ambient temperature is applied to a group of identical test structures. The experimental temperature and current density exceed normal circuit operating conditions and are necessary in order to ensure that failures occur within a reasonable period of time. Figure 9 shows a test structure which was proposed by the National Institute of STandards (NIST) for electromigration testing (62). The design of

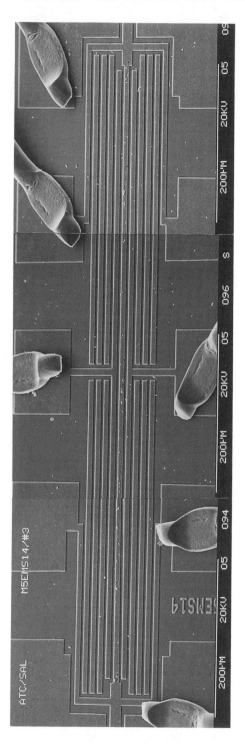

Figure 9: National Institute of Standards proposed electromigration test structure. Voltage taps are 8 µm from the edge of the current tap and current taps are a minimum of 80 µm long. Extrusion sensors are placed on adjacent sides of the test-line.

the structure minimizes thermal gradients, based on thermal modeling, and abrupt current density changes at the current taps. The test-line is the center conductor, connected to current taps on either end. The current tap width is twice the actual test-line width and should extend for a minimum of 80 μm. The voltage taps are symmetrically placed on either side of the current taps to accommodate multiple packaging configurations and should be located approximately 8 μm from the end of the test-line. An extrusion monitor is placed on adjacent sides of the test-line to detect short circuits during the experiment. The unconnected stripes are to ensure uniform etching during a reactive ion etch process. The length of the test-line should be experimentally determined. In general, the length should be several integer multiples of the critical length (section 6.3). The test structures are monitored periodically using an automated test system until a predefined failure condition is reached. Failure is usually defined in terms of the time required for accumulated electromigration damage to produce a specified change in the structure's resistance or until an electrical open circuit occurs. Because of the statistical nature of electromigration failure times, the life test results are typically described in terms of an appropriate distribution law.

6.1   Statistical Nature of Electromigration Failure Times

There have been a number of investigations to determine the appropriate distribution law to describe the failure times. Schoen (63) investigated the statistical nature of the failure distribution using computer simulations. When the grain size was less than the conductor linewidth the failure times were well represented by a lognormal distribution. For linewidths less than the median grain size the logarithmic extreme value distribution provided the most adequate fit, indicating that for linewidths less than the median grain size the lognormal distribution does not accurately characterize the failure distribution.

Lacombe and Parks (64) measured the failure distribution to low cumulative percent rank ($\approx 0.33\%$) on conductors with linewidths in the range 3.4 μm to 5 μm and a median grain size of 1.5μm. The failure times followed a lognormal distribution law, which supports Schoen's results for $W > D_{50}$. Towner (65) fabricated conductors from films with different median grain sizes and measured the failure distributions for the case when $W > D_{50}$ and $W < D_{50}$. The lognormal distribution could be used to adequately represent the failure times for $W > D_{50}$, in agreement with (63, 64). For $W < D_{50}$ the distribution could be represented by either a lognormal or logarithmic extreme value distribution, the latter providing only minimal improvement over the lognormal distribution law in representing the failure distribution. Subsequent investigations for single mode grain size distributions with abnormally large grains resulting in $W \ll D_{50}$ also show lognormal failure distributions, supporting a lognormal distribution law for electromigration induced failure times (66).

Several key factors may contribute to the lognormal distribution law. Empirically, electromigration failure times follow a generalized temperature dependence of the form $A\exp(\Delta H/kT)$, where A is a constant for a given test condition and conductor geometry, T is the film temperature, k is Boltzmann's constant and ΔH is the activation energy for failure. In a typical life test there is a distribution of film temperatures resulting from a normal distribution of film resistance and heat conduction, assuming the applied current exceeds the Joule heating threshold. Based on the temperature dependence of the failure times, Bobbio and Saracco (67) proposed that variations in conductor temperature, ΔT, less than 20°C can result in a lognormal distribution assuming a normal distribution for 1/ΔT. This assumption was verified by Lloyd (68) using computer simulation. By selecting samples which produced the same film temperature when a stress current was applied, the distribution of ΔT was reduced, yielding a 50% smaller spread in the failure distribution compared with the distribution obtained from randomly selected samples.

Schwarz et al. (69) measured the electromigration activation energy of individual conductors using a temperature ramping method and found the activation energies (ΔH) for pure Al films were distributed normally. A normal distribution of ΔH will result in a lognormal distribution of failure times, for the same reason as described previously for the temperature distribution. Similar experiments on Al-4%Cu films also show a distribution of ΔH, but in this case the distribution was multi-modal, with a larger range of activation energies. Unfortunately, the effects of this type of ΔH distribution on the failure distribution were not provided. More recently Longworth and Thompson (67) used bicrystals to investigate the effects of grain boundary orientation relative to the direction of electron current flow on the failure distribution. For small angles between the grain boundary and the electron current vector the spread of the failure distribution was small. As the angle between the grain boundary and current vector increased the magnitude of the failure times increased and the distribution broadened. This is consistent with earlier reports (70) which showed more uniform failure distributions and increased failure times when the Al grains followed a predominant orientation. For films with a distribution of orientations the failure times were consistently shorter and the spread of the failure distribution was larger.

Photolithographic, etch or deposition processes introducing dimensional variation in conductor linewidth and thickness during sample fabrication will also influence the failure distribution, since failure times are a strong function of linewidth and thickness (section 6.3). These variations can be minimized by screening samples using electrical measurements of linewidth and resistance (Chapter 9). Samples that do not fall within a tight distribution of these parameters should be rejected for testing.

Lognormal Distribution Law: The lognormal probability density function includes the parameters μ, which is the mean of the log of the failure times, and σ, which is the standard deviation of the log of the failure times. The latter term is also referred to as the shape parameter (or distribution 'spread' in the preceding discussion). The lognormal density function is given by the equation

$$[22] \qquad f(t) = \frac{1}{\sqrt{2\pi}\sigma t} \exp\left(-\frac{\left(Log(t) - \mu\right)^2}{2\sigma^2}\right)$$

The lognormal cumulative probability density function, F(t), describes the percentage of the population that has failed by time t and is given by

$$[23] \qquad F(t) = \int_{-\infty}^{z_p} f(t)dt$$

The percentile of the distribution (t_p) is the time at which a percentage (p) of the population has failed. The percentile may be estimated by solving the equation $p=F(t_p)$ and can be approximated by $t_p=$antilog$(\mu+z_p\sigma)$. The quantity z_p is the standard normal percentile corresponding to $F(t_p)$ as determined from Eq. (23). The standard normal percentile is negative for $F(t_p)<0.5$ and positive for $F(t_p)>0.5$. The most common percentile is the 50% point which corresponds to the median time to failure, t_{50} (or MTF). In this case Eq. 23 gives F(t)=0.5, and $z_p=0$; the median failure time is given by the equation $t_{50}=$antilog(μ) (71-74). The shape parameter can be estimated using the equation $\sigma\approx$Log(t_{50}/t_{16}).

The parameters t_{50} and σ are usually determined by graphing the data on a lognormal probability plot. The data is 'ranked' in order from shortest to longest failure time. The failure data (logarithmic scale) is plotted versus cumulative percent rank on a lognormal probability scale. There are a number of methods for calculating the rank order, with the Hazen algorithm being the most common. The cumulative percent rank is given by R=(i-0.5)/N, where i=1, 2, ..., N and N is the sample population; i=1 corresponds to the shortest failure time, i=2 to the next shortest failure time, etc. An example of ranking is shown in Table 3 for failure data measured using 5.0 μm wide Al-1.5%Cu films. Data was obtained using T=205°C and J=2.0X10^6 A cm^{-2}. The probability plot for the ranked data is shown in Figure 10 along with the calculated values of t_{50} and σ.

Table 3: Failure times and rank for unpassivated Al-1.5%Cu films.

Time To Failure (Hours)	Cumulative Percent Rank (%)
17.316	3.1250
17.599	9.3750
18.619	15.625
19.479	21.875
21.734	28.125
37.468	34.375
40.630	40.625
41.214	46.875
43.300	53.125
53.756	59.375
54.227	65.625
59.918	71.875
64.151	78.125
67.256	84.375
113.09	90.625
119.21	96.875

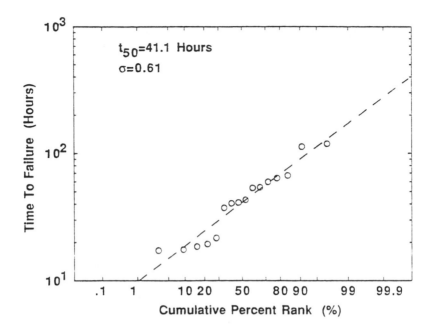

Figure 10: Probability plot for a 5.0 μm wide Al-1.5%Cu conductor (unpassivated). The current density was $J=2X10^6$ A cm^{-2} and the failure data was normalized to a film temperature of T=205°C.

6.2 Extrapolating Life Test Data To Circuit Operating Conditions

The measured failure statistics are typically extrapolated from test conditions to circuit operating conditions using an appropriate failure model. The purpose of this is to establish electromigration design rules that ensure the metallization will not fail before a certain period of time has elapsed, usually 10 years. This assumes that the dependence of the failure mechanism(s) on temperature, current density, conductor geometry and microstructure are understood well enough so that extrapolations to circuit operating conditions can be made from test conditions. For electromigration tests where the film temperature is kept below 50% of the melting point and the film width and thickness are much greater than the median grain size this is a valid assumption. Conductors which fall into this category have microstructures which are polycrystalline; on average it requires several grains in parallel to span the width dimension. In advanced MLM systems utilizing submicron conductor geometries and/or 'hot metal' deposition procedures the median grain size is in the range of the film width and thickness, giving rise to conductors that are 'bamboo' like in structure; a single grain spans the width of the conductor and the line is comprised of single grains connected serially. For these types of conductors there is mounting evidence that the diffusional flux contains both lattice and grain boundary components over the temperature range used to characterize electromigration. Activation energies have been measured in aluminum based films which are near the values for bulk diffusion (1.2-1.4 eV) when the median grain size is greater than the conductor width. These higher activation energies imply a lattice diffusion component in the net diffusional flux which should disappear at lower temperatures since lattice diffusivity decreases rapidly with decreasing temperature. Therefore, extrapolation of t_{50} data measured in thin films of this type and over this temperature range must consider the change in the failure rate due to changes in the components of the diffusional flux with temperature. Recently, a model has been proposed to explain the effects of conductor geometry, grain size distribution and temperature on the grain boundary and lattice diffusion components of electromigration and the impact on life test extrapolations (75).

For $W \gg D_{50}$ the failure times are typically modeled using Black's equation (76), which describes the median time to failure (t_{50}) of a number of identically processed interconnects,

[24] $$t_{50} = Aj^{-2} \exp\left(\frac{\Delta H}{kT}\right) \cdot f(W)$$

or alternatively

[25] $$t_{50} = ATj^{-2} \exp\left(\frac{\Delta H}{kT}\right) \cdot f(W)$$

where j is current density, T is temperature, ΔH is the activation energy and A is a term referred to as the 'process constant'. Typically 'A' is the intercept of the Arrhenius plot at infinite temperature (section 6.3). The term f(W) is an empirically derived function which describes the linewidth dependence of t_{50}. The simplest method for determining f(W) is by measuring t_{50} at the same test conditions for a range of linewidths. For $W \gg D_{50}$ most reports in the literature support $f(W) \approx W$. When $W \le D_{50}$ the linewidth dependence of t_{50} becomes more complex because of structural changes in the conductor. The linewidth dependence is discussed in greater detail in section 6.3.

Equation (25) includes an additional temperature term in the pre-exponential which is the temperature dependence resulting from the Einstein relation. The drawback to Black's equation is that it does not model electromigration failure in terms of the microstructural characteristics of the interconnects. This includes the effects of triple point divergence and structural inhomogeneities. In addition, Eq. (26) does not include a provision for determining the relative change in the diffusivities, and therefore an accurate failure rate, as a function of temperature as mentioned previously. Despite these limitations, it has been the industry standard method for extrapolating experimentally measured failure rate data to circuit operating conditions for over twenty years.

Once f(W) and ΔH have been measured, t_{50} can be extrapolated to circuit operating conditions by taking the ratio of Equation (24) at test conditions to those at operational conditions, and is given by Equation (26)

$$[26] \qquad t_{50_{use}} = t_{50_{test}} \left(\frac{J_{test}}{J_{use}} \right)^2 \exp\left(\frac{\Delta H}{k} \left(\frac{1}{T_{use}} - \frac{1}{T_{test}} \right) \right)$$

The subscripts 'Use' and 'Test' correspond to maximum circuit operating conditions and experimental conditions, respectively. Using the data from Table 3 and $J_{use}=10^5$ A cm^{-2}, $T_{use}=373$ K (100°C) and ΔH=0.6 eV Eq. (26) gives $t_{50use} \approx 10^6$ hours.

Eq. (26) describes the time when 50% of the conductors in a circuit will have failed under operational conditions. Generally, the reliability engineer is concerned with failure rates that are much lower than those which can be measured practically (typically for cumulative percent rank in the range 0.001% to 0.01%), and must extrapolate the experimental data using the appropriate distribution law. The selection of the correct distribution is critical because for cumulative percent rank less than 1% there can be an order of magnitude difference between projections made using a Weibull, logarithmic extreme value or lognormal distribution. As discussed in the previous sections the lognormal distribution appears to be most appropriate. Based on reliability projections for advanced VLSI and ULSI circuits (section 1) the percentile of interest is in the range $t_{0.01}$% to $t_{0.001}$% due to the large number of interconnections required for these technologies. Once the median percentile (t_{50}) and σ are measured experimentally any other percentile can be calculated using the relationship

$$[27] \qquad t_p = t_{50} \exp(z_p \sigma)$$

By using Eq. (27), the time when the first electromigration failures occur can be estimated and provides a more meaningful comparison to the expected life of the metallization than t_{50}. For the example from Table 3, at the use conditions specified in the preceding discussion, if the percentile of interest is $t_{0.01}$ then z_p=-3.719 and Eq. (27) gives $t_{0.01} \approx 10^5$ hours. The use of Eq. (27) assumes the shape parameter is independent of temperature and that the failure distribution is single mode even for very low cumulative percent rank.

6.3   Electromigration Characteristics

A number of investigations have focused on quantifying the failure distribution parameters t_{50} and σ in terms of the microstructural characteristics (orientation, grain size distribution, etc.) of the conductor. The impact of these characteristics on the distribution parameters are difficult to quantify because of the random nature of the microstructure; orientation and grain size depend on the deposition conditions (rate and temperature) and the characteristics of the surface on which the film is deposited. In addition, the distribution parameters are dependent on the conductor dimensions relative to the local film microstructure. Finally, variations in film microstructure between conductors as well as measurement conditions influence the distribution parameters. In the following sections, the electromigration characteristics are discussed in terms of conductor geometry (length and thickness), linewidth in relation to grain size, and activation energy. The effect of crystal orientation was briefly described in section 6.1 in terms of its affect on the failure distribution.

Conductor Length: The effect of conductor length on electromigration has been the subject of a number of experimental investigations and computer simulations. The focus of these investigations has been to understand how the median time to failure and shape parameter depend on the length of the conductor, and the microstructure of the film. Agarwala et al. (77) investigated the length dependence of electromigration for Al stripes, using conductor lengths ranging from 25 μm to 2000 μm, and linewidths in the range 5 μm $\leq$ W $\leq$ 15 μm. The median grain size was estimated to be in the 1.5 μm to 3 μm range. For the 10 μm and 15 μm linewidths they observed an exponential decrease in t_{50}, of the form $t_{50} \approx \exp(\alpha/l)$, where l is conductor length and α is defined as the critical length. For W=10 μm, α=62.5 μm and for W=15 μm, α=82.5 μm (Figure 11). For line lengths greater than 2-3 times the critical length t_{50} was independent of further increases in line length. The shape parameter of the failure distribution showed a similar dependence on line length, decreasing as a function of increasing line length until the length was 500 μm and was essentially independent of length beyond 500 μm.

When a metallization is deposited globally across a wafer, there is a uniform distribution of defects contained in the film. The term 'defect' refers to any structural or compositional element that gives rise to a flux divergence. When the conductor is patterned, the defect distribution is partitioned, such that the probability of incorporating or 'capturing' a defect in the patterned conductor is a function of its width and length. For the shortest conductor length (and fixed linewidth), l=25 μm, the probability of incorporating a critical defect in the patterned conductor is small relative to the longest length, l=2000 μm. Therefore, when the conductor length exceeds 2-3 times the critical length the probability of capturing a defect saturates. Intuitively, this is consistent with a length dependent t_{50} since for short conductor lengths the number of captured defects is small compared to a long conductor. The small number of defects gives rise to a large t_{50} for short lines and a proportionally decreasing t_{50} as conductor length increases. If a conductor is considered as N serially connected elements of length, l, each element should have a randomly occurring failure time, independent of the other elements. If the probability of failure of an element is

[28] $$F(t) = \int_{0}^{t} f(t)dt$$

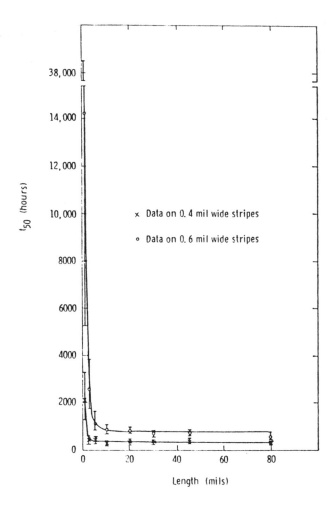

Figure 11: Conductor length dependence of t_{50} for 0.4 mil (10 μm) and 0.6 mil (15 μm) wide lines of aluminum.

then the probability of failure of the weakest element, G(t), of an ensemble of r elements comprising a conductor can be calculated using the equation

$$[29]   G(t) = 1 - \left[1 - F(t) \right]^{r}$$

where r=L/l. If a conductor of arbitrary length L is made up of a single element of length l and has a failure time, t_f, then r=1 and F(t)=G(t)=0.5. If a second element is connected in series with the first element, forming a conductor of length L=2l (r=2), the probability of the 2-element conductor having a failure time greater than that of the single element is 0.25. The probability of the 2 element failure time being less than t_f for the single element is 0.75. Further increases in the number of elements results in decreasing conductor lifetimes, in qualitative agreement with the results in (77).

Attardo et al. (79) used a complex computer simulation to calculate failure time in terms of the distribution of grain boundary mobilities, grain boundary intersections, and grain size divergence induced by variations in the conductor structure. The effects of grain orientation were incorporated by modeling grain boundary mobility in terms of the crystallographic misorientation, θ, between grains using the dislocation core model for a grain boundary (80). Using this approach the flux divergence due to variations in grain boundary mobility (Eq. (16)) could be quantified where boundaries with different degrees of misorientation intersect. The intersection between grain boundaries with identical mobilities and random orientations form an angle, ϕ, between the electric field vector and grain boundary which is proportional to the flux divergence. By partitioning the conductor into sections (Figure 12) and summing grain boundary transport within each section, the time to failure of the conductor was calculated assuming failure occurs when two adjacent sections have a maximum flux divergence compared to other sections. In the computer simulation the number of elements (r) in Eq. (29) is given by L/D_{50}, where L is the conductor length and D_{50} is the median grain size. In Figure 13 the dependence of t_{50} on conductor length, failure time, and shape parameter are shown for pure Al films deposited at 300° C, 400° C, and 500°C. For these films the median grain size (D_{50}) increased with increasing film deposition temperature. The length dependence predicted by the computer simulation is similar to the results in (77). Additionally, for a given conductor length, both t_{50} and the critical length increased with increasing D_{50}. The increase in critical length with increasing D_{50} is consistent with the defect model since the number of elements scales with D_{50}. For a given conductor length the number of defects (proportional to L/D_{50}) will be smaller since a larger D_{50} will in general mean fewer defects.

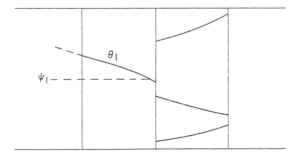

Figure 12:   Model for grain boundary partitions used in ref. (79) for calculating the flux in each region.

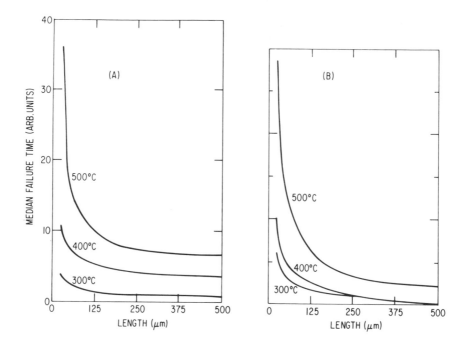

Figure 13: Computer simulation results from ref. (79) showing the length dependence of median time to failure, t_{50}, (left graph) and shape parameter of the failure distribution (right graph) for Al conductors deposited at 300°C, 400°C and 500°C.

Using the method of Attardo et al., Schoen (63) used Monte Carlo computer simulation to model the line length dependence of t_{50}. Rather than using the serial element failure model employed in (79), Schoen simulated the hypothetical failure of test lines and statistically analyzed the 'data' using conventional statistical analysis, assuming a lognormal failure distribution. For conductors with W=1.0 μm and D_{50} = 0.75 μm, and W=2.0 μm with D_{50}=1.25 μm the results of the computer simulations showed that t_{50} decreased exponentially for conductor lengths less than the critical length, in qualitative agreement with the experimental results of Agarwala et al. (77) and Attardo et al. (79). Unfortunately, simulations for lengths longer than the critical length and $W<D_{50}$ were not reported. The simulation results also indicate the shape parameter decreases with increasing line length. The shape parameter is expected to decrease with increasing conductor length because the probability of capturing a defect in a short conductor is small relative to a long conductor. The defect density in short lines will vary over a wider range compared to long conductors since defects will not be found in all short conductors. Therefore, the failure times in short segments will vary over a wider range, resulting in a larger value for σ. As conductor length increases the probability of including the critical

defect increases, increasing the uniformity of the failure times since most lines will contain a larger number of defects.

Linewidth and Grain Size Dependence: The global microstructure (grain size distribution and orientation) of a deposited metal film is dependent on the deposition conditions (rate and temperature) and the characteristics of the surface on which the film is deposited. The grain sizes follow a lognormal distribution law described by D_{50} and the shape parameter, σ (81, 82). By geometrically constraining the global microstructure, i.e. patterning the film to form an interconnect, the resulting interconnect structure can be described by a segment model. A segment may be composed of a single- grain or an element containing several grains. Analytical expressions for the length of single- and poly- grain segments were first described by Fu (83) in his attempt to analytically model the line width dependent electromigration lifetime. The formation of these segments due to the constraint of line width on the grain size distribution can be described in the following way. A linear interconnect pattern defines a rectangular boundary enclosing the local microstructure of the deposited film. If the individual grains which comprise the global distribution are considered the basic unit for the patterned interconnect structure, then we can partition the range of line widths into three regions in terms of D_{50}. For line widths in the range $W \gg D_{50}$, the line is composed of a continuous poly-grain segment whose length is greater than the electromigration threshold length (Figure 14). In this region, the median time to failure (t_{50}) increases monotonically with increasing W for a fixed temperature and current density. Figure 15 shows an example of the increase in t_{50} of an Al conductor with increasing linewidth for $W > D_{50}$ from ref. (79). Scoggin et al. (90) have shown that for $W > D_{50}$ and a single-mode grain size distribution, increasing the conductor line width produces an approximately linear increase in t_{50}.

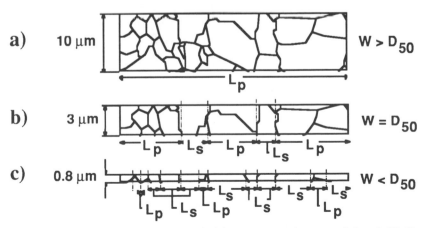

Figure 14: Schematic diagram showing interconnect 'structure' for a) $W > D_{50}$, b) $W \approx D_{50}$ and c) $W < D_{50}$. L_p and L_s refer to the lengths of the single grain and poly grain segments, respectively.

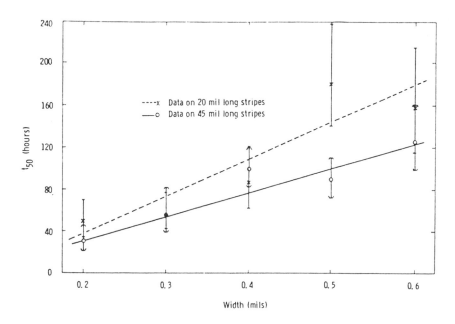

Figure 15: Linewidth dependence of t_{50} for (X) 508 μm and (o) 1143 μm long conductors where $D_{50} > W$. Test conditions were T=130°C and J=0.75X10^6 A cm^{-2} (From ref. (79)).

As linewidth is decreased to the range $W \approx D_{50}$, the interconnect is comprised of poly-grain and single-grain segments in which the corresponding segment lengths are dependent primarily on the grain size distribution parameter, σ. In this range t_{50} approaches a minimum value relative to t_{50} for $W > D_{50}$. The precise line width at which this minima occurs is a function of film thickness and composition (84). Therefore, the critical line width is more aptly described by $W \approx N*D_{50}$, where N is a factor between 1 and 5. When $W << D_{50}$ the line is comprised primarily of single-grain segments having a length greater than the poly-grain segments. In this region t_{50} typically increases with further decreases in line width.

For $W \geq D_{50}$, Black (85), Oliver and Bower (86), Learn (87) and Blair et al. (88) reported that when the conductor width is fixed and the median grain size is increased, t_{50} increases proportionally. Agarwala et al. (89) found t_{50} was also sensitive to changes in the grain size distribution and morphology. For $W > D_{50}$ the increase in t_{50} with increasing line width has been explained in terms of the decreasing probability of aligning a critical 'defect' across the width of the conductor. The term 'defect' refers to any structural inhomogeneity in the conductor producing a flux divergence. As the line width increases, the probability of aligning sufficient critical defects across the conductor width decreases. The decrease in defect probability is manifested as an increase in t_{50}.

Using an argument similar to the 'serial element' model for the dependence of t_{50} on conductor length, the linewidth dependence of t_{50} can be described by a 'parallel element' model. For this model Eq. (29) can be used to describe the probability of failure for an ensemble of $r=N_p$ elements in parallel, where $N_p=W/D_{50}$. In actuality, depending on the linewidth relative to D_{50}, a conductor is composed of both serial and parallel elements connected randomly, if a grain is considered a unit element (see Fig. 14). Based on this, Cho and Thompson (91) proposed a model for the linewidth dependence of t_{50} which includes structural components for both parallel ($N_p=W/D_{50}$, for $W \le D_{50}$) and serial elements ($N_s=L/D_{50}$, for $W > D_{50}$).

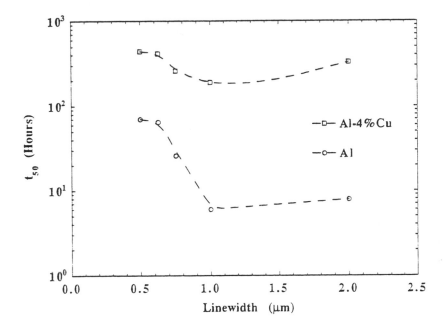

Figure 16: Linewidth dependence of t_{50} for conductors with linewidth in the range of the median grain size for Al-4%Cu and pure Al films. Test conditions were $T=182°C$ and $J=10^6 A\ cm^{-2}$. Data reproduced from ref. 84.

For line widths in the range $W < D_{50}$ (and a single mode grain size distribution) Kinsbron (92) and also Vaidya et al. (93) reported that t_{50} increases with further reductions in line width. Numerous investigators have substantiated these reports for various deposition techniques and film compositions, and the phenomenon reported in these investigations appears to be universal. Figure 16 shows the increase in t_{50} for Al and Al-4%Cu films for W in the range of D_{50} from ref. (84). The increase in t_{50} with decreasing line width has been described statistically using the serial/parallel element model discussed previously; physically it has been attributed to the decrease in poly-grain segment length relative to the electromigration threshold length (92, 93). The concept of threshold length is discussed in section 7.0. As linewidth is decreased there is a corresponding increase in the single-grain segment length relative to the width dimension. This gives rise to a 'bamboo' structure since the line is comprised of single-grains connected serially. Historically, the dependence of t_{50} on decreasing polycrystalline segment length has been explained assuming that diffusion occurs along grain boundaries only and is independent of the film structure or test conditions (4, 91). Dreyer et al. (75) have recently proposed a model for the linewidth dependence based on the relative lengths of single and poly grain segments which supports lattice and grain boundary diffusion at temperatures typically used to characterize electromigration. The results from this model indicate that the linewidth dependence for $W \leq N*D_{50}$ results from the relative change in the grain boundary and lattice components in the diffusional flux. N is an integer and is a function of the electromigration thickness dependence discussed in the next section.

Thickness: The dependence of electromigration lifetime on film thickness is in general a function of the film grain size and linewidth. In general, t_{50} is proportional to the conductor cross sectional area, $A=W*d$, where W is the linewidth and d is the film thickness. The cross sectional area dependence arises because an open circuit failure requires a certain volume of ions migrate away from a localized area, forming a void across the width of the film and through the thickness. Therefore, for a fixed linewidth, t_{50} should in general increase with increasing film thickness. Learn (94) investigated the effects of film thickness on electromigration using 15 μm wide Al-4%Cu-1.7%Si lines that were 1270 μm long. The conductors were tested using $J=8X10^5$ A cm^{-2} and T=227°C. The t_{50} data is shown in Figure 17 as a function of film thickness. With increasing film thickness t_{50} increases and appears to saturate, although the number of data points is limited. The saturation may be due to increased Joule heating in the thicker film since the power required to maintain a constant value of current density in all films increases with increasing film thickness. Whether the saturation effect is real or an artifact of the measurements, the data indicates that t_{50} increases with increasing film thickness. Interestingly, Learn also reported that the activation energy decreases with increasing film thickness, from $\Delta H=0.84$ eV for the thinnest film to $\Delta H=0.73$ eV for the thickest film. Unfortunately, no information on microstructural characteristics of the films were reported.

Kwok investigated the thickness dependence of t_{50} and σ using Al and Al-Cu conductors. In one study (95) t_{50} and σ were measured in Al-Cu films 0.54 μm and 0.86 μm thick. The test structures were 500 μm long and 2 μm wide. The results indicate t_{50} and σ increase with decreasing film thickness, contrary to the results reported in (94). In a second study (84) conductors with Al film thickness of 0.35 μm, 0.75 μm and 0.9 μm and Al-Cu films with thickness of 0.5 μm and 0.86 μm were used for linewidths in the range 0.5 μm to 2 μm. The failure times followed a general trend in which linewidths greater than a critical linewidth show either no change in t_{50} or an increase in t_{50} with increasing linewidth. For linewidths less than the critical linewidth t_{50} increases

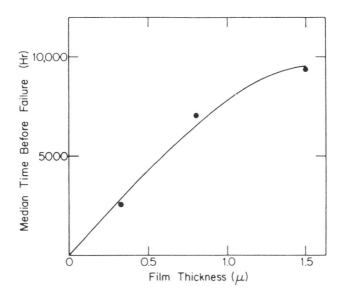

Figure 17: The effect of film thickness on t_{50} for Al-4%Cu-1.7%Si conductors with 15 μm linewidth and 381 μm length. Test conditions were T=227°C and J=8X10⁵ A cm⁻². The activation energy for electromigration decreased from 0.84eV to 0.73 eV as film thickness increased. Reproduced from ref. 94.

sharply (Figure 18). The significant feature of this behavior is that the critical linewidth depends on the thickness of the film. In Figure 18, for Al conductors with W=0.5 μm (W less than the critical linewidth) t_{50} increases from approximately 75-100 hours for the 0.35 μm and 0.75 μm thick films to approximately 200 hours for the 0.9 μm films. For W=0.75 μm t_{50} increases as thickness increases from 0.35 μm to 0.9 μm. This is consistent with the results obtained in (94). However, for W=2 μm t_{50} decreases as the film thickness increases from 0.35 μm to 0.9 μm. These results strongly suggest that the critical linewidth is a function of the film thickness and that the dependence of t_{50} on the film thickness is dependent on the linewidth at which it is measured. For a given grain size, the type of microstructure (i.e. columnar or non-columnar) depends on the film thickness and linewidth. When the film thickness is smaller than the grain size the film is columnar, but not necessarily bamboo. If the linewidth is also less than the grain size, on the order of the film thickness or less, the structure becomes bamboo and t_{50} increased rapidly. Therefore, the effect of film thickness on t_{50} depends on how the film grain size is constrained by the linewidth and thickness.

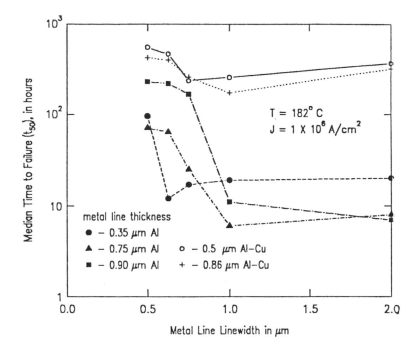

Figure 18: The effect of film thickness on the linewidth dependence of t_{50} for conductors with linewidths in the range of the median grain size for Al and Al-Cu films. Test conditions were T=182°C and J=10^6A cm^{-2}. Reproduced from ref. 84.

Electromigration Activation Energy: Electromigration induced failure times generally follow an exponential temperature dependence. This type of general temperature dependence is common in many thermally activated systems and is referred to as the Arrhenius law. For electromigration the temperature dependence is due to the diffusivity term in Eq. (11) and (12). Using life test data the activation energy can be estimated by plotting the median failure time versus inverse temperature. By taking the logarithm of Eq. (24) the activation energy , ΔH, can be computed from the equation

$$[30] \qquad Log(t_{50}) = \frac{\Delta H}{k} \cdot \frac{1}{T} + Log(A)$$

A linear regression of the form given by Eq. (30) to the experimental life test data can be used to estimate the slope, ΔH/k, from which the activation energy can be calculated. An example Arrhenius plot is shown in Figure 19 for a 10 μm wide Al-1.5%Cu conductor with D_{50}=3 μm. The activation energy for this linewidth is ΔH=0.61 eV. For conductors with W ≥ D_{50} the values obtained using this method yield activation energies which are in reasonable agreement with direct measurements of mass transport for grain boundary diffusion processes. Table 4 shows the activation energies for pure Al and Al alloys determined from mass transport and life test measurements. The activation energies measured in these films are similar to the reported value of grain boundary diffusion, and has been verified using a number of techniques in conductors for which $W \gg D_{50}$ (57-61).

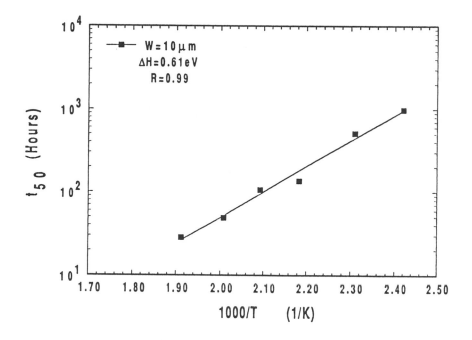

Figure 19: Arrhenius plot for a 10 μm wide Al-1.5%Cu film. Current density was $J=2X10^6$ A cm^{-2}. The resulting activation energy was $\Delta H=0.61$ eV.

For a given metal composition the activation energies obtained from life test results tend to have somewhat greater scatter than energies obtained using direct measurement of mass transport. The probable cause for this result is due to imprecise measurements of peak film temperature due to Joule heating. d'Heurle and Ho (31) and also Lloyd et al. (41) have pointed out that the observed variations can be attributed to improper use of initial film temperature or ambient temperature rather than a meaningful average temperature which includes the progressively increasing Joule heating contribution resulting from damage formation during the test.

The difference between the peak temperature at localized damage sites and the average film temperature is the experimental error and should be proportional to the square of the current density. Venables and Lye (96) have demonstrated by means of computer simulation that ΔH calculated over a range of current densities show a monotonic decrease in ΔH with increasing current density due to the increase in the temperature error. In Figure 20 the activation energy for grain boundary diffusion is assumed to be $\Delta H=0.7$ eV for $J=5X10^4$ A cm^{-2}. Increasing the current density to $J= 5X10^5$ A cm^{-2} increases the Joule heating contribution $\Delta T_j=10°$ C, and yields an activation energy $\Delta H=0.68$ eV. For $J=2X10^6$ A cm^{-2} the Joule heating contribution is $\Delta T_j \approx 20°$ C resulting in an apparent activation energy of $\Delta H=0.65$ eV.

Table 4: Summary of electromigration activation energies for Al based thin film

Composition	Temperature Range (°C)	$D_{50}(\mu m)$	ΔH (eV)	Method	Reference
Al	180-350	0.5 to 2-3	0.7 ± 0.2	TEM[a]	52
Al	320-400	4.0	0.7 ± 0.1	TEM[a]	123
Al	175-350	2.0	0.63	TEM[a]	53
Al	240-550	1.0	0.7	SEM[b]	54
Al	225-400	1.0-1.5	0.55-1.3	Drift velocity	124
Al	220-360	0.3	0.53-1.2	ΔR[c]	49
Al	60-150		0.5-0.6	ΔR[c]	125
Al	100-200		0.5-0.6	Resistance	126
Al	109-260	1.2-8.0	0.48-0.84	Life test	127
Al	140-300	2.0-8.0	0.51-0.73	Life test	102
Al	75-380	1.0	0.41	Life test	128
Al	125-175		0.5-0.7	Life test	129
Al	100		0.3-1.2[g]	Life test	131
Al-(wt%)Cu[d] 0.5% to 8%	70-230		0.45-0.73[e]	ΔR[c]	76
Al-(wt%)Cu[d] 0.3% to 9%	105-230		0.63-0.83[e]	Life test	130
Al-(wt%)Cu[d] 0 to 12%	230-280		0.43-0.78[e]	Life test	132
Al-6%Cu	200-330		1.1	TEM[b]	5
Al-1.5%Cu	140-300	3.0	0.56-1.04[f]	Life test	101
Al-2%Cu-0.3%Cr	175-225	Single crystal	1.22	TEM	99

a. Transported volume (void formation)
b. Mass depletion
c. Resistance change
d. Weight percent Cu intentionally varied over this range
e. Activation energy dependent on concentration
f. Activation energy dependent on linewidth/temperature

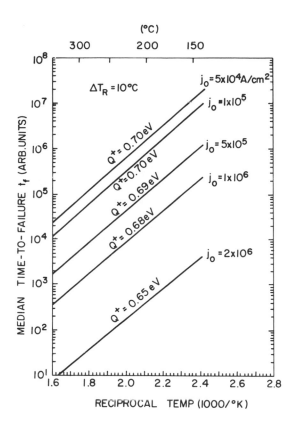

Figure 20: The effective (or apparent) activation energy as a function of reciprocal temperature with current density as a parameter. Initial conditions assume that at J=5x10⁴ A cm⁻² and ΔH=0.7 eV. Reproduced from ref. 96.

There has been some controversy concerning whether or not lattice diffusion contributes significantly to failure over the range of experimental test conditions typically used to characterize electromigration. At circuit operating conditions (typically less than 150° C) lattice diffusion should be nearly non-existent. This has been confirmed experimentally by several investigators in films with $W \gg D_{50}$. For $W < D_{50}$ there is mounting evidence for a lattice diffusion component at test conditions used for electromigration studies. Lattice diffusion has been measured in single grain segments when the segment length exceeds the threshold length in drift velocity structures (97) and in single crystal thin film specimens of Al-Mg (98) and Al-2%Cu (99) at temperatures near 175°C. Moreover, there have been several reports on the grain size dependence of electromigration activation energy at temperatures below one half the melting point of Al (100-106).

For narrow lines with a 'bamboo' or 'near bamboo' structure, mass transport leading to failure often occurs either at triple points or along boundaries separating single grain regions. While the former process is expected to occur, it does not exclude other transport processes. Moreover, for $W < D_{50}$ mass transport will be rate limited by lattice diffusion (107-109). In addition there are a large number of reports in the literature for activation energies, ΔH, with values in the range 0.4 eV $\leq \Delta H \leq 1.4$ eV, which also suggests a mixed mode transport mechanism. The magnitude of the activation energy appears to be a function of the film grain size and distribution, linewidth, and to a lesser extent, grain orientation. Black (100) measured the electromigration activation energy in two groups of Al conductors with W=12.5 μm. One group was fabricated using a 400° C substrate temperature during deposition resulting in large grains ($D_{50} \approx 8.0$ μm); the other group was deposited on a cold substrate and had small grains ($D_{50} \approx 1.2$ μm). The film temperature range was $120°C \leq T \leq 250°$ C. The activation energy for the small and large grain films was $\Delta H = 0.48$ eV and $\Delta H = 0.84$ eV, respectively, which Black interpreted as in increase in the lattice diffusion component. Attardo and Rosenberg (102) used pure Al films to replicate the experimental work in (100). In their case they used conductors with $W = 10$ μm and W=15 μm. By varying the substrate temperature two groups of films were produced having grain sizes of $D_{50} = 2$ μm and $D_{50} = 8$ μm. They reported $\Delta H = 0.51e$ V and $\Delta H = 0.73$ eV for the small and large grain films, respectively, and attributed the difference in activation energy to intrinsic differences between the grain boundary transport properties of large grain and small grain films, rather than to a lattice diffusion component. Recently, Dreyer et al. have reported that the electromigration activation energy increases with decreasing linewidth for the range 0.8μm $\leq W \leq 10$ μm (101) (Figure 21). This data was measured using a film temperature range $140°$ C $\leq T \leq 300°$ C for each linewidth, with $D_{50} = 3.0$ μm. As linewidth decreased ΔH was found to increase, providing strong support for a lattice component of diffusion in 'bamboo' or 'near bamboo' type lines with test conditions commonly used to characterize electromigration. This result suggests a direct relationship among grain size distribution, linewidth, temperature and the diffusion components of electromigration. Based on these results, Attardo and Rosenberg' s argument requires that the grain boundary transport properties be linewidth dependent, i.e. the grain boundary transport properties for narrow conductors should differ from those of wide conductors, which seems unlikely. Recently a model was proposed to account for the change in activation energy with linewidth (75). This model includes the effects of constraining the film' s grain structure in a rectangular interconnect pattern on the diffusional components and includes the temperature dependence of the diffusivities. The proposed model seems to resolve the wide variation in values of activation energy reported for conductors of the same composition, with varying microstructures and measured at different test conditions.

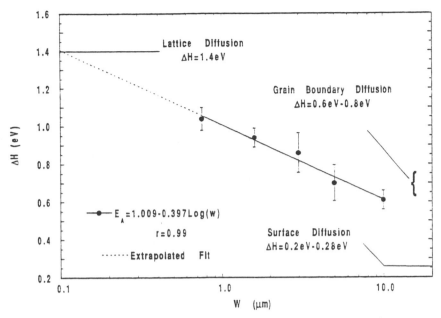

Figure 21: Activation energy measured in Al-1.5%Cu films shown as a function of linewidth. All data was obtained over the same film temperature range, 140° C to 300° C, for each linewidth. For reference the activation energy for surface, grain boundary and lattice diffusion are shown. The error bars represent the 90% confidence intervals.

Several investigators have reported a current density dependent activation energy and proposed a model based on the transfer of energy from incident electrons to activated ions (41, 110, 111). This seems unlikely even under high field conditions since the energy transferred by an electron to an un-activated ion would have to be substantial in order to reduce the apparent height of the ions potential well. Lloyd et al. (41) pointed out that even at a current density of $J=10^7$ A cm^{-2} the apparent energy change would only be on the order of 10^{-4} eV. The source of the apparent current density dependence more likely lies in errors in the measurement of temperature as discussed previously.

In general, measurements of the activation energy should be made under conditions which minimize Joule heating effects. In particular, larger values of current density should only be used if test structures are designed to minimize thermal gradients and the dielectric material underlying the test structure has a sufficiently high heat dissipation factor.

6.4   AC/Pulsed DC Electromigration

Most electromigration characterization is performed using a direct current to provide the driving force. The desired outcome for a life test characterization of a newly developed MLM system are the peak operating current densities or 'design rules'. The design rules for current density are usually expressed in terms of DC conditions and assumptions are generally made for derating the design rules for AC or pulsed DC operation. Over the last several years there has been significant focus on electromigration under AC or pulsed DC conditions since in VLSI or ULSI circuits both types of waveforms may be present. These waveforms may be quite complex with varying duty cycles and waveforms. In general the waveform will be dictated by the technology application (e.g. digital or analog). Experimentally it is impractical to generate complex waveforms. Therefore, more simplistic waveforms such as AC or unipolar square waveforms are used as the driving force. One reason electromigration has not been characterized more extensively using AC/pulse waveforms is that measurements under high frequency AC or pulse conditions are significantly more complex compared to low frequency or DC conditions because of the parasitics associated with the test fixtures, cabling, and packages. Equipment is also generally more expensive. Wafer level testing using specially designed probes can be used to eliminate the problems associated with package testing; wire bond inductance and package capacitance. However, unless the test is highly accelerated, probe stability over long periods becomes an issue. A number of new approaches have been proposed to circumvent these difficulties. These include incorporation of 'on chip' waveform generators to drive the test structure. These chips employ poly-silicon heaters fabricated underneath the test structure to increase the test structure ambient temperature instead of ovens. On-chip drivers and heaters eliminate the need for expensive external function generators and the aforementioned parasitics as well as ovens. The drawback to this testing approach is that the drivers must be fabricated along with the test structures. For advanced silicon technologies this increases the lead time for obtaining samples because of the large number of processing steps required to generate active devices.

The formation of electromigration damage under AC or pulse DC conditions is a complex function of duty cycle and frequency. For pulse DC waveforms the duty cycle (d_c) is defined as the ratio $d_c = t_{on}/t_{tot}$, where t_{on} is the 'on time' during which the current pulse is applied to the conductor and t_{tot} is the waveform period. During the 'on' portion of the waveform vacancies are swept along in the direction of current flow and may coalesce to form voids. During the same time the temperature of the film will rise by an amount over the ambient depending on the peak current density, the thermal time constant of the surrounding dielectric and the frequency of the waveform. Generally, under AC or pulsed DC conditions the amount of Joule heating generated is less than the Joule heating generated under DC conditions if the frequency is much greater than the inverse of the dielectric thermal time constant.

Electromigration lifetime under pulsed conditions generally increases compared to similar DC conditions. Two models for median time to failure under pulsed DC conditions have been proposed. The 'on time' concept assumes that the damage accumulates during the on portion of the pulse only. McPherson and Ghate (112) have proposed that the failure kinetics for the on-time concept are inversely proportional to d_c. This assumes that during the 'on-time' of a current pulse electromigration damage is induced and that in the absence of thermal gradients due to Joule heating no atomic back flow occurs to heal the induced damage. This implies, for example, that for a 10% duty cycle the rate of mass transport would be 1/10 of what it would be compared to DC conditions. Moreover this reduction on mass transport rate should be reflected in the failure time of the conductor. The median time to failure can therefore be expressed using a modified form of Black's equation;

$$[31] \qquad t_{50} = A\left(\frac{1}{d_c}\right) j^{-2} \exp(\Delta H / kT)$$

and the improvement in t_{50} predicted by the on-time model is conservative. Lin and Guan (113) and also Towner and van de Ven (114) have proposed a 'time average' current density model which provides a less conservative estimate of t_{50}. This model assumes that activated ions are subject to an average current density defined by the product of the duty cycle and the peak current density, $j_{ave} = d_c j$. Black's equation, modified as in Eq. 31, predicts the median time to failure is inversely proportional to d_c^2 and can be expressed as

$$[32] \qquad t_{50} = A(d_c j)^{-2} \exp(\Delta H / kT)$$

There are conflicting reports in the literature for the dependence of t_{50} on duty cycle and frequency and little is known about the exact nature of the wearout process under non-DC conditions.

A number of investigations have provided conflicting data on the duty cycle dependence of t_{50}. Over a wide range of conditions Kinsbron et al. (115, 116) found that the duty cycle dependence was proportional to d_c^{-1}, supporting the on-time model given by Eq. (31). On the other hand Liew et al. (117, 118) found that the duty cycle dependence was proportional to d_c^{-2}, supporting the time average model given by Eq. (32).

Not surprisingly, there are also conflicting reports in the literature for the dependence of t_{50} on frequency. Liew et al. (117, 118) measured the dependence of t_{50} on frequency with peak current densities ranging from 10^7 A cm^{-2} to 5×10^7 A cm^{-2}. No variation in t_{50} was observed for frequencies in the range 35 kHz to 14 MHz. Towner and van de Ven (114) used unpassivated Al films with W=4.5 μm with j=2×10^6 A cm^{-2} peak current density and frequencies of 10 Hz, 10^3 Hz and 2×10^4 Hz to investigate electromigration under pulsed DC conditions. The duty cycles investigated were d_c=25% and d_c=75%. For the 2 test temperatures used, 175°C and 250°C, no dependence of t_{50} on frequency was observed. Grain size apparently has no effect on the frequency dependence. Lin et al. (113) used 2 groups of Al-1%Si films to investigate the grain size dependence. One group had a fine grain structure ($D_{50} \approx 1.0$ μm), the other had larger grains ($D_{50} \approx 2.2$ μm). Using frequencies of 0.1 MHz and 1 MHz and a peak current density of 1.2×10^6 A cm^{-2}, no dependence of t_{50} on frequency was observed. English and Kinsbron (119), using an edge displacement technique for

Al-0.5%Cu films, found the rate of mass transport appears to be independent of frequency, for frequencies in the range 0.01 Hz to 10^5 Hz.

An almost equal number of report in the literature indicate that electromigration under pulse conditions is dependent on frequency. Davis (120) measured the increase in t_{50} with increasing frequency, which conflicts with the results obtained by Lin and Guan and also Towner and van de Ven. Suehle and Schafft (121) measured the frequency dependence of t_{50} using unipolar current pulses with a 50% duty cycle for frequencies ranging from 1 mHz to 50 kHz. For frequencies below 10 Hz t_{50} is only weakly dependent on frequency, increasing slightly with increasing frequency. Between 10 Hz and 100 Hz t_{50} increases more abruptly and becomes invariant with frequencies exceeding 100 Hz. The peak current density was 1.5×10^6 A/cm^2. Noguchi et al. (122), using very large peak current densities (1.5×10^7 A/cm^2 to 2.0×10^7 A/cm^2) measured the frequency dependence of t_{50}. The range of frequencies measured was 1 kHz to nearly 1 MHz with a 50% duty cycle. For frequencies below 12 kHz t_{50} was independent of frequency. Above this critical frequency a rapid monotonic increase in t_{50} was observed with no saturation apparent even as the frequency approached 1 MHz. In addition, for a pulse width of 1ms t_{50} was inversely proportional to duty cycle, and for a 20 μs pulse width t_{50} was proportional to d_c^{-2}.

The conflicting reports concerning duty cycle and frequency dependence may be related to the vacancy relaxation time for a given metallization. Liew et al. (117, 118) have proposed a vacancy relaxation model which suggests that for frequencies greater than several multiples of $1/\tau$, where τ is the vacancy relaxation time (on the order of 0.1ms-1ms), the vacancy concentration reaches an steady state value. In this frequency range the median time to failure is independent of frequency since $f \gg 1/\tau$ and is proportional to $1/d_c^2$ relative to similar DC conditions. For frequencies less than $1/t$ there may be a decrease in t_{50} and the duty cycle dependence may be proportional to $1/d_c$.

7.0 DAMAGE FORMATION IN MULTILEVEL METALLIZATION

In general, the rate of damage formation is controlled by the magnitude of the atomic flux and the local flux divergence, which depend on the driving force, the rate of mass transport and the microstructure/morphology of the interconnect (133). Damage formation, as a result of the local change of the atomic concentration, is related directly to the flux divergence. Depending on the test temperature, the flux divergence is determined by the driving force, the diffusion characteristics and the microstructure of the interconnect. The interconnect morphology, such as the line/stud interface, is also important in determining the flux divergence, where the local diffusion flux can change abruptly. The damage formation process of the multilevel interconnect can be significantly different from the single-level conductor lines due to changes in the driving force and the flux divergence for the mass transport. For single-level conductor lines and under normal device operating condition, mass transport is dominated by grain boundary diffusion. As discussed in section 3.1, the grain-boundary triple point is a common site of flux divergence, which causes damage by crack formation in the line. For submicron multilevel interconnects, the reduction in the line dimension and the use of complex multilevel metallization have changed the basic nature of the flux divergence. While the grain structure evolves to assume the bamboo morphology, the use of refractory metal, e.g. Ti, induces compound formation. In addition, the use of TiN barrier and W stud introduces interfaces formed by materials with distinct mass transport and thermal/mechanical properties. In such a structure, mass transport occurs through coupling between grain boundaries and interfaces and flux divergence can ready take place at an interface formed by dissimilar materials, e.g. the Al(Cu)/W stud/line interface. These factors can lead to damage formation processes fundamentally different from single-level line structures (134). This will be illustrated by results from electromigration studies on an Al(Cu)/W two-level interconnect structure in the next section.

In addition to the effects on flux divergence due to the interconnect microstructure and morphology, device scaling impacts the interconnect reliability by increasing the current density, the number of thermal processing cycles and the amount of power to be dissipated. These factors change the driving force and increase the kinetics of damage formation with a decrease of the lifetime of the interconnect. Most of these effects described here have been recognized, however, their quantitative study for submicron interconnects is still at an early stage. The scaling impact on stress-induced voiding is not well understood, although one can attribute the problem directly to device scaling as a result of a reduction in line dimension and an increase in dielectric confinement. Particularly lacking at this time are experimental results to quantity the effect on void formation. The basic mechanism and the present understanding of this problem will be discussed in a later section.

7.1   Electromigration Damage in Multilevel Interconnects

The structure chosen for the electromigration study is a high-performance interconnect employing planarized wiring levels of Al(Cu) alloys in SiO_2 with W interlevel studs. The processing steps of the test structure have been described previously (135), so they are summarized briefly here. The M1-W-M2 line/stud/line test structures were processed by a fully planarized two-level interconnection process. The first level metal line M1 was formed using a Ti(25nm)/Al(2%cu)(0.7μm)/Ti(25nm) structure with Al(Cu)(50nm) and Si(20nm) capping layers. Following M1 patterning, a PECVD silicon oxide interlevel dielectric was deposited and planarized using a chemical-mechanical polishing process. The studs were processed by sequential blanket sputtered Ti/TiN and CVD W, followed by a second planarization process. All samples were annealed at 450°C for 40 mins. in argon to induce a complete reaction of

Ti/Al to form the intermetallic compound TiAl$_3$. The formation of the intermetallic compound is essential for the improvement of the electromigration performance of this metallization (136). The M2 level is processed identically as M1, except that the thickness of the Al(Cu) layer is 0.8μm and a thin TiN capping level was used, instead of a thin Si layer as in M1.

The electromigration performance of this metallization has been investigated recently by Rathore et al. (137). These authors first carried out measurements comparing the lifetime of single-level conductor lines formed using this metallization with the standard Al(Cu) metallization with 2% Cu. For the single-level structure, a meandering metal stripe (790μm long) with Kelvin contacts and extrusion monitor was made with line width ranging from 0.7 to 1.6μm. The result of this study is shown in Figure 22a where it is seen that with a current density of 10^6A/cm^2 and at 250°C, the lifetime of the Ti-Al(Cu)-Ti metallization is better than Al(Cu) by more than 100x over a range of line width from 0.5 to 2μm. Lifetime measurements were subsequently carried out to evaluate the electromigration performance of a two-level M1-W-M2 line/stud/line structure formed using this metallization. The two-level line-stud chain test structure had fixed M1 length (at 10μm) but with variable M2 length ranging from 3 to 300μm. The stud diameter and line width were variable in this study and an extrusion monitor was incorporated in the test structure. The results from one of the tests are plotted in Figure 22b together with the lifetime measured for the single-level conductor line formed by the same metallization. Under an identical test condition, the lifetime of the two-level structure with 1.3μm wide line is about a factor of 50 smaller than that of the single-level line. These results suggest a distinct process of damage formation for the two-level structure. A separate study was carried out to investigate several basic aspects of the electromigration behavior and damage formation in two-level interconnects using Al(Cu) and Ti-Al(Cu)-Ti metallizations (138). In this study, the mass transport was measured using a drift-velocity technique (124) as a function of line length and width. In this experiment, the structure used was a line/stud chain with the M2 line segments of variable length while the M1 line segment is kept constant below a threshold length (defined later). Under the electromigration driving force, materials accumulate in each segment toward the anode and generate a stress, resulting in back-flow of material. The consequence of electromigration and back-flow gives rise to a net drift velocity for materials depleted from the cathode edge

$$[33] \qquad v_d = v_e - v_b = \left(\frac{D}{kT}\right)\left(Z^*e\rho j - \sigma\frac{W}{L}\right)$$

Where V_b is the average velocity of the atoms caused by the back-flow stress σ, W is the atomic volume and L the line segment length. The remaining terms were defined in section 3.1. As seen from Eq.33, the mass transport is reduced by the back-flow stress, resulting in a drift velocity inversely proportional to the line length. Accordingly, one can define a threshold value for the product of the current density J, and a critical line length L_c, below which mass transport vanishes as

[34] $$j \cdot L_c = \frac{\sigma W}{Z^* e \rho}$$

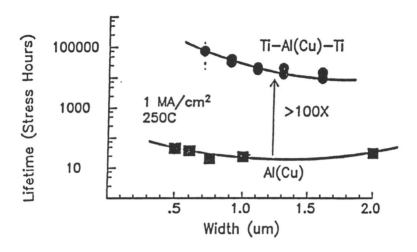

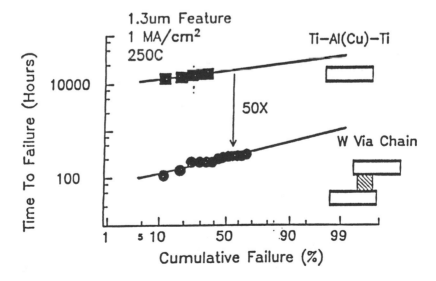

Figure 22: Comparison of electromigration lifetime for Al(Cu) vs. Ti-Al(Cu)-Ti (top) and for Ti Al(Cu)-Ti single level and two-level (with W stud) structures (from Ref. 144).

Results of the drift velocity experiment on Al(Cu) and Ti-Al(Cu)-Ti revealed that the expected linear time dependence of the edge displacement was observed only after an initial incubation period, as shown in Figure 23a. This was traced to a kinetic process involving the dissolution of the Al_2Cu precipitates and depletion of Cu by electromigration from the cathode end. Essentially, Cu has to be depleted beyond a critical length before Al can be depleted from the cathode edge since the electromigration of Al is retarded by the presence of Cu. The dissolution of the precipitates is required because they act as sources to supply Cu.

Resistometric measurements were carried out concurrently with drift velocity experiments in line/stud chain structures to monitor electromigration-induced damage formation. The results of one such measurement correlate with the drift velocity measurement in Figure 23a and is shown in Figure 23b. An initial period of about 20 hrs. was observed with no change in resistance, in agreement with the incubation time for Al depletion found in the drift velocity experiment. Afterwards, the resistance increases linearly with time, coinciding with a linear depletion of Al. This process continues until an open failure occurs with Al completely depleted from the line/stud contact. This indicates that in the Al(Cu)/W two-level interconnect, failure occurs as a result of flux divergence at the line/stud contact. This process is different from the failure due to crack formation at grain boundaries for single-level lines.

With the mechanism established for damage formation, we proceed to investigate the kinetics for electromigration-induced failure in the two-level interconnect. The result was found to follow the standard expression for lifetime:

[35] $$t_{50} = Aj^{-n} \exp(\Delta H / kT)$$

where n is the current density exponent and ΔH is the activation energy. The value of n was determined to be close to 2, being 1.9 ± 0.1, up to a current density of $10^6 A/cm^2$. The activation energy was determined by measuring the lifetime as a function of temperature at $10^6 A/cm^2$. Results are summarized in Figure 24 for 0, 0.5 and 2% of Cu. As expected, there is a general trend for the failure time to increase in proportion to Cu concentration. This reflects the increase in time required to sweep Cu beyond the threshold distance before Al depletion can occur. Significantly, the activation energy is different for pure Al, 0.58eV, compared with 0.80 and 0.83eV for the two Al(Cu) alloys. The value of 0.58eV is typical for grain boundary diffusion in Al (139), demonstrating that grain boundary diffusion of Al is responsible for electromigration   failure in pure Al interconnect. The activation energy for the Al(Cu) alloy is considerably higher, suggesting that the migration of Al does not control the kinetics for failure in the Al(Cu) two-level interconnect. It is interesting to note that the activation energy for electromigration of the alloy is almost the same as the Cu grain boundary diffusion in Al(2%Cu), 0.81eV (35) but higher than the formation energy for Al_2Cu, 0.42eV (140). This suggests that the kinetics for electromigration failure in Al(Cu)/W two-level structure is controlled by the slow process of Cu electromigration along the grain boundary, and not by the fast process of dissolution of the θ phase precipitates.

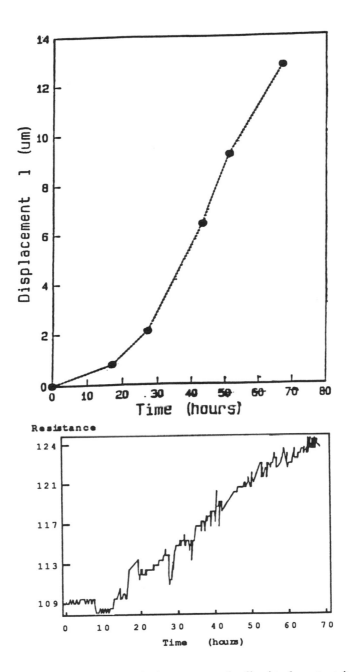

Figure 23. Correlation of the displacement at the line/stud contact (top) and resistance increase of line/stud chain (bottom) measured for Al(2%Cu, 3%Si)/W two-level interconnect at 205°C and $1.5x10^6$ A/cm^2 as a function of time. (From Ref. 138).

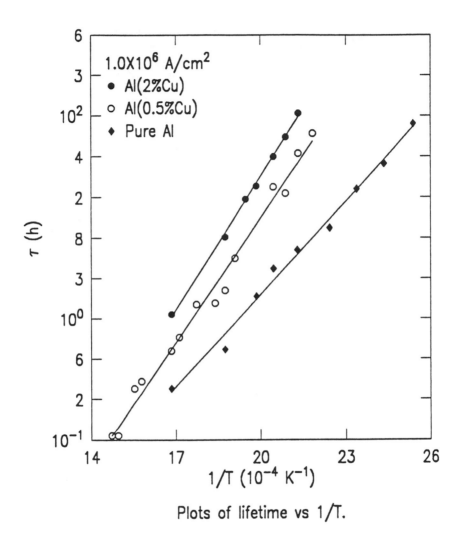

Plots of lifetime vs 1/T.

Figure 24. Results of time-to-failure for Al(Cu)/W two-level interconnects with 0, 0.5 and 2% Cu measured at 10^6 A/cm^2 as a function of temperature. (From Ref. 138).

8.0   STRESS-INDUCED VOID FORMATION

Since its first report in 1984-85 (141, 144), stress-induced void formation has attracted great interest and is being intensively investigated (142). From these studies, several salient features have emerged. First, the stress is induced by the mismatch in thermal expansion between the metal and the silicon substrate. While void formation is induced by the tensile stress generated during cool down in the metal line, other stress-related phenomena have been observed, in particular, interfacial cracking of the dielectric layer. Second, the presence of the dielectric passivation layer is important for void formation so far most of the voids observed occur under a passivation layer. Third, the line width is another important parameter. Reports in the literature indicate void formation occurs mostly in lines with a width less than about 2μm. And fourth, peak void formation occurs at an intermediate temperature range from about 150°C to 180°C. This means that the chip processing temperature and test condition are important in controlling the amount of void formation. Although the investigation of the phenomenon is at a relatively early stage, these basic characteristics can now be explained.

In reviewing stress-induced void formation, we focus on the characteristics of the thermal stress in multilevel structures and the behavior of stress relaxation leading to void formation. At this time, only limited experimental results on void formation for line structures have been reported, so they will be reviewed briefly here. To discuss stress-induced void formation, we can follow the formalism presented above by examining the driving force, the flux divergence in mass transport, and the effect of interconnect geometry and morphology.

Thermal stress is induced by the mismatch in thermal expansion between metal and Si. For Al metallization, the thermal mismatch at 20°C is about 20.5 ppm/°K, taking α for Al 23.5 ppm/°K and for Si 3 ppm/°K. The stress in a layered film on a substrate can be measured by x-ray diffraction or wafer bending technique (143). In Figure 25, we show a typical result for an Al(Cu) film deposited on Si obtained by a bending beam technique (145). One can clearly see the compressive and tensile stresses generated during the heating period and the cooling period in a thermal cycle. During heating, plastic yielding starts at about 200°C to relieve the compressive stress, which continues to the maximum temperature of 400°C. Yielding also occurs during cooling, starting at about 350°C but with a different plastic flow characteristics to reflect the different temperature, strain rate and relaxation history. Subject to the plastic yield, the film reaches a maximum tensile stress of about 300 MPa at the end of the cooling period. The stress behavior during the second thermal cycle is different, showing a higher level of compressive stress, which indicates a reduction in stress relaxation, probably due to the presence of a higher dislocation density at this stage. With the change in the dislocation structure stabilized, the stress behavior of the film becomes invariant although the measured behavior for subsequent cycle is not shown.

In this case, the thermal stress is generated as a result of the film confinement to the Si substrate, which prohibits it from expanding or shrinking freely. Since the confinement applies only at the film/substrate interface, it leaves the other surface free for deformation processes, such as hillock growth, to occur and relieve the stress. If the film is passivated on the top surface by an oxide layer, such as in a bilayered oxide/Al/Si structure, both of the film surfaces are confined, then a different stress behavior is to be expected. With the top surface of the film no longer free, an increase in the yield strength of the film is to be expected. Thus the confinement from the substrate and the dielectric layer plays a central role in controlling the stress level and the deformation characteristics of the metal film in a layered structure.

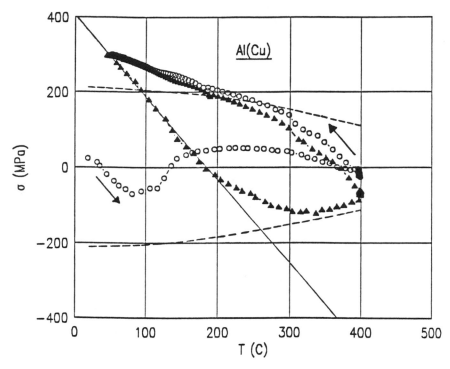

Figure 25: Stress versus annealing temperature of a 1.0μm thick Al(2at%Cu) film. Circles: heating cycle, triangles: heating cycle. Straight line: thermal mismatch from literature values, dashed line: yield stress at 0.2% strain. (From Ref. 145).

A similar effect of dielectric confinement is expected for metal lines in a multilevel interconnect. In this case, the confinement effect can be estimated by comparing the measured stresses with those calculated based on an elastic approximation. For a thin film on a substrate, corresponding to an aspect ratio of h/w=0, the confinement is 2-dimensional and the principal stress components in the elastic limit with no plastic yield can be expressed as (147):

[36] $$\sigma_1 = \sigma_2 = \frac{E\Delta\alpha\Delta T}{1-\nu}, \qquad \sigma_3 = 0$$

where E is the Young's modulus, ν the Poisson's ratio, $\Delta\alpha$ the difference of the thermal expansion coefficient and ΔT the temperature excursion. For a 3-dimensional confinement of a circular cylinder embedded in a substrate, corresponding to an aspect ratio of h/w=1, the elastic stress components are (151):

[37a] $$\sigma_1 = \frac{E\Delta\alpha\Delta T(1+r)}{1+r(1-2v)}$$

[37b] $$\sigma_2 = \sigma_3 = \frac{\sigma_1 r}{1+r}$$

where σ_1, σ_2 and σ_3 are the principal stresses along the axial, width and thickness directions respectively, and $r=E_s/E_m$ is the ratio of the Young's moduli of the metal and the substrate. The thermal stresses for metal lines with aspect ratios between these two extreme cases can be deduced using Eshelby theory of inclusions for embedded lines with elliptical cross sections (148, 149).

For interconnect structures with the metal lines embedded in dielectric layers on a silicon substrate, there is no analytical solution and the thermal stresses have to be calculated numerically. Using a finite element analysis, Jones et al. (150) have computed the principal stresses generated during thermal cycling from room temperature to 400°C for Al lines embedded in oxide dielectrics and the results are summarized in Figure 26. Compared with the Al film, the stress levels in the line structure are about double, indicating a significant stress enhancement due to dielectric confinement. In addition, this work shows a large stress concentration existing at line corners and interfaces, as high as 200% above the average stress. This can be attributed to the additional confinement due to the interconnect geometry and explains why void formation is usually observed at corners and interfaces of stress concentration points. Another interesting result is that a compressive stress of magnitude comparable to the tensile stress in the line is generated during cool down in the dielectric layer. A similar stress state but with opposite stress signs is present during heating, which can induce fracture or delamination to occur in the dielectric layer during heating. This mode of stress-induced damage formation has indeed been reported (151).

The significance of these results is the high value of the thermal stress in the confined lines, 600MPa vs. the value of 300 MPa measured for the Al(Cu) film. Since these stress values are derived using the elastic approximation, it is interesting to compare these results with the stress level and stress relaxation behavior in the actual interconnect structure. A number of experimental studies have been carried out using x-ray diffraction (152-154) and bending beam technique (145, 155) to measure thermal stresses and stress relaxation behavior in narrow Al lines. Results from an x-ray measurement (152) for Al(Si) lines of 2μm wide and 0.8μm thick with dielectrics of phosphorus silicate glass are shown in Figure 27. For comparison, we show in Figure 28 results from a bending beam measurement of Al(Cu) lines of 11μm wide and 1μm thick with sputtered quartz passivation (145). Although the materials and dimensions of the line structures are different, the results are consistent, showing that the stress behavior within the measured temperature range is almost elastic, revealing a large stress difference between 20°C and 400°C with insignificant plastic yield. Indeed, the magnitude of the total stress difference, except along the normal to the thickness direction, is about 600 to 800 MPa, very close to the calculated elastic \ stresses.

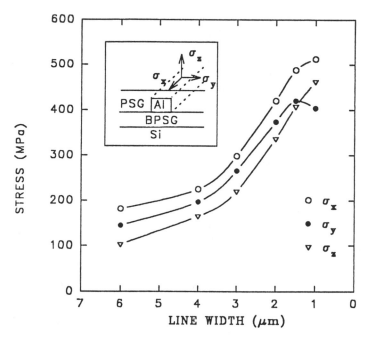

Figure 26: The calculated average stresses in Al confined lines as a function of line width. The insert shows the line configuration and the stress orientations. (From Ref. 150)

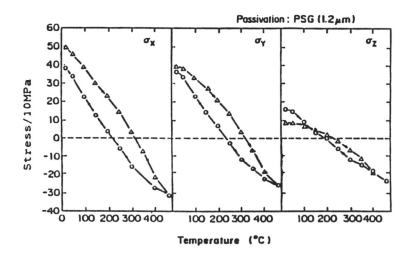

Figure 27: X ray measurement of stress for PSG passivated Al-Si interconnection in a temperature cycle. 0.8μm thick, 2μm width. (From Ref. 152).

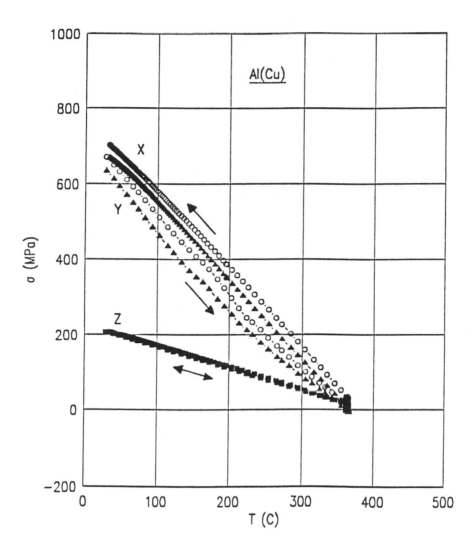

Figure 28: Triaxial stress components (for Al(Cu) lines as a function of annealing temperature in a SiO$_2$/Al(Cu) line structure on Si (line width: 11.35μm, thickness: 1.0μm, thickness of passivation: 1.5μm, width x, length y and height z). (From Ref. 145).

With the stress level raised by dielectric confinement, the elastic energy density $\left(\sigma_x^2 + \sigma_y^2 + \sigma_z^2\right)/2E$ can increase by an order of magnitude over that of the unpassivated film. Korhonen et al. (149) suggest that this provides an energy density exceeding the critical level required for void nucleation. These observations on thermal stresses clarify the basic mechanism for void formation. Essentially, most of the thermal stresses generated remain in the metal lines due to the confinement of the dielectric layer. The confinement increases the yield strength of the line which gives rise to a high level of near-hydrostatic stress. This corresponds to a volumetric strain which, depending on the interconnect microstructure and morphology, is most effectively relieved by void formation distributing throughout the interconnect structure.

The stress relaxation process responsible for void formation has been investigated by x-ray diffraction and bending beam techniques; the latter is particularly suited for such studies since the stress variation can be rapidly and in-situ monitored during thermal cycle. Applying this technique to periodic line structures, Moske et al. have investigated the stress relaxation behavior of Al(Cu) line structures on silicon substrates between 150°C to 250°C following a cool down from 400°C (155). The observed relaxation of the line structure is compared with corresponding unpassivated and passivated layered film structures. The overall behavior of all structures are similar, showing an initial plastic deformation, then a fast relaxation in sequence with a log(time) slow relaxation. The initial fast relaxation under a constant temperature is associated with a dislocation climb controlled process while the subsequent log(time) behavior is attributed to diffusional creep. The kinetics of these relaxation processes are found to be affected by dielectric confinement, decreasing due to a higher degree of dielectric layer confinement, as expected. The results obtained at 200°C for passivated Al(2at% Cu) lines of 3 um wide and 1 um thick are shown in Figure 29. One can see that the relaxation behavior of the principal stress components are anisotropic, with the stress component along the width direction being relaxed at a faster rate. Together with the results obtained at 150°C and 250°C (not shown), the relaxation rate was found not to vary monotonically with the annealing temperature, rather it is maximized at about 200°C.

The two initial relaxation processes due to plastic yield and diffusion flow do not contribute to a volume change, and only the slow process due to dislocation climb is effective in relaxing the near hydrostatic stress and responsible for void formation. For this process, the large increase observed in the relaxation rate perpendicular to the line at 200°C, as compared with 250°C, suggests void formation during cool down below 250°C. Preferred sites for void nucleation are line edges and stud corners and possibly triple point of the grain structure. This is facilitated by the presence of local processing defects and the agglomeration of thermal vacancies and/or volume changes associated with the precipitation of the Al_2Cu particles. Once voids have nucleated, stress relaxation continues with void growth in the directions perpendicular to the lines, especially at junctions where grain boundaries intersecting at 90° with the line/dielectric interface. The void surface near the edge is subject to high local stress, thus acting as efficient defect sources to facilitate the stress relaxation. Such a relaxation behavior is expected to be even more pronounced in submicron lines with a bamboo-like grain morphology although no observation has been reported.

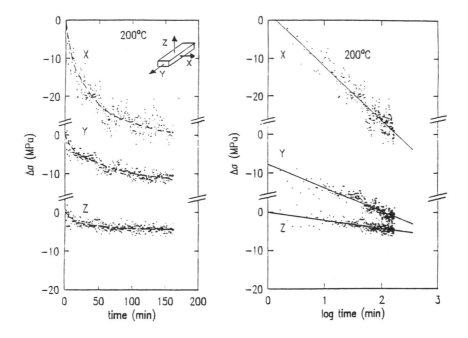

Figure 29: Stress change $\Delta\sigma$ a. versus time t and b. versus log(t) of a SiO$_2$(1.5μm)/Al(Cu)(1.0μm) line structure with 3μm linewidth during annealing at 200°C perpendicular (x), parallel (y) and vertical (z) to the line. (From Ref. 155).

The peak temperature for stress relaxation observed at about 200°C is consistent with a peak void formation temperature. This can be explained by a balance of the driving force against the kinetics of diffusion creep in the stress relaxation process. While the driving force coming from the thermal stress during cool down increases with decreasing temperature, the diffusion kinetics become faster with increasing temperature. This gives rise to an optimum temperature for void formation, which is determined by a complex interaction among the processing conditions, the interconnect morphology/structure and the material composition of the metallization.

An X-ray diffraction technique has been employed to measure stress relaxation in Al line structures at room temperature for longer time periods up to about 5,000 minutes (156). The results are consistent with independent void volume measurements, which are shown in Figure 30. As shown, there is a fast initial growth, following with an exponential growth and a logarithmic growth stage, and the total void volume accounts for most of the thermal mismatch in the structure. A number of kinetic models have been proposed to account for the void growth observed. These include an early model (157) considering grain boundary diffusion driven by thermal stresses be primarily responsible for void growth. Subsequent models extend the treatment to include specific grain boundary relaxation processes and growth by migration and coalescense (158-

160). While most of these models seem able to account for the void growth as a result of thermal mismatch, the development is at an early stage. Future efforts are needed to build on the present framework to fully understand void formation in submicron multilevel interconnect.

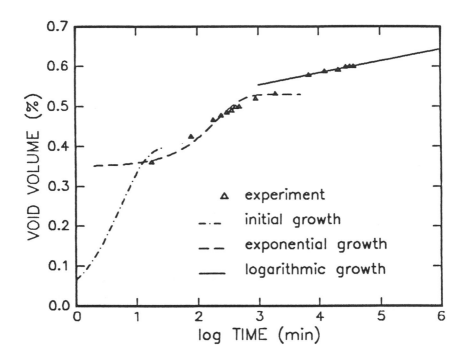

Figure 30: Schematic of the three stages of void growth and stress relaxation. (From Ref. 156).

Acknowledgment

The authors gratefully acknowledge the collaboration and discussion with their colleagues. M. L. Dreyer would like to express thanks to the following Motorola organizations; Core Technologies particularly C. J. Varker and B. Vasquez; the Materials Technology Center; in particular J. Mohr, C. J. Tracy and S. R. Wilson; the Advanced Reliability Engineering Group and also T. Zirkle and D. M. Dreyer. P.S. Ho gratefully acknowledges IBM and the University of Texas at Austin where some of the studies reported in this paper were carried out, in particular, C.K. Hu, M.A. Moske, M.B. Small, D.J. Mikalsen, R. Rosenberg and T. Kwok. He also wishes to express thanks for the support from S. Srikrishnan and M. Haley from the IBM Technology Products Division in Fishkill, New York.

REFERENCES

1. D. L. Crook, Evolution of VLSI Reliability Engineering, Proceedings of the 28th International Reliability Physics Syposium, p. 2 (1990).
2. R. H. Dennard, F. H. Gaensslen, H. N. Yu, V. L. Rideout, E. Bassous and A. R. LeBlanc, IEEE J. Solid State Circuits, SC-9, 256 (1974).
3. D. D. Tang and P. M. Solomon, IEEE J. Solid State Circuits, SC-14, 679 (1979).
4. J. L. Prince, Very Large Scale Integration: Fundamentals and Applications, ed. D.F. Barbe (Berlin: Springer), chapter 2.
5. J.R. Black, Proceedings of the 6th IEEE Symposium on Reliability in Physics (New York: IEEE) p148 (1967).
6. I. A. Blech and E. S. Meieran, Appl. Phys. Lett., 11, 263 (1967).
7. H. A. Schafft, D. A. Baglee and P. T. Kennedy, Proceedings of the 29th International Reliability Physics Syposium, p. 1 (1991) .
8. P.S. Ho, Basic Problems for Electromigration in VLSI Applications, Proceedings of the 20th International Reliability Physics Symposium, p. 288 (1982).
9. P.S. Ho and T. Kwok, Report on Progress in Phys. 52, 3 (1989).
10. D. Moy, M. Schadt, C.K. Hu, F. Kaufman, A. Ray, N. Mazzeo, E. Baran, and D.J. Pearson, IEEE Proc. Int'l VLSI Multilevel Interconnect Conf., (IEEE, New York, 1991), p. 20.
11. J.K. Howard, P.S. Ho and J.F. White, J. Appl. Phys. 49, 4083 (1978).
12. J.J. Estabil, H.S. Rathore and E.N. Lavine, Electromigration Improvements with Titanium Underlay and Overlay in Al(Cu) Metallurgy, IEEE Proc. 8th Int'l VMIC Conf. Proceedings, p. 242 (1991).
13. V. B. Fiks, Fiz. Tverd. Tela (Leningrad) 1, 16 (1959) Sov. Phys. Solid State, 1, 14 (1959).
14. H. B. Huntington and A. R. Grone, J. Phys. Chem. Solids, 20, 76 (1961).
15. C. Bosvieux and J. Friedel, J. Phys. Chem. Solids, 23, 123 (1962).
16. R. Landauer and J. W. F. Woo, Phys. Rev. B, 10, 1266 (1974).
17. R. S. Sorbello, Comment. Solid State Phys., 6, 117 (1975).
18. P. Kumar and R. S. Sorbello, Thin Solid Films, 25, 25 (1975).
19. L. J. Sham, Phys. Rev. B, 12, 3142 (1975).
20. A. K. Das and R. Peierls, J. Phys. C: Solid State Phys., 8, 3348 (1975).
21. H. B. Huntington, in Diffusion in Solids, eds. A. S. Nowick and J. J. Burton (Academic, New York, 1975), p. 303.
22. W. L. Schaich, Phys. Rev. B, 13, 3350 (1976).
23. R. Landauer, Phys. Rev. B, 14, 1474 (1976).
24. R. S. Sorbello and B. Dasgupta, Phys. Rev. B, 16, 5193 (1977).
25. R. S. Sorbello, in Electro- and Thermo-Transport in Metals and Alloys, eds. R. E. Hummel and H. B. Huntington (The Metallurgical Society of AIME, New York, 1977), p. 2.
26. W. L. Schaich, Phys. Rev. B, 19, 620 (1979).
27. P. R. Rimby and R. S. Sorbello, Phys. Rev. B, 21, 2150 (1980).

28. A. H. Verbruggan, IBM J. Res. Dev., 32, 93 (1988).
29. T. Kwok and P. S. Ho, in Diffusion Phenomena in Thin Films and Microelectronic Materials, eds. D. Gupta and P. S. Ho, (Noyes publications, Park Ridge, New Jersey, 1988), p. 369.
30. R. S. Sorbello, Basic Concepts in Electromigration, in Materials Reliability Issues in Microelectronics, eds. J. R. Lloyd, F. G. Yost, and P. S. Ho, (Materials Research Society, Pittsburgh, 1991), p. 3.
31. F. M. d'Heurle and P. S. Ho, in Thin Films- Interdiffusion and Reactions, eds. J. M. Poate, K. N. Tu and J. W. Mayer (Wiley and Sons, New York, 1978) p. 243.
32. J. R. Lloyd, Proceedings of the 21st International Reliability Physics Syposium, p208 (1983).
33. J. R. Lloyd, Proceedings of the 22nd International Reliability Physics Syposium, p48 (1984).
34. J. M. Towner, Proceedings of the 23rd International Reliability Physics Syposium, p81 (1985).
35. R. Rosenberg, M. J. Sullivan, and J. K. Howard, in Thin Films-Interdiffusion and Reactions, eds. J. M. Poate, K. N. Tu and J. W. Mayer (Wiley and Sons, New York, 1978) p. 48.
36. See for example pg. 339, P.S. Ho and T. Kwok, Report on Progress in Phys. 52, 3 (1989).
37. P. B. Ghate, Proceedings of the 19th International Reliability Physics Syposium, p. 243 (1981).
38. J. Chern, W. G. Oldham and N. Cheung, IEEE Trans. Electron. Devices, ED-32, 1341 (1985).
39. S. Vaidya and A. K. Sinha, Thin Solid Films, 75, 253 (1981).
40. J. Cho and C. V. Thompson, Appl. Phys. Lett., 54, 2577 (1989).
41. J. R. Lloyd, M. Shatzkes, and D.C. Challener, "Kinetic Study of Electromigration Failure in Cr/Al-Cu Thin Film Conductors Covered with Polyimide and the Problem of Stress Dependent Activation Energy" Proceedings of the 26th International Reliability Physics Syposium, p216 (1988).
42. H. Schafft, IEEE Trans. Elec. Dev., ED-42 , 664 (1987).
43. W. E. Beadle, J.C.C Tsai and R. D. Plummer, in Quick Reference Manual for Silicon Integrated Circuit Performance, eds. W. E. Beadle, J.C.C Tsai and R. D. Plummer (Wiley and Sons, New York,1985),chapters 1 and 2.
44. J. T. Milek, in Silicon Nitride for Microelectronic Applications: Part 1, Preparation and Properties, (IFI/Plenum, New York, 1971), pg. 63.
45. T. Kwok, Proc. 4th IEEE Int'l VLSI Multilevel Interconnection Conf., (New York: IEEE), p.456 (1987).
46. T. Kwok, Proc. 1st ECS Symposium on ULSI Science and Technology, (New York: The Electrochemical Society), p. 593 (1987).
47. W. Allegreto, A. Nathan, K. Chau and H. P. Baltes, Can. J. Phys., 67, 212 (1989).
48. F. Matsuoka, H. Iwai, K. Hama, H. Itoh, R. Nakata, T. Nakakubo, K. Maeguchi and K. Kanzaki, IEEE Trans. Electron Devices, 37, 562 (1990).
49. R. E. Hummel, R. T. DeHoff and H. J. Geier, J. Phys. Chem. Solids, 37, 73 (1976).

50. R. W. Pasco and J. A. Schwarz, Solid State Electronics, 26, 445 (1983).
51. P. Dutta and P. M. Horn, Rev. Mod. Phys., 53, 256 (1981).
52. I. A. Blech and E. S. Meieran, J. Appl. Phys., 40, 485 (1969).
53. S. J. Horowitz and I. A. Blech, Mater. Sci. Eng., 10, 169 (1972)
54. J. Weiss, Thin Solid Films, 13, 169 (1972)
55. F. M. d'Heurle and A. Gangulee, in Nature and Behavior of Grain Boundary Electromigration and Diffusion, ed. H. Hu (Plenum Press, New York, 1972) p. 339.
56. E. O. Kirkendall, Trans. AIME, 147, 104 (1942).
57. G.G. Mahr von Staszewski and N. E. Walsoe de Reca, Phys. Stat. Sol. 90, 191 (1985).
58. K. L. Tai, P. H. Sun, and M. Ohring, Thin Solid Films, 25, 343 (1975).
59. P. Singh and M. Ohring, J. Appl. Phys., 56, 899 (1984).
60. F. Beniere, K. V. Reddy, D. Kostopoulos and J. Y. Le Traon, J. Appl. Phys., 49, 2743 (1978).
61. K. V. Reddy, F. Beniere, D. Kostopoulos and J. Y. Le Traon, J. Appl. Phys., 50, 2782 (1979).
62. H. Schafft,T. Staton, J. Mandel, and J. Shott, IEEE Trans. Electron Devices ED-34, 673 (1987).
63. J. M. Schoen, J. Appl. Phys., 51, 513 (1980).
64. D. J. LaComb and E. L. Parks, Proceedings of the 24th International Reliability Physics Syposium, p1 (1986).
65. J. Towner, Proceedings of the 28th International Reliability Physics Syposium, p100 (1990).
66. J. Cho and C. V. Thompson, Appl. Phys. Lett, 54, 2577 (1989).
67. A. Bobbio and O. Sarroco, Thin Solid Films, 17, S13 (1973).
68. J. Lloyd, J. Appl. Phys., 50, 5062 (1979).
69. J. A. Schwarz and L. E. Felton, Solid State Electronics, 28, 669 (1985).
70. S. Vaidya and A. K. Sinha, Thin Solid Films, 75, 253 (1981).
71. W. Nelson, in Applied Life Data Analysis, ed. W. Shewhart and S. Wilks (Wiley and Sons, New York, 1982). p 32.
72. J. Aitchison and J. A. C. Brown, in The Lognormal Distribution, (Cambridge University Press, New York, 1957).
73. G. J. Hahn and S. S. Shapiro, in Statistical Models in Engineering, (Wiley and Sons, New York, 1967).
74. N. L. Johnson and S. Kotz, in Distributions in Statistics: Continuous Univariate Distributions, (Houoghton-Miflin, Boston, 1970).
75. M. L. Dreyer, K. Y. Fu and C. J. Varker, To be published in May 15th J. Appl. Phys. (1993); see also Proceedings of the 31st International Reliability Physics Syposium, p304 (1993).
76. J. R. Black, IEEE Trans. Electron. Devices, ED-16, 338 (1969).
77. B. N. Agarwala, M. J. Attardo and A. P. Ingraham, J. Appl. Phys., 41, 3954 (1970).
78. I. Guttman and S. S. Wilks, in Introductory Engineering Statistics (Wiley, New York, 1965), p. 210
79. M. J. Attardo, R. Rutledge and R. C. Jack, J. Appl. Phys., 42, 4343 (1971).
80. D. Turnbull and R. E. Hoffamn, Acta Metallurgica, 2, 419 (1954).
81. P. Feltham, Acta. Met., 5 97 (1957).

82. S. Luby and I. Vavra, Proc. 7th Intern. Vac. Congr. and 3rd Intern. Conf. Solid Surfaces, eds. R. Dobrozemsky, F. Rudenauer, F. P. Viehbock and A. Breth, (Vienna, 1977), pg. 2167.

83. K. Y. Fu, J. Appl. Phys., 69 (14), 2656 (1991).

84. T. Kwok, Proc. 1st ECS Symp. on ULSI Science and Technology, (New York: The Electrochemical Society), p. 593 (1987)

85. J.R. Black, Proceedings of the 6th Reliability Physics Syposium (New York: IEEE) p148 (1967).

86. C. B. Oliver and D. E. Bower, Proceedings of the 8th Reliability Physics Symposium (New York: IEEE) p116 (1970).

87. A. J. Learn, Appl. Phys. Lett., 19 292 (1971)

88. J. C. Blair, P. B. Ghate, and C. T. Haywood, Appl. Phys. Lett., 17 281 (1970)

89. B.N. Agarwala, B. Patnaik, and R. Schnitzel, J. Appl. Phys., 43 1487 (1972)

90. G. A. Scoggin, B. N. Agarwala, P. P. Peressini and A. Brouillard, Proceedings of the 13th International Reliability Physics Syposium (New York: IEEE) p151 (1975).

91. J. Cho and C. V. Thompson, Appl. Pys. Lett., 54, 2577 (1989).

92. E. Kinsbron, Appl. Phys. Lett. 36 968 (1980)

93. S. Vaidya, T. T. Sheng, and A. K. Sinha, Appl. Phys. Lett, 36, 465 (1980)

94. A. J. Learn, J. electrochem. Soc., "Solid State Science and Technology", 123, 894 (1976).

95. T. Kwok, Proceedings of the 26th International Reliability Physics Symposium, p. 187, (1988).

96. J. D. Venables and R. G. Lye, Proceedings of the 10th International Reliability Physics Syposium, p159 (1972).

97. H. Schreiber, Solid-St. Electron. 28, 1153 (1985).

98. A. Gangulee and F. M. D'Huerle, Thin Solid Films, 25, 317 (1975).

99. A. Gangulee and F. M. D'Huerle, Thin Solid Films, 16, 227 (1973).

100. J. R. Black, IEEE Trans. Electron. Devices, ED-16, 338 (1969).

101. M. L. Dreyer and C. J. Varker, Appl. Phys. Lett., 60, 1860 (1992).

102. J. Attardo and R. Rosenberg, J. Appl. Phys., 41, 2381 (1970).

103. S. P. Sim, Microelectronics Reliability, 19, 207, (1979).

104. K. P. Rodbell and S. R. Shatynski, Thin Solid Films, 108, 95 (1983).

105. K. P. Rodbell and S. R. Shatynski, IEEE Trans. Electron Devices, ED-31, 232 (1984).

106. J. Cho and C. V. Thompson, J. Electronic Materials, 19, 1207 (1990).

107. D. Turnbull and R. Hoffman, Acta. Met., 2, 419   (1954).

108. R. Reed-Hill, in *Physical Metallurgy Principles*, ed. W. W. Hagerty, D. Van Nostrand Co., Princeton (1964).

109. P. G. Shewmon, in *Diffusion In Solids*, J. Williams Book Co. (formerly published by McGraw Hill Book Co.), Jenks, OK. (1983).

110. J. Partridge and G. Littlefield, Proceedings of the 23rd International Reliability Physics Syposium, p119 (1985).

111. J. W. McPherson, Proceedings of the 24th International Reliability Physics Syposium, p12 (1986).

112. J. W. McPherson and P. B. Ghate, Proc. of the Symp. on Electromigration of Metals, eds. J. R. Lloyd, R. A. Levy, J. Pierce, R. G. Frieser (The Electrochemical Society: 1985) p. 64.
113. R. W. Lin and D. Y. Guan, 1985 Proc. ECS Eletromigration of metals and the first ECS Multilevel Metallization and Packaging Symposium, p. 75.
114. J. M. Towner and E. P. van de Ven, Proceedings of the 21st International Reliability Physics Syposium (New York: IEEE) p.36 (1983).
115. Kinsbron, Melliar-Smith, and English, 1979 IEEE Trans. ED-26, 22.
116. Kinsbron, Melliar-Smith, English, and Chynoweth, Proceedings of the 16th International Reliability Physics Syposium, p. 248 (1978).
117. B. K. Liew, N. W. Cheung, and C. Hu, Proceedings of the 29th International Reliability Physics Syposium, p. 215 (1989).
118. B. K. Liew, N. W. Cheung, and C. Hu, IEEE Trans. Electron Devices, 37, 1343 (1990).
119. A. T. English and E. Kinsbron, J. Appl. Phys., 54, 275 (1983).
120. J. R. Davis, Proc. IEEE, 123 (1209) and R. J. Miller, Proc. 16th IEEE International Reliability Physics Syposium (New York: IEEE) p.241 (1978).
121. J. S. Suehle and H. A. Schafft, Proceedings of the 28th International Reliability Physics Syposium, p. 229 (1990).
122. T. Noguchi, K. Hatanaka and K. Maeguchi, Proc. VLSI Multilevel Interconnect Conf., p. 183 (1989).
123. L. Berenbaum, Electromigration damage of grain boundary triple points in Al thin films, J. Appl. Phys., 42, 880 (1971)
124. I. A. Blech, Electromigration in thin aluminum films on titanium nitride, J. Appl. Phys., 47, 1203 (1976).
125. R. Rosenberg and L. Berenbaum, Resistance monitoring and effects of non-adhesion during electromigration in aluminum films, Appl. Phys. Lett., 12, 20 (1968).
126. M. C. Shine and F. M. d'Heurle, Activation energy for electromigration in aluminum films alloyed with copper, IBM J. Res. Dev., 15, 378 (1971).
127. J. R. Black, Electromigration failure modes in aluminum metallization for semiconductor devices, Proc. IEEE, 57, 1587 (1969).
128. T. Satake, K. Yokoyama, S. Shirakawa and K. Sawaguchi, Electromigratin in aluminum film stripes coated with anodic aluminum oxide films, Jap. J. Appl. Phys., 12, 518 (1973).
129. I. Ames, F. d'Heurle and R. Horstmann, Reduction of electromigration in aluminum films by copper doping, IBM J. Res. Dev., 14, 461 (1970).
130. F. M. d'Heurle, N. G. Ainslee, A. Gangulee, and M.C. Shine, Activation energy for Electromigration failure in Aluminum Films Containing Copper, J. Vac. Sci. Tech. 9, 289 (1972)
131. S. M. Spitzer and S. Schwartz, J. Electrochemical Society, 116, 1368 (1969).
132. A. J. Learn, J. Electron. Mater., 3, 531 (1974)
133. P. S. Ho and T. Kwok, Report on Progress in Phys. 52, 3 (1989).
134. P. S. Ho, *Basic Problems for Electromigration in VLSI Applications*, Proceedings of the 20th Reliability Physics Syposium, p. 288 (1982).

135. D. Moy, M. Schadt, C. K. Hu, F. Kaufman, A. Ray, N. Mazzeo, E. Baran, and D. J. Pearson, Proc. Int. VLSI Multilevel Interconnect Conf., (IEEE, New York, 1991), p. 20.
136. J. K. Howard, P. S. Ho and J. F. White, J. Appl. Phys. 49, 4083 (1978).
137. J. J. Estabil, H. S. Rathore and E. N. Lavine, *Electromigration Improvements with Titanium Underlay and Overlay in Al(Cu) Metallurgy*, IEEE 8th Int'l VMIC Conf. Proceedings, p. 242 (1991).
138. C. K. Hu, P. S. Ho and M. B. Small, J. Appl. Phys. 72, 291 (1992).
139. See, for example, D. Gupta, D. R. Campbell and P. S. Ho, Chp. 7 in *Thin Films: Interdiffusion and Reactions*, ed. by J. Poate, K.N. Tu and J. Mayer, (John Wiley Publ., New York, 1978).
140. Smithells Metals Reference Book, ed. by E. A. Brandes, (Butterworths, London, 1983), pp. 8-
141. J. Kelma, R. Pyle, E. Domangue, *Reliability Implications of Nitrogen Contamination During Deposition of Sputtered Aluminum/Silicon Metal Films*, IEEE/IRPS Proceedings, p. 1, (1984).
144. J. T. Yue, W. P. Funsten, R. V. Taylor, *Stress Induced Voids in Aluminum Interconnects During IC Processing*, Proceedings of the 23rd Reliability Physics Syposium, p. 126, (1985).
142. See, for example, *Stress-Induced Phenomena in Metallization*, ed. by C. Y. Li, P. A. Totta and P. S. Ho, Am. Inst. Phys., New York (1992).
143. M. F. Doemer and W. D. Nix, J. Materials Res. 1, 601 (1986).
145. M. A. Moske, P. S. Ho, D.J . Mikalsen, J. J. Cuomo and R. Rosenberg, to appear in J. Appl. Phys. (1993).
147. S. Timoshenko, *Strength of Materials*, Van Nostrand, New York (1958), p. 228.
148. H. Niwa, H. Yagi, H. Tsuchikawa and M. Kato, J. Appl. Phys. 68, 328 (1990).
149. M. A. Korhonen, P. Børgesen and C. Y. Li, MRS Bulletin, July (1992), p. 61.
150. R. E. Jones, Jr., *Line Width Dependence of Stresses in Aluminum Interconnect*, Proceedings of the 25th Reliability Physics Syposium, p. 9, (1987).
151. J. L. Freeman, G. Grivna and C.J . Tracy, Proc. of SPIE Conf. on Microelectronics Processing '92, San Jose, CA (1992).
152. A. Tezaki, T. Mineta, H. Egawa and T. Noguchi, Proceedings of the 28th Reliability Physics Syposium, p. 221 (1990).
153. C. A. Paszkiet, M. A. Korhonen, C. Y. Li, *X-Ray Stress Studies of Passivated and Unpassivated Narrow Aluminum Metallizations*, MRS Spring Meeting, Vol. 188, p. 153, (1990).
154. P. A. Flinn, in *Stress-Induced Phenomena in Metallization*, ed. by C. Y. Li, P. A. Totta and P. S. Ho, Am. Inst. Phys., New York (1992), p. 73.
155. M. A. Moske, P. S. Ho, C. K. Hu and M. B. Small in *Stress-Induced Phenomena in Metallization*, ed. by C.Y. Li, P.A. Totta and P.S. Ho, Am. Inst. Phys., New York (1992), p. 195.
156. M. A. Korhonen, C. A. Paszkiet and C. Y. Li, J. Appl. Phys. 69, 8083 (1991).

157. F. G. Yost and F. E. Campbell, IEEE Circuits and Devices <u>6</u>, 40 (1990).

158. C. Y. Li, P. Børgesen and T. Sullivan, Appl. Phys. Lett. <u>59</u>, 1464 (1991).

159. W. D. Nix and A. I. Sauter in *Stress-Induced Phenomena in Metallization*, ed. by C.Y. Li, P.A. Totta and P.S. Ho, Am. Inst. Phys., New York (1992), p. 89.

160. T. D. Sullivan, Appl. Phys. Lett. <u>55</u>, 2399 (1989).

9

MULTILEVEL METALLIZATION TEST VEHICLE

SYD R. WILSON, CHARLES J. VARKER, And JOHN L. FREEMAN, JR.
Motorola Semiconductor Products Sector
Mesa, Arizona

1.0   INTRODUCTION

The primary goals of an interconnect development program for VLSI/ULSI applications are to establish a robust process flow and an associated set of design rules for building an MLM system.   The MLM test vehicle is a mask set designed to provide electrically testable structures for the key elements of an MLM system (metal linewidth, metal thickness, ILD thickness, via and contact resistance, etc.) and is used to both develop and monitor the MLM process. In the development phase these structures are used to: [1] define the necessary processes; [2] validate design rules; [3] determine the manufacturibility of new or special structures; [4] identify design weaknesses or marginal processes which may impact yield, performance, or reliability; [5] provide a database which can be used to predict circuit yield or compare capabilities of different wafer fabs; and [6] compare to the previous generation and potentially look to the next generation to see if the process can be extended. The test structures can be used to establish the upper and lower process limits required to meet product design specifications and establish a data base for product yield improvements. In the manufacturing phase these structures are used to monitor process stability.

The ideal structures for the test vehicle will unambiguously measure, with an automated tester, all of the key metallization parameters of the final circuit or product family. These key items fall into three classes: [1] process control parameters such as linewidth, sheet resistance, via resistance, and other items that are part of the design rules as discussed in section 5 of Chapter 1; [2] defect densities for each of the elements of the MLM system such as metal shorts or opens, via opens, etc.; and [3] reliability parameters for determining the reasonably expected lifetime of the MLM elements. A different tester is usually used to measure the reliability lifetime since this measurement may take days to months and may involve testing at elevated temperatures (see section 4.0 and Chapter 8 of this book.) The structures for measuring each of these classes are discussed in detail in sections 2-4. The use of an automated tester makes it possible to maximize the amount of data that can be obtained from each lot so that sufficient statistics can be obtained to predict yields and generate reasonable trend charts. Where necessary structures can be included that can only be tested on a bench with high precision test equipment. In some cases, structures are used in combination to provide unique information. As an example, a sheet resistance test is used in combination with a split bridge to measure metal linewidths. The ideal test structures for yield and reliability should simulate as closely as possible conditions that may occur in the real circuit in terms of topography, structural density and in the case of the reliability structures, the testing should be done at conditions that can be related to operating conditions of the circuit.

Designs must include groups of discrete test structures and test structure arrays with a range of linear dimensions over varied topography. Parameter extraction for optimizing product design can often be achieved using discrete process control devices. Examples of discrete process control test structures are linear bridges for metal sheet resistance measurements, split bridges for linewidth and space determination, and alignment sensitive test structures for the measurement of pattern registration. Test structure arrays which emulate the layout pattern of the interconnect system provide information on the effects of system integration. Examples of integrated test structures includes snakes and combs patterned over the topography of a preceding layer and via and contact test structures configured in a variety of linear chains and arrays.

Multiple test structures with varied linear dimensions are often included to measure process capability for reduced dimensions and to provide a reference to an existing technology. These structures are designed to include the nominal target dimensions specified in the new design rules and to extend the dimensions above and below the nominal design value. Test results on identical test structures having a range of critical dimensions can be used to provide evidence of fundamental process limitations.

In the early phases of development it is difficult to emulate all possible cases that may occur on a circuit. Therefore, conditions are used that simulate the worst cases as well as the nominal conditions until specific processes and product design rules are finalized and a baseline established.

The engineer designing the test chip must decide how large to make some of the yield structures, whether to use structures over topography and on flat surfaces, and how many variations in sizes (i. e. linewidth, via dimension, etc.) for each of the different test structures can be included in the test vehicle. The yield structures should be as large as possible to make the appropriate yield calculations. The maximum test chip size is set by the field size of the lithography tool. However, to obtain sufficient statistics for each lot of wafers and each test, the number of available die must be taken into account. The number of testable die will depend upon the die size, the wafer size and the number of wafers per lot. The number of different structures or size of structures may have to be reduced in order to reduce the die size and increase the number of available die per wafer. Large die, i.e. few die per wafer, also limit the ability to map the across wafer uniformity of a particular parameter.

In addition to electrically testable structures, it is valuable to have regions of the test chip that are designed for ease of characterization by analytical tools such as those described in Chapter 11. This is especially true for some of the depth profiling techniques that require large areas for analysis due to large spot sizes relative to the test structures discussed in this chapter. The structures are very specific to the particular analytical tool and beyond the scope of this chapter. However, the engineer designing the test mask should discuss these requirements with the analysts at his facility and design in the required structures as necessary.

Several articles in the literature (1, 2) have briefly discussed the desirable features and test structures that may be used in an MLM test vehicle. In this chapter we will discuss in detail structures that can be grouped into three classes; [1] process control, [2] yield and defectivity [3] reliability. For each we will discuss what the structure is, what it is used to measure, how it is built, how it is tested, and what are its limitations.

2.0   PROCESS CONTROL

As discussed in section 1.0, process control (PC) structures are discrete devices that are used alone or in combination to measure the key parameters of an MLM system. Essentially all of these can be measured in a DC mode by forcing a current and measuring a voltage. Most can be used at the silicided polysilicon and silicided monosilicon layers as well as at the metal layers. However due to the low sheet resistance of the metal layers the sheet resistance tester may become quite large. This forces the sheet resistance to be measured independent of the linear, split and alignment bridges. However for the polysilicon and monosilicon layers these can be combined as has been discussed in the literature. These issues are discussed in detail in section 2.2. The size and layout requirements of the process control structures assume they will be tested on a conventional parametric tester and that the desired measurement voltage will be millivolts or greater. This is achievable if they are properly designed. The exceptions are the alignment bridges for metal layers where the voltage resolution of most testers is not adequate to measure the <0.25µm misalignments that are of interest in modern circuits. These alignment measurements must be made using high resolution equipment with precisely designed structures. As discussed in the section on alignment bridges the physical dimensions must be as small as possible and known accurately. Therefore the normal photolithography and metal etch bias may also confound the alignment data if it is not accounted for properly.

2.1   Test Chip Layout Considerations

Typical PC elements are designed for wafer level testing using an automated wafer prober and DC parametric tester. Small area test structures are typically configured within a rectangular area bordered by a 2xn probe pad array. An example is shown in Figure 1 for a rectangular area of $\approx$ 200x4000 µm^2 which is bordered by 44 probe pads having an area of (100x100) µm^2 with a pad spacing of 100 µm. The resulting 2x22 array of pads (shown in the dashed rectangle corresponding to row 1) is repeated as required to form a column of 2x22 test pads. Small area structures such as; planar electromigration test structures, relatively short contact and via chains, linear bridges and discrete test devices are generally located within the basic 2x22 pad array along with the metal runs from the test structures to the pads. Large area structures are accommodated by increasing the distance between the 2x22 pad array and placing the test structures in the intervening space.

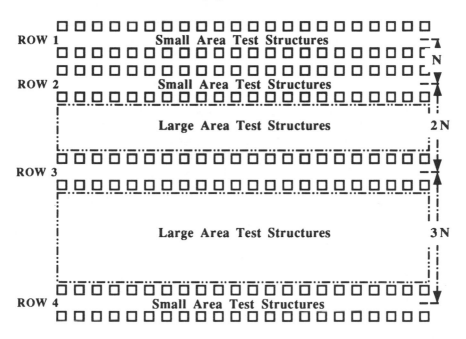

Figure 1. Test chip layout for standardized 2 x 22 probe card

The diagram in Figure 1 shows an example of a configuration in which the 2x22 pad array spacing has been adjusted to accommodate large area test structures such as long contact and via chains, snakes, combs and large area capacitors. The spacing between adjacent pad arrays is generally incremented by an integral multiple of the minimum pad array spacing N, measured from the center of row 1 to the center of row 2, as shown in Figure 1. The spacing between rows 2 and 3 has been increased from 1N to 2N to provide space for large area structures. This type of layout permits the use of a standard probe card independent of the layout requirements of the test chip and it provides a stepping distance which is an integer multiple of the minimum pad array spacing.

2.2   Process Control Structures

Process control test structures are commonly used to monitor process capability and to extract critical process and design parameters which have a direct impact on product yields and performance. The primary consideration for the measurement of process control is the design of test structures which provide test data with the required measurement accuracy and sensitivity to the process parameters of interest. Interconnect, contact and via test structures are normally characterized in terms of their electrical resistance to current flow. Resistance values are calculated from the ratio of the measured voltage V to source current I. This ratio, V/I, is dependent upon the geometry of the test device and its resistivity. Test parameters such as sheet resistance (R_s), contact and via resistance, and dimensional parameters such as line width (W), space (S), and pattern shift are calculated from the voltage measurements, the dimensions of the specific test structure, and the appropriate physical constants.

Resistor structures for sheet resistance measurement: The resistivities of metal films are typically 3 orders of magnitude smaller than the resistivities of ion implanted layers in polycrystalline and monocrystalline silicon. As a result, voltage measurements on thin metal film resistors may require additional design considerations which are often not as important when using high resistance films with comparable geometries. Electrical measurements on MLM test structures may require high resolution because of the relatively small voltage drops in thin metal film test structures. Voltage measurements on resistor bridges may require microvolt measurement resolution unless the structures are designed with large length to width ratios (L/W) or designed for operation at relatively high current densities ($\geq 10^6 A/cm^2$). Although large L/W ratios are easily achieved during layout, operation at high current test conditions can introduce significant Joule heating effects which must be considered. Commercial DC parametric testers offer adequate test capability using standard voltage and current sources when test structures are properly designed.

Simple linear resistor structures normally contain a single pair of terminals to force current and measure voltage. This type of test device provides a measurement of the total load resistance $R_T = V_s/I_s$ at the source terminals, where I_s and V_s are the source current and voltage respectively. The total resistance R_T includes; cable resistance, contact resistances (probes, metal-silicon, and vias), interconnect resistance between pads and test structure, and the resistance of the test structure itself.

The basic electrical parameter which characterizes thin metal film interconnects is the sheet resistance (R_s). Sheet resistance is related to the resistivity of the metal film and it can be derived from the measured voltage drop V on a properly designed thin film linear bridge resistor. The line resistance $R = V/I = \rho * L/W * t_m$, where the proportionality constant ρ is the resistivity of the metal film, L is the conductor length, w is the conductor width, and t_m is the conductor thickness. The linear interconnect or metal film resistor is normally expressed in terms of its L/W ratio. If we consider a unit square of metal film, L=W, we define the sheet resistance $R_s = \rho/t_m$. This value is typically used to characterize thin metal film interconnects having a constant thickness t_m and it is expressed in units of ohms / square.

Although there are many test structure configurations which can be used for the calculation of R_s, the two most commonly used structures are the linear 4 terminal (4-T) bridge resistor and the 4-T van der Pauw bridge resistor. The linear bridge resistor is a four terminal resistor structure which contains a pair of voltage terminals and a pair of current terminals. The voltage V_{lin} measured at the taps is given by

[1] $V_{lin} = V_4 - V_3 = IR_{lin} = I(L/W)R_s.$

An example of a linear bridge resistor is shown in Figure 2. The voltage V_{lin} is measured by sourcing a current from terminals 1 to 2 and measuring the voltage drop (V_4-V_3) between terminals 3 and 4. V_{lin} can be scaled to the desired voltage range consistent with the measurement capability of the test system by scaling the L/W ratio of the bridge. For a fixed deposition technology and resistor design, the parameters R_s and L/W are essentially constant and the voltage is dependent on the source current I.

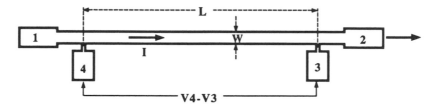

Figure 2. Linear bridge resistor used for sheet resistance measurements

The van der Pauw sheet resistor bridge is commonly configured in the form of a Greek cross. There are many variations of this general bridge and correction factors have been developed by conformal mapping for a variety of geometries and boundary conditions (3, 4). The voltage measured at the taps of the Greek cross is given by

[2] $V_{vdp} = IR_{vdp} = I(\ln 2 / \pi)R_s.$

In contrast with the linear bridge, the V_{vdp} is independent of bridge geometry and therefore, V_{vdp} cannot be scaled with bridge layout dimensions. The van der Pauw bridge requires operation at higher source currents than the stretched linear bridge, to achieve comparable voltages. Figure 3 shows a typical 4-T van der Pauw bridge resistor in the configuration of a Greek cross. In this example, the voltage $V_{vdp} = V_4 - V_3$ is measured by sourcing a current from terminals 1 to 2 and measuring the voltage drop between terminals 3 and 4, in a manner similar to the linear bridge structure.

The 4-T bridge resistors shown in Figures 2 and 3 provide a voltage measurement that is relatively free of significant parasitic voltage drops when a high impedance voltmeter is used during testing. The voltage drop between the measurement taps includes the resistance between the taps but excludes external resistances in the current path. Voltage can be measured at any tap position defined by the test structure design. This type of measurement is often referred to as a Kelvin contact measurement.

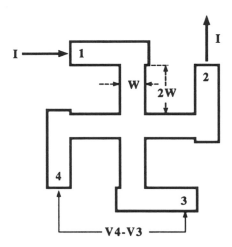

Figure 3. Van der Pauw bridge configured as a Greek cross for sheet resistance measurements

For the structure shown in Figure 2 the length to width ratio L/W represents the number of squares between the voltage taps. The voltage drop V_4-V_3 is equal to the product of the sourcing current I, the L/W ratio and the sheet resistance R_s. For the Greek cross shown in Figure 3, the length to width ratio L/W = 1, independent of its size, and therefore the voltage drop V_4-V_3 depends only on the product of the sourcing current I and the sheet resistance R_s.

A comparison of the essential features of the linear and van der Pauw bridges can be obtained from equations [1] and [2]. If we assume that the voltage taps of both bridges are separated by one square of resistance, L = W, with the same source current I, then, $V_{lin} = (\pi/\ln2)*V_{vdp} = 4.53*V_{vdp}$. Consequently, the test voltage obtained with the linear bridge resistor is ≈ 4.5x larger than the van der Pauw bridge resistor for one square of resistance at an equivalent current. The van der Pauw bridge would therefore require a source current $I_{vdp} = (\pi/\ln2)I_{lin}$ = 4.53 I_{lin} to obtain an equivalent voltage for a one square sheet resistor.

The L/W scaling factor for the linear bridge can be used to increase the voltage for a given source current. Similarly, the area of the van der Pauw bridge can be scaled for operation at higher currents to obtain an equivalent test voltage. Moreover, as a result of the 4-fold rotational symmetry, the van der Pauw bridge voltage is essentially independent of linewidth variations (ΔW/W), which are known to introduce errors in the calculation of resistance. These features offer both advantages and limitations depending on the layout area requirements and the test structure application. The value of ΔW/W for the linear bridge can be made insignificant by using a wide bridge. This increases the structure size and may be helpful for the increased heat dissipation to the substrate.

When testing metal film resistors on thin (< 1.0 μm) silicon dioxide layers the current density (J) should be limited to < 10^6 A/cm^2 to prevent excessive Joule heating which can lead to voltage measurement errors during the test. The temperature rise $\Delta T = \alpha * P$, where $\alpha = t_{SiO2}/k_{SiO2} * A$ is the thermal resistance of the underlying oxide layer, and P is the power dissipation in the metal film. The expression for α (assuming one dimensional heat flow) is analogous to electrical resistance and includes; the thickness t_{SiO2} of the oxide layer, its thermal conductivity k_{SiO2}, and the thermal transport area A (thermal footprint) of the resistor. If we assume that the thickness and the thermal conductivity of the oxide layer are constant so that $t_{SiO2}/k_{SiO2} = K$, we obtain the following expressions for the temperature rise ΔT_{lin} and ΔT_{vdp} for the linear resistor and van der Pauw resistor respectively,

[3] $$\Delta T_{lin} = \alpha_{lin} * P = a_{lin} R_s (L / W) I_{lin}^2,$$

[4] $$\Delta T_{vdp} = \alpha_{vdp} * P = a_{vdp} R_s (\ln 2 / \pi) I_{vdp}^2.$$

The terms a_{lin} and a_{vdp} are the thermal resistances of the oxide layer for the linear and van der Pauw bridges respectively, and are given by the following equations;

[5] $$\alpha_{lin} = t_{SiO_2} / k_{SiO_2} * A_{lin} = K / A_{lin}$$

[6] $$\alpha_{vdp} = t_{SiO_2} / k_{SiO_2} * A_{vdp} = K / A_{vdp}.$$

By equating equations [5] and [6] we satisfy the requirement for equivalent temperature rises in the bridges, and obtain the following equation,

[7] $$A_{vdp}(L / W)I_{lin}^2 = A_{lin}(\ln 2 / \pi)I_{vdp}^2.$$

For equivalent measurement accuracy, the design of these test structures requires that the measurement voltage $V_{vdp} = V_{lin}$ or $I_{vdp}(\ln 2/\pi)R_s = I_{lin}(L/W)R_s$. Using this criteria, the area A_{vdp} of the van der Pauw bridge resistor required for operation at the same voltage and temperature rise as the linear bridge resistor can be expressed by

[8] $$A_{vdp} = (W / L)(\ln 2 / \pi)A_{lin}(\pi / \ln 2)^2(L / W)^2$$

$$= (\pi / \ln 2)L^2$$

The body area of the van der Pauw bridge resistor that is required for equivalent Joule heating [7] and tap voltage as the linear bridge, is proportional to the square of the linear bridge resistor length and therefore the side dimension of the van der Pauw square is proportional to the length of the linear bridge. This analysis neglects the component of thermal transfer contributed by the voltage and current taps external to the resistor body. The current I_{vdp} required for an equivalent voltage is scaled in proportion to the number of squares L/W in the linear resistor, or $I_{vdp} = \pi/\ln2(L/W)I_{lin}$.

As a practical example, consider a van der Pauw resistor structure using a metal film technology having a resistor width of 1μm and sheet resistance of 50 mΩ/square. We assume an underlying oxide thickness of 0.1 μm and further limit the current density to 0.5 MA/cm^2. This represent a sourcing current of 2.5 mA. The measurement accuracy desired is approximately 1% (tester has a limit of 10 μV), therefore, we require a test voltage of ≈ 1mV. To achieve this a linear resistor must have an L/W ratio of 8.0 from [1] or an area = 8.0 μm^2. A van der Pauw will require a sourcing current of $I_{vdp} = \pi/\ln2(L/W) = 32.2x2.5mA = 90.6$ mA. The van der Pauw will require a body area $A_{vdp} = (\pi/\ln2)L^2 = 290$ μm^2. This represents a factor of 36.2 for the van der Pauw body area relative to the area of the linear bridge resistor. Since a van der Pauw resistor in the form of the Greek cross as shown in Figure 3 requires current taps having a width equivalent to the body and a length of approximately 2W the composite area of this structure is approximately 9 times larger than the calculated body area or 2900 μm^2.

The van der Pauw resistor is an area intensive structure relative to the linear bridge when designed to satisfy equivalent voltage measurement conditions. However, it offers a geometry that is symmetrical and therefore independent of line width variations ΔW/W, which can introduce significant voltage errors on narrow linewidth linear bridge resistors. When the van der Pauw bridge is used as an 8-terminal tapped bridge resistor, it offers additional advantages which will be discussed in the section on pattern alignment test structures.

Split-bridge resistors for linewidth, space, and pitch: The split bridge resistor structure consists of a serial combination of linear bridge resistors designed to provide accurate voltage measurements which can be used to calculate the linewidth, space and pitch of a processed test structure. The resulting calculations provide an accurate estimate of the process capability for fabricating structures such as polycrystalline films, implanted and diffused layers and metal film interconnects with the critical dimensions specified in the design rules. Calculations can be compared with optical measurements on the photo mask and optical or SEM measurements on the wafer. These calculations can be used to assess and validate design rules.

Figure 4 shows a diagram of a split-cross bridge resistor (5) designed for measuring the linewidth W, space S, and pitch W+ S for a given process technology. This structure works well for polysilicon and silicides as well as for metal films. For polysilicon and silicides the bridge can be combined with a Greek cross to produce a compact test structure. For metal films such as aluminum the sheet resistance is too low and should be measured independent of the split bridge using one of the structures discussed in the preceding section. The upper resistor and lower resistor pair share a common effective width Wb = 2W+S, where W represents the width of the each parallel resistor in the lower bridge and S represents the space between the resistor pair. The current is assumed to divide equally between these parallel current paths. The basic

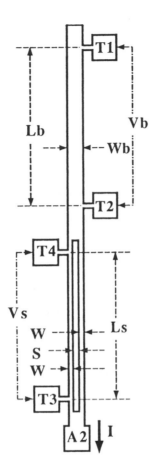

Figure 4. Diagram of a split cross bridge resistor used for electrically measuring the critical dimensions (line and space) of the metal.

principle of operation is the same as the linear bridge resistors discussed in the previous section. The length to width ratio of the upper bridge (Lb/Wb) is equivalent to the number of squares between the voltage taps T1 and T2 and, this ratio is equivalent to the total voltage drop between taps divided by the voltage drop per unit sheet resistance (Vb/IR$_s$). By equating these ratios we obtain an equation for the width of the upper bridge shown in Figure 4, which is given by

[9] $Wb = IR_s(Lb/Vb)$.

A similar equation can be obtained for the lower bridge which is given by

[10] $W = IR_s(Ls/2Vs)$.

These equations can be combined using the expression, Wb = 2W + S to solve for the space S and pitch P respectively, defined by ,

[11] $S = IR_s[(LbVs - LsVb) / VsVb],$

[12] $P = W + S = IR_s[(2LbVsLsVb) / 2VbVs].$

Kelvin resistors for via and contact resistance: In submicron technologies, metal-semiconductor contact resistance is becoming an important factor in the performance of ICs. The resistance of ohmic contacts, vias and plugs increase with decreasing contact area. Contacts to semiconductor layers represent the primary interface between the global metal interconnect system and the underlying active device structures. There are major design rule issues with regard to current crowding at contacts and its effects on performance and reliability. Contact and via resistance provides a measure of the process capability and yield in terms of uniformity, metal step coverage, interface integrity and reliability.

The specific contact resistivity r_c (resistance per unit area) is a basic parameter which is used to characterize the contact interface between the metal and semiconductor layers and the via interface between multilevel metal layers. Specific contact resistivity is a lumped parameter and is often difficult to separate from contact resistance Rc. The contact resistance can be measured using a 4 terminal (4-T) Kelvin contact resistor (6) as shown in Figure 5. The lower contact (shaded) is shown as a metal or semiconductor layer (M1/S). It is drawn with the same layout topology as metal 2 and is rotated 180° to form the 4-T structure. Metal layers are separated by an interlayer dielectric film with a contact opening at the center. Current is sourced from terminal 1 to 3 and voltage is sensed from terminal 2 to 4. The current enters at the lower contact and exits from the upper contact. If we assume uniform current flow within the contact, the contact resistance Rc can be expressed as the ratio of the voltage drop ΔV between taps 2 and 4 to source current I, and is given by

[13] $Rc = \Delta V / I = (V_4 - V_2) / I$

The specific contact resistivity r_c can then be calculated from the contact resistance and the area A of the contact window, $r_c = Rc*A$. This relationship is valid provided that Rc is linear with respect to the contact area. Non-uniform contact area was often a problem for technologies which did not use a barrier layer metal because the contacts were often dominated by silicon precipitates causing contact area non-uniformities. A typical test for uniformity is to obtain a logarithmic scatter plot of Rc vs. A for several Kelvin structures having a range of contact areas. Although the 4-T Kelvin resistor provides a measure of the interfacial contact resistance and provides a test for the uniformity of the contact it does not provide a measurement of contact edge resistance, nor does it allow for corrections due to pattern misalignment.

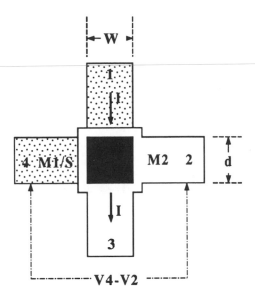

Figure 5. Four terminal contact resistor used for determining via or contact resistance

When the Kelvin structure shown in Figure 5 is used to measure a metal to semiconductor contact, the underlying layer shown as M1/S is represented by a diffused or implanted mono-silicon or polysilicon layer. For these cases, current flow in the underlying film is significantly altered by the presence of the high sheet resistance layer. This results in current spreading within the semiconductor layer and crowding at the entry edge of the contact. (7, 8, 9) Although direct measurements of r_c are not feasible, r_c can be estimated from measurements on properly designed test structures.   On metal to semiconductor contacts the specific contact resistivity r_c requires a measurement of the sheet resistance of the underlying implanted or diffused layer. The current density for most planar contacts to CMOS devices is non-uniform due to current crowding at the contact and is strongly dependent on contact geometry. Therefore, the potential drop is not uniform at the planar interfacial layer of the contact and the voltage measurement is dependent upon the location of the measurement point with respect to the current distribution in the contact. A number of test methods have been developed for the measurement of specific contact resistance based on the transmission line model (TLM) first introduced by Shockley. (10, 11) Front and back end resistance Rf and Rb respectively, for a metal-semiconductor contact, have been calculated using the TLM model, and are given by

[14] $$Rf = V/I = [(Rsd * r_c)^{1/2} / W] \coth[(Rsd/r_c)^{1/2} d],$$

[15] $Rb = V / I = [(Rsd * r_c)^{\frac{1}{2}} / W] \csc h[(Rsd / r_c)^{\frac{1}{2}} d]$,,

where Rsd, W and d are the sheet resistance under the contact, the contact width and length respectively.

Equations [14] and [15] can be combined to give

[16] $Rf = Rb \cosh[(Rsd / r_c)^{\frac{1}{2}} d]$.

Note that these equations neglect the resistance effects of the metal layer. This assumption is valid for most metal-semiconductor contacts due to the large ratio of the semiconductor to metal resistivity. Also, the calculation of Rf will require that values for Rsd and r_c be determined. The 4-T Kelvin structure shown in Figure 5 can be used to measure r_c. For a given contact geometry (W and d known), Rsd can be determined from a measurement of Rb using equation [15]. However, measurements of Rb require a 6 terminal Kelvin structure, shown in Figure 6, which provides an extra pair of voltage taps for this measurement. Rb is measured by sourcing a current from terminals 1 to 6 and measuring the voltage drop between terminals 3 and 5. Note that Rc and Rb must satisfy the linearity test as described above to ensure uniform contact area.

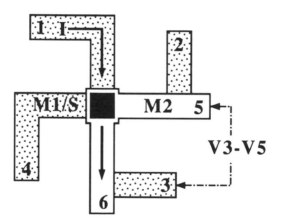

Figure 6. Six terminal Kelvin structure

Note the distinction between the design of the 4-T and that of the 6-T Kelvin resistor. The 2 additional terminals provide voltage taps to the underlying M/S layer. The 6-T Kelvin resistor can be used to obtain a direct measure of r_c and Rb and consequently Rsd and Rf can be calculated using equation [16]. The contact resistance r_c, on the 6-T Kelvin device, is measured by sourcing a current from terminals 1 to 6 and measuring the voltage drop between terminals 4 to 5.

An additional feature of the 6-T resistor is provide by the 4 fold rotational symmetry of the M1/S layer terminals which allows a correction for Rc due to contact misalignment. The test procedure consists of sourcing a current I initially from terminals 1 to 5, and subsequently, from 3 to 5 while measuring the voltage drops for each sourcing current setup from 4 to 6 and 2 to 6. The average value for Rc can then be calculated from the four voltage measurements, $<Rc> = (\sum V_n/4)/I$.

When Kelvin test structures are used for measurements on via structures, very high resolution voltage measurements would be required, and consequently longer test times would be required to obtain sufficient accuracy through averaging. The application of these devices for via measurements is therefore somewhat limited by practical considerations. However, for metal semiconductor contacts the measurements can readily provide test data on design rules governing current crowding and edge effects in sub-micron contacts.

Bridge resistor alignment structures: The precise registration of contacts and vias to their respective silicide and metal interconnects represents an important factor governing the manufacturability of contacts and vias. Interlayer registration errors, which result in poor control of metal/silicide overlap of contacts and metal overlap of single and stacked vias can cause misprocessing resulting in dead structures or nonuniform current flow at the edges of contacts and vias contributing to enhanced current density resulting in poor reliability. The mechanisms which can contribute to reduced reliability at contacts and vias are electromigration and stress voiding which are enhanced locally by high current density and temperature gradients.

A variety of alignment sensitive test structures have been developed for the optical measurement of photomask pattern registration. Microelectronic test structures consisting of alignment sensitive resistors have been developed to extract pattern registration measurements electrically (12, 13). The measurements require a high degree of precision. These measurements are generally obtained with automated parametric testers capable of high measurement precision. Typical structures used for pattern registration measurement include; the tapped bridge resistor (14), the van der Pauw (VDP) resistor bridge (15) and the split resistor bridge (16). As discussed in the introduction to section 2.0, these structures have been applied successfully to diffused and polysilicon layers with or without silicide. However, for metal to metal layers with low resistivities and alignment tolerances ≤0.25 μm, they are virtually impossible to make work in a manufacturing environment with a conventional parametric tester. The reasons they tend not to work are: [1] The required voltage sensitivity for metal layers is beyond the capabilities of conventional testers; [2] The dimensions of the alignment bridges need to be at or near the minimum that can be processed to enhance the voltage sensitivity, but at the same time the dimension of the structure must be known very accurately. This is virtually impossible for a long narrow bridge just due to process variations. [3] Increasing the current to increase the voltage sensitivity results in Joule heating of the narrow structure and thus confounding of the results. We present the drawings and equations for the tapped linear bridge and tapped van der Pauw resistor bridges in the following sections for the reader to use on the polysilicon and monosilicon layers if desired.

The tapped linear bridge resistor: A diagram of a tapped bridge resistor designed for the measurement of via registration is shown in Figure 7.

The bridge consists of a pair of orthogonal tapped linear resistors patterned in metal 1 (M1). The horizontal bridge consists of 2 linear resistors connected in series. Terminals 1 and 9 are the current sourcing terminals and the fixed voltage taps are 2, 3, 5, 6, and 8. The resistor defined by taps 2 and 3, provides a reference voltage drop V1 proportional to (W/L_1). The sheet resistance R_s for the linear bridge is obtained from (1) and is given by

[17] $R_s = (W/L_1)V_1/I.$

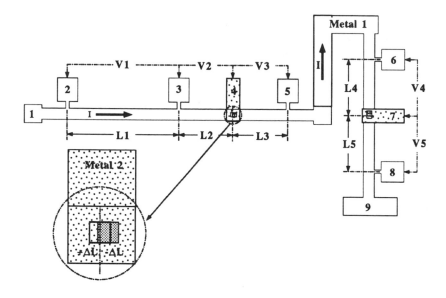

Figure 7. A tapped bridge resistor used for the measurement of via registration

The alignment resistor for the horizontal bridge is defined by M1 taps 3, and 5, with a via located at its midpoint (dark shaded square in inset diagram). The via is contacted by metal 2 (M2) designated as tap 4. The corresponding alignment resistor for the vertical bridge is defined by taps 6 and 8, with a via and M2 tap designated as tap 7.

The inset diagram shows the position of a perfectly aligned via as a dotted square on the center-line of the M2 tap. The via is shown displaced by an amount $+\Delta L$ to simulate a displacement error in the $+X$ direction. A similar tap is located on the vertical resistor bridge to sense displacement errors in the Y direction. When the via is perfectly aligned, $L_2 = L_3 = L_1/2$, and $V_2 = V_3 = V_1/2$. When the via is displaced an incremental distance ΔL, then $L_2 = L_1/2 + \Delta L$, and $L_3 = L_1/2 - \Delta L$. The effect of the via displacement, shown in the inset in Figure 7, is to increase the effective voltage drop on one side of the tap while simultaneously decreasing the voltage drop on the adjacent side. Since (V_1/L_1) is the reference voltage drop per unit length, the voltage drops V_2 and V_3 corresponding to resistors L_2 and L_3 are given respectively by

[18] $V_2 = (V_1 / L_1)(L_2 + \Delta L),$

[19] $V_3 = (V_1 / L_1)(L_3 - \Delta L).$

The voltage resulting from the via displacement ΔL is given by $(V_2 - V_3)$ and the displacement ΔL is given by

[20] $\Delta L = (L_1 / 2)(V_2 - V_3) / V_1.$

Setting $\Delta L = \Delta X$, for the horizontal bridge and $\Delta L = \Delta Y$ for the vertical bridge, we obtain the X and Y components of the via alignment error from the following expressions respectively;

[21] $\Delta X = (V_2 - V_3) / V_1(L_1 / 2),$

[22] $\Delta Y = (V_4 - V_5) / V_1(L_1 / 2).$

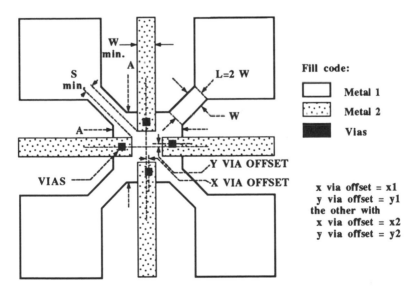

Figure 8. A tapped van der Pauw bridge resistor used for the measurement of via registration

The tapped van der Pauw pattern alignment bridge resistor: The tapped van der Pauw (VDP) resistor bridge combines the sheet resistor and registration sensitive bridge into one unified structure. It uses minimum area and pad count and provides orthogonal registration measurements. Figure 8

shows the layout of the bridge. The body of the bridge (metal 1) consists of a van der Pauw sheet resistor with 4 fixed taps located at the corners. Four alignment sensitive voltage taps (metal 2) contact the sides of the bridge at their midpoints through 4 vias. Alignment calibration is designed into the bridge by offsetting a pair of orthogonal metal 2 taps as shown in Figure 8. The designed offset values are chosen to cover a range of offsets to measure the capabilities of the existing photolithography. These taps provide an internal calibration which is used to measure the precision of the alignment. The current taps at the corner of the bridge were designed to minimize voltage drops at higher forcing currents.

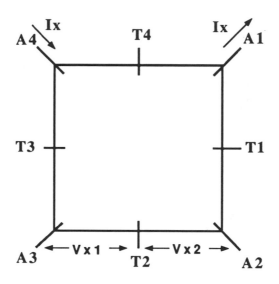

Figure 9. Schematic of the tapped van der Pauw bridge to illustrate the technique

The bridge schematic shown in Figure 9 can be used to illustrate the basic measurement technique. The voltage taps (T1 to T4) contacting the sides of the bridge provide a voltage divider measurement which is sensitive to the orientation of the taps with respect to the body. If a current Ix is forced between corner taps A4 and A1 and the voltages V_{x1} and V_{x2} are measured from A3 to T2 and from T2 to A2 respectively, the component of misalignment is given by the ratio of the difference to the sum of these respective voltage measurements.

[23] $$X = A / 3.1652[(V_{x1} - V_{x2}) / (V_{x1} + V_{x2})]$$

The proportionality constant (3.1652) is derived from conformal mapping of the electrical potentials measured on a thin conductive sheet having a square geometry (15). A is the body dimension of the bridge. The output sensitivity (S) of the bridge in mV/μm provides a figure of merit for a given bridge by expressing the ratio of this measured voltage ($V_{X1}+V_{X2}$) to the body dimension A for a specified forcing current Ix.

[24] $$S = 3.1652(V_{X1} + V_{X2}) / A$$

The sheet resistance (R_s) of the VDP (4, 17) is given by

[25] $$R_s = (\pi / \ln 2)(V_{X1} + V_{X2}) / I_X$$

Inspection of these equations reveals the essential features required for bridge design and measurement accuracy. For a given value of R_s, optimum sensitivity requires a bridge design having minimum body dimension A and maximum test current I. However, as we have discussed above this is contrary to the requirements for measuring R_s. Therefore R_s should be measured independently.

The body dimension cannot be scaled smaller than the space required for movement of the voltage tap contacts over the misalignment range expected and by the finite width of the contacts and taps required to satisfy the minimum design rules. The magnitude of the forcing current Ix is constrained by the current-voltage linearity of the bridge and the thermal dissipation ($I^2 R_s$) induced by the current in the corner taps. The value for R_s is not generally an adjustable parameter because it is dependent upon the structure and process being investigated.

There are several considerations regarding the advantages and limitations of the VDP alignment bridge. The VDP bridge is a compact structure having 4 fold pattern symmetry. Although the structure resembles the van der Pauw sheet resistance tester (and in fact can be used for this purpose on diffused layers), the sheet resistance must be measured independently due to the differences in the requirements on the body dimension. As a result of its symmetry, it does not require the width corrections resulting from image exposure and etch typical of linear structures, however, both structures require a precise knowledge of the physical dimensions A or linear bridge width (W_{lin}) to calculate the displacement X. Unlike the sheet resistance test structures where it is required to make A or W_{lin} large to reduce errors, for alignment bridges it is essential that these dimensions are as small as possible to improve sensitivity. This of course magnifies all of the issues discussed with making the sheet resistance testers as large as possible.

3.0 DEFECTIVITY AND YIELD ENHANCEMENT STRUCTURES

The structures discussed in the following section are those used in conjunction with the process control structures discussed in section 2.0 for developing an MLM process. They are initially used to develop and optimize the various processes. They are also used to verify the design rules. Once the processes are in place, these structures are used to measure yield and defectivity.

With the use of appropriate models, the yield from these structures can be used along with a knowledge of the number of vias, length of metal runs of a particular width and pitch, etc. to predict the effect of the MLM system on circuit yield. However, since these structures are used to predict yield, they tend to be quite large in order to provide a relevant measure of defectivity. To maximize the number of structures within the field of a stepper, it is often judicious to stack these yield enhancement structures on top of each other wherever possible. Many of the test techniques and layout ideas are the same for each of the different structures discussed in this section and therefore they will be discussed only once and referred to from there on.

3.1 Combs, Snakes, and Snakes Combined with Combs

The basic comb structure, shown in Figure 10 consists of interdigitated metal fingers. These structures are used to look for metal shorts between combs on the left and those on the right. The comb can be broken into smaller sections, each of which is wired to a separate probe pad as shown in Figure 10. The test for shorts is performed by applying a voltage to pad 1, grounding pad 2 and measuring the current. As the pattern and etch process improves and defectivity decreases, more combs can be tied together to increase the area of the structure tested. In Figure 10 this would be accomplished by tying pad 1 and 3 together and tying pad 2 and 4 together and then testing. By testing the sections separately and in combination, the area tested can be significantly increased and yet the location of a defect can be determined more easily than just by testing one large structure. Since the yield predicted by models such as Poisson or Murphy decreases exponentially with increasing length or area, it is often advantageous to have the various lengths that are tested cover several decades to facilitate fitting the data to the yield curve. The finger width and space between the fingers should be the minimum for a given layer. Frequently, if there is sufficient area on the test chip, multiple structures with different spacings both greater than and less than the minimum are used to test the robustness of the design rules. Structures with varying spaces between metal lines can also be used to measure the distribution in size of defects that are causing shorts. The width of the metal fingers can be from the minimum value up to a larger more easily manufacturable size. However, increasing the width increases the total area this structure will occupy.

These comb structures are used on each metal layer. The metal 1, metal 2, metal 3, etc. combs may be stacked on top of each other with the associated ILD in between. Testing for shorts between metal layers is accomplished by tying all metal 1 pads together and also tying all metal 2 pads together. A voltage is then applied to all pads on one layer relative to the other and current between layers is measured. This is repeated for each metal layer and its adjacent layers. Typically the combs on each layer are laid out orthogonal to the comb on the prior layer. This will cause some of the worst topography on a given metal layer. To determine the effect of topography under ILD0 on metal 1 processing, stripes of polysilicon or field oxide may be used to produce the maximum topography under ILD0. When chip area is not a problem, one comb for each metal layer is also placed where there is no underlying comb or other topography in order to compare the effects of the planarization scheme on metal shorting.

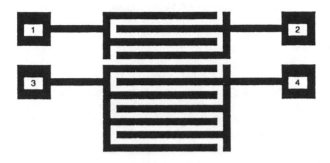

Figure 10. Basic comb structure to test for metal shorts

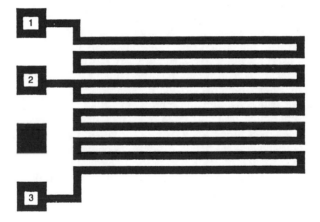

Figure 11. Basic snake or serpentine structure to test for metal continuity

The basic snake or serpentine structure, shown in Figure 11, is a meandering metal line. The width of the line is usually the minimum metal width allowed in the design rules although multiple snakes with different widths may be used if there is enough area on the chip. The length of the snake should approximate the total length of metal with that width that will be on the actual circuit. In many cases this may be more than 1 meter in length. As with the comb structures, the snakes for each metal layer may be stacked on each other to conserve area and to look at the effects of topography. Also like the combs, if the snakes are stacked, they are laid out running orthogonal to the snakes on the preceding layer. As with the combs, when chip area is not a problem, one snake for each metal layer is placed over underlying topography and a second snake of the same length, width etc. is placed where there is no topography in order to compare the effects of the planarization scheme on metal continuity and line resistance.

The snake structures are used to measure metal continuity (resistance of the metal line) by forcing current from one pad to another and measuring the voltage difference. The structure should take into account the minimum number of squares of metal required to produce a voltage drop that is accurately measurable on an automatic tester, typically millivolts, without using currents densities that would cause Joule heating effects. Additionally, on long snakes care should be taken in the choice of currents so as not to induce a high voltage that will cause unwanted breakdown through the ILDs. Multiple taps can be used, as shown by tap 2 in Figure 11, to measure the resistance of the line at intermediate lengths. The resistance of the snake structure shown in Figure 11 can be estimated by knowing the length of the snake, the intended width of the snake and the sheet resistance of the metal. The expected variance from this calculated number can be determined from the variance in the sheet resistance and line widths measured using the structures discussed in section 2.0. If the resistance measurement indicates an open, then the snake is not continuous. If the resistance is substantially lower than expected, this is indicative of metal shorts. Measuring intermediate lengths will indicate if this is a problem across the entire structure or just in one area of the structure. A wide spread shorting problem should also be seen in the comb structures. Shorting problems in the snake structure can be minimized by increasing the space between each length of metal to a value somewhat larger than the design rule. Of course this is done at the expense of increased area used by the snake structure. If the resistance is significantly higher than expected, the cause may be ascertained by combining these results with those from other structures. If the line width is smaller than expected or the metal sheet resistance is high, these problems should be detected by the sheet resistance and line width bridges. High resistance may be a result of the metal thinning as it goes over topography or by changes in the metal linewidth as it passes over topography due to height differences being greater than the depth of focus of the lithography tool. Comparing the line resistance to the resistance of identical structures on flat topography will indicate if topography related issues are a concern. Depth of focus problems generally lead to a narrowing of the metal line over topography and this can usually be observed in either the photoresist pattern or the patterned metal with a high magnification optical microscope or a top view SEM. Thinning of the metal over topography due to low step coverage is more difficult to confirm and usually requires a cross section SEM micrograph.

The snakes that are stacked on top of each other for each metal layer can be used to look at interlayer shorts. This is the same sort of test as was used to test the combs for interlayer shorts. All pads for each layer are tied together and a voltage is applied between the layers. This test will indicate problems in either the ILD, the planarization process, or hillocks in the metal.

The snake and comb structures can be combined into one structure as shown in Figure 12. The snake can be tested for continuity and shorting to the left and right comb. The combs can also be tested for shorting together. A comb to comb short would imply that either the defect was as big as twice the space plus the width of the snake or that the defect density was so high that there were two defects in the same area, one shorting the snake to the left comb and one shorting the snake to the right comb. The tests are performed in the same manner as was discussed in prior paragraphs. This combined structure can conserve area and probe pads when the pad array is laid out on a specific pattern as discussed in section 2.1. However, it does increase the possibility that shorts will confound the continuity data or that opens in the snake will obscure shorts. This is a tradeoff that the engineer designing the test vehicle must make.

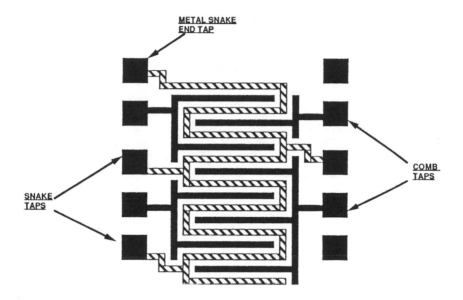

Figure 12. Basic snake and comb combined.

3.2   Via Chains

The basic via chain, shown in Figure 13a in cross section and Figure 13b in top view, is used to test for metal continuity through vias between adjacent metal layers. The chain consists of a series of metal links on the M_i and M_{i+1} layers and vias connecting alternating links. The via dimensions for this structure should be the minimum size in order to evaluate the via filling process. The dimensions of the metal links are usually much wider than the minimum width, so that alignment of the via to the metal link is not an issue and so that the resistance of each link is small relative to the resistance of the via. The number of vias in the chain should be within a factor of 10 of the anticipated number of vias on the circuit so that accurate yield predictions will be possible. For some circuits today this may require several hundred thousand vias or more. If there is room on the test chip it is often helpful to add via chains with vias larger and smaller than the minimum via dimensions to test the robustness of the entire process. There should be at least one via chain for each pair of metal layers that are connected by a via. It should not be assumed that because a via fill process works for one via it will work for another. Changes in ILD thickness, metal film thickness or deposition conditions can change the via filling properties.

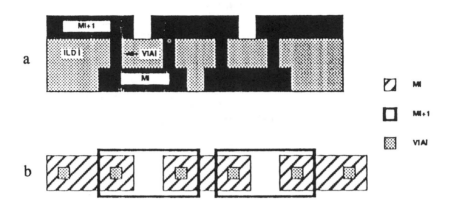

Figure 13. a) Cross section and b) top view of a basic via chain used to test for continuity of vias between adjacent metal layers

The via chain is tested in much the same way as the snakes; a current is forced and the voltage drop is measured. The via chain can also be tapped out (like the snakes) at intermediate lengths to test shorter lengths without having to create multiple chains. This data can also be fit to a yield model to obtain a defectivity for a particular via or to predict the impact of the via on the yield of a particular circuit.

The individual via resistance can be estimated by subtracting the metal link resistances and the lead line resistance from the total measured resistances. These are easily determined by knowing the number of links, the length of the lead lines and the sheet resistance of the metal on each layer. Any variations in the metal width can be accounted for using the information obtained using the split bridges discussed in section 2. This is the reason that wide metal links are used so that the variations in width will have a small effect on the calculated via resistance. The remaining resistance is then divided by the number of vias to give an average via resistance. This average resistance can be compared to the values obtained from the Kelvin structures discussed in section 2.0 Although this resistance number includes more than just the contact resistance, it is often the number circuit designers need.

The two failure modes commonly observed in via chains are opens and high resistance. Opens tend to indicate that the via was not completely etched or there was poor via filling with the metal. High via chain resistance is usually indicative of poor step coverage in the vias resulting in high resistance, via dimensions smaller than the target value, or vias where the etch has partially cleared the via bottom, or a high interfacial resistance metal to metal.

3.3   Contact Chains

Contact chains are similar to via chains except that the bottom layers that form the bottom links are now single crystal silicon or polysilicon rather than a metal layer. The top link is metal 1 for contact chains. The single crystal silicon or polysilicon may have silicide on them depending upon the technology. The chains are designed in the same way as the via chains with the use of wide metal links to minimize the effects of variations in the link width.

The difference in choices for the bottom link layer requires that there be a contact chain for each possibility. This is especially important because of the different contact depths that can occur. Vias for a given layer are essentially all of the same depth especially if the ILD layer has been planarized. Contacts, however, tend to involve three different depths as shown in Figure 14. The three different depths are for contact to single crystal silicon, contact to polysilicon on single crystal silicon or gate oxide, and contact to polysilicon on field oxide. The differences in depth depends upon the degree of planarization achieved at ILD0. For total ILD0 planarization, as shown in Figure 14 the difference in contact depths is a maximum. This increased depth greatly increases the aspect ratio for the contact fill process. In addition, for submicron contacts an etch lag (as discussed in Chapter 7) can result preventing the contacts to single crystal silicon from being completely open if insufficient etch times are not used. Therefore it is necessary to use 2 or 3 different contact chains to evaluate the technologies compared to one via chain. If different size contacts are being evaluated than the number increases that much more. The engineer may choose to use only two of these chains if one type of contact is not allowed, or if space is an issue, the intermediate depth chain may be deleted. In addition, the engineer may choose not to evaluate contacts to single crystal silicon with a multilevel metal test vehicle since this requires additional front end processing and mask layers are to form junctions and isolation. However if this strategy is chosen the mask set used to evaluate the device structures and front end processes must contain contact chains.

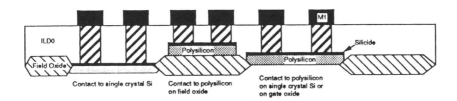

Figure 14. Cross section schematic to show differing contact depths to various areas of the device

Testing contact chains is comparable to testing via chains with two additional considerations. The single crystal or polysilicon (even if they are silicided) links will have a higher sheet resistance and contact resistance than occurs in a via where it is all metal. Therefore the test conditions need to be modified to avoid excessive voltages from being generated. For contact chains it may be necessary to modify the link width relative to a via chain link to keep the voltage low or prevent the link resistance from dominating the resistance of the contact. The second testing issue is preventing the junction in the single crystal silicon from becoming forward biased. If the junction becomes forward biased current can flow into the substrate and invalidate the results. Therefore the polarity of any voltage that will be generated must be considered relative to the type of junctions (n^+/p or p^+/n) that are formed. This is not an issue for vias or contacts to polysilicon on field oxide or gate oxide since these features are isolated from the substrate. For contacts to polysilicon on gate oxide, care must taken to avoid breakdown of the gate oxide

The failure modes for contact chains are essentially the same as for via chains, opens and high resistance. As discussed above the large depths associated with contact chains increases the possibility of incomplete etches, poor step coverage, or interfacial residues.

3.4   Via/Contact Chains at Minimum Pitch

To evaluate the design rules it is usually necessary to build chains at the minimum pitch to determine any issues associated with alignment tolerances and feature dimensions as well as the via/contact etch and fill issues discussed in 3.2 and 3.3. Instead of large links the minimum dimensions are used in conjunction with the minimum overlap. This is shown in Figure 15 for a pair of via chains where adjacent vias are allowed. Usually with this test structure a pair of chains (rather than a single chain) are used to test for shorts between chains. The testing, issues and evaluation of this structure are the same as for standard via and contact chains except for the additional test to look at shorting between structures. If these structure have shorts, the problem should also show up in the comb structures. Opens or high resistance in the chains could also be associated with alignment/overlap issues.

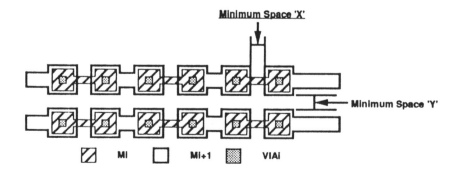

Figure 15. Two via or contact chains built at minimum pitch to look for interactions

The chains and snakes combined with chains presented in the following sections are a few of the more specialized structures that may be used to develop/evaluate an MLM system. Not all of these structures may be useful depending upon the technology design rules. If space is limited the information that can be obtained from these structures must be traded against snakes, combs and chains with varying dimensions.

3.5   Stacked Via Chains or Vias Stacked On Contact Chains

Stacked via chains consist of two or more vias aligned on top of each other as shown in cross section in Figure 16. There is a landing pad at the intermediate layer. The dimension of the landing pad is determined by the larger via and its associated overlap as required by the design rules. Stacked chains can be built with either large links or minimum pitch links, as were used with the standard chains. However in the minimum pitch case, if the design rules change from one layer to another, the largest pitch must be used for all layers. These stacked chains are tested in the same manner as the standard chains.

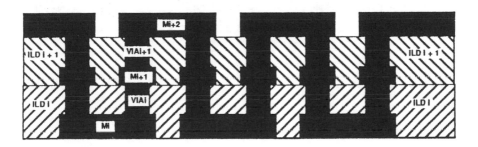

Figure 16. Cross section schematic of a stacked via chain where two vias are placed directly atop one another

3.6   Interdigitated Via Chains and Snakes

There several possible combinations of chains interdigitated with each other or with snakes interdigitated with chains. Figure 17 and 18 show examples of two possible structures. Figure 17 shows two via chains that are interdigitated and Figure 18 shows a via chain with an M_i snake and an M_{i+1} snake weaving through the chain. These test structures are used to look at the effects of topography and to create structures that should closely resemble a real circuit. The testing and evaluation of these structures is just a combination of the ideas used for the individual structures.

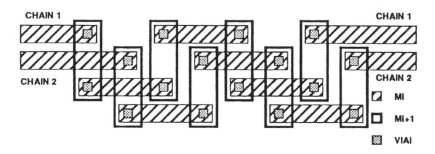

Figure 17. Two via chains that are interdigitated

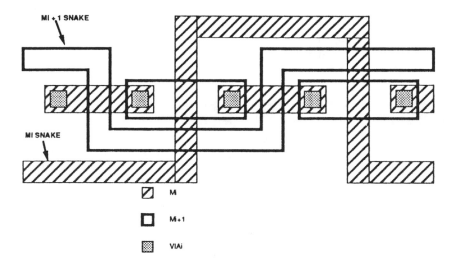

Figure 18. Via chain integrated with metal snakes at the above and below metal layers

3.7 Capacitors

The large plate capacitor is shown in Figure 19 for two layers of metal. The plate capacitors are formed by patterning the Mi metal layer into a large rectangular pattern, depositing the ILD layer, and depositing and patterning a rectangle into Mi+1. These plates are stacked on top of each other for each metal layer. The minimum area of each plate is determined by the area required to produce the minimum capacitance that can be reliably measured with an automatic tester.

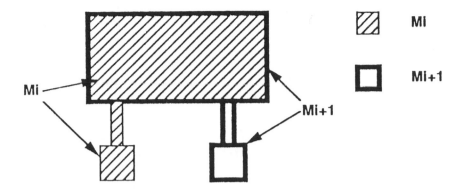

Figure 19. Large metal plate capacitor used to provide information about the ILD

Large plate capacitors are used to measure many of the properties of the various ILD layers. These properties include the ILD thickness (if the dielectric constant is known) or the dielectric constant if the thickness is measured independently, the breakdown voltage of the ILD, the leakage current through the ILD, and mobile ions in the ILD layers. The ILD thickness is monitored by measuring the capacitance and then dividing out the area of the capacitor and the dielectric constant of the ILD which has been measured independently. The dielectric constant can be determined by measuring the capacitance and obtaining the ILD thickness from another method such as cross section SEM. The breakdown voltage of the ILD is measured by ramping the voltage on one tap, grounding the other tap and monitoring the leakage current until the ILD breaks down.

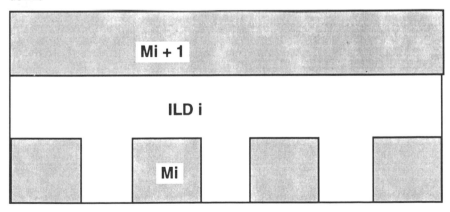

Figure 20. Capacitor structure with the lower plate comprised of fingers

A second type of capacitor structure uses fingers of metal 1 or silicided polysilicon as the lower plate followed by the ILD and a large metal upper plate as shown in Figure 20. This structure is useful in looking at the effects of topography on interlayer shorting as well as ILD step coverage and planarization when combined with SEM. The electrical tests are performed by applying voltage between the two layers and measuring current to look for shorts. The maximum topography can be simulated by stacking two or more layers of fingers on each other followed by the final metal plate.

A final type of capacitor is the one shown in Figure 20 for measuring the capacitance on a signal line with a grounded metal plate above and below the signal line as well as grounded lines adjacent to the signal line. By measuring the capacitance on the signal line some of the modeling discussed in Chapter 1 can be verified. Most parametric testers cannot measure this small of a capacitance and therefore this test is normally done on a bench.

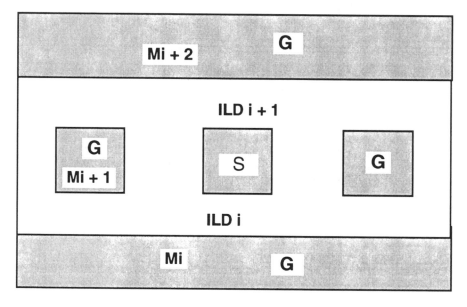

Figure 21. Three layer capacitor where the middle plate is comprised of a signal line surrounded by grounded adjacent metal lines and the top and bottom layers are simple plates.

4.0 TEST STRUCTURES FOR RELIABILITY

4.1 Introduction

Conventional reliability testing relies on the measurement of failure rate (FR) to predict product lifetimes. Reliability is defined in terms of the probability that the performance characteristics of the device or product will remain well within the specified FR limit over its expected lifetime. Reliability measurements provide statistical statements about the average performance characteristic of a test group. The probability value must include a statement of the relevant stress conditions from which the probability was derived, such as: temperature, time, humidity, duty cycle, voltage, and current.

Reliability data is commonly presented in terms of failure rate statistics using a function that describes the failure rate distribution as a function of time. In addition to the 'normal' variation of failure rate for a representative population, a small fraction of the failures may consist of early failures which differ significantly from the FR of the main population. When failure rate distributions are taken collectively the FR vs. time plot resembles a quasi-U shaped curve ('bathtub curve') as shown Figure 22. This hypothetical plot shows three failure rate regions: early failure, steady state failure and intrinsic wearout.

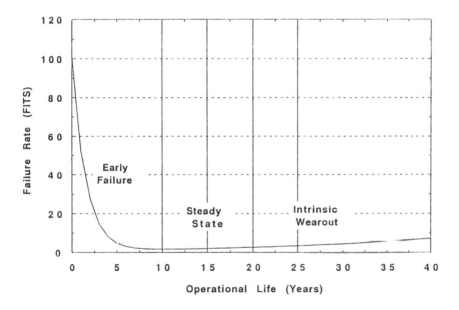

Figure 22. Hypothetical failure rate vs. time curve showing three failure rate regions

For high reliability products such as semiconductor integrated circuits the intrinsic wearout will normally occur at very long times for operational stress conditions. Intrinsic reliability can only be investigated by using accelerated life testing. Reliability tests to measure the FR using accelerated stress conditions provide the data needed to calculate the FR probabilities for accelerated stress, and to estimate by extrapolation, the failure rates for operational stress levels with acceptable statistical confidence. Since most wearout mechanisms are thermally activated, accelerated stress testing for electromigration reliability, which represents the primary intrinsic wearout mechanism for metal interconnect systems, is usually implemented by stressing at temperatures and current densities which are significantly greater than those used in normal operation.

In recent years, the concept of reliability integration (designing reliability into the process and product) has emerged as a viable approach to addressing some of the limitations of conventional reliability testing (18, 19, 20), particularly, when failure rates are extremely small and difficult to measure with a high degree of statistical confidence. This concept relies on the proactive involvement of reliability engineering to develop an understanding of the causes of reliability reduction, as opposed to the measurement of product failure rates. Reliability integration requires involvement in the design and development of device, process and manufacturing technologies and the control and/or elimination of critical reliability degrading factors early in the product development cycle.

Some of the characteristics of reliability integration are: application of wafer level and highly accelerated testing methods, line monitors for in-process measurement and control, development of reliability driven design rules derived from test data, comparative tests to show reliability trends, development of new test methods and tools for fast reliability assessment and novel test structures designed for high sensitivity to critical design and process reliability parameters. However, conventional stress testing will continue to be used for FR calibration and to provide the quantitative FR inputs required for reliability modeling, simulation and prediction.

Implementation of new methods for reliability integration will require a cultural change in the engineering community and must rely heavily on cross functional teams having a shared responsibility for designing reliability into the product. In this environment, acceptable reliability is measured by documented application of reliability driven design rules in process module development, product design and manufacturing. The control of critical design, process and manufacturing parameters which directly impact reliability must be demonstrated and documented.

A detailed discussion of the theory and experimental procedures for electromigration and stress testing are discussed in Chapter 8 of this book.

4.2   Conventional Methods of Reliability Testing and Analysis

The analysis of lifetest data requires the application of statistical methods. Moreover, a range of stress parameters such as temperature, current, etc. will be required to permit failure time measurements obtained under accelerated stress conditions to be extrapolated to operational use conditions. This assumes that the effects of stress can be represented by an acceleration factor which permits extrapolation of the median failure time from accelerated stress conditions to the median failure at use conditions. Since nearly all temperature dependent mechanisms are modeled using the Arrhenius equation, the median lifetime for operational stress can usually be calculated from the temperatures corresponding to the accelerated and operational stress levels and the activation energy which characterizes the failure mechanism.

A major assumption for the application of conventional reliability test methods is that failure rates can be measured accurately. This assumption is dependent on the specified failure rate goal. Using Poisson statistics, a failure rate FR = 100 FIT (1 FIT = 0.01% failure in 10^5 hrs.) requires a minimum sample size of $\approx 10^3$ for an upper confidence limit (UCL) of 60 % with 0 failures. For a FR goal = 10 FIT, a minimum sample size of $\approx 10^4$ is required for a 60 % UCL (18). In other words, as reliability improves, statistical confidence will become progressively poorer, unless we are willing to test an unrealistically large number of samples. The implication of this is that intrinsic failure rate margins will continue to decrease and sample size demands will continue to rise with improved reliability.

Some of the current limitations of conventional reliability testing are: high cost of testing, excessively long test times, data does not provide information on the fundamental or intrinsic causes of reduced reliability, and the test methods do not activate the relevant failure mechanisms. Therefore, a major limitation of conventional testing is that it offers little or no direct impact on process development and product design.

4.3   Multilevel Metallization Reliability Test Structures

A multilevel interconnect system can be divided into three sub-systems; long (cm) horizontal conductors having a global distribution such as power supply and bit lines in logic and memory products, relatively short horizontal conductors (μm) having a local distribution which link circuit components at the cell level and vertical conductors between the multilevel conductor system which consist of contacts, vias, and plugs (CVP).

The design of effective reliability test structures for a multilevel interconnect system must include the following considerations: sensitivity to the anticipated wearout mechanisms, topography of the interconnect system, layout and design rules, stress test and environmental conditions, package and test structure configuration, test equipment and the capabilities of the existing process technology. Design rules which apply to interconnect and contact dimensions and to layout configuration are driven primarily by transistor and circuit performance constraints.

Electromigration tests of planar interconnects and CVP require accurate measurement of stressing conditions including current density and temperature. Other measurements such as; metal resistance, linewidth, film thickness, step coverage and grain size distribution are also required. Testing of submicron contacts, vias and plugs should also include test structures designed to monitor metal-silicon and metal-metal interaction as well as ILD stresses.

The conventional planar electromigration structure (which is described in the subsequent sections) has been used for many years to test wearout on long lines. The theory of electromigration associated with this type of structure has also been explained (see Chapter 8). However, for CVP there is little available theory and data is just beginning to be reported. There is no universally accepted structure, although the Kelvin structures and chains discussed in section 2.0 and 3.0 of this chapter can be used to determine the relative failure rate for one process compared to another. To thoroughly understand these structures and their wearout mechanisms, two or three dimensional modeling of current and heat flow will be required since it is not possible to directly measure them.

Comprehensive reliability test vehicles should also include PC test structures similar to those discussed in section 2.0 to measure parameters such as sheet resistance, critical dimensions (CD) such as line width and space, for comprehensive analysis of reliability test data. Contact and via reliability test structures for MLM technology must include the design and structural features defined by layout rules. Contacts should include the design features of local interconnects, transistors and passive circuit elements which may have a first order impact on contact reliability. Particularly, the non-uniform current flow which is anticipated at the edges of contacts and vias.

4.4   Conventional Planar Electromigration Test Structures

The electromigration reliability of thin metal film interconnects for integrated circuits has traditionally been calculated from failure rate data obtained from simple linear test structures using accelerated stress at constant current density and temperature. Testing is generally conducted on an computer automated tester using wire bonded and packaged test die with temperature controlled ovens. Electromigration lifetime estimates are obtained from log-normal probability plots. Failure rate data is normally extrapolated from accelerated stress test conditions to operational stress levels using Black's equation (section 6.2 in chapter 8). These projected failure rates provide a first order estimate of interconnect lifetime in terms of the median failure rate.

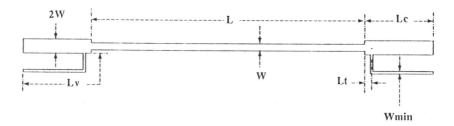

Figure 23. Conventional electromigration test structure developed by NIST

Optimal design for reliability test structures requires high sensitivity to the physical mechanism causing poor reliability without sacrificing measurement accuracy and repeatability. The conventional planar interconnect electromigration test structure is typically a modification of a 4 terminal, Kelvin sensed, linear bridge resistor. A design developed at the National Institute of Standards and Technology (NIST) (21) is shown in Figure 23. The design length and width of the current and voltage taps are based on thermal modeling, to minimize thermal gradients along the line, due to heat transfer at the taps. This structure is designed with a recommended length L of approximately 800 μm. The width W is generally varied within a specified range to provide data on the failure rate dependence on line width. The current sourcing tap of width 2W has a minimum recommended length Lc ≥ 80 μm or approximately 10% of L, and it is designed to prevent electromigration failure in the tap by reducing the tap current density while minimizing the thermal gradient at the end of the test line. The voltage taps are designed with minimum line width and are located a distance Lt from the edge of the test line with a recommended spacing of 8 µm or 1% of L. The minimum length of the test line should be determined experimentally to insure that the failure rate is not dominated by secondary considerations such as a critical length dependence (see Chapter 8).

The voltage taps are Kelvin sensed with a high impedance voltmeter to obtain an accurate measurement of the voltage drop between the current sourcing taps for a specified line current density. Although a constant line current is maintained during stressing, the line current density, which is dependent on the cross sectional area of the test line, is generally assumed to be constant. This assumption is valid for test structures during most of the stressing period but is not necessarily valid in the final stages of wearout when void formation significantly reduces the local cross-sectional area of the line, increasing the local current density and the Joule heating.

Accurate measurement of the relative line temperature is obtained using the measured line resistance ($R=V/I$) and its temperature coefficient of resistance (b), which is given by, $b=\Delta R/R_0\Delta T$, where R_0 is the intercept of a line resistance vs. temperature plot. For typical metal film interconnect test structures on a nominal thickness (0.1 µm) of silicon dioxide (SiO_2), with $J > 1 \times 10^6$ A/cm^2 significant Joule heating can occur and temperature correction may be required for accurate line temperature calculations. During electromigration stressing the test structures are monitored periodically by the test system until a predefined failure condition is reached. Failure is usually defined in terms of the time required for cumulative electromigration damage to produce a specified change in resistance (ΔR) or until an electrical open circuit occurs. Because of the statistical nature of electromigration failure times, the life test results are typically described in terms of an appropriate distribution law (Section 6 in Chapter 8).

Additional sense lines with voltage taps are often located on both sides of the electromigration test structure at minimum space S, as shown in Figure 24, to detect metal extrusions which may occur along the line, resulting from hillock formation. Although, single line EM test structures incorporating extrusion sense lines seldom show extrusion failures on unpassivated structures, extrusion failures on passivated lines can be expected because the metal line is confined by a dielectric capping layer. Under these conditions, electromigration induced hillocks introduce compressive forces which can cause cracking of the passivation layer and extrusion of the compressed metal.

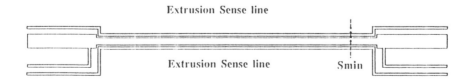

Extrusion Sense line

Extrusion Sense line   Smin

Figure 24. Extrusion monitors added to the conventional NIST electromigration test structure

4.5 Test Structures For Accelerated Wafer Level Reliability Testing

Highly accelerated stressing, $J \geq 10^7$ A/cm² introduces the added requirement for efficient removal of heat during stressing to prevent excessively high line temperatures. This problem is particularly important for highly accelerated wafer level testing, as in the SWEAT test,[22] where self heating is the primary source of thermal stress. Figure 25 illustrates the conceptual transition from conventional electromigration test methods to the highly accelerated methods currently being implemented for wafer level reliability (WLR) testing. Conventional EM testing relies on ambient temperature control, moderate current density, $J \approx 10^6$ A/cm² and minimum Joule heating, whereas, the highly accelerated WLR tests (SWEAT) rely primarily on "controlled" self heating through test structure design, maximum current density and maximum Joule heating. The obvious advantage of highly accelerated stressing is the significant reduction of test time achieved through the increased current density and temperature stress.

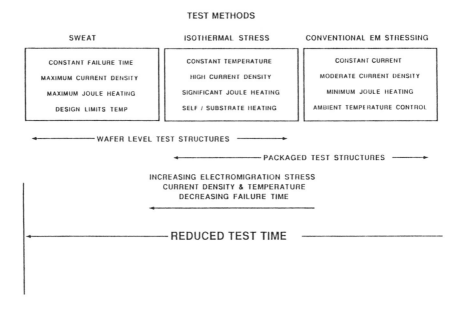

Figure 25. Comparison of test methods which can reduce electromigration test time

The standard wafer electromigration acceleration test (SWEAT) structure shown in Figure 26 has been designed primarily for the qualitative assessment of metal film reliability. It offers the capability for extremely short duration electromigration testing using a conventional wafer probe station. The technique relies on Joule heating at high current densities produced in the test line to accelerate the wearout mechanism. SWEAT structures are designed for operation

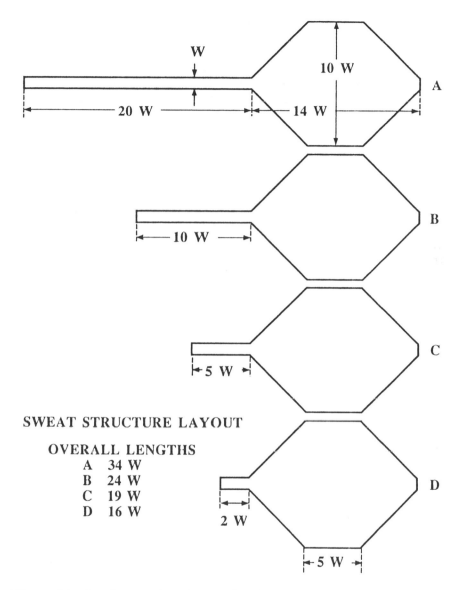

Figure 26. Standard Wafer Electromigration Acceleration Test (SWEAT) structures.

at very high current densities $>10^7$ A/cm^2 to induce very short EM failure times in thin film conductors. Under these test conditions, thin film interconnects on insulating substrates, would be subjected to excessive Joule heating and eventual failure by local melting. In order to permit operation at very high current density the structure must dissipate the excess heat generated in the narrow line segment. To achieve this effect the structure is designed using a 'chain' of relatively short length narrow lines in series with large area links (hexagonal in shape) to form long chain structures having the desired overall length as shown in Figure 27. The surface area of the large area link increases the thermal conductivity of the test structure to the underlying substrate. As a result the temperature of the narrow line is determined by the length of the narrow line segment, its cross sectional area and the effective thermal conductivity of the large area link.

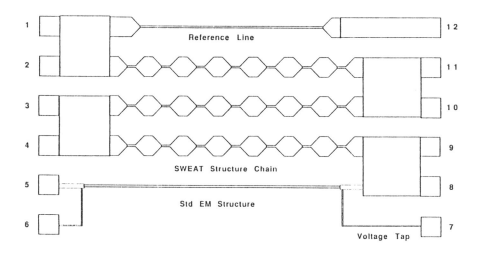

Figure 27. Chain structure composed of the SWEAT elements shown in Figure 26 on which the actual accelerated testing is done

To illustrate a design rule for SWEAT structures, we will consider a requirement to investigate the dependence of failure rate on the test structure linewidth W and its median grain size $D50$. If we assume a $D50 = 2.5\mu m$, we might include three test structures with a W of 1 μm, 2.5 μm and 5 μm, for a $W/D50$ ratio less than, equal to, and greater than unity. Each linewidth could include four structures having different lengths to provide a range of peak line temperatures and thermal gradients at the end of each segment. Examples of four types of structure are shown in Figure 26. Each basic element type (A, B, C, or D) would be used to build a chain. The number of type A, B, C, or D segments would be chosen to keep the total resistance constant. The total resistance is chosen to meet the tester requirements. The reason for varying the line length is to control the peak temperature at the center of the line segment and to vary the temperature gradient across the line segment.

It should be noted that the SWEAT structure is not universally accepted in the reliability community. In some companies it is routinely used to monitor the relative quality of the metal. Others argue that the data is not valid. As time goes on this issue should be resolved. We have also found the structure useful to study stress voiding since voids tend to show up first on the octagonal heat sinks. The test vehicle design engineer should decide for himself based on available data at the time and a discussion with his reliability engineering group whether the SWEAT structure should be included.

In summary, reliability structures should be included in the MLM test vehicle as well as a separate reliability test chip so that this information can be obtained from structures that have been subjected to the complete MLM process flow. In addition much of the information obtained from the structures discussed in sections 2.0 and 3.0 are needed to interpret the reliability data.

5.0   SUMMARY

This chapter has discussed three groups of test structures that are useful in the development of an MLM system. These structures include: [1] PC structures such as bridges for measuring metal sheet resistance, linewidth, etc.; [2] Defectivity and yield enhancement structures such as snakes, combs and chains for developing and monitoring processes; [3] Reliability structures for studying metal line or CVP failures. Additional structures might be included to facilitate physical or chemical analysis of the MLM system as discussed in Chapter 11. These structures are very analytical tool specific and are beyond the scope of this chapter.

REFERENCES

1.   T. E. Wade, "Proposed Comprehensive Test Vehicle for Monitoring Multilevel Interconnection Process Variabilities, Misalignment, Parametrics and Defect Density," Proc. IEEE VLSI Multilevel Interconnection Conf., p. 354, (1986).

2.   D. J. Radack, J. C. Swartz, L. W. Linholm, and M. W. Cresswell, "A Comprehensive Test Chip for the Characterization of Multi-level Interconnect Processes," Proc. IEEE VLSI Multilevel Interconnection Conf., p. 238, (1987).

3.   P. M. Hall, "Resistance Calculations for Thin Film Patterns", Thin Solid Films, Vol. 1, p. 277, (1968).

4.   D.S. Perloff, "Four-point Sheet Resistance Correction Factors for Thin Rectangular Samples," Solid St. Electron., Vol 20, p. 681, (1977).

5.   M. G. Buehler and C. W. Hershey, "The Split-Cross-Bridge Resistor for Measuring the Sheet Resistance, Linewidth, and Line Spacing of Conducting Layers," IEEE Trans. on Electron Dev., Vol., ED-33, No. 10, p. 1572, (1986).

6.   S. J. Proctor, L. W. Lindholm and J.A. Mazer, "Direct Measurement of Interfacial Contact Resistance, End Contact Resistance and Interfacial Contact layer Uniformity," IEEE Trans. on Electron Devices, ED-30, No 11, p. 1535, November 1983.

7.   R. L. Gillenwater, M. J. Hafich, and G. Y. Robinson, "The Effect of Lateral Spreading on the Specific Contact Resistivity in D-Resistor Kelvin Devices", IEEE Trans. on Electron Devices, ED-34, No. 3, p. 537, March 1987.

8.   A. Scorzoni, M. Finetti, K. Grahn, I. Suni and P. Cappelletti, "Current Crowding and Misalignment Effects as Sources of Error in Contact Resistivity Measurements," Part I: Computer Simulation of Conventional CER & CKR Structures, IEEE Trans. on Electron Devices, Vol ED-34, p. 525, (1987).

9.   P. Cappelletti, M. Finetti, A. Scorzoni, K. I. Suni, N. Circelli and G. D. Libera, "Current Crowding and Misalignment Effects as Sources of Error in Contact Resistivity Measurements," Part II: Experimental Results and Computer Simulation of Self-Aligned Test Structures, IEEE Trans. on Electron Devices, Vol ED-34, p. 525, (1987).

10. H. H. Berger, "Models for Contacts to Planar Devices," Solid State Electron., Vol 15, p. 145, (1972).

11. D.A. Scott, W. R. Hunter, and H. Shichijo, "A Transmission Line Model for Silicided Diffusions, Impact on the Performance of VLSI Circuits", IEEE Trans. on Electron Devices, ED-29, No. 4, p. 651, April 1982.

12. D. R. Thomas, and R. D. Pressoni, "An electrical photolithographic alignment monitor", Proceedings of Government Microcircuit Applications Conference, Boulder CO, p. 196, (1974).

13. M. G. Buehler, "Emitter-base electrical alignment test structure", Semiconductor Measurement Technology NBS Special Publication, No.400-25, p. 40, (1975).

14. M. G. Buehler, "The Use of Electrical Test Structure Arrays for Integrated Circuit Process Evaluation", J. Electrochem. Soc., Vol. 127, No. 10, p. 2284, (1980).

15. D. S. Perloff, "A van der Pauw resistor structure for determining mask superposition errors on semiconductor slices", Solid St. Electron., Vol. 21, p. 1013, (1978).

16. K. H. Nicholas, I. J. Stemp, H. E. Brockman, "Measurement and Identification of Distortion, Alignment, and Mask Errors in IC Processing", J. Electrochem. Soc. p. 609, (1981).

17. L.J. van der Pauw, Philips Res. Repts., Vol. 13, No. 1, p. 1, (1958).

18. I.A. Ross, "A Strategic Industry at Risk," A Report to the President and Congress from the National Committee on Semiconductors, Washington, D.C. (1989).

19. D. L Crook, "Evolution of VLSI Reliability Engineering, " Proc. IRPS, p. 2, (1990).

20. H. A. Schafft, D. A. Baglee, and P. E. Kennedy, "Building-In Reliability: Making It Work," Proc. IRPS p. 1, (1991).

21. H. A. Schafft, T. C. Staton, J. Mandel and J. D. Shott, Reproducability of Electromigration Measurements," IEEE Trans. on Electron Devices, Vol. ED-34, 3, p. 673 (1987).

22. B. J. Root and T. Turner, "Wafer Level Electromigration Tests for Production Monitoring," Proc. IRPS, p. 100 (1985).

10

MANUFACTURING AND ANALYTIC METHODS

THOMAS SEIDEL
Sematech
Austin, Texas

1.0   INTRODUCTION (BUSINESS PERSPECTIVES)

As mentioned in earlier chapters, Multilevel Metal (MLM) structures include two or more interconnecting conducting elements on an integrated circuit. For our discussion of the manufacturing aspects, we include: polysilicon (or poly with silicide capping), along with metal ("metal-1"), and higher layers of metal ("metal-2 ... n levels"). Interconnects are separated by various deposited dielectrics. The necessity for interconnect level counts greater than 2 are determined by the complexity of the circuit (see Figure 1). Although the standard number of levels goes from 2 to 3 at 100K gates and from 3 to 4 at 0.5 million gates, the number of levels shown in Figure 1 will be exceeded to achieve leading edge competitive speed performance.(1)

In this introduction, we cover general semiconductor manufacturing strategies and, in particular, those that are applicable to MLM. This introduction contains a discussion of key strategies, along with some technical and operational tactics. They are:

o   Low Cost Strategies
       Consolidated Process Flow
       Tool Extension across Manufacturing Generations
       High Throughput
       Lower Equipment Costs
       Cost of Ownership
o   Technical and Operational High Yield Tactics
       Contamination Control, Defect Reduction
       Monitoring-testing
       Best Shop Practices
       - Metrics
       - Statistical Process Control (SPC)
       - Total Preventive/Productive Maintenance (TPM)
       - Team Solutions
       Communication and Automation

GATE ARRAY/CPU/ULSI LOGIC METAL LEVELS

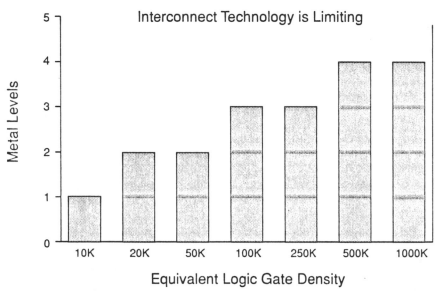

Figure 1. Number of metal levels as a function of on-chip logic gate density (After McCausland, Ref 1.).

Manufacturing strategies are driven by the simultaneous attainment of *low cost and high yield.* To achieve this goal one must consider the product goals: circuit speed and reliability, along with process integration capability including design rules. In the manufacturing environment of a given generation, these elements are held as frozen and "givens" by the manufacturer. Execution to schedule and yield control are key. However, all capabilities must be reviewed regularly to comprehend customer-driven changes. Then the progression to next generation design rules must be planned very carefully.

There are several commonly used approaches to drive cost-to-manufacture down. They include manufacturing and business elements:

i) The simplest, most cost effective integrated process flow (we call this a *"consolidated process flow"*) must be used. This results in the documentation/practice of a set of so-called "unit processes" with the fewest operational steps. These unit processes provide the ability to do "one thing," such as make the dielectric isolation that separates poly from first metal. In fact, the unit process may be composed of several sub-process steps, e.g. clean, sequential film deposition of different compositions and thermal anneals. After completion of these sub-processes, we have the "unit process" isolation composite film we want.

The entire flow is organized into "unit processes" because these are what manufacturing teams are responsible for: the integrity of the entire series of unit processes. As we shall see, it also gives rise to development of original, one step unit process equipment, and ultimately one "machine tool" integrating the sub-processes of the unit operation with pushbutton operation.

ii) *Tool extension across manufacturing generations.* This may be discussed in the context of existing facilities that are undergoing upgrades or expansion and totally new facilities. Manufacturing cultural environments are, by nature, "incremental" and low risk. It is highly unlikely that a totally new technology will be imposed on an already existing operation.

There are a few cultural rules practiced in the name of competitive manufacturing. First, one may run any process sequence with an existing tool set (wafer logistics withstanding), but keep new capitalization to a minimum. The sequence of process steps can be changed to get better performance or higher yields, but the rule "Few New Tools" is practiced to ensure low risk and high profit operation.

The second rule supports the first: if new equipment is required, be sure that it extends across several generations, or that it has the flexibility to run newly developed processes. In addition, a new fab must come up quickly to justify a return on investment, so reliable, high productivity tools are a necessity.

iii) Attain *high throughput.* We drive cost-per-wafer down by improving the net throughput performance of the equipment. This is typically done by improving the deposition or etch rates or the handling time of the tools. Raw or instantaneous throughput may be increased. Reliability and utilization capability should also be increased. (Utilization capability is the percent time the tool can be used to make product under the condition that wafers are always available to run the tool; there are different definitions of this term which we will discuss below.)

The ability for a tool to be fully utilized (uptime/total time) depends on both the tool reliability through the Mean-Time-Between-Failure (MTBF), the Mean-Time-To-Repair (MTTR) and, specifically, their ratio. One must bear in mind that short duration high frequency repairs are not acceptable either, even though the ratio MTTR/MTBF is low and would appear favorable.

A tool can have high throughput by pursuing a batch wafer strategy or a multi-chamber parallel cluster strategy. There are tradeoffs in performance for batch vs. single wafer tool designs and these will be discussed below.

iv) Drive to *lower equipment costs.* Although this is an admirable goal, it is perhaps unrealistic in the face of increasingly complex requirements: higher tool utilization, lower defects, higher degrees of sensor control with self-contained and open communication, automation, flexible recipe capabilities, and in situ cleans with long time to intrusive maintenance (hard clean). All these increasingly sophisticated requirements are driving up the cost of semiconductor manufacturing equipment. It is very difficult to contain the cost, let alone reduce it, although the challenge is not impossible. Strategies to reduce costs include designing the equipment/process with low cost as a specification in the early design stages as well as planning for volume equipment production and partnerships with suppliers.

v) *Cost of Ownership (CoO)* modeling is an approach that calculates the cost per wafer-step, for a given piece of equipment. CoO and equipment cost are not the same thing. Equipment that has a high cost but very good reliability and throughput can be more valuable than a lower cost tool with poor throughput and reliability metrics. Typical key CoO parameters are:

o   Equipment total cost rate ($/hr) which, in turn, depends on:
       capital-depreciation,
       labor,
       plant (footprint, facilities),
       service, and
       materials.
o   Net Throughput (# wafers/hr):
       This comprehends the gross or instantaneous rate and the idle time:
       scheduled clean, maintenance time.
o   Utilization Capability (U.C. %):
       This is the % time for the tool to run *product*. It depends on the
       mean time between failure (MTBF) and the mean time to repair
       (MTTR), and in particular the ratio: MTBF/(MTBF+MTTR). The %
       time the tool is unavailable due to running monitors has to be
       included as a multiplier.

This definition is a user's or customer's definition. The "standard" definition SEMI E-10, 90 (2) was driven by suppliers and includes the % time to run test wafers, as well as product wafers.

The importance of the emphasis on "*product*" wafers is that it drives down the dependencies on "send ahead" or monitor wafers. These overheads, if allowed to be kept in the metrics, will tend to sustain higher costs in the operation. The user is trying to get the equipment to be so predictable and stable, that monitor wafers can be eliminated. Elimination of monitors is an important initiative and is one of the first lessons learned from the Cost of Ownership analysis. (Engineering test wafers are another matter and need to be differentiated against monitors needed to sustain manufacturing. When a tool is brought under engineering control, it is considered removed from manufacturing.)

The simple formulation for the cost per wafer of a tool in manufacture is:

[1]   CoO = Total cost rate / [U.C. x Throughput]

For example, for a $2M (total) investment depreciated in 5 years, with 80% utilization and 30 wafers per hour net, we will have about $2/wafer-step. Other cases are illustrated in Figure 2.

This simple formulation ignored cost of labor, facilities and materials and also assumes that the yield is 100%. If the killer defect density and the chip area are limiting, then the CoO must be increased appropriately. A commonly used factor for the yield goes as $1/(1 + DA)$, where D is the killer density of defects and A the active area in the chip. More sophisticated models have been developed, which may include wafer capacity requirements, time dependent parameters, cost of scrap, cost of money, prove-in costs, training, etc.

Yield control and engineering today require a "total systems" approach because the only satisfactory yield goal is 100% on every product line. Whether one accepts 80-90% yield as a competitive tradeoff is a practical matter, but all competitive manufacturing fabs in the world today must plan to provide 100%

yield capability. The only way to achieve this in deep sub-micron environments is to establish defect- and contamination-free manufacturing.

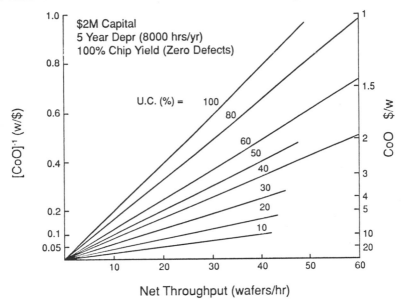

Net Throughput (wafers/hr)

Figure 2. The reciprocal Cost of Ownership (left) and CoO in $/wafer (right) plotted against the net throughput (wafers/hr) for various equipment utilization capabilities.

Regarding technical and operational high yield tactical elements, we have:

i) *Contamination control, defect reduction.* There are two classes of contamination: molecular (e.g. water vapor) and particulate. They both play roles in limiting the yield by causing defects. The need to eliminate contamination from process modules is more important for deep submicron technologies than for large dimension technologies. A current approach is to have "in situ" cleaning that allows a process module to maintain suitably low particle levels <u>and</u> strive for an infinite time before hard cleans. Tools are starting to emerge with this capability today.

ii) *Monitoring-testing.* Control of a fab line is carried out by monitoring equipment set points and parametric testing of discrete and integrated test structures and, ultimately, yield on product. The hierarchy of monitors and test vehicles supports diagnostic capability, and insures very high yield on product, or if yield is less than satisfactory serves to aid root cause diagnostics and recovery. Since monitoring and testing are non-value added steps, one must develop a strategy for only necessary tests and use those sparingly.

iii) Establish *Best Shop Practices.* The monitoring of deposition and etch rates, line widths, selectivities, up times, and other *metrics* are carried out using *statistical process control (SPC) methodology*. This technique affords the fab engineers the opportunity to monitor and react to out of specification values and even small changes that are within control. This activity is the basis for equipment stabilization and health.

However, SPC is costly and is a "reactive operation," wherein we find out what is wrong after something goes wrong. A more attractive approach is to react to small changes before one is out of control and stabilize the change. Beyond this, one can place inexpensive sensors on the equipment designed to forecast when a maintenance is required before the tool goes out of spec. These proactive approaches are known as *Total Preventive/Productive Maintenance or (TPM)*. The best of judgement is required to skillfully execute SPC/TPM.

An element of "Best Shop Practices" is a *team solution* to problems. Regular meetings with factual data, a sense of respect among competent team members and a fierce will to close on issues along with enabling management are among the requirements for success.   Opportunities for continuous improvement using honest quality feedback methodology is the culture.(3)

iv) Implement the right level of *communication and automation*. Computer Integrated Manufacture (CIM) is viewed as essential for successful modern manufacture. Meaningful data entry and access are important for manufacturing operators, engineers, managers and business administration. At the bottom of the communications hierarchy is equipment performance data collection and analysis, at the top is business administrative information such as order in and accounts receivable/closed. The current advanced vision of the utilization of CIM includes the co-design of new products by customer and supplier.(4)

We now proceed with Section 2.0 that describes the implications of the "RC" delay, 3.0 MLM architecture, and 4.0 descriptions of certain unit processes, along with their options that represent opportunities for process consolidation. We discuss special manufacturing issues and performance metrics; and, finally, 5.0 review key design rules. This leads naturally to a discussion of "Design for Manufacture" in Section 6.0. In Section 7.0, we discuss the emerging field of cluster tools and its role in advanced manufacturing (8.0). In Section 9.0 we discuss control and analytic topics: SPC/PM, Analytic Diagnostic methods, test vehicles; and, finally, make some summary comments about assembly and packaging.

2.0   PERFORMANCE DRIVERS

As covered in detail in earlier chapters, MLM technology has been driven by the need for ever faster switching speed. This need has been described by the simple RC time constant model applied to the metal interconnect runner with resistance and parallel plate capacitance:

$$[2] \qquad RC = (\rho \frac{l}{wt_m})(\varepsilon \frac{lw}{t})$$

where ρ is the resistivity of the metal, l the length, w the width, ε the dielectric constant, and t, t_m are the dielectric and metal film thicknesses.

In manufacturing, one must evaluate whether they are meeting the "RC" design needs by statistically monitoring the access time (signal propagation time) of a standard or baseline product. Alternately, one may routinely track the parametric test data on a ring oscillator test circuit. The statistical monitoring of these test parameters will provide an overall metric for MLM performance, but will not give root causes if the results are poor.

More advanced modeling (5) (than simple "RC") includes the parallel plate capacitance, taken in parallel with the fringe and substrate capacitance and in parallel with the interelectrode. See Figure 3. This suggests the need for low interlevel dielectric constant materials to reduce interelectrode capacitance. If the model were generalized to the use of Silicon on Insulator (SOI), substrate parasitic capacitances would be reduced. However, interelectrode capacitance dominates at deep submicron spacing for interconnects. Signal cross talk is an important issue. The use of lower resistivity metal (e.g. Cu) interconnects and/or low dielectric constant insulator materials could benefit performance.

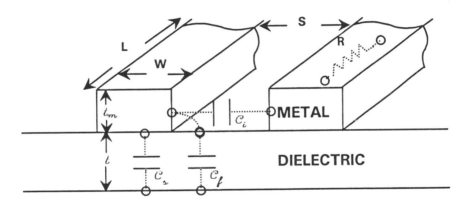

$$RC = \frac{\rho \varepsilon L}{W \cdot t_m} \cdot (C_s + C_f + C_I)$$

$$- \frac{\rho \varepsilon L^2}{x^2} \; ; \text{where } t_m, \ t, \ W, S$$
$$\text{scale as } x.$$

Figure 3. Schematic of metal interconnects over isolating dielectric over substrate with key geometries, resistance and capacitances illustrated. Simple RC model scales as the product of dielectric constant and resistivity, and as the ratio of interconnect length to critical dimension squared.

Manufacturing communities must be prepared to respond rapidly to technology changes, especially if their introduction simultaneously reduces cost and improves performance. For sure, current MLM technologies and tool sets will not be transparent to the use of different insulators or conductors.

In addition to general circuit considerations, there is an evolving need to consider electromigration limitations. As circuits are scaled to smaller design rules, the length of interconnects must be limited locally to keep pace with speed requirements of the logic. Relaxed design rules at higher levels of MLM must be used to connect macro or blocks of circuitry.

As the circuits require ever higher switching rates (dV/dt), the charging current density for constant capacitance per unit area (C/A) must increase (J = C/A x dV/dt). The current density set by the electromigration limit is now recognized as a fundamental limit to progress in the multilevel metal field.(6) Typical values for limiting operation are ~1E5 amps/cm^2. Even this value is obtained by alloying a few % Cu with Al; microstructure is important (fewer grain boundaries are desired). It is also believed that N_2 ambient in film sputter deposition plays a role; lower base pressure has been reported to give higher electromigration values. Progress requires development of materials and processes with higher current density capabilities.(7) Stress migration is also a serious fundamental limit.

It is the responsibility of manufacturing to carry out and insure the reliability of interconnects and dielectrics by performing generic stress and electromigration tests on the test vehicles or product. This includes a statistically significant number of devices, fully packaged with "over-value" testing in voltage, current, frequency, temperature and ambient. Failure rates with single digits in 1E9 device-hours at operating temperatures are required.(8)

3.0   ARCHITECTURE

MLM architecture(s) are impacted by: topography of the underlying front-end-of-line (FEOL) process sequence, design rules (issues such as stacked vias), lithography/etch capabilities, and integration capability of one MLM process with others. These issues will be discussed more in detail in section 6.0 on "Design For Manufacture". In the next section, we will review - by unit process - the architecture of a typical MLM structure. (See Figure 4.) There are simpler and more complex architectural approaches, but it would be unwieldy to try to treat them all in the scope of this chapter.(9)

Successful implementation of modern back-end-of-line (BEOL) processing is more and more interdependent on the front-end-of-line process architecture. It is no longer possible to engineer and manufacture the interconnect levels without careful interactive integration with the front end of the process sectors (isolation and device).

The field oxide is the first major opportunity to limit BEOL topography in the process flow.(10) Depending on the approach, the topography developed by CMOS twin tub processes can also be a factor. These designs not only set the depth of focus requirement for poly lithography, but establish the character of planarization required for poly-to-metal dielectric. Note that in Figure 4, the dielectric (poly-to-metal) is fully planarized. Alternatives include near fully recessed field oxide and the possible use of elevated source/drain films. These approaches undoubtedly help make lithography and MLM sequences more robust, but with more complex processes and greater cost.

SUBMICRON MULTILEVEL METAL STRUCTURE

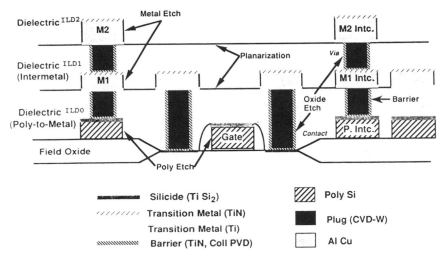

Figure 4. Schematic representation of submicron multilevel metal cross section structure. This architecture is typical of that used in current deep submicron development and manufacture.

Other process issues include the limited thermal budget to provide shallow junctions and compatibility with stress, adhesion, and electromigration. In the most advanced layouts, the critical design rules for contacts, metal and vias may be relaxed relative to the gate dimension, a testimony to the difficulty of achieving the same dimensions in an environment with more complex topography. We discuss implications on MLM for each of the key processes in the flow.

4.0   UNIT PROCESSES

The unit processes we will describe are:

o   Lithography
o   Gate (Poly Etch)
o   Contact (Silicide/Salicide)

o Poly-to-Metal or "First" Dielectric
o Contact/Via
   -        Dielectric Etch
   -        Barrier Films
   -        Fills
o Interconnects
   -        Deposition
   -        Aluminum Etch
o Interlevel or Intermetal Dielectric
o Planarization
o Passivation

In each process, we will review the intent and requirements of the unit process, material options and common material choices, a typical or conventional process sequence used to obtain the unit process, technical issues and future options, parameters for control in manufacturing, metrics (for 0.5-0.35μm technologies) and leading suppliers of equipment used to provide the unit processes.

4.1 Lithography

Lithography is not, strictly speaking, considered a multilevel or interconnect process, but will be discussed here as a unifying concept. Chapter 7 in this book contains much greater detail.

Briefly, lithography (as practiced for semiconductor manufacture) is a printing process that uses an energy source, a lens, and an exposure process with a pattern transferred from a "mask" (a transparent plate with opaque patterns) into "photoresist" (photosensitive film), which is then "developed" (exposed regions of resist are removed); the patterned film is etched, leaving behind a pattern of film that has protective resist. The resist is then stripped, which completes the lithography or pattern transfer.

Although we have not discussed lithography in the context of a MLM technology, it certainly is essential to achieve a system of interconnects. Most of the processes of the back-end-of-the-line are a combination of repeated processes. This repeated sequence is the fundamental concept for modern semiconductor processing. One deposits a film, patterns it (lithography and etch), deposits another and patterns it again. In order to achieve the multilevel system, one must have a capability to carry out lithography processes on poly, oxide and metal films.

The repeated and unifying sequence is:
o Film deposition: either metal or dielectric
      (smoothing or planarization may be added)
o Photolithography: defines film pattern
o Etch of film, transfer the litho pattern into the film
      (Resist strip and clean completes the operation)
Note: Some modules may skip a litho step, such as the blanket etch back of tungsten, which does not use lithography.

Photolithography - in total - is a complex series of processes. It may include: the spin-on of an adhesion promotor (HMDS); the application of an antireflection coating or "ARC" option (this is used especially for high reflectance films such as aluminum, where the optical exposure energy passing through the resist will vary across the high reflectance surface and not pattern well); the application of resist; a stabilization bake of resist; a precise wafer alignment of the litho level of interest to the previous level; exposure of the pattern; post expose bake (option); resist develop; etch image measurement (critical dimension-CD and alignment); and strip-clean.

By performing these operations repeatedly, one obtains the desired system of interconnects which are required to move signals from place to place on the chip (e.g. from logic to memory and vice versa) and ultimately off the chip. The reader is referred to Chapter 7 for more information about technical issues, parameters for control, metrics, and equipment suppliers.

4.2   Gate (Poly Etch)

In general, etch sources have evolved from plasma, to reactive ion or sputter etch, to reactive ion beam, to Electron-Cyclotron Resonance (ECR) and, more recently, to inductively coupled high density. Still others (e.g., helicons) are in development. The latter three sources are high density-low energy sources. They provide higher etch rates, better uniformity and process control and higher selectivity, low damage and independent control of ion energy, respectively. Lower pressure operation provides better CD and defect control in the latter three source types.

Still, technologists are concerned about gate damage during gate and contact etch. A longer range etch option is neutral beams with a possibility for much reduced damage. Operational options for advanced etching include lower wafer temperature, time modulation of gases and potentials, and simpler chemistries.

The details of etching gate structures depends on the nature of the media to be etched: materials may be undoped polycrystalline or amorphous silicon, doped polysilicon, and poly capped with a metal silicide. We will not address all the permutations which may be required for a particular product or technology, but will address the case of undoped poly to illustrate the general considerations.

(However, heavily doped n-type poly etches differently than undoped or p-type poly, which suggests that deep submicron technologies would etch undoped poly before doping gates to maintain similar gate width control. In addition, $TiSi_2$ can be formed on poly and in the source-drain regions after gate and sidewall spacer definition. If it is required to etch $TiSi_2$ on top of polysilicon a two step etch process can be used which calls for optimum etch of silicide metal first, which is not unlike that for aluminum ($BCl_3 + Cl_2$), followed by etch of the poly.)

The gate length is one of the most critically controlled parameters in all of semiconductor manufacture. The uniformity controls the distribution of values of source drain current drive of the device onstate, and the offstate leakage currents. It is important to etch the feature with great line width (critical dimension - CD) control. This is best done with an *anisotropic etch process* (one which etches vertically much faster than laterally, while isotropic etches have vertical and lateral rates that are equal). The etch profile may be tapered, but only if the risk of variability is low. Clearly the taper cannot be large because layout tolerances would have to be increased and the ability of the gate to mask the source-drain implant would be frustrated. It is best to seek ideal anisotropic etches with vertical profiles.

Since gates usually pass over the higher topography of isolating field oxides to serve as poly interconnects (see Figure 4), it is essential to etch this locally thicker poly until all the material is removed. If "over etching" is not carried out, one is left with "stringers" of silicon along the edges of the isolation field oxide. This typically requires an etch time that would remove twice the thickness of the poly, and therefore requires an overetch of ~100%. On the other hand, the poly etch has to stop on thin gate oxide (not shown in Figure 4). If the ratio of poly to oxide etch is not very high ("selectivity" of order or greater than 10:1), the etch will break through the oxide and then continue to etch the underlying bulk silicon. Generally, in order to achieve poly CD uniformity over the entire wafer, very high selectivity is required.

Gas mixtures containing Cl and/or Br are used for anisotropic etching of polysilicon. Although Cl_2, F_2, HF, $CFCl_3$, CF_2Cl_2, CCl_4, $CF_4 + O_2$, SF_6 and NF_3 are all etchants for poly, $Cl_2 + Ar$, $Cl_2 + BCl_3$, and HBr are commonly used. Loading effects (i.e., changes in etch rate with pattern density and wafer lot sizes in batch tools) are minimal with the Cl and Br containing gases. C-containing chemistries are to be avoided, in general, if alternative simple chemistries are available because of the possibility of the formation of polymer residues in the reactor. If an oxide (native or otherwise) is present on the surface, a two-step process which removes the oxide using, for example, C_2F_6, is used as an initiation step. Selectivity with respect to SiO_2 is good for both Cl and Br containing gases.

The basic chemical reaction of the Si/Cl etch system is:

[3] $$Si + 2Cl_2 \rightarrow SiCl_4$$

Typical technical metrics for poly etch are: etch rate of > 2000 Å/min; nonuniformity within wafer and wafer-to-wafer of < 2% 1s; selectivity to resist of > 4:1, to oxide > 50:1, profiles > 88°, residues (none), and CD control of 0.05μm for deep submicron line widths. Monitoring the etch rate and linewidths are key to good manufacturing capabilities. Scanning Electron Microscope (SEM) tools are used at resist and after etch to sample performance of the process and in cross section in development and occasionally in manufacture.

4.3   Contact (Silicide/Salicide)

Metal-silicon compound alloys (e.g., $TiSi_2$) are used in the source, drain, and gate regions to reduce device resistance. When the silicide is processed in a way which makes the silicide "self-aligned" to the gate, it is called "salicide." Requirements for these films are: low sheet and contact resistance (quality junctions) for contact and gate connections; robust reaction selectivity; suitable stress; refractory character for thermal processes that follow; a suitable dopant redistribution capability; and the absence of reaction with films deposited on top of this film (e.g.with barriers, TiN and interconnects, Al). Current materials of use are TiSi$_2$ and CoSi$_2$. Chapter 2 in this book gives more in depth discussions about silicides.

The processes used to create these films are sputter deposition followed by thermal reaction at temperatures that form the "monosilicide" of first phase; followed by a selective etch of the unreacted metal that is covering oxide; and then a final anneal at a somewhat higher temperature, for the formation of the final "disilicide" phase. Although details vary with the choice of material, the general concepts and flow are similar for Ti or Co.

Important issues are: surface microtopology and the physical and chemical state of the silicon surface prior to reaction.

The basic reactions (allowing for some nonstoichiometry) are:

[4] $$Co + Si(400 - 500°C) \rightarrow CoSi$$

[5] $$CoSi + Si(600 - 800°C) \rightarrow CoSi_2$$

The material of most common use is Ti, which reduces native oxides and provides good margin for contact resistance. Interface quality is an important issue.(11) There are, however, generic limitations for Ti such as the formation of dopant precipitation (12) and selectivity limitations on the sidewall oxide.(13) As a result, Co has received increasing attention as an alternate contact and salicide material; it is used in some manufacturing.

Manufacturing must control the cleanliness of the surface, the thickness of the deposited metal film, the integrity of the ambient of evaporation (base pressure and moisture content), the ambient after deposition and thermal reaction, the thermal ambient, temperature and time of the thermally selective process. Sheet resistance is one monitor, but Transmission Electron Microscope (TEM) analysis of the microstructure and the absence of shorts between source-drain and gate or yield/perfection of the salicide process must be monitored.

Typical manufacturing metrics for the Ti sputter chamber (module) and the platform that delivers wafers include a target life of 2000µm cumulative film deposition, wafers between clean > 5,000, and particles < 0.05/cm^2, > 0.3µm for the system, and within wafer uniformity of +/-1.5%(3s). Equipment for the deposition of Ti (for the formation of thermally reacted silicides) is in use in manufacture including the Varian M2000 and the Applied 5500. These tools are evolving to greater performance levels under continuous improvement programs.

4.4 Poly-to-Metal or ILD0

This dielectric (actually composed of a "film stack") is required to fill between closely spaced poly runners without voids, provide a reliable dielectric with insulating properties, and planarize ("smooth") to whatever degree possible. The planarization enables good focus during lithography. The application requires temperatures of about 700-850°C. Lower temperatures might not give "gettering" effects which simultaneously provide planarization achieved by viscous flow of PSG or BPSG, or dopant activation (depending on the process flow) yet higher temperatures may drive shallow junctions to unwanted depths.

The dielectric film stack used in typical CMOS application requires an undoped/doped layer structure to avoid unwanted doping of the junctions. Following the deposition of the doped layer, one anneals - densifying and flowing the doped film - to obtain smooth topology. Finally an undoped layer is deposited to seal the surface from ambient and to prepare the surface for standard oxide lithography. These four operations typically require several tools and wafer transfers between tools.

The industry standard doped dielectric is "BPSG," borophosphosilicate glass with typically 4-5% B and P by weight in an SiO_2 matrix. An older alternate is "PSG," phosphorus-silicate glass usually bounded by 5.5% P, above which the surface becomes highly hygroscopic.(14) Higher temperatures are required to flow the PSG, which drove the boron addition. The approaches to deposit these films are multiple: Atmospheric CVD, subatmospheric CVD, low pressure CVD (LPCVD), and plasma enhanced CVD (PECVD). Chemistries are multiple: silane, TEOS, TEOS/O_3, diethylsilane (DES), and others. The choices are narrowed when cost of process and safety are taken as factors. General chemistry for the deposition of oxide is:

$$[6] \qquad SiH_4 + O_2 \rightarrow SiO_2 + H_2 + \text{other products}$$

The future will require even lower flow/reflow temperatures with better gap-fill and planarization. There are likely to be advances in new material. BPSGG, or borophosphogermansilicate, is a candidate.(15) It is also possible that future technologies will use physical or chemical-mechanical planarization to enable the lithography focus capability.

Manufacturing must control the net and individual results of the film stack. The deposition rates, thickness, uniformity, stress, doping levels, temperature/time/ambient of the flow, topology, stability, particles, gap fill at given design rules, breakdown, dielectric constant, shrinkage and mobile ion content must all be monitored and controlled. Example metrics for the first dielectric are void-free fill at the design rule spacing and poly aspect ratio of the technology, within wafer uniformity and wafer to wafer uniformity of better than 5% (3s), a total film thickness of 1.0 to 3.0μm, dielectric constant < 4.1, and particles < 0.1/cm^2.

Tools commonly used for CVD dielectric deposition are provided by AMAT 5000, Novellus Concept-1 and Watkins Johnson. Flows are carried out in standard furnaces, such as those provided by SVG-Thermco. The recently developed Lam Integrity provides a single tool "unit process" for this application. The Integrity uses CVD deposition and high temperature flow processes in the same chamber.

4.5 Contact/Via - Dielectric Etch

The ILD0 dielectric etch is focused on etch technology for a material that is substantially "oxide." (The material may include undoped oxides and doped oxides as described above, and in addition some applications have nitride films in the stack.) The chemistries commonly used for etching oxides are: C_2F_6, CHF_3, $CF_4 + O_2$, $CF_4 + H_2$, and $SiCl_4$. Sources/systems are usually described in context of ion/electron plasma source type, frequency, chemistry, mix of physical and chemical effects that impact key parameters etch rate, anisotropy and selectivity.(16) Great anisotropy is required, with aspect ratios of order 2:1.

A basic reaction typically involves:

$$[7] \qquad C_2F_6 + \text{ion energy} + SiO_2 \rightarrow SiF_4 + O_2 + \text{other products}$$

SEM measurements are made to insure and confirm that contacts are open, and that they are open across the wafer.

Metrics are: Etch rate: > 5000Å/min, non-uniformity of etch rate, CD control wafer to wafer and across wafer < 2%, 1s, selectivity to resist: > 5:1, to poly and silicide > 25:1, profile > 88°. Selectivity to nitride is key for certain front end and transistor application, but may not impact MLM applications. Microloading and particle requirements are key for advanced MLM applications. Key concerns are the formation of polymers from the CF chemistry and the need for high selectivity at the bottom of the contact, which makes the use of multi-step processes common. Suppliers of oxide etch equipment are Drytek, AMAT, Lam, and Tegal.

4.6   Contact/Via - Barrier Films

After low contact resistance is achieved with stable silicide contacts, the first dielectric applied and contact windows etched, the contact is "ready" for interconnect deposition. Unfortunately, for deep submicron contact sizes, there is no simple way to make reliable, low resistance contact by directly depositing aluminum into the contact. Aluminum sputter-deposited directly into the contact leaves a "bread-loaf" thinning on the sidewalls and there is high risk of contact failure.

If the aluminum were flowed or reflowed using fast (laser) melting or deposited by CVD techniques, the thin sidewalls could be avoided, but it could react with the silicide and penetrate the shallow junction. Alternately, if a tungsten CVD film which fills the contact nicely is deposited, fluorine initiated effects (from WF_6 chemistry) could damage the junction, giving "worm holes" and cause leaky junctions.(17)

A requirement for submicron technology is the use of a barrier layer which will conformally cover the contact to avoid Al and F penetrations. The barrier film must have good adherence to underlying silicide, dielectric, and overlying interconnect metal, if it is to have good thermal and electrical conductance (for its thickness), and not react with the under- or over-layer. "Barriers" may have a dual use: to act as a diffusion barrier for interconnect metals, plugs and chemical species present during deposition of the interconnect and plugs; and to act as adhesion layers for interconnect metals or plugs that are to be applied to the oxide surfaces, which otherwise would not stick to the surfaces.

Candidates for barrier/adhesion layers are TiN, TiW, TiB_2, TiC, and Ti_2N.(18) TiN and TiW have been widely studied, and TiN is most commonly used. Rutherford scattering results show that the TiN is substantially stable at 450°C, but Ti moves to the surface at 500°C in one hour.(19) Above this temperature, TiN and Al react thermodynamically.(20) The films are commonly sputtered and thermally reacted in a nitrogen environment.

The barrier may be deposited by physical vapor deposition (PVD), reactive sputtering, or by CVD techniques. The CVD and collimated PVD give better coverage. (See Figure 5.) Advanced chemistries using MOCVD may lower deposition temperatures to the point where the CVD barrier film can be used for contacts and via applications.(21) In either case, the clean prior to deposition and contact resistance must be engineered carefully. Etching of the barrier and interconnect metal (Al or W) must be compatible. General chemistry is:

[8] $$MO(Ti) + N_2 \rightarrow TiN + \text{Organics}$$

Typically required metrics for a TiN barrier are: 200 μohm-cm resistivity; sheet resistivity uniformity of < 10% (3s); stress < 1E10 dynes/cm^2; and bottom coverage in the contact of > 20% of nominal film thickness with a contact aspect-ratio specified by the technology (a height-to-width aspect ratio of 2.5 is typical). Good step coverage can be addressed by using collimated sputter sources or by using CVD technologies.

The ultimate measure of the barrier effectiveness is low leakage currents of underlying junctions and low contact resistance measured after metal 1 processing. In-line control can be monitored by measuring the resistivity and uniformity of blanket depositions with an automated 4-point probe after deposition and anneals.

Equipment commonly used for barrier deposition are the Varian M2000, Applied Materials 5500, and Novellus Concept 2. No one knows how far sputter technology will extend. Advanced CVD equipment may emerge in the near future that allows conformal coverage down to very small contact sizes and high aspect ratios.

Note: Diffusion barriers may possibly be avoided as they are expensive and if not controlled can be troublesome. One approach is development of an epitaxial silicide or epitaxial metal that acts as an intrinsic barrier because there is no grain boundary diffusion.(22,23) These novel approaches may require much development before they are used in manufacture and may not, in fact, find general application.

BARRIER COVERAGE FOR TiN
(SCHEMATIC)

PVD   CVD

UNCOLLIMATED ⟶ COLLIMATED

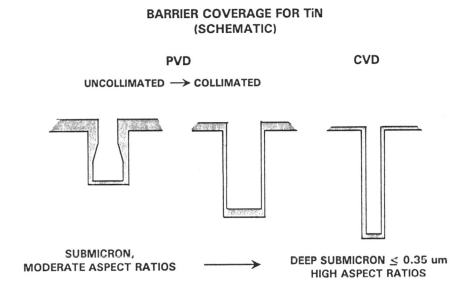

SUBMICRON,   ⟶   DEEP SUBMICRON ≤ 0.35 um
MODERATE ASPECT RATIOS   HIGH ASPECT RATIOS

Figure 5. Illustration of barrier coverage of contacts, typically for TiN. The progression is from PVD uncollimated to PVD collimated to CVD.

4.7   Contact/Via - Fills

Although a conformal barrier contact (e.g. TiN) will provide electrical continuity, direct deposition of aluminum by sputter technology will lead to variable reliability. This difficulty is only made worse for smaller feature sizes, hence, the need for filling the contacts.

Tungsten has been used to fill contact holes using CVD technology. This provides the capability to fill large aspect-very small contacts. Standard chemistry uses WF_6 precursor with H_2. The CVD process was improved with the use of silane to avoid worm hole generation in underlying silicon. The blanket W process with etchback is used for the unit process for contact fills.

[9] $$WF_6 + H_2 + SiH_4 \rightarrow W + H_2 + SiH_4 + F \text{ products}$$

A selective tungsten process is still not used in widespread manufacture. Nonselective nucleation of W on oxide has been an issue for a long time. W, being highly refractory, has a very high electromigration tolerance and does not react with Al at ~400°C.

Some IC's use blanket W as a plug and first metal interconnect, especially in shorter distance, repetitive memory cells. Longer distance interconnects are done with second level, high conducting Al.(24)

Key specifications are thickness uniformity at 3000 to 6000Å with uniformity of ~6% (3s), stress < 1.6E 10 - tensile, and no adhesion loss to a suitably prepared surface and through the process. (Tungsten is not thermodynamically stable on an oxide surface and TiN is required as an adhesion film for tungsten adherence.) Small particles at low density levels are particularly difficult to measure because the rough surface of tungsten provides a high background of high scattered light.

Like the barrier process, for "contact yield" testing for open contacts at M1 electrical test can be carried out. Chains of 100's to 100,000's of contacts (or vias at M2) can be measured for continuity. In the development or prove-in stages, different sizes of contacts are explored - running from larger to smaller than the design rule - which allows the capability to be established.

Unfortunately, in-line monitoring of the contact resistance for submicron sized contacts is very difficult. High resolution optical microscope and the use of SEM measurements of the contact after resist and etch strip help assure open contact control. They also help to assure uniform barrier and plug processes after etch back of plugs.

Equipment for W film deposition includes the Genus 372, Applied 5000 and Novellus Concept 1 and 2. The manufacturing performance of the batch Genus tool was greatly enhanced through an Equipment Improvement Program. (See data in Table I.) The AMAT 5000 provides a unit process capability, wherein deposition and etch back is provided with serial clustering modules. The Novellus Concept 1 provides a continuous batch process, wherein wafers are placed on a "Lazy Susan" reactor plate to serve throughput and uniformity, and the Concept 2 offers integrated etch back.

TABLE I: GENUS BATCH CHEMICAL VAPOR DEPOSITION
BLANKET TUNGSTEN TOOL ACCOMPLISHMENTS

SEMATECH EIP Accomplishments	BASELINE (1989)	CURRENT (1991)
Uniformity (1 sigma)	6%	2.5%
Utilization	40%	82%
Relative Cost of Ownership	1	0.3
Chemical Utilization	14%	50%
Net Throughput (wafers/hour)	9	30

4.8   Interconnects - Deposition

Interconnect lines represent the "R" of the "RC" in the delay considerations mentioned in the introduction of Section 2.0. It is easy to see why aluminum has become the standard for silicon technology.(25) Al has excellent bonding to SiO_2, relatively low resistivity, self-limiting, self-passivating (with Al_2O_3) surfaces and there is considerable experience etching with BCl_3 and Cl chemistries. The bonding means that no underlying TiN adhesion layer is necessary; the self-limiting coating means that no cladding technology has to be applied. Al can be deposited simply by sputter-physical deposition (PVD) systems, which are rather mature.

Although there are a few disadvantages to Al (Si-Al interactions have been overcome with barrier films, stress and hillock formation), these effects would not by themselves lead to the adoption of new materials and new technology.

The thickness of metal interconnects is measured with a mechanical device that steps from and etched region to an unetched region of the wafer. The step height is used to determine the etch rate and uniformity of the metal. Uniformity is also monitored by measurements of the resistivity over the wafer's area. Dimensional control is achieved by measuring the line width after etch with SEM. Line width control is crucial to achieving proper RC time constant performance.

Typical metrics for aluminum deposition are: resistivity of less than 3.5 uohm-cm, uniformity of < 5% (3s), bottom or step coverage with an aspect ratio of < 1.0 is > 30% of nominal film thickness, and less than 0.05 particles/cm^2 down to 0.3µm. Leading suppliers for aluminum deposition are Applied Materials 5500, Varian M2000 and Sputtered Films.

A mature advanced CVD aluminum process (23) would extend the use of Al a little more but, ultimately, would not overcome fundamental resistivity limits. The appeal of an advanced CVD process is the promise of improvement of electromigration, with fill capability and potential elimination of TiN. Any of these attributes would give extension to the use of Al. Similarly, a manufacturable "flowed" Al process could provide extension of the current metallurgy. In the flowed case, barrier technology must be especially robust. (See Figure 6.(26))

Cu is the material of choice for interconnect beyond Al. However, there is no easy way to etch the material; there is no established way to get adherence to some (as of yet undefined compatible) low dielectric constant materials; there is no established surface passivation technology. Fundamental electromigration data is lacking. Challenges and opportunities abound.(27, 28)

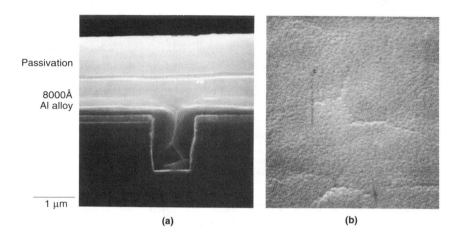

Passivation

8000Å
Al alloy

1 µm

(a)   (b)

Figure 6. Enhanced aluminum step coverage using Multi-step metallization processes. SEM cross sections of planarized Al alloy using 180 sec, 540°C reflow after deposition. (After Kim, et al. Ref 26)

4.9   Interconnects - Aluminum Etch

Aluminum metallization is composed of a top cap film to provide antireflection during patterning (usually TiN) and to limit growth of hillocks, the Al alloy film proper (usually Al-Cu, 0.5-4% or Al-Cu-Si if the metal is to come in contact with the silicon substrate directly as is practiced for technologies above 1um) and an under film (usually Ti) to provide adherence and low contact resistance to materials such as tungsten plugs.

The basic chemistry for etching Al is the formation of volatile chlorides; either BCl_3 or CCl_3 have been used,(29) typically:

[10] $Al + BCl_3 + Cl_2 \rightarrow AlCl_3 +$ by products

This process would be isotropic unless considerable RIE energy were applied. In addition, some believe that there are recombinant sidewall passivation chemistries at work that helps produce an anisotropic etching effect.

With BCl_3, polymer residues do not form. With CCl_3, polymer reactions take place at the reactor walls and, in the presence of residual water, capture moisture. When the reactor is started up, a burst of H_2O comes from the walls, resulting in start-up variations in the process. Processes are migrating to BCl_3 with systems that contain very low levels of residual H_2O.

The basic chemistry with proper electrical bias will provide substantially anisotropic etching for the entire film stack (Ti/Al alloy/TiN), although there may be modifications of bias and chemistries as part of an overall recipe.

It is particularly difficult to etch the Cu in Al alloy because the CuCl is not volatile. RIE mechanisms must be applied to sputter the residual Cu away. Selectivity to SiO_2 is good, but selectivity to resist is poor - although just tolerable - while selectivity to poly is basically incompatible since Cl etches Si as well. This has an impact on the design rules requiring "nailheads" of overlapping Al contacts to poly and silicon.

Post-etch corrosion is an issue, because any moisture will create HCl from the residual Cl, and HCl will attack Al. One practice is the use of an O_2 plasma to strip the PR and passivate the Al to form stable Al_2O_3 on the etched sidewalls.

Typical metrics are: 8500Å/min etch rate, wafer to wafer and within wafer non-uniformity of 2%, 1s, selectivity to resist > 3:1, to oxide > 10:1, profiles > 88°, 0.05µm CD control and no corrosion. Equipment is provided by Lam, AMAT, Drytek and Tegal.

4.10   Interlevel or Intermetal Dielectric

The application here centers on the need to fill the narrow spaces between metal interconnects with a capability for aspect ratios that are greater than one. Interlevel dielectric differs from ILD0 in that it must conform to a low temperature deposition of less than about 450°C to be compatible with aluminum used as first metal. Processes which are standard consist of CVD deposition and etchback followed, in turn, by deposition again. Although the gap fill capability is technically sound, there is a trade-off in throughput - the higher the aspect ratio of the gap, the lower the throughput. Technologies that have etchback (sputtering depositions) running concurrently with deposition appear to have a high throughput at small gap spacing. These systems use ECR depositions with silane chemistry. Even higher deposition rates can be obtained using advanced chemistry and a concurrent chemical doping appears to offer the opportunity for using the same tool for first dielectric.(30)

4.11   Planarization

Planarization of low temperature dielectrics can be carried out with etchback approaches; however, this does not offer global planarization. Global planarization is desired for use with high NA wide field lenses used in modern exposure tools. A standard expectation is that the surface of the dielectric is globally planarized over a 20mm square, the field of a modern exposure tool. There is no need to achieve global planarization over an entire wafer because modern steppers have local levelling capability. A candidate for this requirement is the so-called "Chemical Mechanical Polishing" technology.(31) (See Figure

7.) In this technology, presumably the chemical action of the slurry reacts with the surface providing a ~100Å thick H-rich surface layer.(32) One may speculate that a hydration mechanism is at work wherein -H- bridges itself between oxygen atoms and creates a network which is easily dissolved. The mechanical action sweeps away loose molecules that are -H- rich material with each revolution. Fundamental modeling of this technology is still incomplete, but phenomenological knowledge is rapidly emerging.(33)

Typical metrics for chemical mechanical planarization are: planarity of better than +/- 1200Å over 1cm^2, uniformity across the wafer and from wafer to wafer of less than 5%, 1s and particles less than 0.1/cm^2 at 0.3µm. Suppliers include Westech Systems, Strausbaugh and Bridgestone (NY).

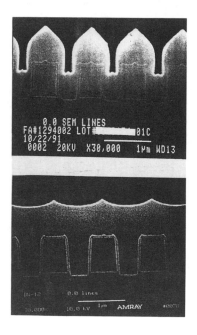
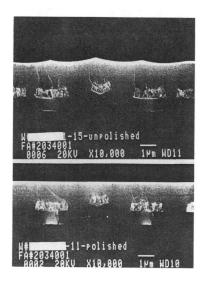

Figure 7. Intermetal dielectric fill between narrow interconnects. On the left photomicrographs, the effect of the ratio of etch to deposition rate - 10% (upper) and 35% (lower) - is apparent. On the lower right, the dielectric has been globally planarized using Chemical Mechanical Planarization, CMP. (After Webb, et al., ref 31.)

4.12   Passivation

Once a circuit is integrated with the last metal level, we are ready for passivating dielectric. The need here is to cover aluminum metal with a dielectric that provides scratch resistance, protects from water/moisture and sodium penetration. The material best suited for this application is silicon nitride. The fact that it must be applied to aluminum and the contact system below, without impacting its physical properties, requires the temperature of deposition to be less than 400°C. Thus, one uses plasma assisted Si_3N_4 deposition, typically between 300-350°C.

The basic reaction involves:

[11] $$SiH_4 + NH_3 \text{ or } N_2 \rightarrow SiN_xH_y + H_2$$

5.0   KEY DESIGN RULES

Design rules are the dimensions of (and between) features on an integrated circuit. They comprise the layout rules by which a circuit is designed. The rules are divided into two groups: the larger dimensions and smaller dimensions (so-called critical design rules or "CD's"). Of course, it is the design rules that scale from generation to generation, and also within a generation, from prototype to volume manufacture.(34)

The critical design rules are most difficult to satisfy. For our purposes we will discuss just those critical rules that impact the ability to meet MLM requirements. When poly line widths were ~2 microns in the early 80's, it was common for all the critical design rules to be essentially the same size: poly, minimum width oxide (tub ties), contact, and metal.

As design rules progressed to deep submicron, the poly size continues to set the standard as it limits transistor switching speed, but contact and metal are often relaxed to somewhat larger sizes. This is because lithography in difficult topography is always more difficult to implement and because the impact of reaching critical contact sizes may not be as important as for poly or metal. It may even be considered a useful trade-off to have larger contact sizes to allow lower contact resistance. However, at some point it will become essential to reach a minimum size capability. Table II shows typical critical rules for deep submicron MLM features:

TABLE II: SELECTED DESIGN RULES

	0.5 µm	0.35 µm	0.25 µm
poly	0.5	0.35	0.25
contact	0.7	0.42	0.30
metal-1	0.6	0.40	0.30
M1 space	0.6	0.45	0.35
metal-2	0.8	0.6	0.40

These rules are not canonical, and some designers may have other values. In this case, the relaxed M1 spacing gives wider interelectrode spacing (lower capacitance) and easier fill capability.

6.0 DESIGN FOR MANUFACTURE

The overriding consideration for design for manufacture is the reduction in cost to manufacture that is afforded by considering **consolidation in the process integration** that allows the same product function to be manufactured, but with fewer steps and thus, lower cost. This consolidation approach goes beyond the obvious initiatives of reducing operations like metrology and cleaning -two "non-value added steps" - to the bare necessities.

As the design of integrated circuits becomes more complex, the number of steps to complete a product will rise. The device architecture and the number of levels of interconnect alone confirm this trend. If no proactive manufacturing design initiatives are undertaken, operational costs and manufacturing costs will balloon.(35)

The number of steps for a typical CMOS process will rise from about 200 steps for two-level metal and single doped poly with n+ lightly doped drains to more than 300 for four-level metal and double doped poly with n+ and p+ lightly doped drains and recessed/ low encroachment field oxides. If these operations were pursued with stand-alone tools, the footprint of the tool set would also escalate. Factory costs can be contained by the use of novel consolidation of process architecture, and then, in addition, by using cluster tools to reduce the number of operational steps.

A prime example of this would be the use of MeV implantation technology to reduce the number of lithography operations.(36) It also reduces the area needed for advanced isolation and reduces the need for expensive epi. The barriers to accepting this kind of thinking include the apparent higher cost of individual equipment (e.g., the apparent higher cost of an MeV implanter vs. a conventional implanter).

Another example would be the use of staircase sidewall oxides to form lightly doped source and drain regions with fewer operations. In this case, four lithographic operations are reduced to two and four separate implant operations reduced to two operations with a high/low two energy implants carried out in one implant operation.(37)

These are documented cases for streamlining the process flow in the front-end-of-line. In the area of multilevel metal, an example would be the use of aluminum which is flowed into contacts that have suitable diffusion barriers.(38) Such an operation would avoid the need to carry out contact fill and etch back operations, and obviously reduce cost by eliminating two process operations.

This is an opportunity that is not easily comprehended by manufacturers who operate only by "freezing" the process flow and then (put blinders on) focus on cost reduction of individual, compartmentalized steps. The integration engineer is the first partner in using the concept of process flow consolidation to achieve optimum "design for manufacture." It is very important that the integration engineer be credible. The ingredients for success include a world class competitive spirit: diligence and persistence. Like so many elements in manufacturing today, this is a continuous process.

6.1 Metrics for Design for Manufacturing

Once a consolidated process flow (a cost effective process flow) is engineered, one must introduce the concept that everything in the process will work with suitable margin. An example of this is the use of the "Process Capability Index" or C_{pk}/C_p methodology and the related 6-sigma approach pioneered by Motorola. The C_{pk} and C_p are manufacturing control indices that are used to measure the stability of a parameter under statistical changes.(39) The index is calculated by reference to control limits (CL) that would cause the product to fail.

It is assumed that a characteristic parameter (e.g. the overlay capability of a stepper or the dimensional control of an etcher, has a mean value (x bar), and a standard distribution or sigma (σ) following a normal or standard distribution. The distribution is established by repeated measurements, with a gauge capability which specifies how precise and with what tolerance the measurement can be made itself. (See Figure 8.) For purposes of our initial discussion we assume that the gauge is infinitely precise and not a factor. C_{pk} is defined as:

[12] $$C_{pk} = (CL - xbar)/3\sigma$$

A $C_{pk} > 1.3$ sets a minimum standard for good practice or good quality, it sets CL - Xbar = 4σ, but it represents a statistical condition that the parameter will be out spec or exceed the control limit for 1 case in about a 1000. That is, reference to a table of statistical values for normal distributions predicts a probability of only 3-nines. The term: C_p is simply a special case that assumes xbar is zero, i.e., we have a parameter that is centered, and has "no bias." Again, the relationships between parameters are diagrammed in Figure 8.

If we change the standard - as Motorola has done for processes and call for a 6-sigma process - the C_p becomes = (CL - xbar) / 3σ > or ~2. This states that only one in a million or so values of parameters will be out of spec. It is common practice to assume that xbar is typically 1.5σ , in this case a 6σ control limit gives a $C_{pk}=1.5$.

Other institutions have adopted a 6-sigma control quality strategy. NASA has practiced a 6-sigma quality control for space vehicle systems. The U.S. commercial aviation carriers have operated with a probability that a commercial flight makes an unsuccessful take-off (once check out is complete and the crew is committed) of only about one in a million. In contrast, the IRS is known to give incorrect information on the phone to taxpayers about 50% of the time. They have practiced a "one-sigma" process; this is recognized and improvements are underway. These examples should give the reader a better idea of the C_{pk} index and its use.

Semiconductor manufacturing requires the highest quality in order to be competitive, and high C_{pk} values insure high process and product yields. There is always tension between C_{pk} capability and the ability to introduce new technology with smaller dimensions. This is why, in particular, it is important to introduce the C_{pk} methodology into engineering practices. The vehicle most commonly used is the Statistical Process Control or "SPC" method, examples of which will be discussed below.

When properly practiced, we get every process to yield with a high probability so that the electrical circuit performance is well controlled and also tight. For example, the on-currents of an MOS transistor (ID-sat) are dependent on the combined control of gate oxide thickness, doping and its distribution in the silicon, and gate length (which in turn depends on lithographic exposure conditions, resist development, etch control, deposition of sidewall oxides and their etch back). In summary, the electrical performance depends on the combined success at control of multi-parameters and processes. It is essential to design the process for success and "for manufacture," as we say, and this means engineered with great tolerances.

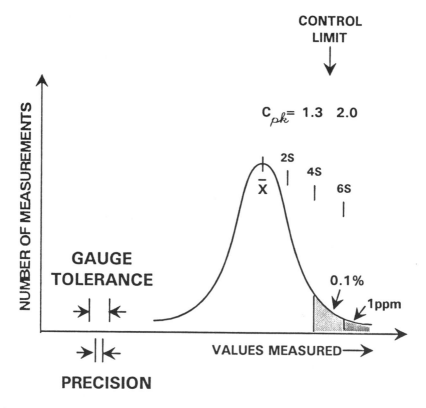

Figure 8. Illustration of relationship between C_p, a manufacturing control index and a normal distribution. A C_p of 1.3 and 2.0 correspond to a 4 and 6-sigma process, respectively.

6.2   MLM Examples of DFM

The *junction contact penetration issue* is key to the manufacturing engineer for the back-end-of-line. This is a temperature-time technical issue. Sufficient thermal budget must be provided to anneal implanted damage to obtain high quality, low leakage junctions, and yet not overdiffuse the junctions so that shorts between source and drain are obtained.(40)

The thermal budget may include steps from the silicide formation and the ILD0 thermal densification. The particular sequence and architecture is usually closely held and considered company proprietary, and may even be customized to a particular technology. In any case, it is the responsibility of the manufacturing engineers to co-characterize the process window with integration engineers for robustness of the allowed temperature-time window.

One of the possible yield limitations comes from the potential for *source and drain to gate shorting*. This can occur if the processes for removing silicide material between the source/drain regions and the gate are not well developed. The development engineers must establish the process windows for zero defects with regard to these shorts.

The *barrier process* (e.g., TiN) must be designed with thickness, diffusion - penetration, and contact resistance properties in mind. Design of experiments are often carried out to define a robust barrier process. This is a rather difficult characterization because of the multi-variable nature and the fact that a complex integrated process must be carried out to gain all the information. Hence, fundamental information as well as engineering results need to be obtained for success.

Open contacts are essential for high-performing, high-yield integrated processes. It is essential that lithography, etch, and dry clean processes be suitably developed to provide clean, critically dimensioned contacts. It is traditional to "relax" the size of the contacts (relative to the gate length dimension) to achieve early success at integration, but it is also a parameter that can be negotiated with designers to obtain lower contact resistance and higher device drive.

The *etch selectivities* (oxide etch rate relative to the silicon etch and silicide etch rate, and the relative rates for oxide and nitride) are a related element for manufacturing consideration. High selectivities of order > 50:1 are essential to provide robust uniformity goals on large wafers. One can - to some extent - trade etch rate uniformity specifications against selectivity if the selectivity is high enough. For example, etching contacts through oxide material of different heights (down to poly and device silicon) can result in essentially the same contact size if selectivity is high enough.

Metal continuity is key to high yield. This, in turn, depends on the planarization and the metal deposition capabilities. Metal sputter source depositions that provide a broad range of incident angles will fill any cusps (see Figure 9 (40)) that otherwise may be the root cause of opens. Alternately void-free dielectric fill and a high degree of planarization allows for ideal metal continuity. Planarization of first and higher level dielectrics is desirable. The metal integrity must be tested for opens and shorts to insure high yield.

Planarization not only enables high yield for metal layers, it enables lithography to expose uniformly across large areas. The limited depth of focus of modern exposure tools is perhaps only 0.5μm, so it is necessary to planarize at least over large chip areas.

Design with contact borders enables subsequent levels to completely cover contact landing areas. This is good practice design for manufacture. The *microstructure of the films* is key to controlling interdiffusion and electromigration.

Ultimately, there is a relationship between the device design rules and the equipment tolerance or C_{pk} values. If, for example, the critical dimension uniformity control of the poly is "X%" then this must be added to the line and spacing design consideration of the poly interconnect runners. This may result in spacings being larger than lines.

There are other issues in the business of "Design for Manufacture," but these examples are typical of the considerations one needs to address.

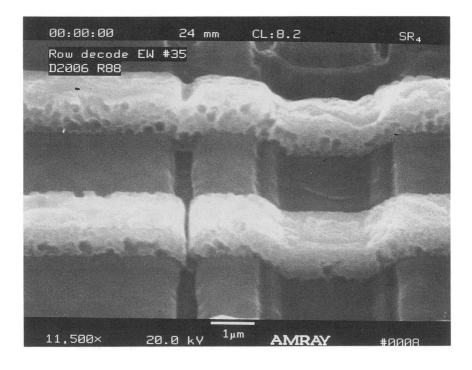

Figure 9. SEM of sputter deposited aluminum running transverse to gaps in dielectrics. This is a test structure that illustrates the problems associated with metal coverage of gap-like topology.

7.0   EQUIPMENT INTEGRATION

Future semiconductor manufacture imposes demanding requirements to minimize contamination and maintain film interface quality. There will also be a drive to reduce costs by reducing the number of operational steps and increase throughput. Processes driven by yield considerations are metal stacks, poly-metal dielectric (cluster tools exist today) and, in the future perhaps, salicide contacts. We address the requirements to achieve a cluster tool environment in the future factory.

7.1   Background

There are many arrangements and definitions of integrated tools.(41,42) For the purposes of this chapter, an integrated tool is a system in which process modules (different or the same) are linked together by a wafer transfer mechanism. Cluster tools emerged from early multi-process tools.(43) One can run two or more different processes served by a common transfer mechanism or wafer handler controlled by a (micro)processor.(44) Examples of applications are deposition or etch of different films and creating or removing a film stack. One process configuration for clustering is the serial linkage of two or more process operations; *i.e., multi-processes with different chambers.* General advantages are maintenance of critical interfaces, lower defect density, reduced handling (possibly increased throughput) and lower overall cost of ownership when compared to several serial stand-alone processes.

In another variation, *multiprocess in the same process chamber*, several different processes are run that require different chemistry. This approach promises even further economy of handling, but requires compatible chemistries and careful purge process recipes. This has been popular in research environments.

Alternately, one can run the same process in multiple (four in the case of the Quad (44)) single wafer process chambers. These batch loaded tools are *multichamber, using the same process configuration.* They have the advantage of single wafer uniformity and better process control and may have moderate cost compared to batch stand-alone tools.

The varied process configurations are:
o   Multiprocess using different chambers;
o   Multiprocess in the same chamber.
o   Multichamber using the same process.

The high interest in cluster tools is in response for increased *yields* in the face of smaller feature sizes with more stringent requirements on particulates and contamination. There are many advocates of the cluster approach, so it is somewhat surprising that there is not more widespread acceptance of the approach by the manufacturing community. Although the vision of serial processing in a self-contained cluster environment is promising, limitations are quickly recognized and there is considerable tension in the initiative, especially regarding the use in volume manufacturing.(41)

For example, the *reliability* of different process chambers must be higher than the redundant multi-process tool case in order to obtain a high equipment utilization. There is the issue of the residence time in each process chamber being balanced with the rate limiting process. In practice, exact balance or commensurate ratioing is rarely possible, so efficiencies are lost just by derating the faster chamber in trying to balance throughput rate.

Throughput is reduced in proportion to adding serial processes but with improved cycle time. Cluster tool configurations will naturally reduce cycle time, which is great for prototyping, but this may be obtained at reduced throughput. Hence, careful analysis and design of insertion to a manufacturing environment is needed. There are many variations of the cluster tool, and it is becoming popular to use mini-batch operations attached as a clustered process chamber.

The *cost* of the tool (and its development) require detailed modeling, analysis and comprehension.

There are many other challenges and requirements:
o   The use of clusters as a testbed for:
        -contamination control,
        -in situ metrology,
        -CIM, through the commonality of control architecture.(41)
o   Rationalization between standards and proprietary platforms; the practicality of a standards vehicle to obtain "best in class" modules.
o   Process consolidation as a driver for operational control of cost as future generations go to higher levels of process complexity.

7.2   Current Tool Status

The emergence of cluster tools (45,46,47,48) suggests that we classify the tools into groups according to their existing architectural configurations. All these tools accept batch cassettes of wafers. Example tools of the groups are described below.

Current Tool Status - Drytek "Quad". The Drytek "Quad" has wafers fed from their cassette in a vacuum-load-locked area into four single wafer chambers.(44) A simple application is to etch the TiN anti-reflection coating, Al alloy, and the Ti glue layer in one chamber. This is the subtractive cluster process for the metal film stack. This is an example of multi-process in the same chamber with multi-chamber arrangements. Some applications in volume manufacture involve three common chambers (multichamber-same process) with the same process, with a fourth that serially strips resist and passivates the stack. Common processes make the tool easier to maintain.

Applied Materials has large market share with the Precision 5000 tool with similar mechanical architecture but supports both CVD deposition and etch back in a single wafer serial cluster.(43) This is a multi-process with different chambers arrangement. However, both Drytek and Applied tools are four chamber mechanical architecture. (See Figure 10a.)

Drytek

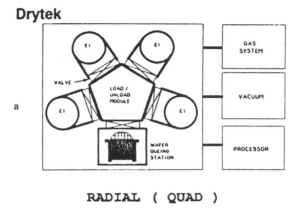

RADIAL (QUAD)

Varian 2000

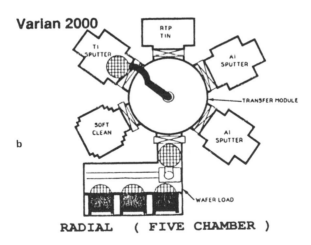

RADIAL (FIVE CHAMBER)

Figure 10. Schematic of two commercial cluster tools, the Drytek Quad (used for etch applications) and the Varian M2000 (used for metal deposition). The Drytek illustrates parallel increased throughput while the Varian is designed for serial processes that provide interface control.

Current Tool Status - Varian M2000. A cluster tool that offers modules that insure the quality of the interfaces during deposition is the Varian

M2000.(49,50) Wafers move serially from chamber-to-chamber in a vacuum environment that insures contamination-free interfaces. The tool is a serial, multi-processing cluster process. Wafers are robotically transported a single wafer at a time through a series of up to five different process modules. It has sputter-etch/clean, metal sputter deposition chambers and recently added an RTP chamber that allows thermally reacted TiN or silicide applications. These tools are radial with up to five process chambers. (See Figure 10b.)

   Current Tool Status - AMAT 5500.   Another cluster tool that offers chambers to guarantee the quality of the interfaces during deposition is the Applied Materials 5500.(51) The tool supports serial, multi-processing cluster processes. Its architecture is one of the first to link a cluster to a cluster. It has a dual radial arrangement that dedicates one handler platform to relatively low pressure PVD applications and the other moderate pressure central handler to serve CVD process modules. The low pressure cluster is for serial processing for the production of the interconnect metal film stacks. (See Figure 11a.) The combination addresses applications for plug contact and via fills.

   Current Tool Status - Rapro 3000.   The Rapro 3000 is a radial/ linear configuration that is chainable in an x and y factory layout scheme. The core central handler is basically radial, but reaches to a set of platforms (52) and then another robot transfer can place the wafers into one of three multi-process chambers. Epi and RTP process modules are available,(52) allowing single wafer gate stacks to be integrated. There is a high degree of compliance to industry standard SEMI/Modular Equipment Standards Committee (MESC) mechanical and controlling interfaces. Brooks offers a platform that adheres to SEMI/MESC standards.   These tools, based on radial architecture, are distinguished by focusing on open architecture.(53)

   Current Tool Status - Novellus Concept I.   The Novellus C-I is a "rotating linear configuration," wherein wafers are robotically fed from their cassette in a load-locked area onto a rotating plate (Lazy Susan) with mini-batch capacity. Vacuum process conditions are established and a CVD process is then run. Novellus has emphasized the mini-batch as an approach to high throughput and good uniformity. Lam has entered the field with a similar architecture ("Integrity" tool), but has vacuum cassette environment and process capability that is used for higher temperature applications, e.g., CVD SiO_2-BPSG. These tools do not fit our definition of a true cluster tool, although *wafers are clustered together* on the rotating plate. Lam refers to this as "intratool clustering". An alternate radial distribution of orifices allows for microlayering. We have also grouped a few sequential processing tools having the chambers in this category since they do not fit our definition of other cluster tools.

   Current Tool Status - Novellus Concept II.   The Novellus C-II is a hybrid/combination architecture with a radial cluster platform that breaks out to a batch environment. Note the Concept-I is the module on the right. The radial central handler serves a single wafer multi-chamber etch back process, while the Concept-II is used as a batch module for CVD tungsten. (See Figure 11b.)

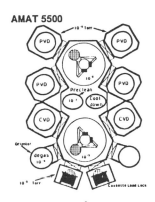

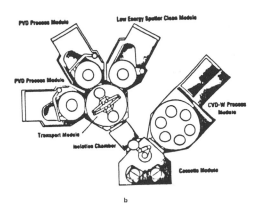

Figure 11. Clustered modules for CVD and PVD: AMAT 5500 and Novellus Concept II (right).

Roadmap Driven by Process Consolidation. Today the tool set used in most integrated circuit factories consists of nonclustered, stand-alone tools. Exceptions to this are interconnect metal stacks of Ti/Al alloy/TiN, which are deposited in low pressure cluster tools.(54) The M2000 and the AMAT 5500 are examples.(50) This film stack can also be etched in a clustered tool. There are other examples shown in Table III, "Cluster Tool Progression Roadmap." The roadmap is not exhaustive but is an example of some of the tools that may emerge in the future.

TABLE III: CLUSTER TOOL PROGRESSION ROADMAP

0.7 μm	0.5 μm	0.35 μm	0.25 μm	0.18 μm	0.12 μm
Al Stk*---→	X------→	X-------→	X---?---→	Adv Stk→ (CVD TiN/Al)	X
W Plug: D/E	X------→	B+X-----→	X---?---→		
ILD: D/E (PECVD)	X-------]	ILD ------→ (ECR)	X+Plan*-→	X--------]	B/LoK
		Salicide→	X+El Drains→	X-------→	X

   B   Indicates barrier metal
   *   Indicates that cleaning steps are included in the cluster
   X   Indicates that the previous generation tool should extend to the next,
sometimes added (+) processes are included in the cluster.
   -------→   indicates extension to the next generation
   -------]   indicates no extension
   LoK indicates relatively low dielectric constant, e.g. 2-2.4

8.0   CHALLENGES AND REQUIREMENTS

8.1   Contamination

     Early experience with vacuum self-contained systems resulted in high
particle counts. When the process module is a vacuum process tool, the issue of
pumpdown to a vacuum environment and transfer from one vacuum environment
must be solved in any case. Much learning has already taken place through the
use of soft roughing techniques,(55) attention to residual levels of moisture and
the gettering of water.(56) Both particulate and molecular contamination can be
monitored using in situ techniques that may, with time, be an essential part of
the process module itself. If send-ahead (monitor) wafers are eliminated and the
conditions leading to maintenance cleaning can be delayed, there will be an
increase in the utilization of the tool.(56) Thus, the need for further
understanding of contamination is key to the success of cluster tools. Handling
is also a big issue, with wafers processed in an upsidedown or vertical
position.(57,58)

8.2   Metrology Testbed

     Cluster tools are going to incorporate as much in situ metrology as needed
to eventually eliminate separate stand-alone metrology tools on the factory floor.
One key aspect is the use of relatively simple, cheap and fast metrology,
ultimately sampling product parameters in "real time", that may be non-unique
or indirect. It may be that relative measurements are all that is necessary in

manufacture. Expensive metrology that gives unique data will still be needed in development and is necessary to substantiate the nonunique measurements by sampling strategies.(59)

Every successful CIM system must have a suitable set of data bases. The data at the cluster tool and chamber level serves as the foundation for such a system. That data would include equipment performance data such as tool power, chemical flow rates, operating pressures, in situ particle and contamination, maintenance history, predictive model for maintenance, and parametric performance parameters (e.g., film thickness, etc.). The data would be fed from the chamber controller and made available to the cluster controller at a local level, possibly stored in memory at the microcontroller of the cluster. There, data could be analyzed and truncated to be fed to a sector control and storage area. In this manner, results, not detailed data, would be communicated to the sector level, reducing the need for large memory and unnecessary detail at the sector and factory levels. Further analyzed data could be communicated to the factory and corporate levels.

The cluster controller itself is a "model" for factory controlling architecture. The connections from cluster controller to the modules of the cluster are similar to, but not necessarily the same as, the connections from the sector controller to individual stand-alone tools in the sector.(56)

A vision of one segment of the industry is to provide an open architecture cluster system that has a central platform with standard mechanical and control interfaces.(49,60) The possibility exists of different suppliers making modules integrated on a "standard" cluster platform. Such a system would enable "best of breed" performance for the user since it is *assumed* that no single supplier would have the capability and dynamic resources to provide the best process modules at all times. It is argued that costs would be reduced since a user would pay for a platform once and be able to add new or upgraded modules at minimal cost. Key to the implementation is the ability to form widely accepted standards. Will all or most suppliers cooperate with the standards approach? Some industry analysts believe that the largest growth will take place with cluster tools that adhere to partial standards or are "semi-flexible." But present market share figures indicate that the answer is far from clear.

An analysis of the market share shows proprietary cluster tools now at $700M/year revenues, $250M from systems that use partial standards, and $4M/year for those that adhere to present SEMI/MESC standards. The greatest growth out to 1995 is expected from the groups using and adhering to standards. Markets of about $1.1B (prop), $1.4B (part standards) and $0.5B (standards) are predicted for the three groups by mid 90's.(53) Some argue with the quantative degree of predicted growth for the standards group.

Most supplier companies that offer open standards are small. It is believed that they do not have extensive process module expertise to provide a wide variety of complete integrated tool capabilities. Those that use partial standards have started to do so by partnering between supplier companies and often use sufficient standardization to provide the current capability to be open to those modules they want to be open to, but may not be open to others. This is a "limited openness," driven by practicality and customer pull. The concept of open architecture alone has driven these limited open architectures. The following issues (and tensions) arise.

8.3   Integrator Ownership

An open cluster system prompts the question: how do we define system-integration-ownership. Usually the platform owner only has access to his own process modules. The key issue is how one can tie to an unfamiliar system without transfer of or loss of proprietary and intellectual property. Some suppliers have worked this out, but their solutions may not allow the addition of even a third party's module to the cluster in the future. The need for standards is thus reinforced. If the user is asking for open-ended process module application (he wants to leave some ports capped), there is almost no way to ensure that an unidentified third party can participate later unless standards are capable of accommodating a third party.

8.4   Cost of a Second Party's Selling Price

If a second party's module is added to an open system, his selling price may be entered as an OEM subsystem by the platform integrator, and there may be another middleman cost incurred. This situation, unless handled by the supplier partnership, may price the tool above alternatives as recognized or perceived by a cost of ownership analysis.

8.5   Limited Flexibility of a Cluster Platform

It can be argued that costs would be reduced if a user would pay for a platform once and be able to add new or upgraded modules. Once metal is cut, the interface ports are sized, and the wafer handling robot is designed with end effector, the number of options available for the open cluster tool is immediately limited. The next larger wafer size may not be compatible. For example, it would also be unwise to build a large platform to handle a 12-inch wafer and then only use it for 8-inch or smaller systems for 3-5 years. The platform is, however, flexible within a given wafer size generation,and this has value.

Thus, there is considerable tension between the standards initiative and the use of proprietary platforms. It is likely that SEMI/MESC standards will evolve and even suppliers who practice proprietary architecture will find ways to open up to second party modules in order to maintain their competitiveness of "best in class." Best in class systems are the common goal of both suppliers and users.

The initiatives in the cluster tool industry will - through users working with their customers - anticipate the sequences that require self-containment for yield and contamination control and cluster these sequences to also take advantage of the desirability of fewer operational steps. They will also cluster steps that are used with great frequency although they can be done with stand-alone tools in an open fab environment, such as lithography exposure tools clustered with a track that applies resist. Fewer operational steps will reduce overall cost of ownership. Much of this concept was covered under Design for Manufacture, Section 2, above.

8.6   Summary of Equipment Integration

It is clear that the initiative toward cluster tools is quite strong, and clusters will play an important role in the drive to a fully automated semiconductor fab. The requirements that must take place include:

TABLE IV: REQUIREMENTS (VALUE) FOR CLUSTER TOOLS

Reliability	3-10X improvement/generation
Contamination	2X improvement/generation
In Situ Metrology	(Key to cost reduction)
CIM	(An enabler)
Standards	(A cost reducer)

9.0   CONTROL AND ANALYTIC CAPABILITY

9.1   Statistical Process Control (SPC)

SPC ensures consistency of production by monitoring key process parameters. If there are out-of-control values or serious deviations from typical or expected values, then action plans are put into place with diagnostics and corrective action to center the parameters.

It is a prerequisite to establish precision and tolerance gauges on all metrology tools used to run the SPC program. The SPC methodology is crucial for driving communication between engineering and manufacturing. In a typical MLM SPC program, one would have key parameters monitored on each unit process described above. In the examples shown here, it will be seen that both support systems and metrology capabilities impact the SPC records.

The particles reported on a commercial oxide etcher were observed to be large and variable; however analysis showed the problem to be due to a mass flow controller which was generating particle bursts. When a filter/valve was placed in series with the controller, the particles were greatly reduced and stabilized.

In Figure 12, the etch rate of a poly silicon etcher was recorded to be highly variable, but the metrology tool was found to have unstable software control. In the case where the software was upgraded or an alternative metrology tool was used, the recordings were stabilized.

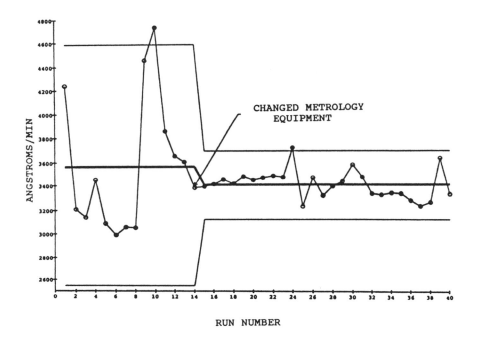

Figure 12. SPC chart exhibiting control for a polysilicon etch tool. Apparent control was improved after a metrology upgrade.

In Figure 13, the uniformity of a CVD oxide deposition tool for interlevel dielectric applications was found to drift towards tighter (better) values, not out of spec. The control limits are set inside the spec limits by the methodology of the SPC process, but clearly the phenomena that is responsible for the trend is a candidate for diagnostics, with the intent of improving the control. If one could understand the root causes for the trend, the opportunity to refine the process toward even tighter values exists.

The typical SPC program system could include considerations of: frequency of measurement of each lot, e.g. from once-a-lot to once-a-month, depending on the stability of the tool; the minimum number of sites set in collaboration with a statistician and with 15-30 observations being typical as a minimum sample size; and the number of wafers in a lot to sample, this may be 1-2. Of course the SPC charts would be linear with expanded scales, but could be logarithmic if the process is unstable.

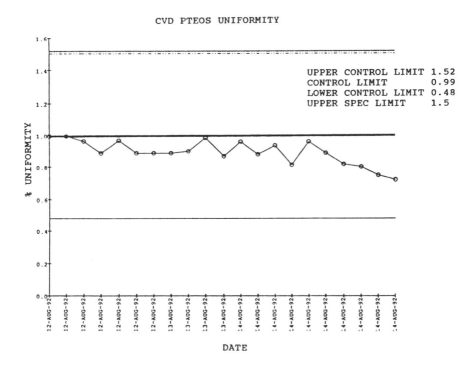

CVD PTEOS UNIFORMITY

UPPER CONTROL LIMIT 1.52
CONTROL LIMIT   0.99
LOWER CONTROL LIMIT 0.48
UPPER SPEC LIMIT   1.5

DATE

Figure 13. Uniformity of a CVD tool showing steady improvements (for undetermined reason), the control limits are inside the spec limits.

9.2   Total Preventive Maintenance (TPM)

In a manufacturer's vigilance to control his line and have the very highest utilization capabilities, he soon comes to the realization that SPC is not enough. It is a reactive methodology and has served us well, but can only respond to problems after the SPC data signals a problem. The proactive approach is to map the performance of tools, with respect to maintenance history, and to measure or predict the timing of maintenance which allows the tool to perform without unscheduled downtime events.

A good familiar analogy is what one does with an auto after 50-60,000 miles of use. A proactive diagnostics is done, and even without new data, one will "change all the rubber under the hood," because based upon history the belt and hoses are known to wear out in the next 40,000 miles. This is an example of TPM that most of us have practiced. In the case of process equipment in the fab, one could anticipate changes of mass flow controllers, plastic containers that may become corroded through use, and sources for sputter targets that are near end of life. The value of being proactive for these kinds of things is that the Utilization Capability will go up. It is not unreasonable to expect a 20% improvement in U.C. by running a disciplined TPM program.

9.3   Analytic Tools

Line width, defect detection, failure and yield analysis are integral parts of a successful manufacturing operation. In general, tools may be considered lab tools or in-the-fab metrology tools dedicated to a particular metric.(61)

For example, the rapid evaluation of poly and metal line width values can be done by lab SEM tools or by newly emerging fab on-line high throughput SEM tools. The value of an automated linewidth capability ~30 w/hr, small footprint, very low cost of ownership fab tool cannot be overemphasized. Such a tool is available from Angstrom. Otherwise the wafers have to be moved to a general purpose lab tool, which is a low throughput, high foot print operation. Ellipsometry is used to determine dielectric film thickness and in certain cases, poly and very thin metal film thicknesses. An emerging technology of great interest for atomic level topology measurements is the atomic force microscope (AFM).

Defect detection tools use scattered light. Laser scan systems are used with high data rates. They have capability to detect down to $0.1\mu m$ with a capture probability of 90% and inspection speeds of several cm^2/sec. These can function on bare or patterned wafers. Example systems are the Tencor P1 for bare wafers, and the KLA 20xx for patterned wafers. An alternative scattered light design uses the principle of broad area exposure (spatial filtering/holographic technology) for very rapid capture of anomalies that stand out against a standard repetitive pattern. Such a tool is the Insystem 8600. Once the coordinates of defects are determined by optical scattering, they are ready for further analysis by high resolution imaging or materials' analysis. Such tools could be clustered.

High resolution failure analysis is carried out using: a defect review scanning electron microscope (SEM) or standard SEM, voltage contrast - electrical probe, defect review tool SEM (DRT), and focussed ion beam imaging (FIB) tools. The resolution and contrast must be such as to reveal anomalies of the order of 10-100Å. Once a defect is observed, the physical and chemical nature of the defect is crucial in understanding its origin and root cause. A defect review tool such as the Amray 18xx can be used to quickly characterize the physical nature of the defect. If round and smooth, the particle is likely to result from a chemical nucleation or thermal (droplet) process, if rough and with jagged surfaces, it is likely to have originated from a secondary source such as a surface wall.

The voltage contrast tool, a SEM with bias applied to a circuit is particularly useful for determining electrical anomalies: voids, thins, compositional variations, etc. In particular, defects in metal test arrays, such as serpentine and comb structures are easily detected. In the early stages of technology and mask development, the VC-SEM will detect any ill-conceived design rules or process flaws when bright spots occur under high bias.

FIB imaging tools are especially useful for determining grain structure of aluminum and so are used in electromigration studies.

Material or chemical compositional analysis is then carried out to further define the nature of the defects. For these analyses, Auger electron spectroscopy (AES) is used to determine composition for films like TiN and TiSi. Rutherford backscattering is used for the same purpose, but without the need for standards. Fourier transform IR spectrometry (FTIR), is used to characterize the composition of BPSG used in first dielectric.

Transmission X-ray fluorescence (TXRF) is used to determine secondary trace layers of material from the inside of process modules onto wafers, atomic force microscopy (AFM) is used to determine micro-roughness from plasma damaged layers, and transmission electron microscopy (TEM) is used extensively to determine the nature of interfaces at contacts in MLM systems.

9.4   Test Vehicles

Electrical test vehicles are used extensively to characterize the defect density of structures in MLM systems. Structures such as closely placed interpenetrating arrays of metal lines (combs) and long meandering metal lines (serpentines), Kelvin bridge contact testers are used.

These test vehicles should be designed with logarithmic defect density sensitivity capability, i.e. be able to detect 100, 10, 1 defects/cm^2.

A more lengthy description is found in Chapter 9.

9.5   Assembly and Packaging

Successful manufacturing of semiconductor systems depend on the close coordination of front end processing which includes "MLM," with *test, assembly* and *packaging*.

Prediction of good system performance includes the planning and coordination of die thermal engineering and management of assembly and reliability testing of the packaged IC. This must be done in environments that represent stressed conditions for the product's applications.

In the future, generalists will have a vision that spans all the way from the design to the application of a particular IC.

10.0   SUMMARY

Future manufacturing capability in the multilevel metal area is limited by cost, defect levels and , perhaps, environmental chemical issues. These are issues that must be addressed through joint efforts of manufacturing and engineering . Cost-of-ownership and lower cost process sequences must be developed. Engineering must comprehend electromigration and stress migration limitations. Advanced etch engineering will provide even higher density. Lower damage source systems and their applicability manufacturing application must be carefully demonstrated. Serially linked cluster applications will emerge where cost and self-contained approaches prove competitive.

ACKNOWLEDGMENTS

The author thanks Tad Dabies, Steve Holland, Theresa Foster, Bert Fowler, Dennis Hartman, Ken Maxwell, Shayam Murarka, and Ron Schutz for source material and insights

REFERENCES

1.   Richard McCausland, Electronic News, Vol. 38, No. 1923, Aug. 3, 1992.
2.   E-10 Guideline for Definition and Measurement of Equipment Reliability, Availability and Maintainability (RAM), available from SEMI, 805 E. Middlefield Road, Mountain View, CA.
3.   S. Shingo, The Sayings of Shigeo Shingo - Key Strategies for Plant Improvement, Productivity Press, Cambridge, MA, 1987.
4.   Manufacturing 21 Report, "The Future of Japanese Manufacturing, Company X: A Hypothetical Semiconductor Manufacturer in the Year 2001," (AMA, Wheeling IL, 1990)
5.   A. K. Sinha, et al, IEEE Elec. Dev. Lett., $\underline{3}$ (4) (April 1982), p. 90.
6.   V. Teal, et al, Thin Solid Films, $\underline{136}$, 21 (1986).
7.   S. P. Murarka, "Multilevel Metallization," in Concise Encyclopedia of Semiconducting Materials and Related Technologies, eds S. Mahajan and L. C. Kimmerling (Pergamon Press, NY, 1992) p. 311.
8.   W.J. Bertram, in VLSI Technology, ed. S.M.Sze (McGraw Hill N.Y., 1983), chapter 14.
9.   Proc. of Ninth International Conference (VMIC), Santa Clara, June 1992, Library of Congress No. 89-644090.
10.   E. Bassous, H. N. Yu, and V. Maniscalco, J. Electrochem. Soc., $\underline{123}$, (1729) (1976).
11.   J. Nulman, in Advanced Metallizations in Microelectronics, eds. A. Katz, S. P. Murarka and A. Appelbaum (Mat. Res. Soc. Proc., $\underline{52}$ Pittsburgh, PA, 1985), p. 123.
12.   K. Maex, L. Hobbs and W. Eichhammer, VLSI Sci. and Tech., edited by J. M. Andrews and G. K. Celler, (ECS Pennington, NJ 1991), p. 254.
13.   S. P. Murarka, IEEE Electron Devices, $\underline{34}$ (10) (October, 1987), p. 2108.

14. R. M. Levin and A. C. Adams, "Low Pressure Deposition of Phosphosilicate Glass Films," J. Electrochem. Soc., 129, 1588 (1982).
15. S. Rastani and A. Reisman, J. Electrochem. Soc. 137, 1288 (1990).
16. R. J. Schutz, Reactive Plasma Etching, VLSI Technology, 2nd Ed., S.M. Sze, Editor, 1988., Ch 5
17. R. S. Blewer, T. J. Headley, and M. E. Tracy, in Proc. of Tungsten and other Refractory Metals for VLSI Applications III, ed. V. A. Wells (MRS, Pittsburgh, PA, 1988) p. 155.
18. S. P. Murarka, Lecture notes at Metallization and Multilevel Metallization for VLSI and ULSI, Rensselaer Polytechnic Institute, August, 1991.
19. R. J. Schutz, MRS Symposia Proceedings, Interfaces and Contacts, 18, 1983.
20. J. N. Mayer and S. S. Lau, Electronic Materials Science: For Integrated Circuits in Si and GaAs, MacMillan, New York (1990) p. 332.
21. I. J. Raaijmakers, MOCVD TiN Barrier and Adhesion Layers for ULSI Applications, presented at the 1992 MRS Spring Meeting, San Francisco, CA, 1992 (unpublished).
22. D. B. Fraser and M. L. A. Dass, Growth of Epitaxial Cobalt Disilicide on <100> Silicon, Paper C3.1, 1992 MRS, San Francisco, CA.
23. K. Tsubouchi, K. Masu, K. Saski, and N. Mikoshiba, IEDM Technical Digest (IEEE), 1991, p. 269.
24. C. Kaanta, W. Cote, J. Cronin, P. Lee, T. Wright, and K. Holland, IEDM Tech. Dig., p. 209 (1987).
25. A. J. Learn, J. Electrochem. Soc., 123, 894 (1976).
26. Y. K. Kim, V. Hoffman, and T. M. Pang, in Proc. SEMINCON/Korea Technical Symposium (1990).
27. A. Jain, et al, Kinetics of Chemical Vapor Deposition of Copper from CuL Precursors, 1992 MRS, San Francisco, CA.
28. B. Zheng, et al, Copper CVD for ULSI Applications, 1992 MRS, San Francisco, CA, 1992.
29. C.J. Mogab, Dry Etching, Ch8, VLSI Technology, Ed.S.Sze, McGraw Hill, 1983.,p 340.
30. C. S. Pai, J. C. Schumacher Symposium, San Diego, CA, February 1992.
31. D. Webb, S. Sivaram, D. Stark, H. Bath, J. Draina, R. Leggett, and R. Tolles, Complete Intermetal Planarization using ECR Oxide and Chemical Mechanical Polish, VMIC Proceedings, 6/9-10/92, Lib. of Congress No. 89-644090, VMIC Cat. No. 92ISMIC-101.
32. J. A. Tragolo and K. Rajan, Detection of Surface Response to Chemical/Mechanical Planarization of Planar Films, 1992 MRS, San Francisco, CA, 1992.
33. S. Sivaram, Chemical Polishing of Oxides: Models for Removal Rate and Planarity, 1992 MRS, San Francisco, CA, 1992.
34. C. Meade and L. Conway, "Introduction to VLSI Systems," (Addison-Wesley, 1980).

35. Graydon B. Larrabee, "Cluster Tools for the Semiconductor Industry," STEP/Cluster Tool Communications. Semicon West '90, Technical Conference Proceedings.

36. K. Tsukamoto, S. Komori, S. Sato, and Y. Akasaka, Nikkei Microdevices, Dec. 1991, p. 95.

37. R. Manukonda and T. E. Seidel, "Staircase Sidewall Spacer for Improved Source/Drain Architecture, U.S. Patent 5,102,816; 1992.

38. W. J. Kim, B. J. Kim, C. J. Lee and D. Hanna in Proc. 1990 VMIC Conference (459) 1990.

39. W. Winchell, Continuous Quality Improvement, Soc.Manuf.Engs.; Dearborn, MI, 1991; Lib. Congress Cat Card No: 91-60113 p. 22

40. D.B.Fraser, Metallization, Chpt.9, VLSI Technology, Ed. S.M. Sze, McGraw Hill, (1983) see pp 348,362.
     S.P. Murarka, Metallization, Chpt 9, VLSI Technology, Ed. S.M. Sze, McGraw Hill (1988) see pp 378,400.

41. Albert Bergendahl, "Cluster Tools Part 2: 16mb DRAM Processing," Semiconductor International, Vol. 13(10), p. 94, September 1990.

42. David Richardson, Standards Overview, Semicon/West '91 Technical Proceedings, May 20, 1991. Cluster Tool: An integrated, environmentally isolated manufacturing system consisting of process, transport and cassette modules mechanically linked together.

43. Richard S. Rosler, "The Evolution of Commercial Plasma-Enhanced CVD Systems," Solid State Technology, Vol. 34(6), p. 67, June 1991.

44. U.S. Patent #4,715,921, December 29, 1987, "Quad Processor."

45. Mary Seppala, "Special Report: Sputtering Equipment," Microelectronic Manufacturing and Testing, Vol. 13(12), p. 18, November 1990.

46. MMT Staff, "Chemical Vapor Deposition," Microelectronics Manufacturing Technology, Vol. 14(7), p. 32, July 1991.

47. MMT Staff, "Dry Etching Equipment Selection Guide," Micro-electronic Manufacturing and Testing, p. 12, January 1989.

48. Nikkei Staff Writer, "Multi-Chamber and Super Clean Technologies are Gathered under One Roof at Semicon Japan '89. Aiming to Penetrate the Market in the 4M Era," Nikkei Microdevices, Vol 11(37), p. 188, November 1989.

49. B. Newboe, "Cluster Tools: A Process Solution?", Semiconductor International, Vol. 13(8), p. 82, July 1990.

50. P. Burggraaf, "Sputtering's Task: Metallizing Holes," Semiconductor International, Vol. 13(13), p. 28, December 1990.

51. G. Birkmaier, et al, "Ultra-High Vacuum in Production Applications," Semiconductor International, Vol. 14(5), p. 108, April 1991.

52. A. Kermani, et al, "Single Wafer RTCVD of Polysilicon," Solid State Technology, Vol. 34(5), p. 71, May 1991.

53. Daniel Holden, Electronic News, EN, August 19, 1991.

54. D. Pramanik, et al, "Barrier Metals for VLSI: Processing and Reliability," ibid, p. 97.

55. Degang Chen, T. Seidel, S. Belinski and S. Hackwood, "Dynamic Particulate Characterization of a Vacuum Loackock System," Journal of Vac. Sci. & Tech., A7(5) pp. 3105-3111, September/ October, 1989.

56. Richard J. Markle, D. Brestovanksy, A.A. Sharif, and G. C. Guzzo, J. Theisen, and H.K. Nguyen, "Reduced Particle Additions and Chamber Cleaning Frequency for a Tungsten Etch Process," to be published Microcontamination Conference Proceedings, Canon Communications, Inc., October, 1991.

57. R. A. Bowling and G. B. Larrabee, "Particle Control for Semiconductor Processing in Vacuum Systems," Microcontamination Conf. Proc., p. 161, 1986.

58. M. M. Moslehi, R. A. Chapman, M. Wong, A. Paranjpe, H. N. Najm, J. Kuehme, R. L. Yeakley, and C. J. Davis, IEEE Trans on E.D. 39, No. 1, 1992, p. 4.

59. Terry Turner, "Parametric Response Surface Control - A Perspective on Real Time Process Control in the Semiconductor Industry," Proceedings: SRC Workshop on Real Time Tool Control, Vancouver, CA, February 18, 1991.

60. "Cluster Tool Communications: The Path to an Open Standard," Solid State Technology, Vol. 33(11), p. 85, November 1990.

61. R. B. Marcus, Diagnostic Techniques (Ch. 12) in VLSI Technology, ed. S. M. Sze, McGraw-Hill, 1983.

11

CHARACTERIZATION TECHNIQUES FOR VLSI MULTILEVEL METALLIZATION

SIMON THOMAS
Core Technologies
Motorola, Inc
Mesa, Arizona

1.0   INTRODUCTION

The trend in multilevel metallization (MLM) technology continues to be in integrating increasing numbers of levels of metallization and decreasing physical dimensions. Four levels of metallization and submicron feature sizes are already being used in state-of-the-art high performance VLSI (1, 2). This successful integration of the MLM is derived through impressive developments in material and process technologies, proper understanding of reliability issues and through extensive characterization of thin film material properties and interactions. MLM technology is essentially a thin film technology where thin film properties, and surfaces/ interfaces dominate physical and electrical properties. The challenges in characterizing thin films and interfaces are formidable and the rewards satisfying. Success in characterizing MLM properties has been a result of exceptional advances in analytical technology, both through the emergence of new techniques and the enhancements of existing techniques. A variety of aspects of MLM need to be characterized and the analytical requirements are conflicting in emphasis. Usually multiple techniques are needed and are used in a synergistic manner, exploiting their strengths, for a comprehensive analysis, especially during the process development phase.

The technical and operational capabilities of analytical techniques also determine their use in different phases of technology development and implementation. These different phases include: 1) the technology development phase when material and process properties are characterized in detail and integrated, 2) the manufacturing phase when emphasis is on monitoring processes for controlling them and, 3) in device failure analysis where significant "deprocessing and detective work" are required. The entire arsenal of analytical techniques will be needed in the characterization of materials and processes during the development phase. Higher expertise of analysts is also essential in this phase. Usually analysis is performed off-line from the wafer fabrication area and the analysis "cycle time" is relatively longer. In the manufacturing phase, less basic characterization is required compared to the development phase and more emphasis is placed on analysis for process control and "cycletime" of analysis. The choice of analytical techniques and integration of the same to a wafer fabrication area are driven by these criteria. The tools generally form part of the "in-line" process monitoring function. In post-production device failure analysis a wide range of analytical tools is required. However, often the specimen preparation methods (destructive deprocessing, artifacts) impact the analysis besides the inherent analytical tool capabilities. The choice of an analytical technique in general, will be influenced by the needs of the above broad functional areas, in addition to the basic technical needs/and capabilities of tools discussed below.

The following is a list of needs generally recognized for MLM analysis tools: capability in analyzing physical, chemical, and electrical properties of conductors and insulators, lateral resolution in the nm range, atomic layer depth resolution, ability to probe lateral and vertical interfaces, ppm - ppb level minimum detectable limits, uniform mass sensitivity to all elements, chemical bonding information, quantitation, crystallinity, topography information, lateral and in-depth distribution and imaging, mechanical stress, minimum artifacts induced by the analysis technique, non-destructive analysis and short analysis time. Evidently no single technique meets all the requirements and the obvious approach is to exploit the strengths of multiple techniques, and the coordinated efforts of analysts from diverse fields of science.

In the MLM technology, particular emphasis is placed on surface and interface characterization since most of the typical IC processing involves modification of surfaces and since surfaces and interfaces play a dominant role in the physical and electrical properties of the integrated MLM structure. With the continued lateral dimensional scaling down and minimum feature sizes <0.5µm currently in production, and high aspect ratio structures, the role of lateral and vertical interfaces is increasingly dominant.

While there is a plethora of analytical techniques available today, this chapter will be limited to the more commonly used analytical tools for MLM characterization in a state-of-the-art wafer fabrication facility. The techniques discussed below have been the subject of prior, detailed review articles and books (3-15) and the instrumental details of the techniques will not be repeated here but instead the reader is directed to these references. The emphasis in this chapter will be to show through illustrative examples, areas of applications and strengths and weaknesses of each technique in MLM characterization.

The physical and chemical analyses of MLM are broadly grouped into imaging analysis of microstructural features, surface and interface chemistry, and bulk compositional analysis with depth compositional information. Analysis techniques discussed below are also broadly grouped into these three categories (Figure 1) based on their primary strengths. Choice of a particular technique to a specific situation may not be obvious from the above broad classification. A variety of factors needs to be considered in choosing a technique, including the technical strengths and weaknesses of the technique. Most of the techniques discussed below use electrons, photons or ions to probe the thin films and the emanating electrons, photons or ions which carry specific information about material properties are analyzed.

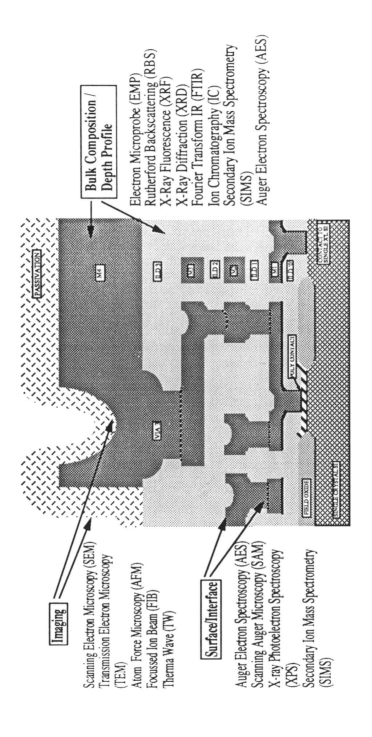

Bulk Composition / Depth Profile

Electron Microprobe (EMP)
Rutherford Backscattering (RBS)
X-Ray Fluorescence (XRF)
X-Ray Diffraction (XRD)
Fourier Transform IR (FTIR)
Ion Chromatography (IC)
Secondary Ion Mass Spectrometry (SIMS)
Auger Electron Spectroscopy (AES)

Imaging

Scanning Electron Microscopy (SEM)
Transmission Electron Microscopy (TEM)
Atom Force Microscopy (AFM)
Focussed Ion Beam (FIB)
Therma Wave (TW)

Surface/Interface

Auger Electron Spectroscopy (AES)
Scanning Auger Microscopy (SAM)
X-ray Photoelectron Spectroscopy (XPS)
Secondary Ion Mass Spectrometry (SIMS)

Figure 1. Schematic of a MLM structure illustrating the various aspects which need characterization. The techniques are grouped into imaging techniques, surface and interface methods and bulk analysis methods.

2.0   IMAGING   TECHNIQUES

2.1   Scanning   Electron   Microscopy   (SEM)

Scanning electron microscopy (SEM) has been the backbone of imaging technology for integrated circuits in general. For MLM evaluation also, SEM has emerged as the most essential analytical tool. The details of the operation of the SEM may be obtained from refs. (5, 14, 15). While SEM offers higher magnification, resolution and depth of focus than optical microscopy, it complements optical microscopy. Optical microscopy is taken for granted as an inspection tool in any wafer fabrication area and the SEM also has achieved that status. High resolution imaging, which is the most often used mode of operation of the SEM, offers invaluable visual information on topography and microstructure of the surfaces and thin film structures. With the continued developments in high brightness electron sources (LaB_6 or field emission), electron optics and image processing, resolution in the range 1 nm is obtainable in commercial SEM. The secondary electron images provide information on the morphology, structural integrity and physical dimensions of the structures which constitute the integrated MLM.

Today's device dimensions are in the submicron range and SEM is the only practical tool available with sufficient resolution, technical maturity and speed of analysis, to measure these dimensions with adequate accuracy. SEMs specifically designed for the measurement of lateral, critical dimensions (CDs) are integral to today's wafer processing areas. The specific thrust of such tools is in high throughput, automated measurement for process monitoring purposes and not so much as analytical microscopes. The requirements of analytical SEMs differ significantly from these monitoring tools. In addition to the routine topography evaluation of MLM structures, a variety of additional information is sought from an SEM. Some of the common applications are given below.

Grain size distribution in metal conductor is an important physical parameter which determines its reliability [see Chapter 8] and is often monitored as a process parameter. SEM is a popular tool used to measure grain sizes encountered in aluminum conductors. Often a wet chemical etch to delineate the grain boundaries will improve the grain identification and measurements. There is a variety of etchants which will delineate grain boundaries of Al and its alloys. A wet etch suitable for Al-Si grain boundary delineation has been reported by Solley et al. (16) which consists of 10ml [7g KNO_3 + 25 ml HNO_3 + 1 l of DI water] + 10 ml NH_3F (40%) + 100 ml DI water + 6 ml KOH (45%). An etching time of 4 minutes for unannealed AlSi samples has been reported. Another wet etch found effective is a dilute Keller's etch consisting of 1ml HF + 1.5ml HCl + 2.5ml HNO_3 + 95ml H_2O (17). Grain boundaries could also be delineated by a very simple solution of 5ml HF and 100ml DI. water. This etchant also delineates the copper precipitates at grain boundaries in AlCu metal. Plasma etching has also been reported to be effective, in selected cases, in grain boundary delineation as in the case of CF_4 + O_2 etch of AlSi films (18). Large grain aluminum films (~1μm) deposited at higher temperature, ~450°C, show thermal grooving and can be easily evaluated without grain boundary delineation. An SEM micrograph of the grain size of an Al-1.5%Cu film, without grain delineation, is shown in Figure 2. Image analysis yields grain size and its distribution and can be used as a process control.

Figure 2. SEM micrograph of the grain structure in Al-1.5%Cu. Grain size and distribution measured from the image analysis are used for process control.

Useful information about the microstructure of interconnect layers can be obtained from the backscattered electron image (BEI)in an SEM. The backscattered electron image contrast due to atomic number differences offers information about the spatial distribution of constituents differing in Z. This is commonly exploited in the evaluation of formation and modification of precipitates in aluminum alloy films. An example of backscattered imaging of copper segregation in an Al-1.5%Cu film is illustrated in Figure 3 a&b. While the secondary electron image in Figure 3.a is relatively featureless, the BEI in Figure 3.b, where the bright areas represent the copper-rich phase, clearly shows the lateral distribution of copper precipitates. The modification of thin film microstructure, specifically, second phase formation and precipitation in alloy films can be characterized *in situ* by backscattered imaging in a SEM equipped with a hot specimen stage. Another practical application of the backscattered image contrast is in evaluating sub-surface voids (for example, stress voids or electromigration induced voids) in aluminum layers (Figure 4 a&b) even when covered with a layer of glass. The voids appear dark in the backscattered image because the number of backscattered electrons are reduced where aluminum is absent.

One of the biggest impacts of SEM in MLM technology development has been in the evaluation of the integrated structure through cross-sectional microscopy. While surface SEM micrographs offer useful information on surface morphology, more detailed and important information is sought on the vertical MLM structure. Invaluable information on the structure of individual layers and their integration can be most effectively derived from the SEM evaluation of cross-sections. The quality of the information is very critically dependent on the quality of the cross-sectional specimens. The cross-sections may be generated by cleaving the wafer (19) or through precision polishing (20-23) to create 90° sections. The cleaved or polished cross sections are usually subjected to a light wet chemical etch to delineate the various layers for better SEM imaging. A brief etch in HF buffered with NH_4F is sufficient to delineate the most commonly used dielectric layers (SiO_2) and Al alloy metal layers. An example of a cross sectional SEM micrograph from a cleaved sample which has been delineated is shown in Figure 5. The various layers of ILD0 (reflowed glass), Al, and the passivation layers can be discerned. The wet etch tends to accentuate the growth seam of the plasma deposited layers through enhanced etching, for example the SiO_2 passivation layer over aluminum in Figure 5. Such artifacts introduced in specimen preparation need to be reckoned with in interpreting chemically delineated cross-section SEM micrographs. Dimensional metrology is often performed on chemically delineated cross sections. While the state of the art SEM offers nm range spatial resolution or better, nm accuracy in dimensional measurement is not assured from delineated cross sections. Electron scattering from delineated interfaces usually results in an apparent broadening of the interface. The accuracy of feature size measurement (which involves two such delineated interfaces) is not much better than ten nm.

While the cleaved cross sections are relatively easy to prepare particularly on repetitive structures and yield outstanding cross-sectional SEMs, the technique lacks the spatial precision required for sectioning through specific locations of micron sized MLM features in VLSI. Also, it is difficult to produce good quality cleaves through MLM involving more than two levels of aluminum layers. Polished cross sectioning techniques have advanced to such a level as to produce exceptionally good quality samples of MLM or defects with a location precision of <1000Å. A SEM micrograph of a four level MLM structure, prepared by the polishing technique and chemical delineation, is shown in Figure 6. All the salient features of the physical structure can be assessed from the micrograph which help process development and failure analysis. A thin cap layer, of thickness in the range 10nm, over aluminum can be distinguished by the delineation. In fact thin (~nm range) interfacial layers of native oxides on Al in vias can be delineated by this technique, though it should be used only as a qualitative gauge of the presence of interfacial layers. Aforementioned enhanced etch artifacts of gaps in glass filling spaces between metal lines (see Figure 6) must be borne in mind when using this method specifically to characterize the gap fill process. Appropriate alternate etchants which do not etch glass can be used in such situations. Of all the imaging techniques used in MLM process characterization, the SEM is the most indispensable tool and is routinely used in process characterization, monitoring and failure analysis.

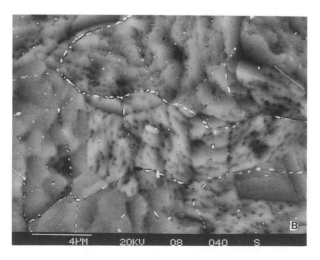

Figure 3.a) Secondary electron image and b) backscattered image from an Al-1.5% Cu film surface. The BEI clearly reveals the lateral distribution of copper precipitates (the bright areas) in the film.

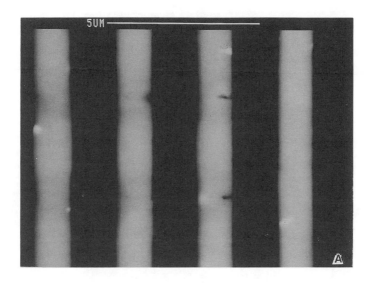

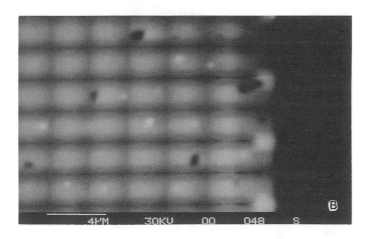

Figure 4. Backscattered images of voids (the dark areas) in AlCu films covered with SiO$_2$ film. a) stress induced voids b) electromigration induced voids.

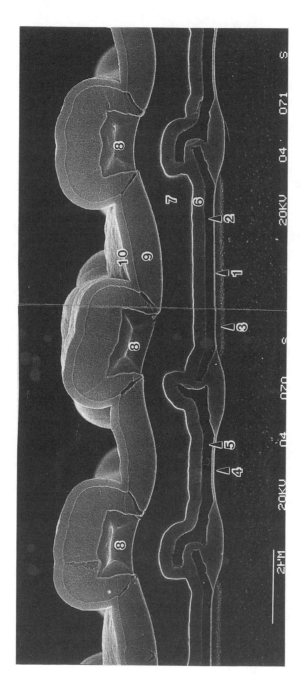

Figure 5. SEM micrograph of cleaved cross section of an integrated structure delineating the various structural layers.

1. Thin SiO$_2$, 2. Polysilicon, 3. N$^+$ Implant, 4. Gate Oxide, 5. Polysilicon gate
6. SiO$_2$, 7.Reflowed glass ILD0, 8. Aluminum , 9&10. Passivation layers

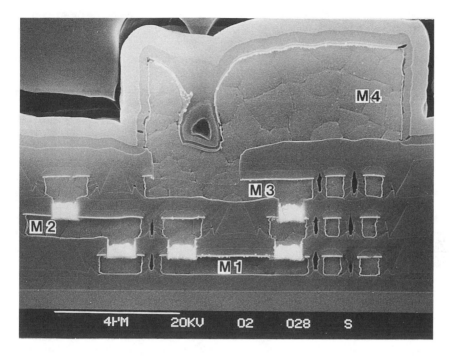

Figure 6. SEM micrograph from polished and delineated cross section of a four level metal structure. The structural layers and their integration are revealed in the micrograph.

2.2. Transmission Electron Microscopy (TEM)

Transmission electron microscopy (TEM) is another electron-beam imaging technique used in materials and process characterization. The most obvious strengths of TEM are its high resolution, and its ability to provide chemical and structural information on a sub-nanometer scale (24-26). Hence for high- resolution microstructural characterization, TEM is the preferred technique. Transmission electron microscopy utilizes a high-energy electron beam, typically 100-300 keV in energy, incident on and through a thin specimen. The beam is diffracted by the various planes of atoms comprising the specimen. An aperture can be used to choose one beam out of the various diffracted beams and the transmitted beam. The image produced by the beam is amplified by the use of a system of electromagnetic lenses and then focussed to produce an image on photographic film. In general, the different layers present in a MLM structure have grains oriented in a variety of directions; these give rise in the TEM to diffracted beams in different directions. When a particular diffracted beam is chosen to form the image, some regions of the specimen are tilted so that planes in that area are in the Bragg condition appropriate for the chosen beam; these regions contribute intensity to the image. By an appropriate choice of diffraction conditions, particular layers or localized regions of the MLM structure can therefore be imaged in the TEM. Image contrast obtained in the manner described above is called 'diffraction contrast'. Other mechanisms of contrast involve phase contrast (due to a change in relative phase of the electron beam as it passes through a particular region of the crystal) or 'Z' contrast due to greater inelastic scattering of electrons by elements with higher atomic numbers. Phase contrast can be used to analyze voids or oxide regions present in MLM structures. Z-contrast is useful for imaging layers with varying atomic numbers, e.g., W/Al or Al/TiW.

Two types of TEM imaging are usually performed: plan-view sections parallel to the sample surface and vertical sections, the former emphasizing microstructural characterization of individual layers and the latter aimed at the structural integrity and defects in multilayers. Specimen preparation for transmission electron microscopy is an art and techniques are different for the plan-view and cross-section imaging. TEM requires specimen areas of interest that are less than a micron in thickness. The starting material must therefore be thinned down to less than a micron in the area of interest. Mechanical thinning coupled with chemical thinning is a possible means of attaining such thicknesses (27-29) and is more suitable for single- phase materials. An alternative, particularly in the case of composite structures comprised of different materials as in the case of MLM, is by a combination of mechanical abrasion and Ar ion beam milling (30-32). Specimens are mechanically ground and polished down to ~5-50 µm. Ion beams are then used to sputter away material till the required thickness is obtained. Composite structures frequently contain 'hard' materials juxtaposed with 'soft' materials. This makes it quite difficult to preserve the whole device structure during the course of ion (or chemical) thinning. The 'soft' materials disappear before 'hard' materials become electron transparent. Use of high-voltage electron microscopy of thicker specimens (2-5 µm), using accelerating voltages 400 kV and above, is one way of circumventing this problem. Preparation of cross sections for TEM evaluation is traditionally a tedious process and often a deterrent to the routine use of TEM for process characterization and monitoring. However, cross-sectioning techniques continue to evolve to overcome this limitation and the information required can be generated by no other technique than TEM. The technique has therefore become an essential part of the materials and process characterization arsenal.

The scanning principles of SEM have been incorporated to TEM to form the Scanning TEM (STEM) (33), which essentially becomes a SEM in the transmission mode. STEM is increasingly being used in IC cross-sectional analysis to supplement SEM analysis, in order to exploit the superior resolution of the former. An example of a cross sectional STEM image of an MLM structure is shown Figure 7. The superior resolution of the image (in comparison to SEM) is evident. Voiding in the ILD glass gap-fill can be seen without concern for sample preparation artifacts (eg.wet etch delineation). In addition to high-resolution imaging, STEM imaging is used to locate the area of interest in a sample for further compositional or structural analysis. Nowadays, TEM and STEM capabilities are built into the same instrument which make it a powerful analytical tool. While STEM offers resolution advantages over SEM, it is not intended to replace the SEM. Considering the time and labor required for STEM specimen preparation and the operator skill level required, STEM should be used only in those selected cases where its capabilities are indispensable.

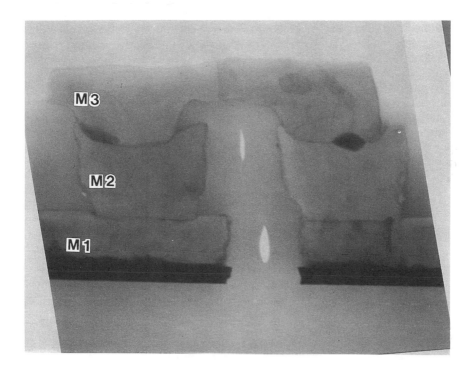

Figure 7. Scanning TEM micrograph from a multilevel metal cross section. Voids in gap-fill glass can be evaluated without concern for etch artifacts seen in SEM cross sections. Due to the superior spatial resolution of the technique, thin interfacial layers (< nm) can be clearly identified.

Plan-view TEM imaging of metallization layers has been used extensively for grain structure evaluation, impurity segregation and defect characterization. For example, AlCu grain structure, size distribution, θ-phase formation nd grain boundary segregation have been the subject of a number of studies (34-37). Grain-size distributions are accurately measured by TEM imaging. Grain-size measurement by optical and SEM microscopies, do generally need grain-delineation techniques (especially for grain sizes <1μm), whereas TEM needs no such surface treatment. In the SEM evaluation a possibility exists that the surface features such as thermal grooves are taken as grain boundaries even after grain boundaries have moved with further thermal processing. TEM analysis on the other hand, unambiguously determines the grain boundaries. Since TEM offers excellent spatial resolution, it is also used in the accurate measurement of physical dimensions where other tools fail. Calibration data is usually generated in such cases using TEM. Thicknesses of very thin films (nm range) and interfacial layers (eg: native oxide in Al via) can be measured by TEM.

Microstructural characterization of interconnect layers is key to the development of reliable metallization. Figures 8 a-d show examples of microstructural changes observed in Al-1.5%Cu during *in situ* annealing in the TEM. As mentioned above, the average grain size and grain size distribution (information relevant to electromigration) can be determined from such plan view TEM micrographs. The AlCu film was deposited at 465°C. It is interesting to note that with progressive annealing, the θ- phase particles coarsened, dissolved and reprecipitated at the grain boundaries (identified by the arrows) (36). In this example, the location and size of the precipitates did not change significantly from the as-deposited sample. TEM is used in characterizing such microstructures and their dependence on process parameters which in turn are controlled in fabricating reliable metallization.

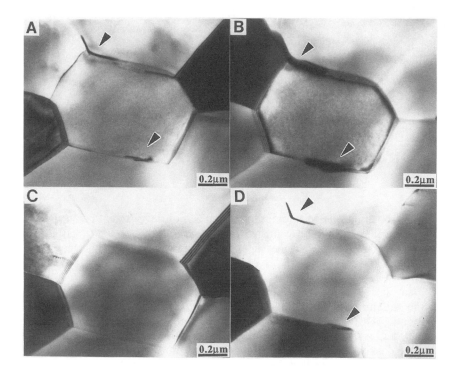

Figure 8. a-d. TEM of AlCu microstructure. In-situ annealing of plan view TEM sample. Al-1.5%Cu as deposited at 465°C. Coarsening, dissolution and reprecipitation of θ particles at the grain boundaries are shown (marked by the arrows). a) 25°C, b) 380°C, 1 min, c) 450°C, 1 min, d) 25°C.

It is important to characterize defect densities and configurations in MLM structures. Extended defects can affect electromigration properties of interconnects and degrade electrical properties of device structures. TEM is very useful for analysis of extended defects. If a crystalline defect is present in the region being imaged, the strain associated with the defect causes local lattice planes to move away from the Bragg condition. There is a loss of intensity contributed to the diffracted beam; locally distorted planes in the vicinity of the defect cause a local drop in diffracted intensity. The defect is therefore delineated in the image. Dislocations can be seen in TEM images as lines of greater or lesser intensity than the matrix. The appearance of the dislocation image changes for images obtained using different diffracted beams. The nature of the change in appearance can be used to derive certain characteristics of the dislocation such as the Burgers vector. Inclined planar faults such as stacking faults produce periodic fringes in the image (38). Precipitates and impurity clusters show local variations in contrast due to local strain or diffraction effects (39). In addition to imaging, diffraction patterns that are obtained from precipitates or second phases can be used to identify their structure (and hence chemical nature). The unique advantage of such micro-diffraction patterns is that the information is specific to a very small area and such information is difficult to obtain using any other technique. The information can lead to better control of the microstructure and phase-structure of MLM materials. This can lead in turn to better control of electrical, mechanical and reliability characteristics.

Analysis of the chemical composition can be performed using energy dispersive X-ray spectroscopy (EDS) (discussed in section 4.1) coupled to the TEM and thus EDS is a common addition to the microscope. This is particularly useful in compositional analysis of small areas typically investigated in the TEM. Highly resolved compositional variation across grain boundaries can be measured by this technique (35), which is quite useful in studying grain boundary segregation phenomena in metal films.

2.3. Scanning Tunneling Microscopy (STM)/ Atom Force Microscopy (AFM)

Scanning Tunneling Microscopy (STM) and Atom Force Microscopy (AFM) are two recently emerged techniques with atomic resolution imaging capability (40-44). The excitement in the capabilities and potential of these tools is reflected by their rapid incorporation into integrated circuit process development and characterization while the tool is still undergoing development. STM operates on the principle of quantum tunneling between a biased metal tip and a conducting surface. The tunneling current depends on the tip to sample separation of the order of 0.1nm. As the tip is scanned over the sample with piezoelectric controls, the height is controlled by a feed back loop to maintain a constant tunneling current and this height variation is processed to give a display of topography of the sample surface with atomic level resolution. Atomic resolution has been reported not only in ultrahigh vacuum but even under atmospheric pressure and liquids. However, one limitation of the technique is that STM is restricted to conducting surfaces which do not form insulating native oxides. Although work has been done in in-situ imaging under a solution, adequate progress has not yet been achieved for routine materials characterization.   This approach might be of use in clean metal surface characterization without concern for gas adsorption from the ambient or in the study of elctrochemical reactions.

The limitation of STM due to the presence of insulating surface film has been overcome by the development of the companion technique of Atom Force Microscopy (AFM) (45). Instead of the tunneling current, AFM measures the interatomic force between atoms at the apex of a fine tip and the atoms on the surface of the sample under study. The tip mounted on a soft cantilever is scanned in the X-Y plane and the vertical deflections of the cantilever due to the above forces are detected and transformed to a three dimensional image, similar to the STM display (43, 44, 46). Similar to STM, the AFM uses a servo loop to lock the tip height maintaining a constant cantilever deflection. Since the AFM cantilever deflections depend on interatomic forces, both conducting and insulating surfaces can be imaged, and this has been found to be more applicable in evaluating surfaces encountered in typical MLM fabrication. Since the tip is small, the forces are extremely small, of the order of 10^{-9} Newtons. At these forces the tip is capable of tracking individual atoms without damage to the surface even though the tip can actually be in contact with the surface. In this contact mode, the tool is like a conventional stylus profilometer with at least five orders of magnitude lower loading force and no damage to the surface. Typical area of AFM imaging ranges from a few nm^2 to μm^2 and depth resolution of atomic dimensions.

An example of an AFM image of AlCu surface is shown in Figure 9. Grain size information can be very easily derived from such micrographs. No grain delineation is needed in sample preparation and the technique is very fast. Grain size and surface profile monitoring for wafer process control are two obvious applications of AFM in MLM, for which tool capabilities already exist and methods are mature. The true uniqueness of AFM is evidently its atomic dimension resolution. Ultrahigh spatial resolution images of surface features and microstructure and grain boundary effects can be obtained with this technique. It must be kept in mind that the image obtained is always that of the interaction between cantilever tip and the net force on the sample surface. Images of AlCu surfaces, for instance, show the surface of the native oxide with a perturbation caused by the interaction between the tip and any residual forces such as electrostatic, capillary, etc. Some samples have a highly repulsive electrostatic force which makes scanning difficult because of the extreme bending of the cantilever away from the sample. In general, such perturbations tend to be small such that much information can still be derived from micrographs of Al surfaces. Moisture on the surface can also affect the resultant attractive force. Any chemical differences along the surface and grain boundaries will also affect the sensed image.

While the AFM images on typical metal surfaces have yielded superb high resolution images, much more work lies ahead in gaining more confidence in quantitative structural characterization of thin films, growth of thin films and surfaces encountered in MLM processes. While the tool is being used with confidence in process development areas, some further tool and method development is needed for it to be usable in a manufacturing environment. Tip technology for metrology of high aspect ratio structures, as in high density MLM structures, still needs significant development. Potential for STM/AFM in surface imaging and critical dimension metrology is immense and it is expected that the tool development will continue at a fast pace and new applications in integrated circuit fabrication will vigorously be pursued. Strengths of these techniques are: microscopy with atomic resolution, surface specificity, no special sample preparation, operation in air or vacuum, analysis of conductors and insulators, and speed of analysis.

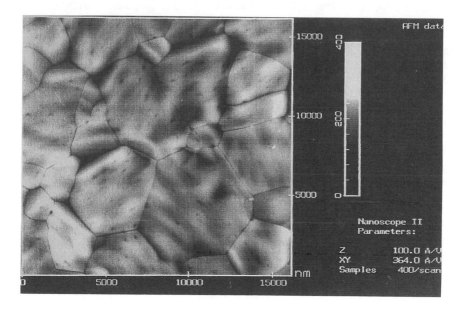

Figure 9. AFM image of AlCu surface. Grain size measurement can be routinely performed.

2.4. Thermal Wave (TW) Imaging

Imaging of subsurface defects in MLM in order to characterize and monitor them non-destructively and in 'real-time' is a requirement in wafer processing. One group of defects which has drawn intense attention because of its severity and impact on reliability of fine line Al metallization is stress-induced voiding.

Optical microscopy and SEM have been used to evaluate these metal voids. However, significant limitations exist for these tools in subsurface void evaluation for process monitoring. A technique which has been adapted for this specific application is the thermal wave imaging (47-53). Thermal waves exist when there is heat generation and heat flow in a medium. Common methods of generating thermal waves involve absorption of energy from a modulated laser beam or electron beam, though for metal film evaluation a laser beam is used. Once thermal waves are generated, they propagate and interact with regions differing in thermal properties eg: defects such as voids or precipitates. This contrasting interaction can be detected to provide an image. In the commercial TW imager, (Therma Wave, Inc., Fremont, CA) specifically designed for metallization defect studies, a modulated Ar ion pump laser generates locally, non-damaging heat waves or thermal waves which propagate and interact with defects/voids. This laser raises the local temperature typically <10°C but which is dependent on the presence of localized defects. A second He-Ne probe laser detects the changes in the wafer surface temperature properties from its reflectivity modulated by the presence of these defects. This modulated optical reflectance of the probe forms the signal for imaging the defect. The area of interest on the wafer is rastered under the laser spot and a composite thermal image is formed from the above type signal, through established video image processing methods.

Although the metal films are opaque to the pump laser, the thermal waves propagate ~3 microns into the metal and can still modulate the reflectivity at the surface and hence voids ~3 microns below the surface can be detected. Voids can be inspected through passivation layers or even multiple levels of metallization. Defects as small as 1kÅ can be distinguished by this imaging technique. The advantages of the technique are its non-destructive nature and relative speed of analysis (~2 min for typical image acquisition). The former makes it a suitable technique for studying void growth dynamics. An example of TW imaging is illustrated in Figure 10 which shows the void size evolution in Al metallization subjected to heating. The bright areas in the TW images (Figure 10. b,c,d) represent voids in the metal. Relatively small size stress voids are seen in the sample heated to 412°C for one hour while the voids have coalesced to form larger voids in the sample heated for eleven hours.

While the imaging of voids has received the most attention in TW imaging applications in MLM, there appear to be many other potential applications largely unexplored. The mapping of precipitates in Al alloy metallization is required in process characterization. An example of a TW image of silicon precipitates in 1μm wide AlSi metal lines, imaged through the passivation layer, is shown in Figure 11. Copper precipitates also have been similarly mapped with the TW. Other extensions to TW applications include quantitative hillock height measurement and grain size distribution (53-55) and W plug and Al via fill quality evaluation. While new applications of TW in the characterization of structural morphology and defects in metallization continue to be recognized, the technique is still qualitative in the fingerprinting of defects. Nevertheless the technique is receiving acceptance as an in-process monitoring tool for metallization defects. Other applications in MLM process development will be explored as more studies are done on this emerging imaging technique.

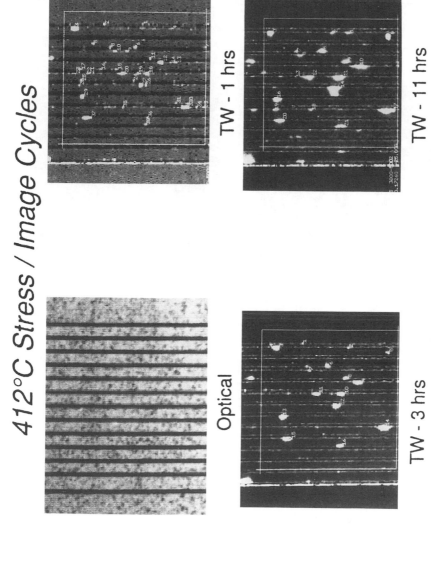

Figure 10. a-d. Thermawave images of stress void evolution in aluminum lines with heating.

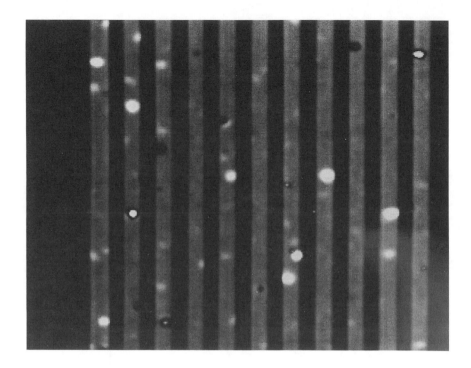

Figure 11. Thermawave image of silicon precipitates (bright areas) in AlSi metal lines.

2.5.   Focused Ion Beam (FIB)

The focused ion beam (FIB) tool is a relative newcomer to the MLM process development and materials characterization (56) tool-set although it has been extensively used in the past for mask and device repairs. While FIB has capabilities in imaging, it is deemed more a technique to prepare cross section samples with high precision using ion milling. The FIB has been applied to substantially simplify and/or improve existing techniques for preparing cross-sections of MLM structures without causing deformation artifacts. It is especially useful when studying specific or low density defects which would be impossible or very tedious to cross section, using existing polishing methods. Imaging of the cross sections can be done in the FIB tool or by transferring to a SEM. Observation of ion channeling contrast in FIB ion microscope mode allows observation of crystallographic orientation changes in metal grains and grain size evaluation.

Development and understanding of the liquid metal ion source (LMIS) (57-60) led to realization of commercial FIB instruments. The liquid metal ion source in its most basic form is a high melting point metal needle coated with the "liquid" metal which will be ionized. When the needle is sufficiently biased as the anode, the liquid metal flows down the needle, balancing the electrostatic force against the liquid metal surface tension, to form a cone with an extremely small tip radius of ~150 Å. When the electric field between the anode and cathode, reaches ~1.5-2 V/Å, field evaporation produces ions of the liquid metal which are subsequently focused to a very fine beam using electrostatic lenses. Gallium is the most widely used "liquid" metal. It is the only source available which can produce the high beam current densities (~5 A/cm^2) and the low energy spread required to achieve finely focused ion beams with reasonable milling rates. The ions leaving the LMIS are extracted and focussed through an electrostatic system with multiple apertures to generate a variety of ion spot sizes and beam currents best suited for milling or imaging. Typical Ga$^+$ ion beams between 0.1 to 1.5 μm diameter with currents of 250 to 4000 pA respectively, are used for milling (61) while a minimum beam diameter of ~350Å at 15 pA is used for imaging.

When the ion beam impinges on a sample surface, secondary ions and electrons are produced in addition to sputtered neutral atoms. A variety of detectors are employed to produce images from these secondary electrons or ions, which often have contrast mechanisms different from those in electron microscopy. Unlike SEM secondary electron micrographs in which the predominant cause of contrast is topography related, FIB images show chemical contrast. For example, using ion induced secondary electrons, silicon dioxide appears dark and aluminum bright, while silicon nitride and silicon appear gray. Interestingly, this contrast is reversed when the FIB positive secondary ion signal is examined.

FIB applications to MLM defect and structure analysis have been reviewed by Nikawa (62) and can be divided in two main groups, cross section sample preparation, and direct imaging of metal grain structure. Today's high density ICs require investigation of defects having very low spatial density (~1 D/cm^2) and submicron size (<0.5 μm). This presents a special challenge to the analyst who must locate, cross-section or otherwise prepare the defects for study, typically by electron microscopy. The traditional techniques usually include defect location using optical microscopy and cross-sectioning either by cleaving or metallurgical polishing which lack the needed precision. Additionally, MLM structures usually contain aluminum which is easily deformed by cleaving or even the polishing technique and there is always concern about sample preparation artifacts. This concern is particularly acute when attempting to section through subsurface voids in aluminum.

Sample preparation by focused ion beam reduces and sometimes eliminates these problems. Applications of and procedures for precision FIB cross-sectioning for IC failure analysis have been reported by many authors recently (63-67). The tool has become an integral part of many failure analysis laboratories. The defect is located using the FIB imaging capabilities. A relatively large (10 x 20μm) "rough cut" box is milled a sufficient distance from the defect to allow a second "clean-up" cut using a smaller beam diameter providing a precise, high quality cut. The cross section of interest is on one of the walls of this 'box'. Well behaved non-charging samples allow beam placement accuracy of ≈0.1 μm. The cross-section sample may be examined in the FIB using secondary ions, or for better resolution examined as usual in the SEM.

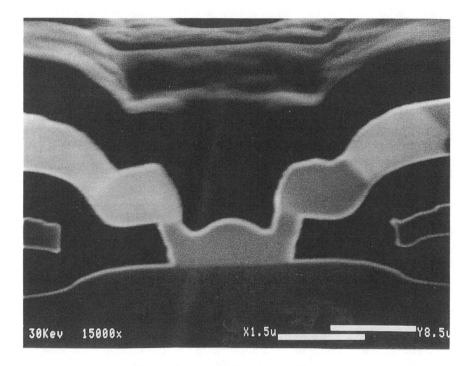

30Kev   15000x   X1.5u   Y8.5

Figure 12. FIB induced electron image from a FIB generated cross section of Al contact to silicon. The grain structure of aluminum is easily discernible.

Information on the structural integrity of metal films in contacts and vias, the grain structure of sheet films and that in vias, is often sought for modeling and understanding failure mechanisms. Direct imaging of the precision sectioned via in the FIB often provides such information. Ion channeling contrast which varies as the crystal orientation changes, is shown in the secondary electron image in Figure 12. This image from a FIB generated cross section is that of an Al contact to silicon and shows the contact as consisting of a single grain and the metal in general has columnar grain structure. Grain size distribution of aluminum sheet films and its variation with thermal treatment have been studied using FIB imaging (64,68). An example of grain size imaging is shown in Figure 13, which is an ion channeling image from an AlCu film. Contrast in the grain structure image is due to the grain orientation. The θ−particles can also be discerned at grain boundaries. The advantage of FIB method in measuring grain size lies in the fact that no special sample preparation is needed unlike SEM or TEM. Another clever variation of the precision milling capability of FIB cross sectioning is in preparing TEM cross sections through specific locations for failure analysis (69-71).

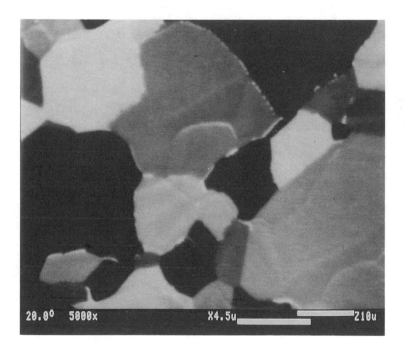

Figure 13. Ion channeling image from an AlCu film. Contrast in the grain structure image is due to the grain orientation. The θ- particles can also be discerned at grain boundaries.

A summary of the strengths and weaknesses of the above listed imaging techniques is given in Table I. These techniques used in a synergistic manner provide the microanalytical capabilities required for IC fabrication.

Table I. A summary of key strengths and weaknesses of the imaging techniques used in MLM characterization.

	Strengths	Weaknesses
SEM	• Most indispensable tool for microscopic observations of MLM topography and structural details. • Excellent spatial resolution ~3-10nm. • Contrast from morphology, atomic weight, electric potential. • In-line dimensional metrology with 10nm accuracy. • Rapid analysis. • Compositional information when coupled with EDS. • Moderate operator skills for routine operations.	• Electron beam induced modification/contamination complicate further surface analysis. • Charging of uncoated dielectric material. • Samples compatible with vacuum.
TEM	• Imaging with atomic resolution. • Planview+cross-sectional microscopy. • Coupled with EDS, compositional analysis from area ~100nm. • Crystal microstructure /defect analysis.	• Time-consuming sample preparation. • Destructive sample preparation. • High vacuum compatible samples. • Operator skill demands.
STM/ AFM	• Surface microscopy with <0.01nm vertical resolution. • Non -destructive • Minimal sample preparation. Sample in vacuum, air or liquid. • Relatively simple instrumentation. • Growing technique with many potential applications in local probing.	• Primary information on surface microroughness. • Imaging susceptible to artifacts • Sensitive to vibration and temperature drift. • Image dependence on tip condition.
T.W.	• Non-destructive imaging of voids and precipitates in metal • In-line process monitoring. • No vacuum restrictions. • Relatively rapid.	• Contrast not specific to material identity. • Spatial resolution ~1μm. • Interpretation of contrast not simple for MLM structures.
FIB	• High precision cross section for microscopy. • Ion beam imaging complements SEM imaging. • Ion image resolution ~10nm. • Metal line cut or repair.	• Long term ion beam stability to section through <0.1μm.

3.0. SURFACE ANALYSIS

3.1. Auger Electron Spectroscopy (AES)

MLM processing involves modification of thin film surfaces. Characterization of the outermost layer of ~10Å of a thin film is imperative in understanding this modification due to processing steps. The physical and electrical properties of integrated multiple layers and multiple levels of thin film metallization are significantly influenced by these altered surface layers. A variety of analytical techniques specifically sensitive to surface layers ~10Å have evolved during the last two decades. Some of the advances in these techniques have been evidently driven by the demands of integrated circuit fabrication.

Auger electron spectroscopy (AES) has emerged as perhaps the most popular surface analytical tool for surface characterization and this holds true for MLM characterization also. The technique has been the subject of exhaustive review articles and books (4, 7, 8,13, 72, 73, 74). Auger electron spectroscopy in its basic form of operation is used to identify elemental composition of surfaces. The principle of the technique is that the sample is bombarded by a finely focussed electron beam of energy in the range of 1-10 keV and electrons of characteristic energy - Auger electrons- [named after Pierre Auger (75) who discovered the phenomenon] are emitted from the sample. The predictable energy of Auger electrons is characteristic of the element and is used to identify elemental composition (76, 77). Most elements except hydrogen and helium have characteristic Auger electrons in the energy range 0-2000 eV. Characteristic spectra of all elements and many compounds have been documented in handbooks (78,79). The reason why AES is surface sensitive is the following. Auger electrons with energy in the above range suffer intense inelastic scattering on their way to escaping from the solid. Only those electrons escaping from the solid without any energy loss will be detected as characteristic Auger electrons. These Auger electrons originate from a surface layer of about 5-30Å (80), depending on their energy. In its simplest form, AES is used to identify elemental composition, however, in selected cases (oxides, nitrides, carbides, some silicides, etc.) chemical bonding information can be obtained from the Auger peak energy, peak shape and fine structure (81 -87). Such chemical effects are extremely valuable in quickly and routinely fingerprinting the chemical state of the surface species. Good examples are the spectra from pure Al vs. Al_2O_3 , adsorbed carbon vs. metal carbides, Si vs. SiO_2 and Si_3N_4. Such fingerprinting is routinely performed in process monitoring.

Auger spectroscopy can be made quantitative from the Auger peak to peak amplitude, and previously established "sensitivity factors" for each element (78). Surface concentration

[1] $$C_x = (I_x / S_x) / \left(\sum I_i / S_i \right)$$

where C_x is the atomic fraction

I is the amplitude of Auger peak

S is the relative sensitivity factor

This method can yield accuracy of ~ 10% in determining atomic concentration. Better accuracy may be obtained by generating sensitivity factors specific to the instrumental set up from good reference samples and calibration with other techniques. The method can be extended to compounds by generating sensitivity factors (if not available from handbook) for reference samples. This is especially true where significant Auger peak shape change exists between elemental and compound form. Minimum detectable limits of the Auger technique range from 0.1-1.0 at %.

The analytical capability of AES is expanded substantially by coupling it with noble gas (argon or xenon) ion sputtering to remove surface layers sequentially and analyzing composition as a function of depth (88). This provides the depth-compositional profile (89). In thin film analysis using AES, such depth profiling is a standard practice. The sputtering time can be converted into depth from known sputter rate data, however, this is not always straightforward in multi-layer structures where non- linearity of sputter rate with depth complicates such conversion. However, depth profiling has been the most extensively used mode in MLM material characterization, particularly in studying thin film interactions as a function of processing. Many publications exist on the study of thin film interdiffusion and rate constants for lattice and grain boundary diffusion from such Auger profiles (90-92). An example of sputter profiling is shown in Figure 14. a-b. These depth profiles are from 7.5kÅ thick Al-1.5%Cu film as deposited on a SiO_2 layer and after an annealing step. The interest has been in the distribution of copper in the film. While copper is homogeneously distributed in the film as deposited, segregation occurs with annealing. This type of profiling is only aimed at the compositional distribution in one dimension -in depth- with no reference to lateral inhomogeneity.

Sputter depth profiling, which is essentially a destructive analysis technique, is not without artifacts due to sputter induced effects (93-97) such as surface roughening, interfacial broadening, ion mixing, preferential sputtering, etc. These artifacts need to be taken into consideration in quantitative analysis of thin film interactions. The problem of sputter induced surface roughening is severe in aluminum metallization especially when using fixed ion beam incidence angle. The roughening effect can be minimized by sample rotation during sputtering (98-101), choice of proper angle of incidence (more grazing) of ion beam (98), choice of ion species Xe^+ vs. Ar^+ or lower ion beam energy.

In the continual quest for improved spatial resolution in surface analysis, Auger electron spectroscopy has been further developed with the incorporation of a scanned primary beam instead of a stationary spot and the Scanning Auger Microprobe (SAM) emerged. Significant advances in spatial resolution, high brightness electron sources, signal detection and processing have been made in scanning Auger microprobe technology. These have opened up much improved tool capabilities and wider applications in surface microanalysis, particularly suitable for microelectronics fabrication. 'State -of -the -art' SAM [eg: PHI model 670, Perkin Elmer Corp, Minnesota] has the capability of analyzing areas as small as ~ 50 nm.   The SAM has secondary electron image capabilities similar to a SEM. The SEM image is particularly useful in locating the area of analysis and identifying the microstructural features and as a starting point for further compositional analysis. SAM is used in multiple forms of operation. In addition to the normal small area analysis, it has the capability to map two dimensional distribution of elements at the surface (Auger map) The Auger map coupled with the secondary electron image (SEI) offers excellent visualization of lateral surface compositional variation.   An example of the SEI and the corresponding Auger map of copper, is given in Figure 15. a-b from an Al-1.5% Cu film taken from a PHI 670 scanning Auger 'nanoprobe'. The familiar θ-phase segregates at grain boundaries are very elegantly mapped in the Auger map. Point analysis capabilities are demonstrated by the spectra taken on the θ-phase vs. the middle of the grain and are shown in Figure 16. a-b. With the continual reduction in device feature size, such high resolution Auger analysis is required in process development and failure analysis. In summary, the strengths of AES are, surface specificity, comparable sensitivity to all elements in the periodic table with minimum detectable limits ~0.1 at %, excellent spatial resolution, depth profiling and imaging capability.

3.2.   X-ray Photoelectron Spectroscopy (XPS)

XPS is another member of the family of surface sensitive techniques used in thin film characterization with particular emphasis on surface and interfacial chemistry (8, 13, 102, 103, 104).   It is a complementary tool to Auger spectroscopy, particularly suited for the analysis of insulating samples and those easily damaged by an electron beam. It is also known by another name, ESCA (Electron Spectroscopy for Chemical Analysis) coined by the Nobel Laureate K. Siegbahn and his group in Uppsala who developed the technique for chemical analysis (105).   In this technique the sample surface is bombarded with monoenergetic X-rays to cause the emission of photoelectrons from the core levels or the valence band, which are energy analyzed to precisely determine the binding energy of the electron in the atom.   The binding energies of photoelectrons are characteristic of the element and of the chemical or oxidation states of atoms from which they are emitted. The binding energy (B.E) of the electron is given by the following simple relation:

$$[2] \qquad B.E.= h\nu - K.E - \phi$$

where, $h\nu$ = Incident X-ray photon energy
K.E.   = Kinetic energy of the emitted electron
ϕ = Work function of spectrometer

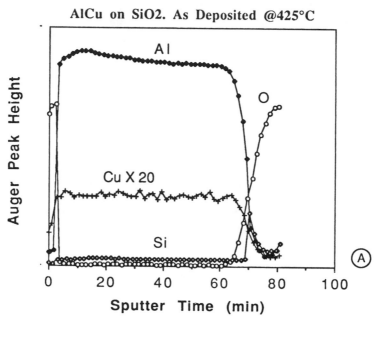

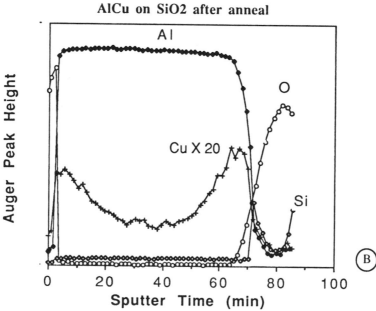

Figure 14. a&b. Auger depth profiles from AlCu film, as deposited and after an anneal step. While the in-depth distribution of copper is homogeneous in the as-deposited film, segregation occurs with the anneal step.

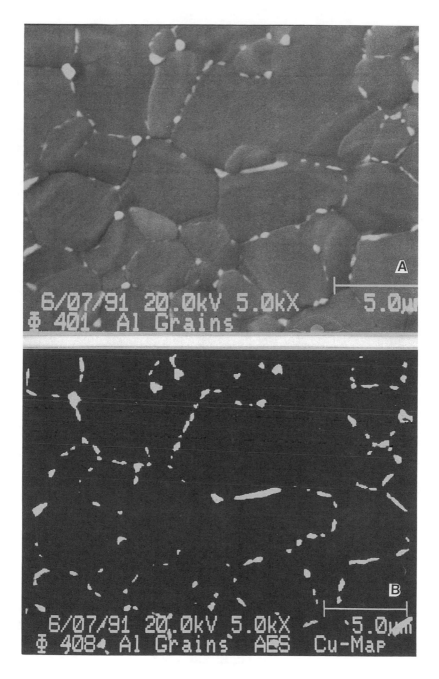

Figure 15 a&b. a) Secondary electron image and b) the corresponding Auger map of copper from an Al-1.5% Cu film taken from a PHI 670 scanning Auger 'nanoprobe'.

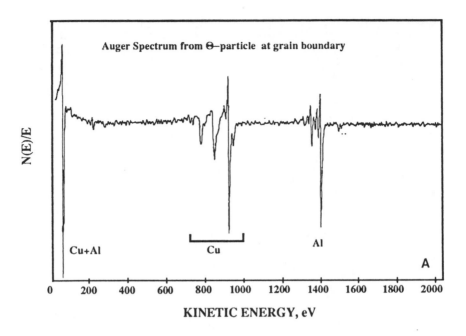

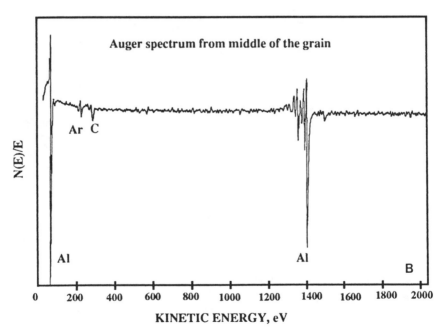

Figure 16 a&b. Auger spectra taken on the θ-phase particle (in Figure 15) vs. the middle of the grain. These spectra illustrate the point analysis capability of state of the art high resolution Auger spectrometers.

In XPS, normally Al K_α or Mg K_α X-rays of energy 1486.6 eV or 1253.6 eV respectively are used as the excitation source. With these excitation sources, photoelectrons from all elements in the periodic table can be excited. These photoelectrons are energy analyzed by a spectrometer from which the B.E of the electrons are calculated using the relation in equation (2). A plot of the photoelectron intensity as a function of the binding energy yields the characteristic XPS spectrum. The binding energy identifies the elemental composition at the surface. In addition to the photoelectrons, Auger electrons are also emitted during the relaxation process occurring in the photoionized atom. The kinetic energy of the Auger electrons provides supplementary information on the surface composition. Figure 17 illustrates the characteristic features in a typical XPS spectrum from an aluminum surface with some etch residues present. Various photoelectron peaks characteristic of the surface elemental components and Auger electron peaks from fluorine and oxygen are seen in this 'survey spectrum'. Such survey spectra are used for the cursory identification of the major elemental constituents. The technique is ideally suited for surface contaminants or process induced residues which may be susceptible to damage/desorption by electron beam bombardment techniques.

The unique strength of XPS lies not just in its ability to detect elemental composition but in its capability to provide chemical bonding information from the so called 'chemical shifts' (105). The chemical shift arises because chemical bonding changes the valence electron distribution which in turn changes the core level binding energy. Correspondingly there will be a shift - chemical shift - in the photoelectron binding energy. This chemical shift is extensively used in characterizing chemical bonding. An example of the chemical shift is illustrated in Figure 18.a in the spectra from elemental Ti and TiN where a shift of 1.1eV is seen in the Ti 2p level binding energy. Chemical shifts can easily distinguish TiN and TiO_2 as shown in Figure 18.b and such shifts are used to fingerprint chemical bonding at the surface. The chemical shifts in Auger transition evident in the XPS spectrum also have been used to complement the photoelectron chemical shift.

Since photoelectrons have only a few hundred eV energy, similar to the case of Auger electron attenuation, these photoelectrons can escape into vacuum, without energy loss, only from a shallow region ~20-40Å, which is approximately the sampling depth of XPS for most metals and insulators. Quantitative evaluation of the surface concentration can be made, similar to AES, from the generalized expression, $C_x = (I_x / S_x)/ \Sigma I_i/S_i$.
Typical minimum detectable limit for the technique is in the range 0.1 at% - 1.0 at%. In combination with in-situ Ar^+ sputter etching, depth profiles of thin films can also be obtained. Depth profile and thickness of thin layers (<50Å) can be obtained non-destructively by angle resolved XPS (106-108). The thickness of a thin overlayer can be estimated from the intensity ratio using the relation:

[3] $$\ln[R / R^\infty +1] = d / \lambda \sin \theta$$

where R and R^∞ are the overlayer to substrate intensity ratios for overlayer thicknesses of (d) and infinite ($>3\lambda$) respectively, where λ is the mean free path of electrons of the specific energy and θ is the take-off angle (the angle between the analyzer axis and the sample surface) of photoelectrons.

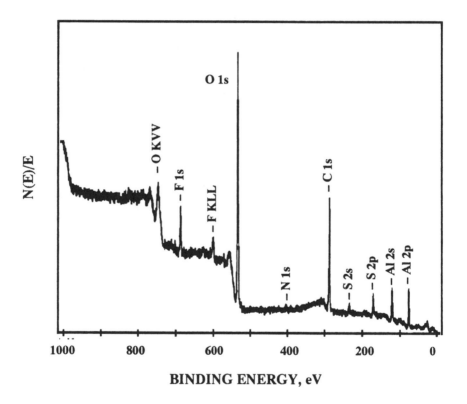

Figure 17. XPS spectrum from an aluminum surface with etch residues present. Various photoelectron peaks characteristic of the surface elemental components and Auger electron peaks from fluorine and oxygen are seen in this 'survey spectrum'. Survey spectra are used for the cursory identification of the major elemental constituents.

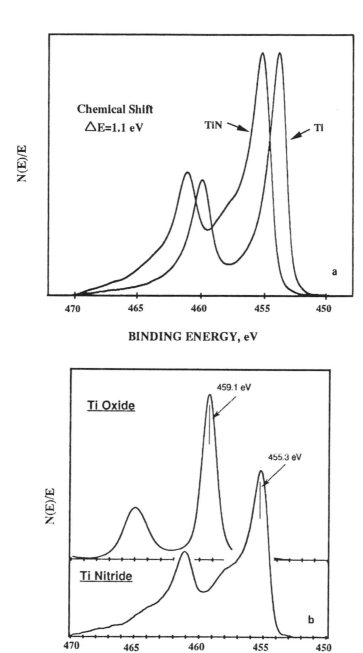

Figure 18. a&b. a) XPS spectra from elemental Ti and TiN where a chemical shift of 1.1eV is seen in the Ti 2p level binding energy. b) Characteristic chemical shifts easily distinguish TiN and TiO_2.

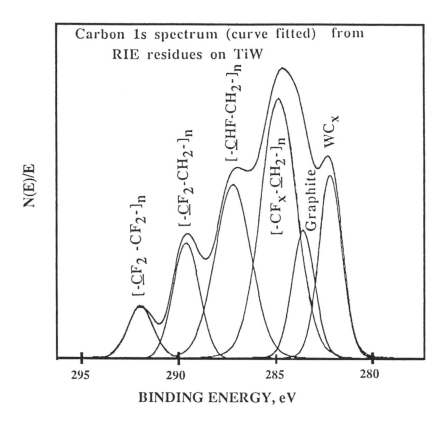

Figure 19. Curve fitted carbon 1s XPS spectrum from surface residues due to a CHF_3 RIE step. The presence of WC_x and $-CF_x$ types of polymeric residues is identified.

One of the limitations of the technique from the stand point of MLM characterization, is the spatial resolution. By and large the technique uses a broad area (order of ~1mm) for analysis, though continual advances are being made in reducing the area requirement. Spatial resolution of the order of 10μm has been demonstrated. XPS is especially suited for evaluating MLM dielectric materials such as SiO_2, Si_3N_4, spin-on glass, PSG, BPSG, etc because of its minimal charging problem with x-rays and minimal damage for organic dielectric materials such as photoresist, polyimides etc. XPS offers the capability to study RIE induced surface modification prevalent in IC processing. An example of TiW surface residues from a CHF_3 RIE step is shown in the carbon XPS spectrum in Figure 19. The presence of graphitic carbon, WC_x and $-CF_x$ types of polymeric residues on the TiW surface is clearly seen. Such surface residues formed by RIE on a variety of surfaces have been extensively studied by XPS. The interpretation of chemical shift data to correlate to the surface chemistry demands scientific expertise and as such, the tool is less accepted for in-line manufacturing evaluation.

3.3.   Secondary Ion Mass Spectrometry (SIMS)

There are very few thin film analytical techniques capable of determining trace level impurities (ppb or less) with a depth resolution of a few nanometers and spatial resolution of ~1 micron. Secondary ion mass spectrometry (SIMS) is the leading technique capable of meeting these demands in integrated circuits fabrication. The ultra-high sensitivity of SIMS is exploited in the analysis of MLM processes and materials to complement other analytical tools which emphasize microscopy and compositional analysis. The operational principles and capabilities of SIMS have been extensively reviewed (12, 13, 109-114) to which the reader is referred for details. The technique is based on sputtering. A primary beam of ions with energy in the range 1-20 keV, bombards the sample surface whereby atoms from the surface are sputtered off. The sputtered species leave this surface as atoms or molecules either in neutral or ionized state, with the ionized species being only a few percent. In SIMS, the secondary ions which may be positively or negatively charged, are detected by a mass spectrometer and sorted according to their mass to charge ratio. Such a mass spectrum gives an identification of the instantaneous elemental composition of the surface. In principle, it is straightforward to identify the elemental composition from the mass spectrum. However, in reality, the mass spectrum is complex and its interpretation nontrivial, since it contains mass peaks from elements, their isotopes and molecules, either singly charged or multiply charged. The large number of peaks due to these different types of ions leads to a peak overlap problem and unambiguous identification of elemental species, especially from multicomponent samples, is rendered difficult. Nevertheless mass spectrum analysis is routine for determining the presence of specified impurities in MLM structures.

The secondary ions are ejected from a relatively thin surface layer of the order of nm and hence the technique is by and large surface sensitive. The secondary ion yield and hence the sensitivity and the minimum detectable limits depend on a number of factors and in fact can vary as much as five orders of magnitude depending on the analytical parameters. The ion yield is dependent on the primary ion species, beam energy, elemental concentration and the sample composition (the so-called matrix effect). However, the primary ion parameters are chosen to maximize the secondary ion yield. Commonly O_2^+, Cs^+ or Ar^+ ions are used as primary ion species. Oxygen ions are used to exploit the higher ion yield from electropositive elements whereas Cs^+ ions are used for electronegative ions. Thus in typical analysis, based on the elements of interest, one uses O_2^+ or Cs^+ ion bombardment and detects positive or negative ions. An example of the negative ion spectrum from an AlCuSi film bombarded by a Cs^+ beam is shown in Figure 20. It is interesting to note that the Si-signal is much higher than Al, although the Si content in the film is only 1% compared to the nominal 98.5% Al. It is simply an indication of the fact that Si forms negative ions more readily rather than their relative abundance. Despite the difficulties associated with secondary yield variation due to a variety of factors, the technique can be made quantitative especially for minor constituents in thin films, by comparing to suitable standards. If standards approaching the composition of the unknown sample are used, under identical operating conditions, accuracy of ~10% can be obtained in determining elemental concentration.

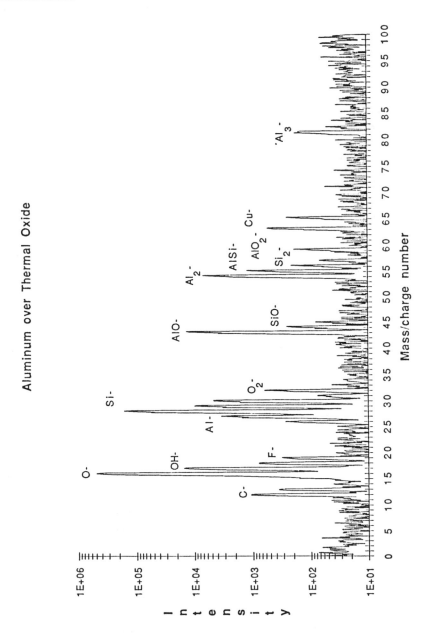

Figure 20. Negative ion spectrum from an AlCuSi film bombarded by a Cs⁺ ion beam. Elemental \ and molecular species are typical of such SIMS spectra.

The more widely recognized application of SIMS in MLM characterization is in depth profiling of minor constituents or trace impurities in thin films. The technique is particularly invaluable where the concentration of the element of interest is 1% or less. The technique supplements other surface analytical techniques such as Auger and XPS which lack its capabilities in minimum detectable limits. As in other sputter profiling techniques the instantaneous composition is monitored as a function of sputter time. In order to discriminate against the signal from the walls of the sputter crater, the primary ion beam is usually rastered over an area and the secondary ions are selected from the flatter crater bottom using electronic gating. Surface profilometry is used to measure the crater depth, subsequent to SIMS measurement from which the sputter time is converted to sputter depth. The method of converting sputter time to a depth is not without error particularly in multilayer thin film structures because of the non- linearity in sputter rate for different films and interfaces. An example of a SIMS depth profile is illustrated in Figure 21, from an AlCuSi film over a TiW film. Positive ion signals from the various elements are monitored. The Ti profile suggests that Ti diffusion from the TiW underlayer into the Al film has occurred at the processing temperature.

Quantitative determination of composition of minor constituents from the depth profile is usually practiced and provides good results. The quantification is achieved by determining the integrated counts from the depth profile and comparing to reference samples (115,116). Such an integrated count approach is used in routine monitoring of processes and impurities in multilayer structures. Minimum detectable limits of most elements of interest in MLM processing lie in the range of $1E13/cm^3$ -$1E16/cm^3$, with the choice of optimum analytical parameters. The depth profiling capability coupled with the ultrahigh sensitivity is used in impurity monitoring in IC processing. The vast majority of examples in depth profiling has been in dopant profiling in silicon. However, the technique is extended to impurity profiling in metallization layers and integrated MLM structures. While there are charging issues associated with SIMS analysis of thick dielectric layers in MLM, various neutralization techniques can be effective in minimizing these. An example is given by the depth profiles of Na and K in a two level metal structure in Figure 22 (117). It is interesting to note that these profiles were obtained by sputtering through conducting Al layers and insulating SiO_2 layers. While there are many ion beam induced effects in profiling through such inhomogenous layers of metals and dielectrics, including ion induced migration of sodium, the technique offers reliable data in process comparison studies. By using the integrated signal intensity and comparing to reference samples the total sodium content could be determined. Such procedures are used in routine monitoring of impurity levels in MLM processing.

SIMS, as other sputter profiling techniques, suffers from the many effects of surface modification due to ion bombardment (93-97) and hence precautions must be taken to account for these artifacts which adversely affect SIMS data. Surface-roughening is significant in metal films and loss of depth resolution is a common occurrence. The same approach of sample rotation used in Auger depth profiling has been used to improve depth resolution in SIMS depth profiling of Al metallization and dielectrics (118-119).

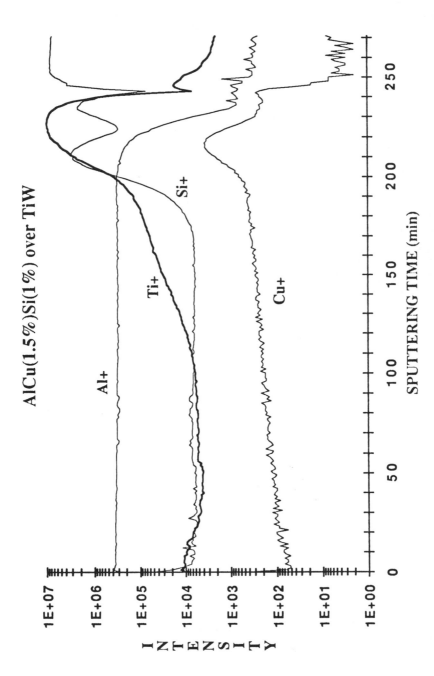

Figure 21. SIMS depth profiles from an AlCuSi film over TiW film after heat treatment. Positive ion signals from various elements are monitored. The Ti profile suggests Ti diffusion from the TiW underlayer into Al.

Na Profile in Two Level Metallization

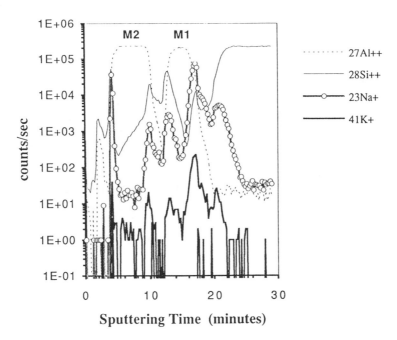

Sputtering Time (minutes)

Figure 22. SIMS depth profiles of Na and K in a two level metal structure. While the apparent segregation of these ions at the various interfaces may be in part due to ion beam effects, the integrated signal counts gives very useful information in comparative analysis.

The secondary ion signal can be used to image lateral distribution of surface elements over the bombarded area. In MLM structures the imaging provides additional visual information on the lateral inhomogeneity of elements as in the case of grain boundary segregation. While, in general, the imaging offers only qualitative information in concentration differences, such information is very valuable in correlating impurity presence to the specific location in the IC structures. A lateral resolution of <1μm is readily achieved in state-of-the-art commercial SIMS instruments. An extension of the two dimensional imaging is to store such two dimensional information and sequentially generate a series of 2D images and through suitable data handling software to generate a composite three dimensional image of the structure and concentration profile. In today's submicron IC structures, which are essentially three dimensional structures, such 3D imaging is becoming valuable and desirable.

The strength of the SIMS technique is in its ultrahigh sensitivity. While there are limitations to this technique such as difficulty in analyzing the widest range of samples, matrix effects, wide range of sensitivity variation, time of analysis etc, its unique strength is exploited in materials and process characterization and SIMS has become an indispensable part of the analytical tool set.

A comparison of the strengths and weaknesses of the above described surface analysis techniques is made in Table II. Such a comparison is merely intended as guide in the selection of a particular technique based its strengths and the information sought from the analysis.

Table II. A summary of strengths and weaknesses of the surface analytical techniques used in MLM characterization.

	Strengths	**Weaknesses**
AES/ SAM	• Most widely used surface/ interface analysis technique • Depth specificity ~1nm • Sensitive to all elements except H_2 & He • Quantitative; Min. Det. Lim 0.1- 10at% (depending on probed area) • Depth profiling with sputtering • Chemical effects in spectra • Excellent spatial resolution in SE imaging (~5nm) and elemental mapping (~25nm)	• Electron beam damage (dielectrics, halogens, etc) • Charging in dielectrics • UHV compatible samples • Sputtering induced artifacts in depth profiling • Limited information on organics
XPS	• Surface composition with chemical bonding information • Sensitive to all elements except H_2 • Non-destructive analysis • Insulators and conductors, Good for organics • Min. Det. Limits ~1at% • Quantitation with ~10% accuracy • Depth profiling with angle resolved XPS and sputtering	• Large area analysis (~mm) (100µm exceptions) • High vacuum compatible samples • Much expertise required in interpreting complex spectra
SIMS	• Ultra high sensitivity (ppm-ppb) • Sensitive to all elements • Isotope information • Good spatial resolution (~1µm) • The primary trace element analysis tool for thin films • Computer generated 3-D information	• Destructive technique • Ion beam effects in interface analysis • 10^3-10^5 variation in sensitivity among elements; Matrix effects • Complex spectra from unknown samples • Quantitation difficulties (without reference samples)

4.0.   THIN FILM 'BULK' ANALYSIS TECHNIQUES

4.1.   Electron Microprobe (EMP)

Electron Microprobe (EMP) is a well established compositional analysis technique which has been practiced for over four decades (120, 6,12, 14,15). The technique is based on the bombardment of the sample with a high energy electron beam, typically (5-30 keV) and the detection of characteristic X-ray emission from the sample. The technique is also known by other synonymous names such as EDS (Energy Dispersive Spectrometry) or WDS (Wavelength Dispersive Spectrometry). The distinction primarily lies in the X-ray detection approach. The former discriminates X-rays by their energy, usually using a solid state detector while the WDS sorts out X-rays by their wavelengths using diffracting crystals. The technique now-a-days is integrated with most analytical SEM and TEM, while stand alone dedicated EMPs are also common especially in more in-depth quantitative compositional analysis. EDS is generally used for more rapid analysis and where its limited energy resolution does not limit the identification of elements due to X-ray energy overlap or proximity.

The energy or the wavelength of characteristic X-rays fingerprints the elemental composition. The X-ray signal originates from a depth of the order of one micron and hence the information is characteristic of a volume of the order of one cubic micron. The spatial resolution of the technique is not governed so much by the primary beam diameter but rather by the spreading and energy loss of the beam inside the sample. So while the electron beam in the SEM might have a beam diameter of ~10nm, it does not translate to comparable spatial resolution in compositional analysis. The intensity of X-rays is a measure of the concentration of the corresponding element. The analysis can be either qualitative or quantitative. While minimum detectable limits of 0.01 wt% are achieved for higher atomic number elements ($Z>20$), for low atomic elements the corresponding limits are at least an order of magnitude worse. With effort, boron can be detected down to $\approx 1wt\%$ (for example in BPSG). In MLM materials and process characterization, qualitative elemental identification is simply performed by fixing the beam over the point of interest (eg: on a particle) or over a rastered area if the film is homogeneous and detecting the X-rays. An example of the local compositional analysis is shown by the X-ray spectra from a Al_2Cu θ-phaseparticle located at the grain boundary in AlCu metallization and off the particle (Figure 23. a&b).

Quantitation of composition of homogeneous samples is fairly routinely performed, while recognizing that a variety of correction factors need to be used. Such correction procedures are well established and are incorporated into the computer software. Quantitative analysis of multilayer thin films is also possible by comparing to suitable standards with similar layered structure. While the X-rays emanate from a depth of $\approx 1\mu m$, it is possible to measure thicknesses of thin layers forming a multilayer structure. For example, the thickness of a TiW cap layer <100Å over an Al layer can be quantified, using proper calibration standards. While electron beam induced charging can be a problem in the analysis of thick ($\approx 1\mu m$) interlevel dielectric layers, a thin layer (~100Å) overcoat of conducting layer of carbon effectively eliminates the problem without significantly impacting the compositional analysis. Phosphorus content in PSG and B and P in BPSG films can be measured by this method and it is a commonly used as a monitoring method.

X-RAY SPECTRUM FROM THE PARTICLE

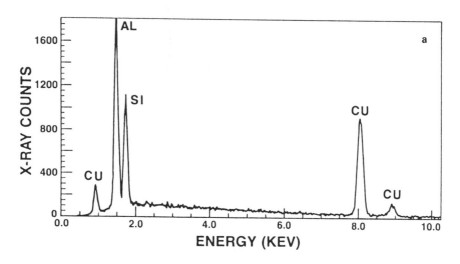

X-RAY SPECTRUM OFF THE PARTICLE

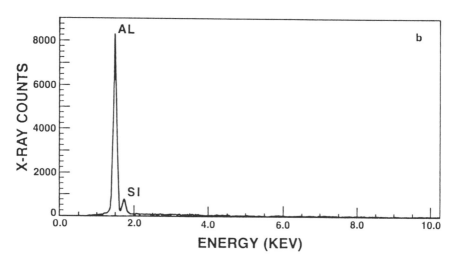

Figure 23 a&b. X-ray spectra from a θ-phase particle in AlCu film and off the particle. The silicon X-ray signal has its origin in the substrate.

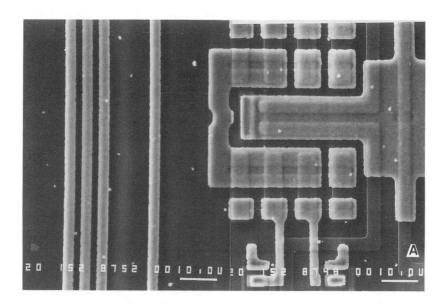

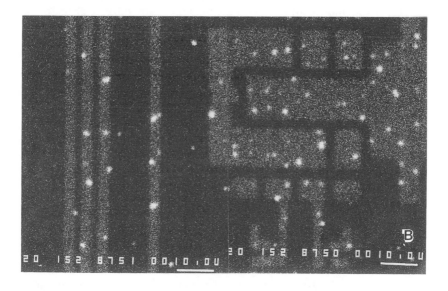

Figure 24. X-ray map of copper distribution in metal pattern. a). SEM image of the area of analysis. Metal lines are covered by a layer of SiO_2. b) X-ray map of copper from the area.

As the primary electron beam is rastered over the area of analysis, X-ray signal from a selected element can be detected and used to image the lateral distribution of the selected element. Such an X-ray image (the X-ray dot map) though qualitative, is very useful in correlating local compositional variations with structural variations and complements the electron images from a SEM or TEM. An example of the X-ray map of copper distribution in AlCu metal pattern (from an area in the SEM image in Figure 24.a) is displayed in Figure 24.b. The metal lines are covered by a layer of SiO_2 which is no hindrance to the imaging. Typical applications of EMP in MLM fabrication include the bulk compositional analysis of films, small volume analysis (such as in the identification of particles), thickness of multilayer films and lateral compositional variation (eg: segregation) through X-ray mapping.

4.2.   Rutherford Backscattering Spectrometry (RBS)

Rutherford backscattering spectrometry (RBS) is an ion beam analytical method that is simple in concept and widely used in the fundamental studies of thin film interactions. While the technique is not readily available in many semiconductor analytical laboratories, its use is pervasive in characterizing thin films, particularly metal films. The principles of the technique and its applications in characterizing thin film metallization have been extensively reviewed [10,121-123]. A monoenergetic helium ion beam, typically in the energy range 1-3 MeV, bombards the thin film sample. The impinging ions encounter Coulomb interaction with the nuclei of target atoms at or below the surface. A transfer of some kinetic energy occurs and a change in direction or scattering of the particles is the result. A few of the projectile ions undergo large-angle scattering, or "backscattering". If the incident ion of mass M_1, atomic number Z_1, and energy E_0, scatters back from a surface atom of mass M_2 and atomic number Z_2 with an energy E_1, then E_1/E_0 is a given by a kinematic factor K_{M2}. For scattering angles $\approx 180°$,

[4] $$K_{M2} = (M_2 - M_1)^2 / (M_1 + M_2)^2.$$

The K factor provides mass identification in the RBS analysis if the energy of the backscattered ions is measured.

The probability that a projected ion will elastically scatter is expressed by the Rutherford scattering cross-section of each target atom and is further described for RBS as a differential scattering cross-section, $d\sigma/d\Omega$, which is defined in the direction of the detector as a ratio of the number of particles scattered into the solid angle $d\Omega$ to the number of incident particles per unit area. The differential scattering cross-section is calculated for any target element. It allows the direct quantitation of atoms of a particular mass in a thin film.

As a penetrating ion moves through the target medium, it is subjected to energy-dissipating, low-angle interactions with surrounding nuclei along the incident path to the scattering atom. The rate of energy loss is determined by the stopping cross-section, which depends on the mass of the incident ion, its velocity, and the composition and density of the target. Extensive tables are available on the stopping cross sections. After backscattering, this process continues along the outward path, causing further energy loss. The energy loss through some distance x, dE/dx, is known and allows depth perception in the analysis. RBS then utilizes kinematics, scattering cross-section, and energy loss to produce a spectrum showing backscattering yield as a function of energy. From this spectrum it is possible to identify atomic masses present in a thin film and determine their atomic concentrations as a function of depth. Since the information is acquired from first principles, there is no reliance upon secondary standards. This aspect makes RBS unique among the commonly used particle beam analytical techniques such as AES, SIMS, and XPS and in fact RBS is used to generate quantitative data on reference samples for these techniques.

Resolution of lighter masses is limited with RBS, making the technique best suited for evaluating heavier-mass constituents on (or in) lighter-mass substrates, an example of which might be a transition metal film on silicon. Figure 25.a depicts a RBS spectrum of such a structure consisting of a platinum film on a silicon wafer. The Pt spectrum has uniform yield over a continuum of energies resulting from backscattering events at the surface and throughout the film to its back interface. The ΔE at FWHM of the Pt spectrum can be converted to a depth scale for Pt. The Si spectrum results from backscattering at the interface and through to 'infinite' depth, which for RBS is generally beyond 1μm. Pure Pt scatters energetic ions far more efficiently than pure Si, a result of the higher scattering cross-section for Pt. This is reflected by the vastly different spectrum heights for Pt and Si in Figure 25.a.

If the Pt/Si sample represented in Figure 25.a is heated to 350°C, an interfacial reaction occurs resulting in the formation of a Pt_xSi_y layer between the pure Pt on the surface and the substrate Si. A RBS spectrum of this system is shown in Figure 25.b. The overall shape of the Pt spectrum has changed, becoming "stepped" on the lower-energy side coupled with a corresponding change in the Si spectrum. The lower height of each element's spectrum reflects the lower atom concentration of each of the constituents in the reacted layer relative to the elemental Pt and Si. The relative concentration ratio of Pt to Si in the reacted layer is derived from a product of the ratio of spectrum heights and the inverse ratio of scattering cross-sections. The result of this analysis is that the reacted layer in Figure 25.b is an intermediate Pt_2Si phase. If the thermal cycle proceeds, the remaining Pt is eventually consumed to form Pt_2Si, followed by the conversion of Pt_2Si to PtSi. RBS analysis at some point during the formation of PtSi produces the spectrum in Figure 25.c which shows Pt_2Si/PtSi layers on Si.

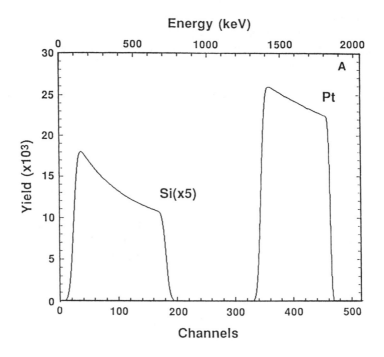

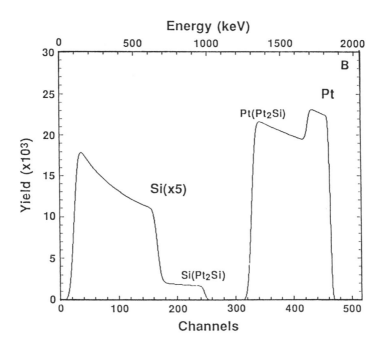

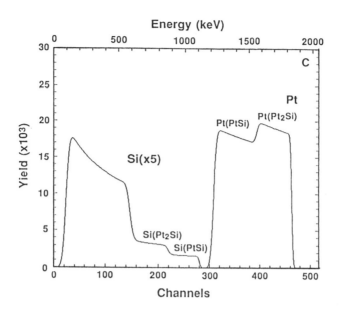

Figure 25.a). RBS spectrum from a platinum film, as deposited, on a silicon wafer. b) On heating to 350°C, an interfacial reaction occurs resulting in the formation of a Pt_2Si layer between the pure Pt on the surface and the substrate Si. c) Spectrum obtained with further heating indicating the formation of $Pt_2Si/PtSi$ over Si.

Applications of RBS to a variety of thin film materials studies are evident in the literature. The principal thrust or strength of the technique is not so much in searching for unknown trace impurities in multilayer structures but rather in the basic understanding of the interactions between multilayer thin films and to aid in developing thin film technology. An example of RBS in the study of thin film interactions for MLM technology is given in the spectra in Figure 26, from a Ti film on Si before and after reaction at elevated temperature. The peak labeled 'unreacted' Ti and the edge labeled 'bulk Si before reaction' represent the 1000Å Ti over lightly doped Si. From the backscattered energy difference between the front and back edges of the Ti peak the film thickness can be easily verified. The spectrum labeled 'after reaction' was from the film after reacting at 650°C for 20 minutes in N_2. The Ti signal consists of two regions. The surface edge of the Ti signal represents the Ti constituent in the TiN layer formed on the surface of film and while the lower energy region represents Ti in $TiSi_2$. The step in the Si signal similarly represents Si in the $TiSi_2$ and the energy difference between the front and back silicon signal edges gives a measure of the thickness of the silicide layer. Stoichiometry of 'TiN', $TiSi_2$ and their thicknesses could be calculated from these spectra. Growth rate of the various layers and process parameters which influence them are typically studied from such RBS spectra. Specific to MLM technology, the largest volume of papers appears to be on silicide characterization. The technique while viewed as a research tool in the past, in recent years has become an integral part of the off-line analytical tool set for MLM characterization.

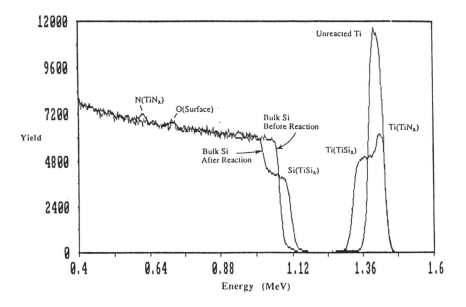

Figure 26.   RBS from a Ti film on Si before and after reaction at 650°C for 20 minutes in N_2. The surface edge of the Ti signal represents the Ti constituent in the TiN layer formed on the surface of film and while the lower energy region represents Ti in $TiSi_2$.

4.3.   X-Ray Diffraction (XRD)

X-ray diffraction (XRD) is an established analytical technique (124-125) for structure analysis of materials and is used in characterizing the crystalline phases formed in thin film growth or due to thin film interactions during device fabrication.   Much of the XRD characterization of metal layers in ICs is performed during the technology development phase in order to establish an understanding of the fundamental properties of thin film interactions. In addition to the identification of crystalline phases, the technique is used to evaluate the strain state and the average grain size in thin films.   The principle of the technique is based on the well known principle where the incident X-ray beam is diffracted by a set of lattice planes (hkl) if the Bragg equation, $2d \sin \theta = \lambda$, is satisfied, where θ is the angle between the incident (or reflection) beam and the planes (hkl) and λ the wavelength of the radiation.   Generally the X-ray intensity is measured as a function of the diffraction angle 2θ between the incident and the diffracted beam.   The phase identification can be done by comparing the measured d-spacing with known standards.

Two different diffractometer designs are used for thin films (125-126). The first one, Bragg-Brentano diffractometer, is commercially available and most widely used. The specimen is mounted in the center of the diffractometer and rotated by an angle θ around an axis in the film plane. Only (hkl) planes parallel to the film plane contribute to the diffracted intensity. The effective thickness, t/sin θ, which a film of thickness t, exposes to the incident beam decreases with increasing diffraction angle and thus give low intensity at high diffraction angles. Preferred orientation of film could also be determined by this design. The second diffractometer design utilizes Seemann-Bohlin ray path. The specimen and the focusing circle remain stationary and the detector tube moves along the circumference of the focusing circle itself. The effective thickness remains constant and can be made very large by using small angles of incidence (as small as 5°) and thus giving higher diffracted intensities than the first design. On the other hand, its arrangement is limited to polycrystalline specimens of random orientation.

An example of the X-ray diffraction pattern from a multilayer thin film structure is given in Figure 27.a-b which shows the thin film diffraction pattern from a Ti/CoSi$_2$/Si structure. The pattern in Figure 27.a is from the structure prior to thermal stressing and shows a CoSi$_2$ pattern plus two additional peaks which are due to the presence of α-Ti with mainly (002) orientation. After thermal nitridation in an NH$_3$ ambient a δ-TiN layer is formed as seen by the diffraction pattern. Such structural determination is performed during process development to characterize thin film interactions during processes. X-ray diffraction methods have not been applied to the study of thin films as extensively as electron-diffraction methods mainly due to the lower diffracted intensities ($\sim 10^3$ times) than that of electrons, which is caused by difference of the physical scattering mechanisms. However, X-ray diffraction methods have advantages of ease and convenience. The diffraction pattern represents the average throughout the film (not localized as electron diffraction) due to the increased penetration, and simultaneous excitation of film and substrate. Identification of fine precipitates (such as CuAl$_2$ phase) with a small volume fraction is very hard with XRD due to the low intensity. However, preferred orientation of deposited films can be easily measured by comparing with the ASTM standard power diffraction data file (127). Especially large diffraction angles offer accurate measurement of plane spacings (accuracy of ~300 ppm) which could be used for homogeneous strain measurement. Line broadening of the peak could be used to estimate crystallite size d, which is given by $d = \lambda/(D \cos \theta)$, where D is the angular width at half the maximum intensity. Recent development of sophisticated procedures [Warren-Averbach Fourier analysis using at least two orders of reflections from the same set of planes, say (111) and (222)] makes it possible to separate each contribution of strain and small crystallite size (128). Improved detection methods for X-rays, the availability of commercial monochromators, and intense microfocus X-ray sources have made X-ray diffraction methods applicable to films as thin as 100 Å (129-130). However, the methods are usually applied to several hundred angstrom thick films.

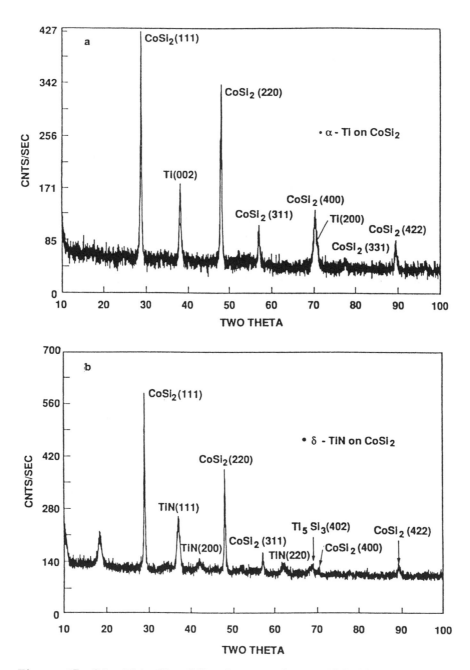

Figure 27.a&b. Thin film diffraction trace from a Ti/CoSi$_2$/Si structure. Fig.a. is from the structure prior to thermal stressing and shows CoSi$_2$ pattern plus α-Ti. b) After thermal nitridation in NH$_3$ ambient, δ-TiN layer is formed as seen by the diffraction pattern.

4.4. X-ray Fluorescence   (XRF)

X-ray Fluorescence is a bulk chemical analytical technique which is used in routine monitoring of composition or thickness of layers used in MLM. In this technique a beam of X-rays is directed onto a sample and core electrons (photoelectrons) are ejected by the X-rays, leaving behind vacancies which are subsequently filled by electrons dropping down from higher electron orbitals. The energy difference between the two orbitals is released in one of two forms; the ejection of Auger electrons or the emission of X-rays. The energies of the emitted X-rays are characteristic of the elements from which they originate. These emitted X-rays are energy discriminated and counted either by allowing them to impact onto a lithium-doped silicon solid state detector (energy dispersive) or by collimating them onto an analyzing crystal which diffracts different energy x-rays at different angles according to Bragg's law and counting the X-rays at appropriate angles with either a gas-filled proportional counter or a scintillation counter (wavelength dispersive). Matrix effects can significantly affect counting intensities, so the calibration of the instrument with matrix-matched standards is essential for accurate analysis results. XRF is typically used to determine the phosphorus and/or boron content of PSG, BSG and BPSG thin films (131-133). It is also used to monitor bulk compositions of TiW and NiCr films, the copper content in AlCu films, and to determine the thickness of metal or dielectric films.

The major advantages of XRF analysis are that it is quite rapid (usually only a few minutes for analysis), automated and it is non-destructive. These attributes make it attractive for in-process monitoring in high volume manufacturing environment. The major disadvantage is that the spatial resolution is very poor, and thus analyses are generally performed on blanket films on test wafers. Small spot XRF systems are available, but the spot size is still much too large to be useful for the analysis of most features on a typical IC.

4.5. Fourier Transform Infrared Spectroscopy (FTIR)

The integrated MLM structure includes not only the metallization layers but also several doped and undoped dielectric films as interlevel dielectric or passivation layers. These materials include doped and undoped films of SiO_2:H, $Si_xO_yN_z$:H deposited using various CVD/ PECVD techniques or various organic layers such as polyimides. These films can have high bonded hydrogen content, high tensile stress, and become unstable with time and/or temperature and can have a negative effect on the integrity of the metal layers when subjected to additional thermal processing. The chemistry of these films deposited at a variety of conditions and subjected to a variety of process conditions needs to be characterized. Fourier Transform Infrared Spectroscopy (FTIR) is widely accepted as the tool of choice to characterize these dielectric films and has been the subject of considerable research (134-138). The nature of the technique allows nondestructive characterization of chemical bonding within the dielectric film.

The ability of IR to characterize chemical bonding is due to the fact that the vibrational and rotational energies of molecules exist in the infrared region of the electromagnetic spectrum (0.8-1000μ). Infrared absorption by a molecule will be observed if there is coupling between the molecular vibration and the incident oscillating electric field of the IR radiation. This change in energy is normally observed as an absorbance band in the infrared spectrum and depends on the atoms present and the type of bonding environment.

The FTIR spectrometer uses a Michelson interferometer instead of conventional monochromators to sort the frequency spectrum. Each frequency of radiation results in an interference pattern that is a modulated cosine wave whose frequency is determined by the velocity of the moving mirror, i.e. the interferometer is taking a Fourier transform of the incoming signal. The resulting signal is the signed summation of all the cosine waves. The detector sees all of these frequencies simultaneously, which results in an interferogram (Intensity vs. Time). The reverse Fourier transform mathematically re-sorts the individual frequencies and transforms them to the frequency domain. This results in a (Intensity vs. Frequency) plot.

For the specific case of analyzing dielectrics, the film normally deposited on a single-side or double-side polished silicon wafer with high resistivity is immediately measured in Transmittance T, where $T=I/I_0$ where I_0 is the energy of the incident light and I is the intensity of the transmitted light through the (film + Si substrate). I_0 is obtained by collecting a background spectrum without a sample in the spectrometer. The Si substrate is measured separately. The ratioed transmittance spectrum of the Si substrate is then converted to Absorbance A, where $A=\log_{10}(1/T)$. This is subtracted from the absorbance spectrum of the (film + Si substrate) resulting in an absorbance spectrum of the film.

An example of FTIR spectrum from a plasma enhanced CVD a-Si_xN_y:H film deposited at 350°C is shown in Figure 28. The different types of bonding configurations can be observed in the spectrum. Since most dielectric films used in this type of application are amorphous, the absorption bands are quite broad with large Full-Width-at-Half-Maximum (FWHM) values. Most a-Si_xN_y:H films are silicon-rich and contain a considerable amount of (Si–H) bonding and (N-H) bonding. A (Si-H) bond stretching absorption is observed at 2175 cm^{-1}. As the more electronegative nitrogen atom replaces the hydrogen originally bonded to silicon, the band shifts from ~2000 cm^{-1} in a-Si:H to 2180 cm^{-1} in a-Si_xN_y:H. Hydrogen bonded to nitrogen is observed at 3334 cm^{-1} (N-H stretching) and near 1190 cm^{-1} (N-H bending). There is also weak evidence of a (SiN-H_2) bending mode at 1539 cm^{-1}. Since hydrogen plays an important role in determining most of the chemical, physical and electrical properties of a-Si_xN_y:H films, quantitative determination of bonded hydrogen is desirable. This is achieved by calibrating to another primary standard technique such as resonant nuclear reaction analysis. FTIR is used routinely for correlating bonded hydrogen content to the physical and electrical properties of these thin films (139).

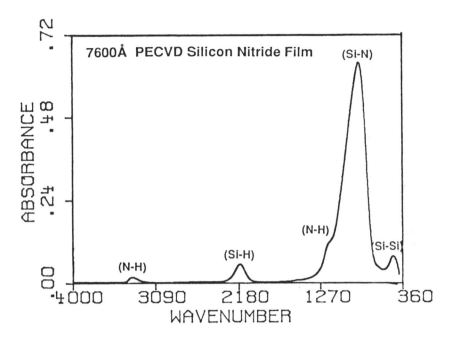

Figure 28. FTIR spectrum from a plasma enhanced CVD a-Si_xN_y:H film deposited at 350°C. The different types of bonding configurations, including hydrogen bonding, arc observed in the spectrum.

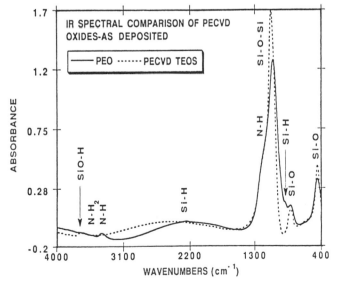

Figure 29. IR spectrum from two plasma enhanced CVD films using SiH_4 and TEOS chemistry.

PECVD a-SiO$_x$:H films (TEOS and SiH$_4$ based) used as ILD are porous and are more susceptible to water absorption. The absorption of water causes drift in the stress of the oxide. This is due to swelling of the glass by moisture and the subsequent hydrolysis of unstable siloxane (Si-O-Si) linkages within the film. When the film is heated during subsequent processing steps, outgassing of the moisture can cause deleterious effects in the interconnect metal films (140-143). Changes with time which occur in the hydroxyl (-OH) stretching region in the spectrum from PECVD TEOS oxide films stored in air have been studied to characterize these films (144). Quantitation of the hydroxyl as water and silanol is possible using FTIR.

Under certain conditions, the use of N$_2$O as the oxidant in the deposition of SiO$_x$:H using SiH$_4$ (PEO) results in the incorporation of a small amount of nitrogen in the film. This can be observed in the IR spectra of two PECVD oxides (TEOS and PEO) deposited using different deposition conditions (Figure 29). The presence of nitrogen is indicated in the figure. The PEO film is stable to moisture due to the incorporation of nitrogen which results in a denser, more relaxed film. In phosphosilicate (PSG) and borophosphosilicate (BPSG) glass films used as ILD, phosphorus and boron incorporation into the films is determined by the deposition parameters and causes film instability due to water absorption. As the amount of phosphorus and/or boron is increased in the film, the tendency toward water absorption increases. FTIR has been used extensively for studying the incorporation of these dopants into silicate glass films and the effect on their resultant physical and chemical properties (134,145-147).

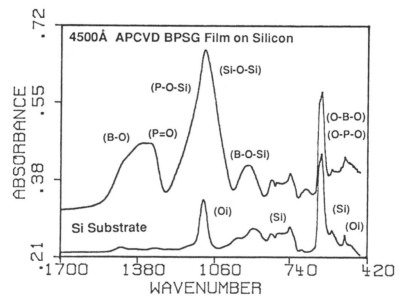

Figure 30. The FTIR spectrum from a CVD BPSG film deposited using SiH$_4$, PH$_3$ and B$_2$H$_6$ in N$_2$/O$_2$ on a Si substrate is compared to the spectrum of the Si substrate. The bands corresponding to the different vibrational absorptions present in the glass are shown in the figure.

The IR spectrum of a CVD BPSG film deposited using SiH_4, PH_3 and B_2H_6 in N_2/O_2 on a Si substrate is shown in Figure 30. This is compared to the spectrum of the Si substrate. The bands corresponding to the different vibrational absorptions present in the glass are shown in the figure. Accurate control of boron and phosphorus is essential for obtaining consistent, reproducible reflow behavior. For process control purposes a rapid, precise and nondestructive technique is necessary. FTIR is ideally suited for the in-process monitoring of these films. Unfortunately, peak-area or band-height measurements, which are sometimes useful for quantitative determination of boron or phosphorus are not successful here due to nonlinear variation of absorptivity with increasing concentration of dopants (147). However, the recent availability of a number of full-spectrum multivariate statistical methods such as Principle Component Regression (PCR) and Partial Least Squares (PLS) have allowed high precision in the analysis of boron and phosphorus in BPSG films to be achieved (148). Careful application of these multivariate statistical techniques can give a standard error of predictive abilities for boron and phosphorus of less than 0.1 wt% and 0.2 wt% respectively.

4.6. Ion Chromatography (IC)

In this technique a dilute acid or base, called the eluent, is pumped through an injector, in which the sample is introduced, then through a separator column, a suppressor column and on to a detector. The separator column is packed with tiny ion exchange resin beads containing fixed charges which have the effect of retarding the flow of any oppositely charged species in the eluent stream. The separation of different ions is possible because some ions have greater affinity for the fixed charge than others, causing different species to elute from the column at different times. Most often the eluent is chemically converted to a less conductive form and the ions of interest are converted to a more conductive form in the suppressor column. The eluting ions are then usually detected by a conductivity detector, although other detectors are useful for some species. Most common anions and cations can be detected at parts per billion levels by direct injection ion chromatography as described above; however, parts per trillion levels can be detected if a preconcentrating ion exchange column is used prior to the injector. For certain very sensitive ions, such as fluoride, chloride, and sodium, detection limits in the low part per trillion range are achievable.

IC is commonly used to determine the ionic contaminant levels in DI water and in semiconductor processing chemicals and materials. Specific to MLM related materials analysis, IC has been used to analyze the boron and phosphorus content in PSG, BSG, and BPSG films (133, 149-150). The films must be stripped off and reacted with HF for a lengthy period of time to convert all of the B_2O_3 in the film to the fluoroborate ion so that it can be detected by IC. The different forms of phosphorus in the film are converted to phosphate, phosphite, fluorophosphate, or polyphosphate, depending on whether the phosphorus was present as P_2O_5, P_2O_3, bonded to silicon as Si–P, etc. Because of the amount of time required for the IC analysis, the analysis of boron and/or phosphorus-containing films is usually only performed to provide information for calibrating other analytical equipment capable of much more rapid analysis, such as XRF or FTIR.

IC is also capable of analyzing the extractable impurities on surfaces. For instance, to determine chloride residues on a wafer surface after an RIE etch such as in metal etch, wafers sampled both before and after the etch can be immersed in ultrapure deionized water and the contaminants extracted from the wafer surface. This can be analyzed for chloride and a wide range of other ions to determine process induced impurities. Similar types of extraction (sometimes involving higher temperatures and pressures) and analysis have been performed on a variety of materials used in wafer processing. The major advantages of IC are that it provides information on speciation; for example, IC can determine whether sulfur is present as sulfate, sulfite, sulfide, etc. The major disadvantages are that the analysis times are generally longer than for the other techniques and that the sample must be converted to a liquid state.

A brief comparison of the strengths and weaknesses of the bulk techniques discussed above are listed in Table III.

5.0   CONCLUSION

The above partial list of the more widely used analytical techniques for MLM materials and process characterization indicates the diversity of tools available to the analysts. Each technique has its own unique strengths and the inevitable weaknesses. No single technique meets all of the complex analytical needs of today's integrated circuit technology and the obvious approach is to use multiple techniques, exploiting their strengths, in a synergistic manner to perform a comprehensive analysis. It is generally recognized that the majority of these analytical techniques have been developed to their present capabilities due to the incessant demands of integrated circuits technology. On the other hand, the successful integration of process technology at the current levels has been in large part, made possible by the availability of such capable techniques for characterization.

While the analytical capabilities of the currently available tools are truly impressive, the demands on them are unabated. With the continued shrinking of device dimensions and the associated MLM features, interfaces, microstructures and trace impurities have a dominant role in determining their electrical properties. Continued demands are seen on higher spatial resolution analysis using many of the existing tools, ppb level detectability in 'nanolayers' and the analysis of subsurface interfaces buried a few microns below the surface. The tradeoff between resolution and sensitivity is generally recognized; one is improved usually at the expense of the other. While it is tacitly assumed that progress will be made in tool and capability developments in the future, these developments by and large, address the needs of the off-line analytical laboratories. Such analyses are usually destructive to the samples and time consuming and required during the process development phase. Adaptation of some of the tools is needed in the manufacturing environment for process monitor and control. The demands in such an environment usually are quite different. Faster and non-destructive analysis is imperative. Automated sample handling, computerized data manipulation and interpretation, and reduced dependence on analytical scientists are typical characteristics sought in such environment. Some of the techniques are expected to be integrated with process equipment for in-situ process monitoring and control.

Table III A summary of key strengths and weaknesses of the above bulk techniques for MLM materials characterization.

	Strengths	Weaknesses
EMP	• Quantitative analysis of bulk elemental composition • Good spatial resolution (~1μm) • High sensitivity to high Z elements; MDL 0.01 wt% for high Z. • SE imaging and elemental mapping • Suitable for in-process monitor	• Low Z elements (Z<11) less sensitive • Charging and e-beam artifacts in dielectrics
RBS	• Quantitative analysis of thin film composition without standards • Non destructive depth profiling (~20nm resolution depending on element) • Thin film thickness measurement • Isotopc sensitivity • High Z element MDL ~0.1at%	• Large area analysis (~mm) and hence laterally integrated .information • Poorer sensitivity for low Z elements. O/Pt signal~1/100 • Off-line research tool
XRD	• Crystal structure determination • Non-destructive • Rapid analysis	• Large area analysis (mm) • Crystalline samples • Lacks depth specificity
XRF	• Rapid, quantitative analysis of thin film composition • Conductors and insulators • Thickness measurement • In -process monitor application	• Large area analysis (mm) • Low sensitivity for low Z elements
FTIR	• Chemical bonding information on dielectrics. Excellent for organics. • Nondestructive • In process monitoring eg: P, B	• Complexity of spectra and interpretation • Min Det Limit ~1wt% • Large area (mm), 100μm possible
IC	• Trace level (ppb-ppt) sensitivity • Liquids and solids analyzed • Calibration of reference samples for other rapid techniques • Chemical bonding information	• Destructive analysis. Films need to be dissolved or impurities extracted • Time consuming sample preparation

Acknowledgements:

The analytical results reported here were generated, by a large group of analysts in Motorola Analytical Laboratories and I would like to thank all these colleagues especially, Messrs. J.Carrejo, J.Christiansen, P.Deal, R.Gregory, R.Hegde, M.Kottke, J.Mohr, T.Remmel, L.Rice, B.Rogers, N.Saha, H.Shin, D.Theodore, H.Tompkins, and D.Werho, and Dr.L.Smith of Therma wave, Inc.

REFERENCES

1. S.R. Wilson, J.L. Freeman and C.J. Tracy, Solid State Technol. 67 (Nov. 1991).
2. R.R. Uttecht and R.M. Geffken, Proc. 8th IEEE -VMIC Conference. 20 (1991).
3. P.F. Kane and G.B Larrabee, eds. Characterization of Solid Surfaces, Plenum Press, (1974).
4. A.W. Czanderna, ed. Methods of Surface Analysis, Elsevier Scientific Publishing, (1975).
5. J.I. Goldstein, H. Yakowitz, Practical Scanning Electron Microscopy, Plenum Press, New York, (1975).
6. K.F.J. Heinrich, Electron Beam X-ray Microanalysis, Van Nostrand-Reinhold, New York (1981).
7. G.B. Larrabee in VLSI Electronics: Microstructure Science, N.G. Einspruch ed., Vol. 2, Academic Press, New York, 37, (1981).
8. D.Briggs and M.P. Seah, Practical Surface Analysis, John Wiley, New York, (1983).
9. R.B. Marcus in VLSI Technology, S.M. Sze ed., McGraw Hill, New York, (1983).
10. L.C.Feldman and J.W.Mayer, Fundamentals of Surface and Thin Film Analysis, North Holland (1986).
11. E.S. Meiran, P.A. Flinn and J.R. Carruthers, Proc. IEEE. **75**, 908 (1987) .
12. R.E. Whan, in Metals Handbook 9th Ed. Vol. 10, Materials Characterizaiton, Am. Soc. Metals, Metals Park, Ohio, (1986).
13. J.M. Walls ed., Methods of Surface Analysis, Cambridge University Press, Cambridge, (1989).
14. J.I. Goldstein, D.E. Newbury, P. Echlin, D.C. Joy, C.E. Fiori, and E. Lifshin. Scanning Electron Microscopy and x-ray Microanalysis, Plenum, New York, (1981).
15. D.E. Newbury, D.C. Joy, P. Echlin, C.E. Fiori, and J.I. Goldstein. Advanced Scanning Electron Micrscopy and X-ray Analysis. Plenum, New York, (1986).
16. E.G. Solley, J.H. Linn, R.W. Belcher and M.G. Shlepr, Solid State Technol. 40 (Jan. 1990).

17. G.L. Kehl, Principles of Metallographic Practice. p424 McGraw Hill, (1949)
18. W. Baerg and P. Jupiter, Proc. of the International Confernece on Materials and Process Characterization for VLSI, World Scientific Publishing Co. N.J. 64 (1988).
19. S. Thomas, Scanning Electron Microscopy, 4, 158 (1983).
20. J.J. Gajda, G. J. Lindstrom, D.J. DeLorenzo, IEEE Trans Vol **CHMT-4**, 509 (1981).
21. J.J. Gajda, T.H. Irish and F.G.Trudeau, Semiconductor International 200 (1984)
22. J.J. Gajda and F.G.Trudeau, Proc. of ISTFA, Los Angeles, 7 (1991).
23. T.Mills and E.H. Sponheimer, 20th Proc. IEEE Reliability Physics Symp, 214 (1982).
24. P.B. Hirsch, A. Howie, R.B. Nicholson, D.W. Pashley and M.J. Whelan, Electron Microscopy of Thin Crystals, Robert E. Krieger Publishing Co, Malabar, Florida, (1977).
25. G. Thomas and M.Goringe, Transmission Electron Microscopy, John Wiley (1979)
26. L. Reimer, Transmission Electron Microscopy, Springer-Verlag, New York, (1984)
27. G.R. Booker and R. Stickler, Brit. J. Appl. Phys., 13, 446 (1962).
28. P.J.Goodfellow, Practical Methods in Electron Microscopy Series, Vol.1, A.M.Glauert, ed., North Holland, (1972)
29. J.C. Bravman, R.M. Anderson and M.L. McDonald, eds., Specimen Preparation for TEM of Materials, Mat. Res. Symp. Proc.vol 115 (1988)
30. T.T. Sheng and R.B. Marcus, J. Electrochem. Soc., 127 (3), 737 (1980).
31. R.B. Marcus and T.T. Sheng, Transmission Electron Microscopy of Silicon VLSI Circuits and Structures, Wiley-Interscience, New York, (1983).
32. T.T. Sheng, Analytical Techniques for Thin Films (K.N. Tu and R. Rosenberg, eds.,), Treatise on Mat. Sci. and Technol. 27, 251 (1988)
33. P.E. Batson, Analytical Techniques for Thin Films (eds. K.N.Tu and R.Rosenberg), Treatise on Mat. Sci and Technol 27: 337 (1988)
34. S. Vaidya, D.B. Fraser and A.K. Sinha, Proc. 18th IEEE Rel. Phys. Symp. 165 (1980)
35. D.R. Frear, J.B. Sanchez, A.D. Romig, Jr and J.W. Morris Jr, Metall. Trans. 21A, 2449 (1990).
36. M. Park, S.J. Krause and S.R. Wilson, Mat. Res. Symp. Proc. 229, 313 (1991)
37. N.D. Theodore, M. Dreyer and C.J. Varker, Proc. 49th EMSA Annual Conf., 884 (1991).
38. R.D. Heidenreich, J. Appl. Phys., 20, 993 (1949).
39. R.F. Egerton, Electron Energy-loss Spectroscopy in the Electron Microscope, Plenum Press, New York (1986).
40. G. Binnig and H. Rohrer, C. Gerber and E. Weibel, Phys. Rev. Lett 49, 57 (1982)
41. G. Binnig and H. Rohrer, Scanning Tunneling Microscopy, Surf Sci. 126, 236 (1983)

42.   G. Binnig and H. Rohrer, Scanning Tunneling Microscope, Sci. Amer. **253**, 50 (1985)

43.   H.K. Wickramasinghe, Scanned-Probe Microscopes, Sci. Amer., **261**, 98 (1989)

44.   R.J. Behm, N. Garcia and H. Rohrer eds., Scanning Tunneling Microscopy and Related Methods. NATO ASI Series, Kluwer Academic Publishers, (1990)

45.   G. Binnig, C.F. Quate and Ch. Gerber, Phys. Rev. Lett, **56**, 930 (1986).

46.   D. Rugar and P. Hansma, Physics Today, 23. 1990

47.   A. Rosencwaig, in VLSI Electronics: Microstructure Science, vol **9**, 227 (1985).

48.   A. Rosencwaig, J. Opsal, W.L. Smith and D. Willenborg, Appl Phys Lett **46**, 1013 (1985)

49.   W.L. Smith, C.G. Welles, D. Willenborg and A. Rosencwaig, Proc. Semicon Osaka, Osaka, Japan (1989).

50.   W.L. Smith, C.G. Welles and A. Bivas, Semiconductor International 92, (Jan 1990)

51.   W.L. Smith, C.G. Welles, A. Bivas, F.G. Yost and J.E. Campbell, Proc. IEEE. Rel. Phys. Symp, 200 (1990).

52.   V. Murali, S. Sachdev, I. Banerjee, S. Casey, P. Gargini, C. Welles and W.L. Smith, Proc. IEEE V-MIC 127 (1990)

53.   W.L. Smith, C.G. Welles, A. Bivas and A. George, Proc. IEEE V-MIC, 279 (1991)

54.   W.L. Smith, Mat. Res. Soc. Symp, **225**, 291 (1991)

55.   J. Opsal, Metallization: Performance and Reliability Issues for VLSI and ULSI, G.S. Gildenblat and G.P. Schwartz, eds. Proc SPIE **1596**, 120 (1991)

56.   J. Melngailis, J. Vac. Sci. Technol. B, **5(2)**, 469 (1987).

57.   D.R. Kingham and L.W. Swanson, Appl. Phys. A**34**, 123 (1984).

58.   D.R. Kingham and L.W. Swanson, Vacuum, **34**, 941 (1984).

59.   A.E. Bell and L.W. Swanson, Nuclear Instruments and Methods in Physics Research B, **10/11**, 783 (1985).

60.   J. Orloff, Sci. American, **265**, 96 (1991).

61.   J. Orloff, J.-Z. Li, and M. Sato, J. Vac Sci. Technol. B, **9(5)**, 2609 (1991).

62.   K. Nikawa, J. Vac. Sci. Technol. B, **9(5)**, 2566 (1991).

63.   R. Boylan, M. Ward, and D. Tuggle, Proc. ISTFA, 249 (1989).

64.   K. Nikawa, K. Nasu, M. Murase, T. Kaito, T. Adachi and S. Inoue, Proc.IEEE. IRPS, 43 (1989).

65.   J.A. Lange and S. Czapski, Proc. 17th ISTFA, Los Angeles, 397 (1991).

66.   L.L. Hsu, R.A. Novo and S.X. Lee, Proc. 17th ISTFA, 409 (1991).

67.   R.G. Lee and J.C. Morgan, Proc. 17th ISTFA, 409 (1991).

68.   D. Pramanik and J. Glanville, Solid State Technol. 77 (May 1990).

69.   J. Szot, R. Hornsey, T. Ohnishi, and S. Minagawa, J. Vac. Sci. Technol. B, **10(2)**, 575 (1992).

70.   S.J. Kirch, R. Anderson and S.J. Klepeis, Proc 49th Ann. Meeting EMSA, 1108 (1991).

71. S. Morris, S. Tatti, E. Black, N. Dickson, H. Mendez, B. Scwiesow and R. Pyle, Proc. 17th ISTFA, 417 (1991).
72. P.H. Holloway, Appl. Surf. Science, **4**, 410 (1980).
73. G.E. McGuire and P.H.Holloway, Electron Spectroscopy; Theory, Techniques and Applications, vol **4**. C.R.Brundle and A.D Baker, eds., Academic Press, (1981).
74. M. Thompson, M.D. Baker, A. Christie and J.F. Tyson, in Chemical Analysis vol. **74**, John Wiley, New York (1985).
75. P. Auger, J. Phys. Radium. **6**, 205 (1925).
76. L.A. Harris, J. Appl. Phys. **39**, 1419 (1968).
77. L.A. Harris J. Appl. Phys. **39**, 1428 (1968).
78. L.E. Davis, N.C. MacDonald, P.W. Palmberg, G.E. Riach and R.E. Weber, Handbook of Auger Electron Spectroscopy , 2nd Ed. Physical Electronics, Perkin- Elmer, Minnesota, (1976).
79. G.E. McGuire, Auger Electron Spectroscopy Reference Manual, Plenum Press, New York, (1979).
80. M. P. Seah and W. A. Dench, Surf. Interface. Anal, **1**, 2 (1979).
81. T.W. Haas, J.T. Grant and G.J. Dooley, J. Appl. Phys. **43**, 1853 (1972).
82. S. Thomas, J. Appl. Phys. **45**, 161 (1974).
83. S. Thomas, J. Appl. Phys. **47**, 301 (1976).
84. M. Salmeron, A.M Baro and J. M. Rojo, Phys. Rev. **B13**, 4348 (1976).
85. S. Thomas, J. Electrochem. Soc. **124**, 1942 (1977).
86. E. Kiny, J.Vac. Sci. Technol. **17**, 658 (1980).
87. H.H. Madden, J. Vac. Sci. Technol. **18**, 677 (1981).
88. P.W. Palmberg, J.Vac. Sci. Technol. **9**, 160 (1972).
89. S. Hofmann in Practical Surface Analysis, D.Briggs and M.P Seah, cds., John Wiley ,(1983).
90. P.M. Hall and J.M. Morabitio, Surf. Sci. **54**, 79 (1976).
91. M.B. Chamberlain and S.L. Lehoczky, Thin Solid Films, **45**, 189 (1977).
92. C.E. Hoge and S. Thomas, Proc. IEEE Reliability Phys. Symp. 301 (1980).
93. S. Hofmann, Appl. Phys, **9**, 59 (1976).
94. J.S. Solomon and V. Meyers, American Laboratory, 31, Mar (1976).
95. J. W. Coburn, Thin Solid Films, **64**, 371 (1979).
96. S. Hofmann and J. M. Sanz, Surf. Interface Anal. **6**, 78 (1984).
97. R. Kelly and D.E. Harrison, Materials Sci and Eng, **69**, 449 (1985).
98. W. Palmer and K. Wangeman, Surf. Interface Anal, **18**, 52 (1992).
99. A. Zalar, Fresenius Z. Anal Chem, **33**, 315 (1989).
100. A. Zalar, Thin Solid Films, **124**, 223 (1985.)
101. A . Zalar, J.Vac. Sci and Technol, **A5**, 2979 (1987).
102. T.A. Carlson, Photoelectron and Auger Electron Spectroscopy, Plenum Press, N.Y. (1975).
103. H. Siegbahn and L. Karlsson, Handbuch der Physik, **31**, 215 (1986)
104. F.J. Grunthaner and P.J. Grunthaner, Materials Sci. Reports, **1**, 65 (1986).
105. K. Siegbahn, C. Norlding, A. Fahlhan, R. Nordberg, K. Hamrin, J. Hedman, G. Johansson, T. Bergmark, S.E. Karlsson, J. Lindgren, and B.

Lindberg, ESCA, Atomic, Molecular and Solid State Structure Studied by Means of Electron Spectroscopy, Almqvistand Wiksells, Uppsala (1967).

106. C.S. Fadley, R.J. Baird, W. Siekhaus, T. Novakov and S. A. L. Bergstrom, J. Electron Spectrosc. **4**, 216 (1974).

107. J.M. Hill, D.G. Royce, C.S. Fadley, L.F. Wagner and F. Grunthaner, Chem. Phy. Lett, **44**, 225 (1976).

108. M. Pijolat and G. Hollinger, Surf. Sci. **105**, 114 (1981).

109. J.A. McHugh, in Methods of Surface Analysis, A.W.Czanderna, ed. Elsevier Science Publishing Co, (1975).

110. R.E. Honig, Thin Solid Films, **31**, 89 (1976).

111. K. Wittmaack, Radiation Effects, **63**, 205 (1982).

112. A. Benninghoven, F.G. Rudenauer and H.H. Werner, Secondary Ion Mass Spectrometry: Basic Concepts, Instrumental Aspects Applications and Trends, John Wiley, (1987).

113. D.E. Sykes, in Methods of Surface Analysis, J.M. Walls, ed., Cambridge University Press, (1989).

114. A.E. Morgan, in Characterization of Solid Materials, vol.1. G.E.McGuire, ed., Noyce Publications (1989).

115. H.W. Werner, Acta Electronica **19**, 53, (1976).

116. D.P. Leta and G.H. Morrison, Anal Chem, **52**, 514 (1980).

117. H. Stevens, Motorola, MOS Analytical Lab (Private Communication)

118. F.A. Stevie and J.L. Moore, Surf. Interface Anal. **18**, 147 (1992).

119. F.A. Stevie, J.L. Moore, P.M. Kahora, and R.G. Wilson, Secondary Ion Mass Spectrometry, SIMS -VIII, A.Benninghoven, ed., John Wiley (1992).

120. R. Castaing, in Adv. in Electronics and Electron Phyiscs, L. Marton, ed., Academic Press, **13**, 317 (1960).

121. W.K. Chu, J.W. Mayer, and M-A. Nicolet, Backscattering Spectrometry, Academic Press, (1978).

122. J.E.E. Baglin and J.S. Williams, in Ion Beams for Materials Analysis, J.R. Bird and J.S. Williams eds., Academic Press, Australia (1989).

123. G. Foti, J.W. Mayer, and E. Rimini, in Ion Beam Handbook for Materials Analysis, J.W. Mayer and E.Rimini eds., Academic Press, New York (1977).

124. B. E. Warren, X-Ray Diffraction, Addison-Wesley, Reading, (1969).

125. B.D. Cullity, Elements of X-Ray Diffraction. Addison-Wesley (1978).

126. A. Semuller and M. Murakami, in Thin Films from Free Atoms and Particles, K.J. Klabunde ed., Academic Press 325, (1985).

127. American Society for Testing Materials, X-ray Department, Philadelphia, PA.

128. B.E Warren , Progr. Metal Phys., **8**, 147 (1959).

129. B. Borie, Acta Cryst., **13**, 542 (1960).

130. B. Borie, C.J. Sparks, and J.V. Cathcart, Acta Met., **10**, 691 (1962).

131. C. Grilletto, Solid State Technol., **20(2)**, 27 (1977).

132. M. Madden, J. N. Cox, and B. Fruechting, Solid State Technol., **32(8)**, 53 (1989).

133. M. C. Hughes and D. R. Wonsidler, J. Electrochem. Soc., **134(6)**, 1488 (1987).

134. W. Pliskin, J. Vac. Sci Technol, **14**, 1064 (1977).

135. J. Theil, D. Tsu, M. Watkins, S. Kim and G. Lucovsky, J. Vac. Sci. Technol. A, **8** (3), 1374 (1990).
136. P. Pai, S. Chao, Y. Takagi and G. Lucovsky, J. Vac. Sci. Technol. A, **4** (3), 689 (1986).
137. G. Lucovsky, J. Non-Cryst. Solids, **141**, 241 (1992).
138. W. Claassen, H. Th v.d. Pol, A. Goemans and A. Kuiper, J. Electrochem. Soc., **133**, 1458 (1986).
139. W. Claassen, W. Valkenburg, M. Willemsen and W. Wijgert, J. Electrochem. Soc., **132**, 893 (1985).
140. A. Shintani, J. Appl. Phys., **51**, 4197 (1980).
141. N. Hirashita, I. Aikawa, T. Ajioka, M. Kobayakawa, F. Yokoyama, and Y. Sakaya, Proc. IEEE Reliability Phys. Symp, 216 (1990).
142. V. Murali, S. Sachdev, I. Banerjee, S. Casey, P. Gargini, C. Welles, and L. Smith, Proc. VMIC, 127 (1990).
143. B. Bhushan, S. Murarka and J. Gerlach, J. Vac. Sci. Technol. B, **8**, 1068 (1990).
144. P. Deal, Motorola, Inc (Personal Communication)
145. J. Wong, J. Electron. Mat, **5**, 113 (1976).
146. W. Kern, RCA Review, **32**, 429 (1971).
147. F. Becker and S. Rohl, J. Electrochem. Soc., **134**, 2923 (1987).
148. D. Haaland, Practical FTIR Spectroscopy Ind. & Lab Chem. Analysis, J. Ferraro and K.Krishnan, eds., Academic Press, 396 (1990).
149. J. E. Tong, K. Schertenleib, and R. A. Carpio, Solid State Technol., **27(1)**, 161 (1984).
150. F. S. Becker, D. Pawlik, J. Schäfer, and B. Staudigl, J. Vac. Sci. Technol. B, **4(3)**, 732 (1986).

12

ELECTRONIC PACKAGING AND ITS INFLUENCES ON INTEGRATED CIRCUIT DESIGN AND PROCESSING

HARRY K. CHARLES, JR. And G. DONALD WAGNER
The Johns Hopkins University
Applied Physics Laboratory
Laurel, Maryland

1.0   OVERVIEW

Electronic packaging technology is one of the major drivers in modern electronic circuit and device design. In fact, the ability to effectively package integrated circuits while still maintaining the designed-in/built-in on-chip performance is a key challenge facing all integrated circuit designers and manufacturers. This challenge can be met by properly understanding and utilizing modern electronic packaging methods and materials as well as by taking consideration of the packaging requirements and constraints (opportunities) during the design process. This chapter is divided into two primary parts: 1) an overview of electronic packaging including basic package structures and the methods of placing, connecting, and qualifying integrated circuits within these packages, and 2) the counter and complementary influences of the integrated circuit on the package and the package on circuit design and performance.

A question that a chip or integrated circuit designer might ask is: "Why do we package at all?". It seems reasonable with today's level of integration (i.e., over 20 million transistors on a single piece of silicon) that it might be possible to put an entire system on a chip - thus why do we need the package. A package structure is needed for several reasons which can be summarized into the four tenets of electronic packaging, i.e. packaging is necessary to provide: 1) input/output; 2) thermal management; 3) mechanical support; and 4) environmental protection.

1.1   Input/Output

It is necessary to get the wanted signals on and off chip. Even if a whole personal computer's central processor and main memory were on a single chip it would still be necessary to connect this element to a display, a key board, a printer, etc. - to allow for human interaction with the device. So the first major tenet of packaging is to get the signals in and out. These signals can be in the form of digital pulses, microwave energy, optical photons, or any other form that conveys information to or from the environment. Similarly, interconnects to chips are extremely small, typically less than the size (diameter) of a human hair, thus humans can not directly wire or probe at this scale without machine help or the intervention of a structure (a package) which enlarges the I/O pattern to a more realistic size.

1.2   Thermal Management

Because integrated circuits dissipate power, proper thermal control must be provided to avoid performance degradation or reliability issues. Proper thermal management involves analysis of the heat loads under the full range of operating conditions as well as providing acceptable heat flow paths, mechanical assemblies which address coefficient of thermal expansion (CTE) mismatches and provide adequate cooling to ensure device performance and reliability.

1.3   Mechanical Support

Silicon as well as other semiconductors is quite brittle and thus fragile. Silicon chips are thin (nominally 250-500 μm) but can be as large as 1.5 cm x 1.5 cm in area for an individual integrated circuit. Thus they can easily break. Some GaAs circuits can be as thin as 100 μm making them extremely fragile. Interconnects cannot be made to free chips - they must be supported thus packaged in some form. In most circumstances the mechanical support also provides help in thermal management and sometimes even in signal input and output (i.e., electrical ground, substrate contact, etc.).

1.4   Environmental Protection

Integrated circuits, because of their fragility, minute patterns, and interconnect structures, typically need protection from the environment including protection from mechanical damage, moisture, chemical attack, etc. Package structures, in addition to their other functions, typically provide protection for the integrated circuits. In certain circumstances such as biomedical electronic human implants, the packaging actually protects the environment from the integrated circuit.

There are many different packaging structures, types of packages and packaging styles, and alternatives available to the integrated circuit and electronic systems packaging industry. In order to provide some order and logic to the myriad of techniques and options open to the potential chip design engineer, we introduce the concept of the electronic packaging level hierarchy. This system consists of 4 levels, Level 1 through Level 4.

1.5   Level 1 - On Chip

This level is associated with anything that is on-chip such as bonding pads, metallization, chip design itself, etc. In the fundamental overview (part 1 (sections 1, 2, and 3)) this level will be assumed to be a given and the chips size, layout, I/O, and performance requirements will become the drivers for subsequent packaging options. In part 2 (section 4) we will address what can be done in chip design (Level 1) to improve overall packaging and package system performance.

1.6   Level 2 - Chip to Package

This is the most important packaging level to the typical integrated circuit designer. Under Level 2 an individual die is placed in a single or multichip package. Multichip packages contain a substrate interconnect structure to interconnect all internal interchip leads. Various Level 2 package types and structures are described in Section 2, below. Once a package is identified, the integrated circuit is mechanically attached, electrically interconnected, protected from the environment, and typically certified to be performing at some rated level as a component piece part. Details of the Level 2 packaging processes are contained in Section 3, below.

1.7   Level 3 - Package to Board

This level influences the choice of package style and board construction. For example, pin packages are either socketed or soldered directly into printed wiring boards (PWBs) made of organic resin materials. Surface mount packages (either leadless or leaded) are mounted on the surface of circuit boards which can be fabricated from either organic or inorganic (primarily ceramic) materials.

1.8   Level 4 - Board to System

This level places the populated board into the system. Although not a direct influence on the initial chip interconnect and mounting, the efficiency by which this operation is done can significantly influence overall system performance and reliability. Items that commonly fall into this area include connectors, cables, flex circuits, and backplanes.

In addition to the above Levels, there are two important technologies which bridge the packaging hierarchy. These are chip-on-board (COB) and wafer-scale integration (WSI). Chip-on-board (1) places the chip directly on a board (with other chips) by passing the mounting of the die in an individual or multichip package (Level 2). Although Level 2 is bypassed, all the operations or processes associated with Level 2 such a die attach and interconnection must be performed at the board level. In addition, assembled board sealing or passivation must be performed prior to inserting the board into the system (Level 4). In wafer scale integration (2), the entire system is interconnected at the chip/wafer level (Level 1), but eventually the wafer level circuit still must be provided with the four tenets of packaging - thus the concepts of packaging and their effect on chip and system performance remain. Further details of the WSI concept and associated packaging is described in Section 2 below.

2.0   PACKAGE TYPES

2.1   Single Chip Packages

The major single chip package types in use today include dual- in-line packages (DIP), quad-flat packages (QFP), and grid arrays. Until recently, the dual-in-line package was the dominant single chip package, but the push for improved performance reliability and cost factors associated with high numbers of inputs and outputs (greater than 64) has forced a switch from the DIP to packages with leads emerging from all four sides (QFPs and chip carriers) and package leads on an area grid emerging from both (grid arrays). Table I summarizes the range of inputs and outputs possible with today's high performance packages. Figure 1 illustrates the number of I/O per unit package area possible as a function of increasing package I/O. Only the grid array structure provides constant I/O density.

Dual-In-Line Packages. The dual-in-line package as shown in Figure 2 has been the standard integrated circuit packaging method for many years with, at times, over 90% of all integrated circuits manufactured shipped in DIPs. DIPs are manufactured using various materials including plastic, glass-ceramic, and ceramic (Al_2O_3). The molded plastic dual-in-line package is typically made of a novalac or B-stage epoxy resin (3). The first step of the assembly process is to attach the integrated circuit die to the lead frame (Figure 2). The bonding pads on the die are then wired to the lead frame using one of the wirebonding methods described in Section 3. Once wired, the wires, die, and the inner portion of the lead frame are placed in a mold for encapsulation. The epoxy is then forced into the evacuated mold under pressure and then the entire unit (epoxy) is cured prior to unmolding. This type of plastic (epoxy) encapsulation, although not truly hermetic, has allowed millions of integrated circuits to function reliably in most consumer, computer, and low humidity industrial environments. The plastic DIP has several advantages including low cost (especially in lead counts below

40); uniform shape ("molded-in" to facilitate automatic handling and insertion); and, according to many reports, has the lowest infant mortality of any packaging type.

Table 1: Package and Application Input/Output Capability and Requirements

PACKAGE TYPE	PIN PITCH (mm)	I/O COUNT

Dual-in Line
   Standard
   High Density
	2.540
	1.270
		4-64
		32-84

Pin-Grid Array
   Standard
   High Density
	2.540
	1.270
		64-256
		64-400+

Leadless Chip
Carriers
   Standard
   High Density
	1.270-1.016
	0.635-0.254
		8-132
		64-240+

Leaded Chip Carriers
   Standard
   High Density
	1.270-1.016
	0.635-0.254
		64-132
		80-200+

Quad Flat Packages Standard High Density	0.635-0.508 0.508-0.254	80-200 100-400
APPLICATION		I/O REQUIREMENTS
Single Device Family	--	4-12
Memory	--	20-80+ ($\leq$32 bit bus) (>32 bit bus or multichips greater)
Microprocessors	--	32-132 (+ multichips)
LED Drivers	--	40-120
Gate Arrays/Custom	--	32-200 (+ multichips)
VHSIC Devices	--	64-320

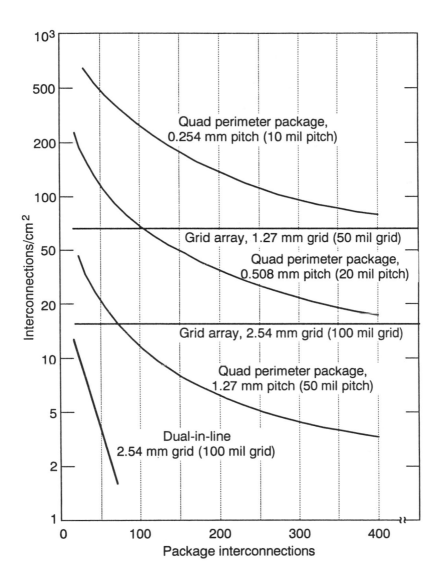

Figure 1. Interconnection Density versus Number of Package Inputs/Outputs for Various Styles

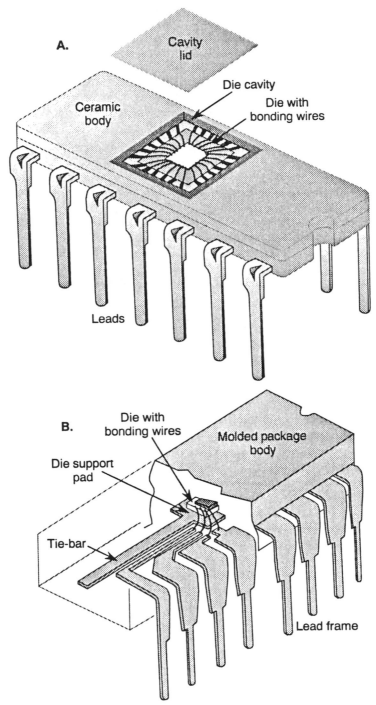

Figure 2. Dual-In-Line Packages: A. Ceramic DIP, B. Plastic or Molded DIP

A second type of DIP is made from a ceramic mixed with glass. These ceramic DIPs or "CERDIPs" come in two versions: a basic unassembled version and an assembled version with an open die cavity and seal ring. To assemble the basic CERDIP, the base is placed on a heater block until the glass seal melts and a lead frame can be embedded into it. Next, the die is attached to a gold pad in the base and the chips' I/O are wired to the lead frame using a conventional wirebonding process. A cap is then sealed to the base in an oven. The assembled open die cavity version is typically used for EPROMS and other light alterable or sensing devices. The open die cavity allows the die to be inserted, attached, and wirebonded. A quartz or sapphire lid (or window) is inserted into a recess in the cap surrounding the cavity and sealed. The main advantage of the CERDIP is that it gives many of the properties of the hermetic high reliability ceramic DIP (described below) of a fraction of the cost (typically five times less). The CERDIP typically costs 1.5 to 2.0 times that of a comparable plastic DIP. It is not as rugged or as uniform as either the plastic or the full ceramic DIP and typically cannot be inserted by automatic means. Because of difficulties in sealing and the nonuniformity of lead planarity, CERDIPS are typically not available with more than 40 leads.

The ceramic DIP is a multilayer alumina technology made by the cofired process. This is a built-up package using green-state or "uncured" alumina tape. Metallization patterns are screened-on in this type and several ceramic-metallized layers are then laminated together and fired at high temperature (nominally >800°C) to create a strong integrated package structure. A lead frame is then brazed to integral metal pads on the sides or top of the package. Typical ceramic DIPs have three layers (top - containing the seal ring, middle - containing the lead frame interconnects, and bottom - containing the chip bonding pad) as shown in Figure 3. The standard ceramic DIP (like its plastic counterpart) has been used up to 64 leads, can be tested in an open mode, is hermetic, and can have a large die cavity (when compared to the CERDIP).

Since DIPs have many desirable features, there has been a great reluctance by many in the packaging industry to move away from the DIP. DIPs, however, cannot support the performance demands of modern circuitry such as speed, electrical behavior (R,L,C), I/O density, thermal conductivity, and the minimization of board level footprint. For these reasons, the DIP has given way to quad flat packages, chip carriers, and grid arrays. In particular, the high density, surface mount packages (chip carrier, quad flat package, etc.) have begun to dominate. Performance parameters for the various package types are given in Table II.

Chip Carriers and Quad Flat Packages. The chip carrier is a square, multilayer package on the bottom (and sides) of which are solder reflow pads located at the periphery. Sometimes leads are connected to the side pads to provide more strength and flexibility in the board level attachment. Usually these leads are wrapped under the package at the end-yielding a "J-like" lead profile. Regardless of whether the chip carrier is leadless or leaded, the external pads or leads are interconnected within the package structure to bonding pads

Table II: Property Comparisons of Dual-In-Line Versus Quad Perimeter

I/O COUNT	PACKAGE AREA (cm²)	PACKAGE WEIGHT (g)	TRACE INDUCTANCE (nH)			LINE-LINE CAPACITANCE			TRACE RESISTANCE (ohms)		
			L	S	R	L	S	R	L	S	R
Dual-In-Line											
16	1.55	1.15	6.4	1.6	4.0	0.074	0.025	2.96	0.242	0.203	1.19
28	5.42	4.02	14.8	1.6	8.1	0.148	0.025	5.92	0.319	0.295	1.08
64	18.1	12.11	49.1	2.3	21.0	0.412	0.033	12.48	1.00	0.89	1.13
Ceramic Chip Carriers											
16	0.21	0.05	1.13	0.73	1.55	0.013	0.009	1.44	0.114	0.108	1.06
28	1.03	0.33	1.80	1.15	1.57	0.019	0.013	1.46	0.139	0.136	1.02
64	4.03	1.63	6.44	4.21	1.53	0.052	0.036	1.44	0.222	0.222	1.00
Quad Flat Packages											
132	6.45	2.60	5.5	4.8	1.15	0.34	0.24	1.41	0.080	0.070	1.14
160	7.84	3.10	8.0	6.5	1.23	0.25	0.19	1.32	0.166	0.143	1.16
208	7.84	3.25	6.1	4.2	1.45	0.40	0.36	1.12	0.025	0.021	1.19

L = Longest Lead   S = Shortest Lead   R = Ratio (Long/Short)

within the die cavity. The die cavity usually has a gold metallized base pad for chip attach and a gold metallized seal frame which allows hermetic lid sealing by a soldering process. The chip carriers are solder reflowed to board-level assemblies using a variety of techniques such as infrared reflow, condensation soldering (vapor phase), and reflow in convection furnaces. A typically chip carrier schematic is shown in Figure 3. There are many chip carrier types (including the leaded varieties) as shown in Figure 4. The number of leads range from 14 to more than 240 I/Os on lead centers from 1.27 mm (50-mils) down to less than 0.32 mm (12.5 mils).

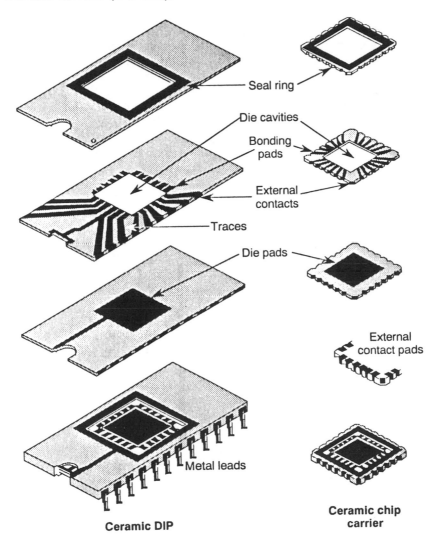

Seal ring

Die cavities

Bonding pads

External contacts

Traces

Die pads

External contact pads

Ceramic DIP

Ceramic chip carrier

Metal leads

Figure 3. Cross Section of a Ceramic DIP and a Leadless Ceramic Chip Carrier of Comparable Size

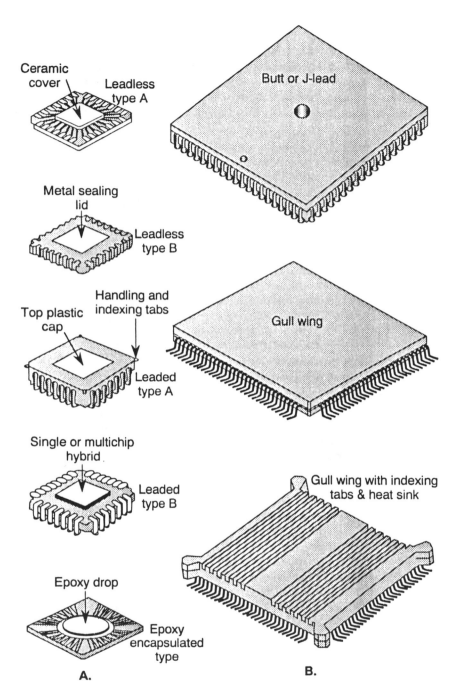

Figure 4. Four Sided Perimeter Packages.   A. JEDEC Style Chip Carriers, B. Quad Flat Packages

The chip carrier may be made from ceramic (single and multilayer versions) with or without leads and from plastic in either pre-molded or post-molded configurations. As shown in Table II, the chip carrier offers a 3-to-1 reduction in size (over the DIP), short and uniform lead length, and a significant weight reduction.

The quad flat package can be all metal in construction with leads exiting the four sides using glass to metal seals. The metal package is typically made of Kovar (an iron, nickel, cobalt alloy with a coefficient of thermal expansion tailor to that of the glass used in the glass-to-metal seals). The leads are also Kovar. The Kovar is typically plated with nickel and then gold to prevent corrosion. The metal packages are sealed with metal lids by either soldering or welding. Welded all metal packages provide excellent hermetic seals. Quad flat packages can also be made of ceramic with ceramic or metal bottoms by techniques similar to chip carriers. Metal or metallized lids can be soldered to the ceramic side frames for hermetic sealing.

Grid Arrays. Grid array structures (pin-grid arrays or pad grid arrays) are similar to ceramic chip carriers in construction except that the I/O's are obtained through pins (or pads in the case of the pad array) that cover the package bottom exterior surface in a grid or array fashion. A typical pin-grid array is a square multilayer cofired ceramic structure with an array of pins on 2.54 mm (100 mil) centers (1.27 mm centers for grid arrays with I/O's exceeding 300). The die cavity can be either on the opposite side of the pins (cavity-up) allowing the full base area for pins or on the same side as the pins (cavity-down) requiring a concentric perimeter array or matrix. Typical pin grid array I/O numbers are compared with those of equivalent size perimeter quad packages in Table III. Low lead count pin grid arrays can be attached to system board level structures using conventional through hole soldering methods (wave, condensation, IR reflow soldering, etc.). To replace the device, a hot-spot air removal gun or other suitable technique must be employed. Care must be exercised to prevent damage to the board and/or surrounding components. As I/O density goes up, pin grid array removal becomes increasingly difficult and a socket or connector is highly desirable. A switch to the pad grid array and surface mounting is another alternative to the repair problem although direct chip mounting by inverted reflow may be preferred to save board real estate. The main advantage of grid array structures over peripheral leaded structure is their ability to support high I/O counts for a given package size (as shown in Table III as well as maintaining constant interconnection density as shown in Figure 1.

2.2   Multichip Packaging

Individually packaged chips attached to circuit boards require larger than optimum board areas because each chip is burdened by its own package structures (spreading lead frames, seal rings, larger I/O footprint (than chip itself), etc.). Denser packaging can be achieved by packaging several bare chips into one large multichip module or hybrid structure. In the multichip module, unencased integrated circuit chips are mounted (via eutectic, solder, or epoxy die attach methods) to host substrates. The substrate can prove single or multilayer interchip interconnect structures and can be fabricated from a wide variety of materials (see Table IV). Chip electrical connection to the substrate is accomplished using wirebonding, tape automated bonding (TAB), and flip chip or inverted reflow (see Section 3 below). The populated multichip substrate is then typically placed-in and interconnected-to a large package structure with either a through hole or surface mount configuration via wirebonding or TAB. Once tested, the multichip module is sealed by soldering or welding if hermetic and by molding or encapsulation if non-hermetic.

Table III: Typical Pin Grid Array I/O Numbers (versus package dimensions) Compared With Equivalent Numbers for Similar Size Quad Packages

PACKAGE STYLE	OVERALL PACKAGE DIMEN- SIONS (cm x cm)	I/O NUMBERS			
		CASE 1	CASE 2	CASE 3	CASE 4
Pin Grid Array	2.54 x 2.54 3.81 x 3.81 5.08 x 5.08	0.38 cm border 0.254 cm grid 64 169 324	0.254 cm border 0.254 cm grid 81 196 363	0.38 cm border 0.127 cm grid 225 576 1156	0.254 cm border 0.127 cm grid 289 676 1296
Chip Carriers	2.54 x 2.54 3.81 x 3.81 5.08 x 5.08	0.127 cm pitch* 64 104 144	0.0635 cm pitch* 128 208 288	0.0508 cm pitch* 160 260 360	0.03175 cm pitch* 256 416 576
Quad Flat Packages	2.54 x 2.54 3.81 x 3.81 5.08 x 5.08	0.116 cm pitch* 80 120 180	0.0635 cm pitch* 128 208 288	0.0508 cm pitch* 160 260 360	0.0254 cm pitch* 320 480 720

*Assumes at least 0.508 cm along each edge not usable for I/O.

Table IV: Properties of Electronic Packaging Materials

MATERIAL	CTE ($\times 10^{-6}$m/m/K)	THERMAL CONDUCTIVITY (W/m_K)
Molybdenum	5.0	140
Kovar*	5.8	17
Cu/Mo/Cu	6.0 (x-y)	166 (Z DIR)
W/Cu	6.5	190
Aluminum	23.0	171
SiC/Al Metal Matrix Composite (70%SiC)	6.2	170
(65% SiC)	6.8	180
(55% SiC)	8.5	150
Alumina (Al_2O_3)	7.1	20-30
Beryllia (BeO)	9.0	250
Aluminum Nitride (AlN)	4.4-4.6	140-200
Silicon	2.6	140-150
GaAs	6.86	46
Silicon Carbide	3.7	270
Diamond	0.8	1000-2000
Parylene	35-69	0.08-0.1
Polyimide	20-60	0.1-0.2
Polyurethanes	200-300	0.0005
Silicones	200-1000	---
Epoxies	60-100	0.01-0.02
Copper	19.7	418
Gold	14.2	297
Aluminum	23.0	240
Tungsten	4.5	200
Nickel	13.3	89-92
Titanium	8.9	22
Platinum	9.0	71-73
Palladium	11.0	70-75
Chromium	6.3	66
Tantalum	6.5	58

*Iron Nickel Cobalt Alloy

Multichip modules (MCMs), in principle, correspond to the classic hybrid (either thick film (4) or thin film (5)) definition as packaged modules containing multiple parts on a common substrate. In today's vocabulary multichip chip modules refer to that subclass of all hybrids where the ratio of the area of the integrated circuits compared to the area of the substrate is greater than 30%. Various parameters and technologies for today's MCM's are given in Table V.

Table V: Multichip Module (MCM) Parameters

PARAMETERS	MCM-C		MCM-D		PRINTED WIRING BOARDS
	COFIRED CERAMIC	LOW-K CERAMIC	SILICON ON SILICON	LOW-K (organic dielectrics on Ceramic, Metal, etc.)	
Line Density (cm/cm²)	20	40	400	200	30
Lines/Spaces (µm)	125/(125-375)	125/(125-375)	10/(10-30)	(15-25)/(35-75)	750
Dielectric Constant	9	5	3.6-4.0	2.4-4.0	3.5-5.0
I/O (x 1000)	1.6-6.4	1.6-6.4	0.8-3.2	0.8-3.2	1.6-3.2
Substrate Area, cm²	225	225	100	100	500+
Terminating Resistors	Built-in	Built-in	Built-in	Surface Mount	Surface Mount
Decoupling Capacitance	Surface Mount	Surface Mount	Built-in	Built-in or Surface Mount	Surface Mount
Transmission Lines	Stripline µstrip	Stripline µstrip	Stripline µstrip 50 Ω	Stripline µstrip 50 Ω	Stripline µstrip

Multichip modules are complex, extremely dense entities which are more difficult to test and repair when compared to individually packaged chips mounted on circuit boards. In the multichip module, individual chips must be removed without damaging the delicate underlying substrate and the surrounding circuitry. In addition, because the multichip module contains several high performance, high density integrated circuits, it is much more difficult to test and determine the cause of failure if the circuit does not function. In fact, it is imperative that circuit fabrication begin with "known good die" and that care must be exercised in the module design to build in testability and self-diagnostics. Since the multichip module is a large area, extremely dense form of packaging, electrical cross talk, thermal loading, and mechanical structural integrity all become major design considerations.

Multichip modules typically require large area substrates (>65 cm^2). Large multilayer areas need low-stress dielectrics. Both organic and inorganic dielectrics can be made stress free over large areas. Different module structures are required depending upon the chip technology and the end use applications. Modules typically fall into three categories: 1) very high density with low speed and/or low power dissipation, e.g., CMOS arrays, RAMs, and low-power application specific integrated circuits (ASICs); 2) high speed with medium power and/or medium density, e.g., VHSIC chips (6) and ECL arrays; and 3) high power and speed at relatively low circuit densities, e.g., gallium arsenide integrated circuits.

A typical high-density multichip module substrate-chip system is shown in Figure 5. In this structure a thermally conductive substrate is used as a base for a multilevel signal structure using polyimide dielectric and either aluminum or copper metallization. Chips are attached directly to the polyimide layers or directly to the substrate in recessed wells. Substrate trace widths can rival the on-chip dimensions. Interconnections can be made by the techniques described above.

Circuits for high speed with medium power and/or density can be made from a combination of organic thin film technology and a cofired ceramic substrate containing the power and ground as shown in Figure 6. This arrangement locates the power and ground planes in the high dielectric constant ceramic (providing intrinsic power supply decoupling), while all the signal layers are in the lower dielectric constant organic material. In other applications where density and speeds are not as stringent, an all cofired ceramic structure can be used for both the signal layers and the power and ground planes.

Figure 7 presents a typical multichip module structure for high power dissipating integrated circuits. In this structure, thermal vias are used to connect the high power dissipating chips to the thermally conductive substrate (usually made of material with thermal conductivity >100 W/mK). Multiple power and ground planes exist with integral decoupling capacitance. Signal lines tend to be wider to control impedance, decrease resistance, and reduce self-heating as described above.

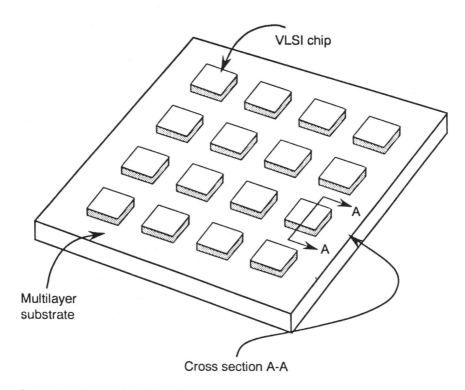

Cross section A-A

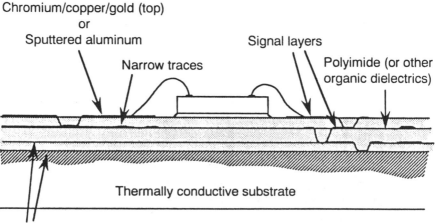

Figure 5. Multichip Module Structure for High Density Applications with Nominally Lower Speed and/or Power Requirements

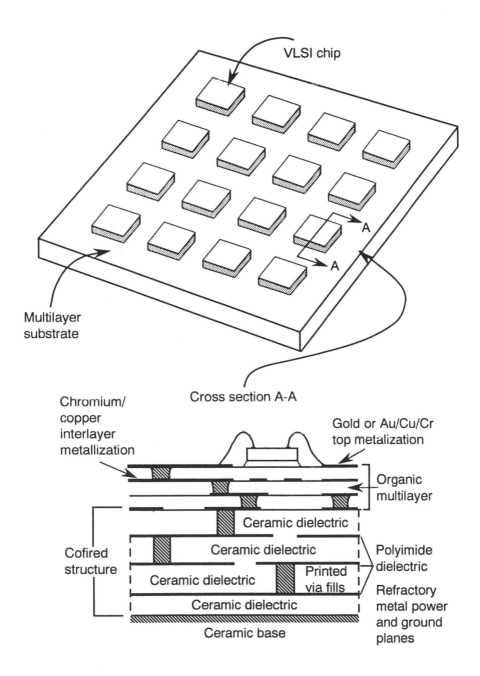

Figure 6. Multichip Module Structure Combining High Performance Organic Signal Layers with Cofired Ceramic Power and Ground Plane Technology

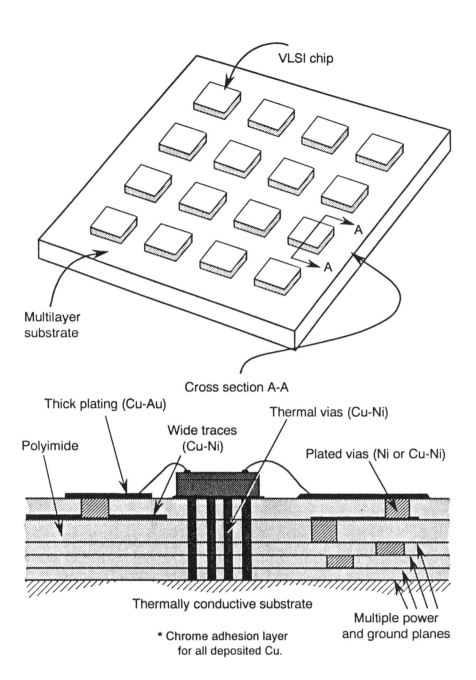

Figure 7. Multichip Module Structure for High Power (and Speed) Applications with Nominally Reduced Density Requirements

2.3   Surface Versus Through-Hole Mounting

There are two major techniques for attaching packaged integrated circuits (either single or multichip packages) to circuit boards: through-hole mounting and surface mounting. In through-hole mounting, packages with pins (bottom or side leads perpendicular to the package bottom) such as PGAs, DIPs, etc., are soldered into plated through-holes in printed wiring boards as shown in Figure 8. These holes are typically placed on 100-mil centers (100-mil grid) and serve, in addition to holding the component leads, as a via for interconnection between circuit board layers. Several through-hole packages are now being fabricated on 50-mil centers to improve circuit density. Through-hole mounting has several disadvantages for high performance integrated circuit applications including reduced board density (due to through-hole via structure); difficulty in repair (as I/O number increases); and increased inductance (due to variable lead frame length in DIPs and round wire leads in PGAs). The repair or removal as mentioned above can be enhanced by the use of a socket that is permanently mounted to the circuit board. Sockets, however, add additional inductance and resistance to the circuit path, which can slow down device performance. In high vibration and thermal excursion environments, the mechanical integrity and, hence, reliability of a socket would also be questioned.

Surface mounting of leadless, beam leaded, gull wing, or "J" leaded components to various circuit board materials is the dominant thrust in electronic assembly today. In this technique, leadless or leaded peripheral I/O packages are soldered to the surface of host circuit boards. No through-hole drilling is required for lead mounting, thus, electrical vias can be formed only where necessary (no through-hole drilling). These staggered vias involve only the necessary levels or layers as shown in Figure 9. Thus, circuit boards can be more dense, with short leads and signal paths, and have higher (faster) performance characteristics. Repair is easier since most joints are accessible and not locked into high aspect ratio holes.

2.4   Chip-on-Board

Direct chip mounting via the inverted solder bump or flip chip technique is becoming increasingly popular. Although pioneered by IBM over twenty years ago (7), this method is finding application in both multichip modules and direct board mounted packageless configurations - i.e., chip-on-board (COB). In this method, a reflow alloy is placed on the chip bonding pads and the chip is turned face-side (active side) down (inverted) and reflowed to mating bonding pads on the underlying substrate. This method can be used to replace wirebonding or TAB in packaged chip configurations or by itself for directly mounting chips to board level structures. In this direct COB style, chips can be mounted on both sides of dense multilayer board structures, achieving densities equal to or greater than wafer scale integration (2) while maintaining repairability and optimum circuit configuration and performance. Once placed, the dies are then overcoated with one or more layers of organic materials to protect the bare die and substrate structures from environmental stresses.

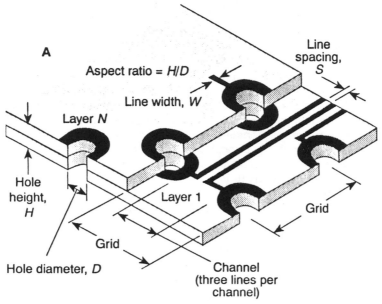

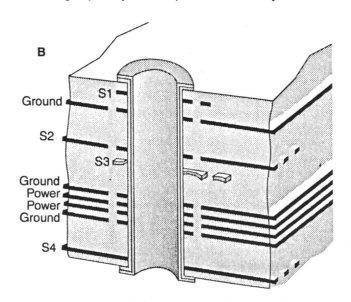

Figure 8. Printed Wiring Board Construction. A. General Lay-up of Layers, B. Detailed Cross-Section of Plated-Through-Hole

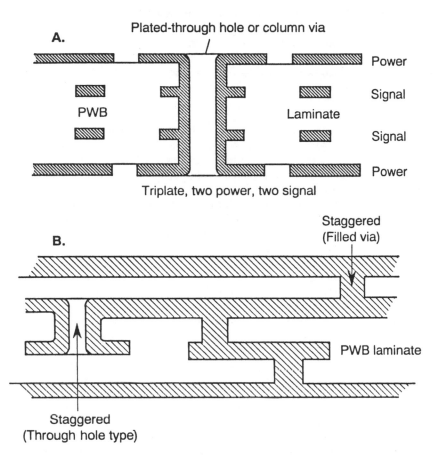

A.

Plated-through hole or column via

Power

Signal

PWB

Laminate

Signal

Power

Triplate, two power, two signal

B.

Staggered
(Filled via)

PWB laminate

Staggered
(Through hole type)

Figure 9. Column Vias (A) or through-hole vias Versus Staggered Vias

2.5   Hermetic vs. Nonhermetic Packaging

From the beginning of the microcircuit industry, the need for hermeticity (the sealing of electronic circuits from the environment) was considered essential for reliability. This view evolved from early reliability tests on both plastic and hermetically sealed (metal can) transistors in which the hermetically sealed units lasted significantly longer. Consequently, the circuits designed for use in high reliability environments (military, space, life support, etc.) were packaged in either metal or ceramic chip packages with soldered, welded, or glass-sealed lids. Such circuits are typically expensive, suffer from low yields (due to extra requirements on die attach, interconnection, and lid sealing), and may actually have sealed-in corrosive or hazardous vapors from some of the adhesives or residual processing chemicals.

For example, some early epoxies used for die attach under thermal aging would produce significant outgassing and sealed package ambient would thus contain large amounts of moisture and ammonia. Both water vapor and ammonia have been correlated with electrical die failures especially with uncapped die, which were prevalent in the early days of the semiconductor industry. Thus, under certain circumstances, the hermetic nature of the package actually accelerated the failure.

In recent times improvements in epoxies, substrates, package materials, and platings have produced highly reliable hermetic packages. Although extremely reliable, these packages are expensive - orders of magnitudes larger than non-hermetic types. The substitution of low melting point glass seals for lid attach can reduce the cost of hermetic packages by a factor of five over welded or soldered lid construction (8).

Even lower costs can be achieved by replacing metal and ceramic package parts with low cost organics; for example, epoxy seals, plastic structural parts, and/or organic overcoats. The use of organic materials, which are in themselves susceptible to moisture penetration as shown in Figure 10, raises serious reliability issues. However, many studies have shown that appropriate inorganic capping (using silicon oxide - nitride combinations), followed by organic junction coatings (e.g., silicone gels) and then plastic (epoxy) encapsulation, have produced extremely reliable integrated circuit packages. The organic junction coatings form excellent chemical bonds with the chip surface precluding moisture interaction with any exposed underlaying chip metallization.

A two-phase organic packaging technique (epoxy encapsulation over a silicone die protective layer) is usually necessary because epoxies typically contain leachable impurities and do not adhere well to the smooth inorganic surfaces of silicon chips. Thus, when moisture finally penetrates through the epoxy to the integrated circuit surface, it will contain impurities leached from the epoxy which, in turn, will cause corrosion at the integrated circuit surface. Silicones, on the other hand, are very pure and, despite their high moisture permeability (Figure 10), bond effectively with the integrated circuit surface (9). This close chemical bonding prevents the formation of liquid water and a galvanic corrosion cell. Corrosion can only occur after the chemical bond is disrupted due to chemicals, mechanical stress, or disbonds by trapped impurity layers.

2.6 Wafer Scale Integration

Many authors have shown the progression of die input/output density (e.g. see Figure 22 Section 4) and increasing die size with time. In fact die size has been increasing (for high performance state of the art die) by greater than 10% per year. Such increases promise to continue at these rates well into the end of this decade (10). Because of this steady upward progression in die size, the concept of wafer-scale integration will be continually advanced as the ultimate system implementation method. Wafer-scale integration or the packaging of an entire system or a single silicon wafer offers the promise of reduced cost, high performance, greater reliability and increased functional density. Random yield

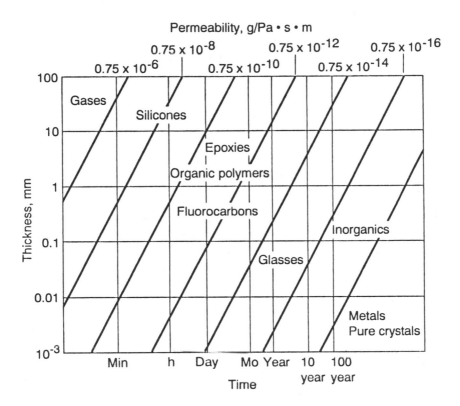

Figure 10. Moisture Penetration Thickness Versus Time for Various Material Classes. Source: C. P. Wong, *Integrated Circuit Device Encapsulants in Polymers for Electronic Application*, J. H. Lau, Ed., CRC Press, 1989, p. 3.

of the system building blocks on the wafer-scale circuit will continue to limit system performance (different interconnection lengths depending on yield of key circuit blocks) as well as produce no standard packaging requirements for each system produced. These drawbacks will be lessened by improving the yield of very large area devices.

Regardless of whether WSI becomes a chip-packaging mainstay - future packaging technologies will have to produce interconnect and support structures capable of handling extremely large silicon chips with a high number of dense I/O. Such complexity will require the use of silicon substrates where line resolution can approach that of the chips themselves as well as incorporating active and passive devices in the substrate - thus allowing chip optimization and perhaps reduction in size.

Thus the formation of a system packaged with high density multichip modules using silicon as the substrate may be the most probable future manifestation of WSI - i.e., the WSI hybrid (11). Such a hybrid structure will allow a standard module footprint and use only pretested die (known good die) thus obviating the non-standard packaging and the redundant circuit building blocks (and non-uniform wiring) to overcome wafer yield. By using the silicon substrate to provide the on and off chip communications drivers as well as the decoupling capacitors and resistors the chip technology can be optimized to perform the system function (e.g. high speed computing, signal processing, etc.). Chip technologies can also be mixed to achieve optimum system performance (i.e. GaAs and silicon, etc.). Optical wave guides can also be formed on the silicon substrate to interface electro-optic based chips and system elements (12).

2.7   Optoelectronic Packaging

Optical interconnections and routing of information both on and off chip by optical signals offers great potential for increased information density, reduced signal line volume, crosstalk and power dissipation. Chips and modules can be optically interconnected in one of three primary techniques: 1) fibers, 2) integrated planar wave guide, or 3) holographic projection through free space on planar guide structures (13).

Fiber-optic cables cannot provide the number and density of connections required for modern chips let alone multichip modules. Integrated planar wave guides are relatively easy to manufacture but they cannot provide the density and propogatoion speed of the holographic approach. They are however inexpensive and can readily be fabricated using standard thin film technology. Their interconnection density is limited by their planar nature and is directly related to the metal pitch. Possible cross talk occurs in most planar guides unless paths cross at exactly 90° angles. Planar optical guides must be separated by relatively large distances thus materially increasing source-to-sink distances. The fibers are made by cladding a high index material (2 to 4) with a material of lower index to produce total internal reflection. The optical signals travel in the high index material which materially slows their propagation speed and increases their transit time, by factors of up to 4 over light in a vacuum.

All these limitations go away with holographic interconnections. Their three dimensional nature provides relatively unlimited interconnect density and optical beams can cross at any angle without cross talk. Direct line-of-sight routing of the optical signals reduces signal path length; and hence, the time of flight. Focusing or aligning the beams is a major problem. In addition to the generation of the holograph, it is the major cost driver. Computer generated holograms may speed the process and reduce costs.

Tradeoff studies have shown that optical interconnects are probably more expensive below 10 GHz because they offer no significant performance advantages. Above 10 GHz (and 10 ps risetimes) signals using conventional paths and interconnects experience significant dispersion and attenuation over typical MCM distances while optical waveforms are basically unaffected.

3.0   FUNDAMENTALS OF PACKAGING

Fundamentals of packaging include the methods for attaching chips to packages or substrates, methods of electrical interconnect, a description of substrate technology for multichip packages, methods of package sealing and/or encapsulation, concepts of package reliability, and methods of accelerated testing and stress screening.

3.1   Die Attach

Separated integrated circuit die may be attached to packages or substrates by three primary techniques: eutectic die bonding, soldering, and organic adhesive attachment.

Eutectic Die Bonding.   Eutectic die bonding refers to using a bi-metallic alloy with a unique melting point, i.e., the eutectic point on the binary alloy phase diagram.   For the alloy Au 94-Si 6 (94 weight percent gold, 6 weight percent silicon), the eutectic temperature or melting point is 370°C.   The source of the gold in the alloy (assuming one wants to attach a silicon chip to a gold metallized package or substrate bonding pad) can be the gold plating on a metal package, the gold thick film ink on a ceramic package, the gold metallization on the bottom or backside of the die, or actual thin sheets of the eutectic alloy (pre-forms) placed between the chip and the underlying bonding pad structures.   The eutectic bond is formed by heating the die and the package to slightly above 370°C and then mechanically scrubbing the die into the underlying metal surface.   This scrubbing (back and forth motion in plane of chip) breaks down any thin oxide layers present and minimizes possible voids under the chip.   The use of the gold-silicon attachment method is what is commonly referred to in most literature as eutectic die attach.   Other common eutectic alloys for die attach and their properties are given in Table VI.   Since these alloys do not rely on the interdiffusion of gold with silicon (and the formation of the gold silicon eutectic), a pre-form of the alloy must be used between the metallized back of the die and the metallized region of the substrate of package bottom.

Solder Attach.   Solder attach refers to the use of soft, lead-based solders (e.g., tin-lead system) for die attach.   In this technique both the die back surface metallization and the substrate metallization are pre-tinned (pre-plated) with the solder alloy.   The components are then fluxed (usually resin based with some level of activation (14)), placed together, and reflowed. Following reflow, the flux must be removed by a vigorous cleaning process. Sometimes an additional solder preform or extra paste solder is used to provide a large volume of attachment alloy that can be "loaded" onto the components during the pretinning operation.   Several common soft-solder compositions are shown in Table VI. Sn 63 is a eutectic alloy, but die attachment using this material would not be termed eutectic die attach but rather solder attach.

Table VI: Common Die Attach Alloys

DIE ATTACH MATERIALS	COMPOSITION Wt. %	MELTING POINT oC (Solidus-Liquidus)
Sn-Pb (tin-lead)		
Sn 63*	63 Pb, 37 Sn	183
Sn 62	62 Pb, 36 Sn, 2 Ag	(179-189)
Sn 60	60 Pb, 40 Sn	(183-191)
Sn 10	80 Pb, 10 Sn, 2 Ag	(268-290)
Sn 5	95 Pb, 5 Sn	(308-312)
Sn-Ag (tin-silver)		
Sn 96*	96 Sn, 4 Ag	221
Au-Sn (gold-tin)		
Sn 90*	90 Sn, 10 Au	217
Sn 20*	20 Sn, 80 Au	280
Au-Si (gold-silicon)		
Au 94*	94 Au, 6 Si	370
Sn-In (tin-indium)		
Sn 50	Sn 50, In 50	(117-125)
Au-Ge (gold-germanium) Au 88*	Au 88, Ge 12	356

*Eutectic Alloys

Both the solder and eutectic attaches provide good mechanical strength, high electrical conductivity, and low thermal resistance. Non-gold-bearing metal alloy attaches perform extremely well, even in harsh radiation environments. Some alloys or solders require flux to aid in the reflow and prevent excessive oxidation. As described above, if flux is needed, it must be removed, thereby introducing extra processing, cleaning, and/or inspection steps.

Organic Adhesive Attach. Organic die attach methods use epoxy, polyimide, and other thermoset or thermoplastic adhesive materials to bond integrated circuit die to packages and/or substrates. Organic adhesives are typically poor thermal conductors and natural electrical insulators. To enhance electrical and thermal performance, organic adhesives are loaded (filled) with conductive particles. Gold and silver are the materials commonly used to provide electrical (and thermal) conductivity. Ceramic particles such as Al_2O_3, BeO, SiC, and AlN are used to provide enhanced thermal conductivity while providing electrical isolation properties. The performance of filled adhesives compared to unfilled materials is given in Figure 11.

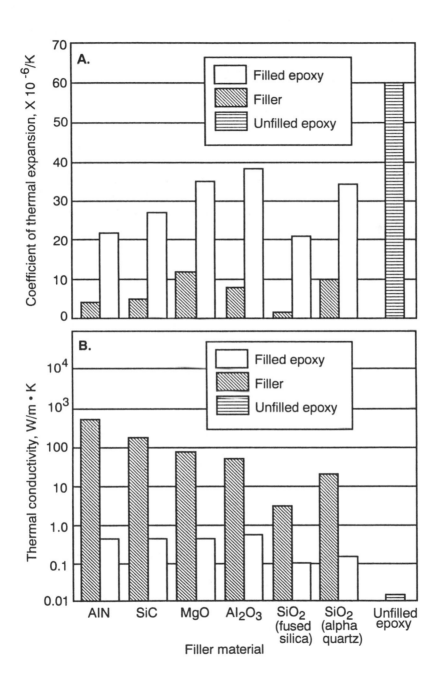

Figure 11. Effect of Filler Materials on the properties of epoxy casting resin. A. Coefficient of Thermal Expansion, B. Thermal Conductivity

A perceived disadvantage of the organic adhesive bond is the potential for outgassing or the subsequent release of absorbed gas or moisture during periods of thermal stress. These outgas products (both known and unknown contaminants) have the potential for adversely affecting device reliability and, hence, product lifetime. The poor thermal stability of organic adhesives (especially above the glass transition temperature), coupled with the outgassing of trapped solvents and reaction by-products, have produced significant concerns for large area die attach reliability, especially in multichip modules.

Low Melting Point Glasses. Silver-filled, low melting point glasses (~400°C) have emerged as a bonding alternative for die attachment. The use of 400°C temperatures and oxidizing ambients (necessary for achieving proper adhesion reactions) has caused major processing concerns. These materials also contain solvents and binders, thus producing conditions similar to the outgassing problems associated with organic adhesives.

3.2   Interconnection

Several chip interconnection methods are being used to meet the performance and high-density interconnection requirements of multichip module. These methods include wafer scale integration, laser pantography, pressure contacts, wirebonding, tape automated bonding, and flip-chip bonding (inverted reflow).

Wafer Scale Interconnect. As mentioned above, in wafer scale integration system building block chips are fabricated (on a single wafer), tested, and then overlaid with an interconnect structure that only links the functional devices. This approach has great potential for high density, high speed, and good thermal management (2), but requires redundancy since the devices and interconnects do not have 100% yield and cannot be tested separately (2). Even more important, different chip technologies cannot be interconnected since all devices must be fabricated on a common wafer. The alternate approach is to fabricate integrated circuits and interconnects separately, then test and assembly them. This approach allows maximum density (no redundancy) since only functional chips and interconnects are assembled - as in today's multichip modules. The five remaining techniques are amenable to this packaging of "known good die."

Laser Pantography. Laser pantography is an experimental technique that utilizes a laser to deposit interconnection metal from gaseous components in a reaction cell (15). Extremely fine lines and, thus, high-density interconnects can be "written" in this manner. Because this is a deposition type process, the lines must be written on an underlying support structure. Thus, to interconnect silicon die, the sides must be beveled (to avoid a vertical step) and passivated after die attach to ensure electrical isolation. Prototype multichip module structures have been fabricated by this technique (16) but no production units have been made.

Pressure Contacts. Pressure contacts in theory provide removable contacts to integrated circuits and mating substrate or package structure. Pressure contacts as shown in Figure 12 are formed or effected by two principle techniques: 1) clamps (mechanical force) and 2) adhesives. Mechanical pressure or clamps force mating surfaces together to make electrical contact. Typically, in a mechanical system chip bonding pads are forced against a bumped substrate or circuit board as shown in Figure 12a. Such a system requires both the circuit board and the chip mounting system to be compliant enough to take up misalignments and variations in bump height. In principle, bumps could be applied to both the chip and the board with a complaint contact strip place in between as shown in Figure 12b. Such compliant interface contact strips have taken various forms as shown in Figure 12c.

Adhesive pressure mounting use adhesives to supply the force necessary to hold chips against a bumped flexible substrate as shown in Figure 12d. The application of force by the adhesive is accomplished in two manners: 1) preloading and 2) controlled shrinkage. In the preloaded mode, the die is compressed against the bumps and flexible substrate by using a known weight or force on the die. This compresses the bumps and circuit structure. Once compressed and held, a rigid epoxy is placed as shown in Figure 12d and cured. The cured epoxy holds the chip in the compressed mode even after the removal of the weight. The controlled shrinkage method uses an adhesive that shrinks as it cures. In this case, uncured adhesive is placed as shown in Figure 12d. As the adhesive cures, it draws the die down closer and in intimate contact with the bumps.

Major Interconnect Methods. The three remaining interconnect methods, wirebonding, tape automated bonding (TAB), and inverted reflow, have all been used in the production of integrated circuits and multichip modules.

Wirebonding. Wirebonding is the most common method in use, but, perhaps, the most limited (see Table VII) for future development of high-density, high-performance multichip modules. There are three wirebonding methods: thermocompression, thermosonic, and ultrasonic. Gold wire can be bonded by all three techniques, while aluminum wire is usually bonded ultrasonically. For multichip modules, thermosonic and ultrasonic bonding (17) are most common. Because of the close pad spacings and high I/O densities of multichip modules, automatic wirebonding is necessary. Even with automatic bonding, however, the relatively long, round wire interconnect necessary in most modules will ultimately limit the performance of packaged integrated circuits and modules because of its high inductance. Electrical performance of the major interconnect schemes is summarized in Table VII. Schematic representatives of the ball (thermosonic) and wedge (ultrasonic) bonding processes are given in Figure 13 and Figure 14, respectively. Figure 15 presents typical bond appearance and geometric factors.

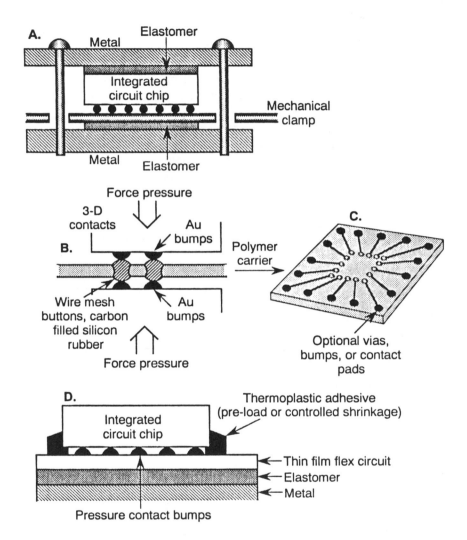

Figure 12. Pressure Contacts. A. Mechanical Clamping, B. Cross Section of Elastomer or Polymer Interconnect, C. Schematic of Polymer Interconnect, D. Adhesive Clamping

Table VII: Current Size and Performance Criteria for Various Chip Interconnect Methods

METHOD	DIAMETER,μm	LENGTH, mm	PITCH, μM	TYPICAL I/O NUMBER	INDUCTANCE, nH	MUTUAL INDUCTANCE, pH
Wirebond	25.4	1	200	256[a]	1-2	100
TAB	50[b]	1	100	400	1	5
Flip-Chip	100	0.1	400	625[c]	0.05-0.1	1
Laser Pantography	5 x 10[-1] [d]	1	25[e]	1600	0.25	1

[a]Ultrasonic wedge chips with I/O's up to 320 have been reported.
[b]Effective ribbon diameter = (thickness + width)/2.
[c]Plus area interconnects up to 16,000.
[d]A 10 μm wide x 1 μm thick laser written line yields a 5 x 10[-3] μm effective diameter.
[e]Laser pantography lines can be written to any practical length because they are entirely supported by the substrate.

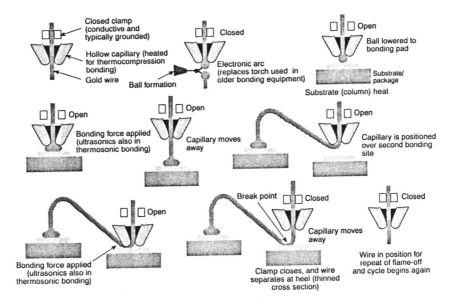

Figure 13. Steps in the Ball Bonding Process

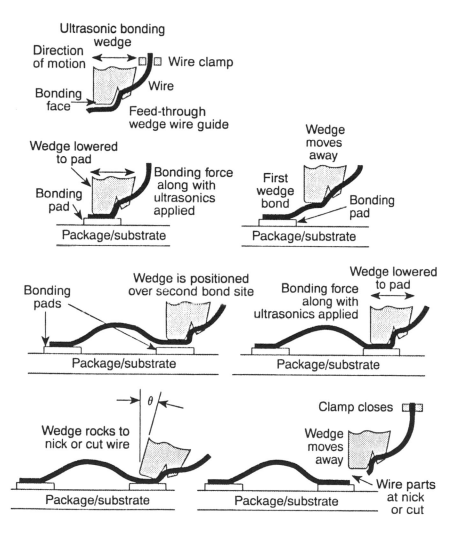

Figure 14. Steps in the Wedge Bonding Process

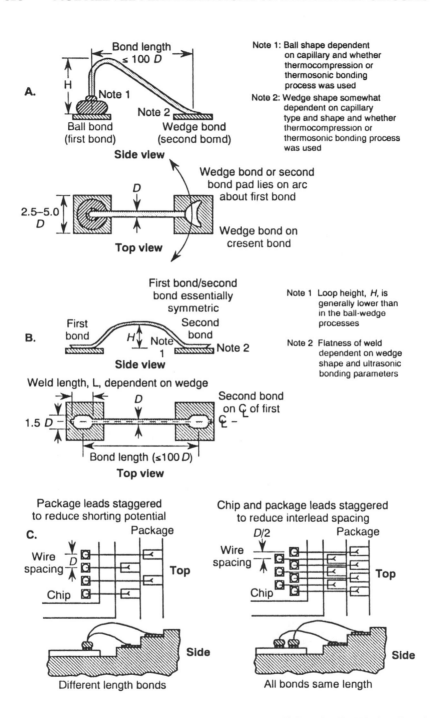

Figure 15. Wire Bond Geometric Factors. A. Ball Bonds, B. Wedge Bonds, C. Package/Die Limitations

Tape Automated Bonding. Tape automated bonding is a planar lead interconnection process that uses prefabricated metallic interconnection patterns (either single or multilevel) on an organic carrier film (typically polyimide). The film is usually one circuit interconnection pattern (or lead frame) wide and hundreds of patterns long, making it suitable for winding on reels or spools for use in automated assembly equipment (18). In order to interconnect a chip to these film-mounted lead frames, the die or tape must undergo additional processing that involves the plating of gold or solder interconnection material, typically in the form of a spheroidal bump, on either the die bonding pads or the tape lead ends. If the integrated circuit has aluminum bonding pads and gold is the interconnection material, a gold diffusion barrier such as titanium-palladium must be used. Since the integrated circuits for multichip modules may come from different vendors who use inherently different processes, it might be difficult to modify the bonding pad and incorporate this integral plating. Thus, plated tape (bumped tape) will be the most attractive module option. Tape automated bonding offers inherent testability since the chip is premounted to the tape lead frame and can easily be handled with a full function tester. It offers improved electrical performance with lower inductance (because of its planar structure and the ability with multilayer tape to bring ground planes close to the chip bond) and high densities (compared to wirebonding), including 50-μm leads on 100-μm centers. The major limitation to the density of these interconnects is the location of the chip bonding pads. Up to 300 inputs/outputs can reasonably be addressed with multilayer peripheral tape structures. Area bonding will have to be used for chips with more than 300 inputs/outputs, requiring the full development of techniques for placing chip bonding pads throughout the active area as well as on the die periphery.

The major disadvantage of tape automated bonding compared to, for example, wirebonding is that high-density, multilayer tape automated bonding systems are expensive and have long production lead times. A specific interconnect pattern is required for each different chip; thus, new masking and tooling are needed for each chip (and/or lead connection pattern) to be mounted. Accordingly, this method has been considered practical only for high-volume production. Because of its performance advantages (see Table VII), it is being considered as an important interconnection method for current and future high-performance chips (19).

Inverted Reflow Bonding. Inverted reflow bonding is the third major chip interconnection scheme in which chips are directly connected by their bonding pads to matching substrate or package pads using plated soldered interconnect metallization, typically in the form of spheroidal bumps. A schematic representation of the bump formation (gold and solder) and attachment process (solder) is given in Figure 16. In this method, solder metallization is attached to the chip bonding pads and/or the substrate lead pattern, as shown in Figure 16b. The chip is then inverted over the substrate pattern, and the bond is formed by solder reflow. The solder metallization melts, and gravity, coupled with surface tension, causes a perfectly shaped and aligned solder joint to be formed, as shown in Figure 16c. Inverted chip solder reflow provides very short, low-resistance interconnections that minimize inductance and capacitance - especially important for a high-frequency operation such as that encountered in multichip modules. Inverted chip solder reflow is also amenable to full area attachment (i.e., bonding pads over the full active device area, not only on the perimeter) since no force is used in the bonding process.   It has three disadvantages: the inverted geometry prevents inspection of the interconnection, heat sinking is poor (if no thermal post is brought into contact with the back of the chip) because the only thermal escape path is through the solder joints, and the devices are difficult to replace if they fail. A full description of the inverted reflow process is given in Reference (20).

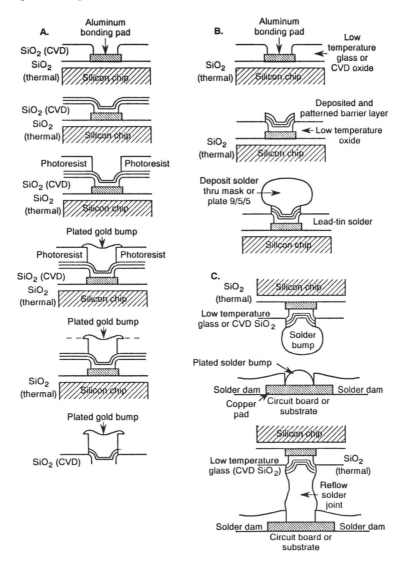

Figure 16. Controlled Collapse (Inverted Reflow) Chip Interconnection Process, A. Gold Bump Formation Process. B. Solder Bump Process, C. Controlled Collapse with Solder Bumps

3.3   Multichip Substrates

Multichip module types are shown in Table V and various substrate materials and their properties are given in Table IV. Multichip module substrates must be strong enough to support multilayer film circuit structures without bowing and should possess a reasonably high thermal conductivity (to ensure proper thermal management and uniform heat distribution) and a coefficient of thermal expansion (CTE) compatible with that of the semiconductor material. Traditionally, thin film multilayer structures have been fabricated on alumina (99.6% Al_2O_3), but more recently silicon has been used (21). Silicon's CTE is totally compatible with that of silicon-based very large scale integrated circuit chips. Silicon also provides a smoother surface than alumina, which allows finer line definition, and has a thermal conductivity many times that of alumina (i.e., 145 W/mK versus 25 W/mK for Al_2O_3). In addition to metal conductors, both active and passive devices can be fabricated in the silicon substrate (22). Because of silicon's low modulus of elasticity when compared to alumina, the number of dielectric and conductor layers must be limited to avoid significant warpage and actual substrate fracture. Another thin film multilayer substrate is sapphire (single crystal Al_2O_3), which has a high thermal conductivity and an extremely smooth surface, allowing greater circuit density. Diamond (both films and substrates) (23) may prove to yield the ultimate in electrical performance while maintaining extremely high thermal conductivity and a compatible CTE.

Cofired ceramic alumina and other ceramic and glass materials substrates in green sheet or tape form have also been used in MCM's as shown in Table V and Figure 6. Such material is useful for its thick dielectric layer possibility, but is limited by its nominally high dielectric constant and high resistance metallizations (molybdenum or tungsten based). It does offer significant potential as a low cost high stability substrate in large volume applications (24).

Organic board materials (epoxy glass and polyimide-glass) can also be used for multichip module substrates. Polyimide-glass provides excellent stability and low dielectric constant, thus ensuring proper signal propagation. Substrate thermal conductivity, high temperature stability, and compatibility with the semiconductor's CTE remain major concerns. Because of electrical design rules and photolithographic limitations on organic boards, circuit density on fiber-reinforced organic materials remains somewhat limited.

Metal substrates such as copper, copper-clad Invar, and copper-clad molybdenum are also suitable for both heat sinks and substrate applications. In a substrate role the metal must be coated with nickel, titanium, chromium, or some other corrosion-resistant material that adheres well to a dielectric material such as polyimide. Metal matrix composites (25) also offer significant promise as substrate materials.

3.4   Package Sealing/Encapsulation

Packages as mentioned in Section 2 fall into two generic types: hermetic and non-hermetic. In polymer sealed or encapsulated packages, moisture will penetrate in a very short time (hours) as shown in Figure 10. Thus, the only true hermetic packages are made of metals, ceramics, and glasses.

Hermetic Package Sealing. Metal packages fall into several types: the transistor outline (TO) (or round header) type, the butterfly (or flat) package, the platform package, and the monolithic (or bathtub) package. These various package styles are shown in Figure 17.

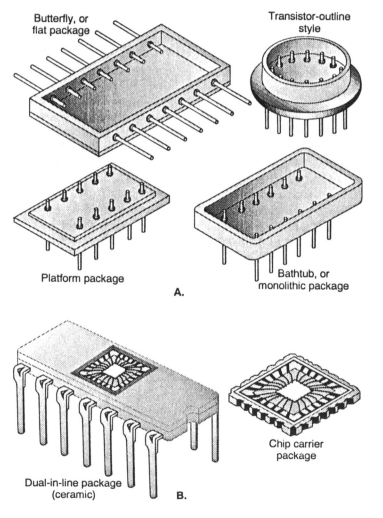

Butterfly, or flat package

Transistor-outline style

Platform package

Bathtub, or monolithic package

A.

Dual-in-line package (ceramic)

Chip carrier package

B.

Figure 17. Common Hermetic Package Styles. A. Metal Packages, B. Ceramic Packages

Metal packages can have large numbers of input and output connections (>200), yet each must be brought out through an isolation glass bead. The lead is sealed to the bead and the bead to the package by a glass-to-metal sealing process (26). Over 75% of all metal packages are welded, using parallel seam sealing, opposed electrode welding, or laser welding. The remaining metal packages are solder sealed using either gold-tin or conventional tin-lead alloys. Solder sealing can be performed in a conveyer furnace, with a seam sealer or a heater platen or cap (27).

There are many types of ceramic packages, including flat packages, ceramic dual-in-line packages (DIPs), chip carriers, grid arrays, and so on. In ceramic packages, a Kovar lead frame is attached to a glazed-alumina package base by temporarily softening the glazing or glass layer. After the die is placed in the package and wirebonded, the glazed-ceramic cap is put on and sealed to the assembly in a furnace or with a hot platen or cap sealer. The glasses used for sealing are high-lead content vitreous or devitrifying glasses with seal temperatures of $400^{\circ}C$ or more. Other ceramic package styles have a Kovar seal ring attached in addition to the lead frame attached to the ceramic body. After plating the lead frame and the sealing with suitable metals, a metal lid is soldered or brazed to the seal ring. A ceramic seal frame is often used instead of the Kovar seal ring. A glassed ceramic lid is typically fused to the ceramic seal ring.

Encapsulation (Nonhermetic Package Sealing). The major chip encapsulation (packaging) techniques are cavity filling, saturation (impregnation), and coating (dip, surface, and conformal). The common cavity-filling processes are potting, casting, and molding.

Potting. In potting, an electronic component (single die, hybrid, and so forth) within a container is filled with a liquid resin that is then cured. The material, container, and circuit become an integral part of the final assembly. Typical resins used for potting include epoxies, silicones, and polyurethanes. Containers are usually made of metal and/or polymeric materials.

Casting. Casting is very similar to potting, except that the container (or outer casing) is removed after the cavity-filled material is cured. Typically, no heat or pressure is used in the process, although some vacuum might be used to allow the filling of remote recesses and to help outgas the polymeric resin.

Molding. Molding involves injecting a premelted polymetric material into a mold containing the electronic circuit or assembly, allowing the resin to set (harden), and then opening or releasing the encapsulated electronic part from the preshaped cavity or mold. The exact control of mold pressure, viscosity of the molding compound, cavity design, and filling mechanism(s) are critical to the success of the molding process. The size and shape of the dies, leads, and cavity determine internal stress buildup in the molded electronic part. Large dies and packages have been known to crack during soldering because of residual stresses induced during the molding process (28). Such stresses must be analyzed and reduced (by the proper choice of package shape, molding resin, lead frame design, and so on) in order to have a high-yield molding system. Finite-element techniques are particularly useful for this type of analysis (29).

Historic molding processes such as injection and pressure-type techniques are being replaced by platen transfer and reactive injection methods using new low-stress molding compounds to minimize the stresses associated with VLSI devices. Details of various molding processes are described in Ref. (27).

Saturation.   Saturation and surface-coating techniques include impregnation, dipping, and conformal coating. Impregnation involves the application of low-viscosity resin to the component which already has a thin layer of the material bonded to the surface (perhaps by surface coating). This is typically used with a cavity-filling process or a conformal-coating technique. Dip coating is performed by dipping the component in the encapsulation resin, withdrawing the component, drying, and, finally, curing the encapsulant. Important parameters for producing effective dip-coated parts (that is, controlled-thickness encapsulation) include resin viscosity, withdrawal rate, and resin temperature. Conformal coating is accomplished by spinning or flow coating. The encapsulant rheology must be tailored, not only to ensure a dense coating with uniform flow, but to accommodate circuit geometries, especially in hybrid applications (30).

3.5 Package Reliability

Designing for reliability means taking reliability and yield figures and single points of failure into account during the system design process such that overall system reliability may be improved (31). Obvious things to be done are elimination of single points of failure (redundancy), wide system performance design margins or robust design (so that components can drift and still allow the system to perform), selection of quality components with improved reliability figures (i.e. come from processes which are under control in the statistical sense), elimination of unnecessary interconnects and interfaces between dissimilar materials, reduce component count, etc. Such measures perhaps take more care and time during the design process but have been shown on numerous occasions to save effort and cost over the system lifetime [32]. System and component reliability can also be significantly impacted by the deployment environment. In addition to humidity, temperature and electrical overstress which can cause failure or accelerated aging as described above, radiation (sun's, nuclear, etc.) can cause component damage and/or hasten the onset of failure. Radiation effects applicable to modern integrated circuits are described in detail by May and Woods (33). Radiation data on various packaging materials has been presented by Bouquet and Winslow (34).

Yield vs. Chip Count.   In multichip modules or hybrids yield concerns naturally increase due to the multiple component nature of the product. Figure 18 (solid curves) shows the system yield for a multicomponent system assuming identical component probabilities of failure.

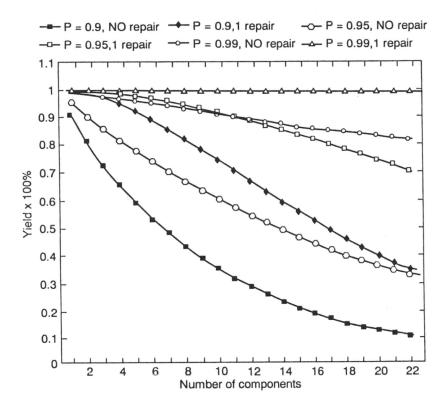

Figure 18. Multichip Module Yield Versus Number of Components With or Without One Repair

Single Points of Failure. In Figure 18 it is assumed that the system will not perform if an individual component does not perform, i.e., all components are single points of failure. System reliability (or yield in this case) can be improved by providing redundancy for each of the single points of failure (or at least those with the highest failure probability in a variable probability system) and by allowing repair. The dashed curves in Figure 18 show the increase in system yield by allowing a single repair in a multicomponent system. Thus if we expect active component yield to fall between 0.9 and 0.95 (which is typical of the dicing and die attach yields of today's known good die), then to achieve 90% module yield the limit on these components in the assembly would fall somewhere between 6 and 11 allowing for 1 repair with a working die. Most of today's MCMs contain less than 10 active die. The yield on passive components is typically, of course, much higher.

Even if an individual component is protected by redundancy (or repair) the effect of the type of component failure on the system must be analyzed. If a redundant component fails open then the redundant element can carry the system. On the other hand, if the element shorts the redundant element might be blocked from becoming operative and the whole system fails. Thus "failure mode" effects analysis (FMEA) must be performed on all elements of a multiple component system.

Fault Trees. A mapping of the single points of failure is known as a fault tree. Fault trees typically contain the effect of the failure on system performance as well as the probability of each single point failure so that risk analysis can be performed. Fault trees, risk assessment, and failure mode effects analysis are beyond the scope of this text but are contained in the book by Raheja (35).

3.6   Quality Assurance and Quality Control

One of the prime drivers in building reliability into products is establishing a consistent, predictable product. The early part of the development-prototyping-production cycle of microcircuit fabrication is aimed in part at identifying failure modes and eliminating or controlling those found. A primary role of quality assurance and quality control is to control the production sufficiently to minimize the instance of failures. Naturally, a first step in establishing consistency is identifying the various factors which do and do not contribute to failure. Subsequently a system can be set up to control those contributing factors to be within acceptable limits. The distinction between quality control and quality assurance is shown, at least in principle, in this sequence. Quality assurance is the system of controls, and quality control is the methods by which that system is carried out. Practicality, however, seldom allows such clean distinctions; the QA system and the QC implementation evolve with time, new materials, and production rate variations. The principle remains, though, that QA is the system and QC is the method.

Structuring a meaningful quality assurance system demands a vision of the goal of the system; significant differences exist between the QA systems for building spacecraft parts and those for making light bulbs, for example. The fundamental question is how good is good enough - for all levels of that system; the answer to that question substantially impacts product cost. For instance, an epoxy used in making military grade microcircuits must meet high and stringent specifications (purity, strength at high temperature, conductivity, etc.), whereas one used on consumer items (such as a "glob-top" epoxy) often is required to meet some minimum - but not optimum - set of requirements; only a select set of properties is specified. The difference is reflected in the tests required of the supplier and of the user, and in the cost of establishing and supporting the QA requirements.

Prime elements of a QA system are specifications for materials and processes, including both quality control and manufacturing processes. For military grade hybrid microcircuits, MIL-H-38534 requires that such specifications be written, maintained, and enforced as part of a Product Assurance Program Plan (PAPP); MIL-STD-1772 gives a checklist of requirements for the quality assurance program for facilities to be certified as able to build high reliability devices. The materials specifications, and the tests or inspections called out in these are just a starting point. These minimize the chance of discrepant materials being introduced into the circuits, by specifying both properties needed for materials and the tests to be carried out to assure those properties are within limits acceptable for the material's use. However, materials specifications by themselves do not assure the materials are used properly.

Proper use of a material, for microcircuits or any other manufactured product, pertains to the manufacturing process. Again, for high reliability products, a consistent product demands a system of documented processes, operating under change control, carried out by trained personnel, and enforced by quality audits. The process documentation spells out the specific way in which processes are to be carried out; process specifications spell out such items as sequence, parameter limits (eg., temperature, time, tools needed) and necessary controls on the process equipment.

Overlooked thus far in the discussion of quality assurance is the need to assure the customer's requirements are met, whether directly specified (i.e., procurement specifications or source control documents) or implied in orders of catalog products; a QA function is to make sure the customer's requirements are allowed for in the product documentation, and that the processes, inspections and testing allow for them to be executed. This is often a matter of end product testing, but can include extra in-process or materials control requirements of a source control document. For high reliability circuits, the various "screening" tests are generally implied, but may include other specific electrical or environmental tests; a discussion of these appears below.

The message of the preceding paragraphs is that quality assurance is itself the *system* of controls over the materials and processes, whose goal is to produce a consistent product. Quality control is the systematic *application* of these controls, whose goal is to assure that the quality assurance specifications are adhered to. It is applied at all stages of production, from raw materials checks, to in-process testing of products and equipment, to tests on the finished product; it is, in essence, a necessary policing function.

A major function of quality assurance is that of material control and certification. Its goal is to assure that the materials used in circuit fabrication meet the requirements demanded by reliability and productability, whether procured (typically epoxies, bonding wire, thick film inks, packages, components) or produced internally (typically substrates). Quality control has the responsibility to assure that any materials used in circuit fabrication have been certified as acceptable and that only those materials so certified are used. A system often used is that of a bonded stockroom for certified parts and materials. In that system, materials go through receiving inspection, are certified and stocked for production use. Certification can involve nondestructive evaluation such as lot sample tests of capacitor sizes and values, package leak testing, or substrate visual inspection; destructive lot sampling evaluations are also often employed, such as substrate wire bondability, or package lead integrity. Materials passing certification tests are often given a different inventory number and stocked. Any materials issued to production must then come from certified stock.

Another function of QC is control of the equipment and training of personnel involved in material control and circuit fabrication. For consistent product to be built, the equipment must operate consistently. For that purpose, another QC function is that of calibration and calibration records. A QA function is to determine what tests and calibrations are sensible and achievable, and at what interval they should be carried out. Loosely analogous, training of personnel in the various operations is needed for establishing a consistent product.

Once material is released to production, QC has functions beyond those of inspections alone. QC should have sign-off of manufacturing aids used in circuit production, when used, as a means of establishing and enforcing product quality standards. QA has the function of providing workmanship standards as guidance to manufacturing personnel; QC inspection criteria are reflected in such standards.

Before departing from the subject of quality assurance and quality control, the subject of quality audits deserves mention. A methodical program of audits done regularly is needed as a check on both performance of the items audited, and on the QA system itself. Discrepancies found in the products, processes, personnel training, or documentation give feedback to both the personnel responsible for action and on the areas needing greater or lesser QC oversight; none of these should be excluded from the audit process.

3.7   Screening and Failure Analysis

Stress Screening. After a circuit is designed properly, allowing for in-process QC tests in addition to its electronic functions, and after it is built with controlled equipment, materials and processes, the user of a device needs assurance that the circuit will work when put in service. Environmental stress screening, or reliability testing, is the testing done to a finished circuit to demonstrate that the circuit will withstand the rigors of service. For military grade circuits, this is a specific and rather complete set of tests; for circuits intended for other uses, cost usually restricts those tests to some lesser set. Theses will be discussed further below.

Reliability tests are a series of tests designed in principle to assure that "flight" grade parts have both a low failure rate and an adequate service life. They are based on the classic "bathtub" curve (see Figure 19).

Figure 19 as previously mentioned is a plot of normally occurring failure rate versus time for a large population of parts, which has three time domains. First is early life, during which failure rates are relatively high but decrease rapidly; these "infant failures" are usually caused by materials or manufacturing related defects. The second phase starts immediately after infant failures have taken place, making the failure rate low; failures in this phase are usually attributed to "random" causes. The third phase is where the devices start succumbing to normal wear-out mechanisms; the failure rate gradually increases above the second phase, and is due to normal physical phenomena such as diffusion spreading in the semiconductors, intermetallic growth in the interconnection metals, embrittlement of adhesives, etc.

Reliability "screening" tests are, in principle, designed to cull out infant failures early in the life of a group of circuits before they are delivered for service. The stress levels dictated by military standards pertaining to electronics are designed to be severe enough to stress the devices beyond what they will encounter in most applications, such that any failures occur in testing rather than in service. This is the series of tests outlined in MIL-STD-883 Method 5004, "Screening Procedures"; Table VIII summarizes the major tests from Method 5004 normally used in military and aerospace circuit screening. These tests are also contained in Method 5008, "Test Procedures for Hybrid and Multichip Microcircuits."   Screening is, then, the test discipline that takes the "sample population," or group of circuits, out of the infant failures phase and places them into the useful life period.

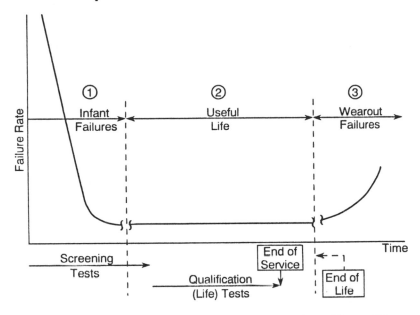

Figure 19. Basic Reliability Curve Illustrating Failure Rate Versus Time

Table VIII: Screening Tests

SCREEN	MIL-STD-883 METHOD
Non-destructive Bond Pull	2023
Internal Visual Inspection	2010/2017
Temperature Cycling	1010
Constant Acceleration	2001
Particle Impact Noise Detection (PIND)	2020
Pre burn-in Electrical Test	Device Specification
Burn-in	1015
Final Electrical Test	Device Specification
Seal (Fine & Gross)	1014

Qualification tests are designed with a different purpose in mind. They are an extended series of tests beyond the screening tests whose purpose, again in principle, is to show that the useful life of a device is greater than its service life; some of these tests are considered destructive. This is done by two types of testing. First, the tests of the screening phase are extended in stress level or duration; the higher levels are considered equivalent to further accelerated aging. Second, some additional tests beyond those in screening are performed. The most noted of these is the 1000-hour life test that is, again in principle, designed to test that a circuit functions after aging equivalent to nominally 20 years; this equivalency is under some assumptions implied in the activation energy assumed and the associated "typical" failure mechanisms. Passing an electrical test after the life test shows that, at least for the circuit configuration built, the wearout mechanisms have not yet affected circuit performance -- that is, that the device is still in its useful life phase at the nominal end of its service life. The specific military requirements are those in MIL-STD-883 Method 5005, "Qualification and Quality Conformance Procedures." Tests on commercial devices aimed at achieving similar qualification are not so standardized, and are considerably more cost constrained. Although similar in goal, they are more tailored to the product; good examples of this are the automobile and communications industries.

No guarantee of absolute reliability (i.e., zero instance of failure) can result from testing. The approach that reliability engineering takes is to invoke a standardized system of testing that allows some consistency in results over the wide variety of electronic configurations and environments seen in high reliability applications.

Failure Analysis. The role of failure analysis in the reliability discipline bears mention. It goes without saying that finding the cause of failures is the first step in correcting them; the failure analysis process is the "eyes" of circuit development. Military specifications (ie., MIL-STD-883, method 5009) give a sequence of steps in destructive physical analysis of a failed microcircuit. Such a method is a good first step, but cannot substitute for a reasoned approach by a trained and experienced failure analyst or analysis team. Many failures have somewhat "generic" causes, such as broken glass seals, mechanical damage, or similar fairly apparent causes. Training and experience are most important in cases where the causes are more subtle, such as chemical action or lot related problems.

Chemical causes are often the most challenging, and often have the greatest impact. Such an example is a lot of units cleaned with a contaminated solvent, which manifests itself in a high failure rate after burn-in. In such a case, the guidance of a good failure analyst usually makes the difference between a focused and a random approach. If the failures are simply sent to an commercial analysis lab without adequate guidance, the results could easily be misleading. Even a well equipped laboratory needs to know the suspected causes to choose the proper tools to apply. In chemical causes in particular, inadequate knowledge of the materials normally found in the failed units can easily mislead interpretation of the various analyses available. The tools are very useful but quite complex (36), with differing applicability whose utility depends on whether the analyst seeks composition, species data, chemical bonding data, or the like.

Regardless of the specific root cause, the message here is simply that failure analysis is a discipline that can be a source of invaluable guidance, but a discipline that must be applied in a controlled scientific manner to yield useful guidance.

4.0 INTEGRATED CIRCUIT DESIGN/STRUCTURE AND PACKAGING INFLUENCES

4.1 Die Size

Integrated circuit chips have traditionally been designed by beginning with either a basic circuit function, high level logic function, or subsystem function to be realized into one or more silicon chips.   More recently with the extraordinary growth of the technology and with the ability to realize millions of transistors on a single chip, designers are required to think in terms of subsystems or even entire systems on a few chips, perhaps even a single chip or in some cases a multichip module.

Initially integrated circuit designers are faced with issues of how big can the chip be since it is usually desired that the design include the maximum amount of functionality that can be realized on a single chip. This size is most often determined by yield or I/O count limitations. The question does not always have a simple unique answer. The allowable size of a chip is determined by several issues.

Defect Density/Process Yield Considerations. Large chips will have low yield dependent upon the process and process defect density. Yield considerations directly translate into issues of cost and allowable die size. Reference 37 provides a basic understanding of many of the issues related to device yield. A simple model for chip yield has the form of

$$[1] \qquad Y = \frac{1}{1 + A_C D_0}$$

where Y = the device or chip yield, A_C = the critical area of the chip and D_O = a process and or technology dependent constant referred to as defect density. It should be recognized that this model is a simplified model, useful primarily for estimating yield relative to chip area, when the process defect density is known. Considerations such as the spatial distribution of defects, packaging yield loss effects, design margin faults, etc. must be treated in a more complex analysis. Typically a designer attempts to realize the maximum amount of circuitry on a single chip, but in order to do this he must decide on a basic architectural approach using major architectural logic building blocks, and decide on how those building blocks might be partitioned into separate chips. Later as the chip designs proceed, each chip is subsequently partitioned into sub-blocks during the chip floor planning phase.

4.2   System Partitioning and its Effect on Costs

As a designer attempts to consider chip cost and ultimately the system costs, he must be aware of the interactions as depicted in Figure 20.

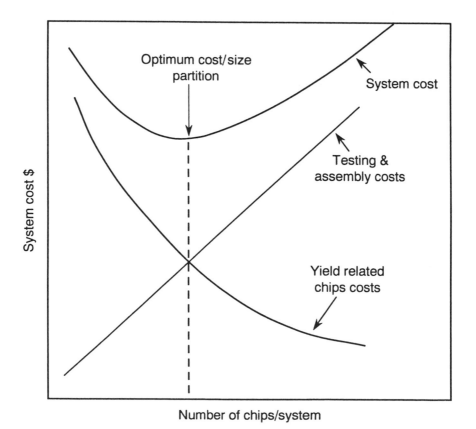

Figure 20. The Effects of Chip Cost on System Partitioning

System cost is most often a driving parameter in the requirements facing the designer. The system must initially be partitioned into functional blocks which ultimately can be realized onto a single chip. Note that if the chips are designed as high yielding (small area) chips, the total system will require many chips, thus driving up assembly and testing costs. Conversely, if only a few large area (likely low yield) chips are required assembly costs and test costs will be low, but basic chip costs will be high. While this model is somewhat simplified, it is clear that optimum system costs are achieved by selection of the right level of integration. It is clear that the use of small scale integration (SSI) will not be appropriate for a system requiring a large number of chips, but to push the state of the art and require very large scale integration (VLSI) or ultra large scale integration (VLSI) can also be inappropriate, if the devices yields are low.

In partitioning the system into single chip functional blocks, major consideration must be given to pin count limits and die size limits. For initial estimation purposes, a relationship known as Rent's Rule can be used (38). The rule, based upon statistical properties of large systems describes the relationship between the number of logic circuits (or gates) contained within a functional block, and the number if I/O signal connections that must be made to that functional block. The equation describing this relationship is

$$[2] \qquad N_C = BN^J$$

where N_C = the number of I/O signal connections that are necessary, N = the number of circuits (or gates) within the functional block, and B and j are constant. Typical data results in B $\approx$ 2 and j $\approx$ _. In actual applications, both B and j will vary depending upon the degree of regularity of the design.

Rent's Rule is applicable to all levels of the design from chip level, multichip module level, card level, and subsystem level, although the constants chosen as best fits to the data may vary for the various levels. The statistical nature of the rule allows it to be useful at a global planning level, but not particularly useful for detailed design. There are examples where Rent's Rule may appear to be broken. Such examples occur where the circuits are very regular and need few interconnects, or perhaps when circuits communicate in serial format instead of requiring numerous independent parallel I/O paths. Large random access memory chips and certain serial digital processors might be examples requiring significantly less I/O pins than predicted by Rent's Rule. More detailed studies (39) have suggested that in these special cases, the rule is still valid, but slightly different coefficients for B and j should be applied. Figure 21 summarizes Rent's modeling for several chip classes.

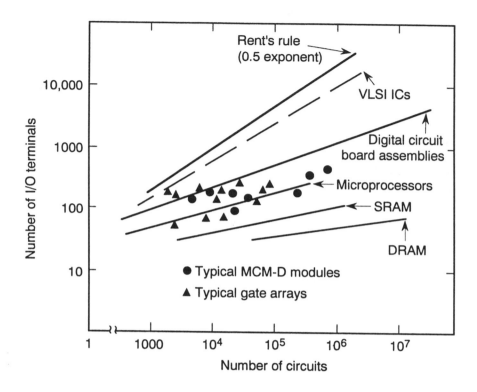

Figure 21. I/O Count Versus Number of Circuits. (Rent's Rule)

A chip design might be considered to be I/O <u>bound</u> or <u>area bound</u>. An I/O bound design is one which could include additional circuit functions, but due to its limited perimeter, no further I/O pin space is available. Conversely, an <u>area bound</u> chip would have available perimeter for I/O pins, but there is no available core area for additional circuit function. An optimum design might be one that is both I/O bound and area bound. In performing the partitioning of the system, the designer might strive for a chip that optimally balances the utilization of peripheral I/O pins and core area.

Partitioning of the system is normally an iterative operation. An initial partition is generated, the expected number of I/O's are tabulated, the chip area is estimated, yield and costs are estimated, and potential packaging schemes are selected. The results of these processes are then reviewed, and the process is iterated until a satisfactory result is achieved.

Technology improvements as the result of improved processes, improved process control or improved photolithography historically have resulted in reducing the area bound chip problem. Figure 22 illustrates the historical and projected increase in die size for microprocessors on random access memories. The increases are the result of process improvements and reduced defect densities.

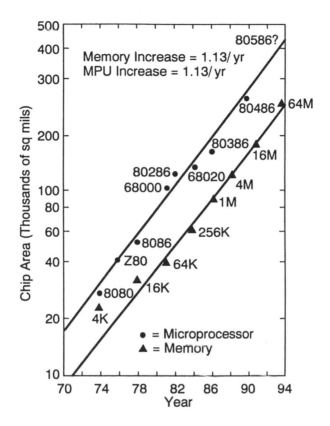

Figure 22. Integrated Circuits Die Size Trends

4.3   Package   Selection

The selection of the actual package is critical to the success of the design. When selecting the actual IC package, consideration must be given to:
⇒   Electrical power/signal needs.
⇒   Next level packaging compatibility.
⇒   Cost goals.
⇒   Thermal management.
⇒   Mechanical support.
⇒   Environmental protection.
Figure 1 illustrates some comparisons of the package area versus the available pin counts for some specific packages currently available in the marketplace. Notice the the rapid increase in package area as the pin count

increases with the traditional DIP package. Those packages such as the quad flat packs are clearly more efficient in the use of area. Consequently, as chip complexity trends continue to increase, the more efficient packages will be utilized. It should also be observed from Table II, that when high I/O counts are needed, those packages which utilize the area more efficiently also have more desirable parasitic resistance, inductance, and capacitance ratios.

4.4   Bonding Pad Layout

Once the system has been partitioned, targets established for die size and I/O pin count, chip floor planning can proceed. Floor planning is an initial step whereby the designer generates a plan for the organization and layout of the building blocks of the chip. This step is somewhat analogous to the system partitioning step. It takes place at the next level down in the hierarchy, but it includes the development of the overall layout concept and the global geometrical and topological placement of the building blocks. During this step, preliminary bonding pad layout locations are selected.

In determining the bonding pad layout, consideration must be given to minimizing the signal path, compatibility with next level package requirements, standardization requirements, manufacturability requirements, and compatibility with the overall floor plan.

Signal lead length is a critical element at all levels of an electronic package design. Robert Noyce, one of the inventors of the integrated circuit, developed the monolithic integrated circuit concept as an improved way to connect transistors. In a 1977 Science article, he stated "The integrated circuit is the component industry's solution to the interconnection problems." (40). While this statement may indicate the major significance of the invention of the integrated circuit, interconnections remain a major challenge at all levels of packaging. Present trends for increased speed and performance require careful design of interconnections at all packaging levels.

At the chip design level the bonding pad layout should be chosen so as to minimize the total system interconnect length. This is desirable in order to minimize signal path delay, but also to minimize space and board area requirements at the next level of packaging. Knausenberger (41) has shown that the cost of interconnection across technology and package level is essentially invariant on a cost per unit length. This emphasizes the need to keep interconnections short, and confirms the work of others. Figure 23 is an attempt to summarize the cost of interconnections versus the density of interconnections for several technologies. In this comparison DSR refers to double sided rigid printed wiring boards, MLB refers to multilayer printed wiring boards, hybrids with 125 μm and 25 μm lines/spaces were studied, and silicon IC's of 2.5 μm were included.

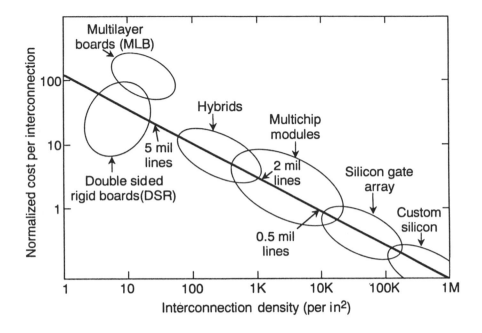

Figure 23. Cost and Density of Interconnections. Intel Corporation.

When considering the compatibility with the next level of packaging, and standardization a bigger perspective is required. Instead of simply trying to optimize the chip design, the chip designer must envision that he is trying to optimize the total design and improve the manufacturability process. Thus the location of bonding pads should consider how to optimize the next level of packaging, conform to standard practices, as well as optimize the chip and package layout. Some examples of these practices exist with:

⇒   General standards for voltage and ground pins on many packages, especially within a logic family of circuits. This feature allows a circuit board designer to standardize many aspects of his layout, and often simplifies test fixtures, and eliminates some catastrophic failures if a device is plugged into the wrong socket.

⇒   Memory devices where input address busses are assigned adjacently and output buses are also assigned adjacently. This feature allows the board designer to interconnect arrays of memory chips together since building large memory arrays requires the parallel routing and connections of output busses as well as the parallel routing and connection of the input address buss. The overall cooperative design of the chip designer and the board designer allows for an efficient design with minimum interconnect length and use of board area or number of layers.

Many of the chips today are I/O pin limited. Traditional wire bonding is most often the chosen interconnection technique at the chip to package level. The bulk of complex IC's utilize die pad pitches of 140 μm (5.0 mils). Active development is ongoing to achieve reduced die pad pitches to further accommodate the density improvements associated with other technology improvements. The usual die shrink process causes many chips to be pad limited. Future die pad pitches as low as 90 μm (3.6 mil) for conventional wire bonding are planned (Ref. 41 and 42). Staggered pads have also been utilized effectively to accommodate greater I/O pin needs. Table IX illustrates the typical wiring pitches generally available to designers.

Table IX: Typical Wiring Pitches for Various Package Levels

TECHNOLOGY	WIRING PITCH
Wire Wrap (Bread Boards/Prototypes)	1250-2500 μm
Printed Wiring Board	200 μm
Multichip Module	
MCM-L	125 μm
MCM-C	125 μm
MCM-D	50 μm
Integrated Circuit (on Chip)	5 μm

4.5 High Speed Considerations

As chip designs continue to be scaled, device performance improves, pushing back the speed barriers, and packaging issues have become more critical. The effects of the interconnection capacitance, inductance and series resistance are no longer negligible and need to be considered in the design, design model and testing of the circuit. Interconnects both on the chip, on the board, and in the rest of the system, need to be understood, carefully modeled, and often need to be treated as transmission lines. Mismatches and the associated reflections from discontinuities at the end of the line may reduce noise margins and degrade high speed performance. Figure 24 illustrates the trends demonstrating that of the total system delay in mainframe computers, the line delay due to level 3 and 4 packaging dominates the gate delay due to levels 1 and 2. Obviously each level of packaging degrades the performance of the device, but as gate level delays continue to improve, the next levels of packaging, e.g., levels 3 and 4, must also improve. Board and system level packaging delays currently dominate the total signal delays in present modern systems. Several trends are emerging as a consequence. (1) Chip designers are doubling and tripling the on-chip clock rates to boost performance on the chip and developing architecture to reduce the amount of data that must go in/out of the chip. Some examples are apparent in the latest microprocessors where on-chip clock rates are higher than off-chip rates, and in the trends to put coprocessors on-chip and cache memory on-chip in order to avoid off-chip data transfers. (2) Package designers are seeking new materials and new design concepts in order to reduce the board and system level delays. Some examples include the use of Multi Chip Modules, Chip on Board and 3 dimensional packaging techniques.

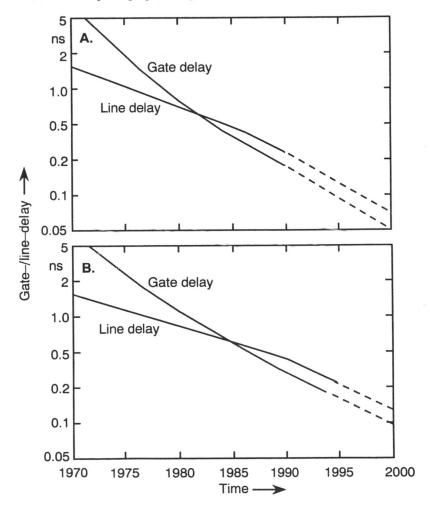

Figure 24. Gate Delay and Line Delay Trends. Wessely, H., Fritz, O. Klimke, P., Koschnick, W., and Schmidt, K. "Electronic Packaging in the 1990's: The Perspective from Europe," in IEEE Transactions on Components, Hybrids, and Manufacturing Technology, Vol. 14, No. 2, June 1991, pp. 272-284.

4.6   Low Voltage Operation

As integrated circuit densities have increased, the fundamental device geometries and parameters have been necessarily scaled to improve or maintain device performance. Recently, scaling has reached the point where gate oxide thickness (typically < 150_) is insufficient to support the traditional 5V standard supply voltage for digital logic circuits. Consequently designers are beginning to adopt a 3.3V or 3V power supply standard (43, 44, 45). The lower supply level will affect both chip design circuit toward package design. Some examples of the effects of changing from a 5 volt standard to a 3 volt standard are:

⇒   Lower power drain, allowing for more applications in battery powered or portable instruments. Figure 25 illustrates how circuit dissipation can be reduced with lower supply voltage.

⇒   Better compatibility with battery technology, resulting in fewer cells and longer operating lifetimes.

⇒   Generally interconnects (line widths and line thickness) can be smaller due to the reduced currents. The series resistance of a conductor (on chip, or on board) might not be as critical, especially if speed is not an issue. This allows designers to shrink many of the conductor dimensions.

⇒   Higher speeds in some circuits, notably CMOS, where the capacitance of the interconnects is critical and since the time to charge a signal line capacitance is proportional to

$$\frac{C\Delta V}{I},$$

higher speed performance can be realized due to the smaller voltage swings.

⇒   Signal noise margins are generally lower. Since the difference between a logical "1" and a logical "0" is less, designers must be careful to reduce circuit noise throughout the system. This will require more careful layouts, reduced line-to-line coupling, reduced signal line length, and better attention to power line decoupling.

⇒   I/O circuitry must be carefully considered in order to achieve compatibility with the more traditional 5 volt designs (46). Often both 3 volt and 5 volt power supplies will be necessary for systems using 3V and 5V chips.

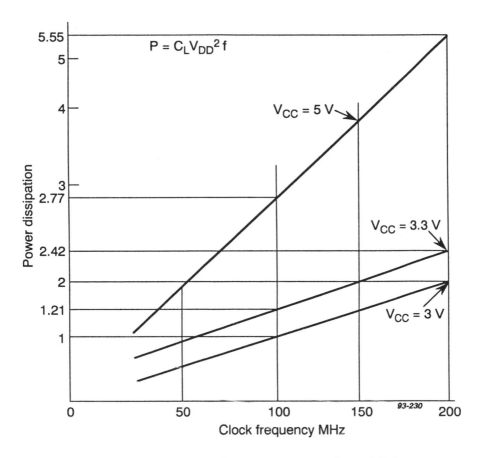

Figure 25. Power Dissipation vs. Clock Frequency at Several Voltages

4.7   End Use Application Effects

To an IC designer, developing a chip intended for use as a standard general purpose product, the end use requirements can be ignored. The end user is responsible for the final levels of the packaging needed in the final application. Therefore, the IC designer attempts to optimize performance costs, size, etc. in order to achieve a broad attraction to a wide market. It is left to the end user to design and package the general purpose devices into a final system or higher level product. In these cases, the end user designers have no influence over the chip design, or package, although they might select from a variety of options. These end users are restricted to Level 3 and Level 4 packaging decisions.

For the IC designer designing a customer application specific integrated circuit (ASIC) chip, the end use might very likely drive the entire design. For example, if speed is the most important parameter, major packaging considerations must be given to interconnections which are short, with least series resistance, low inductance, and low capacitance per unit length. Thus, one must select chip interconnect metallizations for low resistivity, use low dielectric constant insulator materials and keep critical signal levels very short. A special package may be selected to meet unique requirements. A custom-designed package might be created to satisfy the needs. Since high speed operation normally means high power consumption, the package will likely be optimized for high thermal transfer. Efficient thermal transfer will require short thermal paths, good thermal conducting materials, and in some cases forced flow of conductive fluids.

Environmental Effects. Level 1 and Level 2 designers generally design for a broad range of environmental considerations. Some examples might be:

⇒   A full military environment where -55°C to +125°C operation is required. Other requirements might include shock, vibration, salt spray, humidity protection and perhaps nuclear radiation. Applications in military missiles, radars, communications, aircraft instruments, oceanographic instruments, and space satellites are typical.

⇒   A commercial market where 0°C to 70°C operation is more typical and reduced levels of shock, vibration, and humidity might be expected. Applications such as home radio, television, VCR's, personal computers, calculators, appliances, mainframe computers, and industrial controllers are typical. This environment is often well controlled, and consequently is less stressful and allows the designer greater flexibility.

Level 3 and 4 designers generally have a detailed understanding of the environment where their system must function. Hence they resolve final issues of how to deal with the end use environment. For example, they must understand whether a radar will be used:

⇒   on a ground base fixed site - with low vibration, temperature controlled, humidity controlled environment, or

⇒   on an aircraft - with high vibration, little temperature control, little humidity control, or

⇒   on a ship - with high vibration, little temperature control, little humidity control, and possible salt spray conditions, or

⇒   on a spacecraft - with high vibration and some temperature/humidity/salt control during the launch phase, but no vibration, or humidity after orbital insertion.

5.0   SUMMARY

Electronic packaging technology is an important element of modern electronics design and application. The field brings together a variety of disciplines, all critical to the performance and manufacture of electronic devices and systems. This chapter has described the significance of some of the spects of thermal management, mechanical support, and environmental protection. The successful electronic package designer must work with device designers and system designers to achieve a successful total integrated package design. The design must meet the system's electrical performance requirements; while also meeting thermal requirements, mechanical requirements, environmental requirements, reliability and maintainability requirements; and still be capable of being manufactured at reasonable costs. Each specific application requires the designer to weigh the relationship of these requirements accordingly.

Electronic packaging will continue to occupy an important role in the design, development, and application of modern electronics. The ever-increasing density of semiconductor-integrated devices is rapidly approaching fundamental physical limits. This fact, coupled with the system needs for still-greater complexity, will place still greater pressures on improved electronic packaging. In order to meet future system and application needs, packaging engineers will need to:

> Work for improved circuit densities; e.g., smaller line widths, spacings, and apable of higher I/O pin counts.
>
> Work to develop and utilize improved materials and material processes; e.g., to gain better thermal performance, circuit speed, etc.
>
> Work closer with chip and system designers to achieve overall optimal packaging and system performance.
>
> Work with process engineers to improve the packaging process, testability, manufacturability, reliability, and to reduce total system costs.

REFERENCES

1.   Wakamoto, S., COB (Chip-on-Board) Technology, *Proc. 7th Int. Microcircuit Conference*, Yokohama, Japan, 37-44 (1992).

2.   Neugenbauer, C. A., Comparison of VLSI Packaging Approaches to Wafer Scale Integration, *Proc. IEEE 1985 Custom Integrated Circuit Conference*, 32-37 (1985).

3.   Rosler, R. K., Rigid Epoxies in *Packaging*, Vol. 1, *Electronic Materials Handbook*, ASM International, 810-816 (1989).

4.   Borland, W., Thick-Film Hybrids in *Packaging*, Vol. 1, *Electronic Materials Handbook*, ASM International, 332-353 (1989).

5.   Charles, Jr., H. K. and Clatterbaugh, G. V., Thin-Film Hybrids in *Packaging*, Vol. 1, *Electronic Materials Handbook*, ASM International, 313-331 (1989).

6.   Charles, Jr., H. K., VHSIC Packaging for Performance, *APL Technical Review*, 1:47-58 (1989).

7.   Miller, L. F., Controlled Collapse Reflow Chip Joining, *IBM J. Res. Develop.*, 13:239-250 (1969).

8.   Sinnadurai, F. N., High Density Packaging of Chips and Subcircuits in *Handbook of Microelectronics Packaging and Interconnection*

Technologies, Electrochemical Publications, Ltd., Ayr. Scotland, 92-133 (1985).

9.   Sinnaduri, F. N., An Evaluation of Plastic Coating for High Reliability Microcircuits, *Proc. 3rd European Hybrid Microcircuits Conference*, Avignon, France, 482-495 (1981).

10.   Turlik, I. and Messner, Chapter 2: *Background to MCM Technology* in *Thin Film Multichip Modules*, ISHM Reston, VA, 5-44 (1972).

11.   Johnson, R. W., An Introduction to Multichip Module Technology, *Electronic Packaging and Production*, Supplement, 10-14 (October 1992).

12.   Schnachem, S. E., Tessier, T. G., Merkelo, H., Hwang, L.-T., and Turlik, I., Waveguides as Interconnects for High Performance Packaging, *9th IEPS Int. Electronic Packaging Conference*, San Diego, CA, II (1989).

13.   Feldman, M., Optoelectronic Module-to-Module Interconnects, *MCNC Microelectronic Technical Bulletin*, Jan-Feb (1990).

14.   Charles, Jr., H. K., Electronic Packaging Applications for Adhesives and Sealants in *Packaging*, Vol. 1, *Electronic Materials Handbook*, ASM International, 579-603 (1989).

15.   Ehrlich, O. J. and Tsao, J. Y., Laser Direct Writing for VLSI, in *VLSI Electronics: Microstructure Science*, Vol. 7, Academic Press, NY, 129-164 (1983).

16.   Tuckerman, D. B., Ashkenas, D. J., Schmidth, E., and Smith, C., Die Attach and Interconnection Technology for Hybrid VLSI in *Laser Pantography: 1986 Status Report for VHSIC Program*, UCAR-10195, Lawrence Livermore Laboratory, 45 (1986).

17.   Harman, G. G., Wire Bonding Towards 6s Yield and Fine Pitch, *Proc. 42nd Electronic Components and Technology Conference*, San Diego, CA, 903-910 (1992).

18.   Glasser, A. B. and Subak-Sharpe, G. D., Chapter 10: Assembly, in *Integrated Circuit Engineering: Design, Fabrication, and Applications*, Addison-Wesley, Reading, MA, 404-420 (1979).

19.   Brown, S., Kressley, M., Natali, R., Rath, R., and Rima, P., Multilayer Interconnects and TAB for VHSIC Chips, in *Proc. 1987 VHSIC Packaging Conference*, 169-181 (1987).

20.   Charles, Jr., H. K., Electrical Interconnection, in *Packaging*, Vol. 1, *Electronic Materials Handbook*, ASM International, 224-236 (1989).

21.   Hagge, J. K., Ultra-Reliable Packaging for Silicon on Silicon WSI, *Proc. 38th Electronic Components Conference*, Los Angeles, CA, 282-292 (1988).

22.   Bodendorf, D. J., Olson, K. T., Tranko, J. F., and Winnad, J. R., Active Silicon Chip Carrier, *IBM Tech. Disclosure Bulletin*, Vol. 45, 656 (1972).

23.   Burgess, J. F. and Krishnamurthy, V., Use of Diamond in Multichip Packages, *Proc. 1992 Int. Microelectronics Symposium*, San Francisco, CA, 106-111 (1992.

24.   Blodgett, A. J. and Barbour, D. R., Thermal Conduction Module: A High Performance Multilayer Ceramic Package, *IBM J. Res. Dev.*, Vol. 26, 30-36 (1972).

25. Aghajania, K. K., Processing and Properties of Silicon Carbide Reinforced Aluminum Metal Matrix Composites for Electronic Applications, *Proc. 1991 Int. Microelectronics Symposium*, Orlando, FL, 368-371 (1991).

26. Leedecke, C. J., Baird, P. C., and Orphanides, K. D, Glass-to-Metal Seals, in *Packaging*, Vol. 1, *Electronic Materials Handbook*, ASM International, 455-459 (1989).

27. Uy, O. M. and Benson, R. C., Package Sealing and Passivation Coatings, in *Packaging*, Vol. 1, *Electronic Material Handbook*, ASM International, 237-248 (1989).

28. Gee, S. A., van den Bogert, W. F., Akylas, V. R., and Shelton, R. T., Strain Gauge Mapping of Die Surface Stress, *Proc. 39th Electronic Components Conference*, 343-350 (1989).

29. Charles, Jr., H. K. and Clatterbaugh, G. V., Solder Joint Reliability - Design Implications from Finite Element Modeling and Experimental Testing, *Trans. of the ASME, J. Electronic Packaging*, 112:135-146 (1990).

30. Sergent, J. E., Conventional Hybrid Circuits: The Transition to MCMs, *Electronic Packaging and Production*, Supplement, 6-9 (1992).

31. Sinnadurai, F. N., Chapter 6: Reliability of Microelectronics (Packaging and Interconnection), in *Handbook of Microelectronics Packaging and Interconnection Technologies*, Electrochemical Publications, Ayr, Scotland, 134-174 (1985).

32. Dasgupta, A., Verma, S., and Agarwal, R. K., Towards a QML Approach in Electronic Packaging, *Proc. 13th IEEE/CHMT Int. Electronics Manufacturing Technology Symposium*, Baltimore, MD, 19-29 (1992).

33. May, T. C. and Woods, M. H., A New Physical Mechanism for Soft Errors in Dynamic Memories, *IEEE Trans. on Electron Devices*, ED-26:2-9 (1997).

34. Bouquet, F. L. and Winslow, J. W., *IEEE Trans. Nuclear Sci.*, 13:1387 (1984).

35. Raheja, D. G., Assurance Technologies - Principles and Practices, McGraw-Hill, NY (1991).

36. Charles, Jr., H. K., Weiner, J. A., and Blum, N. A., Materials Characterization and Analysis: Applications in Microelectronics, *Johns Hopkins APL Technical Digest*, 6:237-249 (1985).

37. Pimbley, J. M., Ghezzo, M., Parks, H. G., and Brown, D. M., Yield, in *Advanced CMOS Process Technology*, San Diego Academic Press, in *VLSI Electronics Microstructure Science*, 19:227-286 (1989).

38. Fichtner, W., IEEE Journal of Solid State Electronic Devices 2(47) (1978).

39. Einspruch, N. C. and Hilbert, J. C., in Chapter 9: Application Specific Integrated Circuit (ASIC) Technology, San Diego Academic Press, in *Electronic Packaging for ASIC's*, 23:244-272 (1991).

40. Noyce, R. N., Large Scale Integration: What is Yet to Come, *Science*, 195(4283):1102-1106 (1977).

41. Farzaneh, H. and Chanchasem, C., Packaging: The Key to Making Advanced Die Shrinks Work, *ASIC & EDA Technologies for System Design*, 62-65 (August 1992).

42. Shu, B., Fine Pitch Wire Bonding Development using a New Multipurpose Multipad Pitch Test Die, *IEEE Transactions on Components, Hybrids and Manufacturing Technology*, 14(4):680-690 (1991).
43. Morris, B., B. CMOS Buys Back 3V Performance, *ASIC & EDA Technologies for System Design*, 24-28 (July 1992).
44. Pryce, D., 3V Circuits Cut Power and Boost Speed, *EDN*, 73-78 (June 4, 1992).
45. Prince, B., Salters, R.H.W., *IEEE Spectrum*, 22-25 (May 1992).
46. Williams, J., Mixing 3-V and 5-V ICs, *IEEE Spectrum*, 40-42 (March 1993).

13

FUTURE INTERCONNECT SYSTEMS

S. SIMON WONG
Electrical Engineering Department
Stanford University
Stanford, California

1.0 INTRODUCTION

Previous chapters describe the progress during the last decade in the technologies for interconnecting devices on an integrated circuit. In this chapter, we will extrapolate the present trend into the future, examine limitations of existing approaches, explore alternative technologies and discuss potential problems.

As the types of integrated circuits evolve from a few hundred standard analog, memory and logic products of the early seventies to a large variety of standard and custom designs nowadays, the specific requirements for metallization technology also diverge depending on the type of product. It is important to first identify the type of integrated circuit that will be the driver for interconnection technology during the next decade. Although MOS memories, specifically dynamic-random-access-memory (DRAM) and static-random-access-memory (SRAM), are widely considered to have been the technology drivers of the past two decades, they are unlikely to be the type of products that demand the most advanced metallization technology in the future. Figure 1 shows the past and projected evolution of memories. Double level metallization was not widely used until the 1-megabit DRAM and 256-kilobit SRAM generation in the mid-eighties. In future memories, more levels of polysilicon and/or polycide will be used to stack transistors and/or capacitors in order to reduce the cell size. However, two or, at most, three levels of metallization will be sufficient to interconnect the memory arrays and the peripheral circuits. This is evidenced by the fact that all proposed 64-megabit DRAM and 16-megabit SRAM cells have only two levels of metallization. Although the peripheral circuits can benefit from more levels of metallization, it is unlikely that more than three levels will be used because memories are price sensitive products, and process simplicity is essential.

For CMOS logic products, double level metallization is commonly used and triple level metallization is needed for products with 100K or more equivalent gates in order to achieve high packing density (or high utilization in gate arrays) and high speed. Figure 2 shows that the evolution of Motorola's gate array technology has followed this trend. Similar advantages have been observed in microprocessors in which a 20% gain in packing density and performance is observed by going from double to triple level metallization.(1) It is important to note that bipolar logic products have had at least one more level of metallization when compared to CMOS products of the same era, mainly because of the high operational speed and the problem of distributing power and ground. However, bipolar products will not be the driver of interconnection technology due to their small volume and price disadvantage. In the next decade, CMOS logic or microprocessor type products will require the greatest advancement of metallization technology in terms of density, complexity and process yield. These products will therefore be used as a basis for our discussion.

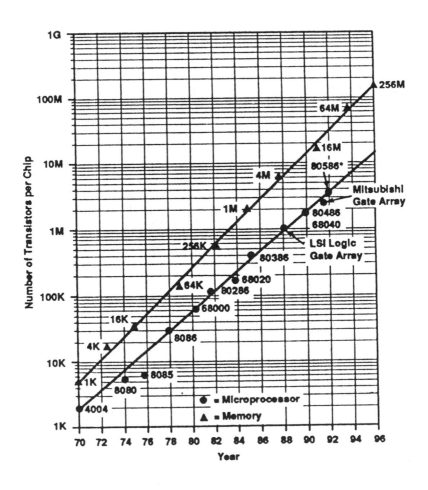

Figure 1. Density trends of microprocessors and memories (DRAM) (Source : W. McClean, Status 1990, a Report on the Integrated Circuit Industry.)

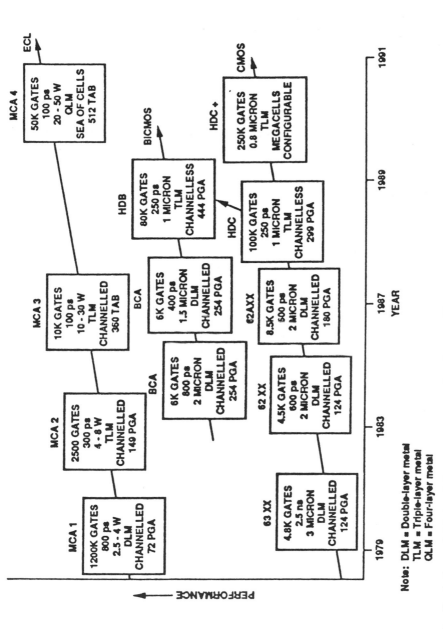

Figure 2. Evolution of Motorola's gate array technology.   (Source : W. McClean, Status 1990, a Report on the Integrated Circuit Industry.)

2.0 METALLIZATION TREND IN THE NEXT DECADE

Since integrated circuit technologies are mostly evolutionary, it is appropriate to project the future trend based on past achievement and present practice. As older approaches encounter fundamental barriers and new technology matures, the industry will adopt the new practices. This changeover does not take place overnight, but rather evolves over a long period of time. In many cases, the old technology continues to be used because of the lower cost, even when the new one is well established. In the following discussion, we will not only extrapolate present practice into the future, but also point out when such projection is inappropriate and outline the reasons. We will also discuss alternatives.

2.1 Metal Feature Size

The feature sizes (line width and line space) of the metallization layer are typically 1.5 to 2 times the minimum feature sizes such as those found in the polysilicon level. This is because the lithography at the interconnection level is complicated by the high reflectivity and the rough surface of the metal layer. More importantly, there is typically more greatly varied topography on the wafer surface during the final stage of processing which limits the lithographic resolution. With the advancement in planarization technology, lithography tools and etching equipment (as described in Chapters 6 and 7), it is increasingly common to find the feature sizes of the lower levels of metallization approaching the critical dimensions. From Figure 3, it is extrapolated that in advanced production, the minimum pitch (width plus space) of metallization layers will be about 0.5 μm by 1995 and 0.2 μm by 2000. Interconnections with these small geometries will experience performance and reliability problems.

The two major parasitic components that contribute to the delay associated with an interconnection are resistance and capacitance. In a typical multilevel metallization, a level of interconnection is sandwiched between two levels of interconnection which run in an orthogonal direction, as depicted in Figure 4. W_1 is the width of the interconnection, W_s is the space between the interconnections, H_1 is the thickness of the interconnection, and H_d is the thickness of the interlayer dielectric. The resistance, R_i, of the interconnection is given by

$$[1] \qquad R_i = \rho l \,/\, H_l W_l$$

where ρ is the resistivity of the metal film, l is the length of the interconnection, and $H_l W_l$ is the cross-sectional area of the interconnection. To the first order, the parasitic capacitances, C_i, of the interconnection can be approximated by

$$[2] \qquad C_i = 2k\varepsilon_o l W_l \,/\, H_d$$

where k is the dielectric constant of the interlayer dielectric and ε_0 is the permittivity of free space. However, as the spaces between interconnections decrease, the two-dimensional distribution of the electric field has to be considered. Some of the electric field lines may terminate at adjacent interconnections instead of the underlying or overlying layer of metallization. In general, a three-dimensional simulation program such as RAPHAEL (2) is required to accurately determine all the parasitic capacitances. If the underlying and overlying metallization layers are approximated by two solid sheets of conductor, the structure as illustrated in Figure 4 can be analyzed with a two-dimensional simulator. The capacitance between two interconnections on the same level, $C_{crosstalk}$, and that between an interconnection and the underlying (or overlying) conductor, C_{plane}, can be determined. The total capacitance, C_{total}, of an interconnection is then twice the sum of $C_{crosstalk}$ and C_{plane}.

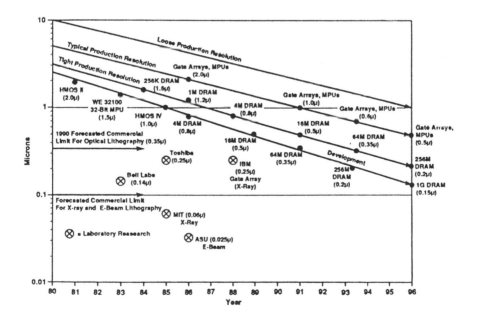

Figure 3. Feature size trends. (Source : W. McClean, Status 1990, a Report on the Integrated Circuit Industry.)

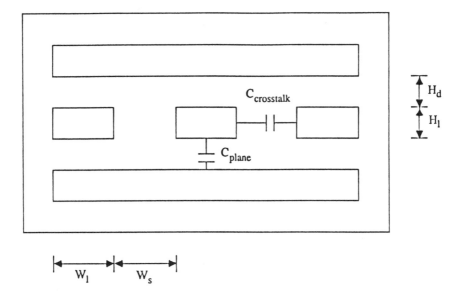

Figure 4. Cross-sectional view of a multilevel interconnection system. (Source : IEEE Transactions on Electron Devices, 39, p. 901 - 907, 1992.)

The variations of the parasitic elements as the interconnections are scaled have been simulated. A set of baseline design rules with $W_1 = W_s = 2\ \mu m$ and $H_1 = H_d = 1\ \mu m$ is chosen to represent the typical feature sizes in the industry. The simplest approach to scaling is to reduce all the geometries, i.e., W_1, W_s, H_1 and H_d, by a factor of K. The resulting resistance and capacitances per unit length, as a function of the metal width, for an aluminum interconnection and oxide dielectric system are illustrated in Figure 5. The $C_{crosstalk}$, C_{plane} and, hence, C_{total}, remain constant as the metal dimensions are scaled. This is because the distribution of the electric field lines is not changed as all the geometries are reduced by the same factor. On the other hand, the resistance per length increases as a function of K^2 because the resistance is inversely proportional to the cross-sectional area, H_1W_1, which is reduced by K^2. The rapid increase in resistance as the metal pitch is scaled results in a severe penalty in the overall performance of the interconnection. In addition, the thinning of the interlayer dielectric increases the pin hole density which leads to more occurrences of accidental shorts between layers of metallization or other failures. Therefore, typical technology reduces the thicknesses of the interconnection and dielectric by a factor of $K^{1/2}$ while the width and space of interconnections are reduced by a factor of K. Figure 6 shows the resistance and capacitances per unit length, as a function of the metal width, in such a scenario. When the vertical dimensions are scaled less aggressively, the resistance increases as a function of $K^{3/2}$. On the other hand, more electric field lines will terminate at adjacent interconnections instead of the overlying or underlying planes. As a consequence, the $C_{crosstalk}$ increases and the C_{plane} decreases, while the C_{total} remains approximately constant at 0.25 fF/μm as the metal pitch is reduced. The fact that C_{total} does not scale with the metal width is in strong disagreement with the overly simplified Equation 2.

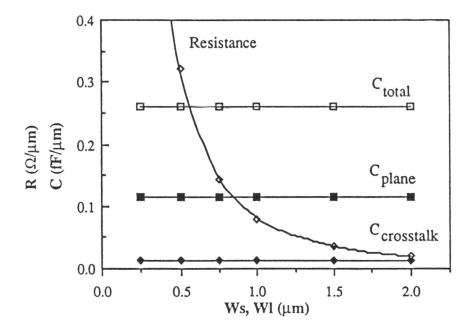

Figure 5. Parasitic capacitances and resistances of an aluminum interconnection and oxide dielectric system when all dimensions are scaled by K. (Source : IEEE Transactions on Electron Devices, 39, p. 901 - 907, 1992.)

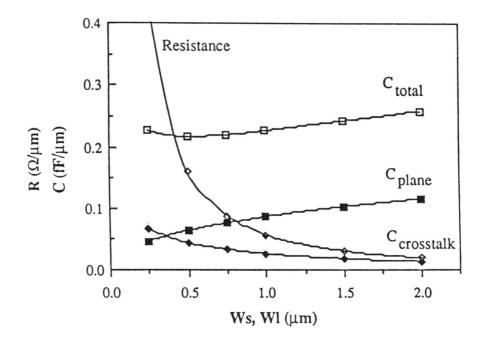

Figure 6. Parasitic capacitances and resistances of an aluminum interconnection and oxide dielectric system when the lateral dimensions are scaled by K and the vertical dimensions are scaled by $\sqrt{K}$. (Source : IEEE Transactions on Electron Devices, 39, p. 901 - 907, 1992.)

The increase in $C_{crosstalk}$ will cause more noise, ΔV, to be coupled into a neighboring interconnection as an interconnection experiences a voltage swing of V (3, 4). In the worst case situation when the neighboring interconnection is floating (e.g., a dynamic node) and other device loadings are small (e.g., a long interconnection), the normalized noise ($\Delta V/V$) is related to the parasitic capacitances with the following relationship:

[3] $\Delta V / V = C_{crosstalk} / C_{total}$

Figure 7 shows that the normalized noise will be larger than 10% when the metal width and space are reduced to below 1 μm. Such a high level of disturbance is unacceptable for reliable operations. Note that the noise, to the first order, does not depend on the permittivity of the dielectric. This study illustrates that in order to avoid the rapid increases in line resistance and crosstalk capacitance, the minimum metal pitch for each level of metallization should not be reduced to much less than 2 μm.

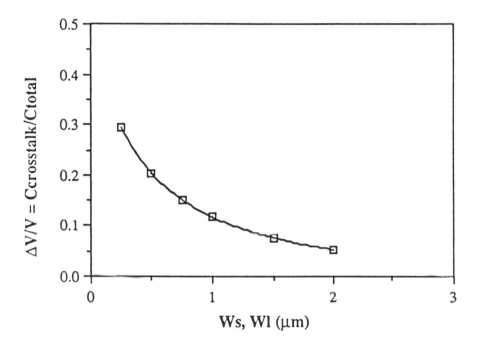

Figure 7 . Cross-talk disturbance as a function of metal width (and space). The thickness of the metal and dielectric are fixed at 1.0 μm. (Source : IEEE Transactions on Electron Devices, 39, p. 901 - 907, April 1992.)

It is important to note that not every interconnection in an integrated circuit, but only a few long ones, are plagued by the problems described above. In custom crafted products such as memories, the problematic interconnections can be identified through experience or by extensive simulation. Either the circuit can be re-designed or the interconnections can be widened accordingly. Hence, it can be expected that metal pitch, at least in the lower levels of metallization, will continue to decrease because this remains the most cost effective means of increasing the interconnection density. Furthermore, the introduction of high conductivity metal and/or low ε_r dielectric will allow the scaling to continue. This will be elaborated in sections 2.3 and 2.4.

For logic or microprocessor type products, the identification and correction of the problematic interconnections cannot be performed manually. Advancement in computer-aided design (CAD) tools will be needed to pinpoint potential delay paths so that appropriate remedies can be implemented. The magnitude of this problem will multiply rapidly as the numbers of devices and interconnections continue to increase. This approach may not be acceptable to fast turnaround, low volume products, typical of ASIC. An alternative to achieving the interconnection density needed in the future is adding more levels of interconnections. It is interesting to note that adding more levels rather than reducing metal pitch is the approach commonly practiced in circuit boards.

2.2 Number of Metallization Levels

As discussed in the introduction, memory products will only have at most three levels of metallization. For ASIC products, the evolution of the number of equivalent gates and metallization levels during the last decade is shown in Figure 2. The trend predicts that MOS ASIC products will have over 1 mega-equivalent gates and five levels of interconnections by 1995, and almost 10 mega-gates and seven levels of metallization by 2000. Such a trend will lead to a dramatic increase in the complexity, cost, and yield loss of the back end process.

Contrary to predictions, the number of interconnection levels will not increase much beyond six. The reduction in area and the gain in performance are about 20% as a third level of metallization is introduced.(1)   The improvement will only be about 10% as a fourth level of metallization is added, and even less for higher levels of metallization. Although the top few levels do not greatly improve the interconnectivity, they will be needed for power and ground distribution, and perhaps major signals and clocks. Beyond the sixth level, the penalty in yield loss will not justify the small improvement in packing density and performance, and hence, it is unlikely that the number of metallization levels will be much more than six. If high conductivity metal and low ε_r dielectric can be successfully implemented, the metal feature sizes can be further scaled. This will further delay the need for additional levels of metallization.

In order to fully utilize the limited density of interconnections, each level in the metallization system will have to be optimized for a particular class of line lengths. Figure 8 is a histogram depicting the occurrence of various lengths of interconnections in a typical integrated circuit. Over half of the interconnections are shorter than 1 mm. These short lines will be implemented in the lower levels which are thinner and have finer pitch because the delay along such a short interconnection is not critically dependent on the line resistance but is determined by the driver transistor, as will be discussed in the next section. The upper levels of interconnections, on the other hand, should be optimized for long lines. The metal will be thicker and wider to minimize the resistance and signal delay. To facilitate the interconnection between levels, stacked and self aligned contacts and vias will be needed.

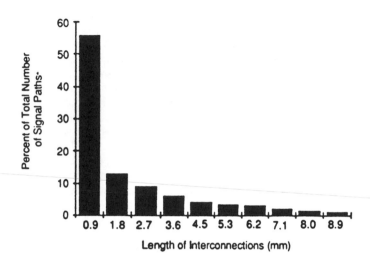

Figure 8. Histogram of interconnection lengths on typical integrated circuits. (Source : IEEE Transactions on Electron Devices, 34, p. 633 - 643, March 1987.)

2.3 Interconnection Metals

Although aluminum alloy, the dominant interconnection material, has many desirable properties and the technology is well developed, it faces some fundamental reliability problems that might render it unacceptable for future devices. In Chapter 8, these issues are discussed in detail. Electromigration in aluminum alloy is a severe problem that has been studied quite extensively. The aluminum alloy lifetime has continued to improve with the addition of impurities such as copper and the use of layered structures.(5) Recent results have shown that the lifetime of aluminum alloy under realistic operating conditions of pulse and AC stress is much improved over that under DC stress.(6) Since only a small number of interconnections in an integrated circuit will experience severe electromigration, they can be properly widened once they are identified. Electromigration, though a critical concern, will not be the sole reason for abandoning aluminum alloy if proper CAD tools are developed to identify and correct the hot spots. Stress-induced-migration, on the other hand, can affect every narrow interconnection and hence is a critical problem. Without any doubt, fundamental studies in this area will lead to a better understanding, more accurate models and innovative solutions. On the other hand, the continuous scaling of line features and the piling of more layers will only aggravate the problem. If proper means cannot be devised to reduce the stress and improve the strength of aluminum alloy, it will have to be replaced.

CVD tungsten has been investigated for over a decade and is becoming a standard contact and via plug technology. Almost all metallization technologies with three or more levels use the blanket deposition and etch back of CVD tungsten for plugs between levels. There are also attempts to use the selective CVD tungsten process because of the potential process simplification and cost savings. As the diameter of the contact and via plugs becomes less than 0.25 μm, a uniform deposition of the seed layer into the contact and via holes by sputtering will be extremely difficult. Selective deposition will become more attractive. It is expected that with the advancement in CVD reactors and the maturing of in-situ deposition-etching processes, or post deposition clean up processes,(7) selective CVD tungsten may become the dominant plug technology. CVD tungsten has also been used as interconnections in the lower levels of metallization. Although it is immune to migration, the high resistivity limits its application to short lines. Whether tungsten will continue to be used for interconnections depends on the development of other metals. If neither aluminum nor some other highly conductive metal can be scaled to the fine pitch required, tungsten will be the only alternative for the lower levels of metallization.

Silver has the lowest resistivity of all metals. However, silver has been reported to suffer from a severe electromigration problem which excludes it from consideration.(8, 9) Copper, which is the second most conductive metal, will be an important metal for the future. With its high conductivity and resistance to electromigration,(10-12) copper will allow the continued scaling of metal pitch, and delay the need for more levels of metallization. Significant progress has been made in addressing the various processing and device compatibility problems associated with copper metallization. These challenges include the development of a low vapor pressure precursor (13) and a CVD technique for blanket and/or selective deposition; an anisotropic dry etching technique (14, 15) for patterning copper interconnections, or a blanket etch-back process such as chem-mechanical polishing;(16) adhesion layers; and metal and dielectric barriers to prevent the diffusion and drift of copper towards devices.(17) In addition, whether Cu offers an improvement in the resistance against stress induced failures has to be verified.(18) Gold has almost all the performance advantages and processing problems associated with copper, but has a resistivity between that of aluminum and copper.(19,20) If substantial effort is required to incorporate either gold or copper onto integrated circuits, it would probably be more appropriate to concentrate on the metal that will deliver the highest level of performance, namely, copper.

In order to develop an understanding of the interplay between conductivity, reliability and performance for the various metals, it is necessary to estimate the signal delay along an interconnection as the geometry is scaled. An interconnection can be modeled as a distributed RLC network.(21-23) For typical interconnections in silicon integrated circuits, the $(LC)^{1/2}$ delay (the signal propagation time) is much smaller than the RC delay (the signal rise time).(24,25) Hence, the inductance can be ignored. For simplicity, the junction capacitance of the MOSFET, C_s, and the load capacitance, C_l, are assumed to be negligible when compared with the interconnection capacitance, C_{total}. With these approximations, the RC circuit as shown in Figure 9(a) is used to model the discharging of an interconnection by a N-MOSFET. A similar analysis can be performed for the charging of an interconnection by a P-MOSFET.

The operations of the N-MOSFET are modeled as a simple resistor of VD_{sat} / ID_{sat} for the linear region (i.e., $VD < VD_{sat}$) and a constant current source ID_{sat} for the saturation region (i.e., $VD > VD_{sat}$), as depicted in Figure 9(b). The input of the MOSFET is driven by a 0V to V_{DD} voltage pulse. The delay for the interconnection to discharge from V_{DD} to V_{ol} can be expressed as

[4]
$$T_d = C_{total} \{ (V_{DD} - VD_{sat} - ID_{sat} R_i) / ID_{sat} \\ + (VD_{sat} / ID_{sat} + R_i) \ln [(VD_{sat} + ID_{sat} R_i) / V_{ol}] \}$$

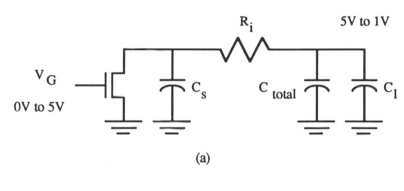

(a)

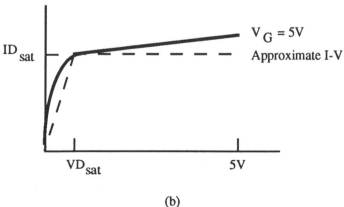

(b)

Figure 9. (a) Simplified model for estimating the signal delay along an interconnection, and (b) approximated I-V characteristic of a N-MOSFET. (Source : IEEE Transactions on Electron Devices, 39, p. 901 - 907, 1992.)

Table I summarizes the expected device performance as the technology is scaled from 5 μm to 0.4 μm. For tungsten interconnections, the width of the N-MOSFET is chosen such that the input gate capacitance is one fifth of the interconnection capacitance, which is a commonly used ground rule to optimize the driver MOSFET. In the case of aluminum interconnections, the current density in the interconnection increases rapidly as the technology is scaled, which may lead to severe electromigration problems. The width of the N-MOSFET is therefore reduced such that the peak current density does not exceed a safe level of 5×10^5 A/cm^2. Although the electromigration lifetime of copper has not been fully evaluated, initial data show that it is much better than Al. For this discussion, we will assume a peak current density of 2×10^6 A/cm^2. Assuming a V_{DD} of 5 V and a V_{ol} of 1 V, the signal delays of aluminum, tungsten and copper interconnections with lengths of 0.25 cm, 0.5 cm, and 1 cm have been modeled. The results are summarized in Figure 10. For long lines, the delay of aluminum and tungsten interconnections are comparable. The inability of aluminum to support a high current density offsets the advantage of a higher conductivity. As expected, copper offers the shortest delay for long lines. Nevertheless, the delay along submicron copper lines may still be unacceptable for high speed products. These long lines will have to be properly widened to minimize the delay. For short lines, the delay is mostly determined by the driver transistor rather than the resistance of the line, and hence copper does not offer any advantage over tungsten. In short aluminum interconnections, because the driver transistors have to be down sized to ensure an acceptable current density, the delay could actually become worse.

Table I: Expected device performance for scaled technology. (Source : IEEE Transactions on Electron Devices, 39, p. 901 - 907, 1992.)

Gate Length (μm)	5	3	2	1.5	1	0.75	0.4
Effective Channel Length (μm)	3.5	2	1.5	1	0.75	0.5	0.25
Gate Oxide Thickness (nm)	90	50	37.5	25	20	17.5	10
VDsat @ VG=5V (volt)	3.2	2.9	2.8	2.7	2.6	2.4	2.3
IDsat @ VG=5V (μA/μm)	42	97	139	217	282	375	647
Metal Line Width (μm)	7.5	4.5	3	2.25	1.5	1.1	0.6
Metal Line Thickness (μm)	2	1.5	1.2	1	0.85	0.75	0.55

Although the results discussed here are only qualitative, the scenario presented accurately reflects the general trend in the industry. The exact line delay will depend on the details of the technology, the device parameters, the interconnection structures and the models used. However, similar conclusions will be reached. The recent trend of reducing the power supply voltage, V_{DD}, from 5 V to 3.3 V or lower values will not greatly increase the signal delay along an interconnection. T_d as expressed in Equation [4] is mostly dominated by the first term which is approximately proportional to V_{DD}/ID_{sat}. Since ID_{sat} almost scales with the maximum gate voltage (i.e., V_{DD}) in today's MOSFET, V_{DD}/ID_{sat}, and hence T_d, are not very sensitive to the reduction in VDD.

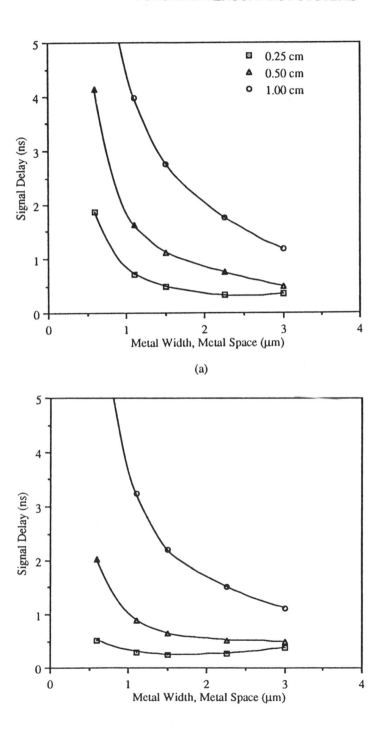

(a)

(b)

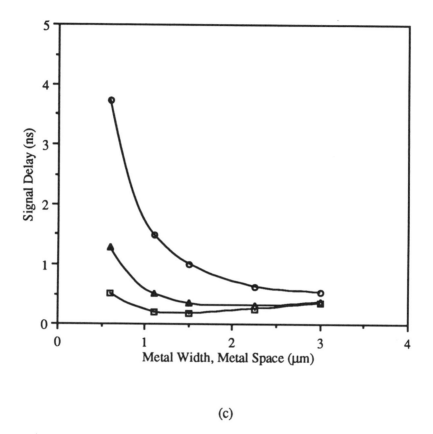

(c)

Figure 10 - Signal delay for (a) aluminum, (b) tungsten, and (c) copper interconnections of various lengths. Peak current density is limited to 5×10^5 A/cm^2 for aluminum and 2×10^6 A/cm^2 for copper (Source : IEEE Transactions on Electron Devices, 39, p. 901 - 907, 1992.)

2.4 Interlayer Dielectrics

Oxide and oxynitride films are the most developed dielectrics. Recent advancements in organometallic precursors and deposition processes have resulted in a significant improvement in the quality and step coverage of oxide films deposited at low temperatures. The availability of cluster tools, which provide reasonable throughput for sequential deposition-etching processes, has simplified the wafer handling procedures. Nevertheless, the advantage of a low ε_r interlayer dielectric in reducing the line capacitance cannot be overlooked.

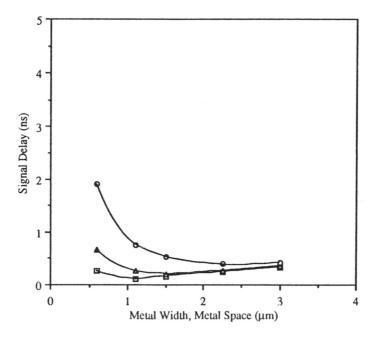

Figure 11. Signal delay for copper interconnections of various lengths when an interlayer dielectric with a ε_r of 2 is adopted. Peak current density is limited to 2×10^6 A/cm^2.

Figure 11 illustrates the significant reduction in signal delay if a dielectric with a ε_r of 2.0 is used in conjunction with copper interconnections. With a reduction in the line capacitance, the interconnections can be scaled to a smaller dimension before the signal delay becomes excessive. This will postpone the need for additional levels of interconnections or new metals. A less obvious benefit of the low ε_r dielectric is the reduction in power dissipation. Figure 12 shows that there is a strong correlation between the power dissipationof a microprocessor versus the product of its chip size and operational frequency. In typical logic circuits, a large portion of the power is dissipated in driving the high capacitances associated with the long clock lines, signal and data buses:

[5] $$P = C_{total}V^2 f / 2$$

where P is the power dissipation, C_{total} is the line capacitance, V is the supply voltage or the switching level and f is the operational frequency. Since the line capacitance is proportional to the chip area, power dissipation is proportional to the product of the chip area and the operational frequency. It is thus obvious that if the line capacitances are reduced by adopting a low ε_r interlayer, the power dissipation will be proportionally reduced. Alternatively, the operational frequency could be increased without greatly affecting the power dissipation.

Polyimide, with a ε_r of about 3.0, is commonly used with copper metallization in multichip modules and circuit boards. However, it is not impervious to the diffusion of copper and moisture and has not been widely accepted in integrated circuits. The development of an alternative, low ε_r dielectric that is also an excellent diffusion barrier will be an important challenge. A deposition technique that provides excellent step coverage, and a compatible planarization technique must be developed concurrently. The latest developments in organic interlayer dielectrics are described in Chapter 5.

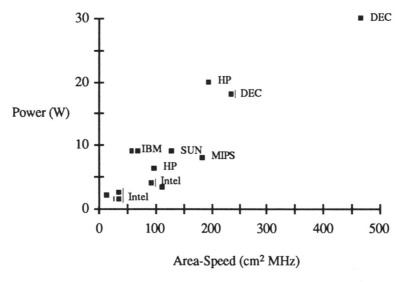

Figure 12 . Power dissipation of various microprocessors versus the product of chip area and operational frequency.

2.5 Interrelationships with Other Technologies

We have compared the performance advantages offered by the various technology options. In practice, the choice is also critically dependent on the development of device and contact structures, deposition tools, etching equipment, planarization technology, packaging requirements and most importantly, cost. Most of these issues have been addressed in the previous chapters. We will only point out in what manner changes in these related areas will impact the metallization technology.

The advancement in device structure will impact the contact technology and the first level metallization. Since strapping of diffusion and polysilicon regions with silicides or refractory metals will be essential for future devices, it can be assumed that contacts to devices will be made to these layers instead of to silicon directly. In addition, local interconnection, which typically involves a modification of silicide or refractory metal strapping technology,(26, 27) will play a role in improving the interconnectivity and reducing the number of contacts to the devices.(28) It may even delay the need for an additional level of metallization. In order to take full advantage of the local interconnection, especially in ASIC type products, proper CAD tools for routing and performance estimation have to be developed.

Advanced packaging technology may require special processing on the bonding pads (e.g., solder bumps). This will complicate the process, which will have to be taken into consideration while designing the top layer of metallization. On the other hand, the advancement of packaging technology may relieve the stringent requirements of metallization. The best example of this is the multichip module. If a large number of bondings (more than 1000) in an array format (e.g., solder bumps) are allowed between the chip and the module, some of the interconnections on the chip (e.g., power, ground and clock distribution) can be implemented on the module instead, which will reduce the number of metallization levels on the chip. Even chips that are packaged individually could benefit from this approach by incorporating the top layers of metallization in the package. It is of utmost importance that the development of multi-level metallization be closely coupled and coordinated with the advancement in packaging technology in order to achieve an overall optimized, low cost solution (see Chapter 12).

2.6 Potential Paradigm Shifts

If a room temperature superconductor that can support a high current density (e.g., $> 10^7$ A/cm^2) is discovered, the interconnections will be greatly simplified. Long, thin and fine pitched interconnections could be implemented without a penalty in signal delay. The super thin conductors also offer a substantial reduction in crosstalk disturbance. The number of interconnection layers could be reduced to two or, at most, three. The recent development of high T_c (< 150 K) superconductors, however, will not impact mainstream electronic systems.

It is extremely unlikely that optical interconnections will play any role in chip level data communication because of the delay and power penalty associated with the optical emitters and detectors. Although optical links have been proposed for data transmission on and off chip because their high data rate allows multiplexing and offers a potential reduction in the number of inputs and outputs, the latest advancements in high density, low parasitic bonding and packaging technology offer a much more cost effective solution.

3.0 SUMMARY PERSPECTIVE ON INTERCONNECTION TECHNOLOGY

In summary, interconnections have been, are and will continue to be a vital determining factor in the advancement and the manufacturing of semiconductor products, especially ASIC and microprocessor type products. It is believed that up to six levels of metallization layers will be used. The lower levels, which are mainly used to provide short routing, will have feature size matching those of the transistors, whereas the upper levels will have relaxed pitch to minimize the performance penalty associated with long interconnections. As the density and levels of interconnections continue to increase, the back end of the process will only become more difficult. Substantial improvements in the deposition, etching and planarization techniques are needed to continue the stacking of metallization layers. In addition, the adoption of low resistivity metal, such as copper, and low ε_r dielectric, such as organic polymer, will allow current practices to be extended into the next decade. It is believed that significant advantages can be derived by closely coupling the development of future metallization technology with the advancement of packages.

REFERENCES

1. J. Schutz, "A CMOS 100MHz microprocessor," International Solid-State Circuits Conference Digest of Technical Papers, p. 90, 1991.
2. RAPHAEL, Technology Modeling Associates, Inc., Palo Alto, CA.
3. S. Seki and H. Hasegawa, "Analysis of Crosstalk in Very High-Speed LSI/VLSI's Using a Coupled Multiconductor MIS Microstrip Line Model," IEEE Transactions on Electron Devices, 31, p. 1948, 1984.
4. V. Tripathi and R. Bucolo, "Analysis and Modeling of Multilevel Parallel and Crossing Interconnection Lines," IEEE Transactions on Electron Devices, 34, 650, (1987).
5. D. Gardner, J. Meindl and K. Saraswat, "Interconnection and Electromigration Scaling Theory," IEEE Transactions on Electron Devices, 34, 633, (1987).
6. B. Liew, N. Cheung and C. Hu, "Projecting interconnect electromigration lifetime for arbitrary current waveforms," IEEE Transactions on Electron Devices, 37, 1343, (1990).
7. R. Uttecht and R. Geffken, "A Four-Level-Metal Fully Planarized Interconnect Technology for Dense High Performance Logic and SRAM Applications," Proceedings of VLSI Multilevel Interconnection Conference, p. 20, 1991.
8. I. Videlo and R. Sutherland, "The Use of Silver as an Alternative to Copper in the Fabrication of Reliable Thick-film Hybrid Microcircuits Requiring High Conductivity Conductor Tracks," Hybrid Circuits, 21, 18 (1990).

9.   H. Naguib and B. MacLaurin, "Silver Migration and the Reliability of Pd/Ag Conductors in Thick-Film Dielectric Crossover Structures," IEEE Trans. CHMT, 2, 196, (1979).

10. C. Park and R. Vook, "Activation Energy for Electromigration in Cu film," Applied Physics Letters, 59, 175, (1991).

11. T. Ohmi, T. Hoshi, T. Yoshie, T. Takewaki, M. Otsuki, T Shibata and T. Nitta, "Large-Electromigration-Resistance Copper Interconnect Technology for Sub-Half-Micron ULSI's," International Electron Devices Meeting Technical Digest, p. 285, 1991.

12. H. Kang, J. Cho and S. Wong, "Electromigration Properties of Electroless Plated Cu Metallization," IEEE Electron Device Letters, 13, 448, (1992).

13. J. Norman, B. Muratore, P. Dyer and D. Roberts, "New OMCVD Precursors for Selective Copper Metallization," Proceedings of VLSI Multilevel Interconnection Conference, p. 123, 1991.

14. G. Schuwartz and P. Schaible, "Reactive Ion Etching of Copper Films," Journal of Electrochemical Society, 130, 1777, (1983).

15. K. Ohno, M. Sato and Y. Arita, "High Rate Reactive Ion Etching of Copper Films in $SiCl_4$, N_2, Cl_2 and NH_3 mixture," Conference on Solid State Devices and Materials Extended Abstracts, p. 215, 1990.

16. A. Krishnan, C. Xie, N. Kumar, J. Curry, D. Duane and S. Muraka, "Copper Metallization for VLSI Applications," Proceedings of VLSI Multilevel Interconnection Conference, p. 226, 1992.

17. J. Cho, H. Kang, I. Asano and S. Wong, "CVD Cu Interconnections for ULSI," International Electron Devices Meeting Technical Digest, p. 297, 1992.

18. P. Borgesen, J. Lee, R. Gleixner, and C-Y. Li, "Thermal Stress Induced Voiding in Narrow, Passivated Cu Lines," Appl. Phys. Lett. 60, 1706 (1992).

19. T. Yuzuriha and S. Early, "Failure Mechanisms in a 4 Micron Pitch 2-Layer Gold IC Metallization Process," Proceedins of VLSI Multilevel Interconnects Conference, p146, 1986.

20. D. Summers, "A Process for Two-Layer Gold IC Metallization," Solid State Technology, p. 137 (1983).

21. I. Ho and S. Mullick, "Analysis of Transmission Lines on Integrated-Circuits Chips," IEEE Journal of Solid-State Circuits, 2, 201, (1967).

22. H. Hasegawa and S. Seki, "Analysis of Interconnection Delay on Very High-Speed LSI/VLSI Chips Using an MIS Microstrip Line Model," IEEE Transactions on Electron Devices, 31, 1954, (1984).

23. T. A. Schreyer, "The Effects of Interconnection Parasitics on VLSI Circuit Performance," Ph.D. Thesis, Stanford University, 1989.

24. C. Mead and M. Rem, "Minimum Propagation Delays in VLSI," IEEE Journal of Solid-State Circuits, 17, 773, (1982).

25. T. Sakurai, "Approximation of Wiring Delay in MOSFET LSI," IEEE Journal of Solid-State Circuits, 18, 418, (1983).

26. S. Wong, D. Chen, P. Merchant, T. Cass, J. Amano and K. Chiu, "HPSAC - a Silicided Amorphous Silicon Contact and Interconnect Technology for VLSI," IEEE Transactions on Electron Devices, 34, 587, (1987).

27.   T. Tang, C. Wei, R. Haken, T. Holloway, L. Hite and T. Blake, "Titanium Nitride Local Interconnect Technology for VLSI," IEEE Transactions on Electron Devices, 34, 682, (1987).
28.   J. Gallia, et al., "High performance BiCMOS 100K-Gate Array," IEEE Journal of Solid State Circuits, 25, 142, (1990).

INDEX

DATE DUE

DEMCO, INC. 38-2931